# *Applied and Numerical Harmonic Analysis*

# Contents

*The first editor dedicates this volume to the second,*
*Professor Časlav V. Stanojević,*
*as mentor, friend, and brother,*
*and on occasion of his $70^{th}$ birthday*

William O. Bray
Department of Mathematics
and Statistics
University of Maine
Orono, ME
USA

Časlav V. Stanojević
Department of Mathematics
and Statistics
University of Missouri
Rolla, MO
USA

**Library of Congress Cataloging in Publication Data**
Analysis of divergence : control and management of divergent processes
/ William O. Bray, Časlav V. Stanojević, [editors].
p. cm. — (Applied and numerical harmonic analysis)
Includes bibliographical references and index.
ISBN 0-8176-4058-4 (alk. paper)
1. Control theory. 2. Divergent series. 3. Asymptotic expansions. I. Bray, William O., 1955– . II. Stanojević, Časlav V., 1928– . III. Series.
QA402.3.A546 1999
629.8′312—dc21 98-44618
CIP

AMS Subject Classifications: #40, #43, #46

Printed on acid-free paper.


*Birkhäuser* 

ISBN 0-8176-4058-4
ISBN 3-7643-4058-4
Typeset by the authors in LaTeX.
Printed and bound by Braun-Brumfield, Inc., Ann Arbor, MI.
Printed in the United States of America

9 8 7 6 5 4 3 2 1

William O. Bray
Časlav V. Stanojević
Editors

# Analysis of Divergence

## Control and Management of Divergent Processes

Birkhäuser
Boston • Basel • Berlin

# IV Asymptotics and Applications 415

# Preface

The $7^{th}$ International Workshop in Analysis and its Applications (IWAA) was held at the University of Maine, June 1-6, 1997 and featured approximately 60 mathematicians. The principal theme of the workshop shares the title of this volume and the latter is a direct outgrowth of the workshop.

IWAA was founded in 1984 by Professor Časlav V. Stanojević. The first meeting was held in the resort complex Kupuri, Yugoslavia, June 1-10, 1986, with two pilot meetings preceding. The Organization Committee together with the Advisory Committee (R. P. Boas, R. R. Goldberg, J. P. Kahne) set forward the format and content of future meetings. A certain number of papers were presented that later appeared individually in such journals as the *Proceedings of the AMS, Bulletin of the AMS, Mathematischen Annalen,* and the *Journal of Mathematical Analysis and its Applications.*

The second meeting took place June 1-10, 1987, at the same location. At the plenary session of this meeting it was decided that future meetings should have a principal theme. The theme for the third meeting (June 1-10, 1989, Kupuri) was Karamata's Regular Variation. The principal theme for the fourth meeting (June 1-10, 1990, Kupuri) was Inner Product and Convexity Structures in Analysis, Mathematical Physics, and Economics. The fifth meeting was to have had the theme, Analysis and Foundations, organized in cooperation with Professor A. Blass (June 1-10, 1991, Kupuri). However, due to unforeseen circumstances, it was postponed. The fifth meeting was later reorganized at the University of Missouri-Rolla, May 17-27, 1995. The proceedings of the third meeting were published by the Mathematical Institute of the Serbian Academy of Sciences and edited by S. Aljancić and Č. V. Stanojević. In spite of all odds, the proceedings of the fourth meeting were published by the University of Novi Sad and edited by O. Hadzić and Č. V. Stanojević. The proceedings of the postponed fifth meeting were published in 1997 in a Special Issue of Mathematica Moravica (University of Cacak) and edited by Č. V. Stanojević, Matt Insall, Andreas R. Blass, and Mališa R. Žižović.

The sixth IWAA was held at the University of Maine, June 15-21, 1992, with the principal theme, Contemporary Aspects in Fourier Analysis. The proceedings of this meeting was published as Vol 157, Lecture Notes in Pure and Applied Mathematics, Marcel Dekker, 1994.

The current volume, based on papers delivered at the seventh meeting provides complete expository and research papers focused on workshop emphasis. Simply put, divergent processes occur throughout analysis and hence papers in this volume touch a large swath of mathematics. As such, this volume addresses a wide mathematical audience, of interest not only to specialists but also graduate students and researchers in related fields.

IWAA is indebted to many institutions and people. Among institutions, we express gratitude to the University of Missouri-Rolla, the University of Maine, the Mathematical Institute of the Serbian Academy of Sciences, the Steklov Mathematics Institute of the Russian Academy of Sciences, the University of Belgrade, the University of Novi Sad, the University of Titograd, and all institutions that supported participants. For the seventh meeting particular gratitude goes out to Mr. B. Stinson and co-workers of the Catering and Conferences Division of the University of Maine; without their work this workshop would not have occurred. Special thanks go to Dr. B. Baishanski, Dr. G. Folland, Dr. O. Hadzić, Dr. M. A. Pinsky, Dr. T. Ostrogorski, Dr. Milan-Brat Tasković, and Dr. M. Žižović. Out of a long list of participants who contributed to the success of IWAA meetings, we would like to express appreciation to R. Askey, O. V. Besov, N. H. Binghan, V. P. Ilyine, B. S. Kasin, W. A. Kirk, M. Mateljevic, B. Milnik, B. Mitjagin, S. M. Nikol'skii, W. Rudin, S. Saitoh, B. S. Stechkin, S. A. Telyakovskii, V. M. Tikhomirov, and B. Walther.

William O. Bray
Časlav V. Stanojević
September, 1998

# Contributors

**Josefina Alvarez** Department of Mathematics
New Mexico State University
LasCruses, NM 88003
alvarez@math.unm.edu

**Bogdan Baishanski** Department of Mathematics
Ohio State University
Colombus, OH 43210-1174
bogdan@math.ohio-state.edu

**Sergey K. Bagdasarov** Department of Mathematics
Ohio State University
Colombus, OH 43210-1174
skbgdsrv@math.ohio-state.edu

**George Benke** Department of Mathematics
Georgetown University
Washington D.C. 20057-0996
gbenke@gumath1.math.georgetown.edu

**William O. Bray** Department of Mathematics and Statistics
University of Maine
Orono, ME 04469
bray@gauss.umemat.maine.edu

**I. Canak** Department of Mathematics and Statistics
University of Missouri-Rolla
Rolla, MO 65401

**Dean A. Carlson** Department of Mathematics
University of Toledo
Toledo, OH 43606
dcarlson@math.utoledo.edu

**Christiane Carton-Lebrun** Mathematique (BATVI)
Universite de Mons-Hainaut
Avenue du Champ de mars 24, B-7000 Mons, Belgium
lebrun@umh.ac.be

**Jeff Connor** Department of Mathematics
Ohio University
Athens, OH 45701
connor@math.sc.edu

**James E. Daly** Department of Mathematics
University of Colorado at Colorado Springs
Colorado Springs, CO 80933
jedaly@math.uccs.edu

**Loukas Grafakos** Department of Mathematics
University of Missouri-Columbia
Columbia, MO 65211
loukas@msindy3.math.missouri.edu

**M. N. Hunter** Department of Mathematics
University of Melbourne
Melbourrne, Australia

**Alex Iosevich** Department of Mathematics
Wright State University
Dayton, OH 45435-0001
iosevich@zara.math.wright.edu

**Nets Hawk Katz** Department of Mathematics
University of Chicago
Chicago, IL

**E. R. Love** Department of Mathematics
University of Melbourne
Melbourrne, Australia

**W. R. Madych** Department of Mathematics
University of Connecticut
Storrs, CT 06269
madych@uconnvm.uconn.edu

**Peter R. Massopust** Department of Mathematics
Sam Houston State University
Houston, TX
prm@galois.shsu.edu

**P. S. Milojević** Department of Mathematics
New Jersey Institute of Technology
Newark, NJ 07102-9938
pemilo@math.njit.edu

**Tatyana Ostrogorski**
Matematicki Institut
Knez Mihajlova 35
11001 Beograd
Serbia

**Harold R. Parks**
Department of Mathematics
Oregon State University
Corvallis, OR 97331
parks@math.orst.edu

**María C. Pereyra**
Department of Mathematics
University of New Mexico
Albuquerque, NM 87131
crisp@math.unm.edu

**Keith L. Phillips**
Department of Mathematics
University of Colorado at Colorado Springs
Colorado Springs, CO 80933

**Mark A. Pinsky**
Department of Mathematics
Northwestern University
Evanston, IL 60208
pinsky@math.nwu.edu

**S. Pilipović**
Institute for Mathematics
University of Novi Sad
Novi Sad, Serbia

**Boris Rubin**
Department of Mathematics
The Hebrew University of Jerusalem
91904 Jerusalem, Israel
boris@math.huji.ac.il

**William H. Ruckle**
Mathematical Sciences
Clemson University
Clemson, SC 29634

**Xiaoping Shen**
Department of Mathematics
University of Wisconsin-Milwaukee
Milwaukee, WI 53201

**Gord Sinnamon**
Department of Mathematics
University of Western Ontario
London, Ontario, Canada N6A5B7
sinnamon@julian.uwo.ca

**Časlav V. Stanojević** Department of Mathematics and Statistics
University ofMissouri, Rolla Bldg 202
Rolla, MO 65409
iwaa@umr.edu

**Vera B. Stanojević** Department of Mathematics
Southwest Missouri State University
Springfield, MO

**A. Stefanov** Department of Mathematics
University of Missouri-Columbia
Columbia, MO 65211

**M. Stojanović** Institute for Mathematics
University of Novi Sad
Novi Sad, Serbia

**Khalifa Trimèche** Faculty of Sciences of Tunis
Department of Mathematics
Campus-1060 Tunis, Tunisia

**W. R. Wade** Department of Mathematics
University of Tennessee
Knoxville, TN 37996

**Gilbert G. Walter** Department of Mathematics
University of Wisconsin-Milwaukee
Milwaukee, WI 53201

**Björn G. Walther** Department of Mathematics
Royal Institute of Technology
SE-10044 Stockholm, Sweden
walther@schrodinger.math.kth.se

**Ahmed I. Zayed** Department of Mathematics
University of Central Florida
Orlando, FL 32816
zayed@pegasus.cc.ucf.edu

# Overview

**William O. Bray**
**Časlav V. Stanojević**

## Introduction

Divergent processes are the *raison d'être* of most of classical and modern analysis. Divergent processes, or rather control and management of such, occur whenever a limit procedure, or combination of, is present. Perhaps the most graphic and at the same time subtle example of this lies in the notion of integration. It is well known that for the Riemann integral of a bounded function on $[a, b]$ to exist, it is necessary as sufficient that the function be continuous almost everywhere. Loosely put, the Riemann integral is basically appropriate for the class of continuous functions on $[a, b]$, $C[a, b]$. Furthermore, the natural notion of convergence for $C[a, b]$ is that of uniform convergence. The latter fact indicates a convergence theorem: one may interchange a sequential limit operation and the Riemann integral, if the limit operation is uniform. Lebesgue's idea, introducing the notion of measurable functions, not only led to a broader class of functions which can be integrated while keeping all operational properties of the Riemann integral, but also led to much richer sequential limit theorems.

Other examples of substantial importance are as follows.

- *Numerical series.* Convergence of an infinite series, $\sum_n a_n$, is meant to be convergence of the sequence of partial sums. In classical analysis, summability methods were introduced to "sum" certain infinite series which do not converge in the ordinary sense or were found useful to speed up convergence of a series which does converge. Examples of summability methods include the Cesáro means or $(C, 1)$ means given by

$$\sigma_N = \frac{1}{N+1} \sum_{n=0}^{N} S_n, \quad S_n = \sum_{k=0}^{n} a_k,$$

  and the Abel means

$$f(r) = \sum_{n=0}^{\infty} a_n r^n, \quad 0 < r < 1.$$

  The series is then said to be Cesáro or Abel summable if

$$\lim_{N\to\infty} \sigma_N = s, \qquad \lim_{r\to 1^-} f(r) = s,$$

respectively. A question of fundamental importance is: How to recover convergence from a summability process? In other words, what condition is needed which when combined with summability yields convergence. Such conditions are known as Tauberian conditions and the following is a well known Tauberian theorem due to Hardy and Littlewood.

If $\sum_n a_n$ is Abel summable to $s$ and $na_n = O(1)$ as $n \to \infty$, then $\sum_n a_n = s$.

The classical perspective of summbility can be found in G. H. Hardy's book *Divergent Series* [6].

- *Fourier series of $L^1$ functions.* On the circle group $\mathbb{T}$, the class of measurable functions which possess Fourier series is $L^1(\mathbb{T})$. Indeed, for $f \in L^1(\mathbb{T})$, we have

$$f \sim \sum_{|n|<\infty} \widehat{f}(n)\, e^{int}, \text{ where}$$
$$\widehat{f}(n) = \frac{1}{2\pi} \int_{\mathbb{T}} f(s)\, e^{-ins} ds.$$

A well known argument based on the uniform boundedness principle shows that there exists $f \in L^1(\mathbb{T})$ whose Fourier series diverges in $L^1$-norm. Moreover, Kolmorgorov provided an example of an $L^1(\mathbb{T})$ function whose Fourier series diverges pointwise everywhere (see [17]). If one replaces convergence in norm or pointwise with $(C, 1)$ summability, then reasonable results can be obtained (see [8]). In this setting, one can then pursue Tauberian theorems to recover convergence in norm or almost everywhere from $(C, 1)$ summability (e.g. [13, 2]). Alternatively, one can replace $L^1(\mathbb{T})$ with a smaller space such as $L^p(\mathbb{T})$, where $p > 1$. Within this framework, we then have convergence almost everywhere and in norm. The latter leads us naturally to the next example.

- *Singular integrals.* Rather than treat the general theory of singular integrals at this point, we wish to mention the classical conjugate function operator

$$f \longrightarrow \widetilde{f}(t) = -\frac{1}{\pi} \int_{-\pi}^{\pi} \frac{f(t+s)}{2\tan\frac{s}{2}}\, ds,$$

where the integral is taken in the principal value sense (provided of course that it exists). A key development in the theory of Fourier series [17] is that the above operator is bounded from $L^p(\mathbb{T})$ to $L^p(\mathbb{T})$ for $p > 1$. This is the basic ingredient in the proof that Fourier series of

$L^p(\mathbb{T})$ functions converge in norm. More generally, for homogeneous Banach spaces satisfying a certain natural condition, convergence in norm is equivalent to the norm boundedness of the conjugate operator. Far reaching generalizations of the conjugate function operator began in the seminal work of Calderón and Zygmund [3].

The above are meant to be key illustrative examples of divergent processes, more appropriately, examples of control and management of divergent processes. Whether the limit process involved be in the theory of infinite series, classical Fourier analysis or its generalizations, singular integrals and their generalization and application, variational or optimal control problems involving singularities, applications of summability ideas to sampling theory, etc., proper control and management of divergence is the heart and soul of obtaining significant results in analysis. This volume is partial testament to this fact.

# Chapter overview

The volume naturally divided into four parts, each with general overall connection with the title and theme. Individual chapters in each part bear relationship to the subtheme as well as to each other. Our exposition here is meant to be brief as an enticement to the reader; more detailed exposition and historical commentary can be found in the introductions to each chapter.

**Part I. Convergence and Summability**

The first chapter authored by Časlav V. Stanojević et. al. gives a new twist to classical Tauberian theory. Specifically, Tauberian conditions for recovery of convergence out of summability, often depend on a parameter and cease to be Tauberian conditions for limiting cases of this parameter. For example, the condition

$$\frac{1}{n}\sum_{k=1}^{n} k^p \left|a_k\right|^p = O(1), \ n \to \infty$$

is a well known Tauberian condition for recovery of convergence of the series $\sum_n a_n$ from Abel summability provided $p > 1$. The limiting case of $p = 1$ is no longer a Tauberian condition. A natural question is: does there exist a regular summability method for which the limiting case is a Tauberian condition? This chapter indicates a positive answer to this question; one application provides a generalized Abel summability method for which the above condition with $p = 1$ is a Tauberian condition.

The second chapter authored by Ruckle takes summability notions into the abstract setting of topological vector spaces, biorthogonal sequences, and natural expansions built out of the latter for functions. Convergence of these expansions is known to fail except in special circumstances. The author's focus is to devise suitable summability methods for the series in which recovery of functions is possible.

Chapter three contributed by W. R. Wade, concerns Cesaro summability of double Fourier-Vilenkin series of unbounded type. In the bounded case, Cesaro summability of such series is well behaved. However, in the unbounded case, examples have been constructed for which divergence of the Cesaro averages at any prescribed point can occur for continuous functions. The focus in this chapter is to provide estimates for the order of growth of Cesaro averages of double Fourier-Vilenkin series of certain unbounded type ($\delta$-quasibounded). The methods center on martingale techniques introduced by Weisz and provide new results in the one dimensional case.

Correction for Gibb's phenomena in wavelet expansions is the subject of Chapter four written by G. Walter and X. Shen. Gibb's phenomena is well known in the theory of Fourier series and orthogonal expansions. Correction of the phenomena in such cases can be achieved via summability processes and other filtering techniques (see [7]). Summability techniques do not work for wavelet expansions because of their structure which comes from multiresolution analysis $\{V_m\}$. In this chapter, the authors overview a scheme using a variation on Abel summability which provides a suitable replacement for the projection operator on $V_m$. The latter has positive kernel, Gibb's phenomena is eliminated, and approximation of positive functions possess the property of being positive. The focus of this paper is the practical implementation of the method.

The chapter written by Love and Hunter (Chapter 5) concerns the classical topic of convergence at a point of generalized Fourier-Legendre expansions on $(-1,1)$ for suitable functions satisfying a Dini condition, i.e. for some $\delta > 0$,

$$\frac{f(x-t)}{x-t} \in L^1(x-\delta, x+\delta).$$

The methods employed do not avail themselves in order to treat convergence at the boundary points $x = \pm 1$. In Chapter six, Pinsky introduces a modified form of the expansions considered by Love and Hunter for which convergence at the end points can be achieved via suitable asymptotic analysis of the Legendre functions.

Based on earlier work of Rubin [12], Chapter seven, coauthored by Bray and Rubin, concerns inversion of the horocycle Radon transform on real hyperbolic space for functions in $L^p$, $p \geq 1$. Contemporary methods for inverting this transform only work for sufficiently smooth functions which decay appropriately at infinity. The basic idea in Chapter seven is to imbed the horocycle transform and its geometric dual transform into an analytic family of operators $R^\alpha$ and $\overset{*}{R}{}^\alpha$ dependent on the parameter $\alpha$ (generalized fractional integral operators) for which a natural inversion formula suggests itself:

$$f = \overset{*}{R}^{n-1-\alpha} R^{\alpha} f.$$

The case in point would correspond to $\alpha = 0$, however the integral defining the dual is in general divergent. Recognizing that the dual above can be re-expressed as a wavelet-like transform and further manipulation leads to and approximate identity (i.e. a summability process) on real hyperbolic space.

The final chapter of Part I, authored by Mark Pinsky, studies the problem of pointwise convergence of eigenfunction expansions corresponding to radial Laplace operator on a ball in Euclidean space and satisfying one of the traditional boundary conditions (Dirichlet, Neumann, or Robin). In contrast to his work in the unbounded domains [11] where the lower bound on the smoothness index for convergence at the origin was $\left[\frac{n-2}{2}\right]$, here the lower bound depends on the type of boundary condition. The front cover of this volume is based on these ideas; it depicts the behavior of the partial Fourier inversion integral of the Fourier transform in three dimensions for the characteristic function of a ball. Note the oscillatory divergence at the origin.

## Part II. Singular Integrals and Multipliers

Chapter nine written by L. Grafakos and A. Stefanov is largely a survey paper on singular integrals with rough kernels. Specifically, given a Calderon-Zygmund convolution singular integral

$$T_{\Omega} f(x) = \lim_{\varepsilon \to 0} \int\limits_{|y|>\varepsilon} \frac{\Omega(y/|y|)}{|y|^n} f(x-y)\, dy,$$

where $\Omega$ is an integrable function on the unit sphere $S^{n-1}$ with mean value zero, a basic question of emphasis is that of finding necessary and sufficient conditions on $\Omega$ so that this operator is bounded from $L^p$ to $L^p$. This chapter briefly considers the classical results laying the foundation for recent results and examples along with some fresh perspectives due to the authors.

Katz and Pereyra authored Chapter 10 which provides a fresh approach to Haar multipliers and dyadic paraproducts based on stopping time arguments. The author's introduce a decaying stopping time and produce new proofs of e.g. Gehring's theorem. After developing and $L^p$ theory of decaying stopping times, they provide weighted inequalities for constant Haar multipliers, dyadic paraproducts, and the dyadic square function. Hence, a new approach to dyadic Littlewood-Paley theory results [5].

Chapter 11 written by Daly and Phillips investigates multipliers and square functions on Vilenkin groups. . The latter includes the additive group of a local field and the $p$-adic numbers ($p = 2$ corresponds with the dyadic group). The chapter begins with a short yet complete historical overview. The author's then focus on the dyadic group and develop a multiplier theorem for $H^p$ which in turn implies that the appropriate square function characterizes $H^p$, resolving a conjecture of Simon.

Alvarez in Chapter 12 considers pseudo-differential operators in the Hormander class, i.e. operators of the form

$$L(f)(x) = \int e^{2\pi i x.\xi} a(x,\xi) \hat{f}(\xi)\, d\xi,$$

where the symbol $a(x,\xi)$ satisfies well known Hormander estimates. The focus of this chapter is on determining the spectrum of such operators as bounded operators on $L^p$. In the translation invariant case, the spectrum is characterized in terms of the symbol. A holomorphic functional calculus based on translation invariant pseudo-differential operators is made precise as an application.

Chapter 13 written by Iosovich investigates singular integrals on curves and surfaces in the context of Orlicz spaces. The methodology is based on scaling techniques and the results proved yield as corollories several known results.

Oscillatory Fourier integrals is the topic of Chapter 14 authored by Walther. These are operators of the form

$$(S^a f)[x](t) = \frac{1}{(2\pi)^n} \int_{\mathbb{R}^n} e^{i(x\xi + t|\xi|^a)} \widehat{f}(\xi)\, d\xi.$$

In the case where $a = 2$, the above are solutions to the time-dependent free Schrödinger equation; when $a = 1$ we have solutions of the classical wave equation. The focus of this paper is on maximal estimates and weighted $L^2$ estimates of $(S^a f)$ using techniques based on spherical harmonic decompositions and asymptotics of Bessel functions. The origins of this work are to be found in the works of Carlson [4], Stein [14], and Strichartz [15].

The final chapter in Part II, co-authored by Alvarez and Carton-Lebrun, investigates natural $n$-dimensional versions of the Hilbert transform in the context of distributions. The focus is to characterize spaces of distributions on which these convolution operators are defined.

**Part III. Integral Operators and Functional Analysis.**

The first paper of Part III is authored by K. Trimèche and is set in the realm of certain hypergroups commonly called Chébli-Trimèche hypergroups. The latter are measure algebras on $\mathbb{R}_+$ on which the convolution structure and the structure of a certain differential operator blend in a natural manner. In this framework there is a natural harmonic analysis which includes e.g. Hankel transforms and Jacobi transforms as special cases. The first part of the paper surveys the underlying theory including various asymptotic expansions. The second half of the paper has focus to apply harmonic analysis on these hypergroups and develop a theory of wavelet like transforms.

Chapters 17 and 18 bear some relation to one another. Sinnamon in Chapter 17 develops inequalities for the operator that integrates a function over a difference of two dilations of a starshaped set in $R^n$. Applications include developing Hardy type inequalities for a variety of integral operators. On the other hand, Ostrogorski in Chapter 18 has focus on the structure of the weights for which Hardy's inequality is valid. While Muckenhoupt [9] has characterized all such weights, in Chapter 18 a broad class of weights is constructed using the notion of regularly bounded functions.

In early work of Landau, Hadamard, Kolmorgorov , and others, the problem of sharp inequalities for intermediate derivatives of function in Sobolev space was studied extensively. Bagdasarov in Chapter 19 pursues such investigations for certain natural generalizations of Sobolev space originally found in the work of Nikol'skii [10].

Massopust in Chapter 20 investigates perturbed Laplace equation where the perturbation is continuous in the angular variables. The focus is characterization of solutions via their relation to hyperfunctions of Lions and Magenes. Key techniques lie in generalized spherical harmonics.

Solvability results for nonresonant semi-abstract semilinear equations involving nonlinear perturbations is the focus of Chapter 21 authored by Milojevic. While these results lie in the framework of abstract functional analysis, the author provides several concrete applications to boundary value problems for the nonlinear parabolic and hyperbolic equations.

Chapter 22 authored by Connor gives a survey of the theory of statistical convergence. The latter notion has its origin in Fourier analysis, ergodic theory, number theory and convergence in density. The latter part of the paper gives new proofs of recent results using techniques from functional analysis.

**Part IV. Asymptotics and Applications.**

Chapters 23 and 24 deal with topics from the calculus of variations. In Chapter 23, Carlson considers variational problems on unbounded intervals which satisfy certain pointwise state constraints. Motivation for such problems lies in mathematical economics. Alternatively, in Chapter 24 Parks considers variational problems with the structure of finding the surface which minimizes an integral with singular integrand. Techniques draw on differential geometry of minimizing surfaces.

Chapter 25 authored by Benke is concerned with the asymptotics of certain exponential sums which arise in modeling the sensitivity of a phased array antenna of omnidirectional sensors. The techniques draw on the theory of generalized Rudin-Shapiro systems developed by the author [1].

Madych in Chapter 26 gives an exposition of a version of the spline summability method for recovering multivariate band limited functions from discrete lattice samples which uses polyharmonic splines. Fourier transform techniques are the primary tool and the author provides computational algorithms and an example in the realm of image processing. Chapter 27 authored by Zayed is also cast in the spline setting. Here, cardinal B-splines are used to construct orthonormal sets and frames in Paley-Wiener space, the space of band limited functions.

Chapter 28 authored by Baishanski is a survey of related problems he has been interested in for many years. Principally, the problem originates in complex analysis and concerns the problem of boundedness in norm of powers of an analytic function on the disk. The twist presented concerns the relationship of this classical problem to generalized central limit theorems.

The final chapter in Part IV and in the volume is authored by Pilipović and Stojanović. It concerns development of a theory of asymptotics for certain classes of generalized functions. The origin of this vein of thought is in the work of Vladimirov [16]. Applications are concerned with the asymptotics of solutions of nonlinear partial differential equations and integral equations.

*References*

[1] G. Benke, *Generalized Rudin-Shapiro systems,* J. of Fourier Anal. and Appl. **1** (1994), 87–101.

[2] W. O. Bray, Časlav V. Stanojević, *Tauberian $L^1$-convergence classes of Fourier series II,* Math. Ann. 269 (1984) p469-486.

[3] A. P. Calderón and A. Zygmund, *On the existence of certain singular integrals*, Acta Math. 88 (1952) p. 85-139.

[4] L. Carleson, *Some Analytic problems to related to Statistical Mechanics,* Euclidean harmonic analysis (Proc. Sem., Univ. Maryland, College Park, Md., 1979), Lecture Notes in Math. 779, Springer, Berlin, 1980 pp. 5–45.

[5] R. E. Edwards, *Littlewood-Paley Theory and Multipliers,* Ergebnisse der Mathematik und ihrer Grenzgebiete Vol 90, Springer Verlag (1977).

[6] G. H. Hardy, *Divergent Series,* Oxford Claredon Press (1949).

[7] A. J. Jerri, *The Gibb's Phenomena in Fourier Analysis, Splines and Wavelet Approximation,* Kluwer Academic Publishers, Boston (1998).

[8] Y. Katznelson, *An Introduction to Harmonic Analysis,* Dover Publ. (1976).

[9] G. Muckenhoupt, *Hardy's inequality with weights*, *Studia Math.* 44 1972, 31-38.

[10] S. M. Nikol'skii, *La série de Fourier d'une fonction dont be module de continuité est donné*, Dokl. Akad. Nauk SSSR 52 (1946), p.p. 191–194.

[11] M. A. Pinsky, *Pointwise Fourier inversion and related eigenfunction expansions,* Communications in Pure and Applied Mathematics, 47(1994), 653-681.

[12] B. Rubin, *Inversion of Radon transforms using wavelet transforms generated by wavelet measures,* Math. Scand. (to appear).

[13] Časlav V. Stanojević, *Tauberian conditions for the $L^1$-convergence of Fourier series,* Trans. Amer. Math. Soc. 271 (1982) p237-244.

[14] E. M. Stein, *Harmonic Analysis: real-variable methods, orthogonality, and oscillatory integrals,* Princeton Mathematical Series, No. 43, Monographs in Harmonic Analysis, III, Princeton University Press, Princeton, NJ, 1993.

[15] R. Strichartz, *Restrictions of Fourier transforms to quadratic surfaces and decay of solutions of wave equations,* Duke Math. J. 44 (1977), pp. 705–714.

[16] Vladimirov V. S., Droshshinov Yu. N., Zavyalov B. I., T*auberian Theorems for Generalized Functions*, Kluwer Academic Publishers, Boston, 1988.

[17] A. Zygmund, *Trigonometric Series,* Cambridge University Press, $2^{nd}$ edition (1959).

# Part I

# Convergence and Summability

# Chapter 1

# Tauberian theorems for generalized Abelian summability methods

**Časlav V. Stanojević**
**I. Canak**
**V. B. Stanojević**

*ABSTRACT. Tauberian conditions for the recovery of convergence out of summability, depending on a parameter cease to be Tauberian conditions for some limiting values of the parameter. This situation motivates the following question. Are there some more general summability methods for which the limiting case would become a Tauberian condition for the behavior of $\{S_n(a)\}$? A general summability method is designed whose special cases lead to generalized Abel's summability methods. The Tauberian theorems for those Abel's summability methods contain as a particular case the answer to the above question.*

## 1.1 Introduction

In the beginning conditions for the recovery of convergence of the series $\left\{\sum_{k=0}^{n} a_k\right\}$ out of its Abel's summability were conditions on the order of magnitude of the Taylor coefficients $\{a_n\}$ of a function

$$f(a) = f(a,x) = \sum_{n=0}^{\infty} a_n x^n, \; |x| < 1$$

analytical in the unit disc. From the original Tauber [1] condition $na_n = o(1)$, $n \to \infty$, via Littlewood's [2] substantial generalization $na_n = O(1)$, $n \to \infty$, all the way up to the condition

$$V_n(|a|,p) = \frac{1}{n}\sum_{k=1}^{n} k^p |a_k|^p = O(1), \; n \to \infty, \; p > 1 \tag{1.1}$$

conjectured by Hardy and Littlewood [3] and proved by Szász [4], the recovery of the convergence of the series $\{S_n(a)\}$ was based on certain information about the behavior of the Taylor coefficients $\{a_n\}$, restricting essentially the growth of $\{S_n(a)\}$. However all above conditions (and other similar) imply that

$$\lim_{\substack{N>M\to\infty \\ N/M\to 1}} (S_N(a) - S_M(a)) = 0, \tag{1.2}$$

i.e. the slow oscillation of $\{S_n(a)\}$, introduced by Schmidt [5]. This led to the generalized Littlewood [6] Tauberian theorem asserting that if the limit

$$\lim_{x\to 1-o} f(a,x)$$

exists and (1.2) holds then the series $\{S_n(a)\}$ converges to

$$\lim_{x\to 1-o} f(a,x).$$

The condition (1.1) is of considerable interest to our study. It does not only imply (1.2) but for $p \in (1,2]$ implies that $g(a) = g(a,x) = \sum_{n=0}^{\infty} \frac{a_n}{n+1} x^n$ belongs to $H^q$, $\frac{1}{p} + \frac{1}{q} = 1$, [7], and as noticed by Rényi [8] for $p = 1$ is no longer a Tauberian condition for the recovery of convergence of the series out of its Abel's summability. Rényi [8] also observed that if

$$V_n(|a|, 1) = O(1), \; n \to \infty$$

is replaced by somewhat stronger condition that

$$\lim_n V_n(|a|, 1) \text{ exists}$$

then one can recover convergence of the series $\{S_n(a)\}$ out of its Abel's summability. This situation motivates the following questions:

(i) Are there some generalized Abel's summability methods of $\{S_n(a)\}$ such that $V_n(|a|, 1) = O(1)$, $n \to \infty$ or its generalization, entail some limiting information about $\{S_n(a)\}$?

(ii) What are the conditions for recovering Abel's summability out of those general summability methods, regular with respect to Abel's summability?

The following variant of Borel [9] summability methods indicates a direction for search to the answers to the questions (i) and (ii). Let

$$P(x) = \sum_{n=0}^{\infty} p_n x^n$$

be an analytical function on $(0,1)$ such that $P$ is not polynomial on $(0,1)$ and $P(x) \to \infty$, $x \to 1 - o$. If

$$\lim_{x\to 1-o} \frac{\sum_{n=0}^{\infty} S_n(a) p_n x^n}{P(x)}$$

exists, then $\{S_n(a)\}$ is $(K,P)$-summable.

Since the convergence of $\{S_n(a)\}$ implies the existence of $\lim_{x\to 1-o} f(a,x)$, method $(K,P)$ is regular. Notice also that for $P(x)=\frac{1}{1-x}=\delta_1(x)$

$$\frac{\sum\limits_{n=0}^{\infty} S_n(a)p_n x^n}{P(x)} = (1-x)\sum_{n=0}^{\infty} S_n(a)x^n = f(a,x).$$

Instead $\{S_n(a)\}$ we may consider

$$\sigma_n^{(1)}(a)=\sigma_n(a)=\sigma_n(S(a))=\frac{1}{n+1}\sum_{k=0}^{n} S_k(a)=\frac{1}{n+1}\sum_{k=0}^{n}\sigma_k^{(0)}(S(a)),$$

or any other sequence $\{A_n(S(a))\}$ generated by $\{S_n(a)\}$. Thus in general we define

$$K(f,P,A(S(a)),x)=\frac{1}{P(x)}\sum_{n=0}^{\infty} A_n(S(a))p_n x^n.$$

For $P(x)=\delta_1$ and $A_n(S(a))=\sigma_n(a)$

$$\begin{aligned} K(f,\delta_1,\sigma,x) &= \frac{\sum\limits_{n=0}^{\infty}\sigma_n(a)x^n}{\delta_1(x)} \\ &= (1-x)\sum_{n=0}^{\infty}\sigma_n(a)x^n \\ &= \sum_{n=0}^{\infty}(\sigma_{n+1}(a)-\sigma_n(a))x^n. \end{aligned}$$

The Taylor coefficients of $K(f,\delta_1,\sigma)$ are $\{\sigma_{n+1}(a)-\sigma_n(a)\}=-\Delta\sigma$.

If $\{S_n(a)\}$ is $(K,\delta_1,\sigma)$-summable, i.e. if

$$\lim_{x\to 1-o}\frac{\sum\limits_{n=0}^{\infty}\sigma_n(a).1.x^n}{\delta_1(x)}$$

exists, and if $V_n(|a|,1)=O(1)$, $n\to\infty$ then the series $\{S_n(a)\}$ is $(C,1)$-summable. Thus $V_n(|a|,1)=O(1)$, $n\to\infty$ is a Tauberian condition for $(C,1)$-convergence of $\{S_n(a)\}$.

For the convergence recovery of $\{S_n(a)\}$ out of $(K,\delta_1,\sigma)$-summability we need Tauberian condition: $\{V_n(a,1)\}$ is a bounded slowly oscillating sequence. (From any slowly oscillating sequence $\{L(n)\}$ one can construct a bounded slowly oscillating sequence: $\{\exp(iL(n))\}$ in the complex case and $\{\sin(L(n))\}$ in the real case.)

However if $\{V_n(a,1)\}$ is slowly oscillating and

$$V_n(|V|,1) = O(1),\ n \to \infty$$

holds we can recover the convergence of $\{S_n(a)\}$ out of its $(K, \delta_1, \sigma)$-summability.

In the next section we shall consider the general situation motivated by the above examples and remarks.

## 1.2 A General Summability Method

The answer to the questions (i) and (ii) in the introduction depends on the choice of a summability method that generalizes the Abel summability method and provides an algorithm for construction of analytical functions in some interval $[\alpha_0, 1)$, $\alpha_0 \in (0,1)$, whose Taylor coefficient generates limiting processes. We need also to establish corresponding Tauberian condition for the recovery of convergence out of that general summability. Hence we have to find a parallel algorithm for those Tauberian conditions.

Our approach depends on a method of integral transformations of the space $\mathcal{A}$ of all functions analytical in the unit disc, or unit interval.

For $f(a) \in \mathcal{A}$, the following denotation will be used

$$f(a) = f(a,x) = \sum_{n=0}^{\infty} a_n x^n,\ |x| < 1$$

$$S_n(a) = \sum_{k=0}^{n} a_k,\ \sigma_n^{(m)}(S(a)) = \sigma_n^{(m)}(a) = \frac{1}{n+1}\sum_{k=0}^{n} \sigma_k^{(m-1)}(a),$$

for integers $m \geq 1$

$$\left(\sigma_n^{(1)}(a) = \sigma_n(a),\ \sigma_n^{(0)}(a) = S_n(a)\right).$$

To describe the class $\Phi$ of kernels of the integral transforms of functions in $\mathcal{A}$, we need the following properties of functions $\varphi$ in $\Phi$:

1. there exists a number $\alpha_0 = \alpha_0(\Phi) \in (0,1)$ such that every $\varphi \in \Phi$ is analytical in $(\alpha_0, 1)$.

2. for every $\varphi \in \Phi$

$$\varphi(x) \to \infty,\ x \to 1 - o.$$

3. every $\varphi \in \Phi$ is zero-free in $[\alpha_0, 1)$.

4. $\frac{\varphi_m(x)}{\varphi_{m-1}(x)} = o(1),\ x \to 1-o,\ m \geq 1$, where $\varphi_0 = \varphi$ and

$$\varphi_m(x) = \int_{\alpha_0}^{x} \varphi_{m-1}(t)\, dt.$$

Then for every $f(a) \in \mathcal{A}$ and $\varphi \in \Phi$ we define

$$\begin{aligned} M_\varphi(f(a)) &= M(f(a), \varphi) = M(f(a), \varphi, x) \\ &= \begin{cases} \dfrac{\int_{\alpha_0}^{x} f(a,t)\varphi(t)dt}{\varphi_1(x)}, & \text{if } x \neq \alpha_0 \\ \lim_{x\to\alpha_0} M(f(a), \varphi, x) = f(\alpha_0), & \text{if } x = \alpha_0 \end{cases}, \end{aligned}$$

or in general

$$M(f(a), \varphi_m, x) = \begin{cases} \dfrac{\int_{\alpha_0}^{x} f(a,t)\varphi_m(t)dt}{\varphi_{m+1}(x)}, & \text{if } x \neq \alpha_0 \\ \lim_{x\to\alpha_0} M(f(a), \varphi_m, x) = f(\alpha_0), & \text{if } x = \alpha_0 \end{cases}$$

for $m \geq 0$.

If for some $\varphi_m \in \Phi$

$$\lim_{x\to 1-o} M(f(a), \varphi_m, x) \tag{1.3}$$

exists, $\{S_n(a)\}$ is $M_{\varphi_m}$-summable to the limit above. Since

$$\varphi'_{m+1}(x) = \varphi_m(x),$$

it is clear that the existence of the limit

$$\lim_{x\to 1-o} f(a,x)$$

implies the existence of the limit (1.3). That is, the summability method $M_\varphi$ is a regular summability method.

An important subclass of $\Phi$ are functions

$$\delta_m(x) = \frac{1}{(1-x)^m},\quad m > 0$$

and $\alpha_0 = 0$.

For $m = 2$, we have

$$\begin{aligned} M(f(a), \delta_2, x) &= \frac{1-x}{x} \int_0^x \frac{f(a,t)}{(1-t)^2} dt = K(f(a), \delta_1, \sigma, x) \\ &= \sum_{n=0}^{\infty} \left(\sigma_{n+1}^{(1)}(a) - \sigma_n^{(1)}(a)\right) x^n \end{aligned}$$

the function $M(f(a), \delta_2)$ is analytical in $(0,1)$ and it is denoted by

$$f(-\Delta\sigma) = f(-\Delta\sigma, x) = A^{(1)}(f(a)) = A^{(1)}(f(a), x).$$

Notice that $A^{(1)}(f(a))$ defines a regular summability method $(A, 1)$, because the Abel summability method $(A, o) = A^{(0)}(f(a)) = f(a) = f(a, x)$ implies $(A, 1)$. An elementary answer to the questions (i) and (ii) is provided by the next theorem.

If the $\lim_{x \to 1-o} A^{(1)}(f(a), x)$ exists then $\{S_n(a)\}$ is $(A, 1)$-summable to that limit.

**Theorem 1.1** *Let $\{S_n(a)\}$ be $(A, 1)$-summable. If*

$$V_n(|a|, 1) = O(1), \; n \to \infty \tag{1.4}$$

*then $\{S_n(a)\}$ is $(A, o)$-summable.*

The Tauberian condition (1.4) (recall that (1.4) is not a Tauberian condition for the recovery of convergence of $\{S_n(a)\}$ out of $(A, o)$-summability) can be generalized all the way up to $\{\sigma_n(a)\}$ being slowly oscillating. Closer to the classical situation is the following theorem.

**Theorem 1.2** *Let $\{S_n(a)\}$ be $(A, 1)$-summable and let $\{V_n(a, 1)\}$ be bounded and slowly oscillating. Then the series $\{S_n(a)\}$ converges to its $(A, 1)$-sum.*

In the next section we shall study more specific $M_\varphi$-summability methods in particular those related to various generalizations of $(A, o)$-summability method.

We close this section with some results regarding $M_\varphi$-summability methods.

**Theorem 1.3** *Let $\{S_n(a)\}$ be $M_\varphi$-summable and let*

$$f'(a, x) = o\left(\frac{\varphi(x)}{\varphi_1(x)}\right), \; x \to 1 - o \tag{1.5}$$

*If $\{V_n(a, 1)\}$ is slowly oscillating then the series $\{S_n(a)\}$ converges to its $M_\varphi$-sum.*

**Proof.** From the identity

$$M(f(a), \varphi, x) = f(a, x) - \frac{\int\limits_{\alpha_0}^{x} f'(a, t)\varphi_1(t)dt}{\varphi_1(x)}$$

and (1.5) it follows that $\{S_n(a)\}$ is $(A, o)$-summable to its $M_\varphi$-sum. The $(A, o)$-summability implies $(A, 1)$-summability. Hence $\{V_n(a, 1)\}$ is $(A, o)$-summable and since $\{V_n(a, 1)\}$ is slowly oscillating we have from the generalized Littlewood theorem that $\{V_n(a, 1)\}$ converges. Therefore $\{\sigma_n^{(1)}(a)\}$

is slowly oscillating and consequently $\{S_n(a)\}$ is slowly oscillating. Recalling again the generalized Littlewood theorem we conclude that the series $\sum_{n=0}^{\infty} a_n$ converges to its $M_\varphi$-sum. ■

A more general result can be obtained by replacing (1.5) with

$$M(f'(a), \varphi_1, x) = o\left(\frac{\varphi_1(x)}{\varphi_2(x)}\right), \quad x \to 1 - o.$$

The composition of two $M_\varphi$-summability methods can be define as follows.

Let $\psi, \varphi \in \Phi$. Then

$$(M_\psi \circ M_\varphi)(f(a), x) = M(M_\varphi(f(a)), \psi, x) = \frac{\int_{\alpha_0}^{x} M(f, \varphi, t)\psi(t)dt}{\psi_1(x)}.$$

Now Theorem 1.3 can be rewritten in the following way.

**Theorem 1.4** *Let $\{S_n(a)\}$ be $M_\psi \circ M_\varphi$-summable and let*

$$f'(a, x) = o\left(\frac{\varphi(x)}{\varphi_1(x)}\right), \quad x \to 1 - o \tag{1.6}$$

$$M'(f(a), \varphi, x) = o\left(\frac{\psi_1(x)}{\psi_2(x)}\right), \quad x \to 1 - o. \tag{1.7}$$

*If $\{V_n(a, 1)\}$ is slowly oscillating then the series $\sum_{n=0}^{\infty} a_n$ converges to its $(M_\psi \circ M_\varphi)$-sum.*

**Proof.** The iterated identity from the proof of Theorem 1.3 yields

$$\begin{aligned}
(M_\psi \circ M_\varphi)(f(a), x) &= M_\varphi(f(a), x) - \frac{1}{\psi_1(x)} \int_{\alpha_0}^{x} M'(f(a), \varphi, t)\psi_1(t)dt \\
&= f(a, x) - \frac{1}{\varphi_1(x)} \int_{\alpha_0}^{x} f'(t)\varphi_1(t)dt \\
&\quad - \frac{1}{\psi_1(x)} \int_{\alpha_0}^{x} M'(f(a), \varphi, t)\psi_1(t)dt.
\end{aligned}$$

The conditions (1.6) and (1.7) imply that $\{S_n(a)\}$ is $(A, o)$-summable. The rest of the proof follows the lines of the proof of Theorem 1.3. ■

## 1.3 Generalized Abel's Summability Methods

The generalizations of $(A, o)$-summability method are based on $M_{\delta_2}$ -summability method or more precciously on $M_{\delta_2} \circ K_{\delta_1}$-summability method. In section 2 we introduced $(A, 1)$-summability of the series $\{S_n(a)\}$ as a generalization of $(A, o)$-summability method. Similarly we can define $(A, 2)$-summability of $\{S_n(a)\}$ as $(A, 1)$-summability of $A^{(1)}(f(a))$. In general, for any integer $m \geq 1$

$$\begin{aligned} A^{(m)}(f(a)) &= A^{(m)}(f(a), x) \\ &= A^{(1)}\left(A^{(m-1)}(f(a), x)\right) = A^{(1)}\left(A^{(m-1)}(f(a))\right). \end{aligned}$$

Notice that $A^{(m)}(f(a))$ defines a regular summability method $(A, m)$.

For the $(C, 1)$-convergence recovery of the series $\{S_n(a)\}$ out of its $(A, 1)$-summability we will introduce Tauberian conditions in the following theorem.

**Theorem 1.5** *Let*

$$\lim_{x \to 1-o} \int_0^x \frac{f(a,t)}{1-t} dt \tag{1.8}$$

*exists. If* $\left\{ \sum_{k=0}^{n} \frac{S_k(a)}{k+1} \right\}$ *is slowly oscillating then the series* $\{S_n(a)\}$ *is* $(C, 1)$-*summable to zero.*

**Proof.** Since $\lim_{x \to 1-o} \int_0^x \frac{f(a,t)}{1-t} dt$ exists, the series $\left\{ \sum_{k=0}^{n} \frac{S_k(a)}{k+1} \right\}$ is $(A, o)$-summable and since $\left\{ \sum_{k=0}^{n} \frac{S_k(a)}{k+1} \right\}$ is slowly oscillating, by the generalized Littlewood theorem, it converges. Set $S_n^*(a) = \sum_{k=0}^{n} \frac{S_k(a)}{k+1}$. It follows that

$$S_n^*(a) - \sigma_n^{(1)}(S^*(a)) = \sigma_n^{(1)}(S(a)). \tag{1.9}$$

This completes the proof. ∎

The next three corollaries are obtained by replacing the slow oscillation of $\left\{ \sum_{k=0}^{n} \frac{S_k(a)}{k+1} \right\}$ by some stronger condition.

**Corollary 1.6** *Let*

$$\lim_{x \to 1-o} \int_0^x \frac{f(a,t)}{1-t} dt$$

*exists. If for some* $p > 1$

$$\frac{1}{n+1}\sum_{k=0}^{n}|S_k(a)|^p = O(1),\ n \to \infty \tag{1.10}$$

*then* $\{S_n(a)\}$ *is* $(C,1)$*-summable to zero.*

**Proof.** (1.10) implies that $\{S_n^*(a)\}$ is slowly oscillating. ∎

**Corollary 1.7** *Let* $\left\{\sum\limits_{k=0}^{n}\frac{S_k(a)}{k+1}\right\}$ *be* $(C,1)$*-summable. Then*
*(i)* $\sigma_n^{(1)}(S(a)) = O\left(\sum\limits_{k=0}^{n}\frac{S_k(a)}{k+1}\right)$, $n \to \infty$
*(ii)* $S_n(a) = O(1),\ n \to \infty$
*implies that* $\{S_n(a)\}$ *is* $(C,1)$*-summable to zero.*

**Proof.** (i) follows from (1.10).
(ii) The series $\{S_n^*(a)\}$ being $(C,1)$-summable is $(A,o)$-summable. The condition $S_n(a) = O(1),\ n \to \infty$ implies that $\{S_n^*(a)\}$ is slowly oscillating. ∎

**Corollary 1.8** *Let*

$$\lim_{x\to 1-o}\int_0^x \frac{f(a,t)}{1-t}dt$$

*exists. If*

$$V_n(S(a),1) = O(1),\ n \to \infty \tag{1.11}$$

*then* $\{S_n(a)\}$ *is* $(C,1)$*-summable to zero.*

**Proof.** (1.11) implies that $\{S_n^*(a)\}$ is slowly oscillating. Recall that

$$f(-\Delta\sigma, x) = A^{(1)}(f(a),x) = \sum_{n=0}^{\infty}\left(\sigma_{n+1}^{(1)}(a) - \sigma_n^{(1)}(a)\right)x^n.$$

From the above identity it is clear that the statements (i) $\{S_n(a)\}$ is $(A,1)$-summable and (ii) $\{\sigma_n^{(1)}(S(a))\}$ is $(A,o)$-summable are equivalent. ∎

For $f(a) \in \mathcal{A}$, the following denotation will be used

$$V_n^{(m)}(a,1) = \frac{1}{n+1}\sum_{k=0}^{n}V_k^{(m-1)}(a,1)$$

for integers $m \geq 1$

$$V_n^{(o)}(a,1) = V_n(a,1) = \frac{1}{n+1}\sum_{k=0}^{n}ka_k.$$

Notice that

$$\sigma_n^{(m)}(S(a)) - \sigma_n^{(m+1)}(S(a)) = V_n^{(m)}(a,1)$$

for integers $m \geq 0$.

From the last identity we obtain that

$$A^{(m)}(f(a),x) - A^{(m+1)}(f(a),x) = f(\Delta V^{(m)}(a,1),x).$$

It is straightforward that if $\{S_n(a)\}$ is $(A, m+1)$-summable then it is $(A,m)$-summable provided that $\{V_n^{(m)}(a,1)\}$ is $(A,o)$-summable.

The analogue to the generalized Littlewood theorem for $(A,m)$-summability method is given in the next theorem. If $\{\sigma_n^{(m)}(S(a))\}$ is slowly oscillating, then $\{S_n(a)\}$ is $(C,m)$ slowly oscillating.

**Theorem 1.9** *Let $\{S_n(a)\}$ be $(A,m)$-summable. If $\{S_n(a)\}$ is $(C,m)$ slowly oscillating then $\{S_n(a)\}$ is $(C,m)$-summable to its $(A,m)$-sum.*

The next three corollaries are obtained by replacing the slow oscillation of $\{\sigma_n^{(m)}(a)\}$ by some stronger conditions.

**Corollary 1.10** *Let $\{S_n(a)\}$ be $(A,m)$-summable. If for some integer $m \geq 1$, $\{V_n^{(m-1)}(a,1)\}$ is bounded then $\{S_n(a)\}$ is $(C,m)$-summable to its $(A,m)$-sum.*

**Proof.** From the identity

$$\sigma_n^{(m)}(S(a)) = \sum_{k=1}^{n} \frac{V_k^{(m-1)}(a,1)}{k}$$

and $V_n^{(m-1)}(a,1) = O(1),\ n \to \infty$ it follows that $\left\{\sigma_n^{(m)}(S(a))\right\}$ is slowly oscillating. ∎

**Corollary 1.11** *Let $\{S_n(a)\}$ be $(A,m)$-summable. If for some $p > 1$*

$$\frac{1}{n+1}\sum_{k=1}^{n} |V_k^{(m-1)}(a,1)|^p = O(1),\ n \to \infty \qquad (1.12)$$

*then the series $\{S_n(a)\}$ is $(C,m)$-summable to its $(A,m)$-sum.*

**Proof.** Rewritten form of (1.12)

$$\frac{1}{n+1}\sum_{k=1}^{n} k^p \left|\frac{V_k^{(m-1)}(a,1)}{k}\right|^p = O(1),\ n \to \infty$$

shows that $\left\{\sum_{k=1}^{n} \frac{V_k^{(m-1)}(a,1)}{k}\right\}$ is slowly oscillating. ∎

**Corollary 1.12** *Let $\{S_n(a)\}$ be $(A,m)$-summable. If for some integer $m \geq 1$*

$$V_n^{(0)}\left(V^{(m-1)}(a,1),1\right) = O(1),\ n \to \infty \tag{1.13}$$

*the series $\{S_n(a)\}$ is $(C,m)$-summable to its $(A,m)$-sum.*

**Proof.** From the condition (1.13) it follows that

$$\sigma_n^{(m)}(S(a)) = \sum_{k=1}^{n} \frac{\beta_k}{k} - \sum_{k=1}^{n} \frac{k+1}{k^2}\beta_{k+1}$$

for some bounded sequence $\{\beta_n\}$. Since $\{\beta_n\}$ is bounded, the last identity implies that $\{\sigma_n^{(m)}(S(a))\}$ is slowly oscillating. ■

For the generalized Abel's summability method $(A,m)$ we have the generalized Littlewood theorem. To recover the convergence of the series $\{S_n(a)\}$ out of its $(A,m)$-summability method we have to assume an extra condition on $\{S_n(a)\}$. As seen in the next theorem weakening the summability method and strengthening the Tauberian condition makes the series $\{S_n(a)\}$ convergent to its $(A,m)$-sum.

**Theorem 1.13** *For some integer $m \geq 1$ let $\{S_n(a)\}$ be $(A,m)$-summable. If $\{S_n(a)\}$ is bounded slowly oscillating then the series $\{S_n(a)\}$ converges to its $(A,m)$-sum.*

**Proof.** If $\{S_n(a)\}$ is $(A,m)$-summable, $\lim\limits_{x\to 1-o} \sum\limits_{n=0}^{\infty} \Delta\sigma_n^{(m)}(a)x^n$ exists. Since $S_n(a) = O(1)$, $n \to \infty$, the series $\{\sigma_n^{(m)}(S(a))\}$ is slowly oscillating. By Theorem 1.9 the series $\{S_n(a)\}$ is $(C,m)$-summable to its $(A,m)$-sum. $(C,m)$-summability implies $(A,o)$-summability. Therefore $\{S_n(a)\}$ converges to its $(A,m)$-sum by the generalized Littlewood theorem. ■

In Theorem 1.13 the Tauberian condition can be replaced by a weaker condition.

**Theorem 1.14** *For some integer $m \geq 1$ let $\{S_n(a)\}$ be $(A,m)$-summable. If $\{V_n^{(o)}(a,1)\}$ is bounded and slowly oscillating then the series $\{S_n(a)\}$ converges to its $(A,m)$-sum.*

**Proof.** Assume that $\{V_n^{(o)}(a,1)\}$ is bounded. Then for each integer $m \geq 1$, $\{V_n^{(m)}(a,1)\}$ is bounded. Therefore $\{\sigma_n^{(m)}(S(a))\}$ is slowly oscillating. Since $\{S_n(a)\}$ is $(A,m)$-summable we obtain that $\{S_n(a)\}$ is $(C,m)$-summable by Theorem 1.9. The condition $V_n^{(o)}(a,1) = O(1)$, $n \to \infty$ implies that $\{\sigma_n^{(1)}(S(a))\}$ is slowly oscillating. Hence $\{S_n(a)\}$ is slowly oscillating. Recalling again the generalized Littlewood theorem we conclude that the series $\{S_n(a)\}$ converges to its $(A,m)$-sum. ■

In the generalized Littlewood theorem the Tauberian condition can be replaced by the condition $\{V_n^{(o)}(a,1)\}$ being slowly oscillating as follows.

**Theorem 1.15** *Let $\{S_n(a)\}$ be $(A, o)$-summable. If $\{V_n^{(o)}(a,1)\}$ is slowly oscillating then the series $\{S_n(a)\}$ converges to its $(A, o)$-sum.*

**Proof.** Since $(A, o)$-summability implics $(A, 1)$-summability. $\{V_n^{(o)}(a,1)\}$ is $(A, o)$-summable. By the generalized Littlewood theorem $\{V_n^{(o)}(a,1)\}$ converges. This implies that $\{S_n(a)\}$ is $(C, 1)$ slowly oscillating. Since $\{V_n^{(o)}(a,1)\}$ is slowly oscillating $\{S_n(a)\}$ is slowly oscillating. Again by the generalized Littlewood theorem we conclude that $\{S_n(a)\}$ converges. ■

The $(C, 1)$-sum of $\left\{V_k^{(m-1)}(a,1)\right\}_{k=0}^{n}$, for some $m \geq 1$, is denoted by $V_n^{(m)}(a,1)$. In the next theorem it will be shown that for any integer $m \geq 0$, assuming that the condition

$$V_n^{(k)}(a,1) = O(1), \ n \to \infty$$

for some $0 \leq k \leq m-1$ is enough to recover $(A, o)$-summability of $\{S_n(a)\}$ out of its $(A, m)$-summability.

**Theorem 1.16** *Let $\{S_n(a)\}$ be $(A, m)$-summable. If for some $k$, $0 \leq k \leq m-1$,*

$$V_n^{(k)}(a,1) = O(1), \ n \to \infty \tag{1.14}$$

*then*
*(i) $\{S_n(a)\}$ is $(A, o)$-summable.*
*(ii) if $\left\{S_n(a) - \sigma_n^{(k+1)}(S(a))\right\}$ is slowly oscillating then $\{S_n(a)\}$ converges to its $(A, m)$-sum.*

**Proof.**

(i) The condition (1.14) implies that $\{S_n(a)\}$ is $(C, m)$ slowly oscillating. By Theorem 1.9, $\{S_n(a)\}$ is $(C, m)$- summable to its $(A, m)$-sum. It then is $(A, 0)$-summable.

(ii) Since $V_n^{(k)}(a,1) = O(1)$, $n \to \infty$ for some $0 \leq k \leq m-1$, $\left\{V_n^{(j)}(a,1)\right\}$ is slowly oscillating for

$$j = k+1, \ldots, m-1.$$

$(A, o)$-summability of $\{S_n(a)\}$ implies that $\left\{V_n^{(j)}(a,1)\right\}$ is $(A, o)$-summable for $j = k+1, k+2, \ldots, m-1$. Combining what we have and observing

$$\begin{aligned} S_n(a) &= \sigma_n^{(m)}(S(a)) + \sum_{j=0}^{m-1} V_n^{(j)}(a,1) \\ &= \sigma_n^{(m)}(S(a)) + \sum_{j=0}^{k} V_n^{(j)}(a,1) + \sum_{j=k+1}^{m-1} V_n^{(j)}(a,1) \end{aligned}$$

and

$$\sum_{j=0}^{k} V_n^{(j)}(a,1) = S_n(a) - \sigma_n^{(k+1)}(S(a))$$

we obtain that $\{S_n(a)\}$ is slowly oscillating. By the generalized Littlewood theorem we conclude that $\{S_n(a)\}$ converges.

■

Notice that Theorem 1.2 is a corollary to Theorem 1.16.

So far we recovered the convergence of the series out of its generalized Abel's summability method and a Tauberian condition corresponding to that method. However the asymptotic behavior of the series $\{S_n(a)\}$ and our generalized Abel's summability method can be related. Appell [10] proved that if $S_n(a) \sim An^\gamma$, $\gamma \geq 0$, $A$ is constant

$$\lim_{x \to 1-o} (1-x)^\gamma f(a,x)$$

exists.

For instance, for $\gamma = 2$ we conclude from Appell's theorem that the limit $\lim\limits_{x \to 1-o} (1-x)^2 f(a,x)$ exists provided that $S_n(a) \sim An^2$, $A$ is constant.

Let $F(a,x) = (1-x)^2 f(a,x)$. Then the series of the partial sums corresponding to $F(a)$ is $(A,1)$-summable. That is, the limit

$$\lim_{x \to 1-o} M(F(a), \delta_2, x) = \lim_{x \to 1-o} \frac{1-x}{x} \int_0^x f(a,t)dt$$

exists. Hence it follows that $\left\{\frac{a_n}{n+1}\right\}$ is $(A,o)$-summable. Assuming the slow oscillation of $\left\{\frac{a_n}{n+1}\right\}$ we obtain that $\left\{\frac{a_n}{n+1}\right\}$ converges.

*References*

[1] A. Tauber, Ein Satz aus der Theorie der Unendlichen Reihen, Monatsh. Math. Phys. 8 (1897) p273-277.

[2] J. E. Littlewood, The converse of Abel's theorem on power series, Proc. London Math. Soc. 9(2) (1911) p434-448

[3] G. H. Hardy and J. E. Littlewood, Tauberian theorems concerning power series and Dirichlet's series whose coefficients are positive, Proc. London Math. Soc. 13(2) (1913/14) p174-191.

[4] O. Szász , Verallgemeinerung eines Littlewoodschen Satzes uber Potenzrielhen, J. Lond. Math. Soc. 3 (1928) p254-262.

[5] R. Schmidt, Uber divergente Folgen und lineare Vittelbildungen, Math. Z. 22 (1924) p89-152.

[6] T. Vijayaraghavan, A Tauberian theorem, J. London Math. Soc.1 (1926) p113-120.

[7] Vera B. Stanojević, Tauberian conditions and structure of Taylor and Fourier coefficients, Publications de L'institut Mathematique Nouvelle Série 58(72) (1995) p101-105 (Slobodan Aljančić memorial volume).

[8] A. Rényi, On a Tauberian theorem of O. Szász, Acta Univ. Szeged Sect. Sci. Math. 11 (1946) p119-123.

[9] G. H. Hardy, Divergent Series, Oxford University Press (1949).

[10] P. Appell, Sur certaines séries ordonnés par rapport aux puissance d'une variable, Comptes Rendus 87 (1878) p689-692.

# Chapter 2

# Series summability of complete biorthogonal sequences

**William H. Ruckle**

*ABSTRACT. A complete biorthogonal sequence in a topological vector space $X$ is a double sequence $(x_i, f_i)$ such that (1) each $x_i \in X$ and each $f_i \in X^*$ the dual space of $X$; (2) the closed linear span of $\{x_i\}$ is dense in $X$, (3) $f_i(x_j) = 1$ if $i = j$ and is 0 otherwise, and (4) $f_i(z) = 0$ for each $i$ only when $z = 0$. If $x \in X$ and $f \in X^*$ the numerical series*

$$\sum_i f_i(x) f(x_i)$$

*will converge to $f(x)$ if (a) $x$ is a finite linear combination of $(x_i)$; (b) $f$ is a finite linear combination of $(f_i)$ or (c) $(x_i)$ is a Schauder basis of $X$. In other cases (1) may not converge, or if it does it may not converge to $f(x)$; but there may be some method of series summability that will associate the "correct" sum, i. e. $f(x)$, to the sequence $(f_i(x) f(x_i))$. In order to determine conditions under which there is such a method of summability we study four sequence spaces associated with the biorthogonal sequence $(x_i, f_i)$, namely $S = \{(f_i(x)) : x \in X\}$, that represents the space $X$, $S^f = \{(f(x_i)) : f \in X^*\}$, that represents the dual space $X^*$ of $X$, $S(S)$, the series space of $S$ that consists of the linear span of all sequences of the form $st$ where $s \in S$ and $t \in S^f$, the multiplier space $M(S)$ consisting of all sequences $u$ such that $us \in S$ whenever $s \in S$. We will discuss how conditions on these four spaces result in summability properties of the biorthogonal sequence $(x_i, f_i)$ and also on the topology of the space $X$.*

## 2.1 Introduction

Suppose $(x_i, f_i)$ is a complete biorthogonal sequence in a topological vector space $X$. This means that (1) each $x_i \in X$ and each $f_i \in X^*$ the dual space of $X$; (2) the closed linear span of $\{x_i\}$ is dense in $X$, (3) $f_i(x_j) = 1$ if $i = j$ and is 0 otherwise, and (4) $f_i(z) = 0$ for each $i$ only when $z = 0$. If

$x \in X$ and $f \in X^*$ the numerical series

$$f(x) \sim \sum_i f_i(x) f(x_i) \tag{2.1}$$

will converge to $f(x)$ if (a) $x$ is a finite linear combination of $(x_i)$; (b) $f$ is a finite linear combination of $(f_i)$ or (c) $(x_i)$ is a Schauder basis of $X$. In other cases (1) may not converge, or if it does it may not converge to $f(x)$; but there may be some method of series summability that will associate the "correct" sum, i. e. $f(x)$, to the sequence $(f_i(x) f(x_i))$. In [4] Dieudonné considered various problems concerning this situation for ordinary convergence. In [5] we introduced the concept of a sum on a topological vector space of sequences and the concept of a sum space and in [16] we showed how the concepts of sums and sum spaces could be used to discuss convergence properties of biorthogonal sequences in Banach spaces. In [7] and other papers we indicated the dual nature of sum spaces and multiplier algebras; see e. g. [4]. Sum spaces were discussed further in the papers [6], [7], [8] and [16]. In the paper [8] we showed how results of Köthe-Toeplitz [1], [2] and Garling [3] could be generalized to a more general and abstract setting. Application of sum spaces to multipliers of sequence spaces, generalized sectional convergence and other topics in summability theory appear in [1], [3], [5], and [6].

The purpose of this paper is to study sum spaces with the $\beta\varphi$ topology and then extend results from [16] for biorthogonal sequences in a Banach space to norming biorthogonal sequences in a locally convex topological vector space. We defined the $\beta\varphi$ topology on a topological vector space of sequences in [8] and studied it further in the monograph [10] and the paper [11].

## 2.2 Preliminaries

This section describes ideas and terminology we shall use in the paper: biorthogonal sequences and their representation by sequence spaces, sequence space terminology, the beta phi topology on a sequence space, sums, sum spaces, multiplier algebras and convergence properties.

### 2.2.1 Biorthogonal Sequences

**Definition 2.1** *Suppose $X$ is a topological vector space with dual space $X^*$. A double sequence $(x_i, f_i)$ with each $x_i \in X$ and each $f_i \in X^*$ is called a* biorthogonal sequence for $X$ *if*

1. *the closed linear span of $\{x_i\}$ is dense in $X$,*
2. *$f_i(x_j) = 1$ if $i = j$ and is $0$ otherwise, and*

*3.* $f_i(z) = 0$ *for each* $i$ *only when* $z = 0$.

**Definition 2.2** *A biorthogonal sequence* $(x_i, f_i)$ *is called* norming *if the topology on* $X$ *is determined by seminorms of the form*

$$p_A(x) = \sup\{|g(x)| : g \in A\} \tag{2.2}$$

*where* $A$ *is an* $X$*-bounded subset of* $[f_i]$*, the linear span of the functionals* $(f_i)$.

If $(x_i, f_i)$ is a norming biorthogonal sequence for the barrelled topological vector space $X$ then the topology of $X$ is determined by the collection of seminorms $p_A$ given by (2.2) as $A$ ranges over all $X$ bounded subsets of $[f_i]$.

### 2.2.2 Sequence Spaces

The K space terminology that we use to describe topological sequence spaces was originally developed by Karl Zeller; see e. g. [15] and elaborated by his successors.

We shall consider a *sequence* to be a real or complex function $s$ on the set $\mathsf{N}$ of natural numbers. The value of the sequence $s$ at the index $i$ will be written $s(i)$ rather than $s_i$ so we can reserve subscripts for sequences of sequences. By a *sequence space* we mean a linear space $s$ of real or complex sequences. For $i \in \mathsf{N}$ we denote by $e_i$ the *ith coordinate sequence*, that is for which $e_i(i) = 1$ and $e_i(j) = 0$ for $i \neq j$. A *K space* is a locally convex space of sequences $s$ such that each *coordinate functional* $E_i$ defined by $E_i(s) = s(i)$ is continuous. In this paper we shall also assume that a K space contains each $e_i$ unless we specifically say otherwise. A K space that is a complete metric linear space is called an *FK* space, and one that is a Banach space is called a *BK* space. A few common K spaces, to which we shall refer, are:

$\omega$ the space of all sequences.

$\varphi$ the space of all sequences which are eventually 0, i. e. $s \in \varphi$ if there is a finite subset $D$ such that $s(i) = 0$ for $i \notin D$.

$l^1$ the space of all sequences $s$ such that $\sum_i |s(i)| < \infty$.

$c_0$ the space of all sequences $s$ which converge to 0.

$c$ the space of all sequences $s$ such that $\lim_i s(i)$ exists.

$l^\infty$ the space of all bounded sequences.

$cs$ the space of all sequences $s$ such that $\sum_{n=1}^{\infty} s(n)$ converges.

If the linear span $\varphi$ of the set $\{e_i\}$ is dense in a K space $s$, then we say $s$ has *AD*. For $s$ a K space $s^f$ denotes the space of all sequences $(f(e_n))$ as $f$ ranges over $s^*$ the dual space of $s$. If $s$ has AD then $s^f$ is a representation of the $s^*$ by means of the correspondence $f \leftrightarrow (f(e_n))$.

### *2.2.3 The Beta-Phi Topology on a Sequence Space*

There is a natural duality between every sequence space $S$ and the space $\varphi$ given by the equation

$$\langle x, y\rangle = \sum_j x(j)y(j), \quad x \in S, y \in \varphi.$$

Since $y \in \varphi$, the series contains only finitely many non-zero terms. A subset $B$ of $\varphi$ is called *S-bounded* if

$$p_B(x) = sup\{|\langle x, y\rangle| : y \in B\} < \infty$$

for each $x \in S$; i. e., if the set

$$B^\circ = \{z \in \omega : |\langle z, y\rangle| \leq 1 \text{ for each } y \in B\}$$

absorbs points of $S$. We denote by $\beta(S)$ the collection of all $S$-bounded subsets of $\varphi$. A sequence space $S$ is said to have a $\varphi$ topology if the topology on $S$ is determined by a family of seminorms $\{p_B\}$ as $B$ ranges over a subcollection of $\beta(S)$.

The $\beta\varphi$ topology is the locally convex topology on a sequence space $S$ determined all of $\beta(S)$. The $\beta\varphi$ topology makes $S$ a K space because each singleton $\{e_i\}$ is an $S$-bounded subset of $\varphi$ and $|E_i(x)| = |\langle x, e_i\rangle|$. Most familiar sequence spaces bear the $\beta\varphi$ topology, e. g., $\varphi$ with the strongest locally convex topology, the $l^p$ spaces $1 \leq p \leq \infty$, and $\omega$ with the topology of coordinatewise convergence; but not $l^p(0 < p < 1)$ with the non-locally convex complete metric linear topology. Every absorbing absolutely convex subset of a K space $S$, that is closed in the relative topology of $\omega$, is a barrel. It follows that if $S$ is barrelled, its topology must be at least as strong as the $\beta\varphi$ topology. If $S$ has its $\beta\varphi$ topology, let $Y$ consist of all sequences $y$ such that $p_B(y) < \infty$ as $B$ ranges over all $S$ bounded subsets of $\varphi$. The $\beta\varphi$ topology on $Y$ is determined by the collection of all semi norms $p_B$, and $Y$ is complete in this topology. See Proposition 3.3 of [6]. If $\widetilde{S}$ denotes the closure of $S$ in the space $Y$ then $\widetilde{S}$ is the unique K space completion of $S$ and the $\beta\varphi$ topology on $\widetilde{S}$ induces the $\beta\varphi$ topology on $S$. See the discussion found in Section 2 of [11].A subspace $T$ of a sequence space $S$ is called a $\beta\varphi$ subspace of $S$ if the space $T$ determines the same bounded subsets of $\varphi$ as does $S$ so that the relative topology on $T$ induced by the $\beta\varphi$ topology on $S$ is the same as the $\beta\varphi$ topology on $T$.

### 2.2.4 Biorthogonal Sequences and Sequence Spaces

If $S$ is an AD sequence space then the double sequence $(e_i, E_i)$ is a complete biorthogonal sequence for $S$. On the other hand, if $(x_i, f_i)$ is a complete biorthogonal sequence for any topological vector space $X$ then the space $S(x_i, f_i)$ consisting of all sequences $s_x = (f_i(x))_i$ is (algebraically) isomorphic to $X$ under the correspondence $\mathsf{T}(x) = (f_i(x))$ because each $f_i$ is linear and each $f_i(x)$ is zero only for $x = 0$. If we define a topology on $S(x_i, f_i)$ by letting $U$ be open in $S(x_i, f_i)$ if and only if $\mathsf{T}^{-1}(U)$ is open in $X$, then the operator $\mathsf{T}$ becomes a topological isomorphism, and $S(x_i, f_i)$ becomes a K space that has AD. Thus we can without loss of generality assume that a complete biorthogonal sequence $(x_i, f_i)$ for a topological vector space is the complete biorthogonal sequence $(e_i, E_i)$ in the associated sequence space $S(x_i, f_i)$. The biorthogonal sequence $(x_i, f_i)$ is norming if and only if $S(x_i, f_i)$ has a $\varphi$ topology. If $(x_i, f_i)$ is norming and $X$ is barrelled then $S(x_i, f_i)$ has its $\beta\varphi$ topology.

### 2.2.5 Multiplier Algebras, Sums, and Sum Spaces

If $S$ is a sequence space then by $M(S)$ we denote the *multiplier algebra* of $S$, the space of all sequences $u$ such that $ux = (u(i)\,x(i))$ is in $S$ whenever $x$ is in $S$. If $S \subset T$ then every $T$-bounded subset of $\varphi$ is $S$-bounded so that the inclusion from $S$ into $T$ is continuous with respect to their $\beta\varphi$ topologies. More generally every row finite matrix between spaces with their $\beta\varphi$ topologies is continuous [11]. In this paper we shall use a special case of this result namely that if $u \in M(S)$ then the linear operator $x \to ux$ is continuous from $S$ into $S$.

**Definition 2.3** *A* sum *on a K space $S$ is a continuous linear functional $E$ on $S$ such that $E(e_n) = 1$ for each $n$.*

Thus the space $S$ has a sum on it if and only if the space $S^f$ contains the sequence $e$ that consists of all $1's$. If $S$ is an FK, BK or $\beta\varphi$ space with a sum on it then $M(S) \subset S^f$ because if $u \in M(S)$ then the functional $f$ defined by $f(x) = E(ux)$ is in $S^*$ and $f(e_n) = u(n)$ for each $n$. If $M(S) = S^f$, that is if the functional $E$ determines the correct "sum" for each sequence $(f(e_n)\,x(n))$, namely $f(x)$ then we make the following definition.

**Definition 2.4** *A K space $S$ is called a* sum space *if $M(S) = S^f$.*

The spaces $\varphi, l^1, cs$ and the space of $C_1$ series summable sequences are all examples of sum spaces; but there are many other examples such as the space of rapidly decreasing sequences, [10]. We shall generally be interested in sums on AD sequence spaces, because these are the sums that truly extend finite summation.

### 2.2.6 Convergence Properties of Sequence Spaces

The following property for sequence spaces is a natural generalization of the property AK of Zeller [15].

**Definition 2.5** *Suppose* $(u_n)$ *is a sequence in* $\varphi$ *such that* $\lim_n u(j) = 1$ *for each* $j$. *A K space* $S$ *is said to have the property* AK$(u_n)$ *if* $\lim_n u_n s = s$ *in the topology of* $S$ *for all* $s \in S$.

When $u_n$ is the sequence $(e_1 + e_2 \ldots + e_n)_{n=1}^{\infty}$ then AK$(u_n)$ is the original property AK. In [11] we studied convergence properties with respect to a sequence of matrices, but in this paper we shall restrict ourselves to diagonal mappings. For a sequence $(u_n)$ in $\varphi$ with $\lim_n u(j) = 1$ for all $j$, the sequence $l^1(u_n)$ consists of all scalar sequence of the form

$$\sum_n t(n) u_n \text{ as } t \text{ ranges over } l^1.$$

The following is a special case of Theorem 5.7 of [11] that we state here for reference.

**Proposition 2.1** *Suppose* $S$ *is a* $\beta\varphi$ *space and* $(u_n)$ *is a sequence in* $\varphi$ *such that* $\lim_n u(j) = 1$ *for all* $j$. *The following statements are equivalent:*

1. *$S$ has AK$(u_n)$;*
2. *$S$ has AD and $l^1(u_n) S \subset \widetilde{S}$;*
3. *$[l^1(u_n) S]$ is a dense $\beta\varphi$ subspace of $\widetilde{S}$;*
4. *$[l^1(u_n) S]$ is a dense barrelled subspace of $\widetilde{S}$ having AK$(u_n)$.*

## 2.3 Sums and Sum Spaces

In this section we shall derive some facts about sums and sum spaces. The main result for a sum space $S$ is Theorem 2.4 that shows that the collection $\beta(S)$ of $S$ bounded subsets of $\varphi$ has a useful algebraic property. In Theorem 2.6 we show how this property implies that the sum on a sum space can be given by a positive series to sequence summability method.

### 2.3.1 Sums

**Proposition 2.2** *A K space* $S$ *with a* $\varphi$ *topology admits a sum if and only if there is an* $S$ *bounded sequence* $(u_n)$ *in* $\varphi$ *such that* $\lim_n u_n(i) = 1$ *for each* $i$.

**Proof.** If $S$ admits a sum then the sequence $e = (1, 1, \dots) \in S^f$ so that by Proposition 3.3 of [8] there is an $S$ bounded subset $B$ of $\varphi$ such that $e \in B^{(\varphi)(\omega)}$, the closure of $B$ in the space $\omega$ with the topology of coordinatewise convergence. Since $\omega$ is a metrizable space there is a sequence $(u_n)$ in $B$ that converges coordinatewise to $e$. The converse also follows from Proposition 3.3 of [8]. ■

The following result was proved for BK spaces in [7] using the Closed Graph Theorem.

**Proposition 2.3** *If a K space $S$ having the $\beta\varphi$ topology admits a sum then $M(S) \subset S^f$.*

**Proof.** If $t \in M(S)$ then the mapping $s \to ts$ is continuous from $S$ into $S$ since it is represented by a diagonal matrix [11]. Thus the functional $f(s) = E(ts)$ is continuous on $S$. Therefore, the sequence $(f(e_i)) = t \in S^f$. ■

The following example shows that admitting a sum does not force much structure on a sequence space.

**Example 2.1** *Suppose $S$ is any BK space (Banach K space) with norm $\| \ \|$. Let $\|E_n\|$ denote the norm of the coordinate functional $E_n$ in the dual space of $S$. The space $T = S / (n^2 \|E_n\|)$ that consists of all sequences $t(n)$ such that $(n^2 \|E_n\| t(n)) \in S$ with the norm $\|t\|' = \|(n^2 \|E_n\| t(n))\|$ is a BK space isometric to $S$. Each member of $T$ is in $l^1$ because for each $n$ we have*

$$\begin{aligned} |t(n)| &= \frac{|E_n((n^2 \|E_n\| t(n)))|}{n^2 \|E_n\|} \leq \frac{\|E_n\| \, \|(n^2 \|E_n\| t(n))\|}{n^2 \|E_n\|} \\ &\leq \frac{\|(n^2 \|E_n\| t(n))\|}{n^2}. \end{aligned}$$

*Since $T \subset l^1$, $T$ has a sum on it, but has properties no different from $S$.*

### *2.3.2 Sum Spaces*

**Theorem 2.4** *For $S$ an AD, K space with the $\beta\varphi$ topology we have 1.⇒ 2.⇒ 3.⇒4. and all of the statements are equivalent if $S$ is complete.*

1. *$S^f \subset M(S)$;*
2. *For each $t \in S^f$ and $A \in \beta(S)$ we have $tA \in \beta(S)$;*
3. *For each $s \in S$ and $A \in \beta(S)$ the set $sA$ is bounded in the space $S$;*
4. *If $A$ and $B$ are in $\beta(S)$ then the set $AB \, (\equiv \{uv : u \in A, \ v \in B\})$ is in $\beta(S)$.*

**Proof.** 1. $\Rightarrow$ 2. If $s \in S$, and $t \in S^f$ then $ts \in S$ so

$$\sup\{|\langle ts, u\rangle| : u \in A\} < \infty$$

because $A \in \beta(S)$. But $\langle ts, u\rangle = \langle s, tu\rangle$ for each $u \in A$ so that

$$\sup\{|\langle s, tu\rangle| : u \in A\} = \sup\{|\langle s, v\rangle| : v \in tA\} < \infty$$

which implies $tA \in \beta(S)$.

2. $\Rightarrow$ 3. If $f$ is a continuous linear functional on $S$ then the sequence $t = (f(e_n))$ is in $S^f$ by definition. Since $tA \in \beta(S)$ we have

$$\begin{aligned} \sup\{|\langle s, v\rangle| : v \in tA\} &= \sup\{|\langle st, u\rangle| : u \in A\} \\ &= \sup\{|\langle su, t\rangle| : u \in A\} < \infty \end{aligned}$$

but since $su \in \varphi$ we have

$$\langle su, t\rangle = \sum_n s(n)\, u(n)\, f(e_n) = f\left(\sum_n s(n)\, u(n)\, e_n\right) = f(su).$$

Therefore, the set $sA$ is weakly bounded, hence strongly bounded in the space $S$.

3. $\Rightarrow$ 4. If $s \in S$ then

$$\begin{aligned} \sup\{|\langle s, uv\rangle| : u \in A,\ v \in B\} &= \sup\{|\langle su, v\rangle| : u \in A,\ v \in B\} \\ &= \sup\{p_B(su) : u \in A\} < \infty \end{aligned}$$

because $sA$ is bounded in $S$.

4. $\Rightarrow$ 1. (when $S$ is complete) If $t = (f(e_n)) \in S^f$ there is a set $A \in \beta(S)$ such that $t$ is in the pointwise closure of $A$. Thus we can find a sequence $(t_n)$ in $A$ such that $\lim_n t_n(i) = t(i)$ for each $i$. If $B \in \beta(S)$ then $AB \in \beta(S)$ so that for $v \in B$ we have

$$|\langle ts, v\rangle| = \lim_n |\langle t_n s, v\rangle| = \lim_n |\langle s, t_n v\rangle| \le p_{AB}(s).$$

This means that

$$p_B(ts) \le p_{AB}(s)$$

implying that $ts$ is in the space $Y = \{w : p_C(w) < \infty \text{ for all } C \in \beta(S)\}$. If we consider the $\beta\varphi$ topology on $Y$, we know $S$ is a $\beta\varphi$ subspace of $Y$ that is closed because $S$ is complete. The diagonal mapping $s \to ts$ takes $\varphi$ into $\varphi$ and is continuous from $S$ into $Y$. Therefore, it takes $S$ into $S$ because $S$ has AD and is complete. ■

**Corollary 2.5** *A complete AD K space S with the $\beta\varphi$ topology is a sum space if and only if (a) there is a sum on S; (b) if $(u_n)$ and $(v_n)$ are S bounded sequences in $\varphi$ then $(u_n s_n)$ is also S bounded.*

**Proof.** Condition (a) is necessary since the sequence $e = (1, 1, 1, \ldots)$ is always part of $M(S)$. Condition (b) is a special case of property 4. of Theorem 2.4.

To show conditions (a) and (b) are sufficient we first observe that if $S$ has a sum on it then $M(S) \subset S^f$ by the observation following Definition 2.3. On the other hand, if $M(S)$ is a proper subspace of $S^f$ then by 4.⇒1 of the previous theorem we know there are $S$ bounded subspaces $A$ and $B$ of $\varphi$ such that $AB$ is not $S$ bounded. Thus there are sequences $(u_n)$ in $A$ and $(v_n)$ in $B$ such that $(u_n v_n)$ is not $S$ bounded. ∎

The following example exhibits numerous proper subspaces $S$ of $l^1$ that are sum spaces with the relative topology of $l^1$ and thus all have $M(S) = S^f = l^\infty$. These spaces are dense in $l^1$ and thus not complete. We also show how to construct a space $S$ that has properties 2., 3. and 4. of Theorem 2.4 but does is not a sum space since $M(S)$ is properly contained in $S^f$.

**Example 2.2** *If $S$ is one of the spaces $l^p$, $0 < p < 1$, $\cap_{p>0} l^p$, or $l^1 \cap \omega_p$, $0 < p \leq 1$ where*

$$\omega_p = \left\{ s : s \in \omega, \ \lim_n \frac{|\{i : i \leq n \text{ and } s(i) \neq 0\}|}{n^p} \right\}$$

*then $S$ is a $\beta\varphi$ subspace of $l^1$ that is barrelled in the relative topology of $l^1$. Moreover, $M(S) = l^\infty = S^f$ so each of the spaces $S$ is a sum space. To construct a space having properties 2. 3. and 4. of the preceding Theorem but not property 1. let $S$ be the linear span of the space $l^{\frac{1}{2}}$ and the sequence $(1/n^{1.1})$. Then $S$ inherits properties 2., 3., and 4. from $l^1$ since $\beta(S) = \beta\left(l^{\frac{1}{2}}\right) = \beta(l^1)$. However, $M(S)$ is smaller than $l^\infty$ because $(1/n^{.1})\ (1/n^{1.1}) = (1/n^{1.2})$ is not in $S$. For suppose it were possible that for each $n$*

$$\frac{1}{n^{1.2}} = c\left(\frac{1}{n^{1.1}}\right) + u(n) \ \text{ with } u \in l^{\frac{1}{2}}.$$

*Then by multiplying each side of the equation by $n^{1.1}$ we obtain*

$$\frac{1}{n^{.1}} = c + n^{1.1} u(n)$$

*Since $(1/n^{1.1})$ is not in $l^{\frac{1}{2}}$ the sequence $(n^{1.1} u(n))$ cannot be bounded away from 0 so it has a subsequence that converges to 0. This means that c has to be 0. But this would imply $(1/n^{1.2})$ is in $l^{\frac{1}{2}}$, a contradiction. This construction shows $l^1$ contains many dense barrelled proper subspaces that are not sum spaces, as well as many that are.*

**Example 2.3** *We can use Theorem 2.4 to help construct an increasing sequence of sum spaces that begins with the space cs. For each $n = 1, 2, \ldots$*

*let*

$$u_n = \sum_{m=1}^{n} e_n.$$

*For each* $k = 1, 2, \ldots$ *let*

$$B_k = \left\{ v : v \in \varphi \text{ and } v = \sum_{n=1}^{\infty} t(n)\, u_{kn},\ \sum_n |t(n)| \leq 1 \right\}.$$

*Since* $u_{kp}u_{kq} = u_{k\min(p,q)}$ *it follows that* $B_k B_k \subset B_k$. *Let* $S(B_k)$ *consist of all sequences* $s$ *such that*

$$p_k(s) = \sup\{|\langle s, t\rangle| : t \in B_k\} < \infty.$$

*With the seminorm* $p_k$ *and the seminorms* $|E_n(s)| = |s(n)|$, $n = 1, 2 \ldots,$ $S(B_k)$ *is an FK space. The space closure of* $\varphi$ *in* $S(B_k)$ *denoted by* $S(B_k)^\circ$ *is a sum space with its* $\beta\varphi$ *topology because every* $S(B_k)^\circ$ *bounded subset of* $\varphi$ *is contained in a finite union of multiples of the sets* $B_k$ *and* $\{e_n\}$, $n = 1, 2, \ldots$ *and these subsets of* $\varphi$ *satisfy property 4. of Theorem 2.4. It is easy to see that* $S(B_1)^\circ = cs$, *and*

$$S(B_k)^\circ = \left\{ v : \sum_{n=1}^{\infty} \sum_{j=1}^{k} v((n-1)k + j) \text{ converges} \right\}.$$

*The spaces* $S(B_{2^k})^\circ$, $k = 0, 1 \ldots$ *form a strictly increasing sequence of sum spaces that begins with* $S(B_1) = cs$. *The union of these spaces consists of all sequences* $s$ *such that*

$$\sum_{n=0}^{\infty} \left( \sum_{i=1}^{2^k} s(n2^k + i) \right)$$

*converges for some* $k$.

**Theorem 2.6** *If* $S$ *is a complete AD sum space with the* $\beta\varphi$ *topology, then there is a sequence* $(u_n)$ *in* $\varphi$ *such that for each* $x \in S$ *we have* $u_n x \to x$ *in the topology of* $S$*; and we can assume that* $u_n(j) \geq 0$ *for each* $n, j$.

**Proof.** By 2.2 there is an $S$ bounded sequence $(v_n)$ in $\varphi$ that converges coordinatewise to the sequence $e$. If $u_n(j) = v_n(j)^2 \geq 0$ then by 4. of Theorem 2.4 $(u_n)$ is also $S$ bounded, and $(u_n)$ converges coordinatewise to $e$. The set $B = \{u_n\}$ is in $\beta(S)$. Let $x$ be any member of $S$, let $A \in \beta(S)$ and let $\varepsilon > 0$ be given. The set $AB$ is also in $\beta(S)$. Let $w \in \varphi$ be such that

$$p_A(x - w) \text{ and } p_{AB}(x - w) \text{ are } < \varepsilon/3.$$

Since $w \in \varphi$ it follows that $u_n w \to w$ in the $\beta\varphi$ topology on $S$. Let $N$ be such that

$$p_A(w - u_n w) < \varepsilon/3$$

for $n > N$. Then we have

$$p_A(x - u_n x) \leq p_A(x - w) + p_A(w - u_n w) + p_A(u_n w - u_n x).$$

We have already arranged for the first of the two summands to be less that $\varepsilon/3$. The third summand is also less than $\varepsilon/3$ because

$$p_A(u_n w - u_n x) \leq p_{AB}(w - x).$$

■

One implication of the preceding theorem is that every sum space is contained in the series to sequence summability field of a non negative matrix. All of the Cesaro series to sequence means determine sum spaces, but there are positive Nörlund means that do not. See [1].

**Corollary 2.7** *If $S$ is an AD sum space with the $\beta\varphi$ topology, then $S$ is barrelled.*

**Proof.** This is an application of Theorem 2.6 and the special case of Theorem 5.7 of [11] described in Proposition 2.1. The space $S$ has AK$(u_n)$ for the diagonal mappings $(u_n)$. Thus by 2.1 it follows that $\left[l^1(u_n) S\right]$ is a dense barrelled subspace of $\widetilde{S}$. Since $S$ is dense in $\widetilde{S}$ we have $M(S) = S^f$ and $M(S) \subset M\left(\widetilde{S}\right)$. Because $\widetilde{S}$ admits a sum it follows that $M\left(\widetilde{S}\right) \subset \widetilde{S}^f = S^f = M(S)$ so that $M(S) = M\left(\widetilde{S}\right)$. By 2.1 applied to $\widetilde{S}$ it follows that $l^1(u_n)\widetilde{S} \subset \widetilde{S}$ so that $l^1(u_n) \subset M(S)$. Therefore, $S$ contains the barrelled space $\left[l^1(u_n) S\right]$ so $S$ itself is barrelled. ■

## 2.4 Inclusion Theorems

In this section we shall derive results pertaining to a norming complete biorthogonal sequence $(x_i, f_i)$ in a barrelled topological vector space $X$. By identifying the space $X$ with the associated sequence space $S_X = \{(f_i(x)) : x \in X\}$ we can assume that we are always dealing with the biorthogonal sequence $(e_i, E_i)$ in a barrelled AD, space that is barrelled in the $\beta\varphi$ topology. The first three results give information about weak convergence while the fourth Theorem 21.5 describes a natural condition under which the space has AK$(u_n)$ for a nonnegative sequence $(u_n)$ in $\varphi$.

**Theorem 2.8** *Suppose $S$ is an AD space with the $\beta\varphi$ topology and $T$ is a $\beta\varphi$ space on which there is a sum $E$. If $t_f = (f(e_n))$ is a sequence in $S^f$ such that $t_f s \in T$ for each $s \in S$ then we have*

$$E(t_f s) = f(s)$$

*for all $s \in S$.*

**Proof.** The diagonal mapping $s \to t_f s$ is continuous from $S$ into $T$ since both spaces have their $\beta\varphi$ topologies. Therefore, the linear functional $s \to E(t_f s)$ is continuous on the AD space $S$. For $s \in \varphi$ we have

$$E(t_f s) = \sum_j t_f(j)\, s(j) = \sum_j s(j)\, f(e_j) = f\left(\sum_j s(j)\, e_j\right) = f(s)$$

since we are dealing with a finite sum. Since $\varphi$ is dense in $S$ the equality must hold for all $s \in S$. ■

We can apply the above theorem along with Theorem 2.6 to obtain the following result.

**Corollary 2.9** *Suppose $S$ is an AD space with the $\beta\varphi$ topology and $T$ is a sum space with the $\beta\varphi$ topology. If $t_f = (f(e_n))$ is a sequence in $S^f$ such that $t_f s \in T$ for each $s \in S$ then there is a sequence $(u_n)$ in $\varphi$ such that $u_n(j) \geq 0$ for each $j$ such that*

$$\lim_n \sum_j u_n(j)\, t_f(j)\, s(j) = f(s)$$

*for each $s \in S$.*

The preceding corollary shows that if for an AD space with the $\beta\varphi$ topology we have $SS^f$ contained in a $\beta\varphi$ sum space $T$ then there is a nonnegative sequence $(u_n)$ in $\varphi$ such that $\lim_n u_n s$ converges weakly to $s$ for each $s \in S$. The following theorem shows we can, in fact, obtain strong convergence.

**Theorem 2.10** *Suppose $S$ is an AD space with the $\beta\varphi$ topology and $T$ is a sum space with the $\beta\varphi$ topology. If for each $s \in S$ and $t \in S^f$ we have $st \in T$ then there is a sequence $(u_n)$ in $\varphi$ such that $u_n(j) \geq 0$ for each $j$ such that*

$$\lim_n u_n s = s$$

*for each $s \in S$.*

**Proof.** Let $(u_n)$ be the sequence determined for $T$ by Theorem 2.6. Then for each $s \in S$, the sequence $(u_n s)$ is weakly bounded, hence strongly bounded in $S$. Thus, if $A$ is an $S$ bounded subset of $\varphi$ then the set $B = \{u_n v : v \in A,\ n = 1, 2, \dots\}$ is also $S$ bounded since

$$\begin{aligned} \sup\{|\langle s, w\rangle| : w \in B\} &= \sup\{|\langle s, u_n v\rangle| : v \in A,\ n = 1, 2, \dots\} \\ &= \sup_n p_A(u_n s) < \infty. \end{aligned}$$

Now let $s \in S$, $A$ an $S$ bounded subset of $\varphi$ and $\varepsilon > 0$ be given. Let $t \in \varphi$ be such that $p_A(s - t)$ and $p_B(t - s)$ are less than $\varepsilon/3$ where $B$ is the set described above. Let $N$ be such that $p_A(u_n t - t) < \varepsilon/3$ when $n > N$. Then if $n > N$ we shall have

$$p_A(s - u_n s) \le p_A(s - t) + p_A(t - u_n t) + p_A(u_n t - u_n s)$$

$$\le \varepsilon/3 + \varepsilon/3 + p_B(t - s) < \varepsilon.$$

■

*References*


[1] J. DeFranza and D. J. Fleming, *Sequence spaces and summability factors*, Math. Z. 199 (1988) 99-108.

[2] J. DeFranza and D. J. Fleming, *Sequence spaces and summability factors*, Math. Z. 199 (1988) 99-108.

[3] ____________, *On the Bohr-Hardy criteria*, Arch. Math. 51(1988), 464-473.

[4] J. Dieudonné, *On biorthogonal systems*, Mich. Math. J. 2(1954), 7 -20.

[5] D. J. Fleming, *Unconditional Toeplitz sections in sequence spaces*, Math. Z. 194 (1987), 405-414.

[6] K. G. Grosse-Erdmann, $T$-solid sequence spaces, Results in Math. 23 (1993), 303-321.

[7] G. Köthe and O. Toeplitz, *Lineare Räume mit unendlich vielen Koordinaten und Ringe unendlicher Matrizen*, J. reine angew. Math. 171(1934), 193-226.

[8] G. Köthe, *Topological Vector Space I*, Springer-Verlag, 1969.

[9] D. J. H. Garling, *On topological sequence spaces*, Proc. Camb. Phil. Soc. 63(1967) 997-1019.

[10] A. B. Latimer and W. H. Ruckle, *Nuclear FK-Sum Spaces*, Math. Ann. 208(1974), 213-220.

[11] R. J. McGivney and W. H. Ruckle, *Multiplier algebras of biorthogonal systems*, Pac. J. Math 29(1969), 375-387.

[12] W. H. Ruckle, *An abstract concept of the sum of a numerical series*, Can. J. Math., 22(1970), 863-874.

[13] ___________, *Lattices of sequence spaces,* Duke Math. J. 35(1968), 491-503.

[14] ___________, *Representation and series summability of complete biorthogonal sequences,* Pac. J. Math. 34(1970), 511-528.

[15] ___________, *Topologies on sequence spaces,* Pac. J. Math. 42(1972), 491-504.

[16] ___________, *On the classification of biorthogonal sequences,* Can. J. Math. 26(1974), 721-763.

[17] ___________, **Sequence Spaces,** London 1981, Pitman.

[18] ___________, and S. Saxon, *Generalized sectional convergence and multipliers,* J. of Math. Anal. and Appl. 193 (1995), 680-705.

[19] S. Saxon, *Non-barrelled dense $\beta\varphi$ subspaces*, J. Math. Anal. and Appl. 196 (1995), 428-441.

[20] H. H. Schaefer, **Topological Vector Spaces**, New York, Macmillan 1966.

[21] A. Wilansky, **Modern Methods in Topological Vector Spaces**, New York, McGraw-Hill, 1978.

[22] K. Zeller, *Abschnittskonvergenz in FK-Räumen*, Math. Z. 55(1951), 55-70.

# Chapter 3

# Growth of Cesàro means of double Vilenkin-Fourier series of unbounded type

**W.R. Wade**

## 3.1 Introduction

Let $\mathbf{N}$ represent the set of natural numbers, and $\mathbf{Q} := [\mathbf{0}, \mathbf{1}) \times [\mathbf{0}, \mathbf{1})$ represent the unit cube. Let $p_0, p_1, p_2, \ldots$ and $q_0, q_1, q_2, \ldots$ be two sequences of natural numbers with $p_n \geq 2$ and $q_n \geq 2$. For each $n \in \mathbf{N}$ set $P_n := p_0 p_1 \ldots p_{n-1}$ and $Q_n := q_0 q_1 \ldots q_{n-1}$, where the empty product is by definition 1. The *double Vilenkin system* associated with these generators is the system $(w_{n,m}; n, m \in \mathbf{N})$ defined on $\mathbf{Q}$ as follows:

$$w_{n,m}(x,y) := w_n(x) w_m(y) := \prod_{k=0}^{\infty} \exp\left(\frac{2\pi i n_k x_k}{p_k}\right) \prod_{k=0}^{\infty} \exp\left(\frac{2\pi i m_k y_k}{q_k}\right),$$

where the coefficients $n_k, m_k, x_k, y_k$ all are integers which satisfy

$$0 \leq n_k < p_k, 0 \leq m_k < q_k, 0 \leq x_k < p_k, 0 \leq y_k < q_k,$$

$$n = \sum_{k=0}^{\infty} n_k P_k, m = \sum_{k=0}^{\infty} m_k Q_k, x = \sum_{k=0}^{\infty} x_k P_{k+1}^{-1}, \text{and } y = \sum_{k=0}^{\infty} y_k Q_{k+1}^{-1}$$

(see Vilenkin [5] for more details). When $p_k \equiv q_k \equiv 2$ for all $k$, the system $w_{n,m}$ is the double Walsh system. When $p_k = O(1)$ and $q_k = O(1)$, the system $w_{n,m}$ is called a double Vilenkin system of *bounded type*.

The *double Vilenkin–Fourier series* of an $f \in L_1(\mathbf{Q})$ is the series whose partial sums are given by

$$S_{n,m} f := \sum_{k=0}^{n-1} \sum_{j=0}^{m-1} \widehat{f}(k,j) w_{k,j},$$

where

$$\widehat{f}(k,j) := \iint_{\mathbf{Q}} f(x,y) w_{k,j}(x,y)\, d(x,y),$$

and the *Cesàro means* of the double Vilenkin–Fourier series of $f$ are given by

$$\sigma_{n,m}f := \sum_{k=0}^{n-1}\sum_{j=0}^{m-1}\left(1-\frac{k}{n}\right)\left(1-\frac{j}{m}\right)\widehat{f}(k,j)w_{k,j}.$$

Summability of one–dimensional Vilenkin–Fourier series for systems of bounded type is fairly well understood. Building on earlier independent work of Fine, Taibleson, and Schipp, Pál and Simon [3] proved that if $f \in L_1[0,1)$ then $\sigma_n f \to f$ almost everywhere on $[0,1)$. This is not the case for Vilenkin systems of unbounded type. Indeed, building on earlier work of Price, Tsutserova [4] has proved that for any unbounded Vilenkin system, there exist continuous functions for which $\sigma_{P_n} f$ diverges at any prescribed point.

We shall obtain an estimate for the growth of $\sigma_{P_N} f$ and $\sigma_{P_n,Q_m} f$ for integrable $f$ valid for a large class of Vilenkin systems of unbounded type.

## 3.2 Fundamental concepts and notation

A *Vilenkin rectangle* of order $n,m$ is a rectangle of the form

$$R := [kP_n^{-1},(k+1)P_n^{-1}) \times [\ell Q_m^{-1},(\ell+1)Q_m^{-1}),$$

where $k$ and $\ell$ are integers which satisfy $0 \le k < P_n$, $0 \le \ell < Q_m$. The $\sigma$–algebra generated by all Vilenkin rectangles of order $n,m$ will be denoted by $\mathcal{F}_{n,m}$. Given a rectangle $R \in \mathcal{F}_{n,m}$ and $r \in \mathbf{N}$, the *r–fold expansion* of $R$ will be the Vilenkin rectangle $R^r \in \mathcal{F}_{n-r,m-r}$ which satisfies $R \subset R^r$.

Let $0 < p < \infty$. A *p–atom* on $\mathbf{Q}$ is a bounded measurable function $a$ on $\mathbf{Q}$ which satisfies

i) $a$ is of *mean zero*, i.e.,

$$\iint_{\mathbf{Q}} a(x,y)d(x,y) = 0,$$

ii) there is a Vilenkin rectangle $R \in \mathcal{F}_{K,K}$ such that $\|a\|_\infty \le |R|^{-1/p}$,

iii) $a$ is supported on $R$.

A *Vilenkin martingale* is a sequence of integrable functions

$$(f_{n,m}; n,m \in \mathbf{N})$$

such that each $f_{n,m}$ is $\mathcal{F}_{n,m}$ measurable, and the conditional expectation operator with respect to these $\sigma$–algebras satisfies $E_{k,\ell}(f_{n,m}) = f_{k,\ell}$ for

$k \le n$ and $\ell \le m$. It is well–known that the partial sums $S_{P_n,Q_m}f$ of the double Vilenkin–Fourier series of any $f \in L_1(\mathbf{Q})$ is a Vilenkin martingale.

Using Vilenkin martingales and quadratic variation instead of distributions and the maximal operator, Weisz [6] introduced Vilenkin Hardy–Lorentz spaces $H_{p,q}(\mathbf{Q})$ for $0 < p,q < \infty$ which reduce to the usual Vilenkin Hardy spaces $H_p(\mathbf{Q})$ when $p = q$. The only thing we need to know about these Hardy–Lorentz spaces is the following interpolation theorem (see Weisz [6], Theorem 2):

**Lemma 3.1** *Suppose $T$ is a sublinear operator defined on Vilenkin martingales and $0 < \delta \le 1$. If for every $\delta \le p \le 1$ there exists an $r \in N$ and a constant $C_{p,r}$, which depends only on $p$ and $r$, such that*

$$\iint\limits_{\mathbf{Q}\setminus\mathbf{R}^r} |Ta|^p \, d(x,y) \le C_{p,r}$$

*holds for all $p$–atoms $a$, and if $T$ is bounded from $L_\infty(Q)$ to $L_\infty(Q)$, then $T$ is of weak type (1,1) on $L_1(Q)$, and for each pair $\delta \le p \le 1$, $0 < q \le \infty$ there is a constant $C_{p,q}$ which depends only on $p$ and $q$ such that*

$$\|Tf\|_{p,q} \le C_{p,q}\|f\|_{H_{p,q}(\mathbf{Q})}.$$

## 3.3 The Vilenkin–Fejér kernel

Let

$$D_n := \sum_{k=0}^{n-1} w_k \qquad \text{and} \qquad K_n := \sum_{k=0}^{n-1}\left(1 - \frac{k}{n}\right) w_k$$

represent the one–dimensional Vilenkin–Dirichlet and one–dimensional Vilenkin–Fejér kernels, respectively. Let $\dot{+}$ and $\dot{-}$ represent addition and subtraction (respectively) which is inherited from the underlying Vilenkin group, e.g., if $x = \sum_{k=0}^{\infty} x_k P_k^{-1}$, and $y = \sum_{k=0}^{\infty} y_k P_k^{-1}$, then

$$x \dot{+} y := \sum_{k=0}^{\infty} (x_k \oplus y_k) P_k^{-1},$$

where $\oplus$ denotes addition modulo $p_k$ (see Vilenkin [5] for details). Recall that

$$D_{P_n}(x) \equiv P_n \chi_{[0,P_n^{-1})}(x) := \begin{cases} P_n & 0 \le x < P_n^{-1} \\ 0 & P_n^{-1} \le x < 1 \end{cases} \tag{3.1}$$

and

$$(\sigma_{n,m}f)(x,y) \equiv \iint\limits_{\mathbf{Q}} f(t,u)K_n(x \stackrel{\bullet}{-} t)K_m(y \stackrel{\bullet}{-} u)\,d(t,u). \qquad (3.2)$$

The following decomposition of the Fejér kernel is crucial to our estimates.

**Lemma 3.2** *Let* $\omega_j := \exp(2\pi i/p_j)$ *for each* $j \in N$. *Then*

$$K_{P_n}(x) = D_{P_n}(x) - \sum_{j=0}^{n-1} \frac{P_j}{P_n} \sum_{k=0}^{p_j-1} \frac{k}{p_j} \sum_{\ell=0}^{p_j-1} \omega_j^{-\ell k} D_{P_n}(x \stackrel{\bullet}{+} \ell P_{j+1}^{-1}).$$

**Proof.** Onneweer [2] introduced the *finite Vilenkin difference* (associated with the sequence $p_0, p_1, \ldots$) of a function $f$ defined on $[0,1)$ to be

$$(d_m f)(x) := \sum_{j=0}^{m-1} P_j \sum_{k=0}^{p_j-1} \frac{k}{p_j} \sum_{\ell=0}^{p_j-1} \omega_j^{-\ell k} f(x \stackrel{\bullet}{+} \ell P_{j+1}^{-1}),$$

and proved that if $k < P_n$, then $d_n(w_k) = k w_k$. Using this, we see that

$$\begin{aligned} K_{P_n}(x) &\equiv \sum_{k=0}^{P_n-1} \left(1 - \frac{k}{P_n}\right) w_k(x) \\ &= D_{P_n}(x) - \frac{1}{P_n} d_n(D_{P_n}(x)) \end{aligned}$$

as required. ■

Given $0 < \delta \le 1$, we shall call a sequence $(p_n; n \in \mathbf{N})$ or the Vilenkin system it generates $\delta$*–quasibounded* if there is a positive number $M$ such that

$$M_n := \sum_{j=0}^{n-1} \left(\frac{p_j}{p_{j+1} \cdots p_n}\right)^{\delta} \le M$$

for all $n \in \mathbf{N}$. A 1–quasibounded system will be called *quasibounded.* Clearly, if $\delta > \mu$ then a $\mu$–quasibounded sequence is $\delta$–quasibounded, but not conversely. In particular, all $\delta$–quasibounded systems are quasibounded.

Notice that if $p_n = O(1)$ as $n \to \infty$, then since $p_n \ge 2$,

$$M_n \le C \sum_{j=0}^{\infty} \frac{1}{2^{\delta j}} < \infty.$$

Thus, for all $0 < \delta \le 1$, any bounded Vilenkin system is $\delta$–quasibounded. Similarly, if $p_n \uparrow \infty$ or if there is an absolute constant $\gamma$ such that $|p_n -$

$p_{n+1}| \leq \gamma$ for all $n \in \mathbf{N}$, then the corresponding Vilenkin system is also $\delta$–quasibounded, because

$$M_n \leq \sum_{j=0}^{n-1} \frac{1+\gamma^\delta}{p_{j+2}\cdots p_n} \leq \sum_{j=0}^{\infty} \frac{1+\gamma^\delta}{2^{\delta j}} < \infty.$$

(Recall the empty product is by definition 1.) On the other hand, if $p_k = 2^n$ for $k = 2^n - 1$ and $p_k = 1$ otherwise, then

$$M_{2^n} \geq \left(\frac{p_{2^n-1}}{p_{2^n}}\right)^\delta = 2^{\delta n} \to \infty$$

as $n \to \infty$. In particular, many unbounded Vilenkin systems are $\delta$–quasibounded but not all.

Our first application of Lemma 3.2 is an estimate of the $L_1[0,1)$ norm of $K_{P_n}$ for $\delta$–quasibounded Vilenkin systems.

**Lemma 3.3** *If $(p_n, n \in N)$ is $\delta$–quasibounded for some $0 < \delta \leq 1$, then*

$$\|K_{P_n}\|_1 = O(p_{n-1})$$

*as $n \to \infty$.*

**Proof.** By Lemma 3.2,

$$|K_{P_n}(x)| \leq |D_{P_n}(x)| + \sum_{j=0}^{n-1} \frac{P_j}{P_n} p_j \sum_{\ell=0}^{p_j-1} |D_{P_n}(x \dot{+} \ell P_{j+1}^{-1})|.$$

Consequently, it follows from (3.1) and the translation invariance of Lebesgue measure with respect to $\dot{+}$ (see Vilenkin [5]) that

$$\begin{aligned} \|K_{P_n}\|_1 &\leq 1 + \sum_{j=0}^{n-1} \frac{P_j}{P_n} p_j^2 \\ &= 1 + p_{n-1} + \sum_{j=0}^{n-2} \frac{p_j}{p_{j+1}\cdots p_{n-1}}. \end{aligned}$$

But this last sum is bounded as $n \to \infty$ since $(p_n; n \in \mathbf{N})$ is $\delta$–quasibounded, hence quasibounded. ∎

## 3.4 The main results

Fix $\alpha > 0$ and choose $r \in \mathbf{N}$ such that $r - 1 < \alpha \leq r$. For each Vilenkin martingale $f = (f_{n,m}; n, m \in \mathbf{N})$ and each $0 < p \leq 1$, let

$$\sigma_{\alpha,p} f := \sup_{|n-m| \leq \alpha} \frac{|\sigma_{P_n,Q_m} f|}{\beta_{n,m}^{1/p} \beta_{n+r,m+r}^{2r/p}},$$

where for each pair $k, j \in \mathbf{N}$, $\beta_{k,j} := \max\{p_0, \ldots, p_{k-1}, q_0, \ldots, q_{j-1}\}$. Maximal operators of this type (without the $\beta$'s) were first discussed for the Walsh case in Móricz, Schipp, and Wade [1].

We shall obtain the following estimates.

**Theorem 3.4** *Suppose $\{p_k\}$ and $\{q_k\}$ are $\delta$–quasibounded for some $0 < \delta \leq 1$. Then $\sigma_{\alpha,\delta}$ is of weak type (1,1) on $L_1(Q)$, and for every $\delta \leq p \leq 1$ and $0 < q \leq \infty$ there is a constant $C_{p,q}$ which depends only on $p$ and $q$ such that*

$$\|\sigma_{\alpha,\delta} f\|_{p,q} \leq C_{p,q} \|f\|_{H_{p,q}(\mathbf{Q})}.$$

**Proof.** Since $r \geq 1$, it is clear by Lemma 3.3 and (3.2) that $\sigma_{\alpha,\delta}$ is of type $(\infty, \infty)$. Hence by Lemma 3.1 it suffices to show $\sigma_{\alpha,\delta}$ is $p$–quasilocal for each $\delta \leq p \leq 1$, i.e., there is a constant $C_{p,r}$ such that if $a$ is a $p$–atom supported on some "square" $R := I \times J$ (i.e., $|I| = P_K^{-1}$ and $|J| = Q_K^{-1}$ for the same $K$), then

$$\iint_{\mathbf{Q}\setminus\mathbf{R}^r} |(\sigma_{\alpha,\delta} a)(x,y)|^p \, d(x,y) \leq C_{p,r}. \tag{3.3}$$

(Here $r$ is the integer chosen above so that $r - 1 < \alpha \leq r$.)
Suppose $|I| = P_K^{-1}$ and $|J| = Q_K^{-1}$. If $n < K$ and $m < K$, then both $w_n$ and $w_m$ are constant on $I$ and $J$. Since $a$ is of mean zero, it follows that $\widehat{a}(n,m) = 0$. Thus we may suppose that either $n$ or $m$ is $\geq K$. If $n \geq K$, then the conditions $|n - m| \leq \alpha$ and $r - 1 < \alpha \leq r$ imply $m \geq n - \alpha \geq n - r \geq K - r$. Hence we may assume that both $m$ and $n$ are $\geq K - r$.
To prove (3.3), notice first that $\mathbf{Q} \setminus \mathbf{R}^r$ can be broken into 8 pieces: two horizontal pieces $([0,1) \setminus I^r) \times J^r$; two vertical pieces $I^r \times ([0,1) \setminus J^r)$; and four rectangular pieces $([0,1) \setminus I^r) \times ([0,1) \setminus J^r)$. Estimates of $\sigma_{\alpha,\delta}$ over the horizontal and vertical pieces are similar, and the estimates over the remaining rectangular pieces are simpler. Consequently, we shall supply the details for the horizontal estimates only.
Fix $(x,y) \in ([0,1) \setminus I^r) \times J^r$. By Lemma 3.3, $\|K_{Q_m}\|_1 \leq C q_{m-1}$. Since $a$ is supported on $R \equiv I \times J$ and $|a| \leq |R|^{-1/p}$, it follows from (3.2) that

$$\begin{aligned} |(\sigma_{P_n,Q_m} a)(x,y)| &\leq \int_J \int_I |a(t,u)| |K_{P_n}(x \stackrel{\bullet}{-} t)| \, |K_{Q_m}(y \stackrel{\bullet}{-} u)| \, dt \, du \\ &\leq C q_{m-1} |R|^{-1/p} \int_I K_{P_n}(x \stackrel{\bullet}{-} t) \, dt. \end{aligned}$$

Using Lemma 3.2 and the fact that $p_j P_j = P_{j+1}$, we can continue this estimate as follows.

$$|(\sigma_{P_n,Q_m}a)(x,y)| \le Cq_{m-1}|R|^{-1/p} \cdot \cdot \left\{ \int_I D_{P_n}(x \overset{\bullet}{-} t)\,dt + \sum_{j=0}^{n-1} \frac{P_{j+1}}{P_n} \sum_{\ell=0}^{p_j-1} \int_I D_{P_n}(x \overset{\bullet}{+} \ell P_{j+1}^{-1} \overset{\bullet}{-} t)\,dt \right\}.$$

At this point it is important to realize that many of the terms of this estimate are zero. First, they are zero if $j$ is too large. Indeed, since $x \notin I^r$ and $D_{P_n}$ is supported on $[0, P_n^{-1})$, if $n \ge j \ge K - r$ then both $D_{P_n}(x \overset{\bullet}{-} t)$ and $D_{P_n}(x \overset{\bullet}{+} \ell P_{j+1}^{-1} \overset{\bullet}{-} t)$ are identically zero for $t \in I$ and $0 \le \ell < p_j$ (because under these conditions, $x$ and $x \overset{\bullet}{+} \ell P_{j+1}^{-1}$ both lie outside $I^r$ so the coefficients of $t$ cannot cancel all "lower order" the coefficients of $x$ or $x \overset{\bullet}{+} \ell P_{j+1}^{-1}$). Thus

$$|(\sigma_{P_n,Q_m}a)(x,y)| \le Cq_{m-1}|R|^{-1/p} \cdot \cdot \sum_{j=0}^{K-r-1} \frac{P_{j+1}}{P_n} \sum_{\ell=0}^{p_j-1} \int_I D_{P_n}(x \overset{\bullet}{+} \ell P_{j+1}^{-1} \overset{\bullet}{-} t)\,dt. \tag{3.4}$$

Next, the terms of this estimate are zero if $x$ is too near to or too far from $I$. To see this, notice that since Lebesgue measure is translation invariant with respect to $\overset{\bullet}{+}$, we may suppose that $I = [0, P_K^{-1})$. Since $n \ge K - r$, and $D_{P_i} \equiv P_i \chi_{[0,P_i^{-1})}$ for all $i \in \mathbf{N}$, we have both

$$\frac{1}{P_n} D_{P_n} \le \frac{1}{P_{K-r}} D_{P_{K-r}},$$

and

$$\int_I D_{P_i}(x \overset{\bullet}{+} \ell P_{j+1}^{-1} \overset{\bullet}{-} t)\,dt = \frac{P_i}{P_K} \chi_{[\xi_{j\ell}, \xi_{j\ell}+P_i^{-1})}(x)$$

for $j \le i \le K-1$ and $x \notin I^r$, where $\xi_{j\ell} := (p_j - \ell)P_{j+1}^{-1}$. Combining these observations with (3.4), we obtain

$$|(\sigma_{P_n,Q_m}a)(x,y)| \le Cq_{m-1}|R|^{-1/p} \cdot \cdot \sum_{j=0}^{K-r-1} \frac{P_{j+1}}{P_{K-r}} \sum_{\ell=0}^{p_j-1} \frac{P_{K-r}}{P_K} \chi_{[\xi_{j\ell}, \xi_{j\ell}+P_{K-r}^{-1})}(x).$$

Since $(a+b)^p \le a^p + b^p$ for all $a, b \ge 0$ and $0 < p \le 1$, it follows that

$$|(\sigma_{P_n,Q_m}a)(x,y)|^p \le C^p q_{m-1}^p |R|^{-1} \cdot \sum_{j=0}^{K-r-1} \frac{P_{j+1}^p}{P_K^p} \sum_{\ell=0}^{p_j-1} \chi_{[\xi_{j\ell}, \xi_{j\ell}+P_{K-r}^{-1})}(x) \tag{3.5}$$

for all $x \in [0,1) \setminus I^r$.
Recall that $|R| = P_K^{-1} Q_K^{-1}$ and $K$ is less than or equal to both $n+r$ and $m+r$. Since

$$q_{m-1}^p p_j = q_{m-1}^p p_j^{1-p+p} \leq \beta_{n,m}^p \beta_{n,m}^{1-p} p_j^p = \beta_{n,m} p_j^p$$

for all $0 \leq j < K-r$,

$$\frac{P_K}{P_{K-r}} \frac{Q_K}{Q_{K-r}} = p_{K-r} \cdots p_{K-1} \cdot q_{K-r} \cdots q_{K-1} \leq \beta_{n+r,m+r}^{2r},$$

and $p/\delta \geq 1$, it follows from (3.5) that

$$\left| \frac{(\sigma_{P_n,Q_m} a)(x,y)}{\beta_{n,m}^{1/\delta} \beta_{n+r,m+r}^{2r/\delta}} \right|^p \leq C^p P_{K-r} Q_{K-r} \sum_{j=0}^{K-r-1} \frac{p_j^p}{p_j} \frac{P_{j+1}^p}{P_K^p} \sum_{\ell=0}^{p_j-1} \chi_{[\xi_{j\ell}, \xi_{j\ell}+P_{K-r}^{-1})}(x).$$

Since $|J^r| = Q_{K-r}^{-1}$ and $|[\xi_{j\ell}, \xi_{j\ell} + P_{K-r}^{-1})| = P_{K-r}^{-1}$ for each $j$ and $\ell$, and $\beta_{n,m}^{p/\delta} > \beta_{n,m}$, we conclude that

$$\int_{[0,1)\setminus I^r} \int_{J^r} |(\sigma_{\alpha,\delta} a)(x,y)|^p \, dy \, dx \leq C^p \sum_{j=0}^{K-r-1} \left( \frac{p_j P_{j+1}}{P_K} \right)^p$$

which is bounded in $K$ because the Vilenkin system is $p$–quasibounded and $p \geq \delta$. ■

Notice that if the Vilenkin systems are of bounded type, then the coefficients $\beta_{n,m}$ are also bounded. Hence the operators $\sigma_{\alpha,\delta}$ defined above reduce to the operator $\sigma_\alpha$ Weisz used in the bounded case (see Theorem 4 in [6]). Along the same lines, if one of the systems is bounded but the other is not, similar modifications in the definition of $\sigma_{\alpha,\delta}$ can be made so that $\beta$ depends only on the $p_k$'s or the $q_k$'s.

Theorem 3.4 can be used to obtain pointwise estimates of Cesàro means of Vilenkin–Fourier series in the quasibounded case.

**Theorem 3.5** *Let $0 < p \leq 1$, $\alpha > 0$, and $r \in N$ satisfy $r - 1 < \alpha \leq r$. If $f \in L_1(Q)$ then the Cesàro means of the Fourier series of $f$ with respect to any $p$–quasibounded Vilenkin system satisfy*

$$|\sigma_{P_n,Q_m} f(x,y) - f(x,y)| = o(\beta_{n,m} \beta_{n+r,m+r}^{2r}), \tag{3.6}$$

*as $n, m \to \infty$ provided $|n-m| < \alpha$, for almost every $(x,y)$ in $Q$.*

**Proof.** In view of Weisz [6], we may suppose that at least one of the sequences $p_n$ or $q_m$ is unbounded, i.e., that $\gamma_{n,m} := \beta_{n,m} \beta_{n+r,m+r}^{2r} \to \infty$ as $n, m \to \infty$. For each $g \in L_1(\mathbf{Q})$ let

$$Tg = \limsup \frac{\sigma_{P_n,Q_m} g}{\gamma_{n,m}} - \liminf \frac{\sigma_{P_n,Q_m} g}{\gamma_{n,m}},$$

where the limits supremum/infimum are taken as $n, m \to \infty$ with $|n - m| < \alpha$. Let $\epsilon > 0$ and choose a double Vilenkin polynomial $\phi$ such that $\|f - \phi\|_1 < \epsilon$. Since $\sigma_{P_n,Q_m}\phi \equiv \phi$ for $n, m$ sufficiently large and $\gamma_{n,m} \to \infty$, it is clear that $Tf = T(f - \phi)$. Since $\sigma_{\alpha,1}$ is of weak type $(1,1)$, it follows that

$$|\{Tf > \lambda\}| = |\{T(f - \phi) > \lambda\}| \leq \frac{2}{\lambda}\|f - \phi\|_1 < \frac{2}{\lambda}\epsilon$$

for each $\lambda > 0$. In particular, $Tf = 0$ almost everywhere on $\mathbf{Q}$. Since $f$ is almost everywhere finite–valued and $\gamma_{n,m} \to \infty$ as $n, m \to \infty$, we conclude that (3.6) holds almost everywhere on $\mathbf{Q}$. ∎

The estimates above can be used to yield new information in the one–dimensional case too. Indeed, since in this case $n = m$, we can take $\alpha = r = 0$. Thus Theorem 3.4 contains the following result.

**Theorem 3.6** *Let $0 < p \leq 1$. If $f \in L_1[0,1)$ then the Cesàro means of the Fourier series of $f$ with respect to any $p$–quasibounded Vilenkin system satisfy*

$$|\sigma_{P_n} f(x) - f(x)| = o(\delta_n), \tag{3.7}$$

*as $n \to \infty$, for almost every $x$ in $[0,1)$, where $\delta_n :=\max\{p_0, p_1, \ldots, p_{n-1}\}$.*

We close with several remarks. First, in the bounded case, the estimates in Theorems 3.5 and 3.6 reduce to $o(1)$. Hence in the bounded case, our results show that Vilenkin–Fourier series are almost everywhere Cesàro summable (with suitable "cone" conditions in the multidimensional case). Next, Theorem 3.6 implies that $|\sigma_{P_n} f(x) - f(x)| = o(P_n^{1/n})$, as $n \to \infty$, in the quasibounded case. We do not know whether this or the estimate in (3.6) are sharp. Finally, we have used "lacunary" Cesàro means (i.e., used the indices $P_n$ and $Q_m$) for all estimates. It is an open question what happens when these indices are replaced with $n$ and $m$ respectively.

*References*

[1] F. Móricz, F. Schipp, and W.R. Wade, *Cesàro summability of double Walsh–Fourier series*, Trans. Amer. Math. Soc. **329**(1992), 131–140.

[2] C.W. Onneweer, *Differentiability of Rademacher series on groups*, Acta Sci. Math. Szeged **39**(1977), 121–128.

[3] J. Pál and P. Simon, *On a generalization of the concept of derivative*, Acta Math. Acad. Sci. Hungar. **29**(1977), 155–164.

[4] N.I. Tsutserova, *On $(C,1)$–summability of Fourier series in a multiplicative system of functions*, Mat. Zametki **43**(1988), 808–848.

[5] N. Ya. Vilenkin, *On a class of complete orthonormal systems*, Izv. Akad. Nauk. SSSR, Ser. Mat. **11**(1947), 363–400.

[6] F. Weisz, *Hardy spaces and Cesàro means of two–dimensional Fourier series*, Bolyai Soc. Math. Studies **5**(1996), 353–367.

[7] F.Weisz, *Cesàro summability of two–dimensional Walsh–Fourier series*, Trans. Amer. Math. Soc. **348**(1996), 2169–2181.

# Chapter 4

# A substitute for summability in wavelet expansions

**Gilbert G. Walter**
**Xiaoping Shen**

*ABSTRACT. Gibbs'phenomenon almost always appears in the expansions using classical orthogonal systems. Various summability methods are used to get rid of unwanted properties of these expansions. Similar problems arise in wavelet expansions, but cannot be solved by the same methods. In a previous work [10], two alternative procedures for wavelet expansions were introduced for dealing with this problem. In this article, we are concerned with the details of the implementation of one of the procedures, which works for the wavelets with compact support in the time domain. Estimates based on this method remove the excess oscillations. We show that the dilation equations which arise, though they contain an infinite number of terms, have coefficients which decrease exponnetially. In addition, an iteration relation for the positive estimation function is derived to reduce the amount of calculation in the approximation. Numerical experiments are given to illustrate the theoretical results.*

## 4.1 Introduction

Gibbs' phenomenon, which occurs at points of jump discontinuities of a function, has been known for over a century [11]. It deals with the "overshoot" of partial sums of orthogonal series in the limit at these points. It may be avoided by using a summability method in place of partial sums. For example, with trigonometric Fourier series, the Abel summability is given by

$$f_r(x) := \sum_{n=-\infty}^{\infty} r^{|n|} c_n e^{inx}, \qquad 0 < r < 1. \tag{4.1}$$

and the limit is taken as $r \longrightarrow 1$.

In this case, we have uniform convergence of the approximations for continuous functions, the upper and lower bounds are the same as those of

$f(x)$, and there is no Gibbs' phenomenon. These properties hold because of the summability kernel associated with $f_r(x)$ is a positive kernel (or delta sequence) [12, p.88].

If we turn to wavelet expansions, some of the shortcomings of Fourier series disappear. Their convergence, for example, is uniform for continuous function [4, 19, 6]. However, Gibbs' phenomenon almost always appears [8] and approximates to positive functions are not necessarily positive. The techniques that work for Fourier series cannot be used because of the peculiar nature of the wavelet expansions.

However two alternative methods have been introduced to replace Abel or Cesàro summability in a previous work [10]. These methods are not, strictly speaking, summability methods as used with classical orthogonal systems. But they achieve the same ends, and for lack of a better term, we will continue to use summability to describe our process.

Unfortunately, the summability function whose translates form a basis of $V_0$, is no longer compactly supported even for a compactly supported scaling function. In order to work with such a summability function in real world calculation, we will need to truncate the infinite series used to define it.

In this article, we will study asymptotic properties of one of the procedures, which works for the wavelets with compact support in the time domain. It is related to Abel summability and gives a positive "summability" kernel. We will show that the exponential decreasing rate is achieved by the dilation coefficients of the summability function in the final section. In addition, an iteration relation for the resulting positive estimation function in terms of the usual estimation function is found. This reduces the amount of calculation required for the approximation. Numerical experiments are given in the last section to illustrate the theoretical results.

## 4.2 Background

We present here a few elements of orthogonal wavelet theory, in which an orthonormal basis $\{\psi_{mn}\}$ of $L^2(\mathbb{R})$ is constructed having the form

$$\psi_{mn}(t) = 2^{m/2}\psi_{mn}(2^m t - n), \quad n, m \in \mathbb{Z} \times \mathbb{Z},$$

where $\psi(t)$ is the "mother wavelet". Usually it is not constructed directly but rather from another function called the "scaling function" $\phi(t) \in$

$L^2(\mathbb{R})$. The scaling function $\phi$, is chosen in such a way that

$$\begin{cases} (i) & \int \phi(t)\phi(t-n)dt = \delta_{0,n}, \quad n \in \mathbb{Z}, \\ (ii) & \phi(t) = \sum_k \sqrt{2} c_k \phi(2t-k), \quad \{c_k\}_{k\in\mathbb{Z}} \in l^2, \\ (iii) & \text{For each } f \in L^2(R), \epsilon > 0, \text{there is a function} \\ & f_m(t) = \sum_n a_{mn}\phi(2^m t - n) \text{ such that } \|f_m - f\| < \epsilon. \end{cases} \tag{4.2}$$

These conditions lead to a "multiresolution approximation" $\{V_m\}_{m\in\mathbb{Z}}$, consisting of closed subspaces of $L^2(\mathbb{R})$. The space $V_m$ is taken to be the closed linear span of $\{\phi(2^m t - n)\}_{n\in\mathbb{Z}}$. Because of (4.2)(ii), the $V_m$ are nested, i.e. $V_m \subseteq V_{m+1}$ and because of (4.2)(iii), $\cup_m V_m$ is dense in $L^2(\mathbb{R})$.

A standard prototype is the Haar system in which $\phi(t) = \chi_{[0,1]}(t)$. It is an easy exercise to show that (4.2) is satisfied. This prototype has poor frequency localization but good time localization. Most of the other examples found, e.g., in [2], attempt to get fairly good time and frequency localization at the same time.

Frequently conditions (4.2) are expressed in terms of their Fourier transforms. We give a sufficient condition for (4.2) as

$$\begin{cases} (i) & \sum_k |\widehat{\phi}(\omega+k)|^2 = 1, \\ (ii) & \widehat{\phi}(\omega) = (\frac{1}{\sqrt{2}}\sum_k c_k e^{ik\omega/2})\widehat{\phi}(\frac{\omega}{2}) = m_0(\frac{\omega}{2})\widehat{\phi}(\frac{\omega}{2}), \\ \text{where} & m_0(\frac{\omega}{2}) = \frac{1}{\sqrt{2}}\sum_k c_k e^{ik\omega/2} \in L^2(-2\pi, 2\pi), \\ (iii) & \widehat{\phi}(\omega) \text{ is continuous at } \omega = 0 \text{ and } \widehat{\phi}(0) = 1. \end{cases} \tag{4.3}$$

The mother wavelet comes from $\phi(t)$ via (4.2)(ii) or (4.3)(ii).

$$(i) \quad \psi(t) = \sum_k (-1)^k \sqrt{2} c_{1-k}\phi(2t-k),$$

$$(ii) \quad \widehat{\psi}(\omega) = e^{i\pi\omega}\overline{m_0(\tfrac{\omega}{2}+\pi)}\widehat{\phi}(\tfrac{\omega}{2}).$$

## 4.3 Summability for Wavelets With Compact Support

The Haar orthogonal system is based on a scaling function which has compact support and is non-negative. Many other wavelets with compact support beginning with those of Daubechies [2] have been constructed. These are continuous, which the Haar system is not, but are never non negative [5]. It is shown in [11] that the sampling function $s(t)$ for the Daubechies wavelet with 4 taps is non negative for all values of $t$. However this appears to be a lucky accident, since this property is not shared by other Daubechies wavelets. To find non-negative estimation in $V_0$, we first ask if there are any non-negative functions in $V_0$, and then if there is a basis of such functions. We start with partitions of unity.

Let $\theta(t)$ be any continuous function on $\mathbb{R}$ with support in an interval $[M, N]$. We say that $\theta(t)$ generates a partition of unity if

$$\sum_{n \in Z} \theta(t-n) \equiv 1, \quad t \in \mathbb{R}. \tag{4.4}$$

This is a property shared by all orthogonal scaling functions of compact support, in particular those due to Daubechies as well as the "Coiflets" of Coifman [2]. We define the Abel means of their series (4.4) to be

$$\rho_r(t) := \sum_{n=-\infty}^{\infty} r^{|n|} \theta(t-n), \quad 0 < r \leq 1, \quad t \in \mathbb{R}. \tag{4.5}$$

This series converges uniformly on $[0,1]$ since it is locally finite. Furthermore $\rho_r(t) \longrightarrow 1$ for $t \in [0,1]$ as $r \longrightarrow 1$ by the regular summability property of Abel means. The series also converges uniformly on any finite interval $[M, N]$ (Since if a series converges to $S$, it is also Abel summable to $S$). Consequently, there exists a real number $r_0$, $0 < r_0 < 1$ such that $\rho_r(t) \geq 1/2$ for $1 > r \geq r_0$, $t \in [M, N]$.

We denote by $V$ the closed linear span of $\{\theta(t-n)\}_{n \in Z}$ in $L^2(\mathbb{R})$. Then if in addition $\{\theta(t-n)\}$ is a Riesz basis of V and $a_n \in l^2$, we have

$$f(t) = \sum_n a_n \theta(t-n) \in V.$$

This requirement will be met if $\widehat{\theta}(\omega)$ satisfies

$$0 < A \leq \sum_k |\widehat{\theta}(\omega + 2k\pi)|^2 \leq B < \infty. \tag{4.6}$$

The properties of $\rho_r(t)$ are given by the next lemma.

**Lemma 4.1** *Let $\theta(t)$ be a continuous function on $\mathbb{R}$ with compact support satisfying (4.4) and (4.6); let $V = CLS\{\theta(t-n), n \in \mathbb{Z}\}$; then there is an $0 < r_0 < 1$, such that $\rho_r$ given by (4.5) for $r_0 \leq r < 1$, satisfies*

(i) $\rho_r(t) \geq 0$, $t \in \mathbb{R}$,

(ii) $\rho_r \in V$.

We now replace $\theta(t)$ by a continuous orthonormal scaling function $\phi(t)$ with compact support, and suppose that $\{V_m\}$ is the associated multiresolution of $L^2(\mathbb{R})$. From Lemma 4.1, we have an answer to the question posed at the start of this sections.

**Corollary 4.2** *Let $\{V_m\}$ be a multiresolution of $L^2(\mathbb{R})$ associate with the scaling function $\phi(t)$. Then there is an element $\rho \in V_0$ such that $\rho(t) \geq 0$.*

We can say much more than this. In fact, we have

**Theorem 4.3** *Let $\rho_r(t) = \sum_n r^{|n|}\phi(t-n)$, where $r$ is chosen so large that $\rho_r(t) \geq 0$, for $t \in \mathbb{R}$; then $\{2^{\frac{m}{2}}\rho_r(2^m t - n)\}_{n\in Z}$ is a Riesz basis of $V_m$; its dual basis is generated by $\rho_r^*$, where*

$$\rho_r^*(t) = \frac{1}{2\pi(1-r^2)}[(1+r^2)\phi(t) - r\{\phi(t+1) + \phi(t-1)\}].$$

The proof of Lemma 4.1 and Theorem 4.3 can be found in [10].

However, this biorthogonal system does not give us the positive kernel we need. Rather we use a modification which gives the desired properties. The kernel that gives us the approximation in $V_0$ to $f \in L^2$ is given by

$$k_r(t,s) = \left(\frac{1-r}{1+r}\right)^2 \sum_n \rho_r(t-n)\rho_r(s-n), \tag{4.7}$$

i.e.

$$f_0^r(t) = \int_{-\infty}^{\infty} k_r(s,t)f(s)ds.$$

This kernel satisfies the conditions [4, p.117] need to generate a positive delta sequence $\{k_{r,m}\}$ where

$$k_{r,m}(s,t) = 2^m k_r(2^m s, 2^m t), \quad m \in Z. \tag{4.8}$$

We summarize some of the properties of $k_{r,m}$ in next proposition. The proof is a direct consequence of the fact that$\{k_{r,m}\}$is a positive delta sequence [12].

**Proposition 4.4** *Let $k_{r,m}$ be as in (4.7), let $f \in L^1 \cap L^2(\mathbb{R})$; let*

$$f_m^r(t) = \int_{-\infty}^{\infty} k_{r,m}(t,s)f(s)ds; \tag{4.9}$$

*then $f_m^r \in V_m$ and*

(i) if $M_1 \leq f(t) \leq M_2$ for $t \in \mathbb{R}$, then $M_1 \leq f_m^r(t) \leq M_2$ for $t \in \mathbb{R}, m \in \mathbb{Z}$,

(ii) if $M_3 \leq f(t) \leq M_4$ for $t \in [a,b]$, then for each $\epsilon > 0, \delta > 0$, there is an $m_0$, such that for $t \in (a+\delta, b-\delta), M_3 - \epsilon \leq f_m^r(t) \leq M_4 + \epsilon$, for $m \geq m_0$.

Clearly, (ii) guarantees that $f_m^r(t)$ will not exhibit Gibbs' phenomenon and oscillations will be controlled by the given $\epsilon$. That is, the overshoot at discontinuities can be no more than $\epsilon$, which is arbitrarily small. Other properties of $f_m^r$ also arise from this positivity of $k_{r,m}(t,s)$. We note that it may also be represented by the series

$$f_m^r(t) = \sum_n c_{mn}^r \rho_r(2^m t - n). \tag{4.10}$$

## 4.4 The Properties Of The Summability Function

### *4.4.1 The rate of decrease of the filter coefficients*

The summability function $\rho_r(t)$, as defined in Theorem 4.3, is no longer compactly supported. A dilation equation relating $\rho_r(t)$ and a series of $\{\rho_r(2t-n)\}$ may be found as an infinite sum. However, the series is convergent rapidly, indeed we have,

$$\begin{aligned}
\widehat{\rho}_r(\omega) &= \frac{1-r^2}{1-2r\cos\omega + r^2}\widehat{\phi}(\omega) \\
&= \frac{1-r^2}{1-2r\cos\omega + r^2} m_0(\frac{\omega}{2}) \, \frac{1-2r\cos\frac{\omega}{2}+r^2}{1-r^2}\widehat{\rho}_r(\frac{\omega}{2}) \\
&= \frac{1-2r\cos\frac{\omega}{2}+r^2}{1-2r\cos\omega + r^2} m_0(\frac{\omega}{2})\widehat{\rho}_r(\frac{\omega}{2}).
\end{aligned}$$

Here, we use the identity $\widehat{\phi}(\omega) = m_0(\frac{\omega}{2})\widehat{\phi}(\frac{\omega}{2})$.

This is the Fourier transformed version of the dilation equation for $\rho_r(t)$. By expanding the periodic factor in a Fourier series and taking the inverse Fourier transform, we can find the coefficients of the dilation equation in the time domain as follows ($h_n = \sqrt{2}c_n$ of (4.2)):

$$\begin{aligned}
a_k &= \sum_{n=-\infty}^{\infty} \left( h_{k-2n} r^{|n|} \frac{1+r^2}{1-r^2} - h_{k-1-2n}\frac{r^{|n|+1}}{1-r^2} - h_{k+1-2n}\frac{r^{|n|-1}}{1-r^2} \right) \qquad (4.11) \\
&= a_k^{(1)} + a_k^{(2)} + a_k^{(3)}.
\end{aligned}$$

Notice that, for every fixed $k$ (4.11) only contains a finite number of terms. We then have the dilation equation:

$$\rho_r(t) = \sqrt{2} \sum_{k=-\infty}^{\infty} a_k \rho_r(2t-k).$$

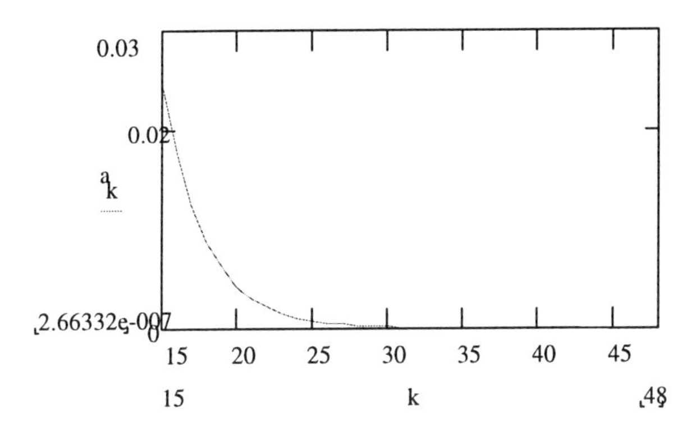


**FIGURE 4.1. Rate of decay of filter coefficients $a_k[8, 0.42]$**

Moreover, the filter coefficients $a_k$ are exponentially decreasing as $k \to \pm\infty$. In fact, in the case of Daubechies scaling functions, $h_n = 0$, for $n < 0$ or $n > 2N$. For $k > 2N - 1$, we have ,

$$|a_k^{(1)}| \leq \frac{1 + r^2}{1 - r^2} \cdot \frac{1 - r^{-N}}{1 - r^{-1}} \cdot \frac{1}{\sqrt{r}} r^{k/2}, k \geq 2N.$$

Similarly, we have the estimation:

$$|a_k^{(2)}| \leq \frac{r}{1 - r^2} \cdot \frac{1 - r^{-N}}{1 - r^{-1}} \cdot \frac{1}{\sqrt{r}} r^{k/2}, \; k \geq 2N,$$

$$|a_k^{(3)}| \leq \frac{r^{-2}}{1 - r^2} \cdot \frac{1 - r^{-N}}{1 - r^{-1}} \cdot \frac{1}{\sqrt{r}} r^{k/2}, \; k \geq 2N.$$

If we denote $C(r, N) = \frac{1+r^{-2}+r+r^2}{(1-r^2)\sqrt{r}} \cdot \frac{1-r^{-N}}{1-r^{-1}}$ , then

$$|a_k| \leq C(r, N) \cdot r^{k/2}, \qquad k \geq 2N.$$

This estimate is also true if $k < 0$. For a typical case, say ${}_4\phi$, if we choose $r = \frac{1}{2}$ , then $C(\frac{1}{2}, 4) = 115\sqrt{2}$. The coefficients corresponding to different $N$ and $r$ (denoted as $a_k[N, r]$) arc list in the Table 4.1. Table 4.1 together with Figure 4.1 illustrate the rate we have just obtained.

**Remark 4.1** *To lessen the amount of calculation in $\rho_r(t)$, the parameter $r$ should be chosen as small as possible, while keeping $\rho_r(t) \geq 0$. The optimum value satisfying this requirement is dependent on the length of support interval of the associated Daubechies scaling functions. Some results on numerical experiments to find an appropriate $r$ can be found in [10].*

| $k$ | $a_k[4, 0.5]$ | $a_k[5, 0.5]$ | $a_k[6, 0.48]$ | $a_k[8, 0.42]$ |
|---|---|---|---|---|
| 15 | 0.02481696 | 0.02637668 | 0.02074980 | 0.00912566 |
| 16 | 0.01745430 | 0.01869251 | 0.01435488 | 0.00590334 |
| 17 | 0.01240848 | 0.01318834 | 0.00995990 | 0.00383278 |
| 18 | 0.00872715 | 0.00934625 | 0.00689034 | 0.00247940 |
| 19 | 0.00620424 | 0.00659417 | 0.00478075 | 0.00160977 |
| 20 | 0.00436358 | 0.00467313 | 0.00330736 | 0.00104135 |
| 21 | 0.00310212 | 0.00329709 | 0.00229476 | 6.76101655E-4 |
| 22 | 0.00218179 | 0.00233656 | 0.00158753 | 4.37366994E-4 |
| 23 | 0.00155106 | 0.00164854 | 0.00110149 | 2.83962695E-4 |
| 24 | 0.00109089 | 0.00116828 | 7.62016543E-4 | 1.83694138E-4 |
| 25 | 7.75529995E-4 | 8.24271388E-4 | 5.28713050E-4 | 1.19264332E-4 |
| 26 | 5.45447020E-4 | 5.84140910E-4 | 3.65767940E-4 | 7.71515378E-5 |
| 27 | 3.87764998E-4 | 4.12135694E-4 | 2.53782264E-4 | 5.00910194E-5 |
| 28 | 2.72723510E-4 | 2.92070455E-4 | 1.75568611E-4 | 3.24036459E-5 |
| 29 | 1.93882499E-4 | 2.06067847E-4 | 1.21815487E-4 | 2.10382282E-5 |
| 30 | 1.36361755E-4 | 1.46035227E-4 | 8.42729335E-5 | 1.36095313E-5 |
| 31 | 9.69412494E-5 | 1.03033924E-4 | 5.84714337E-5 | 8.83605582E-6 |
| 32 | 6.81808775E-5 | 7.30176137E-5 | 4.04510081E-5 | 5.71600313E-6 |
| 33 | 4.84706247E-5 | 5.15169618E-5 | 2.80662882E-5 | 3.71114345E-6 |
| 34 | 3.40904387E-5 | 3.65088069E-5 | 1.94164839E-5 | 2.40072132E-6 |
| 35 | 2.42353123E-5 | 2.57584809E-5 | 1.34718183E-5 | 1.55868025E-6 |
| 36 | 1.70452194E-5 | 1.82544034E-5 | 9.31991226E-6 | 1.00830295E-6 |
| 37 | 1.21176562E-5 | 1.28792404E-5 | 6.46647279E-6 | 6.54645704E-7 |
| 38 | 8.52260969E-6 | 9.12720172E-6 | 4.47355788E-6 | 4.23487240E-7 |
| 39 | 6.05882809E-6 | 6.43962022E-6 | 3.10390694E-6 | 2.74951196E-7 |
| 40 | 4.26130484E-6 | 4.56360086E-6 | 2.14730778E-6 | 1.77864641E-7 |
| 41 | 3.02941404E-6 | 3.21981011E-6 | 1.48987533E-6 | 1.15479502E-7 |
| 42 | 2.13065242E-6 | 2.28180043E-6 | 1.03070774E-6 | 7.47031492E-8 |
| 43 | 1.51470702E-6 | 1.60990506E-6 | 7.15140159E-7 | 4.85013909E-8 |
| 44 | 1.06532621E-6 | 1.14090021E-6 | 4.94739714E-7 | 3.13753227E-8 |
| 45 | 7.57353511E-7 | 8.04952528E-7 | 3.43267276E-7 | 2.03705842E-8 |
| 46 | 5.32663105E-7 | 5.70450107E-7 | 2.37475063E-7 | 1.31776355E-8 |

**TABLE 4.1. Filter Coefficients $a_k$**

### *4.4.2 The calculation of the positive estimation $f_m^r(t)$*

One way to calculate the positive estimation $f_m^r(t)$ approximately is by using the definition (4.9) and then numerically approximating the integral. This is not very efficient, however. A better procedure is to use the series (4.10) instead together with the dilation equation of $\rho_r(t)$ whose coefficients are given by (4.11). But this involves truncating the dilation equation since there are an infinite number of terms in (4.11) which are non zero. Still another procedure involves the calculation of $f_m^r(t)$ in terms of the projection onto $V_m$ given by

$$f_m(t) = \int_{-\infty}^{\infty} q_m(t,s) f(s) ds.$$

where $q_m(t,s) = \sum_{-\infty}^{\infty} \phi_{m,n}(t)\phi_{m,n}(s)$ is the reproducing kernel of $V_m$ .
In order to do so, we first express $f_m^r$ by its Fourier transform,

$$\begin{aligned}
\widehat{f_m^r}(\omega) &= \int_{-\infty}^{\infty} \widehat{k}_m(\omega,s) f(s) ds \\
&= \left(\frac{1-r}{1+r}\right)^2 \int_{-\infty}^{\infty} \widehat{\rho}_r(2^{-m}\omega) \overline{n \sum e^{-i\omega n/2^m}} \rho_r(2^m s - n) f(s) ds \\
&= \left(\frac{1-r}{1+r}\right)^2 \widehat{\rho}_r(2^{-m}\omega) \\
&\qquad \times \frac{1}{2\pi} \int_{-\infty}^{\infty} 2^{-m} n \sum e^{-i\omega n/2^m} \overline{e^{-i\xi n/2^m} \widehat{\rho}_r(2^{-m}\xi)} \widehat{f}(\xi) d\xi \\
&= \left(\frac{1-r}{1+r}\right)^2 \widehat{\rho}_r(2^{-m}\omega) \\
&\qquad \times \frac{1}{2\pi} \int_{-\infty}^{\infty} k \sum \delta((k-\xi)2^{-m} + 2\pi k) \widehat{\rho}_r(2^{-m}\xi) \widehat{f}(\xi) d\xi \\
&= \left(\frac{1-r}{1+r}\right)^2 \widehat{\rho}_r(2^{-m}\omega) \frac{1}{2\pi} k \sum \widehat{\rho}_r(2^{-m}\omega + 2\pi k) \widehat{f}(\omega + 2^{m+1}\pi k),
\end{aligned}$$

where $\widehat{k}_m(\omega, s)$ denote the Fourier transform of $k_m(t,s)$ with respect to the first variable.

We first find $f_0^r(t)$ in terms of $f_0(t)$. To obtain the expression for $f_m^r(t)$, we need merely change the scale. The Fourier transform of $f_0(t)$ is

$$\widehat{f_0^r}(\omega) = \int_{-\infty}^{\infty} \widehat{k_0^r}(\omega, s) f(s) ds,$$

where

$$\begin{aligned}\widehat{k_0^r}(\omega, s) &= \left(\frac{1-r}{1+r}\right)^2 \widehat{\rho}_r(\omega) n \sum e^{-i\omega n} \rho_r(s-n) \\ &= \left(\frac{1-r}{1+r}\right)^2 G_r(\omega) \widehat{\phi}(\omega) n \sum e^{-i\omega n} \rho_r(s-n).\end{aligned}$$

We now use Parseval equality for the Fourier transform (in a generalized sense) to find

$$\begin{aligned}\widehat{f_0^r}(\omega) &= \frac{1}{2\pi} \int_{-\infty}^{\infty} \overline{\widehat{\widehat{k_0^r}}(\omega, \xi)} \widehat{f}(\xi) d\xi \\ &= \left(\frac{1-r}{1+r}\right)^2 \frac{G_r(\omega)\widehat{\phi}(\omega)}{2\pi} \int_{-\infty}^{\infty} n \sum e^{-i\omega n} \overline{\widehat{\rho}_r(\xi) e^{-i\xi n}} \widehat{f}(\xi) d\xi \\ &= \left(\frac{1-r}{1+r}\right)^2 G_r(\omega)\widehat{\phi}(\omega) \int_{-\infty}^{\infty} k \sum \delta(\omega - \xi + 2\pi k) \overline{\widehat{\rho}_r(\xi)} \widehat{f}(\xi) d\xi \\ &= \left(\frac{1-r}{1+r}\right)^2 G_r(\omega)\widehat{\phi}(\omega) k \sum \overline{\widehat{\rho}_r(\omega + 2\pi k)} \widehat{f}(\omega + 2\pi k) \\ &= \left(\frac{1-r}{1+r}\right)^2 G_r(\omega)\widehat{\phi}(\omega) \overline{G_r(\omega)} k \sum \overline{\widehat{\phi}(\omega + 2\pi k)} \widehat{f}(\omega + 2\pi k) \\ &= \left(\frac{1-r}{1+r}\right)^2 |G_r(\omega)|^2 \widehat{\phi}(\omega) k \sum \overline{\widehat{\phi}(\omega + 2\pi k)} \widehat{f}(\omega + 2\pi k).\end{aligned}$$

Since $f_0(t)$ has its Fourier transform giving by

$$\begin{aligned}\widehat{f_0}(\omega) &= \int_{-\infty}^{\infty} \widehat{q}(\omega, t) f(t) dt \\ &= \widehat{\phi}(\omega) k \sum \overline{\widehat{\phi}(\omega + 2\pi k)} \widehat{f}(\omega + 2\pi k).\end{aligned}$$

we find that

$$\widehat{f_0^r}(\omega) = \left(\frac{1-r}{1+r}\right)^2 |G_r(\omega)|^2 \widehat{f_0}(\omega) \tag{4.12}$$
$$= \left(\frac{1-r}{1+r}\right)^2 \frac{(1-r^2)^2}{(1-2r\cos\omega + r^2)^2} \widehat{f_0}(\omega)$$
$$= \frac{(1-r)^4}{(1-2r\cos\omega + r^2)^2} \widehat{f_0}(\omega).$$

This nicely avoids the need to use the truncated decomposition algorithm to find $f_m^r$ as given by the series (4.10). This would have involved first finding the approximate coefficients at the finest scale of interest, and then using the truncated dilation equation with coefficients given by (4.11) to approximate the coefficients in the series (4.10) at all coarser scales. But now we can use the decomposition algorithm for $f_m$ instead. This involves only a finite number of terms in the dilation equation, and hence does not introduce truncation error. We then convert to $f_m^r$ at the final stage by using (4.12).

This also can be phrased directly in terms of the coefficients of $f_0$ and $f_0^r$. The form is particularly simple for the inverse transformation since

$$\widehat{f_0}(\omega) = \frac{1}{(1-r)^4}(1 - re^{i\omega} - re^{-i\omega} + r^2)^2 \widehat{f_0^r}(\omega). \tag{4.13}$$

By taking Fourier transform, we have

$$f_0(t) = \frac{1}{(1-r)^4}[((1+r^2)^2 + 2r^2) f_0^r(t) - 2r(1+r^2)(f_0^r(t+1) + f_0^r(t-1)) + r^2(f_0^r(t+2) + f_0^r(t-2))].$$

Since both $f_0$ and $f_0^r \in V_0$, they may be expanded as a series of $\{\phi(t-k)\}$, say,

$$f_0\,(t) = \sum b_k \phi(t-k), \tag{4.14}$$

and

$$f_0^r(t) = \sum \beta_k^r \phi(t-k). \tag{4.15}$$

Therefore (4.13) is equivalent to

$$(1-r)^4 b_k = r^2 \beta_{k-2}^r - 2r(1+r^2)\beta_{k-1}^r + [(1+r^2)^2 + 2r^2]\beta_k^r - 2r(1+r^2)\beta_{k+1}^r + r^2 \beta_{k+2}^r\,, \tag{4.16}$$

| $k$ | $b_k$ | $\beta_k^{0.5}$ |
|---|---|---|
| -8 | .000000 | 0.002851 |
| -7 | .000000 | 0.005080 |
| -6 | .000002 | 0.008913 |
| -5 | -.000252 | 0.015332 |
| -4 | -.002932 | 0.025676 |
| -3 | .025969 | 0.042319 |
| -2 | -.099477 | 0.061755 |
| -1 | .386723 | 0.087451 |
| 0 | .189967 | 0.083924 |
| 1 | .000000 | 0.062088 |
| 2 | .000000 | 0.041181 |

**TABLE 4.2. The conversion between $b_k$ and $\beta_k^0.5$**

which can be rewritten as

$$\beta_{k+2}^r = C_r b_k - (\beta_{k-2}^r + A_r \beta_{k-1}^r + B_r \beta_k^r + A_r \beta_{k+1}^r) \, , \qquad (4.17)$$

where $A_r = -\frac{2(1+r^2)}{r}$, $B_r = \frac{(1+r^2)^2+2r^2}{r^2}$, and $C_r = \frac{(1-r)^4}{r^2}$.

The series (4.15) has only a finite number of terms for each fixed $t$, since the scaling function $\phi$ has compact support. In fact, if we are only interested in the expansion on a finite interval, the conversion from the conventional expansion (4.14) can be accomplished by using the iteration formula (4.17) in a finite number of steps.

To illustrate the idea, we consider the expansions of the characteristic function of $[0, 1]$ :

$$\chi(t) = \begin{cases} 1, \, t \in [0,1] \\ 0, \, t \notin [0,1] \end{cases} . \qquad (4.18)$$

as an example. To be specific, we use the scaling function of the Daubechies wavelet with 8 taps as a basis, and choose $r = 0.5$. Correspondingly, we have $A_r = -0.5$, $B_r = 8.25$, and $C_r = 0.25$.

We assume that all the coefficients of the conventional expansion (4.14) are known (Notice that in this case, the first and the last non zero coefficients for (4.14) are $b_{-6}$ and $b_0$, respectively). If we are only interested the expansion on an interval $[-1, 2]$, the first non zero coefficient for (4.15) is $\beta_{-8}^{0.5}$, while the last coefficient we need on this interval is $\beta_1^{0.5}$. However, for this example, $\beta_2^{0.5} \neq 0$, but $\phi(t-2) \equiv 0$ for $t \in [-1, 2]$. These coefficients are enough for the expansion on this interval. For other intervals, we have to go on with (4.15). We use (4.16) to calculate $\beta_k^{0.5}$, $k = -8, -7, \ldots, -5$ as initial values, and then employ (4.17) to calculate the rest of the coefficients (See Table 4.2).

**Conclusion.** A substitute for classical summability methods suitable for wavelet expansions has been introduced. If the wavelets have compact support, this becomes a positive summability method which avoids Gibbs's phenomenon. The associated approximation $f_m^r$ can be found in a number of ways: directly from (4.9) or (4.10), in terms of $f_m$ by using (4.12), or by using its series expansion (4.14) as well as (4.17).

*References*


[1] S. Bochner and K. Chandrasekharan, Fourier Transforms, Princeton University, Princeton, 1949.

[2] I. Daubechies, Ten Lectures On Wavelets, CBMS-NSF Series in Appl. Math., SIAM Publ., 1992.

[3] J. W. Gibbs, Letter To The Editor, Nature (London) 59 (1899), 606.

[4] A.J.E.M. Janssen, The Smith - Barnwell Condition And Non - Negative Scaling Functions, IEEE Trans. Info. The., 38, 884-885 (1994).

[5] S. E. Kelly, Gibbs' Phenomenon For Wavelets, Appl. Comput. Harmon. Anal. 3(1996), 72-81.

[6] S. E. Kelly, M. A. Kon and L. A. Raphael, Local Convergence For Wavelet Expansions, J. Funct. Anal 126(1994), 102-138.

[7] Y. Meyer, Ondelettes et Opèrateurs I, Herman, Paris 1990.

[8] H. T. Shim and H. Volkmer, On Gibbs' Phenomenon For Wavelet Expansions, J. Approx. The. 84(1996), 74-95.

[9] G. G. Walter, Wavelets And Other Orthogonal Wavelets With Applications, CRC Press, 1994.

[10] G. G. Walter and X. Shen, Positive Estimation With Wavelets, Cont. Math. 216 (1998), 63-79.

[11] G. G. Walter and H. T. Shim, Gibbs' Phenomenon For Sampling Series And What To Do About It. J Fourier Anal & Appl, 4 (1998) 355-373.

[12] S. Zygmund, Trigonometric Series, Cambridge, 1959.

# Chapter 5

# Expansions in series of Legendre functions

**E. R. Love**
**M. N. Hunter**

## 5.1 Introduction

The Legendre functions $P_\nu^\mu(x)$ occurring in this paper are sometimes called modified Legendre functions, or Legendre functions on the cut. They are defined in [2, p.143, eq. (6)] as

$$P_\nu^\mu(x) = \frac{1}{\Gamma(1-\mu)}\left(\frac{1+x}{1-x}\right)^{\frac{1}{2}\mu} F\left(-\nu; 1+\nu; 1-\mu; \frac{1-x}{2}\right) \tag{5.1}$$

for $-1 < x < 1$, where $F$ is Gauss's hypergeometric function and $\mu$ and $\nu$ are real or complex parameters. They satisfy the recurrence relation

$$(\nu-\mu+1)P_{\nu+1}^\mu(x) + (\nu+\mu)P_{\nu-1}^\mu(x) = (2\nu+1)xP_\nu^\mu(x). \tag{5.2}$$

The Legendre polynomials are $P_n^0(x)$ (usually written $P_n(x)$), the case $\mu = 0$ and $\nu = n$ a positive integer or zero; they will not often appear in this paper.

The main result obtained in [5], Theorem 8, runs as follows.

*If* $(1-t^2)^{-\frac{1}{4}}f(t) \in L(-1,1)$, $f$ *is Dini (see below) at a certain* $x \in (-1,1)$, $|\operatorname{Re}\mu| < \frac{1}{2}$ *and* $\nu$ *is not half an odd integer, then*

$$f(x) = \sum_{n=-\infty}^{\infty} a_n P_{\nu+n}^\mu(x),$$

*where*

$$a_n = (-1)^n \frac{\nu+n+\frac{1}{2}}{2\cos\nu\pi} \int_{-1}^{1} f(t)P_{\nu+n}^{-\mu}(-t)\,dt;$$

*and the two "halves" of the series are separately convergent.*

The Dini condition on $f$ at $x$ is that for some $\delta > 0$

$$\frac{f(x)-f(t)}{x-t} \in L(x-\delta, x+\delta).$$

It allows $f$ to be differentiable at $x$, and more generally Hölder-continuous of any order in $(0, 1]$ at $x$; but it does not allow $f$ to be discontinuous at $x$.

The purpose of the paper is to present several improvements and tidyings of the work in [5] that have since become apparent. Perhaps the most notable of these is one of M. N. Hunter's identities; this appears in Lemma 5.5.

## 5.2 Preliminaries and known results

Several results which can be found in [5] are needed in the sequel, as follows.

### 5.2.1 *Christoffel Summation Formula*

**Theorem 5.1** *If $n$ is a positive integer, $x$ and $t$ are in $(-1, 1)$ and $x \neq t$, then*

$$\sum_{r=0}^{n-1}(-1)^r(\nu + r + \tfrac{1}{2})P^{\mu}_{\nu+r}(x)P^{-\mu}_{\nu+r}(-t) = \frac{(-1)^{n-1}D(\nu + n, t) + D(\nu, t)}{x - t}, \tag{5.3}$$

*where*

$$\begin{aligned} D(\nu, t) &= D(\mu, \nu; x, t) \\ &= \frac{1}{2}\{(\nu - \mu)P^{\mu}_{\nu}(x)P^{-\mu}_{\nu-1}(-t) + (\nu + \mu)P^{\mu}_{\nu-1}(x)P^{-\mu}_{\nu}(-t)\}. \end{aligned} \tag{5.4}$$

**Proof.** The proof is essentially the same as in [5, p.581], with $m$ replaced by 1 and $n$ by $n-1$. It involves nothing more than elementary algebra from (5.2). ■

Christoffel's formula was the case $\mu = 0 = \nu$; it involved only Legendre polynomials.

### 5.2.2 *Stieltjes's Inequality*

As in [5, pp. 582–585]. This includes the following results.

**Theorem 5.2** *If $(1 - t^2)^{-\frac{1}{4}} f(t) \in L(-1, 1)$ and $|\operatorname{Re} \mu| < \frac{1}{2}$, then*

$$\int_{-1}^{1} f(t)P^{\mu}_{\nu}(\pm t)\, dt$$

*exist as L-integrals.*

**Theorem 5.3** *If* $\mu$ *and* $\operatorname{Im}\nu$ *are fixed with* $|\operatorname{Re}\mu| < \frac{1}{2}$, *then*

$$P_\nu^\mu(\cos\theta) = O\left(\frac{(\operatorname{Re}\nu)^{\operatorname{Re}\mu - \frac{1}{2}}}{(\sin\theta)^{\frac{1}{2}}}\right)$$

*as* $\operatorname{Re}\nu \to +\infty$, *uniformly on* $0 < \theta < \pi$.

### *5.2.3 Riemann-Lebesgue-type Theorem*

**Theorem 5.4** *If* $(1-t^2)^{-\frac{1}{4}} f(t) \in L(-1,1)$, $\mu$ *and* $\operatorname{Im}\nu$ *are fixed and* $|\operatorname{Re}\mu| < \frac{1}{2}$, *then*

$$\int_{-1}^{1} f(t) P_\nu^\mu(\pm t)\, dt = o\left((\operatorname{Re}\nu)^{\operatorname{Re}\mu - \frac{1}{2}}\right)$$

*as* $\operatorname{Re}\nu \to +\infty$.

### *5.2.4 Singular Integrals*

**Lemma 5.5** *If* $-\infty < a < b < \infty$, $\psi \in L(a,b)$ *and* $\psi$ *is Dini at a certain* $x \in (a,b)$, *then the singular (that is, Cauchy principal value) integral*

$$p.v.\int_a^b \frac{\psi(t)}{x-t}\, dt := \lim_{y \to 0+} \left( \int_a^{x-y} + \int_{x+y}^{b} \right) \frac{\psi(t)}{x-t}\, dt$$

*exists.*

**Lemma 5.6** *If* $\phi$ *is differentiable at* $x$ *and* $\psi$ *is Dini at* $x$, *then the product* $\phi\psi$ *is Dini at* $x$.

**Theorem 5.7** *(Dirichlet's Integral subdivided) If* $(1-t^2)^{-\frac{1}{4}} f(t) \in L(-1,1)$, $f$ *is Dini at a certain* $x \in (-1,1)$ *and* $|\operatorname{Re}\mu| < \frac{1}{2}$, *then*

$$\sum_{r=0}^{n-1} b_r P_{\nu+r}^\mu(x) = (-1)^{n-1} I_n + I_0\,, \tag{5.5}$$

*where*

$$b_r = (-1)^r(\nu + r + \frac{1}{2}) \int_{-1}^{1} f(t) P_{\nu+r}^{-\mu}(-t)\, dt \tag{5.6}$$

*and*

$$I_n = p.v.\int_{-1}^{1} \frac{f(t)}{x-t} D(\nu+n, t)\, dt. \tag{5.7}$$

**Proof.** Integrating (5.3) in Theorem 5.1 with respect to $f(t)\,dt$ over $(-1,1)$,

$$\sum_{r=0}^{n-1}(-1)^r(\nu+r+\frac{1}{2})P^{\mu}_{\nu+r}(x)\int_{-1}^{1}P^{-\mu}_{\nu+r}(-t)\,f(t)\,dt$$
$$=\int_{-1}^{1}\frac{(-1)^{n-1}D(\nu+n,t)+D(\nu,t)}{x-t}\,f(t)\,dt,$$

the integrals on the left existing by Theorem 5.2, and consequently that on the right by linearity. By (5.4) $D(\nu+n,t)$ is differentiable at $x$ and by hypothesis $f$ is Dini at $x$. So by Lemma 5.6 $D(\nu+n,t)f(t)$ is Dini at $x$. By (5.4) and Theorem 5.2, $D(\nu+n,t)f(t)$ is in $L(-1,1)$. So by Lemma 5.5 the singular integral

$$\int_{-1}^{1}\frac{D(\nu+n,t)\,f(t))}{x-t}\,dt$$

exists; in particular it exists for $n=0$. By linearity of singular integrals, and (5.7),

$$(-1)^{n-1}I_n+I_0=\int_{-1}^{1}\frac{(-1)^{n-1}D(\nu+n,t)+D(\nu,t)}{x-t}\,f(t)\,dt.$$

The integral on the right is equal to the corresponding L-integral, since this exists as remarked in connection with (5.4). With (5.4) and (5.6), this proves (5.5). ■

## 5.3 Neumann's Integral and consequences

Franz Neumann proved [2, 3.6 (29) on p.154] that $Q_n$, a Legendre function of the second kind, is related to the corresponding Legendre polynomial $P_n$ by

$$Q_n(z)=\frac{1}{2}\int_{-1}^{1}\frac{P_n(t)}{z-t}\,dt$$

for all $z$ in the complex plane cut along the real axis from $-1$ to $1$. A generalization of this was given by Love [3, pp. 450–453]; and limit processes indicated by [2, 3.4 on p.143] applied to the generalization give

**Theorem 5.8** *If* $\operatorname{Re}\mu < 1$, $\operatorname{Re}\nu > -1$, $\operatorname{Re}(\mu+\nu) > -1$, $n \geq 0$ *and*

$$\theta(t) := \frac{(1+t)^{\nu+\frac{1}{2}\mu}}{(1-t)^{\frac{1}{2}\mu}} \text{ for } -1 < t < 1,$$

*then* $\theta(t)\, P^{\mu}_{\nu+n}(t) \in L(-1,1)$ *and*

$$\theta(x)\, Q^{\mu}_{\nu+n}(x) = \frac{1}{2} p.v. \int_{-1}^{1} \frac{\theta(t)\, P^{\mu}_{\nu+n}(t)}{x-t}\, dt \text{ for } -1 < x < 1,$$

*where* $Q^{\mu}_{\nu}(x)$ *may be defined by [2, 3.4 (10) on p.144].*

**Proof.** The proof is as in [5, pp. 589–590], with $k = 0$ and $p(t) = 1$. ∎
The key consequences of this result are the following two lemmas.

**Lemma 5.9** *If* $(1-t^2)^{-\frac{1}{4}} f(t) \in L(-1,1)$, $f$ *is Dini at a certain* $x \in (-1,1)$ *and* $|\operatorname{Re}\mu| < \frac{1}{2}$, $\operatorname{Re}\nu \geq -\frac{1}{2}$ *and* $n \geq 0$, *then*

$$\frac{1}{2}\, p.v. \int_{-1}^{1} \frac{f(t)}{x-t} P^{\mu}_{\nu+n}(t)\, dt = f(x)\, Q^{\mu}_{\nu+n}(x) - \frac{1}{2} \int_{-1}^{1} \frac{h(x)-h(t)}{x-t} \frac{(1+t)^{\nu+\frac{1}{2}\mu}}{(1-t)^{\frac{1}{2}\mu}} P^{\mu}_{\nu+n}(t)\, dt,$$

*where*

$$h(t) = \frac{(1-t)^{\frac{1}{2}\mu}}{(1+t)^{\nu+\frac{1}{2}\mu}} f(t)$$

*and the last integral is Lebesgue.*

**Proof.** Let

$$\theta(t) = \frac{(1+t)^{\nu+\frac{1}{2}\mu}}{(1-t)^{\frac{1}{2}\mu}} \quad \text{and} \quad \rho(t) = \frac{h(x)-h(t))}{x-t}. \tag{5.8}$$

Then $f(t) = h(t)\theta(t)$ and

$$\begin{aligned}
f(x)\, Q^{\mu}_{\nu+n}(x) &= h(x)\, \theta(x)\, Q^{\mu}_{\nu+n}(x) \\
&= h(x) \frac{1}{2} p.v. \int_{-1}^{1} \frac{\theta(t)\, P^{\mu}_{\nu+n}(t)}{x-t}\, dt \quad \text{(by Theorem 5.8)} \\
&= \frac{1}{2}\, p.v. \int_{-1}^{1} \frac{h(x)-h(t)+h(t)}{x-t}\, \theta(t)\, P^{\mu}_{\nu+n}(t)\, dt \\
&= \frac{1}{2}\, p.v. \int_{-1}^{1} \rho(t)\, \theta(t)\, P^{\mu}_{\nu+n}(t)\, dt \\
&\quad + \frac{1}{2}\, p.v. \int_{-1}^{1} \frac{f(t)}{x-t} P^{\mu}_{\nu+n}(t)\, dt,
\end{aligned} \tag{5.9}$$

which gives the required equation, provided that the separation into two integrals in the last step is correct and the former of these integrals is Lebesgue.

To justify these provisos, observe first that $h$ is Dini at $x$, by its definition and Lemma 5.6. So there is $\delta > 0$ such that $\rho \in L(x-\delta, x+\delta)$, and $\delta$ can be chosen small enough for $[x-\delta, x+\delta] \subset (-1,1)$. Thus $\rho(t)\,\theta(t)\,(1-t^2)^{-1/4} \in L(x-\delta, x+\delta)$. Outside $(x-\delta, x+\delta)$ but inside $(-1,1)$,

$$\begin{aligned}\delta|(1-t^2)^{-1/4}\theta(t)\rho(t)| &\le (1-t^2)^{-1/4}|\theta(t)|\{|h(x)|+|h(t)|\}\\ &= |h(x)|(1+t)^{\mathrm{Re}(\nu+\frac{1}{2}(\mu-\frac{1}{2}))}(1-t)^{-\frac{1}{2}(\mathrm{Re}\ \mu+\frac{1}{2})}\\ &\qquad + |(1-t^2)^{-1/4}\, f(t)|.\end{aligned}$$

The last two terms are in $L(-1,1)$. Consequently $(1-t^2)^{-1/4}\;\theta(t)\;\rho(t)$ is in $L\{(-1, x-\delta) \cup (x+\delta, 1)\}$.

Having thus established that $(1-t^2)^{-1/4}\,\theta(t)\,\rho(t) \in L(-1,1)$, it follows from Theorem 5.2 that the first integral on the right of (5.9) exists; this justifies the separation of the previous line into the sum of the two integrals in (5.9), and completes the proof of Lemma 5.9. ■

**Lemma 5.10** *If $(1-t^2)^{-1/4} f(t) \in L(-1,1)$, $f$ is Dini at a certain $x \in (-1,1)$, $\mu$ and $\nu$ are fixed with $|\,\mathrm{Re}\ \mu| < \frac{1}{2}$, then as $n \to \infty$*

$$\frac{1}{2}\,p.v.\int_{-1}^{1}\frac{f(t)}{x-t}\,P^{\mu}_{\nu+n}(-t)\,dt = -f(x)\,Q^{\mu}_{\nu+n}(-x) + o\left(n^{\mathrm{Re}\,\mu-\frac{1}{2}}\right).$$

**Proof.** Suppose temporarily that $\mathrm{Re}\ \nu \ge -\frac{1}{2}$. For $n \ge 0$, Lemma 5.9 gives

$$\frac{1}{2}\int_{-1}^{1}\frac{f(t)}{x-t}\,P^{\mu}_{\nu+n}(t)\,dt = f(x)\,Q^{\mu}_{\nu+n}(x) - \frac{1}{2}\int_{-1}^{1}\theta(t)\rho(t)\,P^{\mu}_{\nu+n}(t)\,dt$$

with $\theta$ and $\rho$ defined as in (5.8). Since by Lemma 5.9

$$(1-t^2)^{-1/4}\theta(t)\rho(t) \in L(-1,1),$$

Theorem 5.4 gives that the last integral is $o\left((\mathrm{Re}\ \nu+n)^{\mathrm{Re}\ \mu-\frac{1}{2}}\right)$ as $n \to \infty$, and therefore $o\left(n^{\mathrm{Re}\ \mu-\frac{1}{2}}\right)$.

The temporary hypothesis that $\mathrm{Re}\ \nu \ge -\frac{1}{2}$ can now be omitted. For, for any given $\nu$, there is positive $m$ such that $\mathrm{Re}\,(\nu+m) \ge -\frac{1}{2}$. Replacing $\nu$ by $\nu+m$ and $n$ by $n-m$,

$$\frac{1}{2}\,p.v.\int_{-1}^{1}\frac{f(t)}{x-t}\,P^{\mu}_{\nu+n}(t)\,dt = f(x)\,Q^{\mu}_{\nu+n}(x) + o\left(n^{\mathrm{Re}\ \mu-\frac{1}{2}}\right) \quad \text{as } n \to \infty. \tag{5.10}$$

Now let $y=-x$, $u=-t$ and $g(t)=f(u)=f(-t)$. Then

$$(1-u^2)^{-1/4}g(u)=(1-t^2)^{-1/4}f(t)\in L(-1,1),$$

and also $g$ is Dini at $y$. So the left side of the desired equation is equal, by (5.10), to

$$\begin{aligned}-\frac{1}{2}\,p.v.\int_{-1}^{1}\frac{g(u)}{y-u}\,P^{\mu}_{\nu+n}(u)\,du &= -g(y)\,Q^{\mu}_{\nu+n}(y)+o\left(n^{\mathrm{Re}\,\mu-\frac{1}{2}}\right)\\ &= -f(x)\,Q^{\mu}_{\nu+n}(-x)+o\left(n^{\mathrm{Re}\,\mu-\frac{1}{2}}\right),\end{aligned}$$

as desired. ■

## 5.4 Hunter's Identities

A typical one of several identities due to M. N. Hunter, and the only one that we require in this paper, is as follows. It replaces the asymptotic result given in [5, Lemma 11, p.592].

**Lemma 5.11** *For all real and complex $\mu$ and $\nu$ for which the functions are defined, and for $-1<x<1$,*

$$(\nu-\mu)P^{\mu}_{\nu}(x)\,Q^{-\mu}_{\nu-1}(-x)+(\nu+\mu)P^{\mu}_{\nu-1}(x)\,Q^{-\mu}_{\nu}(-x)=\cos\,\nu\pi.$$

**Proof.** Denote the left hand side by $J(x)$. Using [2, p. 144, eqs. (15), (18) and (17)], we perform the following set of manipulations.

$$\begin{aligned}J(x)=(\nu-\mu)P^{\mu}_{\nu}(x)&\left\{-Q^{-\mu}_{\nu-1}(x)\cos(\nu-\mu-1)\pi\right.\\ &\left.-\frac{1}{2}\pi P^{-\mu}_{\nu-1}(x)\sin(\nu-\mu-1)\pi\right\}\\ +(\nu+\mu)P^{\mu}_{\nu-1}(x)&\left\{-Q^{-\mu}_{\nu}(x)\cos(\nu-\mu)\pi\right.\\ &\left.-\frac{1}{2}\pi P^{-\mu}_{\nu}(x)\sin(\nu-\mu)\pi\right\}\end{aligned}$$

$$\begin{aligned}
J(x) &= (\nu-\mu)P_\nu^\mu(x)\cos(\nu-\mu)\pi\,\tfrac{\Gamma(\nu-\mu)}{\Gamma(\nu+\mu)} \\
&\qquad \times\{Q_{\nu-1}^\mu(x)\cos\mu\pi+\frac{1}{2}\pi P_{\nu-1}^\mu(x)\sin\mu\pi\} \\
&\quad +\frac{1}{2}\pi(\nu-\mu)P_\nu^\mu(x)\sin(\nu-\mu)\pi\,\tfrac{\Gamma(\nu-\mu)}{\Gamma(\nu+\mu)} \\
&\qquad \times\{P_{\nu-1}^\mu(x)\cos\mu\pi-\frac{2}{\pi}Q_{\nu-1}^\mu(x)\sin\mu\pi\} \\
&\quad -(\nu+\mu)P_{\nu-1}^\mu(x)\cos(\nu-\mu)\pi\,\tfrac{\Gamma(\nu-\mu+1)}{\Gamma(\nu+\mu+1)} \\
&\qquad \times\{Q_\nu^\mu(x)\cos\mu\pi+\frac{1}{2}\pi P_\nu^\mu(x)\sin\mu\pi\} \\
&\quad -\frac{1}{2}\pi(\nu+\mu)P_{\nu-1}^\mu(x)\sin(\nu-\mu)\pi\,\tfrac{\Gamma(\nu-\mu+1)}{\Gamma(\nu+\mu+1)} \\
&\qquad \times\{P_\nu^\mu(x)\cos\mu\pi-\frac{2}{\pi}Q_\nu^\mu(x)\sin\mu\pi\}
\end{aligned}$$

$$\begin{aligned}
J(x) &= \tfrac{\Gamma(\nu-\mu+1)}{\Gamma(\nu+\mu)}[\{P_\nu^\mu(x)Q_{\nu-1}^\mu(x)-P_{\nu-1}^\mu(x)Q_\nu^\mu(x)\} \\
&\qquad \times\{\cos(\nu-\mu)\pi\cos\mu\pi-\sin(\nu-\mu)\pi\sin\mu\pi\} \\
&\quad +\frac{1}{2}\pi\{P_\nu^\mu(x)\,P_{\nu-1}^\mu(x)-P_{\nu-1}^\mu(x)P_\nu^\mu(x)\} \\
&\qquad \times\{\cos(\nu-\mu)\pi\sin\mu\pi-\sin(\nu-\mu)\pi\cos\mu\pi\}]
\end{aligned}$$

Finally, we have:

$$J(x)=\tfrac{\Gamma(\nu-\mu+1)}{\Gamma(\nu+\mu)}\{P_\nu^\mu(x)\,Q_{\nu-1}^\mu(x)-P_{\nu-1}^\mu(x)\,Q_\nu^\mu(x)\}\cos\nu\pi. \tag{5.11}$$

Set

$$I(x)=\frac{2}{\pi}\sin\mu\pi(1-x^2)\frac{d}{dx}Q_\nu^\mu(x).$$

Then by [2, p.144, eq. (13)] and [2, p.161, eq. (19)],

$$\begin{aligned}
I(x) &= (1-x^2)\left[\cos\mu\pi\,\frac{d}{dx}P_\nu^\mu(x)-\frac{\Gamma(\nu+\mu+1)}{\Gamma(\nu-\mu+1)}\frac{d}{dx}P_\nu^{-\mu}(x)\right] \\
&= \cos\mu\pi\{-\nu xP_\nu^\mu(x)+(\nu+\mu)P_{\nu-1}^\mu(x)\} \\
&\quad -\frac{\Gamma(\nu+\mu+1)}{\Gamma(\nu-\mu+1)}\{-\nu xP_\nu^{-\mu}(x)+(\nu-\mu)P_{\nu-1}^{-\mu}(x)\} \\
&= -\nu x\frac{2}{\pi}\sin\mu\pi\,Q_\nu^\mu(x)+(\nu+\mu)\left\{\cos\mu\pi P_{\nu-1}^\mu(x)-\frac{\Gamma(\nu+\mu)}{\Gamma(\nu-\mu)}P_{\nu-1}^{-\mu}(x)\right\} \\
&= \frac{2}{\pi}\sin\mu\pi\left\{-\nu\,x\,Q_\nu^\mu(x)+(\nu+\mu)Q_{\nu-1}^\mu(x)\right\};
\end{aligned}$$

thus if $\mu$ is not an integer, and by continuity if it is,

$$(1-x^2)\frac{d}{dx}Q_\nu^\mu(x)=-\nu\,x\,Q_\nu^\mu(x)+(\nu+\mu)Q_{\nu-1}^\mu(x). \tag{5.12}$$

By (5.12) and [2, p. 161, eq. (19)],

$$\begin{aligned}
&P_\nu^\mu(x)\,Q_{\nu-1}^\mu(x) - P_{\nu-1}^\mu(x)\,Q_\nu^\mu(x)\\
&= \frac{1}{\nu+\mu}\left[P_\nu^\mu(x)\left\{(1-x^2)\frac{d}{dx}Q_\nu^\mu(x) + \nu\, xQ_\nu^\mu(x)\right\}\right.\\
&\qquad \left.-Q_\nu^\mu(x)\left\{(1-x^2)\frac{d}{dx}P_\nu^\mu(x) + \nu\, xP_\nu^\mu(x)\right\}\right]\\
&= \frac{1-x^2}{\nu+\mu}\left[P_\nu^\mu(x)\frac{d}{dx}Q_\nu^\mu(x) - Q_\nu^\mu(x)\frac{d}{dx}P_\nu^\mu(x)\right]\\
&= \frac{2^{2\mu}}{\nu+\mu}\,\frac{\Gamma(1+\frac12\nu+\frac12\mu)\Gamma(\frac12+\frac12\nu+\frac12\mu)}{\Gamma(1+\frac12\nu-\frac12\mu)\Gamma(\frac12+\frac12\nu-\frac12\mu)} \qquad (5.13)\\
&= \frac{2^{2\mu}}{\nu+\mu}\,\frac{\Gamma(\frac12)\Gamma(1+\nu+\mu)}{2^{\nu+\mu}}\,\frac{2^{\nu-\mu}}{\Gamma(\frac12)\Gamma(1+\nu-\mu)}; \qquad (5.14)
\end{aligned}$$

here (5.13) follows from [2, p. 146, eq. (25)], and (5.14) from Legendre's duplication formula [2, p.5, eq. (15)]. We thus obtain

$$P_\nu^\mu(x)\,Q_{\nu-1}^\mu(x) - P_{\nu-1}^\mu(x)\,Q_\nu^\mu(x) = \frac{\Gamma(\nu+\mu)}{\Gamma(1+\nu-\mu)}. \qquad (5.15)$$

This shows that (5.11) is equal to cos $\nu\pi$, and so proves Lemma 5.11. ■

**Remark 5.1** *Besides (5.14) and Lemma 5.11 there are several similar identities; for instance*

$$(\nu-\mu)P_\nu^\mu(x)\,Q_{\nu-1}^{-\mu}(x) - (\nu+\mu)P_{\nu-1}^\mu(x)\,Q_\nu^{-\mu}(x) = \cos\,\mu\pi$$

*and*

$$(\nu-\mu)P_\nu^\mu(x)\,P_{\nu-1}^{-\mu}(-x) + (\nu+\mu)P_{\nu-1}^\mu(x)\,P_\nu^{-\mu}(-x) = \frac{2}{\pi}\sin\,\nu\pi;$$

*but they are not needed in this paper.*

**Lemma 5.12** *If* $(1-t^2)^{-1/4}f(t) \in L(-1,1)$, $f$ *is Dini at a certain* $x \in (-1,1)$, *and* $\mu$ *and* $\nu$ *are fixed with* $|\,\mathrm{Re}\,\mu| < \frac12$, *then as* $n \to \infty$

$$(-1)^{n-1}p.v.\int_{-1}^{1}\frac{f(t)}{x-t}D(\nu+n,t)\,dt \to f(x)\cos\,\nu\pi.$$

**Proof.** Replacing $\mu$ in Lemma 5.10 by $-\mu$,

$$\frac12 p.v.\int_{-1}^{1}\frac{f(t)}{x-t}P_{\nu+n}^{-\mu}(-t)\,dt = -f(x)\,Q_{\nu+n}^{-\mu}(-x) + o\left(n^{-\,\mathrm{Re}\,\mu-\frac12}\right). \qquad (5.16)$$

Let $\nu_n = \nu + n$ ( this differs from the $\nu_n$ used in [5]; compare (2.3) therein); thus $\nu_{n-1} = \nu_n - 1$. By (5.4),

$$D(\nu_n, t) = \frac{1}{2}\{(\nu_n - \mu)\{P^{\mu}_{\nu_n}(x)\, P^{-\mu}_{\nu_n - 1}(-t) + (\nu_n + \mu) P^{\mu}_{\nu_n - 1}(x)\, P^{-\mu}_{\nu_n}(-t)\}.$$

By Theorem 5.2, $f(t)P^{-\mu}_{\nu_n}(-t) \in L(-1,1)$; and by Lemma 5.6, $f(t)P^{-\mu}_{\nu_n}(-t)$ is Dini at $t = x$. So by Lemma 5.5 the singular integrals below exist. Thus we have

$$\begin{aligned}
&p.v. \int_{-1}^{1} \frac{f(t)}{x-t} D(\nu_n, t)\, dt \\
&\quad = (\nu_n - \mu) P^{\mu}_{\nu_n}(x) \times \frac{1}{2} p.v. \int_{-1}^{1} \frac{f(t)}{x-t} P^{-\mu}_{\nu_n - 1}(-t)\, dt \\
&\quad\quad + (\nu_n + \mu) P^{\mu}_{\nu_n - 1}(x) \times \frac{1}{2} p.v. \int_{-1}^{1} \frac{f(t)}{x-t} P^{-\mu}_{\nu_n}(-t)\, dt \\
&\quad = (\nu_n - \mu) P^{\mu}_{\nu_n}(x) \{-f(x)\, Q^{-\mu}_{\nu_n - 1}(-x) \\
&\quad\quad\quad + o\left(n^{-\operatorname{Re}\mu - \frac{1}{2}}\right)\} \\
&\quad\quad + (\nu_n + \mu) P^{\mu}_{\nu_n - 1}(x) \{-f(x)\, Q^{-\mu}_{\nu_n}(-x) \\
&\quad\quad\quad + o\left(n^{-\operatorname{Re}\mu - \frac{1}{2}}\right)\} \\
&\quad = -f(x)\{(\nu_n - \mu) P^{\mu}_{\nu_n}(x) Q^{-\mu}_{\nu_n - 1}(-x) \\
&\quad\quad + (\nu_n + \mu) P^{\mu}_{\nu_n - 1}(x) Q^{-\mu}_{\nu_n}(-x)\} \\
&\quad\quad\quad + O(n)\, O\left(n^{\operatorname{Re}\mu - \frac{1}{2}}\right) o\left(n^{-\operatorname{Re}\mu - \frac{1}{2}}\right),
\end{aligned}$$

using (5.15) and Theorem 5.3. So by Lemma 5.11,

$$\begin{aligned}
\int_{-1}^{1} \frac{f(t)}{x-t} D(\nu + n, t)\, dt &= -f(x) \cos(\nu + n)\pi + o(1) \\
&= (-1)^{n+1} f(x) \cos \nu\pi + o(1),
\end{aligned}$$

and this is equivalent to the stated result. ■

**Lemma 5.13** *If* $(1 - t^2)^{-\frac{1}{4}} f(t) \in L(-1,1)$, $f$ *is Dini at a certain* $x \in (-1,1)$, $|\operatorname{Re}\mu| < \frac{1}{2}$, $\nu$ *is not half an odd integer, and for integers* $n$

$$a_n = (-1)^n \frac{\nu + n + \frac{1}{2}}{2 \cos \nu\pi} \int_{-1}^{1} f(t) P^{-\mu}_{\nu+n}(-t)\, dt,$$

*then*

$$\sum_{n=0}^{\infty} a_n P^{\mu}_{\nu+n}(x) = \frac{1}{2} f(x) + \frac{1}{2} \sec \nu\pi \int_{-1}^{1} \frac{f(t)}{x-t} D(\nu, t)\, dt.$$

**Proof.** Using Theorem 5.7 and Lemma 5.12,

$$\begin{aligned}
\sum_{n=0}^{\infty} a_n P^{\mu}_{\nu+n}(x) &= \frac{1}{2}\sec \nu\pi \sum_{n=0}^{\infty} b_n P^{\mu}_{\nu+n}(x) \\
&= \frac{1}{2}\sec \nu\pi \lim_{n\to\infty}\{(-1)^{n-1}I_n + I_0\} \\
&= \frac{1}{2}\sec \nu\pi \{\lim_{n\to\infty}(-1)^{n-1}\int_{-1}^{1}\frac{f(t)}{x-t}D(\nu+n,t)\,dt \\
&\qquad + \int_{-1}^{1}\frac{f(t)}{x-t}D(\nu,t)\,dt\} \\
&= \frac{1}{2}\sec \nu\pi \{f(x)\cos \nu\pi + \int_{-1}^{1}\frac{f(t)}{x-t}D(\nu,t)\,dt\},
\end{aligned}$$

and this gives the stated equation. ∎

**Lemma 5.14** *Under the hypotheses of Lemma 5.13,*

$$\sum_{n=0}^{\infty} a_{-n}P^{\mu}_{\nu-n}(x) = \frac{1}{2}f(x) - \frac{1}{2}\sec \nu\pi \int_{-1}^{1}\frac{f(t)}{x-t}D(\nu,t)\,dt.$$

**Proof.** Using the symmetry property $P^{\mu}_{\nu}(x) = P^{\mu}_{-\nu-1}(x)$ [2, p. 144, eq. (7)], the left side is formally equal to

$$\sum_{n=0}^{\infty} a_{-n-1}P^{\mu}_{\nu-n-1}(x) = \sum_{n=0}^{\infty} a_{-n-1}P^{\mu}_{n-\nu}(x),$$

and

$$\begin{aligned}
a_{-n-1} &= (-1)^{-n-1}\frac{\nu-n-\frac{1}{2}}{2\cos \nu\pi}\int_{-1}^{1} f(t)P^{-\mu}_{\nu-n-1}(-t)\,dt \\
&= (-1)^{n}\frac{-\nu+n+\frac{1}{2}}{2\cos(-\nu\pi)}\int_{-1}^{1} f(t)P^{-\mu}_{-\nu+n}(-t)\,dt.
\end{aligned}$$

The last expression is $a_n$ with $\nu$ replaced by $-\nu$. Further

$$\begin{aligned}
D(-\nu,t) &= \frac{1}{2}\{(-\nu-\mu)P^{\mu}_{-\nu}(x)P^{-\mu}_{-\nu-1}(-t) \\
&\qquad +(-\nu+\mu)P^{\mu}_{-\nu-1}(x)P^{-\mu}_{-\nu}(-t)\} \\
&= -\frac{1}{2}\{(\nu+\mu)P^{\mu}_{\nu-1}(x)P^{-\mu}_{\nu}(-t) + (\nu-\mu)P^{\mu}_{\nu}(x)P^{-\mu}_{\nu-1}(-t)\} \\
&= -D(\nu,t). \qquad (5.17)
\end{aligned}$$

So by Lemma 5.13 with $\nu$ replaced by $-\nu$,

$$\begin{aligned}\sum_{n=0}^{\infty} a_{-n-1} P^{\mu}_{\nu-n-1}(x) &= \sum_{n=0}^{\infty} a_{-n-1} P^{\mu}_{-\nu+n}(x) \\ &= \frac{1}{2} f(x) - \frac{1}{2} \sec \nu\pi \int_{-1}^{1} \frac{f(t)}{x-t} D(\nu, t)\, dt\end{aligned}$$

and the series on the left is convergent.This justifies the formal operations; and replacing $n$ by $n-1$ on the left gives the stated result. ■

**Theorem 5.15** *If* $(1-t^2)^{-\frac{1}{4}} f(t) \in L(-1,1)$, $f$ *is Dini at a certain* $x \in (-1,1)$, $|\operatorname{Re} \mu| < \frac{1}{2}$, $\nu$ *is not half an odd integer, and for integers* $n$

$$a_n = (-1)^n \frac{\nu + n + \frac{1}{2}}{2 \cos \nu\pi} \int_{-1}^{1} f(t) P^{-\mu}_{\nu+n}(-t)\, dt,$$

*then*

$$f(x) = \sum_{n=-\infty}^{\infty} a_n P^{\mu}_{\nu+n}(x)$$

*and the two "halves" of the series are separately convergent.*

**Proof.** This is immediate from Lemmas 5.13 and 5.14. ■

**Corollary 5.16** *If* $(1-t^2)^{-\frac{1}{4}} f(t) \in L(-1,1)$, $f$ *is Dini at a certain* $x \in (-1,1)$, $|\operatorname{Re} \mu| < \frac{1}{2}$, *and for integers* $n$

$$b_n = (-1)^n (n + \frac{1}{2}) \int_{-1}^{1} f(t) P_n^{-\mu}(-t)\, dt,$$

*then*

$$f(x) = \sum_{n=0}^{\infty} b_n P_n^{\mu}(x).$$

**Proof.** By (5.16), $D(0,t) = 0$. So Lemma 5.13 with $\nu = 0$ gives

$$\sum_{n=0}^{\infty} (-1)^n \frac{n + \frac{1}{2}}{2} \int_{-1}^{1} f(t) P_n^{-\mu}(-t)\, dt\, P_n^{\mu}(x) = \frac{1}{2} f(x),$$

from which the stated equation follows. ■

**Remark 5.2** *This corollary generalizes the classical Laplace's expansion in Legendre polynomials $P_n(x)$. For, taking $\mu = 0$, the corollary gives*

$$f(x) = \sum_{n=0}^{\infty} b_n P_n(x)$$

*where*

$$b_n = (n + \frac{1}{2}) \int_{-1}^{1} f(t)(-1)^n P_n(-t)\, dt = (n + \frac{1}{2}) \int_{-1}^{1} f(t) P_n(t)\, dt.$$

**Remark 5.3** *(AN EXTENSION) The Dini condition permits $f$ to be Hölder-continuous of any order in $(0,1]$, at $x$; but it does not permit $f$ to be discontinuous there. However, by a totally different method we have proved the following theorem, which does permit $f$ to have ordinary discontinuity at $x$:*
*If $(1-t^2)^{-\frac{1}{4}} f(t) \in L(-1,1)$, $f$ has bounded variation on a neighbourhood of a certain $x \in (-1,1)$, $|\operatorname{Re} \mu| < \frac{1}{2}$ and $\nu$ is not half an odd integer, then*

$$\lim_{N\to\infty} \sum_{n=-N}^{N} a_n P^{\mu}_{\nu+n}(x) = \frac{1}{2}\{f(x+0) + f(x-0)\}.$$

*But the two "halves" of the series may not be separately convergent; this is shown by the example*

$$f(t) = (1-t^2)^{\frac{1}{2}\mu} \text{ for } t < x, \qquad f(t) = 0 \text{ for } t > x.$$

*This extension is fully described in [4], particularly p. 651, Theorem 4 and p. 670, Theorem 6.*

## *References*


[1] E. T. Copson, Functions of a Complex Variable (Oxford, 1935).

[2] A. Erdélyi, W. Magnus, F. Oberhettinger and F. G. Tricomi, Higher Transcendental Functions, vol. 1 (Bateman Manuscript Project, McGraw-Hill, New York, 1953).

[3] E. R. Love, Franz Neumann's Integral of 1848, Proc. Cambridge Philos. Soc. 61 (1965), 445–456.

[4] E. R. Love, Abel summability of certain series of Legendre functions, Proc. London Math. Soc. (3) 69 (1994), 629–672.

[5] E. R. Love and M. N. Hunter, Expansions in series of Legendre functions, Proc. London Math. Soc. (3) 64 (1992), 579–601.

# Chapter 6

# Endpoint convergence of Legendre series

**Mark A. Pinsky**

*ABSTRACT. We show that the results of Love and Hunter can be reformulated to obtain convergence results at the endpoints of the interval* $-1 \leq x \leq 1$.

## 6.1 Statement of results

In another paper in this volume and elsewhere, Love and Hunter [1,4] have discussed the convergence of series expansions of the form

$$f(x) \sim \sum_{n=-\infty}^{\infty} a_n P^{\mu}_{\nu+n}(x) \qquad -1 < x < 1 \tag{6.1}$$

where $-\frac{1}{2} < \mu < \frac{1}{2}$ and $\nu$ is not half an odd integer; $P^{\mu}_{\nu}(x)$ is defined in terms of the Gauss hypergeometric function by

$$P^{\mu}_{\nu}(x) = \frac{1}{\Gamma(1-\mu)} \left(\frac{1+x}{1-x}\right)^{\mu/2} F(-\nu, 1+\nu; 1-\mu; \frac{1-x}{2}) \tag{6.2}$$

and the Fourier coefficient is defined by

$$a_n = (-1)^n \frac{\nu+n+\frac{1}{2}}{2\cos\nu\pi} \int_{1}^{1} f(t) P^{-\mu}_{\nu+n}(-t)\, dt. \tag{6.3}$$

They prove that the series converges at $x \in (-1,1)$ provided that $f$ satisfies a Dini condition. In another paper, Love[2, p. 651, theorem 4] showed that if $f$ is of bounded variation the series converges to $\frac{1}{2}f(x+0) + \frac{1}{2}f(x-0)$ at an interior point of discontinuity. In particular convergence holds if $f$ has a simple jump at $x$ and is otherwise piecewise smooth.

The question of convergence at the endpoints cannot be directly treated in this framework, since the terms of the orthogonal series are undefined at the endpoints $x = \pm 1$. The purpose of this note is to discuss the endpoint convergence of the closely related series for the function

$$\tilde{f}(x) := \left(\frac{1-x}{1+x}\right)^{\mu/2} f(x) \sim \sum_{n=-\infty}^{\infty} a_n \tilde{P}^{\mu}_{\nu+n}(x) \tag{6.4}$$

where

$$\tilde{P}^{\mu}_{\nu+n}(x) := \left(\frac{1-x}{1+x}\right)^{\mu/2} P^{\mu}_{\nu+n}(x) = \frac{1}{\Gamma(1-\mu)} F(-\nu, 1+\nu; 1-\mu; \frac{1-x}{2}) \tag{6.5}$$

The convergence behavior at interior points is identical to that of the original series, but now we have the possibility of discussing the endpoint behavior, since from the definition of $F(a,b;c;z)$

$$\lim_{x\to 1-0} \tilde{P}^{\mu}_{\nu+n}(x) = \frac{1}{\Gamma(1-\mu)}$$

and it will be shown that

$$\lim_{x\to -1+0} \tilde{P}^{\mu}_{\nu+n}(x) = \frac{\Gamma(-\mu)}{\Gamma(1-\mu+\nu)\Gamma(-\mu-\nu)} \qquad (\mu < 0)$$

$$\lim_{x\to -1+0} \tilde{P}^{\mu}_{\nu+n}(x) = \frac{\Gamma(2-\mu)}{\Gamma(2-\mu+\nu)} \frac{\Gamma(\mu+\nu)\sin \pi(\mu+\nu)}{\pi} \qquad (\mu > 0)$$

We prove the following theorem.

**Theorem 6.1** *Suppose that $\widehat{f}(x), -1 \le x \le 1$ is a piecewise smooth function. Then the series (6.4) converges at both endpoints $x = \pm 1$*

## 6.2 Asymptotic estimates

The hypergeometric function $u = F(a,b;c;z)$ is known to be a solution of the differential equation

$$z(1-z)u'' + [c-(a+b+1)z]u' - ab\,u = 0$$

Making the substitution $z = (1-x)/2$ and $a = -\nu$, $b = 1+\nu$, $c = 1-\mu$ we have

$$(1-x^2)u'' - 2(x-\mu)u' + \nu(1+\nu)u = 0. \tag{6.6}$$

This can be written in the self-adjoint Sturm-Liouville form

$$(su')' + \nu(1+\nu)\rho u = 0 \tag{6.7}$$

by taking

$$s(x) = \frac{(1+x)^{\mu+1}}{(1-x)^{\mu-1}}, \quad \rho(x) = \frac{(1+x)^{\mu}}{(1-x)^{\mu}}.$$

The integral (6.3) defining the Fourier coefficient is transformed from the $(f, P_\nu^\mu)$ notation to the $(\tilde{f}, \tilde{P}_\nu^\mu)$ notation as follows:

$$\int_{-1}^{1} f(t) P_{\nu+n}^{-\mu}(-t)\, dt = \int_{-1}^{1} \left(\frac{1+t}{1-t}\right)^{\mu/2} \tilde{f}(t) \left(\frac{1-t}{1+t}\right)^{-\mu/2} \tilde{P}_{\nu+n}^{-\mu}(-t)\, dt$$

$$= \int_{-1}^{1} \tilde{f}(t)\rho(t)\tilde{P}_{\nu+n}^{-\mu}(-t)\, dt.$$

We use the abbreviation $\phi_n(t) = P_{\nu+n}^{-\mu}(-t)$. When we change $t \to -t$ and $\mu \to -\mu$ the differential equation (6.6) remains unchanged. Therefore $\phi_n$ is also a solution of the second-order differential equation (6.7), written

$$(s\phi_n')' + (\nu+n)(\nu+n+1)\rho\phi_n = 0 \tag{6.8}$$

The following lemma uses partial integration to find an asymptotic formula for the Fourier coefficient.

**Lemma 6.2** *Suppose that $\widehat{f}(t), -1 \le t \le 1$ is a piecewise smooth function with possible jumps at the points $a_1 < \cdots < a_K$. Let*

$$\delta\widehat{f}(a_i) := f(a_i+0) - f(a_i-0), \qquad 1 \le i \le K$$

$$\delta\widehat{f}'(a_i) := f'(a_i+0) - f'(a_i-0), \qquad 1 \le i \le K.$$

*Then we have the identity*

$$(\nu+n)(\nu+n+1)\int_{-1}^{1} \tilde{f}(t)\rho(t)\tilde{P}_{\nu+n}^{-\mu}(-t)\, dt$$

$$= \sum_{i=1}^{K} s(a_i)\left[\phi_n(a_i)\delta\tilde{f}'(a_i) - \phi_n'(a_i)\delta\tilde{f}(a_i)\right]$$

$$- \int_{-1}^{1} \left[s(t)\tilde{f}'(t)\right]'\phi_n(t)\, dt.$$

**Proof.** The differential equation (8) for $\phi_n$ is written

$$\frac{d}{dt}[s(t)\phi_n'(t)] + (\nu+n)(\nu+n+1)\rho(t)\phi_n(t) = 0.$$

We multiply by $\tilde{f}(t)$ and integrate on the interval $(a_{i-1}, a_i)$. Performing a partial integration on the first integral yields

$$(\nu+n)(\nu+n+1)\int_{a_{i-1}}^{a_i} \rho(t)\phi_n(t)\tilde{f}(t)\,dt$$

$$= -s(t)\phi_n'(t)\tilde{f}(t)|_{a_{i-1}}^{a_i} + \int_{a_{i-1}}^{a_i} s(t)\phi_n'(t)\tilde{f}'(t)\,dt.$$

Performing a second partial integration on the final integral yields

$$\int_{a_{i-1}}^{a_i} s(t)\tilde{f}'(t)\phi_n'(t)\,dt = s(t)\tilde{f}'(t)\phi_n(t)|_{a_{i-1}}^{a_i} - \int_{a_{i-1}}^{a_i} \phi_n(t)\left[s(t)\tilde{f}'(t)\right]'\,dt$$

When we combine this with the previous line and sum for $1 \le i \le K$ we obtain the statement of the lemma, with the lower limit replaced by $a_1$ and the upper limit replaced by $a_K$. Finally, noting that $s(t) \to 0$ when $t \to 1$ or $t \to -1$, we can extend the final computation to the interval $[a_K, 1]$ and to the interval $[-1, a_1]$, which completes the proof. ■

The asymptotic behavior of the coefficients is then obtained from the lemma as

$$a_n = (-1)^n \frac{\nu+n+\frac{1}{2}}{2\cos\nu\pi}\frac{1}{(\nu+n)(\nu+n+1)}\left[-\int_{-1}^{1}\phi_n(t)\big(s(t)\tilde{f}'(t)\big)'\,dt\right.$$

$$\left.+\sum_{i=1}^{K} s(a_i)\big(\phi_n(a_i)\delta\tilde{f}'(a_i) - \phi_n'(a_i)\delta\tilde{f}(a_i)\big)\right]$$

The asymptotic ($\nu \to \infty$) behavior of the Legendre function is given by [3, p.162, eq.(2)]

$$P_\nu^\mu(\cos\theta) = \frac{\Gamma(\nu+\mu+1)}{\Gamma(\nu+\frac{3}{2})}\sqrt{\frac{2}{\pi\sin\theta}}\left(\cos\left[(\nu+\frac{1}{2})\theta - \frac{\pi}{4} + \frac{\mu\pi}{2}\right] + O(\frac{1}{\nu})\right)$$

with a corresponding expression for the derivative. Replacing $\nu$ by $\nu+n$ and applying Stirling's formula, we have

$$\frac{\Gamma(\nu+n+\mu+1)}{\Gamma(\nu+n+\frac{3}{2})} \sim n^{\mu-\frac{1}{2}}, \qquad n \to +\infty$$

Hence we have the following estimates when $n \to \infty$:

$$P^{\mu}_{\nu+n}(\cos\theta) = n^{\mu-\frac{1}{2}}\sqrt{\frac{2}{\pi\sin\theta}}\left(\cos\left[(\nu+n+\frac{1}{2})\theta - \frac{\pi}{4} + \frac{\mu\pi}{2}\right] + O(\frac{1}{n})\right)$$

$$\phi_n(t) = P^{-\mu}_{\nu+n}(t) = n^{-\mu-\frac{1}{2}}\sqrt{\frac{2}{\pi\sin\theta}}\left(\cos\left[(\nu+n+\frac{1}{2})\theta - \frac{\pi}{4} - \frac{\mu\pi}{2}\right] + O(\frac{1}{n})\right)$$

$$\phi'_n(t) = P^{-\mu}_{\nu+n}(t) = -n^{-\mu+\frac{1}{2}}\sqrt{\frac{2}{\pi\sin\theta}}\left(\sin\left[(\nu+n+\frac{1}{2})\theta - \frac{\pi}{4} - \frac{\mu\pi}{2}\right] + O(\frac{1}{n})\right)$$

To analyze the behavior for $n \to -\infty$, we use the relation $\Gamma(x)\Gamma(1-x) = \pi/\sin(\pi x)$ to write

$$\frac{\Gamma(\nu+n+\mu+1)}{\Gamma(\nu+n+\frac{3}{2})} = \frac{\Gamma(-\nu-n-\frac{1}{2})\sin\pi(n-\nu-\frac{1}{2})}{\Gamma(-\nu-n-\mu)\sin\pi(n-\nu-\mu)}$$

Another application of Stirling's formula shows that the first factor $\sim (-n)^{\mu-\frac{1}{2}}$ when $n \to -\infty$, so that the same estimates apply to the terms of the series with $n < 0$.

## 6.3 Convergence at the endpoints

### 6.3.1 *Convergence at $x = 1$*

Combining the above asymptotic estimates, we set $\theta_i = \cos^{-1}(a_i)$ and obtain, to within lower-order terms

$$\begin{aligned} a_n P^{\mu}_{n+\nu}(1) &= \frac{(-1)^n}{n\Gamma(1-\mu)}\sum_{i=1}^{K} s(a_i)\frac{\delta\tilde{f}(a_i)}{n^{\mu-1/2}}\sin\left[(\nu+n+\frac{1}{2})\theta_i - \frac{\pi}{4} - \frac{\mu\pi}{2}\right] \\ &= \frac{n^{-\mu-\frac{1}{2}}}{\Gamma(1-\mu)}\sum_{i=1}^{K} s(a_i)\delta\tilde{f}(a_i)\sin\left[(\nu+n+\frac{1}{2})\theta_i - \frac{\pi}{4} - \frac{\mu\pi}{2}\right] \end{aligned}$$

Since $-\frac{1}{2} < \mu < \frac{1}{2}$, we have $\mu + \frac{1}{2} > 0$ so that this is the general term of a convergent series.

### 6.3.2 *Convergence at $x = -1$*

To discuss the convergence at the endpoint $x = -1$, we first consider the case $\mu < 0$. Then we can apply[3, p. 61, eq. (14)] to obtain

$$F(-\nu, 1+\nu; 1-\mu; 1) = \frac{\Gamma(1-\mu)\Gamma(-\mu)}{\Gamma(1-\mu+\nu)\Gamma(-\mu-\nu)}.$$

We use the identity $\Gamma(x)\Gamma(1-x) = \pi/\sin \pi x$ to obtain

$$\tilde{P}_\nu^\mu(-1) = \frac{\Gamma(-\mu)\Gamma(\mu+\nu+1)\sin\pi(\mu+\nu+1)}{\pi\Gamma(1-\mu+\nu)}.$$

By Stirling's formula,

$$\frac{\Gamma(\mu+\nu)}{\Gamma(1-\mu+\nu)} \sim \nu^{2\mu-1}, \qquad \nu \to \infty$$

Combining the above computations and simplifying, we can make the evaluation

$$a_n \tilde{P}^\mu_{n+\nu}(-1) = C_\mu n^{3\mu-\frac{3}{2}} \sum_{i=1}^{K} \cos(n\theta+\alpha)\delta f(a_i) + O(n^{3\mu-\frac{5}{2}}).$$

Since $\mu < \frac{1}{2}$, the final exponent is negative and we have the general term of a convergent series.

If $\mu > 0$ the above formulas cannot be directly applied. However we first employ the reduction formula [3, p.61]

$$F(a, b; c; 1) = \frac{(c-a)(c-b)}{c(c-a-b)} F(a, b; c+1; 1)$$

Applying this with $a = -\nu, b = 1+\nu, c = 1-\mu$ yields

$$\phi_n(-1) = \frac{(1-\mu+\nu)(-\mu-\nu)}{(1-\mu)(-\mu)} F(-\nu, 1+\nu; 2-\mu; 1)$$

Now we can apply [3, p.61, eq.(14)] in the form

$$F(-\nu, 1+\nu; 2-\mu; 1) = \frac{\Gamma(2-\mu)\Gamma(1-\mu)}{\Gamma(2-\mu+\nu)\Gamma(1-\mu-\nu)}$$

$$= \frac{\Gamma(2-\mu)\Gamma(1-\mu)}{\Gamma(2-\mu+\nu)} \frac{\Gamma(\mu+\nu)\sin\pi(\mu+\nu)}{\pi}$$

Stirling's formula provides the evaluation $\Gamma(\mu+\nu)/\Gamma(2-\mu+\nu) \sim \nu^{2\mu-2}$ so that we have to within lower-order terms

$$\phi_n(-1) \sim \nu^{2\mu-1} \frac{\Gamma(-\mu)\Gamma(2-\mu)}{\pi(\mu+\nu)} \sin \pi(\mu+\nu)$$

Again, the exponent $2\mu - 1 < 0$ so that the series converges also at $x = 1$. The terms with $n < 0$ are handled exactly as in the case $x = 1$.

**Acknowledgement.** We would like to thank Professor E.R. Love for a helpful initial correspondence.

*REFERENCES*

[1] E. R. Love and M.N. Hunter, Expansions in series of Legendre functions, Proc. London Math. Soc. 64( 1992), 579-601 .

[2] E.R. Love, Abel summability of certain Legendre series, Proc. London Math. Soc. 69(1994), 629-672 .

[3] A. Erdelyi, W. Magnus, F. Oberhettinger and F.G. Tricomi, Higher Transcendental Functions, vol. 1, Bateman Manuscript Project, McGraw Hill, 1953.

[4] E. R. Love and M.N. Hunter, Expansions in series of Legendre functions, Chapter 5 in this volume.

# Chapter 7

# Inversion of the horocycle transform on real hyperbolic spaces via a wavelet-like transform

**William O. Bray**
**Boris Rubin**

*ABSTRACT. It is proved that the horocycle transform $Rf$ on real n-dimensional real hyperbolic space $H$ is well-defined for $f \in L^p(H)$ if and only if $1 \le p < 2$. Further, the function $f$ can be recovered explicitly in $L^p$-norm and a.e. via a suitably defined wavelet like transform on the space of horocycles.*

## 7.1 Introduction

Let $\mathbb{E}^{n,1}$ be $(n+1)$-dimensional pseudo-Euclidean space endowed with the inner product $[x,y] = x_{n+1}y_{n+1} - x_1 y_1 - \cdots - x_n y_n$. Real n-dimensional hyperbolic space $\mathbb{H}$ can be regarded as the upper sheet of the two sheeted hyperboloid $\mathbb{H} = \{x \in \mathbb{E}^{n,1} | \, [x,x] = 1, \, x_{n+1} > 0\}$. Let $\Gamma$ be the upper part of the light cone in $\mathbb{E}^{n,1}$, i.e. $\Gamma = \{\xi \in \mathbb{E}^{n,1} \, | \, [\xi,\xi] = 0, \, \xi_{n+1} > 0\}$. Geometrically, horocycles are planar sections of $\mathbb{H}$ by hyperplanes of the form $[x,\xi] = 1$ ([8]). Each hyperplane of this type is parallel to a certain generatrix of $\Gamma$. The horocycle transform $Rf(\xi)$ assigns to each sufficiently nice function $f$ on $\mathbb{H}$ the integrals of $f$ over horocycles. For a compactly supported smooth function $f$ one can write

$$Rf(\xi) = \int\limits_{\mathbb{E}^{n,1}} f(x)\, \delta([x,\xi] - 1)\, dx, \tag{7.1}$$

$$f(x) = \begin{cases} \dfrac{(-1)^m}{2(2\pi)^{2m}} \int_\Gamma \delta^{(2m)}([x,\xi] - 1)\, Rf(\xi)\, d\xi, & \text{if } n = 2m+1, \\ \dfrac{(-1)^m \Gamma(2m)}{(2\pi)^{2m}} \int_\Gamma ([x,\xi] - 1)^{-2m} Rf(\xi)\, d\xi, & \text{if } n = 2m. \end{cases} \tag{7.2}$$

Here $\delta$ is the Dirac delta-function and the integrals are interpreted in a suitable sense ([8], [20, p. 162]). In later works, other inversion formulas were obtained using different methods (e.g. [1], [5], [6], [6], [21]). Typically, these methods are based on the use of the dual transform $\overset{*}{R}$ (which integrates $Rf$ over all horocycles passing through a fixed point $x \in \mathbb{H}$) and/or techniques from harmonic analysis. For functions in more general function classes, e.g. $L^p(\mathbb{H})$ or $C(\mathbb{H})$, the methods do not apply.

In the recent papers ([2], [13] – [16]) explicit inversion formulas for Radon transforms in various settings were obtained in the framework of $L^p$-space. The basic idea is to include $R$ and $\overset{*}{R}$ into suitable analytic families $\{R^\alpha\}$ and $\{\overset{*}{R}{}^\alpha\}$ of fractional integrals in such a way that the inverse operator $R^{-1}$ belongs to the family $\{\overset{*}{R}{}^\alpha\}$. The operators $\overset{*}{R}{}^\alpha$ give rise to generalized wavelet transforms. In terms of these transforms it is possible to write out the analytic continuation of $\overset{*}{R}{}^\alpha Rf$ in the form which enables us to work with $L^p$-functions and with continuous functions.

In the present paper we apply this method to the horocycle transform. The required families of fractional integrals were discovered by examining the proof of (7.2). The general idea is as follows (cf. [16]): since the delta function $\delta(t)$ is a member of the analytic family of distributions $t_+^{\lambda-1}/\Gamma(\lambda)$ (see [9]) and these distributions generate Riemann-Liouville fractional integrals (and corresponding wavelet transforms [12]), it is natural to expect that the delta function $\delta([x,\xi]-1)$ can be associated with some fractional integrals and wavelet transforms.

The paper is organized as follows. In the second section, we give basic definitions and auxiliary results, and establish the Solmon type estimate for the horocycle transform (cf. [10], [2]). The third section introduces continuous wavelet transforms associated with $R$ and the inversion formula mentioned in the Abstract is proved; the main result is given by Theorem 7.8. The argument explores the connection between our fractional integrals and wavelet transforms with harmonic analysis on $\mathbb{H}$.

## 7.2 Preliminaries

### 7.2.1 Algebraic and geometric notions

References to this subsection are [3] and [20]. In addition to $\mathbb{E}^{n,1}$, $\mathbb{H}$, $\Gamma$, we define the spaces $\mathbb{R}^n = \{x \in \mathbb{E}^{n,1} \,|\, x = (x_1, \dots, x_n, 0)\}$ and $\mathbb{R}^{n-1} = \{x \in \mathbb{E}^{n,1} \,|\, x = (x_1, \dots, x_{n-1}, 0, 0)\}$ with the corresponding rotation groups $K = SO(n)$ and $M = SO(n-1)$. The coordinate unit vectors are denoted by $e_1, \dots, e_{n+1}$. We write $dk$ for the normalized Haar measure on $K$ so that $\int_K dk = 1$. Let $\mathbf{S}^{n-1} = K/M$ be the unit sphere in $\mathbb{R}^n$; $\omega_{n-1} = |\mathbf{S}^{n-1}| = 2\pi^{n/2}/\Gamma(n/2)$.

The geodesic distance between points $x, y \in \mathbb{H}$ is defined by $\cosh d(x,y) = [x,y]$. The isometry group of $\mathbb{H}$ is $G = SO_o(n,1)$, the locally compact connected group of pseudo-rotations of $\mathbb{E}^{n,1}$ which preserve the bilinear form $[x,y]$. The subgroup $K$ is the isotropy subgroup of the point $O = (0,\dots,0,1) \in \mathbb{H}$, called the origin in $\mathbb{H}$. We then have the homogeneous space identification $\mathbb{H}=G/K$.

The group $G$ possesses a Cartan decomposition $G = KAK$ and an Iwasawa decomposition $G = KAN$ where $A$ is an Abelian subgroup of the form

$$A = \left\{ a = a_t = \begin{bmatrix} I_{n-1} & 0 & 0 \\ 0 & \cosh t & \sinh t \\ 0 & \sinh t & \cosh t \end{bmatrix} \mid t \in \mathbb{R} \right\},$$

and $N$ is a nilpotent subgroup of $G$ given by

$$N = \left\{ n = n_v = \begin{bmatrix} I_{n-1} & -v^{tr} & v^{tr} \\ v & 1-|v|^2/2 & |v|^2/2 \\ v & -|v|^2/2 & 1+|v|^2/2 \end{bmatrix} \mid v \in \mathbb{R}^{n-1}(row) \right\}.$$

Then $A$ normalizes $N$, i.e. $a_t^{-1} n_v a_t = n_{e^{-t}v}$. The Haar measure $dn$ on $N$ is given by Lebesgue measure $dv$ on $\mathbb{R}^{n-1}$, so that $\int_N f(n)dn = \int_{\mathbb{R}^{n-1}} f(n_v)dv$. Any $g \in G$ has a unique expression $g = kan$ up to the centralizer $M$ of $A$ in $K$. Any $x \in \mathbb{H}$ can be written uniquely in the form $x = n_v a_t \circ O$. This leads to the horocycle coordinates on $\mathbb{H}$ given by:

$$x = n_v a_t \circ O = a_t n_{e^{-t}v} \circ O = (e^{-t}v, \sinh t + \frac{|v|^2}{2} e^{-t}, \cosh t + \frac{|v|^2}{2} e^{-t}). \tag{7.3}$$

In terms of the decomposition $x = n_v a_t \circ O$, the invariant Riemannian measure $dx$ on $\mathbb{H}$ has the form $dx = e^{-2\rho t}dtdv$, where $\rho = (n-1)/2$ (the letter $\rho$ has this meaning everywhere throughout the paper).

Horocycles in $\mathbb{H}$ can be defined as translates of the orbit $N \circ O$ under $G$. Any horocycle has the form $ka_t N \circ O$ for some $k \in K$ and $t \in \mathbb{R}$ ($t$ gives the signed distance of the horocycle to the origin $O$). We denote the space of horocycles by $\Xi$. The group $G$ is transitive on $\Xi$, and the subgroup $MN$ of $G$ leaves fixed the "basic horocycle" $N \circ O$. Hence we have the homogeneous space identification $\Xi = G/MN$. Let $\xi_0 = (0,\dots,0,1,1)$. Each horocycle $ka_t N \circ O$ is identified uniquely with the point $\xi \in \Gamma$ according to

$$\xi = ka_t \circ \xi_0 = e^t \, k \circ \xi_0 = e^t b(\omega), \tag{7.4}$$

where $\omega = k \circ e_n \in \mathbf{S}^{n-1}$, $b(\omega) = k \circ \xi_0 = \omega + e_{n+1} \in \Gamma$. In accordance with (4), the invariant measure on $\Gamma$ is defined by $d\xi = e^{2\rho t} dt d\omega$, $dt$ being Lebesgue measure on $\mathbb{R}$ and $d\omega$ the usual surface measure on $\mathbf{S}^{n-1}$ ([20], p. 24).

Finally notice that for each $x \in \mathbb{H}$ and each $\omega \in \mathbf{S}^{n-1}$ there is a unique horocycle passing through $x$ and given by the point $e^t b(\omega) \in \Gamma$ with

$$t = \langle x, \omega \rangle = -\log[x, b(\omega)] \tag{7.5}$$

(cf. [3], p. 80). This quantity is usually called the *horocycle distance function.*

### 7.2.2 The horocycle transform and its dual

Given $\xi \in \Gamma$, let $\hat{\xi}$ be the horocycle defined by $\hat{\xi} = \{x \in \mathbb{H} \,|\, [x, \xi] = 1\}$, and, given $x \in \mathbb{H}$, let $\check{x} = \{\xi \in \Gamma \,|\, [x, \xi] = 1\}$ be the set of points of the cone $\Gamma$ corresponding to all horocycles passing through $x$. We denote by $d_\xi x$ and $d_x \xi$ the induced Lebesgue measures on $\hat{\xi}$ and $\check{x}$ respectively. According to (7.1), for sufficiently nice functions $f : \mathbb{H} \to \mathbf{C}$ and $\varphi : \Gamma \to \mathbf{C}$ the horocycle transform and its dual are defined by

$$Rf(\xi) = \int\limits_{\hat{\xi}} f(x) d_\xi x \quad \text{and} \quad \overset{*}{R} \varphi(x) = \int\limits_{\check{x}} \varphi(\xi) d_x \xi,$$

respectively. If $\xi = e^t b(\omega) = e^t k \circ \xi_0$ and $x = g \circ O,\ g \in G$, then in group theoretic terms these transforms read as follows

$$Rf(\xi) = R_\omega f(t) = \int\limits_N f(k a_t n \circ O)\, dn \quad \left( = \int\limits_{\mathbb{R}^{n-1}} f(k a_t n_v \circ O)\, dv \right), \tag{7.6}$$

$$R^* \varphi(x) = \omega_{n-1} \int\limits_K \varphi(gk \circ \xi_o)\, dk, \tag{7.7}$$

The following statement gives another representation of the dual transform.

**Proposition 7.1** *For each $g \in G$ and each $\omega \in \mathbf{S}^{n-1}$,*

$$\int\limits_K \varphi(gk \circ \xi_o)\, dk = \int\limits_K e^{2\rho \langle g \circ O, k \circ \omega \rangle} \varphi(e^{\langle g \circ O, k \circ \omega \rangle} k \circ \xi_0)\, dk \tag{7.8}$$

*provided that one of these integrals exists.*

**Proof.** We write (7.8) in the equivalent form $I_1 \varphi(g) = I_2 \varphi(g)$, where

$$I_1 \varphi(g) = \int\limits_{\mathbf{S}^{n-1}} \varphi(g \circ b(\omega))\, d\omega,$$

$$I_2 \varphi(g) = \int\limits_{\mathbf{S}^{n-1}} e^{2\rho \langle g \circ O, \omega \rangle} \varphi(e^{\langle g \circ O, \omega \rangle} b(\omega))\, d\omega.$$

Set $g = k'a_r k''$ $(k', k'' \in K, a_r \in A)$. One can readily see that $I_1\varphi(g) = I_2\varphi(g)$ if and only if $I_1\varphi'(a_r) = I_2\varphi'(a_r)$, $\varphi'(\xi) = \varphi(k' \circ \xi)$. Thus it suffices to prove (7.8) for $g = a_r$. By passing to polar coordinates on $\mathbf{S}^{n-1}$ and taking into account the equalities

$$a_r \circ e_n = (\cosh r)e_n + (\sinh r)e_{n+1}, \quad a_r \circ e_{n+1} = (\sinh r)e_n + (\cosh r)e_{n+1},$$

we have

$$I_1\varphi(a_r) = \int_{-1}^{1} (1-\eta^2)^{(n-3)/2} \times \int_{\mathbf{S}^{n-2}} \varphi(\sqrt{1-\eta^2}\theta + (\eta\cosh r + \sinh r)e_n + (\eta \sinh r + \cosh r)e_{n+1})d\theta d\eta,$$

$$I_2\varphi(a_r) = \int_{-1}^{1} \frac{(1-\tau^2)^{(n-3)/2}}{(\cosh r - \tau \sinh r)^{n-1}} \int_{\mathbf{S}^{n-2}} \varphi\Big(\frac{\sqrt{1-\tau^2}\theta + \tau e_n + e_{n+1}}{\cosh r - \tau \sinh r}\Big)\, d\theta d\tau. \tag{7.9}$$

The second expression can be reduced to the first one by changing the variable (put $1/(\cosh r - \tau \sinh r) = \eta \sinh r + \cosh r$). ■

**Corollary 7.2** *For $x \in \mathbb{H}$,*

$$\overset{*}{R}\varphi(x) = \int_{\mathbf{S}^{n-1}} e^{2\rho\langle x,\omega\rangle} \varphi(e^{\langle x,\omega\rangle} b(\omega))\, d\omega. \tag{7.10}$$

Known properties of these transforms are the content of the following lemmas. For convenience of the reader we supply them with simple proofs.

**Lemma 7.3** *(cf. formulas (2.8) and (3.1) from [6]). We assume that $f$ and $\varphi$ are locally integrable on $\mathbb{H}$ and $\Gamma$ such that the integrals below exist a.e.*
*(i) If $f$ is a $K$-invariant function on $\mathbb{H}$, i.e. $f(x) = f_o(x_{n+1})$, then $Rf$ is $K$-invariant, and*

$$e^{\rho t} R_\omega f(t) = 2^{\rho-1}\omega_{n-2} \int_{\cosh t}^{\infty} f_o(s)\,(s-\cosh t)^{\rho-1}\, ds \tag{7.11}$$

$$= 2^{2\rho-1}\omega_{n-2} \int_{u^2}^{\infty} f_o(2\tau - 1)\,(\tau - u^2)^{\rho-1}\, du, \quad u = \cosh\frac{t}{2}. \tag{7.12}$$

*(ii) If $\varphi$ is a $K$-invariant function on $\Gamma$, i.e. $\varphi(\xi) = \varphi_o(\xi_{n+1})$, then $\overset{*}{R}\varphi$ is $K$-invariant, and*

$$\overset{*}{R}\varphi(x) = \frac{2^{\rho-1}\omega_{n-2}}{(\sinh r)^{n-2}} \int_{-r}^{r} \varphi_o(e^s)\,(\cosh r - \cosh s)^{\rho-1} e^{\rho s} ds, \; \cosh r = x_{n+1}. \tag{7.13}$$

**Proof.** (i) Since $f$ is $K$-invariant, then one can ignore $k$ in (7.6), and by (7.3) we have

$$\begin{aligned} R_\omega f(t) &= \int_{\mathbb{R}^{n-1}} f(a_t n_v \circ O)\, dv = e^{-2\rho t} \int_{\mathbb{R}^{n-1}} f(n_v a_t \circ O)\, dv \\ &= e^{-2\rho t} \int_{\mathbb{R}^{n-1}} f_o\Big(\cosh t + \frac{|v|^2}{2} e^{-t}\Big) dv \end{aligned}$$

which gives (7.11). The representation (7.12) can be obtained from (7.11) by putting $\cosh t = 2u^2 - 1$.

(ii) By making use of (7.9), we obtain

$$\begin{aligned} \overset{*}{R}\varphi(x) &= \omega_{n-2} \int_{-1}^{1} (1-\tau^2)^{\rho-1} (\cosh r - \tau \sinh r)^{-2\rho} \varphi_o\Big(\frac{1}{\cosh r - \tau \sinh r}\Big)\, d\tau \\ &= \frac{\omega_{n-2}}{(\sinh r)^{n-2}} \int_{-r}^{r} (\sinh^2 r - (\cosh r - e^{-s})^2)^{\rho-1} e^{s(2\rho-1)} \varphi_o(e^s)\, ds \end{aligned}$$

which coincides with (7.13). ∎

**Lemma 7.4** *(Duality; cf. [6] , p. 103). Let $f$ and $\varphi$ be functions on $\mathbb{H}$ and $\Gamma$, respectively. Then the duality relation*

$$\int_{\Gamma} \varphi(\xi) Rf(\xi)\, d\xi = \int_{\mathbb{H}} \overset{*}{R}\varphi(x) f(x)\, dx \tag{7.14}$$

*holds provided that at least one of the integrals is finite for $\varphi$ and $f$ replaced by $|\varphi|$ and $|f|$, respectively.*

**Proof.** By setting $\xi = e^t b(\omega)$, $\omega = k \circ e_n$, we write the left-hand side of (7.14) in the form (cf. (7.6))

$$\int_{\mathbb{R}} e^{2\rho t}\, dt \int_{\mathbf{S}^{n-1}} \varphi(e^t b(\omega))\, d\omega \int_{\mathbb{R}^{n-1}} f(k a_t n_v \circ O)\, dv.$$

Put $v = e^{-t}u$, $a_t n_v \circ O = n_u a_t \circ O = y$, $[ky, b(\omega)] = e^{-t}$. Then the above expression can be written as

$$\int\limits_{\mathbf{S}^{n-1}} d\omega \int\limits_{\mathbb{H}} \varphi([ky, b(\omega)]^{-1} b(\omega)) f(ky) [ky, b(\omega)]^{-2\rho}\, dy$$

$$= \int\limits_{\mathbf{S}^{n-1}} d\omega \int\limits_{\mathbb{H}} \varphi(e^{-\langle x,\omega\rangle} b(\omega)) f(x) e^{2\rho\langle x,\omega\rangle}\, dx$$

which coincides with the right-hand side of (7.14). ■

We use the above results to characterize the behavior of $R$ on $L^p(\mathbb{H})$ mentioned in the Abstract.

**Proposition 7.5** *If $f \in L^p(\mathbb{H})$, $1 \le p < 2$, then $Rf(\xi)$ exists a.e. and*

$$\int\limits_{\Gamma} |\xi|^{\beta-\rho} \left||\xi| - 1\right|^{-2\beta} |Rf(\xi)|\, d\xi \le C_{n,\beta} \|f\|_p, \tag{7.15}$$

*provided $1 + (n-1)(1/2 - 1/p) < \beta < \min(1, 1/2 + n/p')$, $1/p + 1/p' = 1$. Further, there exists an $\widetilde{f} \in L^p(\mathbb{H})$ such that $Rf(\xi) \equiv \infty$, for every $p \ge 2$.*

**Proof.** Let $\varphi(\xi) = e^{-\rho t}(\cosh t - 1)^{-\beta}$, where $\xi = e^t b(\omega)$. By (7.13),

$$\overset{*}{R}\varphi(x) = const \times (\cosh r - 1)^{1/2-\beta} (\cosh r + 1)^{1-n/2}, \ \cosh r = x_{n+1}.$$

By the conditions on $\beta$, for $f \in L^p(\mathbb{H})$, the right-hand side of (7.14) is an absolutely convergent integral. Consequently (7.14) is valid and it follows that $Rf$ is defined a.e. Substituting into (7.14) and applying Hölder's inequality gives the estimate. The examples in the case when $p \ge 2$ can be constructed easily by considering $K$-invariant functions and using (7.11). ■

**Remark 7.1** *Conceivably, an analog of the above result can be obtained for general rank one symmetric spaces of non-compact type by making use of the formulas (2.8) and (3.1) from [6]) together with the corresponding duality relation. Undoubtedly, the result can be extended to symmetric spaces of any rank, however the computational aspects of the above proof do not seem to generalize easily.*

### *7.2.3 Approximate identities on $\mathbb{H}$*

Approximate identities have been introduced in the symmetric space setting by various authors, e.g. [18] or [4]. Here we introduce a modification appropriate for our needs. Given an integrable function $k_0 : (0, \infty) \to \mathbf{C}$, consider the convolution operator

$$K_\varepsilon f(x) = \int\limits_{\mathbb{H}} k_\varepsilon([x, y])\, f(y)\, dy, \tag{7.16}$$

where $\varepsilon > 0$, and the kernel is given by

$$k_\varepsilon(\tau) = \frac{(\tau^2-1)^{1-n/2}}{\varepsilon} k_0\left(\frac{2(\tau-1)}{\varepsilon}\right).$$

Convolutions of this form arise naturally in the inversion procedure for the horocycle transform given in the next section.

**Lemma 7.6** *Let $f$ be a measurable function on $\mathbb{H}$.*
*(i) If $k_0$ has a decreasing integrable majorant, then*

$$K^* f \leq const \times f^*, \tag{7.17}$$

*where $f^*(x)$ is the Hardy-Littlewood maximal function on $\mathbb{H}$ defined by*

$$f^*(x) = \sup_{r>0} \frac{1}{|B(x,r)|} \int_{B(x,r)} |f(y)|\, dy,$$

*$B(x,r)$ is a geodesic ball of radius $r$ centered at $x$.*
*(ii) Let $f \in L^p(\mathbb{H})$ for some $1 \leq p < \infty$. Then*

$$\lim_{\varepsilon \to 0} K_\varepsilon f = c_0 f, \qquad c_0 = \frac{\omega_{n-1}}{2} \int_0^\infty k_0(s)\, ds, \tag{7.18}$$

*in the $L^p$-norm. Further, if $k_0$ has decreasing integrable majorant, then the limit holds a.e.*
*(iii) If $f \in C_0(\mathbb{H}) = \{f \in C(\mathbb{H}) \mid f(x) \to 0$ as $d(O,x) \to \infty\}$, then the limit holds uniformly on $\mathbb{H}$.*

**Proof.** (i) By passing to polar coordinates on $\mathbb{H}$, we get

$$\begin{aligned} K_\varepsilon f(x) &= \frac{\omega_{n-1}}{\varepsilon} \int_0^\infty k_0\Big(\frac{2(\cosh r - 1)}{\varepsilon}\Big) M_r f(x)\, d(\cosh r) \\ &= \int_0^\infty k_0(s) u(2+\varepsilon s)\, ds \end{aligned} \tag{7.19}$$

where $M_r f(x)$ is the mean value of $f$ over the geodesic sphere of radius $r$ centered at $x$, $u(t) = 2^{-1}\omega_{n-1} M_{\operatorname{arcosh}(t/2)} f$, $t \geq 2$. Properties of $M_r f$ for $f \in L^p(\mathbb{H})$ and $f \in C_0(\mathbb{H})$ were studied, e.g., in [8], [2]. Without loss of generality one can assume $f \geq 0$. Since $k_0$ has a decreasing integrable majorant, then

$$|K_\varepsilon f| \leq c \sup_{h \in \mathbb{R}} \frac{1}{2h} \int_{2-h}^{2+h} u(t)\, dt = \frac{c}{2} \sup_{h>0} \gamma(h) \tag{7.20}$$

(for $t < 0$, $u(t)$ is defined by zero), where $\gamma(h) = h^{-1} \int_1^{1+h} M_{\text{arcosh}\,\tau} f\, d\tau$, $c = \text{const}$. Consider the function

$$\psi(t) = \omega_{n-1} \int_1^t (s^2-1)^{n/2-1} M_{\text{arcosh}\,s} f\, ds = \int_{B(x,\ \text{arcosh}\, t)} f(y)\, dy.$$

Let $\nu(t) = |B(x, \text{arcosh}\, t)|$. Since

$$\nu(t) = \int_{[x,y]<t} dy = c \int_1^t (\tau^2-1)^{n/2-1} d\tau,$$

then $\nu(t) = O((t^2-1)^{-n/2})$ for $t \leq 2$ and $\nu(t) = O(t^{n-1})$ for $t > 2$. Hence $\psi(t) \leq \nu(t) f^*$, and integration by parts yields

$$\begin{aligned}
\gamma(h) &= \frac{\omega_{n-1}^{-1}}{h} \int_1^{1+h} (t^2-1)^{1-n/2} d\psi(t) \\
&\leq \frac{cf^*}{h} \left[ ((1+h)^2-1)^{1-n/2} \nu(1+h) + \int_1^{1+h} \nu(t)(t^2-1)^{-n/2} t dt \right] \\
&\leq Af^*,
\end{aligned}$$

$A$ being independent of $h$. This estimate together with (7.20) implies (7.17).

(ii) Let us prove the limit relation (7.18). By (7.19),

$$K_\varepsilon f - c_0 f = \frac{\omega_{n-1}}{2} \int_0^\infty k_0(s) \left[ M_{\text{arcosh}(1+\varepsilon s/2)} f - f \right] ds \to 0$$

as $\varepsilon \to 0$ in the $L^p$-norm (and uniformly for $f \in C_0(\mathbb{H})$) owing to the properties of $M_r f$ (see Lemma 2.1 from [2]). The a.e. convergence is then a consequence of (7.17) due to the estimate $\|f^*\|_p \leq \|f\|_p$ (for $p = 1$ the corresponding weak estimate holds; see [12], [19]). ■

## 7.3 Inversion of the Horocycle Transform

With the horocycle transform and its dual we associate the following fractional integral operators

$$R^\alpha f(\xi) = c_{n,\alpha} \int_{\mathbb{H}} f(x)\, h_\alpha(x,\xi)\, dx, \tag{7.21}$$

$$\overset{*}{R}{}^{\alpha} \varphi(x) = c_{n,\alpha} \int\limits_{\Gamma} \varphi(\xi)\, h_\alpha(x,\xi)\, d\xi. \tag{7.22}$$

Here $\operatorname{Re}\alpha > 0$, $\alpha \neq 1, 3, \dots,$ $c_{n,\alpha} = 2^{-\alpha}\pi^{-1/2}\Gamma((1-\alpha)/2)/\Gamma(\alpha/2)$,

$$\begin{aligned} h_\alpha(x,\xi) &= |[x,\xi]-1|^{\alpha-1}\,[x,\xi]^{-\rho-\alpha/2} \\ &= \left|[x,\xi]^{1/2} - [x,\xi]^{-1/2}\right|^{\alpha-1} [x,\xi]^{-\rho-1/2}. \end{aligned}$$

These operators share properties with the horocycle transform and its dual, namely, duality

$$\int\limits_{\mathbb{H}} f(x)\, \overset{*}{R}{}^{\alpha} \varphi(x)\, dx = \int\limits_{\Gamma} R^\alpha f(\xi)\, \varphi(\xi)\, d\xi.$$

Furthermore, they intertwine the action of $G$ on $\mathbb{H}$ and on $\Gamma$, i.e. $R^\alpha f(g\circ\xi) = R^\alpha f_g(\xi)$ and $\overset{*}{R}{}^{\alpha} \varphi(g\circ x) = \overset{*}{R}{}^{\alpha} \varphi_g(x)$, where $f_g(x) = f(g\circ x)$ and similarly for $\varphi_g$.

The following lemma links these fractional integrals with the horocycle transform and its dual.

**Lemma 7.7** *Let $f$ and $\varphi$ be smooth compactly supported functions on $\mathbb{H}$ and $\Gamma$ respectively, and let $r_\alpha(t) = 2^{\alpha-1}c_{n,\alpha}e^{-t(\rho+1/2)}\,|\sinh(t/2)|^{\alpha-1}$. Then*

$$R^\alpha f(\xi) = \int\limits_{\mathbb{R}} r_\alpha(t)\, R_\omega f(s-t)\, dt, \quad \xi = e^s b(\omega), \tag{7.23}$$

$$\overset{*}{R}{}^{\alpha} \varphi(x) = \int\limits_{\mathbb{R}} r_\alpha(s) \Big[ \int\limits_{\mathbf{S}^{n-1}} e^{2\rho(s+\langle x,\omega\rangle)} \varphi(e^{s+\langle x,\omega\rangle} b(\omega))\, d\omega \Big] ds, \tag{7.24}$$

*and the following relations hold*

$$\lim_{\alpha\to 0+} R^\alpha f = Rf, \qquad \lim_{\alpha\to 0+} \overset{*}{R}{}^{\alpha} \varphi = \overset{*}{R} \varphi. \tag{7.25}$$

**Proof.** In order to prove (7.23) we write $\xi = e^s k\circ\xi_0$. Then $[x,\xi] = e^s[k^{-1}\circ x, \xi_0)]$ and from (7.21) we have

$$R^\alpha f(\xi) = c_{n,\alpha} \int\limits_{\mathbb{H}} f(k\circ x) \left| e^{s/2}[x,\xi_0)]^{1/2} - e^{-s/2}[x,\xi_0)]^{-1/2} \right|^{\alpha-1}$$
$$\times\, e^{-(\rho+1/2)s}\, [x,\xi_0)]^{-\rho-1/2} dx.$$

We now express the integral using horocycle coordinates (7.3) in the form $x = a_t n\circ O$. Then $[x,\xi_0)] = x_{n+1} - x_n = e^{-t}$, and therefore

$$\begin{aligned} R^\alpha f(\xi) &= c_{n,\alpha} \int\limits_{\mathbb{R}} \int\limits_{N} f(ka_t n\circ O) \left| e^{(s-t)/2} - e^{-(s-t)/2} \right|^{\alpha-1} \\ &\qquad \times e^{-(\rho+1/2)(s-t)} dn\, dt. \end{aligned}$$

This is equivalent to (7.23) by (6). The equality (7.24) can be derived from (7.22) by putting $\xi = e^t b(\omega)$, $d\xi = e^{2\rho t} dt d\omega$, $[x, \xi] = \exp(t - \langle x, \omega \rangle)$ (cf. (5)), and further, $t = s + \langle x, \omega \rangle$. The limit relations (7.25) follow from (7.23) and (7.24) due to normalization (cf. [9], Ch. I, Sec. 3.5). ■

In view of (7.25), the horocycle transform and its dual can be regarded as members of the analytic families of operators $\{R^\alpha\}$ and $\{\overset{*}{R}{}^\alpha\}$, respectively. This is the key observation in our approach. The form of the kernel $h_\alpha$ in (7.21) and (7.22) was discovered taking into account calculations in [20, p. 164].

Following the philosophy given in [2, 14, 16], it is natural to expect that the fractional integral $R^\alpha$ can be inverted by the dual operator via the formula $(R^\alpha)^{-1} = \overset{*}{R}{}^{1-n-\alpha}$. Consequently, the inversion formula for the horocycle transform would take the form ($\alpha = 0$)

$$f = \overset{*}{R}{}^{1-n} Rf. \tag{7.26}$$

However, the integral defining $\overset{*}{R}{}^{1-n}$ is divergent in general, hence the right-hand side of the above formula must be interpreted via analytic continuation in a suitable sense. In what follows, this analytic continuation is carried out via a wavelet-like transform. Specifically, fix a complex-valued function $w \in L^1(0, \infty)$. The *wavelet transform* of a function $\varphi$ living on $\Gamma$ is given for $x \in \mathbb{H}$, and $t > 0$ by

$$W\varphi(x,t) = \frac{1}{t} \int_\Gamma \varphi(\xi)\, w\left( \frac{\left| [x,\xi]^{1/2} - [x,\xi]^{-1/2} \right|}{t} \right) [x,\xi]^{-n/2} d\xi. \tag{7.27}$$

The name wavelet transform is appropriate as the function $w$ (called a wavelet) will be required to satisfy certain growth and moment conditions (specified later). The structure of (7.27) is motivated by the formula

$$\overset{*}{R}{}^\alpha \varphi(x) = \text{const} \times \int_0^\infty W\varphi(x,t) \frac{dt}{t^{1-\alpha}}, \quad 0 < \operatorname{Re} \alpha < 1, \tag{7.28}$$

which can be checked by interchanging the order of integration in the right-hand side. This formula can be used to give the analytic continuation of $\overset{*}{R}{}^\alpha$ for $\operatorname{Re} \alpha \le 0$.

From (7.26) and (7.28) one expects inversion of the horocycle transform via

$$f(x) = c_w \int_0^\infty WRf(x,t) \frac{dt}{t^n},$$

for suitable wavelets $w$. The following makes precise the above formula.

**Theorem 7.8** *Let $f \in L^p(\mathbb{H})$ for some $1 \le p < 2$. Let $w : (0,\infty) \to \mathbf{C}$ satisfy*

$$\operatorname*{ess\,sup}_{s>0}(1+s)^{\mu}\,|w(s)| < \infty, \ \textit{for some } \mu > n; \tag{7.29}$$

$$\int_0^{\infty} s^j w(s)\, ds = 0, \ \textit{for } j = 0, 2, \dots 2[\rho]. \tag{7.30}$$

*Then*

$$f(x) = \lim_{\varepsilon \to 0} c_n^{-1} \int_{\varepsilon}^{\infty} WRf(x,t) \frac{dt}{t^n}, \tag{7.31}$$

*where the limit is understood in the $L^p$-norm and a.e., and*

$$c_n = \frac{\omega_{n-1}\omega_{n-2}}{2} \begin{cases} \Gamma(-\rho) \int_0^{\infty} s^{2\rho} w(s)\, ds, & \textit{if } n \textit{ is even} \\ \frac{(-1)^{\rho+1}}{2\rho!} \int_0^{\infty} s^{2\rho} w(s) \log s\, ds, & \textit{if } n \textit{ is odd} \end{cases}. \tag{7.32}$$

*Moreover, if $f \in C_0(\mathbb{H}) \cap L^p(\mathbb{H})$ for some $1 \le p < 2$, then (7.31) holds uniformly.*

The proof will depend on several preliminaries concerning the wavelet transform (7.27). The growth condition (7.29) and moment conditions (7.30) are the reason we call (7.27) a wavelet transform.

For functions $\varphi$ defined on $\Gamma$, consider the operator

$$B\varphi(x) = \int_{\Gamma} \varphi(\xi)\, b([x,\xi])\, d\xi. \tag{7.33}$$

The wavelet transform given by (7.27) has this structure and we need to determine conditions on the kernel $b$ so that $B\varphi$ is defined when $\varphi = Rf$ for $f \in L^p(H)$ and some $1 \le p < 2$. For this it is useful to express the operator group theoretically as follows. Let $x = g \circ O$ and write $\xi = gka_t \circ \xi_0$. Then

$$\begin{aligned} B\varphi(x) &= \omega_{n-1} \int_{-\infty}^{\infty} \Big[ \int_K \varphi(gka_t \circ \xi_0)\, dk \Big] b([O, a_t \circ \xi_0])\, e^{2\rho t} dt \\ &= \omega_{n-1} \int_{-\infty}^{\infty} \overset{*}{M}{}^{t} \varphi(x)\, b(e^t)\, e^{2\rho t} dt, \end{aligned} \tag{7.34}$$

where $\overset{*}{M}{}^{t} \varphi(x)$ is an averaging operator in brackets. The last formula is justified provided one of the integrals is finite with $\varphi$ and $b$ replaced by $|\varphi|$ and $|b|$, respectively.

**Lemma 7.9** *Let $0 \leq \beta \leq \rho$. Assume that $b : [0, \infty) \to \mathbf{C}$ and $\varphi : \Gamma \to \mathbf{C}$ are non-negative and measurable with*

$$\gamma_\beta = \operatorname*{ess\,sup}_{s>0} s^{\rho-\beta}(1+s)^{2\beta} b(s) < \infty.$$

*Then*

$$B\varphi(x) \leq c_\beta \, \gamma_\beta \, x_{n+1}^{\rho+\beta} \int_\Gamma |\xi|^{\beta-\rho} \, ||\xi|-1|^{-2\beta} \, \varphi(\xi) \, d\xi \ \ a.e., \tag{7.35}$$

*where $c_\beta$ is independent of $b$.*

**Proof.** We start from (7.34) to obtain

$$B\varphi(x) \leq \gamma_\beta \omega_{n-1} \int_{-\infty}^{\infty} \overset{*}{M}{}^{t} \, \varphi(x) \, e^{(\beta-\rho)t}(e^t+1)^{-2\beta} e^{2\rho t} dt,$$

or as an integral over $\Gamma$,

$$B\varphi(x) \leq \gamma_\beta \omega_{n-1} \int_\Gamma \varphi(g \circ \xi) \left[ \xi_{n+1}^{\beta-\rho}(1+\xi_{n+1}^{-2\beta}) \right] d\xi, \qquad x = g \circ O.$$

Changing the variable we see it suffices to estimate the quantity in brackets where $\xi$ is replaced by $g^{-1} \circ \xi$. For this in the case $|\xi| > 1$ we have

$$\begin{aligned} \Omega(x,\xi) &\equiv \frac{(g^{-1}\circ\xi)_{n+1}^{\beta-\rho} \, ((g^{-1}\circ\xi)_{n+1}+1)^{-2\beta}}{|\xi|^{\beta-\rho} \, ||\xi|-1|^{-2\beta}} \\ &= \frac{|\xi|^{\rho-\beta} \, ||\xi|-1|^{2\beta}}{(g^{-1}\circ\xi)_{n+1}^{\rho-\beta} \, ((g^{-1}\circ\xi)_{n+1}+1)^{2\beta}} \\ &\leq \frac{|\xi|^{\rho-\beta} \, ||\xi|+1|^{2\beta}}{(g^{-1}\circ\xi)_{n+1}^{\rho+\beta}}. \end{aligned}$$

Now $(g^{-1} \circ \xi)_{n+1} = [x, \xi]$ and if we write this in coordinate form ($x = (\cosh s, \sinh s \, \omega)$, $\xi = e^t b(\omega')$), then $(g^{-1} \circ \xi)_{n+1} = e^t(\cosh s - \sinh s \, (\omega \cdot \omega'))$, the dot representing ordinary Euclidean inner product. Hence, $(g^{-1} \circ \xi)_{n+1} \geq e^{t-s} = e^{t-d(x,O)}$. The rest of the estimation is straightforward. If $|\xi| < 1$, then $\Omega(x,\xi) \leq (|\xi|/(g^{-1} \circ \xi)_{n+1})^{\rho-\beta}$ which does not exceed $x_{n+1}^{\rho-\beta}$ $(\leq x_{n+1}^{\rho+\beta})$ up to a constant multiple. ∎

For the next lemma we introduce the notation

$$w_1(s) = s^{-1/2} w(s^{1/2})$$

and the associated fractional integral operator

$$I_+^\mu w_1(u) = \frac{1}{\Gamma(\mu)} \int_0^u (u-s)^{\mu-1} w_1(s) \, ds. \tag{7.36}$$

**Lemma 7.10** *Let $f \in L^p(\mathbb{H})$ for some $1 \le p < 2$ and let $w \in L^1(0,\infty)$ satisfy*

$$\operatorname*{ess\,sup}_{s>0}(1+s)^{2\beta+1}\left|w(s)\right| < \infty \tag{7.37}$$

*for some $\beta \in [0, \min(1, \rho, 1/2 + n/p')]$.*

*(i) If $f$ and $w$ are nonnegative, then for each $t > \varepsilon > 0$ and $x \in \mathbb{H}$,*

$$WRf(x,t) \le c_{\beta,\varepsilon} t^{2\beta} x_{n+1}^{\rho+\beta} \left\|f\right\|_p \tag{7.38}$$

*with some constant $c_{\beta,\varepsilon}$, independent of $f$.*

*(ii) Let $f_x(\cosh r) = M^r f(x)$. Then*

$$WRf(x,t) = 2\Gamma(\rho)\omega_{n-1}\omega_{n-2}t^{n-3}\int_1^\infty f_x(2\tau-1)\, I_+^\rho w_1\left(\frac{4(\tau-1)}{t^2}\right) d\tau \tag{7.39}$$

*where $f_x(\cosh r) = M^r f(x)$.*

**Proof.** Set $\varphi = Rf$. Then $W\varphi$ has the form of the operator (7.33) with

$$b(s) = t^{-1}s^{-n/2}w(\left|s^{1/2} - s^{-1/2}\right|/t).$$

We will first demonstrate that $W\varphi$ is well defined for $f \in L^p(\mathbb{H})$ via the previous lemma. For simplicity assume that $w$ and $f$ are non-negative. By Lemma 7.9 and Proposition 7.5 we need to show

$$t^{-1}s^{-\beta-1/2}(1+s)^{2\beta}w(\left|s^{1/2} - s^{-1/2}\right|/t) \le c_{\beta,\varepsilon}t^{2\beta}$$

for $t > \varepsilon$ and $\beta$ satisfying $0 \le \beta \le \rho$ and the hypothesis of Proposition 7.5 (i.e. $\beta \in [0, \min(1, \rho, 1/2 + n/p')]$). This is a straightforward, albeit tedious calculation as follows. Let $s = e^{2u}$ and $v = \left|\sinh u\right|$, then

$$\begin{aligned} e^u &= \cosh u + \sinh u = \begin{cases} \sqrt{1+v^2} + v, & \text{if } u > 0 \\ \sqrt{1+v^2} - v, & \text{if } u < 0 \end{cases} \\ &\ge \sqrt{1+v^2} - v = \frac{1}{\sqrt{1+v^2}+v}. \end{aligned}$$

Consequently,

$$\begin{aligned} \frac{(1+s)^{2\beta}}{t\, s^{\beta+1/2}} w(\left|s^{1/2} - s^{-1/2}\right|/t) &= \frac{(2\cosh u)^{2\beta}}{te^u} w(\frac{2\left|\sinh u\right|}{t}) \\ &\le 4^\beta t^{-1}(1+v^2)^\beta \left(\sqrt{1+v^2}+v\right) w(\frac{2v}{t}) \\ &\le c_\beta t^{-1}(1+v)^{2\beta+1} w(\frac{2v}{t}) \\ &\le \frac{c_\beta}{t}\sup_{s>0}\left(\frac{1+st/2}{1+s}\right)^{2\beta+1} \\ &\le c_{\beta,\varepsilon}t^{2\beta} \end{aligned}$$

provided that $t > \varepsilon$.
Now we will verify (7.39). From (7.34) we have the formula

$$W\varphi(x,t) = t^{-1} \int_{-\infty}^{\infty} \overset{*}{M}{}^{s} \varphi(x)\, w(\frac{2\,|\sinh s/2|}{t})\, e^{(\rho-1/2)s} ds.$$

With $\varphi = Rf$, a simple group theoretic argument shows that

$$\overset{*}{M}{}^{s} \varphi(x) = Rf_x(s).$$

The latter we may compute using (7.12):

$$Rf_x(s) = 2^{2\rho-1}\omega_{n-2} e^{-\rho s} \int_{u^2}^{\infty} f_x(2\tau - 1)\,(\tau - u^2)^{\rho-1} d\tau,$$

where $u = \cosh s/2$. Substituting in the formula for $W\varphi$ we have

$$\begin{aligned} W\varphi(x,t) &= \frac{2^{2\rho-1}\omega_{n-2}}{t} \int_{-\infty}^{\infty} \left[ \int_{\cosh^2 s/2}^{\infty} f_x(2\tau-1)\,(\tau-u^2)^{\rho-1} d\tau \right] \\ &\qquad \times w(\frac{2\,|\sinh s/2|}{t})\, e^{-s/2} ds \\ &= \frac{2^{2\rho}\omega_{n-2}}{t} \int_{0}^{\infty} \left[ \int_{\cosh^2 s/2}^{\infty} f_x(2\tau-1)\,(\tau-u^2)^{\rho-1} d\tau \right] \\ &\qquad \times w(\frac{2\,|\sinh s/2|}{t})\, \cosh s/2\, ds. \end{aligned}$$

Making the change of variables $u = \cosh^2 s/2$ and interchanging the orders of integration we get the following

$$W\varphi(x,t) = \frac{2^{2\rho}\omega_{n-2}}{t} \int_{1}^{\infty} f_x(2\tau-1) \int_{1}^{\tau} w\left(\tfrac{2\sqrt{\tau-1}}{t}\right) \frac{(\tau-u)^{\rho-1}}{\sqrt{u-1}} du\, d\tau.$$

Now the change of variable $\eta = 2\sqrt{\tau-1}/t$ yields:

$$W\varphi(x,t) = 2^{2\rho}\omega_{n-2} \int_{1}^{\infty} f_x(2\tau-1) \int_{0}^{\frac{2\sqrt{\tau-1}}{t}} w(\eta) \left(\tau - 1 - \frac{t^2\eta^2}{4}\right)^{\rho-1} d\eta\, d\tau.$$

This last formula coincides with (7.39). Please note the use of Fubini's theorem is justified by (7.38). ■

The following Corollary relates the integral on the right hand side of (7.31) to a structural form similar to an approximate identity. The proof of Theorem 7.8 will then follow by verifying that indeed we have an approximate identity.

**Corollary 7.11** *Let $f$ and $w$ satisfy the conditions of Lemma 7.10, and let*

$$g(s) = s^{-1} I_+^{\rho+1} w_1(s).$$

*Then*

$$\int\limits_{\varepsilon}^{\infty} \frac{WRf(x,t)}{t^n} dt = \int\limits_{\mathbb{H}} k_\varepsilon([x,y])\, f(y)\, dy, \tag{7.40}$$

*where*

$$k_\varepsilon(\tau) = \frac{\omega_{n-2}\Gamma(\rho)}{2} \frac{(\tau^2-1)^{1-n/2}}{\varepsilon^2} g\left(\frac{2(\tau-1)}{\varepsilon^2}\right). \tag{7.41}$$

**Proof.** Using (7.39) and interchanging the order of integration we have

$$\begin{aligned}
\int\limits_{\varepsilon}^{\infty} \frac{WRf(x,t)}{t^n} dt &= 2\Gamma(\rho)\omega_{n-2} \int\limits_{1}^{\infty} f_x(2\tau-1) \int\limits_{\varepsilon}^{\infty} I_+^{\rho} w_1\left(\frac{4(\tau-1)}{t^2}\right) \frac{dt}{t^3}\, d\tau \\
&= \Gamma(\rho)\omega_{n-2} \int\limits_{1}^{\infty} f_x(2\tau-1) \int\limits_{0}^{4(\tau-1)/\varepsilon^2} I_+^{\rho} w_1(z)\, dz \frac{d\tau}{4(\tau-1)} \\
&= \frac{\Gamma(\rho)\omega_{n-2}}{\varepsilon^2} \int\limits_{1}^{\infty} f_x(2\tau-1)\, g(\frac{4(\tau-1)}{\varepsilon^2})\, d\tau \\
&= \frac{\Gamma(\rho)\omega_{n-2}}{2\varepsilon^2} \int\limits_{0}^{\infty} M^r f(x)\, g(\frac{2(\cosh r-1)}{\varepsilon^2}) \sinh r\, dr.
\end{aligned}$$

This last expression is equivalent to (7.40). Application of Fubini's theorem was possible because of the estimate (7.38). ∎

Finally, we will obtain the proof of Theorem 7.8 by verifying that $k_\varepsilon$ defines an approximate identity. This is accomplished via the following statement which is a consequence of Lemma 2.4 from [14].

**Lemma 7.12** *Let $w \in L^1(0,\infty)$ satisfy (7.29) and (7.30). Set $\gamma = \rho + (\mu - n)/4$. Then*

$$\int\limits_{0}^{\infty} s^{\gamma} |w_1(s)|\, ds \le const \times \int\limits_{0}^{\infty} u^{2\gamma}(1+u)^{-\mu} du < \infty \tag{a}$$

$$g(s) = \begin{cases} O(s^{\rho-1}), & \textit{if } 0 < s \le 1 \\ O(s^{\delta-1}), & \textit{if } s > 1 \end{cases} \tag{b}$$

*where* $\delta = \rho - \min(1 + [\rho], \gamma) < 0$;

$$\int_0^\infty g(s)\, ds = \begin{cases} \Gamma(-\rho) \int_0^\infty s^\rho w_1(s)\, ds, & \textit{if } \rho \notin \mathbf{N} \\ \frac{(-1)^{\rho+1}}{\rho!} \int_0^\infty s^\rho w_1(s) \log s\, ds, & \textit{if } \rho \in \mathbf{N} \end{cases}. \tag{c}$$

Indeed the above Lemma shows that the kernel (7.41) satisfies the conditions of Lemma 7.6 . Hence we deduce that

$$\lim_{\varepsilon \to 0} \int_\varepsilon^\infty \frac{WRf(x,t)}{t^n} dt = c_n f(x)$$

$$c_n = \frac{\omega_{n-2}\Gamma(\rho)}{4} \int_0^\infty g(s)\, ds,$$

with the limit interpreted in the required senses. One can easily check that the constant $c_n$ above coincides with that in (7.32).

**Acknowledgment.** B. Rubin wishes to express his gratitude to his colleague, W. O. Bray, for the hospitality he enjoyed during his visit at the University of Maine in June 1997.

*References*

[1] C. A. Berenstein, C. Tarabusi, *An inversion formula for the horocycle transform on the real hyperbolic space,* Lectures in Appl. Mathematics 30 (1994) p1-6.

[2] C. A. Berenstein, B. Rubin, *Radon transform of $L^p$-functions on the Lobachevsky space and hyperbolic wavelet transforms,* Forum Math. (to appear).

[3] W. O. Bray, *Aspects of harmonic analysis on real hyperbolic space,* Fourier Analysis: analytic and geometric aspects, Vol 157 Lect. Notes Pure Appl. Math. Marcel Dekker (1994) eds. W. O. Bray, P. S. Milojevic, Časlav V. Stanojević, p77-102.

[4] W. O. Bray, *Generalized spectral projections on symmetric spaces of non-compact type: Paley-Wiener theorems,* Jour. Func. Anal. 135 (1996) p206-232.

[5] W. O. Bray, D. C. Solmon, *The horocycle transform and harmonic analysis on the Poincaré disk,* Preprint (1988).

[6] W. O. Bray, D. C. Solmon, *Paley-Wiener theorems on rank one symmetric spaces of non-compact type,* Contemporary Math. 113 (1990) p17-29.

[7] J. L. Clerc, E. M. Stein, *$L^p$-multipliers for non-compact symmetric spaces,* Proc. Nat. Acad. Sci. USA 71 (1974) p3911-3912.

[8] I. M. Gelfand, M. I. Graev, N. Ja. Vilenkin, *Generalized Functions, Vol 5 Integral Geometry and Representation Theory,* Academic Press (1966).

[9] I. M. Gelfand, G. E. Shilov, *Generalized Functions, Vol 1 Properties and Operations,* Academic Press (1964).

[10] S. Helgason, *Geometric Analysis on Symmetric Spaces,* Amer. Math. Soc. (1994).

[11] P. I. Lizorkin, *Direct and inverse theorems of approximation theory for functions on Lobachevsky space,* Proc. of the Steklov Inst. of Math. Issue 4 (1993) p125-151.

[12] B. Rubin, *Fractional Integrals and Potentials,* Addison Wesley Longman (1996).

[13] B. Rubin, *Inversion of Radon transforms using wavelet transforms generated by wavelet measures,* Math. Scand. (to appear).

[14] B. Rubin, *Spherical Radon transforms and related wavelet transforms,* Applied and Computational Harmonic Analysis 5 (1998) p202-215.

[15] B. Rubin, *Inversion of $k$-plane transforms via continuous wavelet transforms,* Jour. Math. Anal. Appl. 220 (1998) p187-203.

[16] B. Rubin, *Fractional calculus and wavelet transforms in integral geometry,* Fractional Calculus and Applied Analysis 1 (1998), No. 2, p193-219.

[17] D. C. Solmon, *A note on $k$-plane integral transforms,* Jour. Math. Anal. Appl. 71 (1979) p351-358.

[18] R. J. Stanton, P. A. Tomas, *Pointwise inversion of the spherical transform on $L^p(G/K)$, $1 \leq p < 2$,* Proc. Amer. Math. Soc. 73 (1979) p398-404.

[19] J. O. Strömberg, *Weak type $L^1$ estimates for maximal functions on non-compact symmetric spaces,* Ann. Math. 114 (1981) p115-126.

[20] N. Ja. Vilenkin, A. U. Klimyk, *Representation of Lie Groups and Special Functions,* Vol 2, Kluwer Academic Publ. (1993).

[21] A. Zorich, *Inversion of horospherical integral transform on real semisimple Lie groups,* International Journal of Modern Physics A, Vol 7, Suppl. 1B (1992) p1047-1071, Proceedings of the RIMS Reseach Project 1991, "Infinite Analysis", World Sci. Publishing Company.

# Chapter 8

# Fourier-Bessel expansions with general boundary conditions

**Mark A. Pinsky**

*ABSTRACT. We find necessary and sufficient conditions for the pointwise convergence of the radial eigenfunctions expansion of the p-dimensional Laplace operator in a ball, where we prescribe either Dirichlet, Neumann or Robin conditions on the boundary.*

## 8.1 Introduction

This paper is a continuation of our program to obtain sharp conditions for pointwise convergence of Fourier expansions of piecewise smooth functions in several variables. The most basic case—the Fourier transform of Euclidean space— has been dealt with in [2], together with extensions to the sphere and the hyperbolic space. More recently [7], we have extended these results to an arbitrary rank one symmetric space of the non-compact type. In all of these problems we have an expansion on the whole space, in the absence of boundary conditions.

When we turn to expansions related to boundary conditions, new phenomena arise. In the simplest case [1], we have the Fourier-Bessel expansion of $f(r) \equiv 1$ in the unit ball of Euclidean space with Dirichlet boundary conditions, which diverges at $r = 0$ in the case the dimension is three or greater, and is otherwise convergent.

In this note we examine the corresponding problem in the presence of general boundary conditions. For Dirichlet boundary conditions, the expansion converges in dimension one and two, while in dimension three or greater we find a necessary and sufficient condition for convergence at $r = 0$.

For non-Dirichlet boundary conditions (Neumann or Robin conditions), it is found that the only obstructions to convergence take place in dimension five and higher. For simplicity of exposition, the computations are carried out for functions which are smooth in the interior, so that all of the "non-smoothness" occurs at the boundary. In a subsequent work we plan to deal with the case of interior discontinuities and more general eigenfunction expansions.

Since our treatment is based on radial functions, we assume that the dimension parameter is any real number $p \geq 1$. Identification of the sum of the series is carried out in Section 4 in the case of integer values of $p$.

The reader should be alerted that our use of the term "Fourier-Bessel series" is different from the prevaling use in the literature, where authors deal with the series

$$\sum_{n=1}^{\infty} A_n J_\alpha(r z_n)$$

where $z_n$ are determined by the boundary conditions. Such a series is identically zero at $r = 0$ whenever $\alpha > 0$, so the convergence question is trivial there. In our treatment we study the series obtained from the radial Laplace operator, namely

$$\sum_{n=1}^{\infty} A_n \frac{J_\alpha(r z_n)}{r^\alpha}$$

which has identical convergence properties for $r > 0$ but radically different behavior at $r = 0$ when $\alpha > 0$. (When $\alpha = 0$ both series coincide)

## 8.2 Statement of Results

Let $\phi_n(r)$ be the radial eigenfunctions of the $p$-dimensional Laplace operator:

$$0 = \Delta\phi + \lambda\phi = \phi''(r) + \frac{p-1}{r}\phi'(r) + \lambda\phi(r), \qquad 0 \leq r \leq a \tag{8.1}$$

with the boundary condition

$$\cos\beta\,\phi(a) + a\sin\beta\,\phi'(a) = 0 \tag{8.2}$$

where $0 \leq \beta < \pi$ ($\beta = 0$ corresponds to the Dirichlet boundary condition $\phi(a) = 0$ and $\beta = \pi/2$ corresponds to the Neumann boundary condition $\phi'(a) = 0$). Let $f(r)$, $0 \leq r \leq a$ be a smooth function with Fourier coefficients

$$A_n = \frac{\int_0^a f(r)\phi_n(r)\, r^{p-1}\, dr}{\int_0^a \phi_n(r)^2\, r^{p-1}\, dr}. \tag{8.3}$$

The formal Fourier expansion is written

$$f(r) \sim \sum_{n=1}^{\infty} A_n \phi_n(r). \tag{8.4}$$

**Theorem 8.1**

- *If $\beta = 0$ and $0 < r \leq a$, then the expansion converges without any further conditions.*

- *If $\beta = 0$ and $r = 0$, then the expansion converges without further conditions if $1 \leq p < 3$. If $p \geq 3$ the expansion converges at $r = 0$ if and only if $f$ satisfies the boundary conditions that*

$$f(a) = 0, (\Delta f)(a) = 0, \ldots, (\Delta^J f)(a) = 0$$

*where $J = [(p-3)/4]$ and $[\ ]$ denotes the integer part.*

- *If $0 < \beta < \pi$ and $0 < r \leq a$, then the expansion converges without any further conditions*

- *If $0 < \beta < \pi$ and $r = 0$, then the expansion converges without further conditions if $1 \leq p < 5$. If $p \geq 5$, the expansion converges at $r = 0$ if and only if $f$ satisfies the boundary conditions that*

$\cos\beta f_j(a) + a \sin\beta f_j'(a) = 0, \ldots$ *for* $j = 0, \ldots, [(p-5)/4]$, *where* $f_0 = f$ *and* $f_j = \Delta^j f$ .

## 8.3 Proofs

Explicitly, we have for the (un-normalized) eigenfunctions

$$\phi_n(r) = \frac{J_\alpha(r\sqrt{\lambda_n})}{r^\alpha},$$

where $\alpha = (p-2)/2$. The Fourier coefficients can be obtained by integrating the Bessel equation (1); beginning with

$$(r^{p-1}\phi_n')' + \lambda_n r^{p-1}\phi_n = 0,$$

we multiply by $f(r)$ and integrate by parts twice to obtain the following version of Lagrange's identity:

$$
\begin{aligned}
\lambda_n \int_0^a r^{p-1}\phi_n(r) f(r)\, dr &= -\int_0^a f(r)(r^{p-1}\phi_n'(r))'\, dr \\
&= -f(a)a^{p-1}\phi_n'(a) + \int_0^a f'(r) r^{p-1}\phi_n'(r)\, dr \\
&= a^{p-1}\left(f'(a)\phi_n(a) - f(a)\phi_n'(a)\right) \\
&\qquad - \int_0^a \phi_n(r)\left(r^{p-1}f'\right)'\, dr \\
&= a^{p-1}\left(f'(a)\phi_n(a) - f(a)\phi_n'(a)\right) \\
&\qquad - \int_0^a r^{p-1}\phi_n(r)\,\Delta f(r)\, dr.
\end{aligned}
$$

This can be iterated to obtain the following finite expansion.

**Proposition 8.2** *For any* $N = 0, 1, 2, \ldots$

$$
\begin{aligned}
\int_0^a r^{p-1}\phi_n(r) f(r)\, dr &= a^{p-1}\sum_{j=0}^{N} \frac{(-1)^j}{\lambda_n^{j+1}}\left(f_j'(a)\phi_n(a) - f_j(a)\phi_n'(a)\right) \\
&\quad + \frac{(-1)^{N+1}}{\lambda_n^{N+1}} \int_0^a r^{p-1}\phi_n(r)\, f_{N+1}(r)\, dr,
\end{aligned}
$$

*where* $f_j = \Delta^j f$.

On the other hand, the normalization coefficients which occur in the denominator of (8.3) are well-known [3]:

- $\beta = 0$

$$
\int_0^a \phi_n(r)^2 r^{p-1}\, dr = \int_0^a J_\alpha(r\sqrt{\lambda_n})^2\, r\, dr = \frac{1}{2} J_{1+\alpha}(a\sqrt{\lambda_n})^2.
$$

- $0 < \beta < \pi$

$$
\begin{aligned}
\int_0^a \phi_n(r)^2 r^{p-1}\, dr &= \int_0^a J_\alpha(r\sqrt{\lambda_n})^2 r\, dr \\
&= \frac{(\lambda_n - \alpha^2 + \cot^2\beta) J_\alpha(a\sqrt{\lambda_n})^2}{2\lambda_n}.
\end{aligned}
$$

In either case, the asymptotic behavior of the Bessel functions [4] yields the result

$$\int_0^a \phi_n(r)^2 r^{p-1}\,dr = \frac{C_{\alpha,\beta}}{n}\left(1+O(\frac{1}{n})\right), n\to\infty \quad (C_{\alpha,\beta}>0)$$

The asymptotic behavior of the eigenfunctions is summarized as follows:

- If $r>0$,

$$\begin{aligned}\phi_n(r) &= \frac{J_\alpha(r\sqrt{\lambda_n})}{r^\alpha}\\ &= \frac{C_\alpha}{r^\alpha\sqrt{r\sqrt{\lambda_n}}}\left(\cos(r\sqrt{\lambda_n}-\theta_\alpha)+O(\frac{1}{\sqrt{\lambda_n}})\right)\\ &= \frac{C_\alpha}{r^{\alpha+1/2}\sqrt{n\pi/a}}\left(\cos(r\sqrt{\lambda_n}-\theta_\alpha)+O(\frac{1}{n})\right).\end{aligned}$$

$$\begin{aligned}\phi_n'(r) &= \frac{d}{dr}\frac{J_\alpha(r\sqrt{\lambda_n})}{r^\alpha}\\ &= -\frac{C_\alpha\sqrt{\lambda_n}}{r^\alpha\sqrt{r\sqrt{\lambda_n}}}\left(\sin(r\sqrt{\lambda_n}-\theta_\alpha)+O(\frac{1}{\sqrt{\lambda_n}})\right)\\ &= -\frac{C_\alpha\sqrt{n\pi/a}}{r^{\alpha+1/2}}\left(\sin(r\sqrt{\lambda_n}-\theta_\alpha)+O(\frac{1}{n})\right).\end{aligned}$$

- If $r=0$,

$$\phi_n(0) = c_\alpha(\sqrt{\lambda_n})^\alpha \sim c_\alpha(\frac{n}{pi})^\alpha$$

where $c_\alpha, C_\alpha, \theta_\alpha$ depend only on $\alpha$.
The eigenvalues $\lambda_n$ have a well-known asymptotic behavior:

$$\sqrt{\lambda_n} \sim \frac{n\pi}{a}, \qquad n\to\infty$$

In more detail, we have

$$\begin{aligned}\beta=0: \quad a\sqrt{\lambda_n} &= (n-1/2)\pi+\theta_\alpha+O(\frac{1}{n}),\\ \beta>0: \quad a\sqrt{\lambda_n} &= n\pi+\theta_\alpha+O(\frac{1}{n}).\end{aligned}$$

Combining the above information, we have the following asymptotic behavior of the $n^{\text{th}}$ term of the series (8.4), where we analyze each of the four cases separately:

- If $\beta = 0$ and $0 < r \leq a$, then $\phi_n(a) = 0$ and we can take $N = 0$ in the asymptotic expansion (5) to obtain

$$A_n\phi_n(r) = \frac{n}{C_{\alpha,\beta}}\left(\frac{a}{n\pi}\right)^2 \frac{C_\alpha^2}{(ra)^{\alpha+1/2}} f_0(a) \cos(r\sqrt{\lambda_n} - \theta_\alpha) \times \sin(a\sqrt{\lambda_n} - \theta_\alpha) + O(\frac{1}{n^2})$$

  which is the general term of a convergent series.

- If $\beta = 0$ and $r = 0$, then $\phi_n(a) = 0$ and we take $N = 1 + [(p-3)/4]$ to obtain the asymptotic behavior

$$A_n\phi_n(0) = \frac{n}{C_{\alpha,\beta}}(\frac{n\pi}{a})^\alpha \sum_{j=0}^{N} \frac{(-1)^j f_j(a)}{(n\pi/a)^{2+2j}} \sqrt{\frac{n\pi}{a}} \sin(a\sqrt{\lambda_n} - \theta_\alpha) + O(\frac{1}{n^2})$$

  The first term is asymptotic to $n^{(p-3)/2} f_0(a) \sin(a\sqrt{\lambda_n} - \theta_\alpha)$, which is the general term of a convergent series if $1 \leq p < 3$. If $p \geq 3$, convergence requires that the first term tend to zero, which is possible if and only if $f_0(a) = 0$. The next term in the expansion of $A_n\phi_n(0)$ is $n^{(p-7)/2} f_1(a) \sin(a\sqrt{\lambda_n} - \theta_\alpha)$, which is the general term of a convergent series if $p < 7$. Continuing inductively, we find that if $4k + 3 \leq p \leq 4k + 6$, the convergence takes place if and only if $f_0(a) = 0, \ldots, f_k(a) = 0$, as required.

- If $0 < \beta < \pi$ and $0 < r \leq a$, then $a\phi_n'(a) = -\cot\beta\,\phi_n(a)$; we take $N = 0$ to obtain the asymptotic behavior

$$A_n\phi_n(r) = \frac{n}{C_{\alpha,\beta}}(\frac{a}{n\pi})^2 a^{p-1} (af'(a) + \cot\beta f(a))\, a^{p-1}\phi_n(a) + O(\frac{1}{n^2})$$

  which is the general term of a convergent series.

- If $0 < \beta < \pi$ and $r = 0$, then we take $N = 1 + [(p-5)/4]$ in the proposition and use the boundary condition $a\,\phi_n'(a) = -\cot\beta\,\phi_n(a)$ to obtain the asymptotic behavior

$$A_n\phi_n(0) = \frac{n}{C_{\alpha,\beta}}(\frac{n\pi}{a})^\alpha \sum_{j=0}^{N} (-1)^j \frac{af_j'(a) + \cot\beta f_j(a)}{(n\pi/a)^{2+2j}} \sqrt{\frac{a}{n\pi}} \times \cos(a\sqrt{\lambda_n} - \theta_\alpha) + O(\frac{1}{n^2})$$

The first term is $n^{(p-5)/2}[af_0'(a)+\cot\beta f_0(a)]\cos(a\sqrt{\lambda_n}-\theta_\alpha)$, which is the general term of a convergent series if $1\le p<5$. If $p\ge 5$ convergence requires that the first term tend to zero, which is possible if and only if $af_0'(a)+\cot\beta f_0(a)=0$. The next term in the expansion of $A_n\phi_n(0)$ is $n^{(p-9)/2}[f_1'(a)+a\cot\beta f_1(a)]\cos(a\sqrt{\lambda_n}-\theta_\alpha)$, which is the general term of a convergent series if $p<9$. Continuing inductively, we find that if $4k+5\le p\le 4k+8$, the convergence takes place if and only if $af_0'(a)+\cot\beta f_0(a)=0,\dots,f_k'(a)+a\cot\beta f_k(a)=0$, as required.

The proof is complete.

## 8.4 Identifying the limit

Once we know that the series converges, we can identify the limit by a summability procedure based on the heat equation. We restrict the discussion to integral values of $p$.

The heat kernel (relative to the general boundary conditions) is the function $K(t,x,y)$ defined for $t>0$, $0\le|x|<a, 0\le|y|<a$ and satisfying the partial differential equation

$$\frac{\partial K}{\partial t}=\Delta_x p=\frac{\partial^2 K}{\partial r^2}+\frac{p-1}{r}\frac{\partial K}{\partial r}+\text{angular terms}\qquad t>0, 0\le r\le a$$

with initial and boundary conditions given by

$$K(0+,x,y)=\delta_y(x)$$
$$\left(\cos\beta\, K+a\sin\beta\frac{\partial K}{\partial r}\right)\|_{|y|=a}=0$$

It is proved in [5] Chapter 3, that there exists a unique smooth function with these properties, and that for $t\to 0$, $K(t,x,y)=e^{-|x-y|^2/4t}/(4\pi t)^{p/2}+$ a function which tends uniformly to zero when $t\to 0$. Furthermore it was shown by Hsu [6] that there is a uniformly convergent eigenfunction expansion,

$$K(t,x,y)=\sum_{n=1}^{\infty}e^{-\lambda_n t}\Phi_n(x)\Phi_n(y)$$

where $\Phi_n$ are the (normalized) eigenfunctions of the Laplace operator satisfying the general boundary conditions. These are expressed as the product of the Bessel functions $\phi_n(r)$ in the radial variable with Gegenbauer polynomials in the angular variables.

Now suppose that $f$ is a piecewise smooth function on $[0, a]$. We multiply the series by $f\,dV$ and integrate term by term: the angular terms contribute zero, and we obtain

$$\int_{|y|\leq a} f(|y|)\, K(t, x, y)\, dV = \sum_{n=1}^{\infty} e^{-\lambda_n t} A_n(r)\phi_n(r).$$

When $t \to 0$ and $0 \leq r < a$ the left side tends to the normalized value $\frac{1}{2}f(r+0) + \frac{1}{2}f(r-0)$. If, in addition, the Fourier-Bessel series converges then by an Abelian lemma, proved below, it follows that

$$\lim_{t\to 0} \sum_{n=1}^{\infty} e^{-\lambda_n t} A_n(r)\phi_n(r) = \frac{1}{2}\left(f(r+0) + f(r-0)\right).$$

This completes the identification of the limit. We summarize the result as follows.

**Proposition 8.3** *If the Fourier-Bessel series (8.4) converges, then the sum of the series is*

$$\sum_{n=1}^{\infty} A_n(r)\phi_n(r) = \frac{1}{2}f(r+0) + \frac{1}{2}f(r-0) \qquad p = 1, 2, \ldots, 0 \leq r < a$$

### *8.4.1 An Abelian lemma*

**Lemma 8.4** *Suppose that $\mu(x)$ is a function of bounded variation for $x > 0$, with $\mu(0) = 0$, $L = \lim_{x\to\infty} \mu(x)$. Then $\lim_{t\to 0} \int_0^\infty e^{-tx} d\mu(x) = L$.*

**Proof.** Define $I(t) = \int_0^\infty e^{-tx} d\mu(x)$. Then we can write

$$\int_0^N e^{-tx} d\mu(x) = e^{-Nt}\mu(N) + \int_0^N te^{-tx}\mu(x)\, dx$$

When $N \to \infty$ the first term tends to zero, so we have

$$I(t) = \int_0^\infty te^{-tx}\mu(x)\, dx$$

$$I(t) - L = \int_0^\infty te^{-tx}[\mu(x) - L]\, dx$$

Given $\epsilon > 0$, choose $M$ so that $|\mu(x) - L| < \epsilon$ for $x > M$. Then

$$|\int_M^\infty te^{-tx}[\mu(x) - L]\,dx| < \epsilon$$

On the other hand $\int_0^M te^{-tx}[\mu(x) - L]\,dx \to 0$ when $t \to 0$. This proves that $\limsup_{t\to 0} |I(t) - L| < \epsilon$ for any $\epsilon > 0$, which completes the proof. ■

This is applied above to the jump function $\mu(x) = \sum_{\lambda_n \le x} A_n$ to conclude that if $\sum_n A_n = L$ and $\lim_{t\to 0} \sum_n e^{-\lambda_n t} A_n = M$, then $L = M$.

*References*

[1] A. Gray and M. Pinsky, *Gibbs' phenomenon for Fourier-Bessel series*, Expositiones Mathematicae, 11(1993), 123-135.

[2] M. Pinsky, *Pointwise Fourier inversion and related eigenfunction expansions,* Communications in Pure and Applied Mathematics, 47(1994), 653-681.

[3] M. Pinsky, *"Partial Differential Equations and Boundary-Value Problems with Applications"*, Second Edition, Mc Graw Hill, Inc. 1991.

[4] G. N. Watson, *"A Treatise on the Theory of Bessel Functions"*, Second Edition, Cambridge University Press, 1958.

[5] A. Friedman, *Partial Differential Equations of Parabolic Type*, Prentice Hall, 1964.

[6] P. Hsu, *Seminar on Stochastic Processes*, Birkhauser Boston 1986, pp. 108-116

[7] W. O. Bray and M. Pinsky, *Pointwise Fourier inversion on rank-one symmetric spaces and related topics,* Journal of Functional Analysis 15(1997) 306-334.

# Part II

# Singular Integrals and Multipliers

# Chapter 9

# Convolution Calderón-Zygmund singular integral operators with rough kernels

L. Grafakos[1]
A. Stefanov

*ABSTRACT. A survey of known results in the theory of convolution type Calderón-Zygmund singular integral operators with rough kernels is given. Some recent progress is discussed. A list of remaining open questions is presented.*

## 9.1 Introduction

Throughout this article, $\Omega$ will be a complex-valued integrable function over the sphere $\mathbf{S}^{n-1}$, with mean value zero with respect to surface measure. Define a tempered distribution $K_\Omega$ on $\mathbb{R}^n$ by setting

$$K_\Omega(f) = \lim_{\varepsilon \to 0} \int_{|x|>\varepsilon} \frac{\Omega(x/|x|)}{|x|^n} f(x)\, dx = p.v. \int_{\mathbb{R}^n} \frac{\Omega(x/|x|)}{|x|^n} f(x)\, dx, \qquad (9.1)$$

for $f$ in the Schwartz class $\mathcal{S}(\mathbb{R}^n)$. The limit in (9.1) can be easily shown to exist for any $f$ $C^1$ function on $\mathbb{R}^n$ which satisfies $|f(x)| \leq C|x|^{-\delta}$ for some $C, \delta > 0$ and all $|x|$ large.

We will denote by $T_\Omega$ the operator given by convolution with $\Omega$ initially defined on the set of Schwartz functions $\mathcal{S}(\mathbb{R}^n)$. The operators $T_\Omega$ were introduced by Calderón and Zygmund in [1] and today are referred to as Calderón-Zygmund singular integral operators (of convolution type).

In this article we shall be concerned with the following questions: What conditions on $\Omega$ imply $L^p$ boundedness for $T_\Omega$ and other related operators? It is a classical result, that if $\Omega$ has some smoothness on $\mathbf{S}^{n-1}$, say Lipschitz of order $\alpha > 0$, then $T_\Omega$ is a bounded operator on $L^p(\mathbb{R}^n)$ for $1 < p < \infty$.

---

[1]Research partially supported by NSF grant DMS 9623120 and by the University of Missouri Research Board

In fact, for such $\Omega$'s we have that $K_\Omega$ satisfies Hörmander's condition

$$\int_{|x|\geq 2|y|} |K_\Omega(x-y) - K_\Omega(x)|\,dx \leq B, \tag{9.2}$$

for some $B = B(n,\Omega) > 0$. Condition (9.2) implies that $T_\Omega$ is of weak type $(1,1)$, a property which will be discussed in section 4. This property, together with the $L^2$ boundedness of $\Omega$ (which follows from a Fourier transform calculation), implies that $T_\Omega$ is bounded on $L^p(\mathbb{R}^n)$ for $1 < p < \infty$. See [19] for details.

In 1956 Calderón and Zygmund [2] introduced the method of rotations. The idea is the following: If $\Omega$ is an odd function on $\mathbf{S}^{n-1}$, then it is easy to see that

$$(T_\Omega f)(x) = \frac{1}{2}\int_{\mathbf{S}^{n-1}} \Omega(\theta)(H_\theta f)(x)\,d\theta, \tag{9.3}$$

where $H_\theta f$ is the directional Hilbert transform of $f$ in the direction $\theta \in \mathbf{S}^{n-1}$, defined by

$$(H_\theta f)(x) = p.v.\frac{1}{\pi}\int_{\mathbb{R}^1} \frac{f(x-t\theta)}{t}\,dt = \frac{1}{\pi}T_{\delta_\theta - \delta_{-\theta}}, \tag{9.4}$$

where $\delta_a$ is Dirac mass at $a$. (Of course $\Omega = \delta_\theta - \delta_{-\theta}$ is not in $L^1$, but we can extend the definition of $T_\Omega$ for $\Omega$ bounded Borel measures on $\mathbf{S}^{n-1}$.) Using a rotation, it is easy to show that $H_\theta f$ maps $L^p(\mathbb{R}^n) \to L^p(\mathbb{R}^n)$ with the same norm as the usual Hilbert transform from $L^p(\mathbb{R}^1) \to L^p(\mathbb{R}^1)$. It follows from (9.3) that $T_\Omega$ maps $L^p(\mathbb{R}^n)$ into itself for any $\Omega$ odd in $L^1(\mathbf{S}^{n-1})$.

In the same paper [2], Calderón and Zygmund proved that if

$$\int_{\mathbf{S}^{n-1}} |\Omega(\theta)| Log^+ |\Omega(\theta)|\,d\theta < \infty, \tag{9.5}$$

then $T_\Omega$ is a bounded operator on $L^p$, $1 < p < \infty$. In view of the previous discussion about odd kernels, condition (9.5) is only relevant to even $\Omega$'s.

The general question along these lines is the following:

**Question 9.1** *Let $\Omega$ be an integrable even function on $\mathbf{S}^{n-1}$ with integral zero. Given a $1 < p < \infty$, find a necessary and sufficient condition on $\Omega$ such that $T_\Omega$ extends to a bounded operator from $L^p(R^n) \to L^p(R^n)$.*

It is likely that such a condition will depend on the parameter $p$.

## 9.2 $L^2$ boundedness

$L^2$ is a good starting point to study boundedness of the operators $T_\Omega$ on $L^p$ spaces. We begin with the following natural question: If $\Omega$ is merely an $L^1$ function with integral zero, is $T_\Omega$ a bounded operator on $L^2(\mathbb{R}^n)$?

The answer is known to be negative. More precisely, an example constructed by M. Weiss and A. Zygmund gives a dramatic answer to this question:

**Theorem 9.1** *(M. Weiss and A. Zygmund [21]) Let $\phi(u)$ be a non-negative increasing (nonnecessarily strictly) function defined for $u \geq 0$ which satisfies:*

$$\lim_{u \to \infty} \frac{\phi(u)}{u \log u} = 0.$$

*Then there exists an $\Omega$ in $L^1(\mathbf{S}^{n-1})$ with integral zero which satisfies*

$$\int_{\mathbf{S}^{n-1}} \phi(|\Omega(\theta)|)\, d\theta < +\infty,$$

*and a continuous $f \in L^p(\mathbb{R}^n)$ for all $1 \leq p \leq \infty$, which tends to zero at infinity such that*

$$\limsup_{\varepsilon \to 0} \left| \int_{|y| > \varepsilon} \frac{\Omega(y/|y|)}{|y|^n} f(x-y)\, dy \right| = +\infty$$

*for almost all $x$ in $\mathbb{R}^n$.*

In particular, taking $\phi(u) = u$, we conclude that there exists an $\Omega$ in $L^1(\mathbf{S}^{n-1})$ such that $T_\Omega$ is not a bounded operator on all $L^p$ spaces. Taking $\phi(u) = u(\log u)^{1-\varepsilon}$ we obtain that $\Omega \in LLog^{1-\varepsilon}L$ is not a strong enough condition to imply $L^p$ boundedness for $T_\Omega$.

However, the question is far from over. We know precisely when a convolution operator maps $L^2(\mathbb{R}^n)$ into itself. This happens exactly when the Fourier transform of the convolving distribution is a bounded function. Let us compute the Fourier transform of the distribution $K_\Omega$. Fix $f$ in the

Schwartz class. We have

$$\widehat{K_\Omega}(f) = \int\limits_{\mathbb{R}^n} K_\Omega(x)\widehat{f}(x)\,dx$$

$$= \lim_{\substack{\varepsilon\to 0\\ N\to\infty}} \int\limits_{\mathbb{R}^n} f(y) \left[ \int\limits_{\varepsilon\le|x|\le N} \frac{\Omega(x/|x|)}{|x|^n} e^{-2\pi i y\cdot x} dx \right] dy \tag{9.6}$$

$$= \lim_{\substack{\varepsilon\to 0\\ N\to\infty}} \int\limits_{\mathbb{R}^n} f(y) \left[ \int\limits_{n-1} \Omega(\theta) \left\{ \int\limits_{r=\varepsilon/|y|}^{N/|y|} e^{-2\pi i r\, y'\cdot\theta} \frac{dr}{r} \right\} d\theta \right] dy$$

where $y' = y/|y|$. It can be shown (see [19] for details) that the expression inside the curly brackets above converges pointwise to

$$\frac{\pi i}{2} sgn(\theta\cdot y') + \log\frac{1}{|\theta\cdot y'|}\,.$$

Therefore, if we assume that

$$\sup_{y'\in\mathbf{S}^{n-1}} \int\limits_{\mathbf{S}^{n-1}} |\Omega(\theta)| \log\frac{1}{|\theta\cdot y'|}\, d\theta < +\infty, \tag{9.7}$$

it is an easy consequence of the Lebesgue dominated convergence theorem that $\widehat{K_\Omega}$ is the bounded function:

$$\widehat{K_\Omega}(y) = \int\limits_{\mathbf{S}^{n-1}} \Omega(\theta) \left[ \frac{\pi i}{2} sgn(\theta\cdot y') + \log\frac{1}{|\theta\cdot y'|} \right] d\theta. \tag{9.8}$$

More generally, it can be seen from the calculations above that $\widehat{K_\Omega}$ is a function in $L^\infty(\mathbb{R}^n)$ if and only if the limit of the bracketed expression in (9.6) exists and is equal to a bounded function, i.e.

$$\lim_{\substack{\varepsilon\to 0\\ N\to\infty}} \int\limits_{\varepsilon\le|x|\le N} \frac{\Omega(x/|x|)}{|x|^n} e^{-2\pi i y\cdot x} dx = m(y) \in L^\infty(\mathbb{R}^n). \tag{9.9}$$

Condition (9.7), even though not equivalent to (9.9) contains most of its essence.

An easy consequence of the above is the following

**Theorem 9.2** *Suppose that $\Omega$ satisfies (9.7) or more generally (9.9). Then $T_\Omega$ extends to an operator bounded from $L^2(\mathbb{R}^n)$ into itself. In fact condition (9.9) is equivalent to the $L^2$ boundedness of $T_\Omega$.*

**Exercise.** Use Young's inequality in the context of Orlicz spaces to prove directly that condition (9.5) implies condition (9.7).

## 9.3 $L^p$ boundedness, $1 < p < \infty$

It is well known that if a convolution operator maps $L^p \to L^p$ then by duality it also maps $L^{p'} \to L^{p'}$ with the same norm. ($p' = p/(p-1)$ throughout this paper.) It follows that it maps $L^2 \to L^2$ by interpolation. Since condition (9.9) is equivalent to $L^2$ boundedness, it is unlikely to expect that condition (9.9) would imply that $T_\Omega$ is $L^p$ bounded. Condition (9.7) is slightly weaker, and we can pose the following question:

**Question 9.2** *Let $\Omega$ be an integrable function on $\mathbf{S}^{n-1}$ with integral zero satisfying condition (9.7). Does it follow that $T_\Omega$ is a bounded operator on $L^p(R^n)$ for some $p \neq 2$?*

A weaker question is answered in Theorem 9.4.

Let us denote by $H^1(\mathbf{S}^{n-1})$ the 1-Hardy space on the sphere in the sense of Coifman and Weiss [6]. It is a known result that functions $\Omega$ on $\mathbf{S}^{n-1}$ which satisfy (9.5) are in $H^1(\mathbf{S}^{n-1})$. It is natural to ask whether $T_\Omega$ is $L^p$ bounded when $\Omega \in H^1(\mathbf{S}^{n-1})$. With the aid of a theorem in [3] and with a bit of work one can show that the condition $\Omega \in H^1(\mathbf{S}^{n-1})$ is equivalent to

$$\frac{\Omega(x/|x|)}{|x|^n}\chi_{1/2\le|x|\le 2} \in H^1(\mathbb{R}^n), \tag{9.10}$$

where $H^1(\mathbb{R}^n)$ denotes the Hardy space on $\mathbb{R}^n$. See [18] for details.

We now investigate connections between condition (9.10) and $L^2$ boundedness. Take $\Omega$ to be an even function in this discussion. Using polar coordinates and the fact that $\Omega$ has mean value zero, it is easy to see that

$$\log 4 \int_{\mathbf{S}^{n-1}} \Omega(\theta)\log\frac{1}{|\theta\cdot\xi|}\,d\theta = \int_{1/2\le|x|\le 2} \frac{|\Omega(x/|x|)|}{|x|^n}\log\frac{1}{|x\cdot\xi|}\,dx, \tag{9.11}$$

where both integrals in (9.11) are finite for almost all $\xi \in \mathbb{R}^n$ by an easy application of Fubini's theorem. The $H^1$-BMO duality now gives

$$\left|\int_{1/2\le|x|\le 2}\frac{\Omega(x)}{|x|^n}\log\frac{1}{|x\cdot\xi|}dx\right| \le \left\|\log\frac{1}{|x\cdot\xi|}\right\|_{BMO(dx)}\left\|\frac{\Omega(x)}{|x|^n}\chi_{1/2\le|x|\le 2}\right\|_{H^1} \tag{9.12}$$

Since the $BMO$ norm is invariant under rotations, it is easy to see the $BMO$ norms of the functions $x \to -\log|x\cdot\xi|$ are uniformly bounded in $\xi$. It follows from (9.11) and (9.12) that

$$\sup_{|\xi|=1}\left|\int_{\mathbf{S}^{n-1}}\Omega(\theta)\ln\frac{1}{|\theta\cdot\xi|}\,d\theta\right| \le C\left\|\frac{\Omega(x)}{|x|^n}\chi_{1/2\le|x|\le 2}\right\|_{H^1(dx)}. \tag{9.13}$$

Since $\Omega$ is even, the left hand side of (9.13) is equal to $\|\widehat{K_\Omega}\|_{L^\infty}$ in view of (9.8). We conclude that $T_\Omega$ is $L^2$ bounded, and hence condition (9.10) implies $L^2$ boundedness.

We now show that the $H^1(\mathbf{S}^{n-1})$ condition implies that $L^p$ boundedness for $T_\Omega$ for $1 < p < \infty$. The theorem below was independently discovered by Connett [7] and Ricci and Weiss [14]. See also [6] for a proof in dimension $n = 2$. The proof we give below uses the equivalent hypothesis (9.10).

**Theorem 9.3** *(W. Connett, F. Ricci and G. Weiss) Let $\Omega$ be an integrable function on $S^{n-1}$ with mean value zero which satisfies condition (9.10). Then $T_\Omega$ extends to a bounded operator from $L^p(\mathbb{R}^n)$ into itself for $1 < p < \infty$.*

**Proof.** As discussed before, it suffices to consider $\Omega$ even. Denote by $R_j$ the $j^{\text{th}}$ Riesz transform given by convolution with $p.v. \dfrac{\Gamma(\frac{n+1}{2})}{\pi^{\frac{n+1}{2}}} \dfrac{x_j}{|x|^{n+1}}$. Since

$$I = \sum_{i=1}^{n} R_j^2,$$

it follows that

$$T = \sum_{i=1}^{n} R_j T_j, \tag{9.14}$$

where $T_j = R_j T$. Observe that $T_j$ is well defined as an operator on $L^2$. Let $V_j$ be the kernel of $T_j$. Since $T$ has an even kernel and $R_j$ has an odd kernel, $T_j$ has an odd kernel $K_j$ which is also homogeneous of degree $-n$. Write

$$K_j(x) = R_j\left(p.v. \frac{\Omega(\cdot)}{|\cdot|^n}\right)(x) = \frac{V_j(x/|x|)}{|x|^n},$$

where $V_j$ is an odd distribution on the sphere. ( $V_j(x/|x|)$ denotes the distribution $\phi \to \langle V_j, \phi(x/|x|)\rangle$ on $\mathbb{R}^n$). We will show that $V_j$ is a function satisfying

$$\int_{\mathbf{S}^{n-1}} |V_j(\theta)| d\theta < \infty. \tag{9.15}$$

To prove (9.15) write $K_j = K_j^0 + K_j^1 + K_j^\infty$, where

$$\begin{aligned}
K_j^0 &= R_j\left(p.v. \frac{\Omega(\cdot)}{|\cdot|^n} \chi_{|\cdot| < \frac{1}{2}}\right), \\
K_j^1 &= R_j\left(\frac{\Omega(\cdot)}{|\cdot|^n} \chi_{\frac{1}{2} \le |\cdot| \le 2}\right), \\
K_j^\infty &= R_j\left(\frac{\Omega(\cdot)}{|\cdot|^n} \chi_{2 < |\cdot|}\right).
\end{aligned}$$

Fix $x$ in the annulus $3/4 \leq |x| \leq 3/2$. Then

$$\begin{aligned}
\frac{\pi^{\frac{n+1}{2}}}{\Gamma(\frac{n+1}{2})}\left|K_j^0(x)\right| &= \left|\lim_{\varepsilon\to 0}\int\limits_{\varepsilon<|y|<\frac{1}{2}} \frac{x_j-y_j}{|x-y|^{n+1}}\frac{\Omega(y)}{|y|^n}dy\right| \\
&= \left|\int\limits_{|y|<\frac{1}{2}} \left(\frac{x_j-y_j}{|x-y|^{n+1}} - \frac{x_j}{|x|^{n+1}}\right)\frac{\Omega(y)}{|y|^n}dy\right| \\
&= \left|\int_0^{\frac{1}{2}}\int_{\mathbf{S}^{n-1}} \theta\cdot\nabla\left(\frac{x_j}{|x|^{n+1}}\right)(x-\rho\,\theta\, t_{x,\rho\theta})\,\Omega(\theta)\,d\theta d\rho\right| \\
&\leq \frac{1}{2}\|\Omega\|_{L^1}\max_{1/4\leq|x|\leq 7/4}\left|\nabla\left(\frac{x_j}{|x|^{n+1}}\right)\right| = C\|\Omega\|_{L^1},
\end{aligned}$$

for some $t_{x,\rho\theta} \in [0,1]$. Similarly,

$$\begin{aligned}
\frac{\pi^{\frac{n+1}{2}}}{\Gamma(\frac{n+1}{2})}\left|K_j^\infty(x)\right| &= \left|\int\limits_{|y|>2} \frac{x_j-y_j}{|x-y|^{n+1}}\frac{\Omega(y)}{|y|^n}dy\right| \\
&\leq \int\limits_{|y|>2} \frac{1}{|x-y|^n}\frac{|\Omega(y)|}{|y|^n}dy \\
&\leq \int\limits_{|y|>2} \frac{4^n}{|y|^{2n}}|\Omega(y)|\,dy = C\|\Omega\|_{L^1},
\end{aligned}$$

for $3/4 \leq |x| \leq 3/2$. Finally, $K_j^1$ is in $L^1(\mathbb{R}^n)$ since by assumption $(\Omega(x/|x|)/|x|^n)\,\chi_{1/2\leq|x|\leq 2}$ is in the Hardy space $H^1(\mathbb{R}^n)$. See [20, p.114].

It follows that $K_j$ is integrable over the annulus $3/4 \leq |x| \leq 3/2$. Therefore $V_j(x/|x|)/|x|^n$ has to be integrable over a sphere $a\mathbf{S}^{n-1}$, for some $3/4 \leq a \leq 3/2$. By homogeneity $V_j$ is integrable over $\mathbf{S}^{n-1}$. Therefore $T_j = T_{V_j}$ and by identity (23.7) for $\Omega = V_j$ we deduce that $T_j = T_{V_j}$ is bounded on $L^p$. (9.14) now gives that $T$ is bounded on $L^p$. ■

**Remark 9.1** *In the proof of Theorem 9.3, we showed that condition (9.10) implies that $V_j$ is integrable over $\mathbf{S}^{n-1}$. In fact, the converse is also true. It is shown in [14] that $V_j \in L^1(\mathbf{S}^{n-1})$ for all $j = 1,\ldots,n$ if and only if $\Omega \in H^1(\mathbf{S}^{n-1})$. Moreover, condition $\Omega \in H^1(\mathbf{S}^{n-1})$ is equivalent to condition (9.10) as shown in [18]. Therefore all these three conditions on $\Omega$ are equivalent and they all imply that $T_\Omega$ is bounded on $L^p(\mathbb{R}^n)$, $1 < p < \infty$.*

We end this section with a another sufficient condition on $\Omega$ that implies $L^p$ boundedness for $T_\Omega$. The theorem below is proved based on ideas developed in [9]. Littlewood-Paley decomposition and a bootstrapping argument are used in conjunction with the logarithmic decay at infinity of the Fourier transform of the expression in (9.10). For a proof we refer the reader to [12].

**Theorem 9.4** *Let $\alpha > 0$. Let $\Omega$ be an even function in $L^1(\mathbf{S}^{n-1})$ with mean value zero which satisfies:*

$$\sup_{y' \in \mathbf{S}^{n-1}} \int_{\mathbf{S}^{n-1}} |\Omega(\theta)| \left( \log \frac{1}{|\theta \cdot y'|} \right)^{1+\alpha} d\theta < +\infty. \tag{9.16}$$

*Then $T_\Omega$ extends to a bounded operator from $L^p(\mathbb{R}^n)$ into itself for $(2+\alpha)/(1+\alpha) < p < 2+\alpha$.*

**Remark 9.2** *It follows that if condition (9.16) holds for every $\alpha > 0$, then $T_\Omega$ maps $L^p \to L^p$ for all $1 < p < \infty$. It is natural to ask how condition (9.16) for all $\alpha > 0$ compares with condition (9.5) or even with condition $\Omega \in H^1(\mathbf{S}^{n-1})$. The authors have constructed examples of functions $\Omega$ which satisfy condition (9.16) for all $\alpha > 0$ but do not satisfy the $H^1$ condition (9.10). See [12] for details. Conversely, the function*

$$\Omega(\theta) = \sum_{k=2}^{\infty} \frac{e^{ik\theta}}{(\log k)^2}$$

*is in $H^1(\mathbf{S}^1)$ but it behaves like $\theta^{-1} \log^{-2}(\theta^{-1})$ as $\theta \to 0+$ and therefore it fails to satisfy condition (9.16) for any $\alpha > 0$. See [22] p. 189 for a justification of this.*

## 9.4 The $L^1$ theory

We now turn to questions regarding the behavior of $T_\Omega$ on $L^1(\mathbb{R}^n)$. $T_\Omega$ is said to be of weak type $(1,1)$ if there is a constant $C = C(n, \Omega) > 0$ such that for all $f \in L^1(\mathbb{R}^n)$ we have

$$|\{x : |(T_\Omega f)(x)| > \alpha\}| \leq C \|f\|_{L^1} / \alpha.$$

The question of weak type $(1,1)$ boundedness of $T_\Omega$ for $\Omega$ rough has puzzled many authors who obtained partial results. An important question along these lines was whether a condition bearing on the size of $\Omega$ alone sufficed for the weak type $(1,1)$ boundedness of $T_\Omega$. The answer turned out to be positive. See M. Christ [4] and S. Hofmann [13] for the case $\Omega \in L^q(\mathbf{S}^1)$, $q > 1$, and M. Christ and J.-L. Rubio de Francia [5] for $\Omega \in LLog^+L(\mathbf{S}^1)$. The latter authors were able to extend their result to all dimensions $n \leq 7$ (unpublished). Finally A. Seeger [15] proved that $T_\Omega$ is weak type $(1,1)$ bounded when $\Omega \in LLog^+L(\mathbf{S}^{n-1})$ in all dimensions.

**Theorem 9.5** *Let $\Omega$ be in $L^1(\mathbf{S}^{n-1})$ with integral zero. Suppose that $\Omega$ satisfies condition (9.5). Then $T_\Omega$ can be extended to an operator of weak type* $(1,1)$.

At this point it is natural to ask whether the method of rotations can be used to show that $T_\Omega$ is of weak type $(1,1)$. This is known to be false. The following question is therefore more difficult than its $L^p$ counterpart:

**Question 9.3** *Let $\Omega$ be an integrable odd function on $\mathbf{S}^{n-1}$. Is $T_\Omega$ of weak type* $(1,1)$*?*

Outside the context of odd functions, the general question for weak type $(1,1)$ which is analogous to Question 9.1 can be phrased as follows:

**Question 9.4** *Let $\Omega$ be an integrable function on $\mathbf{S}^{n-1}$ with integral zero. Find a necessary and sufficient condition on $\Omega$ such that the associated operator $T_\Omega$ is of weak type* $(1,1)$.

In the context of Question 9.4 posed above, it is not as natural to assume that $\Omega \in L^1(\mathbf{S}^{n-1})$, as it is to assume that $\Omega$ is a general distribution on the sphere. The reason for that it is sometimes easier to handle finite sums of Dirac masses than general $L^1$ functions. In this case, it is conceivably easier to handle a finite sum of directional Hilbert transforms than a general $T_\Omega$ with $\Omega \in L^1(\mathbf{S}^{n-1})$. Furthermore, one sees from (9.8) that certain distributions $\Omega$ give rise to bounded operators on $L^2$.

**Question 9.5** *Let $\Omega$ be a distribution on $\mathbf{S}^{n-1}$ with mean value zero. Find a necessary and sufficient condition on $\Omega$ such that the associated operator $T_\Omega$ is of weak type* $(1,1)$. *Likewise for $T_\Omega$ to be bounded on $L^p$.*

Obtaining weak type $(1,1)$ bounds is usually a more difficult task than proving $L^p$ boundedness, for, the latter bounds follow from the weak type $(1,1)$ bounds by interpolation. In some occasions a more natural aspect of the $L^1$ theory is to prove that the operator in question is bounded from the Hardy space $H^1$ to $L^1$.

It is fairly easy to check that if $K_\Omega$ possesses a certain amount of smoothness then $T_\Omega$ extends to a bounded operator from $H^1 \to L^1$. Here is a precise statement.

**Theorem 9.6** *Suppose that $\Omega \in L^1(\mathbf{S}^{n-1})$ has mean value zero and assume that $K_\Omega$ satisfies (9.2) and $\Omega$ satisfies (9.9). Then $T_\Omega$ extends to a bounded operator from $H^1 \to L^1$.*

**Proof.** The proof is standard. Fix an atom $a_Q$ and prove that $\|T(a_Q)\|_{L^1} \le C$ with $C$ independent of $Q$. For $x \in 2Q$ use the $L^2$ estimate (which is follows from (9.9)) and Hölder's inequality. For $x \notin 2Q$ subtract $K(x)a_Q(x)$ from $T(a_Q)(x)$ and then use condition (9.2). ∎

Even though $H^1 \to L^1$ boundedness holds for $\Omega$ smooth enough, it may fail for $\Omega$ rough. A good starting point to study $H^1 \to L^1$ boundedness is the directional Hilbert transform. Consider the unit vector $e_1 = (1,0)$ in $\mathbb{R}^2$ and the operator $H^{e_1}$. Let $f$ be the $H^1$ function in $\mathbb{R}^2$ defined by $f(x_1,x_2) = \chi_{|x|<1,x_2>0} - \chi_{|x|<1,x_2<0}$. Then it is easy to see that $|(H^{e_1}f)(x)| \geq C|x|^{-1}$ when $|x| \geq 2$ and $|x_2| \leq 1/2$. It follows that $H^{e_1}f$ cannot be in $L^1(\mathbb{R}^2)$.

Other examples can be found in [8]. Below, we give the an example communicated to us by M. Christ.

**Example 9.1** *(M. Christ) There exists an $\Omega$ in $L^2(\mathbf{S}^1)$ such that $T_\Omega$ does not map $H^1(\mathbb{R}^2)$ to $L^1(\mathbb{R}^2)$.*

For $x \in \mathbb{R}^2$, let $\theta_x = Arg x$ denote the argument of $x$. Choose a lacunary sequence $\lambda_j \geq 2^j$ whose properties will be specified later and let $a_j$ be a square summable sequence also to be chosen later. Define

$$\Omega(x) = \sum_{j=1}^{\infty} a_j e^{i\lambda_j \theta_x}.$$

We have that $\Omega$ is in $L^2(\mathbf{S}^1)$ and it has mean value zero. Now take $f$ to be a $C^\infty$ and radial atom which is supported in the unit disc in $^2$. Fix $x \in \mathbb{R}^2$ satisfying $1/2 \leq |x_2|/|x_1| \leq 2$ in the annulus $\lambda_j\mu_j \leq |x| \leq 2\lambda_j\mu_j$ for some $j \geq 1$. When we write $O(\,\cdot\,)$, we are tacitly implying that the constants involved in the bounds are independent of the $\lambda_j$'s and $x$ but may depend on $f$ and the other parameters. For $1 \leq k \leq j$ we calculate

$$\begin{aligned}
&\left(f * \frac{e^{i\lambda k\theta_y}}{|y|^2}\right)(x) \\
&= \frac{1}{|x|^2}\iint\limits_{|y|\leq 1} f(y)e^{i\lambda k\theta_{x-y}}\,dy + \iint\limits_{|y|\leq 1} f(y)e^{i\lambda k\theta_{x-y}}\left(\frac{1}{|x-y|^2} - \frac{1}{|x|^2}\right)dy \\
&= \frac{1}{|x|^2}\int_0^1 f(\rho(1,0))\rho\left[\int_0^{2\pi} e^{i\lambda k Arg(x-\rho\, e^{i\phi})}d\phi\right]d\rho + O\left(\frac{1}{|x|^3}\right) \\
&= \frac{1}{|x|^2}\int_0^1 f(\rho(1,0))\rho e^{i\lambda k\theta_x}\left[\int_0^{2\pi} e^{i\lambda k(Arg(x-\rho\, e^{i\phi})-\theta_x)}d\phi\right]d\rho + O\left(\frac{1}{|x|^3}\right) \\
&= \frac{1}{|x|^2}\int_0^1 f(\rho(1,0))\rho e^{i\lambda k\theta_x}\int_0^{2\pi}\Bigg[1 + \sum_{m=1}^{6}\frac{(i\lambda_k)^m}{m!}(g_\phi(\rho) - g_\phi(0))^m \qquad (9.17)\\
&\qquad + O\left(\lambda_k^7(g_\phi(\rho)-g_\phi(0))^7\right)\Bigg]d\phi\,d\rho + O\left(\frac{1}{|x|^3}\right),
\end{aligned}$$

where $g_\phi(\rho) = \arctan[(x_2-\rho\sin\phi)/(x_1-\rho\cos\phi)]$. The mean value theorem and an easy estimate give that

$$g_\phi(\rho) - g_\phi(0) = \frac{\rho}{|x|^2}(-x_1 \sin\phi + x_2 \cos\phi) + O\left(\frac{1}{|x|^2}\right).$$

Pluging in the estimate above in (9.17), calculating, and integrating with respect to $\phi$, we obtain that

$$\begin{aligned}\left(f * \frac{e^{i\lambda k \theta_y}}{|y|^2}\right)(x) &= \frac{1}{|x|^2}\int_0^1 f(\rho(1,0))\rho e^{i\lambda k\theta_x}\left[1 - 4\frac{\lambda_k^2\rho^2}{|x|^2} + c_4\frac{\lambda_k^4\rho^4}{|x|^4} + c_6\frac{\lambda_k^6\rho^6}{|x|^6}\right.\\ &\qquad \left. + O\left(\frac{\lambda_k^7}{|x|^7}\right)\right] d\rho + O\left(\frac{1}{|x|^3}\right).\end{aligned}$$

Since $f$ is an atom we have that $\int_0^1 f(\rho(1,0))\rho\, d\rho = 0$. At this point we select $f$ such that $c_f = \int_0^1 f(\rho(1,0))\rho^3\, d\rho \neq 0$, but $\int_0^1 f(\rho(1,0))\rho^5\, d\rho = \int_0^1 f(\rho(1,0))\rho^7\, d\rho = 0$. It follows that

$$\left(f * \frac{e^{i\lambda k\theta_y}}{|y|^2}\right)(x) = -4c_f\frac{\lambda_k^2}{|x|^4}e^{i\lambda k\theta_x} + O\left(\frac{\lambda_k^7}{|x|^9}\right) + O\left(\frac{1}{|x|^3}\right)$$

and therefore

$$\begin{aligned}\left(f * \sum_{k=1}^{j} a_k\frac{e^{i\lambda k\theta_y}}{|y|^2}\right)(x) &= -4c_f a_j\frac{\lambda_j^2}{|x|^4}e^{i\lambda_j\theta_x} + O\left(\frac{\lambda_{j-1}^2}{\lambda_j^4\mu_j^4}\right)\\ &\qquad + O\left(\frac{\lambda_j^7}{\lambda_j^9\mu_j^9}\right) + O\left(\frac{j}{\lambda_j^3\mu_j^3}\right).\end{aligned}$$

For fixed $x$ as above, let $I_{x,\rho}$ be the set of all $\phi \in [0, 2\pi]$ with $|x - \rho e^{i\phi}| \leq 1$.

We have

$$\left|\left(f * \sum_{k=j+1}^{\infty} a_k \frac{e^{\lambda k \theta_y}}{|y|^2}\right)(x)\right|$$
$$= \left| \int_{|x|-1}^{|x|+1} \int_{I_{x,\rho}} f(x-\rho e^{i\phi}) \sum_{k=j+1}^{\infty} a_k e^{i\lambda_k \phi} d\phi \frac{d\rho}{\rho} \right|$$
$$= \left| \int_{|x|-1}^{|x|+1} \int_{I_{x,\rho}} \frac{d}{d\phi}(f(x-\rho e^{i\phi})) \sum_{k=j+1}^{\infty} a_k \frac{e^{i\lambda_k \phi}}{i\lambda_k} d\phi \frac{d\rho}{\rho} \right|$$
$$\leq \int_{|x|-1}^{|x|+1} \left\| \frac{d}{d\phi}(f(x-\rho e^{i\phi})) \right\|_{L^2(d\phi)} \left\| \sum_{k=j+1}^{\infty} a_k \frac{e^{i\lambda_k \phi}}{i\lambda_k} \right\|_{L^2(d\phi)} \frac{d\rho}{\rho}$$
$$\leq \int_{|x|-1}^{|x|+1} \|\nabla f\|_{L^\infty} \left( \sum_{k=j+1}^{\infty} \frac{1}{\lambda_k^2} \right)^{1/2} d\rho = O\left(\frac{1}{\lambda_{j+1}}\right).$$

Combining this result with the one obtained above for the remaining terms we obtain that

$$|(T_\Omega f)(x)| \geq \frac{c_f |a_j|}{\mu_j^4 \lambda_j^2} - C \left[ \frac{\lambda_{j-1}^2}{\lambda_j^4 \mu_j^4} + \frac{1}{\lambda_j^2 \mu_j^9} + \frac{j}{\lambda_j^3 \mu_j^3} + \frac{1}{\lambda_{j+1}} \right]$$

for $x$ satisfying $\lambda_j \mu_j \leq |x| \leq 2\lambda_j \mu_j$ and $1/2 \leq |x_2|/|x_1| \leq 2$. Estimate the $L^1$ norm of $T_\Omega f$ from below by

$$\|T_\Omega f\|_{L^1} \geq c_f \frac{\pi}{10} \sum_{j=1}^{\infty} \frac{|a_j|}{\mu_j^2} - C \sum_{j=1}^{\infty} \left[ \frac{\lambda_{j-1}^2}{\lambda_j^2 \mu_j^2} + \frac{1}{\mu_j^7} + \frac{j}{\lambda_j \mu_j} + \frac{\lambda_j^2 \mu_j^2}{\lambda_{j+1}} \right]. \quad (9.18)$$

Choose now $\mu_j = j^{1/6}$ and $a_j = j^{-5/8}$. Select also a lacunary sequence $\lambda_j$ such that $\sum_{j=1}^{\infty} \lambda_j^2 \mu_j^2 / \lambda_{j+1} < \infty$. This choice of numbers makes the expression in (9.18) equal to infinity.

**Remark 9.3** *Observe that the proof above gives that for a radial $C^\infty$ function $f$ supported in the unit disc we have*

$$\left(f * \frac{e^{i\lambda\theta_y}}{|y|^2}\right)(x) = c_0 \frac{1}{|x|^2} e^{i\lambda\theta_x} + c_1 \frac{\lambda^2}{|x|^4} e^{i\lambda\theta_x} + O\left(\frac{\lambda^4}{|x|^6}\right) + O\left(\frac{1}{|x|^3}\right)$$

*with bounds independent of $\lambda$ for $|x| \geq \lambda$ satisfying $|x_2|/|x_1| \sim 1$. The constants $c_0$ and $c_1$ are multiples of the integral and of the first moment of $f$ respectively.*

**Remark 9.4** *M. Christ has informed us that his example can be modified so that $\Omega \in L^\infty$.*

**Remark 9.5** *A fundamental result of J. Daly and K. Phillips says that if $T_\Omega$ maps $H^1(\mathbb{R}^n)$ into $L^1(\mathbb{R}^n)$, then the function $\Omega$ has to be in $H^1(\mathbf{S}^{n-1})$. Using this theorem we conclude that for every $\Omega \in L^1(\mathbf{S}^{n-1}) - H^1(\mathbf{S}^{n-1})$ we have that the corresponding operator $T_\Omega$ does not map $H^1$ to $L^1$. See [8] for the case of $n = 2$.*

**Remark 9.6** *Recently A. Seeger and T. Tao [16] have shown that the best possible result on $H^1$ is that $T_\Omega$ maps $H^1$ to $L^{1,q}$ for $q \geq 2$, where $L^{1,q}$ denotes the Lorentz space. This means that for some $\Omega \in L^\infty$, $T_\Omega$ does not map $H^1$ into $L^{1,q}$ for $q < 2$.*

Techniques from the $L^1$ theory can be used to answer some questions about the $L^p$ theory.

**Question 9.6** *Give an example of an $\Omega \in L^1(\mathbf{S}^{n-1})$ such that $T_\Omega$ is bounded on some $L^q$ but not on some other $L^p$.*

If $\Omega$ is allowed to be a distribution on the sphere, such an $\Omega$ is shown to exist by abstract methods. To be more precise, let us introduce the following Banach spaces of distributions on the sphere.

$$\mathcal{S}_p = \{\Omega\colon\ \Omega \text{ distribution on } \mathbf{S}^{n-1} \text{ and } \|\Omega\|_p = \|T_\Omega\|_{L^p \to L^p} < \infty\}.$$

By duality and interpolation we see that $\mathcal{S}_p = \mathcal{S}_{p'}$ and $\mathcal{S}_p \subseteq \mathcal{S}_q$, where $1 < p < q \leq 2$. What is not immediately clear here is that $\mathcal{S}_p, \mathcal{S}_q$ are different spaces.

**Theorem 9.7** *(M. Christ) We have $\mathcal{S}_p \subsetneq \mathcal{S}_q$ whenever $1 < p < q \leq 2$.*

**Proof.** It suffices to prove the theorem in dimension $n = 2$. For $x \in \mathbb{R}^2$, let us denote by $\theta_z = Arg x$ the argument of $x$. Consider the operators $T_N = T_{\Omega_N}$ where $\Omega_N(x) = e^{iN\theta_x}$ for $N = 1, 2 \ldots$. According to ([8], Section 3)

$$\widehat{\left(\frac{e^{iN\theta_x}}{|x|^2}\right)}(\xi) = \frac{2\pi i(isgn(N))^{N+1}}{N} e^{iN\theta_\xi}.$$

Therefore $\|T_N\|_{L^2 \to L^2} \leq CN^{-1}$. Next, we show that $\|T_N\|_{H^1 \to L^1} \leq C$. By dilation invariance, it suffices to consider $f$ to be an atom supported in the unit ball and $\|f\|_\infty \leq 1$. For $|x| \leq N$ we use the Cauchy-Schwarz inequality to deduce that

$$\int\limits_{|x| \leq N} |T_N f(x)|\, dx \leq CN\|T_N f\|_{L^2} \leq CNN^{-1}\|f\|_{L^2} = C.$$

For $|x| \geq N$ we have

$$\begin{aligned} T_N f(x) &= \int\limits_{|y|\leq 1} \frac{e^{iNArg(x-y)}}{|x-y|^2} f(y)dy \\ &= \int\limits_{|y|\leq 1} \left( \frac{e^{iNArg(x-y)}}{|x-y|^2} - \frac{e^{iNArg(x)}}{|x|^2} \right) f(y)dy. \end{aligned}$$

Now use that, for $|x| \geq N \geq 2$ we have $|Arg(x-y) - Arg(x)| \leq C/|x|$, to obtain

$$\begin{aligned} \int\limits_{|x|\geq N} |T_N f(x)|dx &\leq \int\limits_{|y|\leq 1} |f(y)|dy \left( \int\limits_{|x|\geq N} \frac{\left|e^{iN(Arg(x-y)-Arg(x))} - 1\right|}{|x|^2} dx \right. \\ &\qquad \left. + \int\limits_{|x|\geq N} \frac{C}{|x|^3} dx \right) \\ &\leq C \left( \int\limits_{|x|\geq N} \frac{N}{|x|^3} dx + C \right) = C. \end{aligned}$$

Therefore $\|T_N\|_{H^1 \to L^1} \leq C$. By interpolation we see that

$$\|T_N\|_{L^p \to L^p} \leq CN^{2/p-2}$$

for every $1 < p < 2$. On the other hand, as we saw in the remark after the previous example, for suitable $f$ the following is true

$$T_N f(x) = c_f \frac{e^{iNArg(x)}}{|x|^2} + O\left(\frac{N^2}{|x|^4}\right) + O\left(\frac{1}{|x|^3}\right),$$

for $|x| \geq N$ near the diagonal, with bounds independent of $N$. Therefore for $|x| \geq N$ and $N$ very large, the first two terms above are the dominant ones and hence

$$\|T_N f\|_{L^p} \geq C_1 \left( \int\limits_{|x|\geq N} \frac{1}{|x|^{2p}} dx \right)^{1/p} = C_2 N^{2/p-2}.$$

From this we conclude that

$$\|\Omega_N\|_p \sim CN^{2/p-2} \qquad \text{for } N \text{ large.}$$

Suppose now that $\mathcal{S}_p = \mathcal{S}_q$. Then by the open mapping theorem we must have that

$$c\|\Omega\|_p \leq \|\Omega\|_q \qquad \text{for every } T \text{ in } \mathcal{S}_q.$$

In particular, for every $N$ large

$$CN^{2/p-2} \leq c\|\Omega_N\|_p \leq \|\Omega_N\|_q \leq CN^{2/q-2},$$

which is a contradiction as $N \to \infty$ since $p < q$. ∎

## 9.5 Another $H^1$ condition in dimension 2

The following fact is well known. If $f$ is supported in a ball $B$ in $\mathbb{R}^n$, $f$ is in $L^p(B)$ for some $1 < p \leq \infty$ (or more generally in $LLog^+L(B)$), and $f$ has mean value zero, then $f$ is in the Hardy space $H^1(\mathbb{R}^n)$. It is also known that $LLog^+L(B)$ cannot be replaced by $L^1(B)$ nor $LLog^{1-\varepsilon}L(B)$ in this context. The following question is therefore naturally raised:

**Question 9.7** *Let $B$ be a ball in $R^n$. Find a condition bearing on the size of a function such that for all $f$ supported in $B$ we have*

$$\int_B f(x)\,dx = 0 \quad \textit{and} \quad (\textit{size condition on } f) \quad \Longleftrightarrow f \in H^1(\mathbb{R}^n).$$

In dimension 1 an interesting answer was given in [18]. The condition discovered by the author reflects more the oscillation/variation of the function than its size. We state the result below:

**Theorem 9.8** *Let $f$ be supported in $[0,1]$, be integrable, and have integral zero. Define*

$$m_f(y) = \int_0^1 f(x) \log \frac{1}{|x-y|}\,dx, \tag{9.19}$$

*for $A_f \subset [0,1]$, where $A_f$ a full measure subset of $[0,1]$, on which the integral giving $m_f(y)$ converges absolutely. Then $f \in H^1(\mathbf{S}^1)$ if and only if $m_f$ is a function of bounded variation on $A_f$. Quantitatively speaking, there exists a constant $C > 0$ such that for all $f$ supported in $[0,1]$ with integral zero, we have*

$$\|f\|_{H^1} \leq Var_{A_f}(m_f) + C\|f\|_{L^1},$$
$$Var_{A_f}(m_f) \leq C\|f\|_{H^1}.$$

**Remark 9.7** *The Variation of $m_f$ over $A_f$ is defined as*

$$Var_{A_f}(m_f) = \sup_P \{\sum_{j=1}^n |m_f(x_j) - m_f(x_{j-1})| :$$
$$P = \{0 = x_0 < x_1 < \cdots < x_n = 1\},\ x_j \in A_f\}.$$

Let us now try to explain Theorem 9.8 along some heuristic lines. Recall the following: A function is in $H^1(\mathbb{R}^1)$ if and only if its Hilbert transform is in $L^1(\mathbb{R}^1)$. Theorem 9.8 states that $m_f$ is of total variation if and only if $Hf$ is integrable. Formally speaking, to find the derivative of the function $m_f$ we differentiate under the integral sign to obtain the Hilbert transform of the function $f$. Of course this argument cannot be justified for a general $f \in L^1$ since $Hf$ is not necessarily given in a form of a convergent integral. ($Hf$ can be written as a convergent integral for smooth enough $f$.) However, $m_f$ is defined almost everywhere and the condition that $m_f$ has finite variation makes sense for all integrable functions $f$. Theorem 9.8 is first proved for step functions and then by approximation is extended to general functions. The extension to general functions is a little delicate because of the convergence problems indicated above.

We now use the result in Theorem 9.8 to state an alternative characterization of the "$H^1$ condition" for singular integrals in $\mathbb{R}^2$. We have the following:

**Theorem 9.9** *Let $\Omega \in L^1(\mathbf{S}^1)$ have mean value zero. If there exists a $G \subset \mathbf{S}^1$ with measure $|G| = 2\pi$ such that*

$$Var_G(m_\Omega) < +\infty, \tag{9.20}$$

*where*

$$m_\Omega(\xi) = \int_{\mathbf{S}^1} \Omega(\theta) \log \frac{1}{|\theta \cdot \xi|} d\theta,$$

*then $T_\Omega$ maps $L^p(\mathbb{R}^2)$ into itself for $1 < p < \infty$.*

Compare condition (9.20) to condition (9.7) which is essentially required for $L^2$ boundedness of $T_\Omega$.

The idea of the proof of Theorem 9.9 is straightforward. In view of Theorem 9.8 and via a simple transference argument from the interval to the circle, we obtain that condition (9.20) is equivalent to the condition that $\Omega \in H^1(\mathbf{S}^1)$. As observed before, this condition is equivalent to (9.10). Now Theorem 9.3 gives the desired conclusion. We refer the reader to [18] for details.

## 9.6 Maximal functions and maximal singular integrals

In this section we discuss two operators related to $T_\Omega$, the maximal function $M_\Omega$ and the maximal singular integral $T_\Omega^*$. First consider the maximal function

$$(M_\Omega f)(x) = \sup_{r>0} r^{-n} \int_{|y|\leq r} |f(x-y)||\Omega(y/|y|)|\, dy,$$

where $\Omega$ is in $L^1(\mathbf{S}^{n-1})$. The following theorem is a straightforward consequence of the method of rotations See [20] p. 72.

**Theorem 9.10** *If $\Omega$ is in $L^1(\mathbf{S}^{n-1})$ then $M_\Omega$ maps $L^p(\mathbb{R}^n)$ into itself for $1 < p \leq \infty$.*

Note that $M_\Omega$ is a positive operator and no mean value property is imposed on $\Omega$.

It is reasonable to ask if there is an $L^1$ theory for $M_\Omega$. Again the main question here is whether a condition bearing only on the size of $\Omega$ suffices for the weak type $(1,1)$ property. This question was first answered positively by M. Christ ($\Omega \in L^q(\mathbf{S}^{n-1})$, $q > 1$, $n = 2$), and later by M. Christ and J.-L. Rubio de Francia ($\Omega \in LLog^+L(\mathbf{S}^{n-1})$ all $n$).

**Theorem 9.11** *(M. Christ and J.-L. Rubio de Francia) If $\Omega$ satisfies (9.5), then $M_\Omega$ is of weak type $(1,1)$.*

The question still left open is the following:

**Question 9.8** *Is $M_\Omega$ weak type $(1,1)$ when $\Omega$ is merely in $L^1(\mathbf{S}^{n-1})$?*

It is fairly easy to check that the singular integral operator with kernel $K_\Omega$ shares the same mapping properties as its truncated version having kernel $\Omega(x/|x|)|x|^{-n}\chi_{|x|>\varepsilon}$. Obtaining estimates for the supremum of the truncated singular integrals allows us to conclude that the principal value integral in (9.1) is almost everywhere convergent for $f \in L^p(\mathbb{R}^n)$, $1 \leq p < \infty$.

For $\Omega$ an integrable function of the sphere with mean value zero, define

$$(T^*_\Omega f)(x) = \sup_{\varepsilon>0} \left| \int_{|y|>\varepsilon} \frac{\Omega(y/|y|)}{|y|^n} f(x-y)\, dy \right|.$$

We call this operator the maximal singular integral operator associated with $T_\Omega$. The $L^p$ boundedness of $T^*_\Omega$ for $\Omega$ in $LLog^+L$ is due to Calderón and Zygmund [2]. $T^*_\Omega$ is also $L^p$ bounded for $\Omega \in H^1(\mathbf{S}^{n-1})$. The theorem below was proved by the authors and independently by Fan and Pan [10] in a more general context. The proof given combines ideas from [2] and from the proof of Theorem 9.3.

**Theorem 9.12** *Let $\Omega$ be an integrable function on $\mathbf{S}^{n-1}$ with mean value zero which satisfies condition (9.10). Then $T^*_\Omega$ extends to a bounded operator from $L^p(\mathbb{R}^n) \to L^p(\mathbb{R}^n)$ for $1 < p < \infty$.*

**Proof.** For a unit vector $\theta \in \mathbf{S}^{n-1}$ define

$$(M_\theta f)(x) = \sup_{a>0} \frac{1}{2a} \int_{-a}^{a} |f(x - r\theta)| \, dr. \tag{9.21}$$

$$(H^*_\theta f)(x) = \sup_{\varepsilon>0} \left| \int_{|r|>\varepsilon} \frac{f(x - r\theta)}{r} \, dr \right| \tag{9.22}$$

It is a classical result that for some $C_p > 0$ and all $f$ we have

$$\sup_{|\theta|=1} \|M_\theta f\|_{L^p} + \sup_{|\theta|=1} \|H^*_\theta f\|_{L^p} \leq C_p \|f\|_{L^p}.$$

For $\Omega$ odd, the method of rotations gives

$$|(T^*_\Omega f)(x)| \leq \int_{\mathbf{S}^{n-1}} |\Omega(\theta)| (H^*_\theta f)(x) d\theta$$

and therefore $\|T^*_\Omega f\|_{L^p} \leq C_p \|\Omega\|_{L^1(\mathbf{S}^{n-1})} \|f\|_{L^p}$. Let us now consider the case when $\Omega$ is even.

Fix $\Phi$ to be a smooth radial function such that $\Phi(x) = 0$ for $|x| < 1/4$ and $\Phi(x) = 1$ for $|x| > 1/2$. We have that

$$(T^\varepsilon_\Omega f)(x) = \int_{\mathbb{R}^n} \frac{\Omega(x-y)}{|x-y|^n} \Phi\left(\frac{x-y}{\varepsilon}\right) f(y) dy - \int_{|x-y|<\varepsilon} \frac{\Omega(x-y)}{|x-y|^n} \Phi\left(\frac{x-y}{\varepsilon}\right) f(y) dy,$$

where we extended $\Omega$ to be a homogeneous of degree zero function on $\mathbb{R}^n$. Since the pointwise estimate

$$\begin{aligned}
\sup_{\varepsilon>0} \left| \int_{|x-y|<\varepsilon} \frac{\Omega(x-y)}{|x-y|^n} \Phi\left(\frac{x-y}{\varepsilon}\right) f(y) dy \right|
&\leq C \sup_{\varepsilon>0} \int_{\varepsilon/4<|x-y|<\varepsilon} \frac{|\Omega(x-y)|}{|x-y|^n} |f(y)| dy \\
&\leq C \sup_{\varepsilon>0} \int_{\mathbf{S}^{n-1}} |\Omega(\theta)| \frac{1}{\varepsilon} \int_{\varepsilon/4}^{\varepsilon} |f(x - r\theta)| dr d\theta \\
&\leq C \int_{\mathbf{S}^{n-1}} |\Omega(\theta)| (M_\theta f)(x) d\theta
\end{aligned}$$

is valid, and the last term above maps $L^p(\mathbb{R}^n) \to L^p(\mathbb{R}^n)$ for $1 < p < \infty$, it suffices to obtain an $L^p$ bound for the smoothly truncated maximal singular integral operator

$$\begin{aligned} (\widetilde{T}^*_\Omega f)(x) &= \sup_{\varepsilon>0} |(\widetilde{T}^\varepsilon_\Omega f)(x)|, \\ \text{where} \quad (\widetilde{T}^\varepsilon_\Omega f)(x) &= \int_{\mathbb{R}^n} \frac{\Omega(x-y)}{|x-y|^n} \Phi\left(\frac{x-y}{\varepsilon}\right) f(y)\, dy. \end{aligned}$$

As usually, we denote by $R_j$ the $j^{\text{th}}$ Riesz transform.

Let $U_j = R_j\left(\Omega(\,\cdot\,)/|\cdot|^n\right)$.

We can write $U_j(x)$ as $V_j(x/|x|)/|x|^n$, where $V_j$ is an odd distribution on $\mathbf{S}^{n-1}$. Here we use the fact that $\Omega$ is even and that $R_j$ has an odd kernel.

It turns out that the fact $\Omega \in H^1(\mathbf{S}^{n-1})$ is equivalent to the fact that $V_j$ are integrable functions on $\mathbf{S}^{n-1}$ for all $j = 1, \ldots, n$. See [14]. Here we use only that $V_j \in L^1(\mathbf{S}^{n-1})$ a fact proved in Theorem 9.3.

Also let $\widetilde{V}_j(x) = R_j\left(\Omega(\,\cdot\,)\Phi(\,\cdot\,)/|\cdot|^n\right)$. Then

$$(\widetilde{T}^\varepsilon_\Omega f)(x) = \sum_{j=1}^n R_j\left(\frac{\Omega(\,\cdot\,)}{|\cdot|^n}\Phi\left(\frac{\cdot}{\varepsilon}\right)\right) * R_j(f) = \sum_{j=1}^n \frac{1}{\varepsilon^n}\widetilde{V}_j\left(\frac{\cdot}{\varepsilon}\right) * R_j(f) \tag{9.23}$$

We shall need the following lemma whose proof is postponed until the end of this section (see also [2], p.299).

**Lemma 9.13** *There exist $G_j$, homogeneous of degree 0, integrable on $\mathbf{S}^{n-1}$ functions, such that*

$$\begin{aligned} &|\widetilde{V}_j(x)| \le G_j(x) \quad \textit{forevery} \quad |x| \le 1, \\ &|\widetilde{V}_j(x) - U_j(x)| \le C\|\Omega\|_{L^1(\mathbf{S}^{n-1})}|x|^{-n-1} \quad \text{for every} \quad |x| > 1. \end{aligned}$$

Using Lemma 9.13 and (9.23), we obtain

$$\begin{aligned} |(\widetilde{T}^\varepsilon_\Omega f)(x)| &= \left|\sum_{j=1}^n \frac{1}{\varepsilon^n}\int \widetilde{V}_j\left(\frac{x-y}{\varepsilon}\right)(R_j f)(y)dy\right| \\ &\le A_1(f,\varepsilon) + A_2(f,\varepsilon) + A_3(f,\varepsilon), \end{aligned}$$

where

$$A_1(f,\varepsilon) = \sum_{j=1}^{n} \left| \int_{|x-y|>\varepsilon} U_j(x-y)(R_j f)(y)dy \right|,$$

$$A_2(f,\varepsilon) = C\sum_{j=1}^{n} \varepsilon \int_{|x-y|>\varepsilon} \frac{|R_j f(y)|}{|x-y|^{n+1}} dy,$$

$$A_3(f,\varepsilon) = \sum_{j=1}^{n} \frac{1}{\varepsilon^n} \int_{|x-y|<\varepsilon} |G_j(x-y)||(R_j f)(y)|dy.$$

First we observe that the $\sup_{\varepsilon>0} |A_1(f,\varepsilon)|$ is controlled by a sum of maximal singular integral operators associated with odd integrable kernels applied to the Riesz transforms of $f$, hence this term is bounded on $L^p$. The $j^{\text{th}}$ term in $A_2(f,\varepsilon)$ is controlled by

$$\begin{aligned}
\varepsilon \int_{|x-y|>\varepsilon} \frac{|(R_j f)(y)|}{|x-y|^{n+1}} dy &\leq \varepsilon \int_{\mathbf{S}^{n-1}} \int_{\varepsilon}^{\infty} \frac{1}{r^2} |R_j f(x-r\theta)| dr d\theta \\
&\leq C \int_{\mathbf{S}^{n-1}} \sum_{k=0}^{\infty} 2^{-k} \frac{1}{2^k \varepsilon} \int_{2^k\varepsilon}^{2^{k+1}\varepsilon} |(R_j f)(x-r\theta)| dr d\theta \\
&\leq C \int_{\mathbf{S}^{n-1}} M_\theta(R_j f)(x) d\theta,
\end{aligned}$$

hence

$$\|\sup_{\varepsilon>0} |A_2(f,\varepsilon)|\|_{L^p} \leq C \sum_{j=1}^{n} \sup_{\theta} \|M_\theta(R_j f)\|_{L^p} \leq C_p \|f\|_{L^p}.$$

Finally,

$$\begin{aligned}
\frac{1}{\varepsilon^n} \int_{|x-y|<\varepsilon} &|G_j(x-y)||R_j f(y)| dy \\
&\leq \int_{\mathbf{S}^{n-1}} |G_j(\theta)| \frac{1}{\varepsilon} \int_0^{\varepsilon} |R_j f(x-r\theta)| dr d\theta \\
&\leq \int_{\mathbf{S}^{n-1}} |G_j(\theta)| M_\theta(R_j f)(x) d\theta,
\end{aligned}$$

which implies that

$$\|\sup_{\varepsilon>0} |A_3(f,\varepsilon)|\|_{L^p} \leq C \sum_{j=1}^{n} \sup_{\theta} \|M_\theta(R_j f)\|_{L^p} \leq C_p \|f\|_{L^p}.$$

■

Theorem 9.12 is now proved and we turn our attention to the proof of Lemma 9.13 left open.

**Proof.** If $|x| > 1$ since $\Phi(y) = 1$ for every $|y| > 1/2$ , we have

$$\begin{aligned}
|\widetilde{V}_j(x) - U_j(x)| &\leq \left| \int \frac{x_j - y_j}{|x-y|^{n+1}} (\Phi(y) - 1) \frac{\Omega(y)}{|y|^n} dy \right| \\
&\leq C \int\limits_{|y|<1/2} \frac{|\Omega(y)|}{|y|^n} \left| \frac{x_j - y_j}{|x-y|^{n+1}} - \frac{x_j}{|x|^{n+1}} \right| dy \\
&\leq \frac{C}{|x|^{n+1}} \int\limits_{|y|<1/2} \frac{|\Omega(y)|}{|y|^n} dy \\
&\leq C \|\Omega\|_{L^1(\mathbf{S}^{n-1})} |x|^{-n-1}.
\end{aligned}$$

For the case $|x| < 1$, notice first that if $|x| < 1/8$, then $|\widetilde{V}_j(x)| \leq C\|\Omega\|_{L^1}$, because the singularity is away from $x$. If $1/8 \leq |x| \leq 1$, then

$$\begin{aligned}
|\widetilde{V}_j(x) - \Phi(x)U_j(x)| &\leq \int\limits_{|y|>2} \frac{|\Omega(y)|}{|y|^n} |\Phi(y) - \Phi(x)| \frac{|x_j - y_j|}{|x-y|^{n+1}} dy \\
&+ \int\limits_{1/16<|y|<2} \frac{|\Omega(y)|}{|y|^n} |\Phi(y) - \Phi(x)| \frac{|x_j - y_j|}{|x-y|^{n+1}} dy \\
&+ \int\limits_{0<|y|<1/16} \frac{|\Omega(y)|}{|y|^n} |\Phi(y) - \Phi(x)| \left| \frac{|x_j - y_j|}{|x-y|^{n+1}} - \frac{x_j}{|x|^{n+1}} \right| dy \\
&= P_1(x) + P_2(x) + P_3(x).
\end{aligned}$$

The first term is easy:

$$P_1(x) \leq C \int\limits_{|y|>2} \frac{|\Omega(y)|}{|y|^{2n}} dy \leq C\|\Omega\|_{L^1}.$$

For the second term $P_2(x)$, we use that $\Phi$ is a Lipshitz function to obtain

$$\begin{aligned}
P_2(x) \leq C \int\limits_{1/16<|y|<2} \frac{|\Omega(y)|}{|y|^n |y-x|^{n-1}} dy &\leq \int \frac{|y|^{1/2} |\Omega(y)|}{|y|^n |y-x|^{n-1}} dy \\
&\leq C|x|^{n-3/2} \int \frac{|y|^{1/2} |\Omega(y)|}{|y|^n |y-x|^{n-1}} dy.
\end{aligned}$$

For the third term use the elementary inequality

$$\left| \frac{x_j - y_j}{|x-y|^{n+1}} - \frac{x_j}{|x|^{n+1}} \right| \leq C|y|$$

to get

$$P_3(x) \leq C \int\limits_{0<|y|<1/16} \frac{|\Omega(y)|}{|y|^{n-1}} dy \leq C\|\Omega\|_{L^1}.$$

Therefore choose $G_j$ to be

$$G_j(x) = C\left[|V_j(x)| + \|\Omega\|_{L^1} + |x|^{n-3/2} \int \frac{|\Omega(y)|}{|y|^{n-1/2}|y-x|^{n-1}} dy\right].$$

This proves the Lemma. ■

We note that condition (9.16) also implies $L^p$ boundedness for $T^*_\Omega$ for a certain range of $p$'s depending on $\alpha$. We refer the reader to [12] for details.

For $\Omega \in L^1(\mathbf{S}^{n-1})$, let us define three operators

$$(M^*_\Omega f)(x) = \int\limits_{\mathbf{S}^{n-1}} |\Omega(\theta)|(M_\theta f)(x)\, d\theta, \tag{9.24}$$

$$(H_\Omega f)(x) = \int\limits_{\mathbf{S}^{n-1}} |\Omega(\theta)||(H_\theta f)(x)|\, d\theta, \tag{9.25}$$

$$(H^*_\Omega f)(x) = \int\limits_{\mathbf{S}^{n-1}} |\Omega(\theta)|(H^*_\theta f)(x)\, d\theta, \tag{9.26}$$

where $M_\theta$ and $H^*_\theta$ are given in (9.21) and (9.22), and $H_\theta$ in (9.4). As observed in the proof of the previous theorem, $L^p$ estimates for the operators $M^*_\Omega$, $H_\Omega$, and $H^*_\Omega$ were useful in establishing $L^p$ bounds for the operators $M_\Omega$, $T_\Omega$, and $T^*_\Omega$. One may wonder whether (9.24), (9.25), and (9.26) are also of weak type $(1,1)$. The answer turns out to be false. In fact with $\Omega = 1$, there is an example of R. Fefferman [11] which says that $M^*_\Omega$ is not of weak type $(1,1)$. Examples can also be given to show that $H_\Omega$ and $H^*_\Omega$ are also not of weak type $(1,1)$.

The $L^1$ theory of $T^*_\Omega$ for $\Omega$ rough is still open as of this writing. The basic question here is whether $T_\Omega$ is of weak type $(1,1)$ if $\Omega$ does not possess any smoothness. The following problem was posed by A. Seeger.

**Question 9.9** *Is $T^*_\Omega$ of weak type $(1,1)$ when $\Omega$ is in $L^\infty(S^{n-1})$?*

We end this exposition with three tables of known and open questions regarding the operators $M_\Omega$, $T_\Omega$ and $T^*_\Omega$.

Tables 9.1 and 9.2 refer to general functions $\Omega$, while table 9.3 refers to odd functions.

| $\Omega \in$ | $L^q$ | $L\log^+ L$ | (9.16) $\forall\alpha > 0$ | $H^1$ | $L^1$ |
|---|---|---|---|---|---|
| $M_\Omega$ | yes, trivial | yes, trivial | yes, trivial | yes, trivial | yes, trivial |
| $T_\Omega$ | yes, [2] | yes, [2] | yes, [12] | yes, [14], [7] | no, [21] |
| $T^*_\Omega$ | yes, [2] | yes, [2] | yes, [12] | yes, [10][1] | no, [21] |

[1] This result independently proved by the authors (Theorem 9.12).

**TABLE 9.1. Boundedness from $L^p$ to $L^p$**

| $\Omega \in$ | $L^q$, $q > 1$ | $L\log^+ L$ | $H^1$ | $L^1$ |
|---|---|---|---|---|
| $M_\Omega$ | [4] ($n = 2$) | [5] (all $n$) | irrelevant | open[2] |
| $T_\Omega$ | [13] ($n = 2$), [5][3] ($n \le 7$) | [5][3], [15] (all $n$) | open | no, [21][2] |
| $T^*_\Omega$ | open | open | open | no, [21][2] |

[2] true for the subspace of $L^1$ consisting of radial functions [17]

[3] only the case $n = 2$ is given in this reference

**TABLE 9.2. Weak type (1,1) boundedness**

| | $L^p \to L^p$, $1 < p < \infty$ | $L^1 \to$weak $L^1$ |
|---|---|---|
| $M_\Omega$ | yes, by method of rotations | open |
| $T_\Omega$ | yes, by method of rotations | open |
| $T^*_\Omega$ | yes, by method of rotations | open |

**TABLE 9.3. The case $\Omega$ is an odd function in $L^1$**

*References*


[1] A. P. Calderón and A. Zygmund, *On the existence of certain singular integrals*, Acta Math. 88 (1952) p. 85-139.

[2] A. P. Calderón and A. Zygmund, *On singular integrals*, Amer. J. Math. 78 (1956) p.289-309.

[3] D. Chang, S. Krantz, and E. M. Stein, *Hardy spaces and elliptic boundary value problems*, Cont. Math. 137 (1992) p.119-131.

[4] M. Christ, *Weak type (1,1) bounds for rough operators*, Ann. Math. 128 (1988) p.19-42.

[5] M. Christ and J.-L. Rubio de Francia, *Weak type (1,1) bounds for rough operators II*, Invent. Math. 93 (1988) p.225-237.

[6] R. R. Coifman and G. Weiss, *Extensions of Hardy spaces and their use in analysis*, Bull. Amer. Math. Soc. 83 (1977) p.569-645.

[7] W. C. Connett, *Singular integrals near* $L^1$, Proc. Sympos. Pure Math., Amer. Math. Soc. 35 (1979) p.163-165 (S. Wainger and G. Weiss, eds.).

[8] J. E. Daly and K. Phillips, *On the classification of homogeneous multipliers bounded on* $H^1(R^2)$, Proc. Amer. Math. Soc. 106 (1989) p.685-696.

[9] J. Duonandikoetxea and J.-L. Rubio de Francia, *Maximal and singular integral operators via Fourier transform estimates*, Invent. Math. 84 (1986) p.541-561.

[10] D. Fan and Y. Pan, *Singular integral operators with rough kernels supported by subvarieties*, Amer. J. Math. 119 (1997) p.799-839.

[11] R. Fefferman, *A theory of entropy in Fourier analysis*, Adv. Math. 30 (1978) p.171-201.

[12] L. Grafakos and A. Stefanov, $L^p$ *bounds for singular integrals and maximal singular integrals with rough kernels*, (1997), to appear, Indiana Univ. Math. J. (1998).

[13] S. Hofmann, *Weak (1,1) boundedness of singular integrals with nonsmooth kernels*, Proc. Amer. Math. Soc. 103 (1988) p.260-264.

[14] F. Ricci and G. Weiss, *A characterization of* $H^1(\Sigma^{n-1})$, Proc. Sympos. Pure Math, Amer. Math. Soc. 35 (1979) p.289-294.

[15] A. Seeger, *Singular integral operators with rough convolution kernels,* J. Amer. Math. Soc. 9 (1996) p.95-105.

[16] A. Seeger and T. Tao, personal communication.

[17] P. Sjögren and F. Soria, *Rough maximal operators and rough singular integral operators applied to integral radial functions,* Rev. Math. Iber. 13 (1997) p1-18.

[18] A. Stefanov, *Characterizations of $H^1$ and applications to singular integral operators applied to integral radial functions,* submitted.

[19] E. M. Stein, *Singular Integrals and Differnetiability Properties of Functions,* Princeton University Press (1970).

[20] E. M. Stein, *Harmonic Analysis: Real Variable Methods, Orthogonality, and Oscillatory Integrals,* Princeton University Press (1993).

[21] M. Weiss and A. Zygmund, *An example in the theory of singular integrals*, Studia Math. 26 (1965) p.101-111.

[22] A. Zygmund, *Trigonometric Series*, Vol I, Cambridge University Press (1959).

# Chapter 10

# Haar multipliers, paraproducts, and weighted inequalities

**Nets Hawk Katz**[1]
**María Cristina Pereyra**[2]

*ABSTRACT. In this paper we present a brief survey on Haar multipliers, dyadic paraproducts, and recent results on their applications to deduce scalar and vector valued weighted inequalities. We present a new proof of the boundedness of a Haar multiplier in $L^p(\mathbb{R})$. The proof is based on a stopping time argument suggested by P. W. Jones for the case $p = 2$, that it is adapted to the case $1 < p < \infty$ using an new version of Cotlar's Lemma for $L^p$. We then prove some weighted inequalities for simple dyadic operators.*

## 10.1 Introduction

A *Haar multiplier* is an operator of the form:

$$Tf(x) = \sum_{I \in \mathcal{D}} \omega_I(x) \langle f, h_I \rangle h_I(x),$$

where the sum runs over the dyadic intervals $\mathcal{D} = \{(k2^{-j}, (k+1)2^{-j}] : k, j \in \mathbb{Z}\}$; $h_I$ is the *Haar function* associated to $I$; $\langle .,. \rangle$ denotes the $L^2$ inner product; and finally the *symbol* $\omega_I(x)$ is a function of both the variables $x \in \mathbb{R}$ and $I \in \mathcal{D}$. These operators are formally similar to pseudodifferential operators, but the trigonometric functions have been replaced by step functions. When the symbol is a function independent of $x$, $\omega_I(x) = \omega_I$, the corresponding operators are the *constant Haar multipliers*, known to be bounded in $L^p$ if and only if the sequence $\{\omega_I\}_{I \in \mathcal{D}}$ is bounded. We will be concerned with symbols of the form

$$\omega_I^t(x) = \left(\frac{\omega(x)}{m_I \omega}\right)^t,$$

[1] Research supported by EPSCR GR/110024
[2] Research supported by EPSCR Visiting Fellowship GR/16066

where $t$ is a real number, $\omega$ is a weight, and $m_I\omega$ denotes the mean value of $\omega$ over $I$. The corresponding multipliers will be denoted $T^t_\omega$. We prove that under certain conditions on the weight these operators are bounded in $L^p(\mathbb{R})$.

The proof presented here is based on a stopping time argument suggested by P. W. Jones for the case $p = 2$, $t = 1$, that is adapted to the case $p \neq 2$ using a version of Cotlar's Lemma in $L^p$ which could be of interest in its own right.

The non constant Haar multipliers corresponding to $t = 1$ appeared for the first time in [24], in connection with the existence and boundedness of *the resolvent of the dyadic paraproduct*. The ones corresponding to $t = \pm 1/2$ appeared in the work of Treil and Volberg concerning matrix valued weighted inequalities for the Hilbert transform [25].

The paraproducts are a class of operators that have made their way into mainstream harmonic analysis since their discovery in the late 70's. They were first introduced by Bony in the context of nonlinear differential equations [2]. After the proof of the $T(1)$ Theorem by David and Journé [10], it became clear that in the theory of singular integral operators the objects of study where of two very different natures: either similar to translation invariant operators, or to paraproducts, whose boundedness depended on a certain BMO/Carleson condition to be satisfied. The simplest version of the paraproduct is the dyadic one. Given a sequence of real numbers $\{b_I\}_{I\in\mathcal{D}}$, the paraproduct $\pi_b$, acts on locally integrable functions on the line by:

$$\pi_b f(x) = \sum_{I\in\mathcal{D}} m_I f\, b_I h_I(x),$$

where $m_I f$ denotes the mean value of $f$ over $I$. The operator $\pi_b$ is bounded in $L^p$ if and only if the sequence of $b_I^2$'s satisfies a Carleson condition, namely there exists a constant $C$ such that for all $I \in \mathcal{D}$:

$$\sum_{J\in\mathcal{D}(I)} b_J^2 \leq C|I|,$$

which is equivalent to saying that the function $b = \sum_{I\in\mathcal{D}} b_I h_I$ belongs to dyadic $BMO$, see [19], [4].

The boundedness of the paraproducts can be interpreted as an embedding theorem. This approach has been favored by the russian school, see [20], [25]. They refer to these results as *Carleson embedding theorems.*

After the Fourier transform the Hilbert transform is, perhaps, the most important operator in analysis. The Hilbert transform $H$, that acts on functions on the real line, is defined by:

$$Hf(x) = P.V. \int \frac{f(y)}{x-y} dy.$$

A necessary and sufficient condition for a single scalar weight $\omega$ so that $H$ maps $L^2(\omega)$ into itself continuously, that is:

$$\int |Hf|^2\omega \leq C\int |f|^2\omega, \qquad \forall f \in L^2(\omega),$$

was first given (as part of a problem in prediction theory) by Helson and Szegö in 1960 using complex analysis, see [14]. In the late 70's, Cotlar and Sadosky extended these results to two weights, and to matrix and operator valued contexts, see [7],[8].

In 1973, Hunt, Muckenhoupt and Wheeden, see [13], presented a new proof, where for the first time the $A_p$-condition for weights appeared as the necessary and sufficient condition for the boundedness of the Hilbert transform in $L^p(\omega)$:

$$\omega \in A_p \iff \left(\frac{1}{|I|}\int_I \omega\right)\left(\frac{1}{|I|}\int_I \omega^{\frac{-1}{p-1}}\right)^{p-1} < C;$$

for all intervals $I \subset \mathbf{R}$.

This was soon extended to a large class of singular integral operators, see [5]. This work depended heavily on certain maximal functions defined only in the scalar case.

It was not understood how to make a sensible theory in the case of matrix (or operator) valued weights. In 1995 S. Treil and A. Volberg managed to prove that the natural matrix valued generalization of $A_2$ was the correct necessary and sufficient condition for the vector valued problem, with equal weights, $p = 2$, and *when the dimension of the Hilbert space is finite*, see [25]. Their bounds depend on the dimension of the space. We suspect the bounds are not sharp although they should still depend on the dimension. Their proof depends on the existence of a weighted Carleson Embedding Theorem which can be interpreted as the boundedness of certain non constant *Haar multipliers*. They have since extended the results for all $1 < p < \infty$, for a classical approach see [26], for a novel one using Bellman functions see [21].

The difficulties extending these results to the infinite dimensional setting become apparent by the recent discovery by Nazarov, Treil and Volberg [22] of a counterexample for an infinite dimensional Carleson embedding theorem. They construct a sequence of matrix-valued Carleson measures indexed in the dimension $N$ of the underlying space, for which the embedding bounds are of the order $\sqrt{\log N}$. They believe those estimates are sharp. The best known upper bound, $\log N$, is due to Katz [15], in his work on matrix valued paraproducts, and, independently, to Nazarov, Treil and Volberg [22].

The scalar two weights problem for the Hilbert transform is still open. Sufficient conditions for the matrix and operator valued case were given in [18]. The main tool used there is an operator valued Schur's Lemma to deduce, as Treil and Volberg did, the boundedness of the operator from the rate of decay of the off-diagonal coefficients of the standard matrix representation of the operator in the Haar basis. Parts of the operator are treated as if they were constant Haar multipliers and others like paraproducts, very much in the spirit of the T(1) Theorem.

The techniques used, in particular paraproducts, stopping time arguments, Schur's Lemma, and Cotlar's Lemma had been exploited with success by the first author to solve a longstanding problem on maximal operators over arbitrary sets of directions, see [16].

Recently Nazarov, Treil and Volberg solved the scalar two weights problem for a toy model of the Hilbert transform, namely the *signed constant Haar multipliers*:

$$T_\sigma f = \sum_{I\in\mathcal{D}} \pm\langle f, h_I\rangle h_I .$$

where $\sigma$ indicates a choice of signs. The result gives necessary and sufficient conditions on a pair of weights $(u, v)$ so that the $T_\sigma$'s are uniformly bounded from $L^2(u)$ into $L^2(v)$. It is not known what happens for $p \neq 2$, see [23].

The paper is organized as follows. In the next section we present some notation and preliminaries. Then we define a decaying stopping time, we present the main Weight Lemma, and we deduce from it Gehring's Theorem and a dyadic relation between $RH_p^d$ weights and dyadic $BMO$ proved by R. Fefferman, C. Kenig and J. Pipher in the doubling case. We then prove a multilinear version of Schur's Lemma, and some lemmas in the $L^p$ theory for decaying stopping times, in particular one that can be thought as a version of Plancherel's Lemma in $L^p$, and the second being an $L^p$ version of Cotlar's Lemma. Then comes the stopping time proof of the boundedness of the Haar multiplier $T^1_\omega$ in $L^p$, as a corollary we deduce that $T^t_\omega$ is bounded in $L^p$, and that it is of weak type $(1,1)$. Finally we use those results to show weighted inequalities for constant Haar multipliers, dyadic paraproducts and the dyadic square function, this provides an alternative proof for the classical dyadic Littlewood-Paley Theory (some of these results appeared in [18]).

## 10.2 Preliminaries

### *10.2.1 Dyadic intervals and Haar basis*

Let us denote by $\mathcal{D}$ the family of all dyadic intervals in $\mathbf{R}$, i.e. intervals of the form $(j2^{-k}, (j+1)2^{-k}]$, $j, k \in \mathbb{Z}$. $\mathcal{D}_k$ denotes the $k^{th}$ generation of $\mathcal{D}$, consisting of those dyadic intervals of length $2^{-k}$. Given any interval $J$, $\mathcal{D}(J)$ denotes the family of dyadic subintervals of $J$;

$$\mathcal{D}_n(J) = \{I \in \mathcal{D}(J) : |I| = 2^{-n}\,|J|\},$$

$|I|$ denotes the length of the interval $I$. Given an interval $J$ we will denote the right and left halves respectively by $J_r$ and $J_l$; they are the elements of $\mathcal{D}_1(J)$.

The *Haar function* associated to an interval $I$ is given by:

$$h_I(x) = \frac{1}{|I|^{1/2}}\left(\chi_{I_r}(x) - \chi_{I_l}(x)\right),$$

here $\chi_I$ denotes the characteristic function of the interval $I$. The Haar functions indexed on the dyadics, $\{h_I\}_{I\in\mathcal{D}}$, form a basis of $L^2(\mathbb{R})$, see [6].

### *10.2.2 Weights*

A *weight* is a locally integrable function which is almost everywhere strictly positive. Let $1 < p < \infty$.

A weight $\omega$ is in *dyadic reverse Hölder p*, $RH_p^d$, if there exists a constant $C$ such that for every interval $I \in \mathcal{D}$

$$\frac{1}{|I|}\int_I \omega^p \leq C\left(\frac{1}{|I|}\int_I \omega\right)^p.$$

The infimum of such constants is called the $RH_p^d$ *constant of* $\omega$. Eg: $\omega(x) = |x|^\alpha$ for $\alpha > -1/p$.

A weight $\omega$ is in the *dyadic Muckenhoupt's class*, $A_p^d$, if there exists a constant $C$ such that for every interval $I \in \mathcal{D}$

$$\left(\frac{1}{|I|}\int_I \omega\right)\left(\frac{1}{|I|}\int_I \omega^{\frac{-1}{p-1}}\right)^{p-1} < C.$$

Eg: $\omega(x) = |x|^\alpha$ for $-1 < \alpha < p-1$.

We say that a weight $\omega$ is in $A_\infty^d$ if it belongs to $RH_p^d$ for some $1 < p < \infty$.

The main property of these classes of weights is Gehring's Theorem which states that if $\omega \in RH_p^d$ then $\omega \in RH_{p+\epsilon}^d$ for some $\epsilon > 0$, see [11].

This result is well known for the corresponding non dyadic classes, see [GC-RF]. We present a proof in the next section based in the Weight Lemma 10.3.

Corresponding properties for matrix valued weights are less clear. In particular an analogue of Gehring's Theorem would be very useful.

The following lemma will be used later. For non dyadic classes it is due to Strömberg and Wheeden [SW], see also [9]. For dyadic classes it still holds. For completeness we present a proof.

**Lemma 10.1** *Assume $s > 1$. A weight $\omega \in RH_s^d$ if and only if $\omega^s \in A_\infty^d$.*

**Proof.** ($\Rightarrow$) Enough to show that there exists $1 < p < \infty$ such that $\omega^s \in RH_p^d$ by the definition of $A_\infty^d$.

Assuming $\omega \in RH_s^d$ then Gehring's Theorem implies that $\omega \in RH_{s+\epsilon}^d$ for some $\epsilon > 0$. Let $p = \frac{s+\epsilon}{s} > 1$ then,

$$\begin{aligned} \frac{1}{|I|}\int_I \omega^{sp} &= \frac{1}{|I|}\int_I \omega^{s+\epsilon} \leq C\left(\frac{1}{|I|}\int_I \omega\right)^{s+\epsilon} \\ &= C\left(\frac{1}{|I|}\int_I \omega\right)^{sp} \leq C\left(\frac{1}{|I|}\int_I \omega^s\right)^{p} \end{aligned}$$

where we have used the $RH_{s+\epsilon}^d$ in the first inequality and Hölder's inequality in the last step. This implies $\omega^s \in RH_p^d$.

($\Leftarrow$) We assume that there exists $1 < p < \infty$ such that $\omega^s \in RH_p^d$. Let $t = \frac{sp(s-1)}{sp-1}$ and $q = \frac{sp-1}{s-1} > p$, then $tq = sp$ and $(s-t)q' = 1$, where $\frac{1}{q} + \frac{1}{q'} = 1$. Then by Hölder's inequality

$$\begin{aligned} \frac{1}{|I|}\int_I \omega^s &= \frac{1}{|I|}\int_I \omega^t\omega^{s-t} \leq \left(\frac{1}{|I|}\int_I \omega^{tq}\right)^{\frac{1}{q}}\left(\frac{1}{|I|}\int_I \omega^{(s-t)q'}\right)^{\frac{1}{q'}} \\ &= \left(\frac{1}{|I|}\int_I \omega^{sp}\right)^{\frac{1}{q}}\left(\frac{1}{|I|}\int_I \omega\right)^{\frac{1}{q'}}. \end{aligned}$$

Since $\omega^s \in RH_p^d$ we get,

$$\left(\frac{1}{|I|}\int_I \omega^{sp}\right)^{\frac{1}{q}} \leq C\left(\frac{1}{|I|}\int_I \omega^s\right)^{\frac{p}{q}},$$

therefore

$$\left(\frac{1}{|I|}\int_I \omega^s\right)^{1-\frac{p}{q}} \leq C\left(\frac{1}{|I|}\int_I \omega\right)^{\frac{1}{q'}}.$$

But the choice of $t$ and $q$ forces $\frac{q}{q'} = s(q-p)$, hence

$$\frac{1}{|I|}\int_I \omega^s \leq C\left(\frac{1}{|I|}\int_I \omega\right)^{s},$$

and $\omega \in RH_s^d$. ■

It is well known that any doubling $RH_q^d$ weight is in $A_p^d$ for some $p > 1$ and viceversa. Without the doubling condition half of it is still true, namely,

**Lemma 10.2** *If $\omega \in A_p^d$ then $\omega \in RH_q^d$ for some $q > 1$.*

**Proof.** This is a consequence of Gehring's Theorem and the following,

**Claim:** If $\omega \in A_p^d$ then $\omega^{\frac{1}{p}} \in RH_p^d$.

Assuming the claim, $\omega \in A_p^d$ implies $\omega^{\frac{1}{p}} \in RH_p^d$ implies $\omega^{\frac{1}{p}} \in RH_{p+\epsilon}^d$ which by the previous lemma implies $\omega^{1+\frac{\epsilon}{p}} \in A_\infty^d$ and by the same lemma it implies that $\omega \in RH_{1+\frac{\epsilon}{p}}^d$.

**Proof of the claim:** By Hölder's inequality twice and the hypothesis $\omega \in A_p^d$ we get,

$$
\begin{aligned}
1 &\le \left(\frac{1}{|I|}\int_I \omega^{\frac{1}{p}}\right)^p \left(\frac{1}{|I|}\int_I \omega^{\frac{-p'}{p^2}}\right)^{\frac{p^2}{p'}} \\
&\le C\left(\frac{1}{|I|}\int_I \omega^{\frac{1}{p}}\right)^p \left(\frac{1}{|I|}\int_I \omega^{\frac{-1}{p-1}}\right)^{p-1} \\
&\le C\left(\frac{1}{|I|}\int_I \omega^{\frac{1}{p}}\right)^p \left(\frac{1}{|I|}\int_I \omega\right)^{-1},
\end{aligned}
$$

where $\frac{1}{p}+\frac{1}{p'}=1$ implies $\frac{p}{p'}=p-1$. We conclude that

$$
\frac{1}{|I|}\int_I \omega \le C\left(\frac{1}{|I|}\int_I \omega^{\frac{1}{p}}\right)^p,
$$

that is $\omega^{\frac{1}{p}} \in RH_p^d$. ■

If a given weight $\omega$ is bounded away from zero and from infinity, that is: $\lambda^{-1} < \omega(x) < \lambda$, for some $\lambda > 1$; then it is doubling and it belongs to any of the classes. In general belonging to $A_\infty^d$ does not guarantee being bounded or doubling (in the non dyadic setting $A_\infty$ implies doubling), but it says that the set where the weight is close to zero intersected with any interval where the mean is far from zero is small in a very precise way, to be described next.

## 10.3 Weight lemma and decaying stopping times

Let us first define a *stopping time* $\mathcal{J}$ for an interval $I$.

For a given interval $I$, let $\mathcal{J}(I)$ be the collection of dyadic intervals contained in $I$ which are maximal with respect to a given property. Let $\mathcal{F}(I)$ be the collection of dyadic intervals contained in $I$ but not contained in any interval $J \in \mathcal{J}(I)$.

We say the property is *admissible* if for all $J \in \mathcal{D}(I)$, $J \in \mathcal{F}(J)$ and hence $\mathcal{F}(J)$ is not empty.

Given an admissible property, let $\mathcal{J}^o(I) = \{I\}$. For $n > 0$ define now the collections $\mathcal{J}^n(I)$ and $\mathcal{F}^n(I)$ inductively. $\mathcal{J}^n(I)$ is the collection of intervals belonging to $\mathcal{J}(J)$ for some $J$ in $\mathcal{J}^{n-1}(I)$. Similarly, $\mathcal{F}^n(I)$ is the collection of intervals belonging to $\mathcal{F}(J)$ for some $J$ in $\mathcal{J}^{n-1}(I)$.

The family of collections of intervals $(\mathcal{J}^n, \mathcal{F}^n)$ is the stopping time $\mathcal{J}$ for the interval $I$ corresponding to the given admissible property. The intervals in $\mathcal{F}^n$ are "good", those in $\mathcal{J}^n$ are "bad" but not so bad because their parents are "good".

Clearly for each $n > 0$ the intervals in $\mathcal{J}^n(I)$ are pairwise disjoint. By definition the elements of $\mathcal{J}^n(I)$ are subintervals of the elements of $\mathcal{J}^{n-1}(I)$.

Also $\mathcal{D}(I) = \bigcup_{j=0}^{\infty} \mathcal{F}^j(I)$, and the $\mathcal{F}^j$'s are disjoint collections of dyadic subintervals.

We say that $\mathcal{J}$ is a *decaying stopping time* if there exists $0 < c < 1$ so that for every $I \in \mathcal{D}$, one has

$$\sum_{J \in \mathcal{J}(I)} |J| \leq c|I|.$$

Iterating this property we conclude that for decaying stopping times,

$$\sum_{J \in \mathcal{J}^k(I)} |J| \leq c^k|I|. \tag{10.1}$$

We now prove what will be the most fundamental lemma in the theory of weights. It should be viewed as an analogue of the John-Nirenberg theorem for functions in $BMO$.

Given a weight $\omega$, we define the stopping time $\mathcal{J}_\omega$ where $\mathcal{J}_\omega(I)$ denotes the set of pairwise disjoint dyadic subintervals $J$ of $I$ which are maximal with respect to the property that $m_J\omega \geq \lambda m_I\omega$ or $m_J\omega \leq \frac{1}{\lambda} m_I\omega$, where $\lambda \geq 1$ is to be specified in the proof of the following Lemma. It depends only on the $RH_p^d$ constant of $\omega$.

**Lemma 10.3 (Weight Lemma)** *Let $\omega \in RH_p^d$. Then, for $\lambda$ sufficiently large, $\mathcal{J}_\omega$ is a decaying stopping time.*

**Proof.** First let $\lambda > 3$. We may divide up $I$ into three disjoint subsets,

$$I = \bigcup_j I_j^\lambda \bigcup_j I_j^{\frac{1}{\lambda}} \bigcup G,$$

where the intervals $I_j^\lambda, I_j^{\frac{1}{\lambda}}$ are an enumeration of all the different elements of $\mathcal{J}_\omega(I)$ such that for each $j$

$$\lambda\, m_I\omega \le m_{I_j^\lambda}\omega \le 2\lambda\, m_I\omega, \tag{10.2}$$

$$m_{I_j^{\frac{1}{\lambda}}}\omega \le \frac{1}{\lambda} m_I\omega,$$

and that $\omega(x) \le \lambda\, m_I\omega$ a.e. on $G$. Notice that the second inequality in (10.2) is just a consequence of the maximality assumption in the definition of $\mathcal{J}_\omega(I)$.

Suppose the lemma is false. Then $G$ can be arbitrarily small. Suppose $|G| \le \frac{|I|}{3\lambda}$. Thus

$$\int_G \omega \le \frac{1}{3}\int_I \omega,$$

and since $\lambda > 3$,

$$\sum_j \int_{I_j^{\frac{1}{\lambda}}} \omega \le \frac{1}{3}\int_I \omega,$$

so that

$$\sum_j \int_{I_j^\lambda} \omega \ge \frac{1}{3}\int_I \omega.$$

By the second inequality in (10.2), this implies that

$$\sum_j |I_j^\lambda| \ge \frac{1}{6\lambda}|I|. \tag{10.3}$$

We will now use (10.3) and (10.2) to contradict $\omega \in RH_p^d$,

$$\int_I \omega^p \ge \sum_j \int_{I_j^\lambda} \omega^p \ge \sum_j \frac{1}{|I_j^\lambda|^{p-1}}\left(\int_{I_j^\lambda} \omega\right)^p = \sum_j |I_j^\lambda|(m_{I_j^\lambda}\omega)^p,$$

with the second inequality being just an application of Hölder's inequality. But

$$\sum_j |I_j^\lambda|(m_{I_j^\lambda}\omega)^p \ge \lambda^p \sum_j |I_j^\lambda|(m_I\omega)^p \ge \frac{1}{6}\lambda^{p-1}|I|(m_I\omega)^p.$$

This contradicts $RH_p^d$ provided we chose $\lambda \geq (6C)^{\frac{1}{p-1}}$, where $C$ is the $RH_p^d$ constant of $\omega$. If this is the case, then $|G| \geq \frac{1}{3\lambda}|I|$ and hence, we have proved the lemma with $c = 1 - \frac{1}{3\lambda}$. ∎

We will use this lemma to show the classical Gehring's Theorem (see Section 2.2). It can also be used to prove the fact that for doubling weights $\omega$, $\log \omega$ is in dyadic $BMO$ whenever $\omega \in A_\infty^d$, see [17]. First we prove a more "dyadic" relation between $RH_p^d$ weights and dyadic $BMO$ first discovered by Fefferman, Kenig, and Pipher [FKP].

A function $b$ is in dyadic $BMO$ if and only if the sequence $b_I = \langle b, h_I \rangle$ satisfies a Carleson condition, namely,

$$\sum_{J \in \mathcal{D}(I)} b_J^2 \leq C|I|, \quad \forall I \in \mathcal{D}.$$

**Corollary 10.4** *Let $\omega \in RH_p^d$ for some $1 < p < \infty$. We define the function*

$$b(x) = \sum_{I \in \mathcal{D}} \frac{\langle \omega, h_I \rangle}{m_I \omega} h_I.$$

*Then $b(x)$ is in dyadic $BMO$.*

**Proof.** It suffices to show that there exist a constant $C > 0$, so that for every $I \in \mathcal{D}$,

$$\sum_{J \in \mathcal{D}(I)} \left| \frac{\langle \omega, h_J \rangle}{m_J \omega} \right|^2 \leq C|I|.$$

We define $\mathcal{F}_\omega^j(I)$ to be the set of dyadic intervals contained in some interval of $\mathcal{J}_\omega^{j-1}(I)$ but not contained in any interval of $\mathcal{J}_\omega^j(I)$. This is the decaying stopping time $\mathcal{J}_\omega$. For any interval $J$, we define the function $\omega_J(x)$ supported on $J$ to be equal to $\omega(x)$ when $x$ is not contained in any interval of $\mathcal{J}_\omega(J)$ and to be equal to $m_K \omega$ when $x \in K \in \mathcal{J}_\omega(J)$. Then

$$|J| m_J^2 \omega \;+ \sum_{K \in \mathcal{F}_\omega(J)} |\langle \omega, h_K \rangle|^2 = \int_J \omega_J^2(x).$$

However, by definition of $\mathcal{J}_\omega(J)$, in particular by (10.2),

$$\sum_{K \in \mathcal{F}_\omega(J)} |\langle \omega, h_K \rangle|^2 \leq \int_J \omega_J^2(x) \leq 2\lambda m_J \omega \int_J \omega_J(x) = 2\lambda |J| m_J^2 \omega.$$

Now for any $K \in \mathcal{F}_\omega^j(J)$, there is a unique interval $J$ of $\mathcal{J}_\omega^{j-1}(I)$ containing

$K$. Then $K \in \mathcal{F}_\omega(J)$ and $m_K\omega > \frac{1}{\lambda} m_J\omega$. Thus

$$\begin{aligned}
\sum_{K \in \mathcal{F}_\omega^j(I)} \left| \frac{\langle w, h_K \rangle}{m_K \omega} \right|^2 &\le \lambda^2 \sum_{J \in \mathcal{J}_\omega^{j-1}(I)} \sum_{K \in \mathcal{F}_\omega(J)} \left| \frac{\langle \omega, h_K \rangle}{m_J \omega} \right|^2 \\
&\le \lambda^2 \sum_{J \in \mathcal{J}_\omega^{j-1}(I)} \frac{1}{m_J^2 \omega} 2\lambda |J| m_J^2 \omega \\
&\le 2\lambda^3 \sum_{J \in \mathcal{J}_\omega^{j-1}(I)} |J| \le 2\lambda^3 c^{j-1} |I|,
\end{aligned}$$

where we have used (10.1) to get the last inequality. Thus

$$\sum_{J \in \mathcal{D}(I)} \left| \frac{\langle \omega, h_J \rangle}{m_J \omega} \right|^2 \le 2\lambda^3 (1 + c + c^2 + \dots)|I| \le \frac{2\lambda^3 |I|}{1 - c},$$

which was to be shown. ■

**Lemma 10.5 (Gehring's Theorem)** *Suppose $\omega \in RH_p^d$ for some $1 < p < \infty$. Then there is $\epsilon > 0$ depending only on $p$ and the $RH_p^d$ constant of $\omega$ so that $\omega \in RH_{p+\epsilon}^d$.*

**Proof.** For any interval $J$, define

$$\mathcal{G}_1(I) = \left( \bigcup \mathcal{J}_\omega(I) \right)^c,$$

where we are denoting $\bigcup \mathcal{J}_\omega(I) = \bigcup_{J \in \mathcal{J}_\omega(I)} J$. Observe that for almost every $x \in \mathcal{G}_1(I)$, one has $\frac{1}{\lambda} m_I \omega \le \omega(x)$, and one has, for $0 < c < 1$,

$$|\mathcal{G}_1(I)| \ge (1 - c)|I|,$$

by Lemma 10.3. Thus

$$\int_{\mathcal{G}_1(I)} \omega^p \ge (1 - c)|I| (\frac{1}{\lambda})^p (m_I \omega)^p.$$

Since $\omega \in RH_p^d$ this means there exists $a > 0$ depending only on $p$ and the $RH_p^d$ constant for $\omega$ (since $\lambda$ and $c$ depend only on these) so that for every $I$,

$$\int_{\mathcal{G}_1(I)} \omega^p \ge a \int_I \omega^p. \tag{10.4}$$

We define

$$\mathcal{G}_j(I) = \bigcup \mathcal{J}_\omega^{j-1}(I) \setminus \bigcup \mathcal{J}_\omega^j(I).$$

Clearly for $x \in \mathcal{G}_j(I)$ we have that

$$\omega(x) \leq (2\lambda)^j m_I \omega. \tag{10.5}$$

Further, we claim that

$$\int_{\mathcal{G}_j(I)} \omega^p \leq (1-a)^{j-1} \int_I \omega^p. \tag{10.6}$$

We will prove (10.6) by induction.

To prove (10.6), it suffices to show that

$$\int_{\bigcup \mathcal{J}_\omega^j(I)} \omega^p \leq (1-a) \int_{\bigcup \mathcal{J}_\omega^{j-1}(I)} w^p,$$

for every $j$, since

$$\int_{\mathcal{G}_j(I)} \omega^p \leq \int_{\bigcup \mathcal{J}_\omega^{j-1}(I)} \omega^p.$$

But by (10.4)

$$\begin{aligned}
\int_{\bigcup \mathcal{J}_\omega^j(I)} \omega^p &= \sum_{J \in \mathcal{J}_\omega^{j-1}(I)} \Big( \int_J \omega^p - \int_{\mathcal{G}_1(J)} \omega^p \Big) \\
&\leq (1-a) \int_{\bigcup \mathcal{J}_\omega^{j-1}(I)} \omega^p.
\end{aligned}$$

Thus (10.6) is shown.

Now we estimate

$$\begin{aligned}
\int_I \omega^{p+\epsilon} &\leq \sum_{j=1}^{\infty} \int_{\mathcal{G}_j(I)} \omega^{p+\epsilon} \\
&\leq (m_I\omega)^\epsilon \sum_{j=1}^{\infty} (2\lambda)^{j\epsilon} \int_{\mathcal{G}_j(I)} \omega^p \\
&\leq (m_I\omega)^\epsilon \sum_{j=1}^{\infty} (2\lambda)^{j\epsilon} (1-a)^{j-1} \int_I \omega^p.
\end{aligned} \tag{10.7}$$

Here we have obtained the second inequality using (10.5) and the third using (10.6). Now we choose $\epsilon$ sufficiently small so that $(1-a)(2\lambda)^\epsilon < 1$, and the sum in (10.7) converges. There is a $C > 0$ so that for every $I$,

$$\frac{1}{|I|} \int_I \omega^{p+\epsilon} \leq C (m_I\omega)^\epsilon \frac{1}{|I|} \int_I \omega^p.$$

Now applying the $RH_p^d$ condition shows there exists $C > 0$ so that

$$\frac{1}{|I|}\int_I \omega^{p+\epsilon} \leq C(m_I\omega)^{p+\epsilon},$$

which was to be shown. ■

The following result will be used later. For a proof see [19].

**Lemma 10.6 (Carleson's Lemma)** *Assume the sequence of positive numbers $\{a_I\}_{I\in\mathcal{D}}$ satisfies the following Carleson condition:*

$$\sum_{J\in\mathcal{D}(I)} a_J \leq C|I|, \quad \forall I \in \mathcal{D}.$$

*Let $\{\lambda_I\}_{I\in\mathcal{D}}$ be any sequence of positive numbers, then*

$$\sum_{I\in\mathcal{D}} \lambda_I a_I \leq C\int \sup_{I\ni x} \lambda_I \, dx.$$

## 10.4 $L^p$ Lemmas for decaying stopping times

### *10.4.1 $L^p$ Plancherel Lemma*

First we state explicitly a multilinear version of Schur's Lemma that we will need.

**Lemma 10.7** *Suppose $C(j_1, j_2, \dots, j_{n-1})$ are given positive numbers where the $j$'s range over $\mathbf{Z}$ and suppose that for some $C \geq 0$*

$$\sum_{j_1=-\infty}^{\infty} \cdots \sum_{j_{n-1}=-\infty}^{\infty} C(j_1, j_2, \dots, j_{n-1}) \leq C,$$

*then for any sequence $\{x_j\}_{j\in\mathbf{Z}}$ in $l^n$, one has*

$$\sum_{j_1=-\infty}^{\infty} \cdots \sum_{j_n=-\infty}^{\infty} C(j_2 - j_1, j_3 - j_2, \dots, j_n - j_{n-1}) x_{j_1} \dots x_{j_n} \leq C^n \|x\|_n^n.$$

**Proof.** We simply use Hölder's inequality,

$$\begin{aligned}\sum_{j_1=-\infty}^{\infty} \cdots \sum_{j_n=-\infty}^{\infty} & C(j_2 - j_1, j_3 - j_2, \dots, j_n - j_{n-1}) x_{j_1} \dots x_{j_n} \\ &\leq \Big( \sum_{j_1, j_2, \dots, j_n} C(j_2 - j_1, \dots, j_n - j_{n-1}) x_{j_1}^n \Big)^{\frac{1}{n}} \dots \\ &\qquad \Big( \sum_{j_1, j_2, \dots, j_n} C(j_2 - j_1, \dots, j_n - j_{n-1}) x_{j_n}^n \Big)^{\frac{1}{n}} \leq C^n \|x\|_n^n.\end{aligned}$$

■

We now prove a small result in the theory of decaying stopping times, which can be viewed as an $L^p$ Plancherel type lemma.

Let $\mathcal{J}$ be a decaying stopping time. Let us denote $\mathcal{F}^j = \mathcal{F}^j([0,1])$, similarly $\mathcal{J}^j$. We define the operator

$$\Delta_{\mathcal{F}^j} f = \sum_{I \in \mathcal{F}^j} \langle f, h_I \rangle h_I,$$

which is just the projection onto the closed subspace spanned by $\{h_I\}_{\mathcal{F}_j}$. This is a constant Haar multiplier with bounded symbol $c_I^j = \chi_{\mathcal{F}^j}(I)$, hence it is bounded on $L^p$ for $1 < p < \infty$ with constant depending only on $p$, by Lemma 10.11 in the next section.

**Lemma 10.8** *Let $\mathcal{J}$ be a decaying stopping time and the $\mathcal{F}^j$'s be as above. For any $1 < p < \infty$ there is a constant $C(p) > 0$ so that for all $f \in L^p(\mathbb{R})$,*

$$\sum_{j=1}^{\infty} \|\Delta_{\mathcal{F}^j} f\|_p^p \leq C(p) \|f\|_p^p \tag{10.8}$$

**Proof.** Consider the sequence $\Delta_{\mathcal{F}} = \{\Delta_{\mathcal{F}^j}\}$ as a single operator from $L^p(\mathbb{R})$ to $L^p(l^p)$. Then (10.8) asserts the boundedness of this operator. Let $X$ be the subspace of $L^p(\mathbb{R})$ of functions supported on $[0,1]$ and having mean zero on [0,1]. Let $E_X$ be the $L^2$ projection into this space. $E_X$ is a constant Haar multiplier, $E_X f = \sum_{I \in \mathcal{D}} c_I \langle f, h_I \rangle h_I$ where $c_I = \chi_{\mathcal{D}([0,1])}(I) \in l^\infty$ hence, by the classical Littlewood-Paley theory $E_X$ is bounded on $L^p$, see Lemma 10.11. Now, it is easy to see that $\Delta_{\mathcal{F}} E_X = \Delta_{\mathcal{F}}$ so it suffices to show (10.8) for $f$ in $X$. Let $\Delta_{\mathcal{F}}^*$ be the dual mapping acting on sequences $L^q(l^q)$ for $\frac{1}{p} + \frac{1}{q} = 1$, it can be seen that,

$$\Delta_{\mathcal{F}}^*(\{f_j\}) = \sum_j \Delta_{\mathcal{F}_j} f_j.$$

Therefore.

$$\Delta_{\mathcal{F}}^* \Delta_{\mathcal{F}} = E_X,$$

Thus showing the reverse inequality to (10.8), for $f \in X$ and all $1 < p < \infty$, that is,

$$\|f\|_p^p \leq C(p) \sum_{j=1}^{\infty} \|\Delta_{\mathcal{F}^j} f\|_p^p, \tag{10.9}$$

implies that $\Delta_{\mathcal{F}}^*$ is bounded from $L^p(l^p)$ to $L^p(\mathbb{R})$ for all $1 < p < \infty$ which, by duality, implies the boundedness of $\Delta_{\mathcal{F}}$.

Now we assert that there exists $C(p) > 0$ and $0 < c_1 < 1$ so that for any $j, k$ and any $f \in L^p(\mathbf{R})$, one has the inequality

$$\int |\Delta_{\mathcal{F}^j} f|^{\frac{p}{2}} |\Delta_{\mathcal{F}^k} f|^{\frac{p}{2}} \leq C(p) c_1^{|k-j|} \|\Delta_{\mathcal{F}^j} f\|_p^{\frac{p}{2}} \|\Delta_{\mathcal{F}^k} f\|_p^{\frac{p}{2}}. \tag{10.10}$$

To show (10.10), it suffices to let $k > j$. We define $f_j = \Delta_{\mathcal{F}^j} f$ and $f_k = \Delta_{\mathcal{F}^k} f$. We observe that on any $J \in \mathcal{J}^j([0,1])$, the function $f_j$ is constant, and we denote by $f_j(J)$ its value there. We compute

$$\begin{aligned}
\int |f_j|^{p/2} |f_k|^{p/2} &= \int_{\bigcup \mathcal{J}^{k-1}([0,1])} |f_j|^{p/2} |f_k|^{p/2} \\
&\leq \left( \int_{\bigcup \mathcal{J}^{k-1}([0,1])} |f_j|^p \right)^{\frac{1}{2}} \left( \int_{\bigcup \mathcal{J}^{k-1}([0,1])} |f_k|^p \right)^{\frac{1}{2}} \\
&\leq \|f_k\|_p^{\frac{p}{2}} \left( \sum_{J \in \mathcal{J}^j([0,1])} \int_{\bigcup \mathcal{J}^{k-j-1}(J)} |f_j(J)|^p \right)^{\frac{1}{2}} \\
&\leq c^{\frac{k-j-1}{2}} \|f_k\|_p^{\frac{p}{2}} \left( \sum_{J \in \mathcal{J}^j([0,1])} |J| \, |f_j(J)|^p \right)^{\frac{1}{2}} \leq c^{\frac{k-j-1}{2}} \|f_k\|_p^{\frac{p}{2}} \|f_j\|_p^{\frac{p}{2}}.
\end{aligned}$$

As in (10.1), we iterated the fact that $\mathcal{J}$ is a decaying stopping time to conclude,

$$\sum_{I \in \mathcal{J}^{k-j-1}(J)} |I| \leq c^{k-j-1} |J|.$$

This implies (10.10). Fix $n > p$. Let $f \in X$. First, by Hölder's inequality and (10.10) for the same $C(p) > 0$ and $0 < c_1 < 1$, for $j_1 < j_2 < \ldots j_n$, one has

$$\begin{aligned}
\int |f_{j_1}|^{\frac{p}{n}} \ldots |f_{j_n}|^{\frac{p}{n}} &\leq \left( \int |f_{j_1}|^{\frac{p}{2}} |f_{j_2}|^{\frac{p}{2}} \right)^{\frac{1}{n}} \ldots \left( \int |f_{j_n}|^{\frac{p}{2}} |f_{j_1}|^{\frac{p}{2}} \right)^{\frac{1}{n}} \\
&\leq C(p) c_1^{\frac{j_2 - j_1}{n}} \ldots c_1^{\frac{j_n - j_{n-1}}{n}} c_1^{\frac{j_n - j_1}{n}} \|f_{j_1}\|^{\frac{p}{n}} \ldots \|f_{j_n}\|^{\frac{p}{n}} \\
&\leq C(p) c_1^{j_n - j_1} \|f_{j_1}\|^{\frac{p}{n}} \ldots \|f_{j_n}\|^{\frac{p}{n}}.
\end{aligned} \tag{10.11}$$

Furthermore $f = \sum_{j=1}^{\infty} f_j$. Thus by Jensen's inequality for $\frac{p}{n} < 1$

$$\int |f|^p \leq \sum_{j_1, j_2, \ldots, j_n} \int |f_{j_1}|^{\frac{p}{n}} \ldots |f_{j_n}|^{\frac{p}{n}}. \tag{10.12}$$

But (10.11) together with (10.12) and Lemma 10.7 for the choice $x_j = \|f_j\|^{\frac{p}{n}}$, $C(j_1, \ldots, c_{n-1}) = c^{j_1 + \ldots j_{n-1}}$, prove (10.9) and with it the lemma. ■

### 10.4.2 $L^p$ version of Cotlar's Lemma

Much effort goes into the search for good $L^p$ substitutes to Cotlar's lemma, see [3]. The proof we just did to prove the $L^p$ Plancherel Lemma was the source of inspiration to state and prove the following version of Cotlar's Lemma for $L^p$.

**Lemma 10.9** *Let $\mathcal{J}$ be a decaying stopping time, and $\Delta_{\mathcal{F}^j}$ be as above. Let $T$ be a linear operator on functions on $\mathbb{R}$ and write $T_j = T\Delta_{\mathcal{F}^j}$. Suppose $T = \sum_{j=1}^{\infty} T_j$. Let $1 < p < \infty$. Suppose there are $C > 0$ and $0 < c < 1$ so that for every $j, k$ (possibly equal), one has*

$$\int |T_j f|^{\frac{p}{2}} |T_k f|^{\frac{p}{2}} \leq C c^{|k-j|} \|f_j\|_p^{\frac{p}{2}} \|f_k\|_p^{\frac{p}{2}}. \tag{10.13}$$

*Then $T$ is bounded on $L^p(\mathbb{R})$ with constant depending only on $p$, $C$, $c$, and the rate of decay of $\mathcal{J}$.*

**Proof.** Let $n$ be an integer so that $2n > p$. We wish to estimate

$$\int |Tf|^p = \int (|Tf|^{\frac{p}{2n}})^{2n}.$$

Now we observe that by Jensen's inequality,

$$|Tf|^{\frac{p}{2n}} \leq \sum_j |T_j f|^{\frac{p}{2n}}.$$

Thus we observe that

$$\int |Tf|^p \leq C(p) \sum_{j_1 \leq j_2 \leq \cdots \leq j_{2n}} \int |Tf_{j_1}|^{\frac{p}{2n}} \ldots |Tf_{j_{2n}}|^{\frac{p}{2n}}. \tag{10.14}$$

Now applying Hölder's inequality and (10.13), as we did in (10.11), it is readily apparent that

$$\int |Tf_{j_1}|^{\frac{p}{2n}} \ldots |Tf_{j_{2n}}|^{\frac{p}{2n}} \leq A c^{\frac{j_{2n}-j_1}{r}} \|f_{j_1}\|_p^{\frac{p}{2n}} \ldots \|f_{j_{2n}}\|_p^{\frac{p}{2n}}. \tag{10.15}$$

The reader may easily verify that (10.14), (10.15) and Lemma 10.7 imply that there is a constant $C$ depending only on $A$, $c$, and $p$ so that

$$\int |Tf|^p \leq C \sum_k \|f_k\|_p^p = C \sum_k \|\Delta_{\mathcal{F}^k} f\|_p^p, \tag{10.16}$$

we apply now the Littlewood-Paley Lemma to prove the lemma. ∎

## 10.5 Boundedness of $T_\omega^t$

### *10.5.1 Boundedness of $T_\omega$*

We are interested in the boundedness of Haar multipliers. Let us discuss the simplest possible case. Which is of course well known, see [4], where the proof uses the classical Littlewood-Paley Theory. Our proof will use the following lemma, which also holds for $p \neq 2$, $1 < p < \infty$.

**Lemma 10.10** *Let $T$ be a linear or sublinear operator which is of strong type $(2,2)$. Suppose that for every dyadic interval $I$, the function $Th_I$ is supported only on $I$. Then $T$ is of weak type $(1,1)$.*

**Proof.** Suppose we have $f \in L^1(\mathbb{R})$. We pick $\lambda > 0$. We now apply the Calderón-Zygmund decomposition. We write $f = g + b$ with $||g||_\infty \leq 2\lambda$, $||g||_1 \leq 2||f||_1$, and $b$ supported on a disjoint sequence of dyadic intervals $\{I_j\}$, having mean zero on those intervals, and so that

$$\sum_j |I_j| \leq \frac{||f||_1}{\lambda}.$$

Now observe that

$$|\{x : |(Tf)(x)| \geq \lambda\}| \leq |\{x : |(Tg)(x)| \geq \frac{\lambda}{2}\}| + |\{x : |(Tb)(x)| \geq \frac{\lambda}{2}\}|. \tag{10.17}$$

For the first term on the right, we apply the fact that $T$ is of strong type (2,2) and hence of weak type (2,2) and that by Hölder's inequality

$$||g||_2 \leq ||g||_1^{\frac{1}{2}} ||g||_\infty^{\frac{1}{2}} \leq \sqrt{2\lambda ||f||_1}.$$

Thus,

$$|\{x : |(Tg)(x)| \geq \frac{\lambda}{2}\}| \leq \frac{4C||g||_2^2}{\lambda^2} \leq \frac{8C||f||_1}{\lambda}. \tag{10.18}$$

On the other hand $\langle b, h_J \rangle = 0$ for $J$ dyadic unless $J \subset I_j$ for some $j$. Thus by assumption,

$$|\{x : |(Tb)(x)| \geq \frac{\lambda}{2}\}| \leq |\bigcup_j I_j| \leq \frac{||f||_1}{\lambda}. \tag{10.19}$$

Now applying (10.18) and (10.19) to (10.17) shows that $T$ is of weak type (1,1) which was to be shown. ■

**Lemma 10.11** *The constant Haar multiplier $Tf = \sum_{I \in \mathcal{D}} c_I \langle f, h_I \rangle h_I$ is bounded in $L^p$ if and only if the sequence $\{c_I\}_{I \in \mathcal{D}}$ is bounded.*

**Proof.** The necessity follows immediately from applying $T$ to the Haar functions.

The sufficiency is a consequence of the previous lemma. Clearly, the multiplier $T$ satisfies the hypotheses of Lemma 10.10. Thus $T$ is of weak type (1,1). Thus by the Marcinkiewicz Interpolation Theorem $T$ is bounded on $L^p(\mathbf{R})$ for all $p$ with $1 < p \leq 2$. But $T$ is selfadjoint. Thus $T$ is bounded on $L^p(\mathbf{R})$ for all $1 < p < \infty$. ∎

We will now use Cotlar's Lemma in $L^p$ to prove the main theorem of this section. See [24] for a different argument.

**Theorem 10.12** *Let $\omega$ be a weight. Define the operator*

$$T_\omega f(x) = \sum_{I \in \mathcal{D}} \frac{\omega(x)}{m_I \omega} \langle f, h_I \rangle h_I(x) = w(x)(M_\omega f)(x),$$

*where $M_\omega$ is the (possibly unbounded) Haar multiplier with coefficients $\frac{1}{m_I \omega}$. Then $T_\omega$ is bounded on $L^p$ if and only if $\omega \in RH_p^d$.*

**Remark 10.1** *If $\frac{1}{\lambda} < \omega(x) < \lambda$ for a.e. $x$, then there is a trivial bound, far from being sharp. Just observe,*

$$\|T_\omega f\|_p^p = \int \omega^p |M_\omega f|^p \leq \lambda^p \|M_\omega f\|_p^p \leq C\lambda^{2p} \|f\|_p^p,$$

*where the last inequality uses the fact that $M_\omega$ is a constant Haar multiplier with symbol bounded by $\lambda$.*

**Proof.** The necessity of $\omega \in RH_p^d$ follows immediately from applying $T_w$ to the Haar functions.

For the proof of sufficiency, it suffices to prove the theorem for

$$Tf(x) = \sum_{I \in \mathcal{D}([0,1])} \frac{\omega(x)}{m_I \omega} \langle f, h_I \rangle h_I(x).$$

If $\omega \in RH_p^d$, by the Weight Lemma, the stopping time $\mathcal{J} = \mathcal{J}_\omega([0,1])$ is decaying. We will abuse our notation denoting by $\bigcup \mathcal{J}^n(I)$ the set $\bigcup_{J \in \mathcal{J}^n(I)} J$, and $\mathcal{F}^j = \mathcal{F}^j([0,1])$, similarly for $\mathcal{J}^j$.

We define

$$T_j = T\Delta_{\mathcal{F}^j} = \omega M_j,$$

where

$$M_j f(x) = \sum_{J \in \mathcal{F}^j} \frac{1}{m_J \omega} \langle f, h_J \rangle h_J(x).$$

We define for every dyadic $I$, the multiplier

$$M_I f(x) = \sum_{J \in \mathcal{F}_\omega(I)} \frac{1}{m_J \omega} \langle f, h_J \rangle h_J(x).$$

Then

$$M_j = \sum_{I \in \mathcal{J}^{j-1}} M_I.$$

Each $M_I$ is a bounded constant Haar multiplier since by the definition of the stopping time, $(m_J\omega)^{-1} \leq \lambda(m_I\omega)^{-1}$ for all $I \in \mathcal{D}$ and for all $J \in \mathcal{F}_\omega(I)$. Therefore, for any $f \in L^p$ and for each $I$, there is $C(p) > 0$ depending only on $p$ such that

$$\|M_I f\|_p \leq C(p)\frac{\lambda}{m_I\omega}\|f\|_p.$$

Here $\lambda$ is fixed, depending just on $p$, so that $\mathcal{J}$ is decaying. In fact, defining

$$\Delta_I f = \sum_{K \in \mathcal{F}_\omega(I)} \langle f, h_K\rangle h_K,$$

we have $M_I \Delta_I f = M_I f$ hence,

$$\|M_I f\|_p \leq C(p)\frac{\lambda}{m_I\omega}\|\Delta_I f\|_p. \tag{10.20}$$

Also notice that $f_j = \Delta_{\mathcal{F}^j} f = \sum_{I \in \mathcal{J}^{j-1}([0,1])} \Delta_I f$.

We shall prove that the $T_j$'s satisfy the condition (10.13) in Cotlar's Lemma. We will begin by proving that each is bounded on $L^p$. We write

$$\int |T_j f|^p = \int_{\bigcup \mathcal{J}^{j-1}([0,1]) \setminus \bigcup \mathcal{J}^j([0,1])} |T_j f|^p + \int_{\bigcup \mathcal{J}^j([0,1])} |T_j f|^p, \tag{10.21}$$

and we will estimate each term separately.

Observe that for every $I$, on $I \setminus \bigcup \mathcal{J}(I)$ one has almost everywhere that $\omega \leq \lambda m_I\omega$. Thus by the remark right before this proof and (10.20) we conclude that,

$$\begin{aligned}
\int_{\bigcup \mathcal{J}^{j-1}([0,1]) \setminus \bigcup \mathcal{J}^j([0,1])} |T_j f|^p &= \sum_{J \in \mathcal{J}^{j-1}([0,1])} \int_{J \setminus \bigcup \mathcal{J}(J)} |T_j f|^p \\
&\leq \sum_{J \in \mathcal{J}^{j-1}([0,1])} (\lambda m_J\omega)^p \int_{J \setminus \bigcup \mathcal{J}(J)} |M_J f|^p \\
&\leq \sum_{J \in \mathcal{J}^{j-1}([0,1])} C(p)\lambda^{2p}\|\Delta_J f\|_p^p \;\leq\; C(p)\lambda^{2p}\|f_j\|_p^p.
\end{aligned}$$

Here the last line comes from the disjointness of the elements of $\mathcal{J}^{j-1}$. We have completed our estimate of the first term of (10.21).

Observe that for every $I$, we have that $M_I f$ is constant on all $J \in \mathcal{J}(I)$. We denote its value on $J$ by $(M_I f)(J)$. Also observe that for such $J$, we have by the definition of $\mathcal{J}$ that $m_J\omega \leq 2\lambda m_I\omega$. Now we estimate using the $RH_p^d$ condition, and (10.20),

$$\begin{aligned}
\int_{\bigcup \mathcal{J}^j([0,1])} |T_j f|^p &= \sum_{J\in\mathcal{J}^{j-1}([0,1])} \int_{\bigcup \mathcal{J}(J)} |T_j f|^p \\
&= \sum_{J\in\mathcal{J}^{j-1}([0,1])} \sum_{K\in\mathcal{J}(J)} (M_J f(K))^p \int_K \omega^p \\
&\leq C \sum_{J\in\mathcal{J}^{j-1}([0,1])} \sum_{K\in\mathcal{J}(J)} (M_J f(K))^p |K| m_K^p\omega \\
&\leq C \sum_{J\in\mathcal{J}^{j-1}([0,1])} 2^p\lambda^p m_J^p\omega \int_J |M_J f|^p \leq C2^p\lambda^{2p}\|f_j\|_p^p.
\end{aligned}$$

Thus we have shown that there exists $C(p)$ so that

$$\int |T_j f|^p \leq C(p)\|f_j\|^p.$$

We claim further that there exists $0 < c < 1$ so that for any $k > j$, one has that

$$\int_{\bigcup \mathcal{J}^{k-1}([0,1])} |T_j f|^p \leq C(p)c^{k-j}\|f_j\|_p^p. \tag{10.22}$$

We will use Hölder's inequality, the decaying of $\mathcal{J}$ and the fact that $\omega \in RH_{p+\epsilon}^d$ to compute

$$\begin{aligned}
\int_{\bigcup \mathcal{J}^{k-1}([0,1])} |T_j f|^p &= \sum_{J\in\mathcal{J}^j([0,1])} \int_{\bigcup \mathcal{J}^{k-j-1}(J)} |T_j f|^p \\
&= \sum_{J\in\mathcal{J}^j([0,1])} |(M_j f)(J)|^p \int_{\bigcup \mathcal{J}^{k-j-1}(J)} \omega^p \\
&\leq \sum_{J\in\mathcal{J}^j([0,1])} c^{\frac{\epsilon(k-j-1)}{p+\epsilon}} |(M_j f)(J)|^p \left( \int_{\bigcup \mathcal{J}^{k-j-1}(J)} \omega^{p+\epsilon} \right)^{\frac{p}{p+\epsilon}} |J|^{\frac{\epsilon}{p+\epsilon}} \\
&\leq \sum_{J\in\mathcal{J}^j([0,1])} c^{\frac{\epsilon(k-j-1)}{p+\epsilon}} |(M_j f)(J)|^p |J| \left( \frac{1}{|J|} \int_J \omega^{p+\epsilon} \right)^{\frac{p}{p+\epsilon}}
\end{aligned}$$

$$\leq C \sum_{J \in \mathcal{J}^j([0,1])} c^{\frac{\epsilon(k-j-1)}{p+\epsilon}} |J| \left( \frac{1}{|J|} \int_J w \, (M_j f)(J) \right)^p$$
$$\leq \sum_{J \in \mathcal{J}^j([0,1])} C c^{\frac{\epsilon(k-j-1)}{p+\epsilon}} \int_J \omega^p |M_j f|^p$$
$$= C c^{\frac{\epsilon(k-j-1)}{p+\epsilon}} \int |T_j f|^p \leq C(p) c^{\frac{\epsilon(k-j-1)}{p+\epsilon}} \|f_j\|_p^p.$$

This proves (10.22). But (10.22) implies (10.13) because $T_k f$ is supported on $\bigcup \mathcal{J}^{k-1}([0,1])$. Thus the theorem is proven. ■

**Remark 10.2** *Notice that $T_\omega h_J = \frac{\omega(x)}{m_J \omega} h_J(x)$ is supported on $J$, and the previous theorem for $p = 2$ will give the strong type (2,2) for the operator (for this result ordinary Cotlar's Lemma is all that is needed, not the $L^p$ version), hence by Lemma 10.10, $T_\omega$ is of weak type (1,1). Thus by interpolation it is of strong type $(p,p)$ for $1 < p < 2$. It is not true that the adjoint $T_\omega^*$ is localized when acting on Haar functions, hence we cannot repeat the argument and then use duality to get $2 < p < \infty$.*

### *10.5.2 Some corollaries*

We can consider the following one parameter family of Haar multipliers:

$$T_\omega^t f(x) = \sum_{I \in \mathcal{D}} \left( \frac{\omega(x)}{m_I \omega} \right)^t \langle f, h_I \rangle h_I(x) = \omega^t M_\omega^t f(x),$$

where $t \in \mathbf{R}$, and $M_\omega^t$ is the constant Haar multiplier with symbol $(m_I \omega)^{-t}$.

Checking the action on the Haar functions one obtains a necessary condition for the boundedness of $T_\omega^t$ in $L^p(\mathbf{R})$, namely *Condition $C_{tp}$*:

$$m_I \omega^{tp} \leq (m_I \omega)^{tp}, \qquad \forall I \in \mathcal{D}.$$

Clearly condition $C_s$ coincides with $RH_s^d$ when $s > 1$, with $A_{1-1/s}^d$ when $s < 0$, and it always holds when $0 \leq s \leq 1$.

The following lemma will allow us to deduce the boundedness of $T_\omega^t$ as a simple corollary of Theorem 21.1.

**Lemma 10.13** *If $\omega \in C_{tp} \cap A_\infty^d$ then (i) $\omega^t \in RH_p^d$ and (ii) $m_I \omega^t \leq C(m_I \omega)^t$ for all $I \in \mathcal{D}$.*

**Proof.** The proof of (i) follows from Lemma 10.1. Enough to show that $\omega^{tp} \in A_\infty^d$ since that would imply that $\omega^t \in RH_p^d$.

The cases $tp = 0$ and $tp = 1$ are trivial. For $0 < tp < 1$ then $\omega \in A_\infty^d$ implies that $\omega^{tp} \in RH_{1/tp}^d \subset A_\infty^d$. For $1 < tp$ then $\omega \in RH_{tp}^d$ if and only if $\omega^{tp} \in A_\infty^d$. For $tp < 0$ then $\omega \in A_{1-1/tp}^d$ if and only if $\omega^{tp} \in A_{1-tp}^d$ but Lemma 10.2 implies that $\omega^{tp} \in RH_q^d$ for some $q > 1$ therefore $\omega^{tp} \in A_\infty^d$.

The proof of (ii) is a simple application of Hölder's inequality for $1 < p < \infty$ and the use of the $C_{tp}$ condition:

$$(m_I \omega^t)^p \leq m_I(\omega^{tp}) \leq C(m_I \omega)^{tp}.$$

■

**Corollary 10.14** *Assume $\omega \in A_\infty^d$, $t \in \mathbf{R}$, and $1 < p < \infty$. Then $T_\omega^t$ is bounded in $L^p(\mathbf{R})$ if and only if $\omega \in C_{tp}$.*

**Proof.** We already pointed out the necessity of the $C_{tp}$ condition.

As for the sufficiency, notice that we can factorize $T_\omega^t$ as a composition of two bounded operators. In fact:

$$T_\omega^t = T_{\omega^t} S_{\omega,t},$$

where $S_{\omega,t}$ is the constant Haar multiplier with symbol $c_I = \frac{m_I \omega^t}{(m_I \omega)^t}$ which is a bounded sequence by Lemma 10.13, therefore $S_{\omega,t}$ is a bounded operator in $L^p$. Since $\omega \in A_\infty^d \cap C_{tp}$ also by Lemma 10.13, $\omega^t \in RH_p^d$, therefore by the Theorem 11.3, $T_{\omega^t}$ is bounded in $L^p$. ■

**Corollary 10.15** *Assume $\omega \in A_\infty^d \cap C_{tp}$ for some $1 < p < A_\infty$ then $T_\omega^t$ is of weak type $(1,1)$, for all $t \in \mathbf{R}$.*

**Proof.** By Corollary 2 $T_\omega^t$ is of strong type $(p,p)$, and it clearly satisfies the hypothesis of Lemma 9. Therefore it is of weak type $(1,1)$. ■

## 10.6 Haar multipliers and weighted inequalities

As an example on how to prove weighted inequalities with the aid of the Haar multipliers $T_\omega^t$, we show that $A_p^d$ is a sufficient condition for the boundedness in $L^p(\omega)$ of constant Haar multipliers, the dyadic square function $Sf$, and dyadic paraproducts.

Recall that, for $\omega$ some weight, and for $t$ any real number, we denote by $M_\omega^t$, the multiplier given by

$$M_\omega^t f = \sum_{I \in \mathcal{D}} \left(\frac{1}{m_I \omega}\right)^t \langle f, h_I \rangle h_I.$$

**Corollary 10.16** *Let $\omega \in A_p^d$. Then for any $1 < p < \infty$, the operators $\omega^{\frac{1}{p}} M_\omega^{\frac{1}{p}}$ and $M_\omega^{-\frac{1}{p}} \omega^{-\frac{1}{p}}$ are bounded on $L^p(\mathbf{R})$.*

**Proof.** Since, by Lemma 10.2, $A_p^d \subset A_\infty^d$, setting $t = 1/p$, $t = -1/p$ in Corollary 10.14 we get, respectively that $T_\omega^{\frac{1}{p}} = \omega^{\frac{1}{p}} M_\omega^{\frac{1}{p}}$ is bounded in $L^p$ and $\omega^{-\frac{1}{p}} M_\omega^{-\frac{1}{p}}$ is bounded on $L^{\frac{p}{p-1}}(\mathbf{R})$. Since $\frac{p}{p-1}$ is the dual index of $p$ and $M_\omega^{-\frac{1}{p}} \omega^{-\frac{1}{p}}$ is the dual of $\omega^{-\frac{1}{p}} M_\omega^{-\frac{1}{p}}$, we have shown that $\omega^{-\frac{1}{p}} M_\omega^{-\frac{1}{p}}$ is bounded in $L^p$. ∎

This allows us to interchange the weight $\omega$ and the multiplier $M_\omega^{-1}$ when proving weighted norm inequalities. We will give some easy applications.

**Corollary 10.17** *Let $\{\alpha_I\}_{I\in\mathcal{D}}$ be a bounded set of numbers and $T_\alpha$ be the associated Haar multiplier, i.e*

$$T_\alpha f = \sum_{I\in\mathcal{D}} \alpha_I \langle f, h_I\rangle h_I.$$

*Let $1 < p < \infty$ be given and let $\omega \in A_p^d$. Then $T_\alpha$ is bounded from $L^p(\omega)$ to $L^p(\omega)$.*

**Proof.** It suffices to show that $\omega^{\frac{1}{p}} T_\alpha \omega^{-\frac{1}{p}}$ is bounded on $L^p(\mathbf{R})$. Since Haar multipliers commute,

$$T_\alpha = M_\omega^{-\frac{1}{p}} T_\alpha M_\omega^{\frac{1}{p}},$$

and is bounded on $L^p(\mathbf{R})$. Then

$$\omega^{\frac{1}{p}} T_\alpha \omega^{-\frac{1}{p}} = (\omega^{\frac{1}{p}} M_\omega^{\frac{1}{p}})(M_\omega^{-\frac{1}{p}} T_\alpha M_\omega^{\frac{1}{p}})(M_\omega^{-\frac{1}{p}} \omega^{-\frac{1}{p}}).$$

Since all three factor of the right hand side are bounded, then so is the left hand side. ∎

**Corollary 10.18** *Let $\omega \in A_p^d$ and $b$ in dyadic BMO. Then $\pi_b$, the dyadic paraproduct defined by*

$$\pi_b f = \sum_{I\in\mathcal{D}} b_I m_I f\, h_I,$$

*is bounded on $L^p(\omega)$.*

**Proof.** We will prove the result for $p = 2$ and then an extrapolation argument proves it for $p \neq 2$ , see [GC-RF].

It suffices to bound $\omega^{\frac{1}{2}} \pi_b \omega^{-\frac{1}{2}}$ on $L^2$. We may write

$$\omega^{\frac{1}{2}} \pi_b \omega^{-\frac{1}{2}} = (\omega^{\frac{1}{2}} M_\omega^{\frac{1}{2}})(M_\omega^{-\frac{1}{2}} \pi_b \omega^{-\frac{1}{2}}).$$

Thus it suffices to bound $M_\omega^{-\frac{1}{2}} \pi_b \omega^{-\frac{1}{2}}$ on $L^2$. However for any $f \in L^2$,

$$\|M_\omega^{-\frac{1}{2}} \pi_b \omega^{-\frac{1}{2}} f\|_2^2 = \sum_{I\in\mathcal{D}} (m_I\omega) b_I^2 (m_I(\omega^{-\frac{1}{2}} f))^2.$$

We use the fact that $\omega$, $\omega^{-1} \in A_2^d$, which implies that $\omega^{-1} \in RH^d_{\frac{2+\epsilon}{2}}$ for some $\epsilon > 0$, Carleson's Lemma 10.6, boundedness of the maximal function on $L^{\frac{2+2\epsilon}{2+\epsilon}}$, and Hölder's inequality to estimate

$$\begin{aligned}
\|M_\omega^{-1/2}\pi_b\omega^{-1/2}f\|_2^2 &\leq \sum_I (m_I\omega)b_I^2(m_I\omega^{-2+\epsilon/2})^{\frac{2}{2+\epsilon}}(m_I f^{2+\frac{\epsilon}{1+\epsilon}})^{\frac{2+2\epsilon}{2+\epsilon}} \\
&\leq C\sum_I (m_I\omega)b_I^2(m_I(\omega^{-1}))(m_I f^{\frac{2+\epsilon}{1+\epsilon}})^{\frac{2+2\epsilon}{2+\epsilon}} \leq C\sum_I b_I^2(m_I f^{\frac{2+\epsilon}{1+\epsilon}})^{\frac{2+2\epsilon}{2+\epsilon}} \\
&\leq C\int\left(\sup_{I\ni x} m_I f^{\frac{2+\epsilon}{1+\epsilon}}\right)^{\frac{2+2\epsilon}{2+\epsilon}} dx \leq C\|f^{\frac{2+\epsilon}{1+\epsilon}}\|_{\frac{2+2\epsilon}{2+\epsilon}}^{\frac{2+2\epsilon}{2+\epsilon}} = C\|f\|_2^2.
\end{aligned}$$

Which was to be shown. ■

We can also prove weighted inequalities for dyadic square functions. For an alternative proof see [1]. In particular the classical dyadic Littlewood-Paley Theory can be deduced setting $\omega = 1$. The next result can be viewed as weighted Littlewood-Paley Theory.

**Corollary 10.19** *Let $\omega \in A_p^d$ then the dyadic square function*

$$Sf(x) = \Big(\sum_j |\Delta_j f|^2\Big)^{\frac{1}{2}}, \quad \Delta_j f = \sum_{I\in\mathcal{D}_j} \langle f, h_I\rangle h_I,$$

*is bounded in $L^p(\omega)$.*

**Proof.** We prove it for $p = 2$ and use extrapolation for $p \neq 2$ as Steve Buckley did in his proof of the same fact, see [1].

It suffices to show that $\omega^{\frac{1}{2}}S\omega^{-\frac{1}{2}}$ is bounded on $L^2$.

Computing we get

$$\begin{aligned}
\|\omega^{\frac{1}{2}}S\omega^{-\frac{1}{2}}f\|_2^2 &= \sum_j \int \omega(x)|\Delta_j(f\omega^{-\frac{1}{2}})|^2 \\
&= \sum_{I\in\mathcal{D}} |\langle f\omega^{-\frac{1}{2}}, h_I\rangle|^2 m_I\omega = \|M_\omega^{-\frac{1}{2}}w^{-\frac{1}{2}}f\|_2^2 \leq C\|f\|_2^2.
\end{aligned}$$

The last inequality by Corollary 10.16. ■

For operators like the Hilbert transform it is more complicated than these, see for example [TV1], [18]. But the same ideas are used. The following tautology holds: *an operator $T$ is bounded from $L^p(u)$ into $L^p(v)$ if and only if the operator $v^{\frac{1}{p}}Tu^{-\frac{1}{p}}$ is bounded from $L^p(\mathbf{R})$ into itself.* As we mentioned before, and we hope has been highlighted by the examples, the Haar multipliers will allow us to replace multiplication on space side by multiplication on frequency side, whenever it is convenient. For example to show the boundedness of the Hilbert Transform from $L^2(\omega)$ into itself it is enough to check that $S = M_{\omega}^{\frac{1}{2}} H M_{\omega}^{\frac{-1}{2}}$ is bounded in $L^2(\mathbf{R})$. In this case the estimates are more laborious to obtain than in the cases presented here. The strategy followed in [TV1], [KP] follows a method introduced by Coifman and Semmes, widely used in Wavelet Theory. One studies the decay of the matrix $t_{IJ} = \langle Sh_I, h_J \rangle$ using, for example, Schur's Lemma [6]. Some pieces will be analized like if they were constant Haar multipliers, others as if they were paraproducts, very much in the spirit of the $T(1)$ Theorem of Journé and David [10].

*References*


[1] S. Buckley, *Summation conditions on weights.* Michigan Math. J. 40, 153–170 (1993).

[2] J. Bony, *Calcul symbolique et propagation des singularités pour les equations aux dérivées non-linéaires.* Ann. Sci. Ecole Norm. Sup. 14, 209–246 (1981).

[3] A. Carbery, *A version of Cotlar's lemma for $L^p$ spaces and some applications.* Contemp. Math. 189, 117–134 (1995).

[4] M. Christ, *Lectures on singular integral operators.* Regional Conferences Series in Math. AMS. 77, (1990).

[5] R. Coifman, C. Fefferman, *Weighted norm inequalities for maximal functions and singular integrals.* Studia Math. 51, 241–250 (1974).

[6] R. Coifman, P. W. Jones, S. Semmes, *Two elementary proofs of the $L^2$ boundedness of Cauchy integrals on Lipschitz curves.* J. of the AMS 2, 553–564 (1989).

[7] M. Cotlar, C. Sadosky, *On the Helson-Szegö theorem and a related class of modified Toeplitz kernels.* Proc. Symp. Pure Math. AMS. 35 (1979).

[8] M. Cotlar, C. Sadosky, *On some $L^p$ versions of the Helson-Szegö theorem.* In "Conference on harmonic analysis in honor of Antoni Zygmund." Eds. Beckner et al., Wadsworth (1983).

[9] D. Cruz-Uribe (SFO), C. J. Neugebauer, *The structure of the reverse Hölder classes.* Trans. of the AMS. 347 #8, 2941–2960 (1995).

[10] G. David, J.-L. Journé, *A boundedness criteria for generalized Calderón-Zygmund operators.* Ann. of Math. 20, 371–397 (1984).

[11] F. W. Gehring, *The $L^p$-integrability of the partial derivatives of a quasiconformal mapping.* Acta Math. 130, 265-277 (1973).

[12] A. Haar, *Zur theorie der orthogonalen funktionen systems.* Math. Ann. 69, 331–371 (1910).

[13] R. Hunt, B. Muckenhoupt, R. Wheeden, *Weighted norm inequalities for the conjugate function and the Hilbert transform.* Trans. AMS 176, 227–252 (1973).

[14] H. Helson, G. Szegö, *A problem in prediction theory.* Am. Math. Pura Appl. 51, 107–138 (1960).

[15] N. Katz, *Matrix valued Paraproducts.* To appear in the Proceedings of El Escorial Conference, June 1996.

[16] N. Katz, *Remarks on maximal operators over arbitrary sets of directions.* To appear J. Lond. Math. Soc.

[17] N. Katz, Lecture notes. Preprint 1996.

[18] N. Katz, M. C. Pereyra, *On the two weights problem for the Hilbert transform.* Rev. Mat. Iberoamericana 13 # 1, 211-243 (1997).

[19] Y. Meyer, *Ondelettes et Opérateurs II.* Herman (1990).

[20] N. K. Nikolskii, *Treatise on the shift operator.* Springer-Verlag (1985).

[21] F. Nazarov, S. Treil, *The Hunt for Bellman function: applications to estimate of singular integral operators and to other classical problems in harmonic analysis.* Preprint 1996.

[22] F. Nazarov, S. Treil, A. Volberg *Counterexamples to infinite dimensional Carleson embedding theorem.* Preprint 1996.

[23] F. Nazarov, S. Treil, A. Volberg *The solution of the problem on Haar multipliers in two-weighted $L^2$ spaces.* Preprint 1996.

[24] M. C. Pereyra *On the resolvent of dyadic paraproducts.* Rev. Mat. Iberoamericana 10 # 3, 627–664 (1994).

[25] S. Treil, A. Volberg, *Wavelets and the angle between past and future.* J. Funct. Anal. 143, no 2, 269-308 (1997).

[26] A. Volberg, *Matrix $A_p$ weights via $S$-functions.* Preprint 1996, 1-24.

# Chapter 11

# Multipliers and square functions for $H^p$ spaces over Vilenkin groups

**James E. Daly**
**Keith L. Phillips**

*ABSTRACT. This work is a short history of the Hardy Spaces $H^p$ for Vilenkin Groups, beginning with the square function characterizations of the Lebesgue Spaces $L^p$ by Sunouchi in the dyadic case in 1951. For the dyadic group $G$ it is proved that the function with values $\varphi(k) = k2^{-j}$ on the $j^{th}$ dyadic block of $G$ is an $H^p$ – multiplier with respect to the Walsh functions $\{\omega_k : 0 \leq k \leq \infty\}$. This theorem implies that Sunouchi's square function $S_2$ characterizes $H^p$, solving a conjecture of Simon for $H^1$ made in 1985. The inverse of $\varphi$ is also an $H^p$ – multiplier.*

## 11.1 Introduction

Hardy spaces have been studied for certain totally disconnected topological groups and martingales since the introduction of real variable methods for classical Hardy spaces in the 1970s. The paper [2] by John Chao treats conjugate functions and $H^p$ spaces for local fields and is probably the first paper explicitly on $H^p$ theory for totally disconnected spaces.

In this paper the setting will be a locally compact Vilenkin group, say $G$, of bounded order. Thus $G$ contains a decreasing sequence of compact open subgroups $(G_n)_{n=-\infty}^{\infty}$ such that

$$i)\ \cup_{-\infty}^{\infty} G_n = G \text{ and } \cap_{-\infty}^{\infty} G_n = \{0\}$$
$$ii)\ \sup_n \{order(G_n/G_{n+1})\} < \infty.$$

In the case that $G$ is compact, we use the convention that $G_n = G$ if $n \leq 0$. The additive group of a local field is a Vilenkin group, as is its ring of integers. In particular, the the *p-adic* numbers are a Vilenkin group. In the case that $p = 2$, the ring of integers is also called the dyadic group.

Let $\Gamma$ denote the dual group of $G$ and $\Gamma_n = \{\gamma \in \Gamma : \gamma(x) = 1$ for all $x \in G_n\}$. The Haar measures $\mu$ on $G$ and $\lambda$ on $\Gamma$ are chosen so that $\mu(G_0)$

$= \lambda(\Gamma_0) = 1$ and consequently, $\mu(G_n) = (\lambda(\Gamma_n))^{-1} := (m_n)^{-1}$ for each $n \in \mathbb{Z}$. There is a norm on $G$ defined by $|x| = (m_n)^{-1}$ if $x \in G_n \backslash G_{n+1}$. The Fourier transform and inverse Fourier transform respectively are denoted by $\wedge$ and $\vee$ , and satisfy

$$(\xi_{G_n})^\wedge = (\lambda(\Gamma_n))^{-1} \xi_{\Gamma_n}$$

where $\xi_A$ denotes the characteristic function of a set $A$. Consequently,

$$(\xi_{\Gamma_n})^\vee = (\lambda(G_n))^{-1} \xi_{G_n}.$$

## 11.2 Historical Comments

In this section we discuss the history of $H^p$ theory on locally compact Vilenkin groups and closely related martingales. In the third section we give two important new examples of multipliers for $H^p$ for $p \leq 1$.

For $p > 1$, $H^p = L^p$. There are several ways to define and/or characterize the $H^p$ spaces for $p \leq 1$. The buzz words are atomic, square functions, maximal functions, conjugate functions, multipliers, singular integrals. In [2] , Chao uses conjugate systems. The fundamental paper of Coifman and Wiess [9] in 1977 treats the theory in a broad setting including spaces of homogeneous type, in particular local fields. This treatment uses the atomic theory as well as maximal functions. Taibleson's book [28] has a chapter on conjugate systems over local fields, but does not treat $H^p$ per se. This chapter, Chao-Taibleson [8] , and [9] form the foundation for [2] . Taibleson has a survey in [29] .

There are three Paley-Littlewood type square functions, introduced by Sunuchi in the dyadic case in the 1951 paper [27] . Let $\{a_k\}_{k=0}^{\infty}$ be a sequence of complex numbers, $\{\omega_k\}_{k=0}^{\infty}$ the Walsh functions on $[0,1)$, and define

$$\begin{aligned} W_n(x) &= \sum_{k=0}^{n-1} a_k \omega_k(x), \\ D_m(x) &= W_{2^m}(x) - W_{2^{m-1}}(x) = \sum_{k=2^{m-1}}^{2^m-1} a_k \omega_k(x), \\ \sigma_n(x) &= \sum_{k=0}^{n-1} (1 - \frac{k}{n}) a_k \omega_k(x). \end{aligned}$$

The three square functions are

$$S_1^2 = \sum_{n=1}^{\infty} \frac{(W_n - \sigma_n)^2}{n}, \quad S_2^2 = \sum_{m=1}^{\infty} (W_{2^m} - \sigma_{2^m})^2, \quad S_3^2 = \sum_{m=1}^{\infty} D_m^2.$$

Sunuchi proves that they all characterize $L^p$ for $p > 1$. In 1985 Simon in [25] studies $H^1$-continuity of square functions in the dyadic case and conjectures that the square function $S_2$ characterizes $H^1$. This conjecture is proved in [11]. In the fourth section, we extend the result by showing that $S_2$ characterizes $H^p$ for $0 < p \leq 1$. Simon proves that $S_1$ is not even continuous on $H^1$, that $S_2$ is continuous on $H^1$, and that $S_3$ characterizes $H^1$. The book [24] (1990) by Schipp, Wade, and Simon contains a detailed chapter on the case in which $G$ is the dyadic group. In [15] (1993) Gat and in [26](1985) Simon study $H^1$ for Vilenkin groups, in particular proving that the atomic and maximal function definitions do not agree in the case that the orders of the factor groups $G_n/G_{n+1}$ are unbounded.

The essential equivalence of the martingale theory and the local field theory was apparent from the start. The first paper explicitely treating martingales seems to be Janson [16] in 1977, in which he characterizes $H^1$ by sets of matrices analogous to conjugate systems. The dyadic case is not covered by this work but is treated later by Chao in [3]. The individuality of the dyadic, or characteristic 2, case is a persistent phenomenon. Chao published a survey for the martingale case in 1982 [4]. Other works on the martingale case are Chao and Janson [5] (1981), Chao and Long [6], [7] (1992); and a book by Long [20] (1993). This book contains proofs of the equivalence of atomic, maximal function, and square function definitions. The Hungarian mathematician F. Weisz makes a detailed study of conjugate systems in [32] (1992), then in 1994 published the book [33] on martingale Hardy spaces. As in [16], sets of matrices characterize martingale $H^1$. In some cases the results of Weisz give characterizations in terms of diagonal matrices, which are martingale multipliers. As far as we know there are no results comparing these multiplier results to multipliers on Vilenkin groups.

We turn now to multiplier transforms for $H^p$ spaces on Vilenkin groups. They are, of course, preceded by $L^p$ results. The first paper we know of which is explicitly for $H^p$ is Janson [17], in which he proves that certain finite sets of coset constant multipliers characterize $H^1$ over a local field. Janson's martingale paper [16] gives results for homogenous multipliers, but phrased in terms of martingales. In 1983 Daly-Phillips [10] give conditions on a homogeneous multiplier on a local field implying continuity on $H^1$, as well as explicit necessary conditions. Our hypothesis for homogeneous multipliers in [10] is the very same as for any multiplier in [12] for Vilenkin groups. Kitada [18] (1987) studies general $H^p$ multipliers over Vilenkin groups for which the orders of the factor groups are bounded. He obtains results for $0 < p < \infty$. For $0 < p < 1$, Onneweer and Quek [23] (1989) improve [18] and have the best condition that a function $\varphi$ on $\Gamma$ be an $H^p$ multiplier. We will use this condition in our third section. For $p = 1$, our condition in [12] improves the Onneweer-Quek-Kitada results for Vilenkin groups. We state these two theorems. For $\psi \in L^\infty(\Gamma)$ and $j \in \mathbb{Z}$, set $\psi_j = \psi\xi_{\Gamma_j}$ and $\Delta_j = \psi_{j+1} - \psi_j$. The multiplier operator $T_\psi$ is defined by

$[T_\psi(f)]^\wedge = \psi f^\wedge$. It operates on various function spaces. The theorems are as follows.

**Theorem 11.1** (Kitada-Onneweer-QuekTheorem.) *If* $\psi \in L^\infty(\Gamma)$ *and*

$$\sum_{j=-\infty}^{\infty} \|(\Delta_j(\psi))^\vee\|_1 < \infty$$

*then* $T_\psi$ *is continuous from* $H^1 \to H^1$.

**Theorem 11.2** Daly-PhillipsTheorem *If* $\psi \in L^\infty(\Gamma)$ *and*

$$\sup_N\Big( \sum_{j=N+1}^{\infty} \int_{(G_N)^c} |(\Delta_j(\psi))^\vee|\, dx\Big) < \infty$$

*then* $T_\psi$ *is continuous from* $H^1 \to H^1$.

Related results on multipliers for certain spaces (not $H^p$ ) over totally disconnected groups appear in Krotov [19] , Wade [30] , Watari [31] , and Young [34] . We have recently received a preprint indicating that Onneweer has new results on multipliers for $L^1(G)$, $G$ a Vilenkin group.

## 11.3 Multipliers for H$^p$ (0 <p <1)

The (atomic) Hardy spaces on $G$ are given as follows. See Kitada [18] . A function $a : G \to C$ is a $p$-atom, $0 < p \leq 1$, if

$$i)\ \text{support of } a \subset I_n := x + G_n$$

$$ii)\ \|a\|_\infty \leq (\mu(I_n))^{-1/p}$$

$$iii)\ \int_G a(x)dx = 0.$$

A distribution $f \in S'(G)$ belongs to $H^p\,(G)$ if $f$ is given by $f = \sum_{i=1}^\infty \lambda_i a_i$, where each $a_i$ is a $p$-atom, $\sum_{i=1}^\infty |\lambda_i|^p < \infty$, and convergence is in $S'(G)$. We set

$$\|f\|_{H^p} = \inf\left(\sum_{i=1}^\infty |\lambda_i|^p\right)^{1/p}$$

with the infimum taken over all such atomic decompositions of $f$. A function $\phi \in L^\infty\,(\Gamma)$ is a (Fourier) multiplier on $H^p$ if there exists a constant $C > 0$ so that for all $f \in H^p \cap L^2$,

$$\left\|(\phi f^\wedge)^\vee\right\|_{H^p} \leq C\,\|f\|_{H^p}\,.$$

For $p > 1$, the equality $H^p = L^p$ holds. For $p = 1$, the inclusion $H^1 \subset L^1$ holds. Thus the theory breaks naturally into the three cases $0 < p < 1$, $p = 1, 1 < p < \infty$.

Our example is for the dyadic group $G = \prod_{k=0}^{\infty}\{0,1\}_{(k)}$. The group $G$ is mapped to the interval $[0,1)$ by $x \to \sum_{j=1}^{\infty} x_j 2^{-j}$. The subgroups $G_n$ correspond to the dyadic intervals $[0, \frac{1}{2^n})$. The character group $\Gamma$ viewed as functions on [0,1) is the set of Walsh functions $\{\omega_k : 0 \le k\}$. Writing

$$k = \sum_{j=0}^{\infty} k_j 2^j$$

the $k^{th}$ Walsh function is

$$\omega_k(x) = \prod_{j=0}^{\infty} e^{i\pi x_{j+1} k_j} .$$

See [10] or [24] for details. The Onneweer condition [22] that $\psi$ be an $H^p$ multiplier is that

$$\left[\sum_{n=0}^{\infty} \left\| 2^{1-\frac{1}{p}} \sum_{k\in B_j} \psi(k)\omega_k\gamma_n \right\|_1^p \right]^{\frac{1}{p}}$$

is bounded in $j$. The function $\gamma_n$ is the characteristic function of $I_n = [\frac{1}{2^{n+1}}, \frac{1}{2^n})$ and the set $B_j$ is the block

$$B_j = \{k : 2^{j-1} \le k < 2^j\}.$$

In this paper our interest is in the multiplier

$$\varphi(k) = \frac{k}{2^j} \text{ if } k \in B_j$$

and its reciprocal $\frac{1}{\varphi}$. In [11] we prove that both of these functions are multipliers for $H^1$ and use this fact to prove that the square function $S_2$ characterizes $H^1$. In this note we extend the result to $p < 1$ for the dyadic group $G$.

**Theorem 11.3** *Let $G$ be the dyadic group. The operators $T_\varphi$ and $T_{\frac{1}{\varphi}}$ are continuous from $H^p(G) \to H^p(G)$ and are inverses of each other for* $0 < p \le 1$.

**Proof.** Let $0 < p < 1$. The case $p = 1$ will follow from this case by interpolation. First consider $\varphi$. Ignoring the constant $2^{1-\frac{1}{p}}$ because it does not depend on $j$, we need to prove that the function

$$2^{-jp}\sum_{n=0}^{\infty}\left\|\sum_{k\in B_j} k\omega_k\gamma_n\right\|_1^p$$

is bounded in $j$. We have

$$\left\|\sum_{k\in B_j} k\omega_k\gamma_n\right\|_1 = \int_{I_n} |X_j(x)|\,dx$$

and so we need to estimate the character sum $X_j = \sum_{k\in B_j} k\omega_k$ on the interval $I_n$. In [11] we obtain exact expressions for $X_j$ on dyadic subintervals, so the integral can be calculated exactly. We use the notation

$$J(j,m) = [\frac{m}{2^j}, \frac{m+1}{2^j}).$$

The function $X_j$ is 0 on most intervals, it is nonzero only on the intervals

$$J(j,0),\ J(j,1),\ J(j,2^r),\ J(j,2^{r+1}),\ 0 \le r < j.$$

We consider cases for the pair $(n,j)$.

<u>The case $n \geqq j$.</u> Then

$$\frac{1}{2^n} \le \frac{1}{2^j} \Rightarrow I_n \subset J(j,0).$$

The value of $X_j$ on $J(j,0)$ is given in [11, subsection 4.3]. It results in

$$\int_{I_n} |X_j(x)|\,dx = \frac{1}{2^{n+1}}\left[2^{2j-2} + 2^{2j-3} - 2^{j-2}\right].$$

<u>The case $n = j-1$.</u> Then

$$\frac{1}{2^n} = \frac{1}{2^{j-1}} \Rightarrow I_n = J(j,1)$$

and the formula is the same as for $n \geqq j$.

<u>The case $n < j-1$.</u> For the pair $(j-1,n)$, we need to find that $r$ for which

$$I_n \cap J(j,2^r) \ne \emptyset \text{ or } I_n \cap J(j,2^r+1) \ne \emptyset.$$

The first intersection requires that

$$\frac{2^r}{2^j} < \frac{1}{2^n} \text{ and } \frac{2^{r+1}}{2^j} > \frac{1}{2^{n+1}}.$$

The second condition may be written

$$2^{j-n-r-1} < 1 + \frac{1}{2^r}$$

implying that $r = j - n - 1$ and that

$$I_n \cap J(j, 2^r) = \left[\frac{1}{2^{n+1}}, \frac{1}{2^{n+1}} + \frac{1}{2^j}\right).$$

The analysis of $I_n \cap J(j, 2^r + 1) \neq \emptyset$ is similar, yields $r = j - n - 1$, and

$$I_n \cap J(j, 2^r + 1) = \left[\frac{1}{2^{n+1}} + \frac{1}{2^j}, \frac{1}{2^{n+1}} + \frac{1}{2^{j-1}}\right).$$

This results in

$$J(j, 2^r) \cup J(j, 2^r + 1) \subset I_n.$$

On the interval $(j, 2^r) \cup J(j, 2^r + 1)$, we have

$$|X_j(x)| = 2^{2j-r-3} = 2^{j+n-2},$$

and so

$$\int_{I_n} |X_j(x)|\, dx = 2^{n-1}.$$

This gives

$$\begin{aligned}\sum_{n=0}^{\infty} \left\| \sum_{k \in B_j} k\omega_k \gamma_n \right\|_1^p &= \sum_{n=0}^{j-2} 2^{p(n-1)} \\ &\quad + \sum_{n=j-1}^{\infty} \left\{\frac{1}{2^{n+1}}\left[2^{2j-2} + 2^{2j-3} - 2^{j-2}\right]\right\}^p \\ &\leq 2^{p(j-1)} + 2^{(2j-1)p} \sum_{n=j-1}^{\infty} \frac{1}{2^{p(n+1)}} \\ &= 2^{p(j-1)} + 2^{(2j-1)p} \frac{2^{-p(j-1)}}{1 - 2^{-p}} = 2^{p(j-1)} + \frac{2^{jp}}{2^p - 1}.\end{aligned}$$

Hence

$$2^{-jp} \sum_{n=0}^{\infty} \left\| \sum_{k \in B_j} k\omega_k \gamma_n \right\|_1^p \leq \frac{1}{2} + \frac{1}{2^p - 1}.$$

We turn now to the function $\frac{1}{\varphi}$. Arguing as with the function $\varphi$, we must show that the expression

$$2^{jp} \sum_{n=0}^{\infty} \left\| \sum_{k \in B_j} \frac{\omega_k}{k} \gamma_n \right\|_1^p$$

is bounded as a function of $j$. We split the sum on $n$ into $n < j$ and $n \geq j$. For $n \geq j$ the inclusion $I_n \subset J(j,0)$ holds, and so $\omega_k = 1$ on the interval $I_n$. Since

$$\frac{1}{2} < \sum_{k \in B_j} \frac{1}{k} < 1,$$

we have

$$\sum_{k \in B_j} \frac{\omega_k}{k} \gamma_n \leq \gamma_n,$$

and hence

$$\left\| \sum_{k \in B_j} \frac{\omega_k}{k} \gamma_n \right\|_1^p \leq \left[ \frac{1}{2^{n+1}} \right]^p = 2^{-(n+1)p}.$$

Thus

$$2^{jp} \sum_{n=j}^{\infty} \left\| \sum_{k \in B_j} \frac{\omega_k}{k} \gamma_n \right\|_1^p \leq 2^{jp} \sum_{n=j}^{\infty} 2^{-(n+1)p} = \frac{1}{2^p - 1}.$$

For $n < j$ we use equalites and estimates for character sums, Dirichlet kernels, and Fejer kernels appearing in the book [24]. The kernels are

$$\Delta_m = \sum_{k=0}^{m-1} \omega_k \text{ (Dirichlet) and } K_m = \frac{1}{m} \sum_{k=0}^{m-1} \Delta_k \text{ (Fejer).}$$

The Dirichlet kernel at dyadic integers is particularly simple:

$$\Delta_{2^j} = 2^j \varsigma_j$$

in which $\varsigma_j$ is the characteristic function of $[0, 2^{-j})$. Thus $x \in I_n$ and $n < j$ give $\Delta_{2^j}(x) = 0$. It follows from Theorem 17 of Chapter 1 of [24] (p. 48) that

$$\sum_{k \in B_j} \frac{1}{k} \omega_k = \frac{1}{2^{j-1}+1} K_{2^{j-1}} - \frac{1}{2^j+1} K_{2^j} + \sum_{k=2^{j-1}+1}^{2^j} \left(\frac{1}{k-1} - \frac{1}{k+1}\right) K_k. \tag{11.1}$$

(This equality is derived from Theorem 17 in [11].) First consider the terms $K_{2^j}$. By Theorem 16 of [24],

$$K_{2^j} = \frac{1}{2}[2^{-j}\Delta_{2^j} + 2^{-j}\sum_{r=0}^{j} 2^r \tau_{e_r}\Delta_{2^j}],$$

in which $e_r = \frac{1}{2^{r+1}}$ and $\tau_{e_r}$ is translation in $G$ by $e_r$. Again $x \in I_n$ implies that $\Delta_{2^j}(x) = 0$, so

$$K_{2^j}(x) = \frac{1}{2}\sum_{r=0}^{j} 2^r \varsigma_j(x \oplus e_r).$$

For $r \geq n$, $x \oplus e_r \in G_n \backslash G_{n+1} \sim I_n$. Since $n < j$, $\varsigma_j(x \oplus e_r) = 0$. For $r < n-1$,

$$r + 1 \leq n \leq j - 1, \text{ implying that } \varsigma_j(x \oplus e_r\ ) = 0.$$

Hence in the sum for $K_{2^j}(x)$ only the term $r = (n-1)$ is non-zero, giving

$$K_{2^j}(x) \leq 2^{n-2}\varsigma_j(x \oplus e_{n-1}) \quad (x \in I_n, n \leq j-1).$$

Next we analyze

$$K_{2^{j-1}}(x) = 2^{-j}\Delta_{2^{j-1}}(x) + 2^{-j}\sum_{r=0}^{j-1} 2^r \Delta_{2^{j-1}}(x \oplus e_r\ ).$$

Take the case $n = j-1$ first. Since $x \in G_n$, the first term is $\frac{1}{2}\varsigma_n(x)$. The sum in the second term is

$$\sum_{r=0}^{n} 2^r \Delta_{2^n}(x \oplus e_r\ ).$$

For $r = n$ or $r = n-1$, we have $x \oplus e_r\ \in G_n$ and so $\Delta_{2^n}(x \oplus e_r\ ) = 2^n$. For $r \leq n-2$ we have $x \oplus e_r\ \in G_{r+1}\backslash G_n$ and so $\Delta_{2^n}(x \oplus e_r\ ) = 0$. Hence

$$\begin{aligned} 2^{-j}\sum_{r=0}^{n} 2^r \Delta_{2^n}(x \oplus e_r\ ) &\leq 2^{-j}[2^n 2^n \varsigma_n(x \oplus e_n) + 2^{n-1}2^n\varsigma_n(x \oplus e_{n-1})] \\ &= 2^{n-1}\varsigma_n(x \oplus e_n) + 2^{n-2}\varsigma_n(x \oplus e_{n-1}) \leq 3 \cdot 2^{n-2}. \end{aligned}$$

This gives

$$K_{2^n}\gamma_n \leq [\frac{1}{2} + 3 \cdot 2^{n-2}]\gamma_n\ ,\ n = j-1.$$

For $n < j-1$, and $x \in I_n$, $\Delta_{2^{j-1}}(x) = 0$. The only nonzero term of the series for $K_{2^{j-1}}$ is $r = n-1$, giving

$$K_{2^{j-1}}(x)\gamma_n(x) = 2^{n-1}\gamma_n(x)\varsigma_{j-1}(x \oplus e_{n-1}),\ n < j-1.$$

Hence for $0 \le n \le j-1$ we have

$$K_{2^{j-1}}(x)\gamma_n(x) = 2^{n-1}\varsigma_{j-1}(x \oplus e_{n-1})$$

It remains to estimate the third term in (11.1). The unblocked Fejer kernels $K_k$ are estimated in [24], Theorem 16. The result we need is

$$|K_k| \le \sum_{r=0}^{j-1} 2^{r-j} \sum_{q=r}^{j-1} (\Delta_{2^q} + \tau_{e_r}\Delta_{2^q})$$

on $[0,1)$. Reversing the order of the double sum,

$$|K_k| \le 2^{-j} \sum_{q=0}^{j-1} \sum_{r=0}^{q} 2^r (\Delta_{2^q} + \tau_{e_r}\Delta_{2^q})$$

on $[0,1)$. The first term of the double sum is seen to be

$$\sum_{q=0}^{j-1} \sum_{r=0}^{q} 2^r \Delta_{2^q} = \sum_{q=0}^{j-1} 2^q[2^{q+1}-1]\varsigma_q = \frac{2}{3}4^{n+1} - 2^{n+1} + \frac{1}{3} \le 4^{n+1}$$

for $x \in G_n$. Next consider the second double sum for $x \in I_n$ and $n \le j-1$. Recall that $\Delta_{2^q} = 2^q\varsigma_q$. The term $\varsigma_q(x \oplus e_r) = 0$ if $(x \oplus e_r) \notin G_q$. We have

$$x \in G_n \backslash G_{n+1} \text{ and } e_r \in G_{r+1} \backslash G_{r+2}.$$

It follows that

$$(n < r+1) \wedge (n < q) \Rightarrow \varsigma_q(x \oplus e_r) = 0,$$
$$(n > r+1) \wedge (r+1 < q) \Rightarrow \varsigma_q(x \oplus e_r) = 0.$$

Applying these implications to the double sum, it is

$$\sum_{q=0}^{j-1} \sum_{r=0}^{q} 2^r 2^q \varsigma_q(x \oplus e_r) = \sum_{q=n+1}^{j-1} \sum_{r=0}^{n-1} 2^r 2^q \varsigma_q(x \oplus e_r) + \sum_{q=0}^{n} \sum_{r=q-1}^{q} 2^r 2^q \varsigma_q(x \oplus e_r).$$

The first sum of the last term is 0 for $x \in G_n$ because $\varsigma_q(x \oplus e_r) = 0$. The second sum is

$$\sum_{q=0}^{n} \sum_{r=q-1}^{q} 2^r 2^q \varsigma_q(x \oplus e_r) \le \frac{4^{n+1}-1}{2} \le 2 \cdot 4^n.$$

Hence

$$|K_k|\, \gamma_n \le 6 \cdot 2^{-j} \cdot 4^n \gamma_n$$

and this in turn gives

$$\sum_{k=2^{j-1}+1}^{2^j} (\frac{1}{k-1} - \frac{1}{k+1}) K_k \gamma_n \le [6 \cdot 2^{-j} \cdot 4^n] \gamma_n \sum_{k=2^{j-1}+1}^{2^j} (\frac{1}{k-1} - \frac{1}{k+1})$$

$$= [6 \cdot 2^{-j} \cdot 4^n] \frac{1}{2^j} \gamma_n = 6 \cdot 4^{n-j} \gamma_n.$$

Consequently

$$\sum_{k \in B_j} \frac{\omega_k}{k} \gamma_n \le 2^{-j} 2^{n-2} \varsigma_j (x \oplus e_{n-1}) + 2^{-j} 2^n \varsigma_{j-1} (x \oplus e_{n-1})$$

$$+ 6 \cdot 4^{-j} \cdot 4^n \gamma_n,$$

which gives

$$\left\| \sum_{k \in B_j} \frac{\omega_k}{k} \gamma_n \right\|_1 \le 2^{-2j} \cdot 2^n \cdot 7 \text{ for } n < j.$$

Thus

$$\sum_{n=0}^{j-1} \left\| \sum_{k \in B_j} \frac{\omega_k}{k} \gamma_n \right\|_1^p \le 7 \cdot 2^{-jp}$$

and finally

$$2^{jp} \sum_{n=0}^{\infty} \left\| \sum_{k \in B_j} \frac{\omega_k}{k} \gamma_n \right\|_1^p = \frac{1}{2^p - 1} + 7.$$

■

## 11.4 Square Function Characterization of $H^p$.

The multiplier theorems in the previous section yield a characterization of $H^p$ by the square function $S_2$ for $0 < p \le 1$, much as in the case $p = 1$ which was treated in [11].

**Theorem 11.4** *Let $G$ be the dyadic group. There are postive constants $A_p$ and $B_p$ for which*

$$B_p \|f\|_{H^p} \le \left\{ \int_0^1 \|S_2 f(x)\|^p \, dx \right\}^{1/p} \le A_p \|f\|_{H^p}$$

*holds for all $f \in H^p(G)$ with $0 < p \le 1$ .*

**Proof.** We express the operator $S_2$ in terms of a multiplier operator $\varphi$. First write $S_2^2$ as

$$S_2^2 = \sum_{m=1}^{\infty}[\sum_{k=0}^{2^m-1} a_k\omega_k - (1-\frac{k}{2^m})a_k\omega_k]^2 = \sum_{m=1}^{\infty}[2^{-m}\sum_{k=0}^{2^m-1} ka_k\omega_k]^2.$$

Write this in terms of the blocks, namely

$$S_2^2 = \sum_{m=1}^{\infty}[2^{-m}\sum_{j=0}^{m}\sum_{k\in B_j} ka_k\omega_k]^2 = \sum_{m=1}^{\infty}[\sum_{j=0}^{m} 2^{-m+j}\sum_{k\in B_j}\frac{k}{2^j}a_k\omega_k]^2.$$

As $\varphi$ is the sequence

$$\varphi(k) = \frac{k}{2^j} if\ 2^{j-1} \le k \le 2^j - 1,$$

and $T_\varphi$ the corresponding multiplier operator, we have

$$S_2^2 f = \sum_{m=1}^{\infty}[\sum_{j=0}^{m} 2^{-m+j} D_j(T_\varphi(f))]^2.$$

Define sequences $d$ and $b$ on $\mathbb{Z}$ by

$$d_j = D_j(T_\varphi(f)) and\ b_j = 2^{-j} if\ 0 \le j; d_j = b_j = 0 \text{ if } j < 0.$$

The sum over $j$ is the $\mathbb{Z}$-convolution $b * d(m)$. Hence the equality

$$S_2 f = \|b * d\|_{\ell^2}$$

holds. The convolution $b * d$ depends on $x \in [0, 1)$. This dependency is surpressed on both sides of the equation. We have

$$\|b * d\|_{\ell^2} \le \|b\|_{\ell^1} \cdot \|d\|_{\ell^2} = 2[\sum_{j=0}^{\infty} |D_j(T_\varphi(f))|^2]^{1/2} = 2S_3(T_\varphi(f)).$$

Thus we have shown that $S_2 f$ is pointwise bounded by $2S_3(T_\varphi(f))$. By Theorem 11.3 we know that $T_\varphi$ is a bounded operator on $H^p$. It follows that there is a constant $A_p$ for which

$$\|S_2 f\|_p \le A_p \|f\|_{Hp}.$$

We show that the reverse inequality holds, with a different positive constant. The sequence $b$ defines a convolution operator $U_b(c) = b * c$ on $\ell^2(\mathbb{Z})$, where

$$b * c(m) = \sum_{j=0}^{\infty} b(j)c(m-j).$$

The multiplier for $U_b$ is the function $\mu$ defined on the circle by

$$\mu(e^{it}) = \sum_{j=0}^{\infty} 2^{-j} e^{ijt} = [1 - \frac{1}{2} e^{it}]^{-1} = \frac{2}{2 - e^{it}}.$$

The function $\mu$ is bounded away from 0 on the circle, and so $\mu^{-1}$ is a bounded function on the circle. Thus the multiplier operator is invertible as an operator on $L^2$, and so is the convolution operator $U_b$ on $\ell^2$. The convolution inverse is, in fact, $U_b^{-1} = (...0, 0, 1, -1/2, 0, 0...)$ . We apply these facts for $f \in H^p$, namely:

$$\begin{aligned} \|f\|_{H^p}^p &= \int_0^1 \|d(x)\|_{\ell^2}^p dx = \int_0^1 \|U_b^{-1}(U_b(d(x)))\|_{\ell^2}^p dx \\ &\leq C \int_0^1 \|U_b(d(x))\|_{\ell^2}^p dx \\ &= C \int_0^1 \|b * d(x)\|_{\ell^2}^p dx = C \int_0^1 (S_2 f(x))^p dx. \end{aligned}$$

Taking $B_p = C^{-1}$, we have proved the theorem. ■

Note that this proof does not explicitly use the boundedness of the multiplier operator $T_{\frac{1}{\varphi}}$. Theorem 11.4 was motivated by the study of $H^1$ and other spaces using the square functions $S_1$, $S_2$, and $S_3$. The proof using invertible convolutions on $\ell^2(\mathbb{Z})$ is more general. With little modification, the above argument proves the following theorem.

**Theorem 11.5** *Let $G$ be the dyadic group. If $T$ is a bounded invertible operator on $H^p(G)$, $0 < p \leq 1$, and $U$ is a bounded invertible convolution operator on $\ell^2(N)$, then*

$$\left( \int_0^1 \|U(\{D_j(T(f)(x))\})\|_{\ell^2}^p dx \right)^{1/p}$$

*is an equivalent norm for $H^p(G)$.*

*References*


[1] Butzer, P.L. and H.J. Wagner, Walsh-Fourier series and the concept of a derivative, Applicable Analysis, v. 3, 1973, pp. 29-46.

[2] Chao, J.-A. $H^p$ spaces of conjugate systems on local fields, Studia Math. 49, 1974, pp. 267-287.

[3] Chao, J.-A. Conjugate characterizations of $H^1$dyadic martingales, Math. Ann. 240, 1979, pp 63-67.

[4] Chao, J.-A. Hardy space on regular martingales, in Martingale Theory in Harmonic Analysis and Banach Spaces, Proceedings, Cleveland, 1981, ed. J.-A. Chao and W.A. Woyczynski, Lecture Notes in Mathematics 939, Springer-Verlag, 1982, pp 18-28.

[5] Chao, J.-A. and S. Janson, A note on $H^1$ q-martingales, Pac. Journ. Math., v. 97, no. 2, 1981, pp307-317.

[6] Chao, J.-A. and R.-L. Long, Martingale transforms and Hardy spaces, Probability Theory and Related Fields 99, 1992, pp 399-404.

[7] Chao, J.-A. and R.-L. Long, Martingale transforms with unbounded multipliers, Proc. Amer. Math. Soc. 114, no. 3, 1992, pp 831-838.

[8] Chao, J.-A. and M. Taibleson, A subregularity inequality of conjugate systems on local fields, Studia Math. 46 (1973), pp 249-257.

[9] Coifman, R.R. and G. Weiss, Extensions of Hardy spaces and their use in analysis, Bull. Am. Math Soc. 83, 1977, pp 569-645.

[10] Daly, J. and K. Phillips, On singular integrals, multipliers, $H^1$ and Fourier series — a local field phenomenon, Math. Ann. 265, 1983, pp 181-219.

[11] Daly, J. and K. Phillips, Walsh multipliers and square functions for the Hardy space $H^1$, Acta Math. Hung. 79(4), 1998, pp 311-327.

[12] Daly, J. and K. Phillips, A note on $H^1$ multipliers for locally compact Vilenkin groups, to appear in Canadian Math. Bulletin.

[13] Fine, N.J. On the Walsh functions, Trans. Amer. Math Soc. 65, 1949, pp 372-414.

[14] Fridli, S. and P. Simon, On the Dirichlet kernels and a Hardy space with respect to the Vilenkin system, Acta Math. Hung. 45(1-2) 1985, pp 223-234.

[15] Gát, G. Investigations of certain operators with respect to the Vilenkin system, Acta Math. Hung, 61, 1993, pp 131-149.

[16] Janson, S. Characterizations of $H^1$ by singular integral transforms on martingales and $R^n$, Math Scand 41, 1977, pp 140-152.

[17] Janson, S. Non-homogeneous multipliers characterizing $H^1$ on local fields, Mittag-Leffler Report 16, 1978, pp. 1-5.

[18] Kitada, T. $H^p$ multiplier theorems on certain totally disconnected groups, Sci. Rep. Hirosaki Univ. 34, 1987, pp 1-7.

[19] Krotov,V. On the multipliers of Fourier series with respect to the Haar system, Analysis Mathematica 3, 1977, pp. 187-198.

[20] Long, *R*. Martingale spaces and inequalities, Peking University Press, 1993.

[21] Onneweer, C.W. Hormander-type mulitpliers on L.C. Vilenkin groups: the $L^1(G)$-case, preprint, 1997.

[22] Onneweer, C.W. Multipliers for $H^p(G)-$spaces, Revista de la Union Matematica Argentina, v. 37, 1991, pp. 135-141.

[23] Onneweer, C.W. and T.S. Quek, $H^p$ multiplier results on locally compact Vilenkin groups, Quart. J. Math Oxford(2), 40, 1989, pp 313-323.

[24] Schipp, F., W.R. Wade, and P. Simon (with assistance from J. Pal), Walsh Series, Akademiai Kiado, Budapest, 1990; co-edition Adam Hilger, Bristol and New York.

[25] Simon, P. $(L^1, H)$-type estimations for some operators with respect to the Walsh-Paley system, Acta Math. Hung. 46, 1985, 307-310.

[26] Simon, P. Investigations with respect to the Vilenkin system, Annales Univ. Sci. Budapest, Sect. Math. 27, 1985, pp 87-101.

[27] Sunouchi, G.I. On the Walsh-Kaczmarz series, Proc. Amer. Math. Soc. 2, 1951, pp 5-11.

[28] Taibleson, M. Fourier Analysis on Local Fields, Princeton University Press, 1975,

[29] Taibleson, *M*. An introduction to Hardy spaces on local fields, Harmonic Analysis in Euclidean Spaces, Proceedings of Symposia in Pure Mathematics, *v*. 35, part 2, 1979, pp 311-316.

[30] Wade, W., $L^r$ inequalities for Walsh series, $0 < r < 1$, Acta Sci. Math. 46, 1983, pp 233-241.

[31] Watari, C. Multipliers for Walsh Fourier series, Tohoku Math. J. 16, 1964, pp 239-251.

[32] Weisz, F. Conjugate Martingale Transforms, Studia Mahematica 103(2), 1992, pp 207-220.

[33] Weisz, F. Martingale Hardy Spaces and their Applications in Fourier Analysis, Lecture Notes in Mathematics 1568, 1994

[34] Young, W.S. Littlewood-Paley and multiplier theorems for Vilenkin-Fourier series, Can. J. Math. 46(3), 1994, pp 662-772.

[35] Young, W.S. Almost everywhere convergence of Vilenkin-Fourier series of $H^1$ functions, Proc. Amer. Math. Soc. 108, no 2, 1990,pp 433-441.

# Chapter 12

# Spectra of pseudo-differential operators in the Hörmander class

**Josefina Alvarez**

*ABSTRACT. We characterize the spectra of $L^p$-bounded translation invariant pseudo-differential operators with symbols in the Hörmander class $S_\rho^m$. In particular, we obtain for these operators a precise version of their $L^p$-spectral invariance. We also prove a partial result on the spectra of $L^p$-bounded pseudo-differential operators with symbols in the Hörmander class $S_{\rho,\delta}^m$. We use these results to study the holomorphic functional calculus of translation invariant pseudo-differential operators.*

## 12.1 Introduction

We consider pseudo-differential operators in the Hörmander class $L_{\rho,\delta}^m$, ([6], [8], [2]), $m \in \mathbb{R}$, $0 < \rho \leq 1$, $0 \leq \delta < 1$. These are operators $L$ of the form

$$L(f)(x) = \int e^{2\pi i x.\xi} a(x,\xi) \hat{f}(\xi)\, d\xi$$

where $f \in S$, and the function $a(x,\xi)$ belongs to the class $S_{\rho,\delta}^m$. That is to say, $a(x,\xi)$ is a smooth function satisfying the estimates

$$\left|\partial_x^\alpha \partial_\xi^\beta a(x,\xi)\right| \leq C_{\alpha,\beta}(1+|\xi|)^{m+\delta|\alpha|-\rho|\beta|}$$

where $m \in \mathbb{R}$, $0 < \rho \leq 1$, $0 \leq \delta < 1$. The parameter $m$ is usually referred to as the order of the operator, although such name is not always well defined.

The function $a(x,\xi)$ is uniquely determined by the pseudo-differential operator $L$ and it is called the symbol of the operator. We will sometimes denote $Op(a)$ the pseudo-differential operator with symbol $a$.

In this article we study the spectra $\sigma_p(L)$ of $L^p$-bounded pseudo-differential operators $L$, where

$$\sigma_p(L) = \{z \in \mathbb{C} : (L - zI)^{-1} \text{ does not exist as a bounded operator on } L^p(\mathbb{R}^n)\}$$

Sometimes we will denote the spectrum $\sigma_p(a)$.

Our main result is the characterization of the spectra of $L^p$-bounded translation invariant pseudo-differential operators. These are operators with symbols not depending on the variable $x$. Thus, they are operators $L$ of the form

$$\overset{\wedge}{L(f)}(\xi) = a(\xi)\overset{\wedge}{f}(\xi) \tag{12.1}$$

where $a \in S_\rho^m = \{a \in S_{\rho,0}^m : a = a(\xi)\}$.

For these operators we prove in Section 12.3 that $\sigma_p$ coincides with the closure of the image of $a$, for $p$ in some finite interval that is optimal. That is to say, $\sigma_p(a) = \overline{a(\mathbb{R}^n)}$.

This characterization shows explicitly that the spectrum of $L$ does not depend on $p$ for appropriate values of $p$. Thus, we recapture for translation invariant operators the optimal $L^p$-invariance properties obtained in [3].

Operators of the form (12.1) are multiplier operators defined by smooth multipliers. Striking differences can be observed on the spectral invariance properties of multiplier operators.

For example, given a finite Borel measure $\mu$ on the circle $T$, it can be proved that the $L^p(T)$-spectrum of the convolution operator $S_\mu(f) = \mu * f$ depends on $p$ ([15], [7]). However, P. Sarnak has proved in [11] that if the $L^2(T)$-spectrum of $S_\mu$ has capacity zero, then the $L^p(T)$-spectrum of $S_\mu$ does not depend on $p$ for $1 < p < \infty$. H. Widom [15] obtained positive as well as negative results concerning the $L^p$-spectral invariance of the finite Hilbert transform

$$H_E(f)(x) = p.v. \int_E \frac{f(x)}{x-y} dy$$

where $E$ is a measurable subset of $\mathbb{R}$ with positive measure.

Given $a(x,\xi) \in S_{\rho,\delta}^m$, we consider in Section 12.4 the following set ([1])

$$K_a = \bigcap_{M \geq 0} \overline{\{a(x,\xi) : x \in \mathbb{R}^n, |\xi| \geq M\}}$$

This is a compact set when $m \leq 0$.

We prove that if $Op(a)$ is $L^p$-bounded, then $K_a \subset \sigma_p(a)$. This result was proved in [1] under different hypotheses.

The set $K_a$ plays an important role in proving uniqueness properties for functions of operators. We discuss this connection in Sections 12.4 and 12.5. We also study in Section 12.5 a precise version of the classical holomorphic calculus of translation invariant operators ([3]).

Although we state our results within the context of $L^p$ spaces, they can be extended without difficulty to weighted Sobolev spaces (see [12] for the definition).

The notation used in this paper is standard in the subject. The symbols $C_0^\infty, C^\infty, S, S', L^p$, etc. will indicate the usual spaces of distributions or functions defined on $\mathbb{R}^n$ with values in the complex space $\mathbb{C}$. Moreover, $\|f\|_p$ will denote the $L^p$ norm of the function $f$. We will say that a function $f$ is smooth if $f \in C^\infty$. Derivatives will be denoted $\partial_x^\alpha$ or $\partial_{x_j}$ as appropriate. We will denote $H_p^s$ the usual Sobolev space of order $s$ based on $L^p$. When $s = 0$ the space $H_p^s$ is $L^p$. Given a complex Banach space $X$, we will denote $L(X)$ the space of linear and continuous operators from $X$ into $X$ with the operator norm. The identity operator will be denoted $I$. The letter $C$ will indicate an absolute constant, probably different at different occurrences. Other notations will be introduced at the appropriate time.

## 12.2 Preliminary results

We collect in this section several results that will be used later. We do not include proofs for these results, but in all cases we give precise references.

Let $L$ be a pseudo-differential operator as defined in the introduction. We are interested on the spectrum of $L$ as a bounded operator on $L^p$ for some $p$. Thus, we need to single out the values of $m, \rho, \delta$ for which the operator $L$ will be bounded on $L^p$ for a given $p$. The following theorem gives an optimal result.

**Theorem 12.1** *([5] ($\delta < \rho$), [2] ($\delta \geq \rho$)) Let $m \in \mathbb{R}$, $0 < \rho \leq 1$, and $0 \leq \delta < 1$. Then given $1 < p < \infty$, $L_{\rho,\delta}^m \subset L(L^p)$ if and only if*

$$m \leq m_p = -n\left[(1-\rho)\left|\frac{1}{p} - \frac{1}{2}\right| + \lambda\right], \lambda = \max\left(0, \frac{\delta - \rho}{2}\right).$$

We observe that $m_p \leq m_{p'}$ when $p \leq p' \leq 2$, and $m_p = m_{p'}$ when $\frac{1}{p} + \frac{1}{p'} = 1$.

The next three results can be stated within the context of weighted Sobolev spaces (see [12] for the definition). We will restrict our attention to their $L^p$ versions.

The first result is a very interesting extension, due to J. Ueberberg ([13]), of a famous result of R. Beals characterizing pseudo-differential operators ([4]). Before stating this result, we need to introduce two important commutators.

Given a linear and continuous operator $L : S \to S'$ let

$$P_j L = \left[\partial_{x_j}, L\right] = \partial_{x_j} L - L\partial_{x_j}$$
$$Q_k L = [2\pi i x_k, L] = (2\pi i x_k) L - L(2\pi i x_k)$$

where $2\pi i x_k$ denotes the operator of multiplication by $2\pi i x_k$.

When $L = Op\,(a)$ for some $a \in S^m_{\rho,\delta}$, integration by parts shows that

$$P_j L = Op\left(\partial_{x_j} a\right)$$
$$Q_k L = Op\,(-\partial_{\xi^k} a)$$

Given multi-indexes $\alpha, \beta$ we indicate $P^\alpha Q^\beta$ the iteration of the commutators $P_j$ and $Q_k$ for values of $j, k$ given by $\alpha$ and $\beta$ respectively. It is clear from the discussion above that $P^\alpha Q^\beta L$ is a pseudo-differential operator in the class $L^{m+\delta|\alpha|-\rho|\beta|}_{\rho,\delta}$.

**Theorem 12.2**

(i) Let $L : S \to S'$ be a linear and continuous operator. Suppose that for some $m \in \mathbb{R}$, $0 < \rho \leq 1$, $0 \leq \delta < 1$, $\delta \leq \rho$, $1 < p < \infty$,

$$Q^\alpha P^\beta L \in L\left(H_p^{s+m-\rho|\alpha|+\delta|\beta|}, H_p^s\right)$$

for all $s \in \mathbb{R}$ and all multi-indexes $\alpha, \beta$. Then, $L \in L^m_{\rho,\delta}$.

(ii) Given $L \in L^m_{\rho,\delta}$ for some $m \in \mathbb{R}$, $0 < \rho \leq 1$, $0 \leq \delta < 1$, $\delta \leq \rho$, $L \in L\left(H_2^{s+m-\rho|\alpha|+\delta|\beta|}, H_2^s\right)$ for all $s \in \mathbb{R}$ and all multi-indexes $\alpha, \beta$.

(iii) Given $L \in L^m_{1,\delta}$ for some $m \in \mathbb{R}$, $0 \leq \delta < 1$, $\delta \leq \rho$, $L \in L\left(H_p^{s+m-|\alpha|+\delta|\beta|}, H_p^s\right)$ for all $s \in \mathbb{R}$, $1 < p < \infty$, and all multi-indexes $\alpha, \beta$.

It is important to notice that since the hypotheses on $m$ in Theorem 12.1 are sharp, the converse of Theorem 12.2(i) cannot be true when $\rho < 1$ and $p \neq 2$. The case $p = 2$ was proved by Beals in [4] for a larger class of pseudo-differential operators.

**Proposition 12.3** *([12] when $\rho = 1$, [3] when $0 < \rho \leq 1$): Let $L \in L^m_{\rho,\delta}$, $0 < \rho \leq 1$, $0 \leq \delta < 1$, $\delta \leq \rho$, $m \leq m_p = -n\,(1-\rho)\left|\frac{1}{p} - \frac{1}{2}\right|$ for some $1 < p < \infty$.*

*If $(L - zI)$ is invertible in $L\left(H_p^{s_o}\right)$ for some $s_o$, then $(L - zI)$ is invertible in $L\left(H_p^s\right)$ for every $s \in \mathbb{R}$.*

Theorem 12.2 and Proposition 12.3 play a crucial role in the proof of the following result, which will be used later in the proof of Theorem 12.5 and Proposition 12.6.

**Proposition 12.4** *([3]): Let* $L \in L^m_{\rho,\delta}$, $0 < \rho \leq 1$, $0 \leq \delta < 1$, $\delta \leq \rho$, $m \leq m_p = -n(1-\rho)\left|\frac{1}{p} - \frac{1}{2}\right|$ *for some* $1 < p < \infty$. *If* $z \notin \sigma_p$, *then* $(L - zI)^{-1} \in L^0_{\rho,\delta}$.

**Remark 12.1** *It is important to observe that the hypotheses of Proposition 12.4 cannot be satisfied with* $z = 0$ *when* $\rho < 1$ *and* $p \neq 2$. *Indeed, in this case, we would have* $I = L^{-1}L \in L^m_{\rho,\delta}$ *with* $m < 0$.

We are now ready to state and prove our results.

## 12.3 The $L^p$-spectrum of translation invariant pseudo-differential operators

Let $a(\xi) \in S^m_\rho$. We assume that

$$0 < \rho \leq 1, \quad 0 \leq m \leq m_{p_o} = -n(1-\rho)\left|\frac{1}{p_o} - \frac{1}{2}\right|$$

for some $1 < p_o < \infty$. Under these assumptions we can prove the following result:

**Theorem 12.5** *Let* $\sigma_p(a)$ *be the spectrum of* $Op(a)$ *as a bounded operator on* $L^p$ *for* $p$ *between* $p_o$ *and* $p'_o$. *Then,*

$$\sigma_p(a) = \overline{a(\mathbb{R}^n)}$$

**Proof.** Let $z \notin \overline{a(\mathbb{R}^n)}$. Then, there exists $C > 0$ such that $|a(\xi) - z| \geq C$ for every $\xi \in \mathbb{R}^n$. This implies that $\frac{1}{a(\xi)-z} \in S^0_\rho$.
If $\rho = 1$ or $p = 2$, the operator $Op\left(\frac{1}{a(\xi)-z}\right)$ is continuous on $L^p$. Thus, $z \notin \sigma_p(a)$.
If $\rho < 1$ and $p \neq 2$, we observe (Remark 12.1), that $z \neq 0$ because $m_{p_o} < 0$. Thus, we can write,

$$\frac{1}{a(\xi) - z} = -\frac{1}{z}\left(\frac{1}{1 - \frac{a(\xi)}{z}}\right) = -\frac{1}{z}\left(1 + \frac{a(\xi)}{z} + \frac{a^2(\xi)}{z(z - a(\xi))}\right)$$

Since the function $\frac{a^2(\xi)}{z(z-a(\xi))}$ belongs to $S^{2m}_\rho$ we conclude that

$$\frac{1}{a(\xi) - z} \in S^0_1 + S^m_\rho$$

Then, the operator $Op\left(\frac{1}{a(\xi)-z}\right)$ is continuous from $L^p$ into itself for $p$ between $p_0$ and $p'_o$. Thus, $z \notin \sigma_p(a)$.

Conversely, let $z \notin \sigma_p(a)$. Then, by definition, $(Op(a) - zI)^{-1}$exists as a continuous operator from $L^p$ into itself. According to Proposition 12.4, we can conclude that $(Op(a) - zI)^{-1} = Op(b)$ for some $b \in S^0_{\rho,0}$. Thus, we can write,

$$\begin{aligned}(Op(a) - zI)^{-1}(Op(a) - zI)(f)(x) &= \int e^{2\pi i x.\xi} b(x,\xi)(a(\xi) - z)\widehat{f}(\xi)\,d\xi \\ &= f(x)\end{aligned}$$

for every $f \in S$. Let us now fix $x_o, \xi_o \in \mathbb{R}^n$ and let us consider an approximation of the identity about $\xi_o$. That is to say, given $\varphi \in C_o^\infty$ such that $\int \varphi dx = 1$, $supp(\varphi) \subset \{|x| \leq 1\}$, and $\varphi(x) \geq 0$ for very $x \in \mathbb{R}^n$, we set

$$\varphi_j(\xi) = j^n \varphi(j(\xi - \xi_o))$$

for $j = 1, 2, \ldots$ Thus, $\varphi_j = \widehat{f_j}$, with

$$f_j(x) = e^{2\pi i x.\xi_o} \check{\varphi}\left(\frac{x}{j}\right)$$

where $\check{\varphi}$ denotes the inverse Fourier transform. We have

$$\int e^{2\pi i x_o.\xi} b(x_o, \xi)(a(\xi) - z)\varphi_j(\xi)\,d\xi = e^{2\pi i x_o.\xi_o} \check{\varphi}\left(\frac{x_o}{j}\right)$$

Taking limit on both sides as $j \to \infty$ we obtain

$$e^{2\pi i x_o.\xi_o} = e^{2\pi i x_o.\xi_o} b(x_o, \xi_o)(a(\xi_o) - z).$$

Thus, $a(\xi) - z \neq 0$ for every $\xi \in \mathbb{R}^n$ and

$$b(x,\xi) = b(\xi) = \frac{1}{a(\xi) - z}$$

Moreover, since the order of $b$ is zero, there exists $C > 0$ such that

$$\left|\frac{1}{a(\xi) - z}\right| \leq C$$

for every $\xi \in \mathbb{R}^n$. Thus, $z \notin \overline{a(\mathbb{R}^n)}$. This completes the proof of Theorem 12.5. ■

## 12.4 The set $K_a$

Let $a(x,\xi)$ be a function in the Hörmander class $S^m_{\rho,\delta}$ as defined in the introduction.

Associated with the function $a(x,\xi)$ we consider the set ( [1])

$$K_a = \bigcap_{M \geq 0} \overline{\{a(x,\xi) : x \in \mathbb{R}^n,\ |\xi| \geq M\}} \tag{12.2}$$

The set $K_a$ is compact if $m \leq 0$. From the inequality

$$|a(x,\xi)| \leq C(1+|\xi|)^m$$

we deduce that the set $K_a$ reduces to $\{0\}$ when $m < 0$.

The set $K_a$ was used in [1], p. 53 to obtain a uniqueness result for the mapping $f \to f(Op(a))$ module regularizing operators, for a particular class of symbols $a$ in $S^m_{\rho,\delta}$, $0 \leq \delta < \rho < 1$.

More precisely, given a smooth function $f$ vanishing in a neighborhood of $K_a$, the function $f(a(x,\xi))$ vanishes for $x \in \mathbb{R}^n$, $|\xi| \geq M$, for some $M > 0$. Thus, $f(a(x,\xi))$ belongs to $S^{-\infty} = \bigcap_m S^m_{\rho,\delta}$. As a consequence, the pseudo-differential operator with symbol $f(a(x,\xi))$ is a regularizing operator, in the sense that it maps the Sobolev space $H^s_p$ into the Sobolev space $H^t_q$ for any $t, s \in \mathbb{R}$ and for any $1 < p, q < \infty$. We will use this observation in last section.

**Proposition 12.6** *Let $a(x,\xi) \in S^m_{\rho,\delta}$. We assume that $0 < \rho \leq 1$, $0 \leq \delta < 1$, $\delta \leq \rho$, $0 \leq m \leq m_{p_o} = -n(1-\rho)\left|\frac{1}{p_o} - \frac{1}{2}\right|$ for some $1 < p_o < \infty$. We also assume that $m + \delta - \rho < 0$. Let $\sigma_p(a)$ be the spectrum of $Op(a)$ as a continuous operator on $L^p$ for $p$ between $p_o$ and $p'_o$. Then,*

$$K_a \subset \sigma_p(a)$$

**Proof.** Let $z \notin \sigma_p$. Then by definition, $(Op(a) - zI)^{-1}$ exists as a continuous operator on $L^p$. According to Proposition 12.4, we can conclude that $(Op(a) - zI)^{-1} = Op(b)$ for some $b \in S^0_{\rho,\delta}$. The functional calculus of pseudo-differential operators shows that we can write

$$\begin{aligned} 1 &= (a(x,\xi) - z)\, b(x,\xi) \\ &\quad + \frac{i}{2\pi} \sum_{j=1}^{n} \int_0^1 \int_{\mathbb{R}^{2n}} e^{2\pi i (x-y)\cdot(\xi-\eta)} \partial_{\xi_j} a(x, \xi + s(\eta - \xi)) \\ &\qquad \times \partial_{x_j} b(y,\xi)\, dy d\eta ds \end{aligned}$$

Moreover, since $\partial_{\xi_j} a \in S^{m-\rho}_{\rho,\delta}$ and $\partial_{x_j} b \in S^{m+\delta}_{\rho,\delta}$ we have

$$1 = (a(x,\xi) - z)\, b(x,\xi) + c(x,\xi)$$

where $c(x,\xi) \in S_{\rho,\delta}^{m-\rho+\delta}$. The assumption $m+\delta-\rho<0$ implies that $c(x,\xi) \to 0$ as $|\xi| \to \infty$ uniformly on $x \in \mathbb{R}^n$. Thus, for some $M>0$ we can say that $|c(x,\xi)| \leq \frac{1}{2}$ if $x \in \mathbb{R}^n$ and $|\xi| \geq M$. Since $b(x,\xi)$ is a bounded function, we conclude that there exists $C>0$ such that

$$|a(x,\xi) - z| \geq C$$

if $x \in \mathbb{R}^n$ and $|\xi| \geq M$. Thus, $z \notin K_a$. This completes the proof of Proposition 12.6. ∎

**Remark 12.2** **a)** *The condition $m+\delta-\rho<0$ implies that we do not know whether the conclusion of Proposition 12.6 holds when $p_o=2$ and $\delta=\rho<1$.*

**b)** *It is clear that $K_a$ is always included in $\overline{a(\mathbb{R}^n \times \mathbb{R}^n)}$. In general, the inclusion is strict. For instance, if $a(x,\xi) = \left(1+|\xi|^2\right)^{-k}$, $k>0$, then $K_a = \{0\}$, whereas $\overline{a(\mathbb{R}^n)} = [0,1]$.*

**c)** *Proposition 12.6 is proved in [1], p. 39 under the assumptions:*

1. $p_o = 2$
2. $|a(x,\xi)| \leq C$
3. $\left|\partial_{x_j} a(x,\xi)\right| \leq C(1+|\xi|)^{\delta}$
4. $\left|\partial_{\xi}^{\alpha} a(x,\xi)\right| \leq C(1+|\xi|)^{-\rho|\alpha|}$

*where $x,\xi \in \mathbb{R}^n, 0 \leq \delta < \rho \leq 1, |\alpha| \leq \left[\frac{n}{2}\right]+1$*

*The proof uses the following observation: By definition of $K_a$, given $z \in K_a$ we have $z = \lim_{k\to\infty} a(x_k,\xi_k)$ where $x_k \in \mathbb{R}^n$ and $|\xi_k| \to \infty$ as $k \to \infty$. Thus, to show that $z \in \sigma_2(a)$, it suffices to find functions $f_k \in L^2$ such that $\|f_k\|_2 = C$ independent of $k$,*

$$\|Op(a)(f_k) - a(x_k,\xi_k) f_k\|_2 \to 0 \text{ as } k \to \infty.$$

**d)** *When $m<0$, Proposition 12.6 states that $\{0\} \subset \sigma_p(a)$. This was observed already at the end of Section 12.2.*

## 12.5 Applications

Theorem 12.5 provides a characterization of the $L^p$-spectrum of $Op(a)$ that is independent of $p$ between $p_o$ and $p_o'$. As a consequence, we obtain the $L^p$-spectral invariance proved in [3] Theorem 2.5 b), when $Op(a)$ is translation invariant.

Moreover, using Theorem 12.5 and Proposition 12.6, we can make more precise the holomorphic functional calculus discussed in [3], p. 8. Before presenting these results, we will outline the construction of this functional calculus, following the definition given by F. Riesz and B. Sz-Nagy in [10], p. 431.

Let $X$ be a complex Banach space and let $T \in L(X)$. A holomorphic function $f : U \to \mathbb{C}$ is called admissible with respect to $T$, if $U$ is an open subset of $\mathbb{C}$ containing the spectrum $\sigma(T)$ of $T$. We will denote $\mathcal{A}(T)$ the space of admissible functions with respect to $T$.

Given $f \in \mathcal{A}(T)$, let $\Gamma$ be a simple, closed curve such that $\sigma(T)$ is contained in the interior $\overset{\circ}{\Gamma}$ of $\Gamma$. Moreover, we assume that $\Gamma \cup \overset{\circ}{\Gamma} \subset U$.

We consider the integral

$$f(T) = \frac{1}{2\pi i} \int_{\Gamma} f(z) (zI - T)^{-1} \, dz \tag{12.3}$$

where $\Gamma$ is parametrized in the counterclockwise direction. The integral in (12.3) is calculated as the Bochner integral of the operator valued holomorphic function $f(z)(zI - T)^{-1}$, with values in the Banach space $L(X)$. The integral does not depend on the curve $\Gamma$, provided that $\Gamma$ satisfies the conditions stated above. We have to justify the name $f(T)$ given to this integral.

The function $f(z) = z^m$ is admissible with respect to any operator $T$ and we can select $\Gamma = \{|z| = R\}$, with $R > \|T\|$. Thus, we can write

$$z^m (zI - T)^{-1} = \sum_{k=0}^{\infty} z^{m-k-1} T^k$$

where the convergence is uniform on $\{|z| \geq R\}$. We can then integrate the series term by term, to obtain

$$f(T) = \frac{1}{2\pi i} \sum_{k=0}^{\infty} T^k \int_{\Gamma} z^{m-k-1} dz = T^m \tag{12.4}$$

In particular, we have

$$I = \frac{1}{2\pi i} \int_{\Gamma} (zI - T)^{-1} \, dz$$

$$T = \frac{1}{2\pi i} \int_{\Gamma} z (zI - T)^{-1} \, dz$$

It is clear that (12.4) can be extended by linearity to any polynomial $P(z)$. Now, given an admissible function $f$, Mergelyan's theorem ([14])

implies that there is a sequence $\{P_k(z)\}$ of complex polynomials that converges to $f$ uniformly on $\Gamma$. Since the operator valued function $(zI - T)^{-1}$ is bounded on $\Gamma$, we can conclude that

$$f(T) = \lim_{k \to \infty} P_k(T)$$

We can easily obtain the following two consequences.

1. Given $f, g \in \mathcal{A}(T)$, we have

$$(f.g)(T) = f(T) \circ g(T) \tag{12.5}$$

   where $f.g$ denotes pointwise multiplication.

2. The operator $f(T)$ commutes with any operator that commutes with $T$.

Moreover, we can also extend to this setting the Spectral Mapping Theorem. Namely, given $f \in \mathcal{A}(T)$, we have

$$\sigma(f(T)) = f(\sigma(T)) \tag{12.6}$$

In fact, given $z_o \in \sigma(T)$ define

$$g(z) = \begin{cases} \frac{f(z) - f(z_o)}{z - z_o} & z \in U,\ z \neq z_o \\ f'(z_o) & z = z_o \end{cases}$$

The function $g$ is holomorphic on $U$. Thus, $g \in \mathcal{A}(T)$. Moreover,

$$(z - z_o)\, g(z) = f(z) - f(z_o)$$

for every $z \in U$. Thus, using (12.5) we have

$$(T - z_o I) \circ g(T) = f(T) - f(z_o) I$$

If $f(z_o) \notin \sigma(f(T))$, this means that the operator $f(T) - f(z_o) I$ is invertible in $L(X)$. Thus, $(T - z_o I) \circ g(T)$ is invertible in $L(X)$. Let $C$ be its inverse. We have

$$(T - z_o I) \circ g(T) \circ C = C \circ (T - z_o I) \circ g(T) = I$$

According to Property 2 above, the operator $g(T)$ commutes with $(T - z_o I)$. Thus, we can also write

$$C \circ g(T) \circ (T - z_o I) = I$$

Then, $(T - z_o I)$ has a left inverse and a right inverse in $L(X)$. Thus, $(T - z_o I)$ is invertible in $L(X)$. This contradicts the fact that $z_o \in \sigma(T)$. Thus, we have proved that $\sigma(f(T)) \supset f(\sigma(T))$.

Conversely, let $\lambda \in \sigma(f(T))$ and let us assume that $\lambda \notin f(\sigma(T))$. Since $f(\sigma(T))$ is a compact subset of $\mathbb{C}$, there exists an open neighborhood $V$ of $f(\sigma(T))$ such that $\lambda \notin V$. Let $U_1 = f^{-1}(V)$. The set $U_1$ is an open subset of $U$ that contains $\sigma(T)$ . This implies that

$$\frac{1}{f(z)-\lambda} \in \mathcal{A}(T)$$

with domain $U_1$. Thus, $\left(\frac{1}{f(z)-\lambda}\right)(T)$, defined by (12.3), belongs to $L(X)$. Clearly, we have

$$\left(\frac{1}{f(z)-\lambda}\right)(f(z)-\lambda) = (f(z)-\lambda)\left(\frac{1}{f(z)-\lambda}\right) = 1$$

for every $z \in U_1$. Thus, using again (12.5), we can write

$$\left(\frac{1}{f(T)-\lambda I}\right) \circ (f(T)-\lambda I) = (f(T)-\lambda I) \circ \left(\frac{1}{f(T)-\lambda I}\right) = I$$

Then, the operator $(f(T)-\lambda I)$ is invertible in $L(X)$, which contradicts the fact that $\lambda \in \sigma(f(T))$. Thus, $\lambda \in f(\sigma(T))$. This completes the proof of (12.6).

With a similar reasoning, we can also prove that given $f \in \mathcal{A}(T)$ and $g \in \mathcal{A}(f(T))$, we have $g \circ f \in \mathcal{A}(T)$ and

$$(g \circ f)(T) = g(f(T))$$

The above properties justify the notation $f(T)$, as well as the name functional calculus, given to the operation defined by (12.3).

We will now apply this notion of functional calculus to the case of translation invariant pseudo-differential operators. Let $a(\xi) \in S^m_\rho$. We assume that $0 < \rho < 1$, $0 \le m \le m_{p_o} = -n(1-\rho)\left(\frac{1}{p_o} - \frac{1}{2}\right)$ for some $1 < p_o < 2$. Let $f \in \mathcal{A}(Op(a))$. That is to say, $f$ is a holomorphic function on an open neighborhood $U$ of $\sigma_p(a) = \overline{a(\mathbb{R}^n)}$. We observe that $0 \in \sigma_p(a)$. Let $\Gamma$ be a simple, closed curve contained in $U$ and such that $\overline{a(\mathbb{R}^n)} \subset \overset{\circ}{\Gamma}$ . Then, we can define $f(Op(a))$ by means of (12.3).

$$f(Op(a)) = \frac{1}{2\pi i} \int_\Gamma f(z)(zI - Op(a))^{-1} dz \tag{12.7}$$

Since $Op(a)$ is a translation invariant operator, we can write

$$f(Op(a)) = Op\left[\frac{1}{2\pi i} \int_\Gamma f(z)(z-a)^{-1} dz\right]$$

In fact

$$\frac{1}{2\pi i}\int_{\Gamma} f(z)\left(zI - Op(a)\right)^{-1} dz = \sum_{k=0}^{\infty}\frac{1}{2\pi i}\int_{\Gamma} f(z)\left(Op(a)\right)^{k} z^{-k-1}dz$$
$$= \sum_{k=0}^{\infty}\frac{1}{2\pi i}\int_{\Gamma} f(z)\, Op\left(a^{k}\right) z^{-k-1}dz$$
$$= Op\left[\frac{1}{2\pi i}\int_{\Gamma} f(z)\left(\sum_{k=0}^{\infty}(a)^{k} z^{-k-1}\right) dz\right]$$

That is to say, we obtain that the symbol of $f(Op(a))$ is $f(a)$.

We can refine this classical result in the following way. As in the proof of Theorem 12.5, we have

$$\frac{1}{z-a(\xi)} = \frac{1}{z}\left(\frac{1}{1-\frac{a(\xi)}{z}}\right) = \frac{1}{z}\left(1+\frac{a(\xi)}{z}+\frac{a^{2}(\xi)}{z(z-a(\xi))}\right)$$

Thus,

$$f(a) = \frac{1}{2\pi i}\int_{\Gamma} f(z)\frac{dz}{z} + \frac{1}{2\pi i}\int_{\Gamma} f(z)\left(\frac{a(\xi)}{z}+\frac{a^{2}(\xi)}{z(z-a(\xi))}\right)\frac{dz}{z} \qquad (12.8)$$

The first term in (12.8) reduces to $f(0)$. The second term is in the class $S_{\rho}^{m}$. Then, we have obtained an explicit decomposition of $f(a)$ in the class $S_{\rho}^{0}+S_{\rho}^{m}$.

We observe again that since $m<0$, the set $K_a$ given by (12.2) reduces to $\{0\}$. Let us now assume that $f$ vanishes in an open neighborhood $V \subset U$ of zero. Then, $a(\xi) \in V$ when $|\xi| \geq M$, for some $M>0$. Thus, the function $f(a) \in C_{o}^{\infty}$. This implies that $Op(f(a))$ is a convolution operator with a kernel in the Schwartz class $S$. As a consequence, the operator $Op(f(a))$ is a regularizing operator, in the sense that it maps the Sobolev space $H_{p}^{s}$ into the Sobolev space $H_{q}^{t}$ for any $t,s \in \mathbb{R}$ and for any $1<p,q<\infty$.

Thus, we have proved that the values of the function $f$ on a neighborhood of zero determines the operator $f(Op(a))$ module regularizing operators.

**Acknowledgement.** I thank Eugenio Hernández for kindly providing reference [11].

*References*

[1] J. Alvarez, A. P. Calderón: *Functional calculi for pseudo-differential operators, I,* Proc. Sem. Fourier Anal., El Escorial, Spain, (Editors: M. de Guzmán, I. Peral) (1979), 1-61.

[2] J. Alvarez, J. Hounie: *Estimates for the kernel and continuity properties of pseudo-differential operators,* Arkiv för Mat. 28 (1990), 1-22.

[3] J. Alvarez, J. Hounie: *Spectral invariance and tameness of pseudo-differential operators on weighted Sobolev spaces,* J. of Op. Th. 30 (1993), 41-67.

[4] R. Beals: *Characterization of pseudo-differential operators and applications,* Duke Math. J. 44 (1977), 45-57.

[5] C. Fefferman, *$L^p$ bounds for pseudo-differential operators,* Israel J. Math. 14 (1973), 413-417.

[6] L. Hörmander: *Pseudo-differential operators and hypoelliptic equations,* Proc. Symp. Pure Math. 10, Amer. Math. Soc. (1967), 138-183.

[7] S. Igari: *Functions of $L^p$-multipliers,* Tôhoku Math. J. 21 (1969), 304-320.

[8] H. Kumano-go: *Oscillatory integrals of symbols of pseudo-differential operators and the local solvability theorem of Nirenberg and Trèves,* Kakata Symp. on PDE (1972), 166-191.

[9] F. Riesz, B. Sz-Nagy: *Functional analysis,* Dover (1990).

[10] P. Sarnak: *Spectra of singular measures as multipliers on $L^p$,* J. of Funct. Anal. 37 (1980), 302-317.

[11] E. Schrohe: *Boundedness and spectral invariance for standard pseudo-differential operators on anisotropically weighted $L^p$-Sobolev spaces,* Integral Eq. and Op. Th. 13 (1990), 235-242.

[12] J. Ueberberg: *Zur Spektralinvariantz von Algebren von Pseudodifferentialoperatoren in der $L^p$-Theorie,* Manuscripta Math. 61 (1988), 459-475.

[13] J. L. Walsh: *Interpolation and approximation,* Amer. Math. Soc. Coll. Publ. 20 (1969).

[14] H. Widom: *Singular integral equations in $L^p$,* Trans. Amer. Math. Soc. 97 (1960), 131-161.

[15] N. Wiener, A. Wintner: *Fourier Stieltjes transforms and singular infinite convolutions,* Amer. J. of Math. 60 (1938), 513-522.

# Chapter 13

# Scaling properties of infinitely flat curves and surfaces

**Alex Iosevich**[1]

*ABSTRACT. We shall give simple sufficient conditions for the Orlicz type bounds for the averaging operators and restriction operators associated with infinitely flat curves in the plane. Our results, obtained by scaling, can be used to recover, up to the endpoints, the results previously obtained in [4], [1], and [2]. We also prove some three dimensional analogs of those results.*

## 13.1 Introduction

Let $\gamma \in C^\infty([0,\infty))$, $\gamma(0) = \gamma'(0) = 0$, $\gamma''(s) \geq 0$, and $\gamma''(s) = 0$ iff $s = 0$. Let

$$T_\gamma f(x) = \int_0^2 f(x_1 - s, x_2 - \gamma(s))ds = f * \mu(x). \tag{13.1}$$

If $\gamma''(s) > 0$ on $[0,2]$, it is well known that $T_\gamma : L^p(\mathbb{R}^2) \to L^q(\mathbb{R}^2)$ if and only if $(\frac{1}{p}, \frac{1}{q})$ is contained in the triangle with endpoints $(0,0)$, $(1,1)$ and $(\frac{2}{3}, \frac{1}{3})$. (See [10], [7]). If $\gamma''$ vanishes of order $m-2$, $m \geq 2$, then $T_\gamma : L^p(\mathbb{R}^2) \to L^q(\mathbb{R}^2)$ if and only if $(\frac{1}{p}, \frac{1}{q})$ is contained in the trapezoid with the endpoints $(0,0)$, $(1,1)$, $(\frac{2}{m+1}, \frac{1}{m+1})$, and $(\frac{m}{m+1}, \frac{m-1}{m+1})$. (See [8]).

If $\gamma''$ vanishes of infinite order, the estimate $T_\gamma : L^p(\mathbb{R}^2) \to L^q(\mathbb{R}^2)$ may not hold for any $q > p$. However, the Orlicz space estimates may be possible. For example, in [4], the authors showed that for some flat curves in $\mathbb{R}^2$ there exists a Young's function $\Phi$, with $\lim_{t\to\infty} \frac{\Phi(t)}{t^2} = 0$ such that the estimate

$$||T_\gamma f||_{L^2(\mathbb{R}^2)} \leq C||f||_{L^\Phi(\mathbb{R}^2)} \tag{13.2}$$

holds, where $L^\Phi(\mathbb{R}^2)$ denotes the standard Orlicz space, associated to an

[1] research partially supported by NSF grant number DMS97-06825

increasing Young function $\Phi$, equipped with the norm

$$||f||_{\Phi} \equiv \inf\left\{s>0: \int \Phi\left(\frac{|f(x)|}{s}\right)dx \leq 1\right\}. \tag{13.3}$$

More precisely, the result proved in [4] is the following.

**Theorem 13.1** *(Bak, McMichael, and Oberlin [4]) Let $T_\gamma$ be as above. If there exist constants $d \in (0,2]$, $\epsilon > 0$, and $\beta > 0$ such that*

$$\int_0^{\epsilon} D(\beta s)(\gamma(s))^{\frac{2}{d}} s^{3(2/d-1)} ds < \infty, \tag{13.4}$$

*then there exists a constant $C$ and a Young function $\Phi$ with $\Phi(t) \approx \left(tH_+^{-1}(t^{-d})\right)^2$ such that $T_\gamma : L^{\Phi}(\mathbb{R}^2) \to L^2(\mathbb{R}^2)$, where $H_+^{-1}(x) = 1$ for $x > 1$, $H^{-1}(x)$ for $x \leq 1$ with $H(x) = x^3\gamma(x)$, and $D(s) = |\{\xi \in \mathbb{R}^2 : |\widehat{\mu}(\xi)| > s\}|$.*

In [1] the following three dimensional result was proved.

**Theorem 13.2** *(Bak [1]) Let*

$$T_\gamma^n f(x) = \int_{\{0 \leq |y| \leq 2\}} f(x'-y, x_n - \gamma(|y|))dy, \tag{13.5}$$

*where $(x', x_n) \in \mathbb{R}^{n-1} \times \mathbb{R}$, $\gamma(0) = \gamma'(0) = \gamma''(0) = 0$, $\gamma''(s) > 0$ for $s > 0$, and $\frac{\gamma'(s)}{s}$ is non-decreasing for $s > 0$. Let $\gamma_n(s) = t^n\gamma'(s)$. Let $n+1 \leq q < \infty$ and assume that for some $\delta > 0$ and $c = c(q) > 0$ $\frac{\gamma_n(s)}{s^{q+\delta}}$ is non-decreasing for $0 < s \leq c$. Let $\Phi = \Phi_q$ be a Young's function such that $\Phi^{-1}(s) \approx \Psi^{-1}(s) = \frac{s^{\frac{1}{q}}}{\left(\gamma_n^{-1}(1/s)\right)^{n-1}}$ if $s \geq 1$, and $\Psi^{-1}(s) = s^{\frac{1}{q}}$ if $s < 1$. If $n = 2$ or $3$ there exists a constant $C$ such that for every Borel set $E \subset \mathbb{R}^n$ with $|E| < \infty$ $T_\gamma^n : L^{\Phi}(\mathbb{R}^n) \to L^q(\mathbb{R}^n)$, $f = \chi_E$, the characteristic function of $E$.*

In this paper we shall give a simple set of sufficient, and in many cases, necessary conditions, such that $T_\gamma : L^{\Phi}(\mathbb{R}^2) \to L^{\Psi}(\mathbb{R}^2)$. We shall also see that the techniques of this paper can be used to obtain $L^{\Phi}(\mathbb{R}^2) \to L^2(\mathbb{R}^2)$ bounds for the restriction operator $\mathcal{R}f = \hat{f}|_{\{(s,\gamma(s))\}}$. See Theorem 13.6. We will also show that these results generalize, in a straightforward way, to surfaces of rotation in $\mathbb{R}^3$. See the section "Three dimensions" below. Our result for the restriction operator is motivated by the following result due to Bak.

**Theorem 13.3** *(Bak, [2]) Suppose that $\frac{\gamma(s)}{s^3}$ is increasing on $[0,2]$. Suppose that $\frac{\gamma''(s)}{s}$ is increasing on $(0,\delta)$ for some $\delta > 0$. Then for $1 \leq q < \infty$ and*

*$0 < d < 1$ there exists a constant $C = C_{d,q}$ such that for all $f \in \mathcal{S}(\mathbb{R}^2)$, the class of rapidly decreasing functions,*

$$\left( \int_0^{\delta} \left| \hat{f}(s, \gamma(s)) \right|^q ds \right)^{\frac{1}{q}} \leq C + C \int_{\mathbb{R}^2} |f(x)| \cdot \left[ \gamma_q^{-1}(|f(x)|) \right]^d dx, \tag{13.6}$$

*where $\gamma_q(s) = s^{q-1}\gamma(s^q)$.*

The main idea behind our two-dimensional results is the following lemma motivated by the results in [5] where it was observed that even though finite type convex curves (e.g. $\{(s, s^m) : -1 \leq s \leq 1\}$) behave very well under the usual diagonal scaling of the form $(x_1, x_2) \to (2^j x_1, 2^{mj} x_2)$, infinite type curves (e.g. $\{(s, e^{-1/s^2}) : -1 \leq s \leq 1\}$) require a more careful non-diagonal scaling.

**Lemma 13.4** *Let $\tau_j f(x) = f(2^{-j}x_1, \gamma_j x_1 + h_j x_2)$, where $\gamma_j = \gamma(2^{-j})$ and $h_j$ is defined as above. Let*

$$T_j^* f(x) = 2^j \tau_j T_j {\tau_j}^{-1} f(x), \tag{13.7}$$

*where*

$$T_j f(x) = \int_{2^{-j}}^{2^{-j+1}} f(x_1 - s, x_2 - \gamma(s)) ds. \tag{13.8}$$

*Then $T_j^* : L^p(\mathbb{R}^2) \to L^q(\mathbb{R}^2)$ for $\left(\frac{1}{p}, \frac{1}{q}\right) \in \mathcal{T}$ with constants independent of $j$.*

**Proof.** Since $T_j^*$ clearly maps $L^1 \to L^1$ and $L^\infty \to L^\infty$, with constants independent of $j$, it suffices to check that $T_j^* : L^{\frac{3}{2}}(\mathbb{R}^2) \to L^3(\mathbb{R}^2)$ with constants independent of $j$. The classical proof of the $L^p$ improving properties of measures supported on surfaces with non-vanishing Gaussian curvature (see e.g. [10] and the subsection "Simplification" below) shows that it suffices to check that if we write $T_j^* f(x) = f * \nu_j(x)$, then

$$|\hat{\nu_j}(\xi)| \leq C(1 + |\xi|)^{-\frac{1}{2}}, \tag{13.9}$$

with $C$ independent of $j$. Let $T_j f(x) = f * \mu_j(x)$. Let $A_j$ denote the matrix given by the equation $\tau_j f(x) = f(A_j x)$. Then

$$f * \nu_j(x) = 2^j \mu_j * \tau_j^{-1} f(A_j x). \tag{13.10}$$

Using the elementary properties of the Fourier transform,

$$\hat{f}(\xi)\hat{\nu_j}(\xi) = 2^j \hat{f}((A_j^{-1})^t A_j^t \xi)\hat{\mu_j}((A_j^{-1})^t \xi) = 2^j \hat{f}(\xi)\hat{\mu_j}((A_j^{-1})^t \xi), \tag{13.11}$$

since $(A_j^{-1})^t A_j^t = I$, where $A_j^t$ denotes the transpose matrix of $A_j$. By Proposition 4.2 of [5],

$$2^j|\hat{\mu}_j((A_j^{-1})^t\xi)| \leq C(1+|(A_j^{-1})^t A_j^t\xi|)^{-\frac{1}{2}} = C(1+|\xi|)^{-\frac{1}{2}}. \qquad (13.12)$$

Diving both sides of (13.11) by $\hat{f}(\xi)$ and using (13.12) completes the proof of the Lemma. ■

The main idea behind the three dimensional results in this paper is the estimate of the Fourier transform of the measure carried by a family of radial surfaces dependent on a parameter. See Lemma 13.8 below. It turns out that while it is difficult to find a scaling that gives the optimal decay in all directions, even the most naive scaling allows one to get the optimal decay in the direction normal to the surface at the origin. This turns out to be enough to obtain the desired results for the averaging and restriction operators.

Our plan is as follows. In the section titled "Scaling" we shall give sufficient conditions for the $L^\Phi(\mathbb{R}^2) \to L^\Psi(\mathbb{R}^2)$ boundedness of the operator $T_\gamma$ and the $L^\Phi(\mathbb{R}^2) \to L^2(\mathbb{R}^2)$ boundedness of the restriction operator $\mathcal{R}$ in terms of the Orlicz norms of the family of dilation operators of the form $\tau_A f(x) = f(Ax)$ where $A$ is an invertible matrix. The subsection titled "Simplification" is devoted to showing that the assumptions of our main result can be simplified if we are willing to assume that $\sup_{0<a<b} \frac{\gamma''(a)}{\gamma''(b)} \leq C$. In the subsection titled "Three dimensions" we shall extend our results to radial surfaces in $\mathbb{R}^3$.

Finally in the section titled "Orlicz norms of dilation operators" we shall compute upper and lower bounds for the Orlicz norms of the operators $\tau_A$ under various assumptions on the Young functions in question. We shall conclude the paper with the subsection on examples.

## 13.2 Scaling

The following definition is motivated by interpolation results for $L^p$ and Orlicz spaces (see e.g. [12], [13]).

**Definition 13.1** *Let $\mathcal{T}$ denote the triangle with the endpoints $(0,0)$, $(1,1)$, and $(\frac{2}{3}, \frac{1}{3})$. We say that $(\Phi, \Psi) \subset \mathcal{T}$ if every linear operator bounded from $L^p(\mathbb{R}^2) \to L^q(\mathbb{R}^2)$, $\left(\frac{1}{p}, \frac{1}{q}\right) \in \mathcal{T}$, is bounded from $L^\Phi(\mathbb{R}^2) \to L^\Psi(\mathbb{R}^2)$.*
*Let $\mathcal{T}' = \{p : 1 \leq p \leq \frac{6}{5}\}$. We say that $\Phi \subset \mathcal{T}'$ if every linear operator bounded from $L^p(\mathbb{R}^2) \to L^2(\mathbb{R}^2)$, $p \in \mathcal{T}'$, is bounded from $L^\Phi(\mathbb{R}^2) \to L^2(\mathbb{R}^2)$.*

Our main results are the following.

**Theorem 13.5** *Let $T_\gamma$ be as above. Suppose that there exists $\epsilon > 0$ so that $h'(t) > \epsilon \frac{h(t)}{t}$ for every $t > 0$, where $h(t) = t\gamma'(t) - \gamma(t)$. Let $(\Phi, \Psi) \subset \mathcal{T}$. Let $\tau_j f(x) = f(2^{-j}x_1, \gamma_j x_1 + h_j x_2)$ where $\gamma_j = \gamma(2^{-j})$ and $h_j = h(2^{-j})$. Let $N_j(\Phi)$ denote the $(L^\Phi(\mathbb{R}^2), L^\Phi(\mathbb{R}^2))$ norm of the operator $\tau_j$, and let $N_j^{-1}(\Psi)$ denote the $(L^\Psi(\mathbb{R}^2), L^\Psi(\mathbb{R}^2))$ norm of the operator $\tau_j^{-1}$. Suppose that*

$$\sum_{j=0}^{\infty} 2^{-j} N_j(\Phi) N_j^{-1}(\Psi) < \infty. \tag{13.13}$$

*Then $T_\gamma : L^\Phi(\mathbb{R}^2) \to L^\Psi(\mathbb{R}^2)$.*

**Proof.** Note that by duality it suffices to check $T_\gamma : L^{\Psi^*}(\mathbb{R}^2) \to L^{\Phi^*}(\mathbb{R}^2)$. In other words, it suffices to check the condition (13.13) with $N_j(\Phi)$ replaced by $N_j(\Psi^*)$, and $N_j^{-1}(\Psi)$ replaced by $N_j^{-1}(\Phi^*)$. Let $T_j, T_j^*$ be defined as above. Then

$$||T_j f||_\Psi = 2^{-j} ||\tau_j^{-1} T_j^* \tau_j f||_\Psi. \tag{13.14}$$

By Lemma 13.4, $T_j^* : L^p(\mathbb{R}^2) \to L^q(\mathbb{R}^2)$ for $\left(\frac{1}{p}, \frac{1}{q}\right) \in \mathcal{T}$ with constants independent of $j$. By definition of $N_j(\Phi)$ and $N_j^{-1}(\Psi)$ it follows that $T : L^\Phi(\mathbb{R}^2) \to L^\Psi(\mathbb{R}^2)$ if (13.13) holds. ■

**Theorem 13.6** *Let $\mathcal{R}f$ be defined as above. Let $\gamma$ satisfy the assumptions of Theorem 13.5. Let $\sigma_j f(x) = |A_j| f(A_j^t x)$ with $A_j$ defined by $\tau_j f(x) = f(A_j x)$, where $\tau_j$ is as above. Let $N_j^{-1}(\Phi)$ denote the $(L^\Phi(\mathbb{R}^2), L^\Phi(\mathbb{R}^2))$ norm of the operator $\sigma_j^{-1}$. Let $\Phi \in \mathcal{T}'$. Suppose that*

$$\sum_{j=0}^{\infty} 2_j^{-\frac{j}{2}} N_j^{-1}(\Phi) < \infty. \tag{13.15}$$

*Then $\mathcal{R} : L^\Phi(\mathbb{R}^2) \to L^2(\Gamma)$, where $\Gamma = \{(s, \gamma(s)) : 0 \leq s \leq 2\}$.*

**Proof.** The proof is along the same lines as the proof of Theorem 13.5. Let $\mathcal{R}_j f = \hat{f}|_{\Gamma_j}$, where $\Gamma_j = \{(s, \gamma(s)) : 2^{-j} \leq s \leq 2^{-j+1}\}$. Let $\sigma_j f(x) = |A_j| f(A_j^t x)$. Using the elementary properties of the Fourier transform it is not hard to see that $\mathcal{R}_j \sigma_j f = \hat{f}(A_j^{-1}\Gamma_j(s))$. It follows that

$$||\mathcal{R}_j \sigma_j f||_2 = 2^{-\frac{j}{2}} ||\hat{f}(A_j^{-1}\Gamma_0(2^{-j}\cdot)||_2. \tag{13.16}$$

By Proposition 4.2 in [5], the Fourier transform of the Lebesgue measure on $A_j^{-1}\Gamma_0(2^{-j}\cdot)$ is bounded by $C(1 + |\xi|)^{-\frac{1}{2}}$, where $C$ is independent of $j$. The proof of the restriction theorem for curves with non-vanishing Gaussian curvature (see e.g. [9]) implies that

$$||\mathcal{R}_j \sigma_j f||_2 \leq C 2^{-\frac{j}{2}} ||f||_\Phi, \tag{13.17}$$

for $\Phi \subset \mathcal{T}'$. It follows that

$$||\mathcal{R}_j f||_2 \le C2^{-\frac{j}{2}} ||\sigma_j^{-1} f||_{\Phi}. \tag{13.18}$$

■

### 13.2.1 Simplification

It is not hard to see that the statements of Theorems 13.5 and 13.6 can be simplified if we are willing to assume that $\gamma''$ is increasing on $[0,2]$, or, even, that $\sup_{0<a\le b} \frac{\gamma''(b)}{\gamma''(a)} \ge C$. More precisely, under this assumption, everywhere in the statements of these Theorems, we can replace the scaling transformation $\tau_j f(x) = f(2^{-j}x_1, \gamma_j x_1 + h_j x_2)$ by a simpler scaling transformation $\tau'_j f(x) = f(2^{-j}x_1, \gamma_j x_2)$. This is the consequence of the following rephrasing of the classical $L^p$-improving result for curves with everywhere non-vanishing curvature and the Stein-Tomas restriction theorem in $\mathbb{R}^2$.

**Lemma 13.7** *Let $Tf(x) = \int_I f(x_1 - s, x_2 - \phi(s))ds$, $\mathcal{R}f = \hat{f}|_{\{(s,\phi(s)):s\in I\}}$, where $I$ is a compact interval and $\phi$ is a smooth function. Suppose that $\phi''(s) \ge 1$ on $I$. Then*

$$T : L^p(\mathbb{R}^2) \to L^q(\mathbb{R}^2), \left(\frac{1}{p}, \frac{1}{q}\right) \in \mathcal{T},$$

*and*

$$\mathcal{R} : L^p(\mathbb{R}^2) \to L^p(\mathbb{R}^2), p \in \mathcal{T}',$$

*with constants independent of $\phi$.*

Before proving the lemma, let's see how it implies the claim in the preceding paragraph. The analog of the operator $T_j^*$ in the proof of Theorem 13.5 is just the averaging operator over the curve $(s, \phi_j(s))$, $1 \le s \le 2$, where $\phi_j(s) = \frac{\gamma(2^{-j}s)}{\gamma_j}$. By assumption and the mean value theorem $\phi_j''(s) \ge 1$, so Lemma 13.7 implies that the new $T_j^* : L^p(\mathbb{R}^2) \to L^q(\mathbb{R}^2)$, $\left(\frac{1}{p}, \frac{1}{q}\right) \in \mathcal{T}$. The rest of the proof goes through as before and we get the claim. The argument in the restriction theorem is basically the same.

**Proof.** Again, we prove the case of the $L^p$-improving theorem, the restriction proof being similar. Let $K_z(x) = \frac{1}{\Gamma(z)}(x_2 - \phi(x_1))^{z-1}\chi_I(x_1)$, where $\Gamma$ denotes the standard gamma function and $\chi_I$ is the characteristic function of the interval $I$. Let $T_z f(x) = f * K_z(x)$. It is clear that if $Re(z) = 1$, then $T_z : L^1(\mathbb{R}^2) \to L^\infty(\mathbb{R}^2)$ with universal constants. It remains to show that when $Re(z) = -\frac{1}{2}$, then $T_z : L^2(\mathbb{R}^2) \to L^2(\mathbb{R}^2)$ with constants independent of $\phi$. Let's compute $m_z(\xi) = \hat{K_z}(\xi)$. Well,

$$m_z(\xi) = \int e^{-i\langle x,\xi\rangle} K_z(x)dx. \tag{13.19}$$

Let $y_1 = x_1$, $y_2 = x_2 - \phi(x_1)$. We get

$$\frac{1}{\Gamma(z)} \int \int_I e^{-i(y_1\xi_1 + \phi(y_1)\xi_2)} e^{-iy_2\xi_2} y_2^{z-1} dy_1 dy_2. \tag{13.20}$$

When $Re(z) = -\frac{1}{2}$, the expression in (13.20) is controlled by $|\widehat{d\sigma}(\xi)||\xi_2|^{\frac{1}{2}}$, so, by Plancherel's theorem it remains to show that $|\widehat{d\sigma}(\xi)| \leq C|\xi_2|^{-\frac{1}{2}}$ with constants independent of $\phi$. This immediately follows by the van der Corput lemma since $\phi''(s) \geq 1$ on $I$. ■

### *13.2.2 Three dimensions*

In this section we shall see that if $S = \{x = (x', x_3) \in \mathbb{R}^3 : x_3 = \gamma(|x'|)\}$, where $\gamma$ satisfies the assumptions of the previous subsection, then one can easily generalize Theorem 13.5 and Theorem 13.6 to three dimensions by replacing the scaling transformation $\tau_j'$ by its three dimensional version, $\tau_j' f(x) = f(2^{-j}x', \gamma_j x_3)$, using the following three dimensional version of Lemma 13.7.

**Lemma 13.8** *Let*

$$Tf(x) = \int_B f(x' - y, x_n - G(y))dy, \mathcal{R}f(x) = \hat{f}|_{\{(y,G(y)):y\in B\}},$$

*where $B$ is the annulus $\{y : 1 \leq |y| \leq 2\}$, and $G(y) = \phi(|y|)$. Suppose that $\min\{\phi', \phi''\} \geq 1$ on $B$. Then*

$$T : L^p(\mathbb{R}^3) \to L^q(\mathbb{R}^3) for \left(\frac{1}{p}, \frac{1}{q}\right) \in \mathcal{T}_3,$$

*the triangle with the endpoints $(0,0)$, $(1,1)$, and $(\frac{3}{4}, \frac{1}{4})$, and*

$$\mathcal{R} : L^p(\mathbb{R}^3) \to L^2(\mathbb{R}^3), p \in \mathcal{T}_3' = \{p : 1 \leq p \leq \frac{4}{3}\},$$

*with constants independent of $G$.*

As we noted above, we can now prove the obvious analogs of Theorems 13.5 and 13.6. We let $\tau_j f(x) = f(2^{-j}x', \gamma_j x_3)$. The analog of the operator $T_j^*$ is the averaging operator over the hypersurface $\{x : 1 \leq |x'| \leq 2;\ \ x_3 = G_j(x')\}$, where $G_j(x') = \frac{\gamma(2^{-j}|x'|)}{\gamma_j}$. The determinant of the hessian matrix of $G_j$ is $2^{-2j}\gamma''(2^{-j}r)\left(\frac{2^{-j}\gamma'(2^{-j}r)}{r}\right)$, where $r = |x'|$. By assumption and the mean value theorem this quantity is bounded below by 1. The rest of the argument is the same as the proof of Theorem 13.5, with Lemma 13.8 replacing Lemma 13.4. The argument for the restriction operator is similar.

**Proof.** (of Lemma 13.8) By the proof of Lemma 13.7 it suffices to show that

$$|F(\xi)| = \left| \int_B e^{i(\langle x',\xi\rangle + \xi_3 G(x'))} dx' \right| \leq C|\xi_3|^{-1}, \tag{13.21}$$

with $C$ independent of $G$. Going into polar coordinates, applying standard stationary phase, and making a change of variables sending $r \to r\xi_3^{-\frac{1}{2}}$, we get ${\xi_3}^{-1}$ times

$$I(A,t) = \int_{t^{\frac{1}{2}}[1,2]} e^{i(rAt^{-\frac{1}{2}} - t\phi(t^{-\frac{1}{2}}r))} r^1 b(rAt^{-\frac{1}{2}}) dr, \tag{13.22}$$

where $A = |x'|$, $t = \xi_3$, and $b$ is a symbol of order $-\frac{1}{2}$. It suffices to prove that $I$ is uniformly bounded with constants independent of $A$, $t$, and $\phi$. Suppose that either $A \approx |t|$ or $A >> |t|$. Now, by the van der Corput lemma, the integral $I(A,t)$ is bounded by

$$C|t^{\frac{1}{2}} b(A)|, \tag{13.23}$$

where $C$ is a universal constant, since the second derivative of the phase function of $I$, $\phi''(rt^{-\frac{1}{2}})$, is bounded below by 1 on $[t^{\frac{1}{2}}, 2t^{\frac{1}{2}}]$ by assumption. The expression in (13.23) is bounded above by another universal constant $C'$ since $b$ is a symbol of order $-\frac{1}{2}$, and $|t| \leq A$. It remains to handle the case when $A << t$. We undo the change of variables sending $r \to rt^{-\frac{1}{2}}$, and we let $h(r) = rA - t\phi(r)$. It is not hard to see that $|h'(t)| \geq |A - t| \geq |t|$. Van der Corput lemma gives us the decay $\frac{C}{t}$ and the proof of the Lemma is complete. ■

## 13.3 Orlicz norms of dilation operators

**Lemma 13.9** *Let $\Phi$ be a Young function such that*

$$\Phi(a)\Phi(b) \leq \Phi(Cab) \tag{13.24}$$

*for all $a, b > 0$ with some $C > 0$. Let* $\det(A) = t$. *Then*

$$\frac{c}{\Phi^{-1}(t)} ||f||_\Phi \leq ||\tau_A f||_\Phi \leq C\Phi^{-1}\left(\frac{1}{t}\right) ||f||_\Phi. \tag{13.25}$$

**Lemma 13.10** *Let $\Phi$ be a Young function such that*

$$\Phi(a)\Phi(b) \geq \Phi(Cab) \tag{13.26}$$

*for all $a, b > 0$ with some $C > 0$. Then*

$$c\Phi^{-1}\left(\frac{1}{t}\right)||f||_\Phi \le ||\tau_A||_\Phi \le \frac{C}{\Phi^{-1}(t)}||f||_\Phi. \tag{13.27}$$

**Proof of Lemmas 13.9 and 13.10.** We must estimate $\inf\{s > 0\}$ such that

$$\int \Phi\left(\frac{|f(Ax)|}{s}\right) dx \le 1, \tag{13.28}$$

where we may assume that $\int \Phi(|f(x)|)dx = 1$. Making a change of variables and using (13.24) this immediately reduces to $\frac{1}{\Phi(s)} \le t$ which implies that $\Phi(s) \ge \frac{1}{t}$. It follows that $s \ge \Phi^{-1}\left(\frac{1}{t}\right)$. Taking the inf proves the second inequality of (13.25). Replacing $A$ by $A^{-1}$ we see that the second inequality implies the first. This completes the proof of Lemma 13.9.
Making a change of variables and using (13.26) reduces (13.28) to $\Phi\left(\frac{1}{s}\right) \le t$ which implies that $s \ge \frac{1}{\Phi^{-1}(t)}$. This proves the second inequality in (13.27). Replacing $A$ by $A^{-1}$ we see again that the second inequality implies the first. This completes the proof of Lemma 13.10. ■

**Lemma 13.11** *Let $\Phi^*$ denote the conjugate function of $\Phi$ given by the equation*

$$\Phi^*(s) = \inf_t (ts - \Phi(t)). \tag{13.29}$$

Let $N_A(\Phi)$ denote the $(L^\Phi, L^\Phi)$ norm of the operator $\tau_A$. Then

$$\frac{1}{t} \le N_A(\Phi)N_A(\Phi^*). \tag{13.30}$$

**Proof.** By Holder's inequality

$$||\tau_A f||_2^2 \le ||\tau_A f||_\Phi ||\tau_A f||_{\Phi^*}. \tag{13.31}$$

Since $||\tau_A f||_2^2 = \frac{1}{t}||f||_2$, the conclusion follows. ■
The following is a sample result obtained by combining Theorem 13.5, Lemma 13.9 and Lemma 13.10.

**Theorem 13.12** *Let $T_\gamma$ be as above. Suppose that $\gamma$ satisfies the conditions of Theorem 13.5. Suppose that $\Phi$ and $\Psi$ satisfy (13.24). Then $T_\gamma : L^\Phi(\mathbb{R}^2) \to L^\Psi(\mathbb{R}^2)$ if*

$$\sum_{j=0}^{\infty} 2^{-j}\Phi^{-1}\left(\frac{2^j}{h_j}\right)\Psi^{-1}(2^{-j}h_j) < \infty. \tag{13.32}$$

*Suppose that $\Phi$ and $\Psi$ satisfy (13.26). Then $T_\gamma : L^\Phi(\mathbb{R}^2) \to L^\Psi(\mathbb{R}^2)$ if*

$$\sum_{j=0}^{\infty} \frac{2^{-j}}{\Phi^{-1}(2^{-j}h_j)\Psi^{-1}\left(\frac{2^j}{h_j}\right)} < \infty. \tag{13.33}$$

*Suppose that $\Phi$ satisfies (13.24) and $\Psi$ satisfies (13.26). Then $T_\gamma : L^\Phi(\mathbb{R}^2) \to L^\Psi(\mathbb{R}^2)$ if*

$$\sum_{j=0}^{\infty} 2^{-j} \frac{\Phi^{-1}\left(\frac{2^j}{h_j}\right)}{\Psi^{-1}\left(\frac{2^j}{h_j}\right)} < \infty. \tag{13.34}$$

Suppose that $\Phi$ satisfies (13.26) and $\Psi$ satisfies (13.24). Then $T_\gamma : L^\Phi(\mathbb{R}^2) \to L^\Psi(\mathbb{R}^2)$ if

$$\sum_{j=0}^{\infty} 2^{-j} \frac{\Psi^{-1}(2^{-j}h_j)}{\Phi^{-1}(2^{-j}h_j)} < \infty. \tag{13.35}$$

**Remark 13.1** *A three dimensional version of this result can be generated using Lemma 13.8 and the proof of Theorem 13.5 under the assumption that $\gamma''$ is increasing on $[0,2]$ by replacing $h_j$ by $\gamma_j$ and $2^j$ by $2^{2j}$ in Theorem 13.12 above.*

The following is a sample theorem obtained by combining Theorem 13.6, Lemma 13.9 and Lemma 13.10.

**Theorem 13.13** *Let $\mathcal{R}$ be as above. Suppose that $\gamma$ satisfies the conditions of Theorem 13.6. Suppose that $\Phi$ satisfies (13.24). Then $\mathcal{R} : L^\Phi(\mathbb{R}^2) \to L^2(\Gamma)$ if*

$$\sum_{j=0}^{\infty} \frac{2^{\frac{j}{2}}\Phi^{-1}(2^{-j}h_j)}{h_j} < \infty. \tag{13.36}$$

*Suppose that $\Phi$ satisfies (13.26). Then $\mathcal{R} : L^\Phi(\mathbb{R}^2) \to L^2(\Gamma)$ if*

$$\sum_{j=0}^{\infty} \frac{2^{\frac{j}{2}}}{h_j \Phi^{-1}\left(\frac{2^j}{h_j}\right)} < \infty. \tag{13.37}$$

**Remark 13.2** *Three dimensional analogs can be generated using Lemma 13.8 and the proof of Theorem 13.6 by replacing $h_j$ by $\gamma_j$ and $2^j$ by $2^{2j}$.*

## 13.4 Examples

**Example 13.1** *Let $\Phi(s) = \int_0^s \phi(t)dt$, where $\phi(t) = t^{q-1}\log^{-l}(t)$, $l > 0$. Let $\Psi(s) = s^q$. Let $\gamma(s) = e^{-\frac{1}{s^\alpha}}$. A calculation shows that $\Phi(s) \approx$*

*$s^q \log^{-l}(s)$ and that $\Phi$ satisfies the condition (13.24). Since $\Psi$ satisfies any condition you want, Theorem 13.12, along with the subsection "Simplification" implies that $T_\gamma : L^\Phi(\mathbb{R}^2) \to L^\Psi(\mathbb{R}^2)$ if*

$$\sum_{j=0}^{\infty} 2^{-j} 2^{j\frac{\alpha l}{q}} < \infty, \tag{13.38}$$

*since $\Phi^{-1}(s) \leq s^{\frac{1}{q}} \log^{\frac{l}{q}}(s)$ (see [3], Example 1.3).The sum in (13.38) converges if $\alpha l < q$.*

**Example 13.2** *Let $\Phi^{-1}(s) \approx s \log^{-l}(s)$. Let $\gamma$ be as in the previous example. It is not hard to see that $\Phi$ satisfies (13.26). Theorem 13.13 and the subsection "Simplification" imply that $\mathcal{R} : L^\Phi(\mathbb{R}^2) \to L^2(\Gamma)$ if*

$$\sum_{j=0}^{\infty} 2^{-\frac{j}{2}} 2^{j\alpha l} < \infty, \tag{13.39}$$

*which takes place if $\alpha l < \frac{1}{2}$.*

The necessity results in [1], [2], and [4], associated with Theorems 28.1, 28.2, 28.3 in the introduction, show that the above examples give optimal results, at least up to the endpoints. Indeed, to test the sharpness of the first, we just test $T_\gamma$ against a characteristic of a cube with side-lengths $\delta$ and $\gamma(\delta)$, (see [4]), whereas the sharpness of the second follows by a variant of the classical Knapp homogeneity argument. See [1].

**Acknowledgement.** The author wishes to thank J. Vance for many helpful conversations and for pointing out the reference [5].

*References*

[1] J.-G. Bak, *Restrictions of Fourier tranforms to flat curves in* $\mathbb{R}^2$,Illinois J. Math. 38 (1994).

[2] J.-G. Bak, *Sharp convolution estimates for measures on flat surfaces*,(preprint 1994) .

[3] J.-G. Bak, *Averages over surfaces with infinitely flat points,* J. Func. Anal. 129 (1995).

[4] J.-G. Bak, D. McMichael, and D. Oberlin, *Convolution estimates for some measures on flat curves,* J. Func. Anal. 101 (1991).

[5] A. Carbery, M. Christ, J. Vance, S. Wainger, and D. Watson, *Operators associated to flat plane curves: $L^p$ estimates via dilation methods,* Duke Math. J. 59 (1989).

[6] L. Carleson and P. Sjolin, *Oscillatory integrals and the multiplier problem for the disc,* Studia Math. 44 (1972).

[7] W. Littman, $(L^p, L^q)$ *estimates for singular integral operators,* Proc. Symp. Pure Math. 23 (1973).

[8] F. Ricci and E. Stein, *Harmonic analysis on nilpotent groups and singular integrals II,* J. Func. Anal. 78 (1988).

[9] E. Stein, **Harmonic Analysis**, Princeton Univ. Press (1993).

[10] R. Strichartz, *Convolution with kernels having singularities on a sphere,* Trans. Amer. Math. Soc. 148 (1970).

[11] P. Tomas, *A restriction theorem for the Fourier transform,* Bull. Amer. Math. Soc. 81 (1975).

[12] A. Torchinsky, *Interpolation of operators and Orlicz classes,* Studia Math. 59 (1976)

[13] A. Torchinsky, **Real Variable Methods in Harmonic Analysis** Academic Press, Orlando (1986).

# Chapter 14

# Some $L^p(L^\infty)-$ and $L^2(L^2)-$ estimates for oscillatory Fourier transforms

**Björn G. Walther**

*ABSTRACT. Let* $(S^a f)(t)\hat{}= \exp(it|\xi|^a)\widehat{f}(\xi)$. *We discuss some examples of maximal estimates and weighted* $L^2$*-estimates for* $Sf$. *The techniques used include asymptotics for Bessel functions and the complete orthogonal decomposition of* $L^2(\mathbb{R}^n)$ *using spherical harmonics.*

## 14.1 Introduction

The purpose of this paper is to discuss some examples of norm inequalities for oscillatory Fourier integrals. We will consider $L^p(L^\infty)$ and $L^2(L^2)$. $L^p(L^\infty)$ means that we are considering Lebesgue space estimates for maximal functions an d $L^2(L^2)$ means that we take an $L^2$-norm with respect to time and a (weighted) $L^2$-norm with respect to range.

The problems treated here are traced to the work of Carleson [5], Stein [23] and Strichartz [26]. More recent development is treated e.g. in Ben-Artzi, Devinatz [1], Bourgain [4], Craig, Kappeler, Strau ss [8], Heinig, Wang [10, 11], Kolasa [15], Kenig, Ponce, Vega [14], Moyua, Vargas, Vega [16], Prestini [17], Sjölin [19], Walther [29] and Wang [30]. If $u$ is the solution to the time-dependent free Schrödinger equation or some other oscillatory Fourier integral $L^p(L^q)$-estimates of $u$ reveal its regularity and decay nature. The case $q = \infty$ gives us maximal estimates for $u$ which in turn is i nteresting because of the relationship of such estimates with the problem of pointwise convergence. Alternatively, this pointwise convergence problem can be seen as a summabilty problem. In our examples the kernel for the summability process will be non-summable.

**Definition 14.1** *For $f$ in the Schwartz class $S(\mathbb{R}^n)$ of rapidly decreasing $C^\infty$-functions and for $a > 0$ we define*

$$(S^a f)[x](t) = \frac{1}{(2\pi)^n} \int\limits_{\mathbb{R}^n} e^{i(x\xi + t|\xi|^a)}\, \widehat{f}(\xi)\, d\xi, \quad t \in \mathbb{R}.$$

*Here $\widehat{f}$ is the Fourier transform of $f$*

$$\widehat{f}(\xi) = \int\limits_{\mathbb{R}^n} e^{-ix\xi} f(x)\, dx.$$

Let $u(x,t) = (S^a f)[x](t)$. If $a = 1$, then $u$ is a solution to the classical wave equation $\Delta_x u = \partial_t^2 u$ with the initial conditions $u(x,0) = f(x)$ and $[\partial_t u](x,0) = g(x)$, $\widehat{g}(\xi) = |\xi|\, \widehat{f}(\xi)$. If $a = 2$, then $u$ is a solution to the time-dependent free Schrödinger equation $\Delta_x u = i\partial_t u$ with the initial condition $u(x,0) = f(x)$. These cases of $a$ explains some of the relevance for the oscillatory Fourier integrals studied here. Due to the non-vanishing of $\xi \mapsto \exp(it|\xi|^a)$ at infinity it is clear that the kernel of the summability process $t \to 0$ is non-summable. For fixed $t$ however $(Sf)(t)$ is an element in $L^2(\mathbb{R}^n)$ if $f \in L^2(\mathbb{R}^n)$.

**Definition 14.2** *By $H^s(\mathbb{R}^n)$ we will denote the set of all tempered distributions $f$ such that*

$$\|f\|^2_{H^s(\mathbb{R}^n)} = \int\limits_{\mathbb{R}^n} \left(1 + |\xi|^2\right)^s d\xi$$

*is finite.*

We will also need some auxiliary definitions. By $B^n$ and $\Sigma^{n-1}$ we denote the open unit ball and the unit sphere in $\mathbb{R}^n$ respectively. $B^1$ will be denoted by $B$. The symbol $d\sigma$ is used when we integrate with respect to the surface measure induced by the Lebesgue measure on $\mathbb{R}^n$.

Throughout this work we will use auxiliary functions $\chi$ and $\psi$ such that $\chi \in C^\infty(\mathbb{R})$ is even,

$$\chi(\mathbb{R} \setminus 2B) = \{0\}, \quad \chi(\mathbb{R}) \subseteq [0,1], \quad \chi(B) = \{1\}$$

and $\psi = 1 - \chi$. From $\chi$ we obtain new functions as follows: for each positive number $N$ set $\chi_N(x) = \chi(x/N)$.

If $A$ is a normed space, $L^p(X, A)$, $X \subseteq \mathbb{R}^n$ will denote those functions $u : X \to A$, measurable in a suitable sense, such that

$$\int\limits_X \|u\|_A^p\, dx < \infty.$$

Unless otherwise explicitly stated functions $f$ will belong to $\mathcal{S}(\mathbb{R}^n)$.

Numbers denoted by $C$ may different at each occurence even within the same chain of inequalities.

## 14.2 $L^p(L^\infty)$-estimates

**Theorem 14.1** *There is a number $C$ independent of $f$ such that the inequality*

$$\|S^1 f\|_{L^2(\mathbb{R}^n, L^\infty(B))} \leq C \|f\|_{H^s(\mathbb{R}^n)}$$

*holds if $s > 1/2$.*

**Proof.** We will prove a more general result of which the theorem will follow. In the general result we consider conditions on (families of) multipliers reminiscent to those given in Stein [24, § XI.4.1 p. 511] and in the references [18], [20] and [23] cited there. See also Cowling [6] and Cowling, Mauceri [7].

Let us define

$$(S_m f)[x](t) = \int_{\mathbb{R}^n} e^{ix\xi}\, m(t,\xi)\, \widehat{f}(\xi)\, d\xi.$$

The assumptions on $m$ are the following:

$$|m(t,\xi)| \leq C\, w_1(t), \; |[\partial_t m]\,(t,\xi)| \leq C\, (w_2(t) + w_1(t)|\xi|^a)\,, \; a > 0. \quad (14.1)$$

Here $C > 0$ is independent of $(t, \xi)$. $w_1$ and $w_2$ are functions in $L^2(\mathbb{R})$. The general result is that under these assumptions there is a number $C$ independent of $f$ such that

$$\|S_m f\|_{L^2(\mathbb{R}^n, L^\infty(\mathbb{R}))} \leq C \|f\|_{H^s(\mathbb{R}^n)}, \quad s > a/2.$$

It is clear that $m(t,\xi) = \chi(t) \exp(it|\xi|^a)$ satisfies the condition in (14.1) and so the theorem follows from this result.

To prove the general result we use Parseval's formula to get that there is a number $C$ independent of $f$ such that

$$\|S_m f\|_{L^2(\mathbb{R}^n, H^0(\mathbb{R}))} \leq C \|f\|_{H^0(\mathbb{R}^n)}. \quad (14.2)$$

If we differentiate $(S_m f)\,[x](t)$ with resepect to $t$ and use Parseval's formula we get that there is a number $C$ independent of $f$ such that

$$\|\partial_t\, (S_m f)\|_{L^2(\mathbb{R}^n, H^0(\mathbb{R}))} \leq C \|f\|_{H^a(\mathbb{R}^n)}. \quad (14.3)$$

Combining (14.2) and (14.3) and invoking the equivalence of norms in Sobolev spaces (cf. e.g. Stein [22, lemma 3 p. 136]) yields that there is a number $C$ independent of $f$ such that

$$\|S_m f\|_{L^2(\mathbb{R}^n, H^1(\mathbb{R}))} \leq C \|f\|_{H^a(\mathbb{R}^n)}. \quad (14.4)$$

Interpolation of vector valued Lebesgue spaces and Bessel potential spaces (cf. Bergh, Löfström [3, theorem 5.1.2 p. 107 and theorem 5.4.1 (7) p. 153]) applied to (14.2) and (14.4) now yields

$$\|S_m f\|_{L^2(\mathbb{R}^n, H^s(\mathbb{R}))} \leq C \|f\|_{H^{sa}(\mathbb{R}^n)}, \quad 0 < s < 1.$$

The general result now follows form this inequality with $s > 1/2$ in conjunction with the linear embedding $H^s(\mathbb{R}) \subseteq L^\infty(\mathbb{R})$, $s > 1/2$. ■

We now turn our attention to the necessary condition on the regularity in the case $a = 1$. This is very well understood in the case $n = 1$. In the case $n > 1$ we use the asymptotics of one single Bessel function to reduce the problem to the case $n = 1$. To the author of this paper it is not known where the case $n > 1$ is treated in detail and so it is included here.

**Theorem 14.2** *Assume that there exists a number $C$ independent of $f$ such that the inequality*

$$\|S^1 f\|_{L^1(B^n, L^\infty(B))} \leq C \|f\|_{H^s(\mathbb{R}^n)}$$

*holds and that $f$ is radial. Then $s > 1/2$.*

**Proof.** We will use the well-known fact that a necessary condition for the linear embedding $H^s(\mathbb{R}) \subseteq L^\infty(\mathbb{R})$ is that $s > 1/2$. A counterexample is provided for $s = 1/2$ by the function $g_N$ given by

$$\widehat{g_N}(\xi) = |\xi|^{-1/2} (\log|\xi|)^{-3/4} \psi(\xi + |\xi|) \chi_N(\xi). \tag{14.5}$$

Now, let us assume that there is a number $C$ independent of $f$ such that

$$\|S^1 f\|_{L^p(B, L^\infty(B))} \leq C \|f\|_{H^{1/2}(\mathbb{R})}, \quad p > 0. \tag{14.6}$$

Let $\mathcal{J}_s$ be the Bessel potential of order $s$. (Cf. e.g. Stein [22, p. 131].) If $\operatorname{supp} \widehat{f} \subset \mathbb{R}_+$, then

$$|f(0)| = \left( \int_0^1 |(S^1 f)[-x](x)|^p \, dx \right)^{1/p} \leq \|S^1 f\|_{L^p(B, L^\infty(B))} \leq$$

$$\leq C \|f\|_{H^{1/2}(\mathbb{R})} = C \left\|\mathcal{J}_{-1/2} f\right\|_{L^2(\mathbb{R})}$$

which yields

$$\left| \left[\mathcal{J}_{1/2} g\right](0) \right| \leq C \|g\|_{L^2(\mathbb{R})}, \quad g = \mathcal{J}_{-1/2} f.$$

In particular this inequality is valid when we replace $g$ by $g_N$. Now

$$\left[\mathcal{J}_{1/2} g_N\right](0) = \frac{1}{2\pi} \int_{\mathbb{R}} \frac{\psi(\xi + |\xi|) \chi_N(\xi)}{(1+\xi^2)^{1/4} |\xi|^{1/2} (\log|\xi|)^{3/4}} \, d\xi \tag{14.7}$$

which goes to infinity as $N$ goes to infinity. However

$$\sup_N \|g_N\|^2_{L^2(\mathbb{R})} \leq \frac{1}{2\pi} \int_{\mathbb{R}} |\xi|^{-1} (\log|\xi|)^{-3/2}\, \psi(\xi + |\xi|)^2\, d\xi \; < \infty$$

and this contradiction proves the theorem in the case $n = 1$.

We now turn our attention to the case $n > 1$. Assume that there is a number $C$ independent of $f$ such that

$$\left\|S^1 f\right\|_{L^1(B^n, L^\infty(B))} \leq C\, \|f\|_{H^{1/2}(\mathbb{R})}.$$

If $g = \mathcal{J}_{-1/2} f$ this yields

$$\left\|\left[S \circ \mathcal{J}_{1/2}\right] g\right\|_{L^1(B^n, L^\infty)} \leq C\, \|g\|_{L^2(\mathbb{R})}.$$

Consider

$$(\widetilde{S}g)[x](t) = \int_{\mathbb{R}^n} e^{i(x\xi + t|\xi|)}\, (1 + |\xi|^2)^{-1/4}\, \widehat{g}(\xi)\, d\xi.$$

Then $\widetilde{S} = S \circ \mathcal{J}_{1/2}$ and under the assumption made

$$\|\widetilde{S}g\|_{L^1(B^n, L^\infty(B))} \leq C\, \|\widehat{g}\|_{L^2(\mathbb{R}^n)},$$

where $C$ may be chosen to be independent of $g$. Let $J_\lambda$ denote the Bessel function of order $\lambda$. Specialising to real and radial Fourier transforms, using Stein, Weiss [25, theorem 3.10 p. 158] and setting $t = r$ in the lef t hand side gives that

$$\int_0^1 r^{-n/2+1} \left| \int_0^\infty J_{n/2-1}(r\rho)\, e^{ir\rho}\, (1 + \rho^2)^{-1/4}\, \widehat{g}(\rho)\, \rho^{n/2}\, d\rho \right| r^{n-1}\, dr \leq$$

$$\leq C \left( \int_0^\infty \left| \widehat{g}(\rho)\, \rho^{n/2-1/2} \right|^2 d\rho \right)^{1/2}$$

and hence that

$$\int_0^\infty \left| \int_0^\infty r^{1/2} J_{n/2-1}(r\rho) \rho^{1/2} e^{ir\rho} (1 + \rho^2)^{-1/4} \widehat{g}(\rho) d\rho \right| \chi(2r) r^{n/2-1/2} dr$$

$$\leq C\, \|g\|_{L^2(\mathbb{R})}$$

must hold, where $g \in \mathcal{S}(\mathbb{R})$ and $C$ may be chosen to be independent of $g$. In particular

$$\int_0^\infty \left| \int_0^\infty r^{1/2} J_{n/2-1}(r\rho) \rho^{1/2} e^{ir\rho} (1 + \rho^2)^{-1/4} \widehat{g_N}(\rho) d\rho \right| \chi(2r) r^{n/2-1/2} dr \tag{14.8}$$

is a bounded function of $N$ where $\widehat{g_N}$ is even and is given by (14.5) for $\rho > 0$. We shall prove that this does not hold.

Remainder term estimate. It follows from the Poisson representation of Bessel functions (cf. [25, lemma 3.1 p. 153]) and from [25, lemma 3.11 p. 158] that there are numbers $C_1$ and $C_2$ such that

$$\left|\int_0^\infty [r^{1/2} J_{n/2-1}(r\rho)\rho^{1/2} - C_1 e^{-ir\rho} - C_2 e^{ir\rho}] e^{ir\rho}(1+\rho^2)^{-1/4}\widehat{g_N}(\rho)d\rho\right|$$

$$\leq C \int_0^\infty \left\{1 \wedge (r\rho)^{-1}\right\} (1+\rho^2)^{-1/4} \left|\widehat{g_N}(\rho)\right| d\rho \tag{14.9}$$

where $C$ may be chosen to be independent of $N$. Now we multiply by $\chi(2r)r^{n/2-1/2}$, integrate with respect to $r$, reverse the order of integration and apply Cauchy–Schwarz inequality to get

$$C \left[\int_0^\infty \left(\int_0^\infty \left\{1 \wedge (r\rho)^{-1}\right\} \chi(2r)\, r^{n/2-1/2}\, dr\right)^2 (1+\rho^2)^{-1/2}\, d\rho\right]^{1/2} \times$$

$$\times \left\|g_N\right\|_{L^2(\mathbb{R})} \tag{14.10}$$

as upper bound. By dominated convergence the inner integral is majorized by $C/(1+\rho)$ where $C$ may be chosen to be independent of $\rho$. Hence the outer integral is convergent. Also, $\sup_N \left\|g_N\right\|_{L^2(\mathbb{R})}$ is finite.

Oscillatory term estimate. Next, by Cauchy–Schwarz inequality and two applications of Parseval's formula

$$\int_0^\infty \left|\int_0^\infty C_2 e^{2ir\rho}(1+\rho^2)^{-1/4}\widehat{g_N}(\rho)\, d\rho\right| \chi(2r) r^{n/2-1/2}\, dr \leq C \left\|g_N\right\|_{L^2(\mathbb{R})}, \tag{14.11}$$

where $C$ may be chosen to be independent of $N$. Also, $\sup_N \left\|g_N\right\|_{L^2(\mathbb{R})}$ is finite.

Main term estimate. Finally,

$$\int_0^\infty \left|\int_0^\infty C_1 (1+\rho^2)^{-1/4}\, \widehat{g_N}(\rho)\, d\rho\right| \chi(2r)\, r^{n/2-1/2}\, dr =$$

$$= C \left|\left(\mathcal{J}_{1/2} g_N\right)(0)\right| \tag{14.12}$$

tends to infinity as $N$ tends to infinity.

Conclusion. Now, by representing the expression $r^{1/2}\, J_{n/2-1}(r\rho)\, \rho^{1/2}\, e^{ir\rho}$ as

$$\left[r^{1/2}\, J_{n/2-1}(r\rho)\, \rho^{1/2} - C_1\, e^{-ir\rho} - C_2\, e^{ir\rho}\right] e^{ir\rho} + C_2\, e^{2ir\rho} + C_1$$

and by combining (14.9) and (14.12) it is clear that the expression in (14.8) cannot be a bounded function of $N$. ■

In theorem 14.1 we treated maximal estimates for multiplier operators. We imposed regularity and decay conditions on the symbol $m$. Next we will consider maximal estimates for operators which are more of a pseudodifferential operator type. The assumptions on the symbol $m$ will be that it is bounded and radial in the frequency.

**Theorem 14.3** *Let $n \geq 2$ and let $m \in L^\infty(T \times \mathbb{R}^n \times \mathbb{R})$. Then there exists a number $C$ independent of $f$ such that*

$$\int_{\mathbb{R}^n} \left( \sup_{t \in M} \left| \int_{\mathbb{R}^n} e^{ix\xi}\, m(t,x,|\xi|)\, \widehat{f}(\xi)\, d\xi \right|^2 \right) \frac{dx}{(1+|x|)(1+\log^+|x|)^b} \leq$$

$$\leq C\, \|f\|^2_{H^s(\mathbb{R}^n)} \tag{14.13}$$

*if $b > 1$ and $s > 1/2$.*

**Proof.** (Cf. Soljanik [21].) Let us write $\left[\widehat{f}\right]_\rho(\xi) = \widehat{f}(\rho\xi)$, $L^\infty = L^\infty(T \times \mathbb{R}^n \times \mathbb{R}_+)$ and

$$(S_m f)[x](t) = \int_{\mathbb{R}^n} e^{ix\xi}\, m(t,x,|\xi|)\, \widehat{f}(\xi)\, d\xi.$$

By polar coordinates we get

$$(S_m f)[x](t) = \int_0^\infty m(t,x,\rho) \left( \int_{\Sigma^{n-1}} e^{i\rho x\xi} \left[\widehat{f}\right]_\rho(\xi)\, d\sigma(\xi) \right) \rho^{n-1}\, d\rho.$$

Let $N$ be a dyadic integer. Upon integrating with respect to $x$

$$\int_{|x|\leq N} \|(S_m f)[x]\|^2_{L^\infty(T)}\, dx \leq$$

$$\leq \int_{|x|\leq N} \left( \sup_{t\in I} \int_0^\infty |m(t,x,\rho)| \left| \int_{\Sigma^{n-1}} e^{i\rho x\xi} \left[\widehat{f}\right]_\rho(\xi) d\sigma(\xi) \right| \rho^{n-1} d\rho \right)^2 dx \leq$$

$$\leq \|m\|^2_{L^\infty} \int_{|x|\leq N} \left( \int_0^\infty \left| \int_{\Sigma^{n-1}} e^{i\rho x\xi} \left[\widehat{f}\right]_\rho(\xi)\, d\sigma(\xi) \right| \rho^{n-1}\, d\rho \right)^2 dx.$$

Applying Cauchy–Schwarz inequality to the integral with respect to $\rho$ gives that there is a number $C$ independent of $f$ such that

$$\int\limits_{|x|\leq N} \|(Sf)[x]\|^2_{L^\infty(T)}\,dx \leq C\,\|m\|^2_{L^\infty} \times$$
$$\times \int\limits_{|x|\leq N} \left( \int\limits_0^\infty (1+\rho^2)^s \left| \int\limits_{\Sigma^{n-1}} e^{i\rho x\xi} \left[\widehat{f}\right]_\rho (\xi)\,d\sigma(\xi) \right|^2 \rho^{2n-2}\,d\rho \right) dx$$

for $s > 1/2$ and changing the order of integration yields together with a change of variables

$$\int\limits_{|x|\leq N} \|(Sf)[x]\|^2_{L^\infty(T)}\,dx \leq C\,\|m\|^2_{L^\infty} \times$$
$$\times \int\limits_0^\infty (1+\rho^2)^s \left( \int\limits_{|x|\leq \rho N} \left| \int\limits_{\Sigma^{n-1}} e^{ix\xi} \left[\widehat{f}\right]_\rho (\xi)\,d\sigma(\xi) \right|^2 dx \right) \rho^{n-2}\,d\rho.$$

Here we use Hörmander [12, theorem 7.1.26 p. 173] on the integral with respect to $x$ to get

$$\begin{aligned}
&\int\limits_{|x|\leq N} \|(Sf)[x]\|^2_{L^\infty(T)}\,dx \\
&\leq C\,\|m\|^2_{L^\infty} \int\limits_0^\infty (1+\rho^2)^s \left( N \left\| \left[\widehat{f}\right]_\rho \right\|^2_{L^2(\Sigma^{n-1})} \right) \rho^{n-1}\,d\rho \\
&= C\,N\,\|m\|^2_{L^\infty}\,\|f\|^2_{H^s(\mathbb{R}^n)},
\end{aligned} \tag{14.14}$$

where $C$ is independent of $f$ and $N$. The theorem now follows from (14.14) since $\sum_N (1+\log N)^{-b}$ is convergent. ∎

**Corollary 14.4** *(Soljanik [21].) Let $n \geq 2$ and let $m(t,x,\rho)$ be independent of $x$. Then there exists a number $C$ independent of $f$ and $R > 0$ such that*

$$\int\limits_{|x|\leq R} \|(S_m f)[x]\|^2_{L^\infty(T)}\,dx \leq C\,R\,\|m\|^2_{L^\infty}\,\|f\|^2_{H^s(\mathbb{R}^n)},$$

*if $s > 1/2$.*

Our source for combining polar coordinates, Cauchy–Schwarz inequality and [12, theorem 7.1.26 p. 173] is Soljanik [21], whose result (corollary

14.4) is generalized in two respects: the symbol $m$ in theorem 14.3 n ot only depends on $t$ and $|\xi|$ but also on $x$; the estimate of theorem 14.3 is $x$-global.

For $m(t,x,|\xi|) = m_0(t\varphi(|\xi|))$, $m_0 \in L^\infty(\mathbb{R})$, $t \in \mathbb{R}$ Vega [28, teorema 1.23, p. 48] uses the method in the preceeding proof to obtain (14.14).

One of the purposes of including theorem 14.3 here is that it implies the inequality

$$\int\limits_{\mathbb{R}^n} \left( \sup_{t\in\mathbb{R}} \left| \int\limits_{\mathbb{R}^n} e^{it|\xi|^a} e^{ix\xi} \widehat{f}(\xi)\, d\xi \right|^2 \right) \frac{dx}{(1+|x|)^b} \leq C\, \|f\|_{H^s(\mathbb{R}^n)},$$

$$b > 1, \quad s > 1/2 \tag{14.15}$$

stated in Vega [27, p. 874]. $C$ is assumed to be independent of $f$. In [27] the inequality

$$\int\limits_{\mathbb{R}^n} \int\limits_{\mathbb{R}} \left| \int\limits_{\mathbb{R}^n} e^{it|\xi|^a} e^{ix\xi} \widehat{f}(\xi)\, d\xi \right|^2 \frac{dt\, dx}{(1+|x|)^b} \leq C\, \|f\|_{H^{-1+b/2}(\mathbb{R}^n)}, \quad b > 1 \tag{14.16}$$

is used as an argument for (14.15). See [27, p. 874 line 3 and 4 from below]. It is assumed that $b > 1$ may be chosen arbitrarily close to 1. However, for such $b$ Wang [31, theorem 1 p. 88] has given a counterexample disproving (14.16).

One motivation for studying (14.15) is that it has consequences for the almost everywhere convergence problem. For that problem it is however sufficient to consider $x$-local estimates. $x$-global estimates as in theorem 14.3 are of in dependent interest for they reveal global regularity properties of maximal functions. Such properties are also studied in Heinig, Wang [10, 11].

## 14.3 $L^2(L^2)$-estimates

Wang's counterexample to Vega's claim raises the question whether there are any weighted inequalities

$$\int\limits_{\mathbb{R}^n} \|(S^a f)[x]\|^2_{L^2(\mathbb{R})} \frac{dx}{(1+|x|)^b} \leq C\, \|f\|^2_{H^s(\mathbb{R}^n)} \tag{14.17}$$

at all. This problem is of interest indepent of results for the Schrödinger maximal function. In this section we obtain a characterization in the case $1 < a \leq n$. The case $a > n$ will be treated elsewhere.

It is noteworthy that thelow frequencies of $f$ are important when analyzing the behaviour of $\|(S^a f)[x]\|_{L^2(\mathbb{R})}$ at infinity. This is expressed by the fact that it is enough to test the inequality (14.17) for functions whos e Fourier transform has support in $B^n$ to obtain the necessary conditions in the parts (b) of the theorems below.

We begin with a lemma which can be thought of as a counterpart to Guo's uniform $L^{4+\varepsilon}$-property [9, lemma 3.2 p. 1333] for Bessel functions.

**Lemma 14.5** *There is a number $C$ independent of $\lambda \in \mathbb{N}/2$ such that*

$$\sup_{r>0} \frac{1}{r} \int_0^r J_\lambda(\rho)^2 \rho \, d\rho \leq C.$$

**Proof.** Combine [12, theorem 7.1.26 p. 173] with [25, theorem 3.10 p. 158]. ■

**Lemma 14.6** *Let $b > 1$. Then there exist a number $C$ independent of $\rho \geq 1$ and $k$ such that*

$$\int_\rho^\infty J_{\nu(k)}(r)^2 \, r^{1-b} \, dr \leq C \, \rho^{1-b}, \quad \nu(k) = \frac{n}{2} + k - 1.$$

**Proof.** Let $N$ be a dyadic integer and let $\rho \geq 1$. We have

$$\int_\rho^\infty J_{\nu(k)}(r)^2 \frac{dr}{r^{b-1}} = \sum_N (2N\rho) \frac{1}{2N\rho} \int_{N\rho}^{2N\rho} J_{\nu(k)}(r)^2 \, r \frac{dr}{r^b} \leq$$

$$\leq \sum_N \frac{2N\rho}{(N\rho)^b} \frac{1}{2N\rho} \int_0^{2N\rho} J_{\nu(k)}(r)^2 \, r \; dr \leq$$

$$\leq \sum_N \frac{2N\rho}{(N\rho)^b} C \leq \frac{2C}{\rho^{b-1}} \left( \sum_N N^{1-b} \right)$$

where the number $C$ may be chosen to be independent of $k$ by lemma 14.5. ■

**Theorem 14.7** *Assume that $1 < a < n$ and that $n \geq 2$.*

*(a) There is a number $C$ independent of $f$ such that*

$$\int_{\mathbb{R}^n} \|(S^a f)[x]\|^2_{L^2(\mathbb{R})} \frac{dx}{(1+|x|)^a} \leq C \, \|f\|^2_{H^{(1-a)/2}(\mathbb{R}^n)}.$$

*(b) Assume that there is a number $C$ independent of $f$ such that*

$$\int_{\mathbb{R}^n} \|(S^a f)[x]\|^2_{L^2(\mathbb{R})} \frac{dx}{(1+|x|)^b} \leq C\,\|f\|^2_{L^2(\mathbb{R}^n)}, \quad \operatorname{supp}\widehat{f} \subseteq B^n$$

*and that $f$ is radial. Then $b \geq a$.*

**Proof.** The auxiliary operator $\widetilde{S^a}$. Define

$$(\widetilde{S^a} f)[x](t) = \frac{1}{(1+|x|)^{b/2}} \int_{\mathbb{R}^n} e^{i(x\xi + t|\xi|^a)}(1+|\xi|^2)^{-s/2} f(\xi)\,d\xi, \quad t \in \mathbb{R}. \tag{14.18}$$

Part (a) of our theorem follows if we can show that for $b = a$ and $s = (1-a)/2$ there is a number $C$ independent of $f$ such that

$$\|\widetilde{S^a} f\|_{L^2(\mathbb{R}^{n+1})} \leq C\,\|f\|_{L^2(\mathbb{R}^n)}. \tag{14.19}$$

The mapping $f \to \widetilde{f^a}$. For $\rho > 0$ we define

$$\begin{aligned} \widetilde{f^a}[x](\rho) &= \widetilde{f^a}(\rho)[x] \\ &= \frac{(1+\rho^{2/a})^{-s/2}\rho^{(n-a)/a}}{a(1+|x|)^{b/2}} \int_{\Sigma^{n-1}} e^{i\rho^{1/a}x\xi'}\, f(\rho^{1/a}\xi')\,d\sigma(\xi'),\ \rho > 0 \end{aligned}$$

(and by $\widetilde{f}(\rho) = 0$ for $\rho \leq 0$). The formula

$$(\widetilde{S^a} f)[x](t) = \int_0^\infty e^{it\rho}\widetilde{f^a}[x](\rho)\,d\rho \tag{14.20}$$

follows by polar coordinates and change of variables in (14.18) and

$$\left\|\widetilde{S^a} f\right\|_{L^2(\mathbb{R}^{n+1})} = (2\pi)^{1/2}\left\|\widetilde{f^a}\right\|_{L^2(\mathbb{R}^{n+1})}$$

follows from Parseval's formula on $\mathbb{R}$ applied to (14.20). Hence, to prove (14.19) (where $C$ is independent of $f$) it is sufficient to prove that

$$\left\|\widetilde{f^a}\right\|_{L^2(\mathbb{R}^{n+1})} \leq C\,\|f\|_{L^2(\mathbb{R}^n)} \tag{14.21}$$

where $C$ may be chosen to be independent of $f$.

Orthogonality for $\widetilde{f^a}(\rho)$ in $L^2(\Sigma^{n-1})$. For a non-negative integer $k$ we will consider the linear span $\mathfrak{H}_k(\mathbb{R}^n)$ in $L^2(\mathbb{R}^n)$ of $Pf$ where $P$ is a spherical harmonic of degree $k$ and $f$ is a radial function. $\mathfrak{H}_k(\mathbb{R}^n)$ is a closed subspace of $L^2(\mathbb{R}^n)$ of infinite dimension. Cf. Stein, Weiss [25, p. 151].)

If $f_i \in (\mathfrak{H}_{k_i} \cap \mathcal{S})(\mathbb{R}^n)$, $k_1 \neq k_2$ we have

$$\int_{\Sigma^{n-1}} \widetilde{f_1^a}(\rho)[rx']\,\overline{\widetilde{f_2^a}}(\rho)[rx']\,d\sigma(x') = 0$$

for all $r > 0$ and all $\rho > 0$. In fact

$$\int_{\Sigma^{n-1}} \widetilde{f_1^a}(\rho)[rx']\,\overline{\widetilde{f_2^a}}(\rho)[rx']\,d\sigma(x') =$$

$$= \frac{(1+\rho^{2/a})^{-s}\rho^{2(n-a)/a}}{a^2(1+r)^b} \int_{\Sigma^{n-1}} \left( \int_{\Sigma^{n-1}} e^{i\rho^{1/a}rx'\xi'}\, f_1(\rho^{1/a}\xi')\,d\sigma(\xi') \right) \times$$

$$\times \left( \int_{\Sigma^{n-1}} e^{-i\rho^{1/a}rx'\xi'}\, \overline{f_2}(\rho^{1/a}\xi')\,d\sigma(\xi') \right) d\sigma(x').$$

It follows from [25, theorem 3.10 p. 158] that each of the inner integrals in the right hand side will produce a linear combination of products of spherical harmonics of degree $k_i$ times certain functions of $(r, \rho)$. Because of the assumption $k_1 \neq k_2$ the spherical harmonics (or rather thesurface spherical harmonics) will be orthogonal in $L^2(\Sigma^{n-1})$, which proves our assertion.

Orthogonality for $\widetilde{f^a}$ in $L^2(\mathbb{R}^{n+1})$. We have the orthogonality relation

$$\left\langle \widetilde{f_1^a}, \widetilde{f_2^a} \right\rangle_{L^2(\mathbb{R}^{n+1})} = 0, \quad f_i \in \mathfrak{H}_{k_i}(\mathbb{R}^n), \quad k_1 \neq k_2.$$

In fact

$$\int_{\Sigma^{n-1}} \left\langle \widetilde{f_1^a}[rx'], \widetilde{f_2^a}[rx'] \right\rangle_{L^2(\mathbb{R})} d\sigma(x') =$$

$$= \int_0^\infty \int_{\Sigma^{n-1}} \widetilde{f_1^a}[rx'](\rho)\overline{\widetilde{f_2^a}}[rx'](\rho)\,d\sigma(x')\,d\rho = 0$$

where we have applied the preceeding orthogonality assertion to get the last equality. Integrating from 0 to $\infty$ with respect to the measure $r^{n-1}dr$ now proves our assertion.

Orthogonality preserved by $f \mapsto \widetilde{f^a}$. For any element

$$f : \xi \longmapsto \sum P(\xi)\, f_0(|\xi|)\, |\xi|^{-n/2-k+1/2}, \quad f_0 \in C_0^\infty(\mathbb{R}_+) \tag{14.22}$$

in $\mathfrak{H}_k(^n)$ we can assume that thesurface spherical harmonics $P|_{\Sigma^{n-1}}$ are orthogonal in $L^2(\Sigma^{n-1})$ and hence, by polar coordinates, that the terms in

(14.22) are orthogonal in $\mathfrak{H}_k(\mathbb{R}^n)$. Then the terms of $\widetilde{f^a}$ will be orthogonal in $L^2(\mathbb{R}^{n+1})$, since the terms of $x \mapsto \widetilde{f^a}(\rho)[rx]$ restricted to $\Sigma^{n-1}$ will be orthogonal in $L^2(\Sigma^{n-1})$ for fixed $(r, \rho)$. Also

$$\|f\|^2_{L^2(\mathbb{R}^n)} = \sum \|f_0\|^2_{L^2(\mathbb{R}_+)}. \tag{14.23}$$

Key estimate for Bessel functions. It is a consequence of orthogonality that it is sufficient to prove the estimate (14.21) for $f \in (\mathfrak{H}_k \cap \mathcal{S})(\mathbb{R}^n)$ where the number $C$ has to be independent of $f$ and $k$. In turn, it is a consequence of orthogonality that it is sufficient to prove the estimate (14.21) for

$$f : \xi \longmapsto P(\xi)\, f_0(|\xi|)\, |\xi|^{-n/2-k+1/2}, \quad f_0 \in C_0^\infty(\mathbb{R}_+),$$

where the number $C$ has to be independent of $f$and $k$. Straightforward computations using change of variables, [25, theorem 3.10 p. 158] and polar coordinates show that for such $f$

$$\|\widetilde{f^a}\|^2_{L^2(\mathbb{R}^{n+1})} = \\ = \frac{(2\pi)^n}{a} \int_0^\infty \int_0^\infty J_{\nu(k)}(\rho r)^2 (1+\rho^2)^{-s}\, \rho^{2-a}\, |f_0(\rho)|^2\, r \frac{dr\, d\rho}{(1+r)^b}.$$

Hence for $b = a$ it is sufficient to prove the estimate

$$\int_0^\infty \int_0^\infty J_{\nu(k)}(\rho r)^2\, (1+\rho^2)^{-s}\, \rho^{2-a}\, f_0(\rho)\, r\, \frac{dr\, d\rho}{(1+r)^b} = \\ = \int_0^\infty \int_0^\infty J_{\nu(k)}(r)^2 (1+\rho^2)^{-s} f_0(\rho) \rho^{b-a} r \frac{dr\, d\rho}{(\rho+r)^b} \le C\, \|f_0\|_{L^1(\mathbb{R}_+)}, \\ f_0 \ge 0,$$

where $C$ may be chosen to be independent of $f_0 \in C_0^\infty(\mathbb{R}_+)$ and $k$. This estimate will be proved by replacing $f_0(\rho)$ by $\chi(2\rho) f_0(\rho)$ and $\psi(2\rho) f_0(\rho)$.

Estimate for $\chi(2\rho) f_0(\rho)$. We shall derive the estimate

$$\int_0^\infty \int_0^\infty J_{\nu(k)}(r)^2\, \chi(2\rho)\, f_0(\rho)\, \rho^{b-a}\, r\, \frac{dr\, d\rho}{(\rho+r)^b} \le C \int_0^\infty \chi(2\rho)\, f_0(\rho)\, d\rho \tag{14.24}$$

where $C$ may be chosen to be independent of $f_0 \in C_0^\infty(\mathbb{R}_+)$ and $k$. Recall that $b = a$. We have

$$\int_0^\infty \int_0^\infty J_{\nu(k)}(r)^2 \, \chi(2\rho) \, f_0(\rho) \, \rho^{b-a} \, r \, \frac{dr \, d\rho}{(\rho + r)^b} \leq$$

$$\leq \int_0^\infty \chi(2\rho) \, f_0(\rho) \left( \int_0^\infty J_{\nu(k)}(r)^2 \, r^{1-a} \, dr \right) d\rho.$$

Here we split the integration with respect to $r$ into two pieces and use the Poisson representation for $0 \leq r \leq 1$ and lemma 14.6 (with $\rho = 1$) for $r \geq 1$ to conclude that the inner integral is bounded from above by a number $C$ independent of $k$. We have proved (14.24).

Estimate for $\psi(2\rho) f_0(\rho)$. We shall derive the estimate

$$\int_0^\infty \int_0^\infty J_{\nu(k)}(r)^2 \, \psi(2\rho) \, f_0(\rho) \, \rho^{b-a-2s} \, r \, \frac{dr \, d\rho}{(\rho + r)^b}$$

$$\leq C \int_0^\infty \psi(2\rho) \, f_0(\rho) \, d\rho \tag{14.25}$$

where $C$ may be chosen to be independent of $f_0 \in C_0^\infty(\mathbb{R}_+)$ and $k$. Recall that $b = a$ and that $s = (1 - a)/2$. We have

$$\int_0^\infty \int_0^\infty J_{\nu(k)}(r)^2 \, \psi(2\rho) \, f_0(\rho) \, \rho^{b-a-2s} \, r \, \frac{dr \, d\rho}{(\rho + r)^b} \leq$$

$$\leq \int_0^\infty \psi(2\rho) \, f_0(\rho) \, \rho^{b-1} \left( \int_0^\infty J_{\nu(k)}(r)^2 \, r \, \frac{dr}{(\rho + r)^b} \right) d\rho.$$

Here we use $\rho$ to split the integration with respect to $r$ into two pieces and use lemma 14.5 for $0 \leq r \leq \rho$ and lemma 14.6 for $r \geq \rho$ to conclude that the inner integral can be estimated from above by $C \, \rho^{1-b}$ for some number $C$ independent of $\rho \geq 1/2$ and $k$. We have proved (14.25).

We have now concluded the proof of part (a) of our theorem.

Proof of part(b). We have the estimate

$$\int_0^\infty \int_0^\infty J_{\nu(0)}(\rho r)^2\, \chi(\rho) f_0(\rho)\, \rho^{2-a}\, r \,\frac{dr\, d\rho}{(1+r)^b}$$

$$\geq 2^{-b} \int_0^\infty \left( \int_{r\geq 1} J_{\nu(0)}(\rho r)^2\, (\rho r)^{1-b}\, d(\rho r) \right) \rho^{b-a}\, \chi(\rho) f_0(\rho)\, d\rho$$

$$= 2^{-b} \int_0^\infty \left( \int_{r\geq \rho} J_{\nu(0)}(r)^2\, r^{1-b}\, dr \right) \rho^{b-a}\, \chi(\rho) f_0(\rho)\, d\rho. \tag{14.26}$$

Now assume that the estimate in part (b) of the theorem holds with a number $C$ independent of $f$. It follows from the discussion above and from (14.26) that we then have the inequality

$$\int_0^\infty \left( \int_\rho^\infty J_{\nu(0)}(r)^2\, r^{1-b}\, dr \right) \rho^{b-a}\, \chi(\rho) f_0(\rho)\, d\rho \leq C \,\|\chi f_0\|_{L^1(\mathbb{R}_+)}, \tag{14.27}$$

where $C$ may be chosen to be independent of $f_0 \in C_0^\infty(\mathbb{R}_+)$. From this statement it is clear that we cannot allow $b < a$. ■

**Theorem 14.8** *Assume that $a = n \geq 2$.*

*(a) If $b > a$, then there is a number $C$ independent of $f$ such that*

$$\int_{\mathbb{R}^n} \|(S^a f)[x]\|^2_{L^2(\mathbb{R})} \frac{dx}{(1+|x|)^b} \leq C \,\|f\|_{H^{(1-a)/2}(\mathbb{R}^n)}.$$

*(b) Assume that there is a number $C$ independent of $f$ such that*

$$\int_{\mathbb{R}^n} \|(S^a f)\,[x]\|^2_{L^2(\mathbb{R})} \frac{dx}{(1+|x|)^b} \leq C \,\|f\|_{L^2(\mathbb{R})}, \quad \operatorname{supp} \widehat{f} \subseteq B^n$$

*and that $f$ is radial. Then $b > a$.*

**Proof.** We follow the proof of the preceeding theorem; the estimate for $\chi(2\rho) f_0(\rho)$ has to be carried out in a slightly different way when $n = a$.

Estimate for $\chi(2\rho) f_0(\rho)$. We shall derive the estimate

$$\int_0^\infty \int_0^\infty J_{\nu(k)}(\rho r)^2\, \chi(2\rho)\, f_0(\rho)\, \rho^{2-a}\, r \,\frac{dr\, d\rho}{(1+r)^b} \leq C \int_0^\infty \chi(2\rho)\, f_0(\rho)\, d\rho$$

where $C$ may be chosen to be independent of $f_0$ and $k$. Cf. (14.24). Here we split the integration with respect to $r$ into two pieces and use the Poisson representation for $0 \le r \le 1$. For the remaining part we estimate as follows:

$$\int_0^\infty \int_0^\infty J_{\nu(k)}(\rho r)^2 \, \chi(2\rho) \, f_0(\rho) \, \rho^{2-a} \, \psi(2r) \, r \, \frac{dr \, d\rho}{(1+r)^b} \le$$

$$\le \int_0^\infty \left( \int_{r \ge 1/2} J_{\nu(k)}(\rho r)^2 \, (\rho r)^{1-b} d(\rho r) \right) \rho^{b-a} \, \chi(2\rho) f_0(\rho) \, d\rho =$$

$$= \int_0^\infty \left( \int_{\rho/2}^\infty J_{\nu(k)}(r)^2 \, r^{1-b} \, dr \right) \rho^{b-a} \, \chi(2\rho) f_0(\rho) \, d\rho \le$$

$$\le \int_0^\infty \left( C \int_{\rho/2}^1 r^{a-b-1} \, dr + \int_1^\infty J_{\nu(k)}(r)^2 \, r^{1-b} \, dr \right) \rho^{b-a} \chi(2\rho) f_0(\rho) \, d\rho.$$

Here we have used the Poisson representation in the last inequality. The number $C$ may be chosen to be independent of $k$. Recall that $b > a = n$. By lemma 14.6 with $\rho = 1$ the expression in the parentheses can be estimated from above by $C \, \rho^{a-b}$ for some number $C$ independent of $\rho \le 1$ and $k$.

Estimate for $\psi(2\rho) f_0(\rho)$. We replace $b = a$ in the proof of theorem 14.7 by $b > a$ and check the estimates in that paragraph.

Proof of part(b). We follow the proof of part (b) in the preceeding theorem; observe that there is no number $C$ independent of $f_0 \in C_0^\infty(\mathbb{R}_+)$ such that the inequality

$$\int_0^\infty \left( \int_\rho^\infty J_{\nu(0)}(r)^2 \, r^{1-n} \, dr \right) \chi(\rho) f_0(\rho) \, d\rho \le C \, \|\chi f_0\|_{L^1(\mathbb{R}_+)}$$

holds. Cf. (14.27). This means that we cannot allow $b = a$. We cannot allow any smaller $b$ either. ∎

In the case $a = 2$ the cases (a) in theorem 14.7 and 14.8 have been treated in Ben-Artzi, Klainerman [2] and Kato, Yajima [13].

**Acknowledgements.** I would like to thank the organizers of the7th Meeting of International Workshop in Analysis and Its Applications for their kind invitation. My participation was made possible thanks to their generous financial support and fin ancial support from theSwedish Natural Sciences Research Council (contract No. R–RA 11968–300). Part of this paper

was prepared during my guest professorship atDepartment of Applied Mathematics, Comenius University, Bratislava, Slovak Repub lic in May 1997 sponsored by theSlovakian Academy of Sciences and theRoyal Swedish Academy of Science.
I would like to thank Professor Krzysztof Stempak, Wrocaw, Poland for suggestions on how to improve the presentation of the proofs of theorems 14.7 and 14.8.

*References*

[1] M. Ben-Artzi, S. Devinatz, *Local smoothing and convergence properties of Schrödinger type equation,* J. Funct. Anal. 101 (1991), pp. 231–254, MR 92k:35064.

[2] M. Ben-Artzi, S. Klainerman, *Decay and Regularity for the Schrödinger Equation*, J. Anal. Math. 58 (1992), pp. 25–37, MR 94e:35053.

[3] J. Bergh; J. Löfström, *Interpolation Spaces. An Introduction,* Grundlehren der Mathematischen Wissenschaften, No. 223. Springer-Verlag, Berlin-New York, 1976, MR 58 #2349.

[4] J. Bourgain, *A remark on Schrödinger operators*, Israel J. Math. 77 (1992), pp. 1–16, MR 93k:35J101.

[5] L. Carleson, *Some Analytic problems to related to Statistical Mechanics,* Euclidean harmonic analysis (Proc. Sem., Univ. Maryland, College Park, Md., 1979), pp. 5–45, Lecture Notes in Math. 779, Springer, Berlin, 1980, MR 82j:8 2005.

[6] M. Cowling, *Pointwise behavior of solutions to Schrödinger equations*, Harmonic analysis (Cortona, 1982), pp. 83–90, Lecture Notes in Math. 992, Springer, Berlin-New York, 1983, MR 85c:34029.

[7] M. Cowling, G. Mauceri *Inequalities for some maximal functions. I*, Trans. Amer. Math. Soc. 287 (1985), pp. 431–455, MR 86a:42023.

[8] W. Craig, T. Kappeler, W. Strauss, *Microlocal dispersive smoothing for the Schrödinger equation*, Comm. Pure Appl. Math. 48 (1995), 769–860, MR 96m:35057.

[9] K. Guo, *A uniform $L^p$ estimate of Bessel functions and distributions supported on* $S^{n-1}$, Proc. Amer. Math. Soc. 125 (1997), pp. 1329–1340, MR 97g:46047.

[10] H. P. Heinig, Sichun Wang, *Maximal Function Estimates of Solutions to General Dispersive Partial Differential Equations*, Preprint, Department of Mathematics and Statistics, McMaster University, Hamilton, Ontario, L8S 4K1 Canada.

[11] H. P. Heinig, Sichun Wang, *Maximal Function Estimates of Solutions to General Dispersive Partial Differential Equations*, Trans. Amer. Math. Soc. (to appear), CMP 1 458 324.

[12] L. Hörmander, *The Analysis of Linear Partial Differential Operators I. Distribution theory and Fourier analysis.* Second edition, Springer Study Edition, Springer-Verlag, Berlin, 1990, MR 91m:35001b, 85g:35002a.

[13] T. Kato, K. Yajima, *Some Examples of Smooth Operators and the associated Smoothing Effect,* Rev. Math. Phys. 1 (1989), pp. 481–496, MR 91i:47013.

[14] C. E. Kenig, G. Ponce, L. Vega, *Oscillatory Integrals and Regularity of Dispersive Equations*, Indiana Univ. Math. J. 40 (1991), pp. 33–69, MR 92d:35081.

[15] L. Kolasa, *Oscillatory integrals and Schrödinger maximal operators*, Pacific J. Math. 177 (1997), pp. 77–101, MR 98h:35043.

[16] A. Moyua; A. Vargas; L. Vega, *Schrödinger maximal function and restriction properties of the Fourier transform*, Internat. Math. Res. Notices 1996, pp. 793–815, MR 97k:42042.

[17] E. Prestini, *Radial functions and regularity of solutions to the Schrödinger equation*, Monatsh. Math. 109 (1990), pp. 135–143, MR 91j:35035.

[18] J. L. Rubio de Francia, *Maximal functions and Fourier transforms*, Duke Math. J. 53 (1986), pp. 395–404, MR 87j:42046.

[19] P. Sjölin, *Regularity of Solutions to the Schrödinger Equation*, Duke Math. J. 55 (1987), pp. 699–715, MR 88j:35026.

[20] C. Sogge, E. M. Stein, *Averages of functions over hypersurfaces in $R^n$*, Invent. Math. 82 (1985), 543–556, MR 87d:42030.

[21] A. Soljanik, Letter communication, April 1990.

[22] E. M. Stein, *Singular Integrals and Differentiability Properties of Functions*, Princeton Mathematical Series, No. 30, Princeton University Press, Princeton, NJ, 1970, MR 44 #7280.

[23] E. M. Stein, *Maximal functions. I. Spherical means*, Proc. Nat. Acad. Sci. U.S.A. 73 (1976), pp. 2174–2175, MR 54 #8133a.

[24] E. M. Stein, *Harmonic Analysis: real-variable methods, orthogonality, and oscillatory integrals,* Princeton Mathematical Series, No. 43, Monographs in Harmonic Analysis, III, Princeton University Press, Princeton, NJ, 1993, MR 95c:42002.

[25] E. M. Stein, G. Weiss, *Introduction to Fourier Analysis on Euclidean Spaces,* Princeton Mathematical Series, No. 32, Princeton University Press, Princeton, NJ, 1971, MR 46 #4102.

[26] R. Strichartz, *Restrictions of Fourier transforms to quadratic surfaces and decay of solutions of wave equations,* Duke Math. J. 44 (1977), pp. 705–714, MR 57 #23577.

[27] L. Vega, *Schrödinger Equations: Pointwise Convergence to the Initial Data,* Proc. Amer. Math. Soc. 102 (1988), pp. 874–878, MR 89d:35046.

[28] L. Vega, *El multiplicador de Schrödinger, la funcion maximal y los operadores de restricción,* Tesis doctoral dirigida por Antonio Córdoba Barba, Departamento de Matemáticas, Universidad Autónoma de Madrid, Febrero 1.988.

[29] B. G. Walther, *Maximal Estimates for Oscillatory Integrals with Concave Phase,* Contemp. Math. 189 (1995), pp. 485–495, MR 96e:42024.

[30] Sichun Wang, *On the Maximal Operator associated with the Free Schrödinger Equation,* Studia Math. 122 (1997), 167–182, CMP 1 432 167.

[31] Si Lei Wang, *On the Weighted Estimate of the Solution associated with the Schrödinger equation,* Proc. Amer. Math. Soc. 113 (1991), 87–92, MR 91k:35066.

# Chapter 15

# Optimal spaces for the $S'$-convolution with the Marcel Riesz kernels and the $N$-dimensional Hilbert kernel

**Josefina Alvarez**
**Christiane Carton-Lebrun**[1]

*ABSTRACT. We consider in the context of distribution spaces, the convolution operators defined by the vector Riesz kernel $pv\frac{x}{|x|^{n+1}}$, $x = (x_1, \ldots, x_n)$, and $pv\frac{1}{x_1} \otimes \cdots \otimes pv\frac{1}{x_n}$. These operators are natural $n$-dimensional extensions of the classical Hilbert transform. We characterize the distribution spaces where the $S'$-convolution with these kernels exists. We next investigate the $S'$-convolvability with a single Riesz kernel. In this case, necessary conditions are found which are in accordance with the result obtained in the vector case.*

## 15.1 Introduction

In this paper, we investigate several problems related to the convolvability of tempered distributions. The convolutions under consideration are of the form $pv\ k \star f$ where $f$ is a tempered distribution. Depending on the problem we are considering, $pv\ k$ is the vector M. Riesz kernel $pv\frac{x}{|x|^{n+1}}$, one of its components, or the kernel $pv\frac{1}{x_1 \ldots x_n}$ of the $n$-dimensional Hilbert transform. In all three cases, the convolution is understood in a precise sense which is often referred to as the $S'$-convolution in the literature (see [1], [12], for instance).

Beyond the frame of the present work, let us mention that many definitions have been introduced in the literature in relation to the general problem of the convolvability of two distributions $f, g$ in $\mathcal{D}'$ and its analogue in $S'$. A survey on the definitions of convolutions can be found in [8, Section 2]. We also refer to the papers by R. Shiraishi [12] and by P. Dierolf and J. Voigt [1] where many of these definitions are discussed and important

[1] Research partially supported by a FNRS grant(Belgium)

results are established about their equivalence.

Much work has been done too in the study of the convolvability with particular kernels and the mapping properties of the associated operators. Important contributions in this direction have been made by J. Horváth [6] and by N. Ortner and P. Wagner [7], [8], for a large class of finite part kernels. Their work also shows the role of weighted $\mathcal{D}'_{L^p}$ spaces in the study of convolutions (see [8] and the references given there).

The present account of the literature is not exhaustive. Additional references will be given in the next sections, in relation to the specific kernels that are considered there. For a more detailed bibliography on the convolutions of distributions and related topics, we refer to [1], [4], [5], [6], [7], [8], [11], [12] and the references given there.

In the present paper, we study the existence of $pv\, k\star f$ in each of the three cases that have been mentioned at the beginning of this introduction. In Section 19.3, we introduce the main definitions and notation. Section 19.4 is devoted to the study of the $\mathcal{S}'$-convolution with the vector M. Riesz kernel. In this case, we show that the space of all tempered distributions that are $\mathcal{S}'$-convolvable with the kernel is the weighted space $(1+|x|^2)^{n/2}\mathcal{D}'_{L^1}$. This result is close to results contained in [6, Theorem 3] and [7, Satz 4]. In Section 15.4, we study the $n$-dimensional Hilbert transform. We show that the optimal space for the $\mathcal{S}'$-convolution with $pv\frac{1}{x_1} \otimes \cdots \otimes pv\frac{1}{x_n}$ is the weighted space $\left[\prod_{i=1}^n (1+x_i^2)^{1/2}\right]\mathcal{D}'_{L^1}$. For $n \geq 2$, the latter is a proper subspace of $(1+|x|^2)^{n/2}\mathcal{D}'_{L^1}$. In the one-dimensional case, both spaces reduce to the space introduced by L. Schwartz in [10], in his study of the convolution with $pv\frac{1}{x}$. Finally, we investigate in Section 15.5 the $\mathcal{S}'$-convolvability with a single Riesz kernel in $\mathbb{R}^2$. In this case, we show that the tempered distributions that are $\mathcal{S}'$-convolvable with the kernel belong to a space of the form $\mathcal{E}' + \Sigma\alpha_\nu W_\nu \mathcal{D}'_{L^1}$ where $\alpha_\nu$ is an appropriate partition of unity and the $W_\nu$ are some $C^\infty$ positive weights. It is worth noting that the intersection of this space with its analogue for the second Riesz kernel is exactly $(1+|x|^2)\mathcal{D}'_{L^1}(\mathbb{R}^2)$, so that we find again the condition obtained in Section 19.4 for the vector Riesz kernel.

## 15.2 Definitions and notation

The notation we use is standard. The Euclidean norm on $\mathbb{R}^n$ is denoted $|\cdot|$. The symbols $\mathcal{D}$, $\mathcal{S}$, $\mathcal{E}$, $L^p$, $\mathcal{D}'$, $\mathcal{S}'$, $\mathcal{E}'$ indicate the usual topological spaces of distributions or functions defined on $\mathbb{R}^n$ and $\|\cdot\|_p$ is the $L^p$ norm. $C^\infty$ is the set of infinitely differentiable functions. When we need a particular setting we write $\mathcal{D}'(\mathbb{R})$, $\mathcal{S}(\mathbb{R}^2)$, $\|\cdot\|_{L^p(K)}$, for instance. Derivatives are denoted $\partial_j = \frac{\partial}{\partial x_j}$, $\partial^\alpha = \partial_1^{\alpha_1} \ldots \partial_n^{\alpha_n}$, as appropriate. For $\alpha = (\alpha_1, \ldots, \alpha_n)$, $\alpha_j \in \mathbb{N}$, we write $|\alpha| = \sum_{j=1}^n \alpha_j$. Given a function $g$, $\check{g}$ is defined by $\check{g}(x) = g(-x)$.

Analogously, given a distribution $f$ and a relevant test function $\varphi$, we write $\langle \check{f}, g\rangle = \langle f, \check{\varphi}\rangle$. The letter $C$ denotes an absolute constant, which may be different at different occurrences.

### 15.2.1 Function and distribution spaces

In addition to the spaces we have just mentioned, we consider the space of integrable distributions, $\mathcal{D}'_{L^1}$. The latter is the dual of

$$\dot{\mathcal{B}} = \mathcal{B}_0 = \{\varphi \in C^\infty : \partial^\alpha \varphi(x) \to 0 \text{ as } |x| \to \infty, \text{ for each } n\text{-tuple } \alpha\}$$

with respect to the topology defined on $\dot{\mathcal{B}}$ by the family of norms

$$s_k(\varphi) = \sup\{\|\partial^\alpha \varphi\|_\infty : 0 \leq |\alpha| \leq k\}$$

([5], [11], [1]).

$\mathcal{D}'_{L^1}$ is also the dual of the topological space

$$\mathcal{B}_c = \{\varphi \in C^\infty : \partial^\alpha \varphi \text{ is bounded on } \mathbb{R}^n \text{ for each } n\text{-tuple } \alpha\}$$

where the topology on $\mathcal{B}_c$ is that induced on the bounded subsets of $\mathcal{B}$ by the topology of $\mathcal{E}$. In other words, $\psi_j$ converges to zero in $\mathcal{B}_c$ if for each $\alpha \in N^n$, $\sup_j \|\partial^\alpha \psi_j\|_\infty < \infty$ and $\partial^\alpha \psi_j$ converges to zero uniformly on any compact set of $\mathbb{R}^n$ (see [11], [1] and the references given there).

As a set, $\mathcal{B}_c$ coincides with $\mathcal{B} = \mathcal{D}_{L^\infty}$. On the latter space, however, the topology is defined by the uniform convergence on $\mathbb{R}^n$ of the functions and all their derivatives ([11], [1]). $\dot{\mathcal{B}}$ is thus a closed subspace of $\mathcal{B}$. $\mathcal{D}$ is dense in $\dot{\mathcal{B}}$ and in $\mathcal{B}_c$ but it is not dense in $\mathcal{B} = \mathcal{D}_{L^\infty}$.

Let us point out that the topology we consider on $\mathcal{D}'_{L^1}$ is the strong one. We also repeatedly use the fact that the space $\mathcal{D}'_{L^1}$ is closed under multiplication by functions in $\mathcal{B}$ ([11], theorem XXVI). For brevity, we also use the symbol $\mathcal{B}$ when no topology is required on this set. Some known results in the setting of the $\mathcal{D}_{L^q}$, $\mathcal{D}'_{L^p}$ spaces, $1 < q, p < \infty$, will be mentioned at occasion although these spaces do not play a central role in our study. We refer to [11] for the study of these spaces and to [1], [5], [11] for more information on $\dot{\mathcal{B}}$, $\mathcal{B}_c$, $\mathcal{D}'_{L^1}$.

The following weighted distribution spaces are also needed to our purpose :

**Definition 15.1** *(a) For $n \geq 1$, $k \in \mathbb{N}$, $w(x) = (1 + |x|^2)^{1/2}$,*

$$w^k \mathcal{D}'_{L^1} = \left\{ f \in \mathcal{S}' : w^{-k} f \in \mathcal{D}'_{L^1} \right\} \tag{15.1}$$

*The topology on this space is induced by the map :*

$$f \mapsto w^{-k} f : w^k \mathcal{D}'_{L^1} \to \mathcal{D}'_{L^1}.$$

(b) *For* $n \geq 1$, $x = (x_1, \ldots, x_n)$, $w_i(x_i) = (1 + x_i^2)^{1/2}$, $i = 1, 2, \ldots, n$,

$$w_1 \ldots w_n \mathcal{D}'_{L^1} = \{ f \in \mathcal{S}' : w_1^{-1} \ldots w_n^{-1} f \in \mathcal{D}'_{L^1} \} \tag{15.2}$$

*The topology on this space is that induced by the map :*

$$f \mapsto w_1^{-1} \ldots w_n^{-1} f : w_1 \ldots w_n \mathcal{D}'_{L^1} \to \mathcal{D}'_{L^1}.$$

*As it has been mentioned in the introduction, both spaces reduce to the space introduced by Schwartz in [10], when* $n = 1$.

### 15.2.2 The $\mathcal{S}'$-convolution

We use the following definition of the $\mathcal{S}'$-convolution (A detailed bibliography concerning this and related definitions can be found in [1], [12]) :

**Definition 15.2** *Let* $f, g \in \mathcal{S}'$. *We say that* $f$ *and* $g$ *are* $\mathcal{S}'$*-convolvable if* $(\varphi \star \check{g}) f \in \mathcal{D}'_{L^1}$ *for every* $\varphi \in \mathcal{S}$. *In this case, the* $\mathcal{S}'$*-convolution* $f \star g$ *is defined on* $\mathcal{S}$ *by*

$$\langle f \star g, \varphi \rangle = {}_{\mathcal{D}'_{L^1}} \langle (\varphi \star \check{g}) f, 1 \rangle_{\mathcal{B}_c} \quad , \quad \varphi \in \mathcal{S}(\mathbb{R}^n). \tag{15.3}$$

In [12], Shiraishi proved that the above definition coincides with the following one :

$$\langle f \star g, \varphi \rangle = {}_{\mathcal{D}'_{L^1}(\mathbb{R}^{2n})} \langle \varphi^{\Delta} (f \otimes g), 1 \rangle_{\mathcal{B}_c(\mathbb{R}^{2n})} \quad , \qquad \varphi \in \mathcal{S}(\mathbb{R}^n)$$

where $\varphi^{\Delta} : (x, y) \mapsto \varphi^{\Delta}(x, y) = \varphi(x + y)$; for this reason, $f \star g \in \mathcal{S}'$ and the convolution considered in (17.3) is commutative.

Note also that the $\mathcal{S}'$-convolution $f \star g$ coincides with the usual one when $f$ or $g$ has compact support ([5], [11]). For $n = 1$ and $g = pv\frac{1}{x}$, it also coincides with the definition introduced by L. Schwartz in [10].

For more information on the $\mathcal{S}'$-convolvability, we refer to the references given in the introduction.

### 15.2.3 Partition of unity on $\mathbb{R}^n$

The following partitions of unity are used in the paper :

(a) In $\mathbb{R}$, we consider the partition $\xi + \eta = 1$ where $\xi$ denotes an even function in $\mathcal{D}(\mathbb{R})$ such that $\xi(x) = 1$ for $|x| \leq 1$, $\xi(x) = 0$ for $|x| \geq 2$, $0 < \xi(x) \leq 1$ for $|x| < 2$.
When we consider the radial extensions of this partition to $\mathbb{R}^n$, we also write $\xi + \eta = 1$; in this case, $\xi$, $\eta$ stand for $\xi(|\cdot|)$, $\eta(|\cdot|)$, respectively.
In connection with these functions, we introduce the notation $\tilde{\xi}$ (resp. $\tilde{\eta}$) to denote a $C^\infty$ function supported in a neighborhood of the support of $\xi$ (resp. $\eta$) and such that $\tilde{\xi}\xi = \xi$ (resp. $\tilde{\eta}\eta = \eta$).

(b) In some parts of the paper, we need the following partition on $\mathbb{R}^2$ :

$$\begin{aligned} 1 &= (\xi_1 \otimes \xi_2) + (\xi_1 \otimes \eta_2) + (\eta_1 \otimes \xi_2) + (\eta_1 \otimes \eta_2) \\ &= \quad \alpha_0 \quad + \quad \alpha_1 \quad + \quad \alpha_2 \quad + \quad \alpha_3 \end{aligned}$$

Here, the one variable functions $\xi_i$, $\eta_i$ are defined by $\xi_i(x_i) = \xi(x_i)$, $\eta_i(x_i) = \eta(x_i)$, $i = 1, 2$, where $\xi, \eta$ are defined as in (a) above. As in the preceding definition, we use some auxiliary functions $\widetilde{\alpha_\nu} \in C^\infty$ such that $\alpha_\nu = \widetilde{\alpha_\nu} \cdot \alpha_\nu$

Additional definitions and notation will be introduced in the specific context where they are needed.

## 15.3 Optimal space for the $\mathcal{S}'$-convolution with the vector Riesz kernel

The vector M. Riesz kernel $pv\frac{x}{|x|^{n+1}}$, $x = (x_1, \dots, x_n)$, belongs to a more general class of principal value kernels which have been investigated by J. Horváth in [4]. From his work, one can immediately deduce the following facts :

- $pv\frac{x}{|x|^{n+1}}$ exists as a distribution of order 1 and belongs to the subspace $\mathcal{D}'_{L^q}$ of $\mathcal{S}'$ for $1 < q < \infty$ ([4], Theorem 1).
- If $\varphi \in \mathcal{D}$ or $\mathcal{S}$, then $pv\ \frac{x_j}{|x|^{n+1}} \star \varphi = R_j\varphi(x)$, $j = 1, 2, \dots, n$, where $R_j\varphi$ denotes the $j$-th Riesz transform of $\varphi$, i.e.

$$R_j\varphi(x) = \lim_{\epsilon \to 0+} \int\limits_{|x-t|>\epsilon} \frac{x_j - t_j}{|x - t|^{n+1}} \varphi(t) dt. \tag{15.4}$$

Moreover, $R_j\varphi \in C^\infty$ and $\partial^\alpha R_j\varphi = R_j\partial^\alpha\varphi$ ([4], Lemma 1).

- The operator $R_j$ admits a continuous extension on $\mathcal{D}_s$ for all $1 < s < \infty$ ([4], Lemma 2).
- The mapping $f \mapsto pv\frac{x}{|x|^{n+1}} \star f$ is continuous from $\mathcal{D}'_{L^p}$ into $\mathcal{D}'_{L^p}$ for all $1 < p < \infty$ ([4], Theorem 2).

Let us mention here that the class of kernels considered in [4] is related to the theory of Calderon-Zygmund operators. We refer to [3], [6], [13], [14] for a detailed exposition and bibliography on many aspects of this theory.

In what follows, we are concerned with the $\mathcal{S}'$-convolution in the sense of Definition 15.2. Using only simple arguments, we show in Theorem 15.1

below that the optimal space for the $\mathcal{S}'$-convolution with $pv\frac{x}{|x|^{n+1}}$ is $w^n\mathcal{D}'_{L^1}$. As it is indicated in Remark 15.1 (i), closely related results have been obtained previously by J. Horváth [6, Theorem 3] and N. Ortner [7, Satz 4] in the context of hypersingular operators.

**Theorem 15.1** *Let $f \in \mathcal{S}'$. Then, the following statements are equivalent :*

*(a) $f \in w^n\mathcal{D}'_{L^1}$;*

*(b) $f$ is $\mathcal{S}'$-convolvable with the vector Riesz kernel $pv\frac{x}{|x|^{n+1}}$;*

*(c) $w^{-1}f$ is $\mathcal{S}'$-convolvable with $\frac{1}{|x|^{n-1}}$.*

**Proof.** For $\varphi \in \mathcal{S}$, denote

$$I\varphi(x) = \int_{\mathbb{R}^n} \frac{\varphi(y)}{|x-y|^{n-1}} dy \tag{15.5}$$

and let $R_j\varphi$ be defined by (15.4). In view of Definition 15.2, the statements (b) and (c) are respectively equivalent to

(b') $(R_j\varphi)f \in \mathcal{D}'_{L^1}$ for every $j \in \{1, \dots, n\}$ and every $\varphi \in \mathcal{S}$;

(c') $(I\varphi)f \in w\mathcal{D}'_{L^1}$ for every $\varphi \in \mathcal{S}$

In what follows, we show that (a) $\Rightarrow$ (b'), (b') $\Rightarrow$ (c') and (c') $\Rightarrow$ (a).

• *Proof of (a) $\Rightarrow$ (b').* In view of the hypothesis, one can write $f = w^n g$ for some $g \in \mathcal{D}'_{L^1}$. Assertion (b') will thus follow if we show that $R_j\varphi \in w^{-n}\mathcal{B}$ for every $j = 1, \dots, n$ and $\varphi \in \mathcal{S}$.

For $j \in \{1, \dots, n\}$, $0 < \epsilon < 1$, and $\varphi \in \mathcal{S}$ given, denote

$$J_{1,j,\epsilon} = \int_{\epsilon<|y|<1} w(x)^n \frac{y_j}{|y|^{n+1}} \varphi(x-y) dy,$$

$$J_{2,j} = \int_{|y|>1} w(x)^n \frac{y_j}{|y|^{n+1}} \varphi(x-y) dy \quad .$$

In order to estimate $J_{1,j,\epsilon}$, we use the Mean Value Theorem and observe that
$w(x) \leq C\, w(x-sy)$ for $|y| \leq 1$, $0 < s < 1$. This yields

$$|J_{1,j,\epsilon}| \leq C \int_{|y|<1} \int_0^1 \frac{w(x-sy)^n}{|y|^{n-1}} \sum_{i=1}^n |(\partial_i\varphi)(x-sy)|\, ds\, dy \quad .$$

To estimate $J_{2,j}$, we notice that $w(x) \leq C|y|w(x-y)$ for $|y| > 1$. Thus we obtain

$$|J_{2,j}| \leq C \int_{\mathbb{R}^n} w(y)^n |\varphi(y)| dy \quad .$$

To complete the proof of (a) $\Rightarrow$ (b'), we only need to observe that

$$\partial^\alpha [w^n R_j \varphi] = w^n R_j(\partial^\alpha \varphi) + \sum_{0<\beta\leq\alpha} b_\beta w^n (R_j(\partial^{\alpha-\beta}\varphi))$$

where the functions $b_\beta$ belong to $\mathcal{B}$.

• *Proof of (b') $\Rightarrow$ (c').* We remark that for each $j \in \{1, \ldots, n\}$ fixed, the following identity holds for every $\varphi \in \mathcal{S}$ :

$$x_j R_j \varphi(x) = \left( \frac{y_j^2}{|y|^{n+1}} \star \varphi \right)(x) + R_j(y_j \varphi)(x).$$

As a consequence,

$$\sum_{j=1}^n x_j (R_j \varphi)(x) = I\varphi(x) + \sum_{j=1}^n R_j(y_j \varphi)(x) \tag{15.6}$$

for every $\varphi \in \mathcal{S}$. In view of the hypothesis, $\sum_{j=1}^n R_j(y_j\varphi) f \in \mathcal{D}'_{L^1} \subset w\mathcal{D}'_{L^1}$ and $\sum_{j=1}^n x_j (R_j\varphi) f \in w\mathcal{D}'_{L^1}$. From (15.6), it thus follows that $(I\varphi)f \in w\mathcal{D}'_{L^1}$ for every $\varphi \in \mathcal{S}$, which ends the proof of (b') $\Rightarrow$ (c').

• *Proof of (c') $\Rightarrow$ (a).* Let $\eta = 1 - \xi$ be defined as in 15.2.3(a). If (c') holds, then $(I\varphi)\eta f \in w\mathcal{D}'_{L^1}$ for every $\varphi \in \mathcal{S}$. This means that to each $\varphi \in \mathcal{S}$, we can associate $g_{[\varphi]} \in \mathcal{D}'_{L^1}$ such that $(I\varphi)\eta f = w\eta g_{[\varphi]}$. Therefore, if there exists a function $\psi \in \mathcal{D}$ such that $(I\psi)^{-1}\tilde{\eta} \in w^{n-1}\mathcal{B}$, we will obtain $\eta f \in w^n \mathcal{D}'_{L^1}$ and (a) will follow. In fact, any function $\psi \in \mathcal{D}$ such that $\psi(x) > 0$ for $|x| < \frac{1}{2}$, $\psi(x) = 0$ for $|x| \geq \frac{1}{2}$ and $\int \psi dx = 1$, satisfies the above condition. Indeed,

$$|I\psi(x)| = \left| \int_{|y|<\frac{1}{2}} \frac{\psi(y)}{|x-y|^{n-1}} dy \right| \geq \frac{C}{|x|^{n-1}} > 0$$

for $|x| \geq 1$, which shows that $(I\psi)^{-1}\tilde{\eta}$ is a $C^\infty$ function on $\mathbb{R}^n$ whose behaviour as $|x| \to \infty$ is of order $O\left(|x|^{n-1}\right)$. Furthermore, it can be seen that the growth of the partial derivatives $\partial^\alpha \left[(I\psi)^{-1}\right]$ as $|x| \to \infty$ does not exceed this order. This ends the proof of the theorem. ■

The following corollary immediately follows from Theorem 15.1:

**Corollary 15.2** *Let $g \in \mathcal{S}'$. Then, $g$ is $\mathcal{S}'$-convolvable with $\frac{1}{|x|^{n-1}}$ if and only if $g \in w^{n-1}\mathcal{D}'_{L^1}$.*

**Remark 15.1** *(i) A different proof of the sufficiency assertion in Corollary 15.2 and of the assertion (a) $\Rightarrow$ (b) in Theorem 15.1 can be derived from [6, pp 186-189] and [7, p 24]. A similar remark holds as concerns the necessity assertion in Corollary 15.2 (see [7, pp 29-30]).*
*(ii) Necessary conditions will be obtained in Section 15.5 as concerns the $\mathcal{S}'$-convolvability with a single Riesz kernel. From this study, it will follow that the analogue of the assertion (b) $\Rightarrow$ (a) in Theorem 15.1 is no longer true if we replace the vector Riesz kernel by a single Riesz kernel.*

## 15.4 Optimal space for the $\mathcal{S}'$-convolution with $pv\frac{1}{x_1} \otimes \cdots \otimes pv\frac{1}{x_n}$

First, we mention some preliminary facts :

([15], [2]). For $\varphi \in \mathcal{S}$, the $n$-dimensional Hilbert transform is defined by

$$\mathcal{H}\varphi(x_1, \ldots, x_n) = \lim_{\epsilon_1, \ldots, \epsilon_n \to 0} \int_{|y_1|>\epsilon_1} \cdots \int_{|y_n|>\epsilon_n} \frac{\varphi(x_1 - y_1, \ldots, x_n - y_n)}{y_1 \cdots y_n} dy_1 \ldots dy_n \quad .$$

The existence of the limit results from the factorization of the operator $\mathcal{H}$ as $\mathcal{H} = H_1 H_2 \ldots H_n$ where $H_j$ denotes the one-dimensional Hilbert transform in the $j$-th variable. Furthermore, in view of the definition of the tensor product of distributions, $\mathcal{H}\varphi$ coincides with the regularization $\left(\bigotimes_{j=1}^n pv\frac{1}{x_j}\right) \star \varphi$ and it is a $C^\infty$ function ( [5], [11]). Since the factorization of $\mathcal{H}$ holds on $L^q$ for $1 < q < \infty$ ([15], [2]), and $\partial^\alpha \mathcal{H}\varphi = \mathcal{H}\partial^\alpha\varphi$ for $\varphi \in \mathcal{S}$, it follows that $\mathcal{H}$ can be extended as a continuous operator on $\mathcal{D}_{L^q}$, $1 < q < \infty$.

Our purpose now is to show that the optimal space for the $\mathcal{S}'$-convolution with $pv\frac{1}{x_1}\otimes\cdots\otimes pv\frac{1}{x_n}$ is $w_1 \ldots w_n \mathcal{D}'_{L^1}$. The latter space is defined by (15.2). Furthermore,

**Proposition 15.3** *For $n \geq 2$, $w_1 \ldots w_n \mathcal{D}'_{L^1}$ is strictly contained in $w^n \mathcal{D}'_{L^1}$.*

**Proof.** The inclusion $w_1 \ldots w_n \mathcal{D}'_{L^1} \subset w^n \mathcal{D}'_{L^1}$ clearly holds since $w^{-n}(w_1 \ldots w_n) \in \mathcal{B}$. In order to show that the inclusion is strict, we consider the distribution $f = \delta_0 \otimes \cdots \otimes \delta_0 \otimes 1$ where $\delta_0$ is the Dirac measure supported at 0 in $\mathbb{R}$. Then, given $\varphi \in \mathcal{D}$,

$$\begin{aligned} |\langle w_n^{-2} f, \varphi\rangle| &= |\langle f, w_n^{-2}\varphi\rangle| = \left| \int_{\mathbb{R}} \frac{\varphi(0,t_n)}{(1+t_n^2)} dt_n \right| \\ &\leq C\|\varphi\|_\infty \quad . \end{aligned}$$

So, $f \in w_n^2 \mathcal{D}'_{L^1} \subset w^2 \mathcal{D}'_{L^1} \subset w^n \mathcal{D}'_{L^1}$. To prove that $f$ does not belong to $w_1 \ldots w_n \mathcal{D}'_{L^1}$, we now consider the sequence $\varphi_k = 1 \otimes \cdots \otimes 1 \otimes \gamma_k$, $k \in \mathbb{N}$, where $\gamma_k \in \mathcal{D}(\mathbb{R})$ for each $k$, and $\gamma_k$ converges to 1 in $\mathcal{B}_c(\mathbb{R})$ as $k \to \infty$. Then, $\varphi_k$ converges to 1 in $\mathcal{B}_c(\mathbb{R}^n)$ as $k \to \infty$. However,

$$\begin{aligned} \lim_{k\to\infty} \langle w_1^{-1} \ldots w_n^{-1} f, \varphi_k\rangle &= \lim_{k\to\infty} \langle w_n^{-1}, \gamma_k\rangle \\ &= \lim_{k\to\infty} \int_{\mathbb{R}} \frac{\gamma_k(t_n)}{(1+t_n^2)^{1/2}} dt_n \\ &= +\infty \quad , \end{aligned}$$

which shows that $f$ does not belong to $w_1 \ldots w_n \mathcal{D}'_{L^1}$. ∎

We now prove the main result of this section :

**Theorem 15.4** *Let $f \in \mathcal{S}'$. Then, the following statements are equivalent :*

*(a) $f \in w_1 \ldots w_n \mathcal{D}'_{L^1}$;*

*(b) $f$ is $\mathcal{S}'$-convolvable with $pv\frac{1}{x_1} \otimes \cdots \otimes pv\frac{1}{x_n}$.*

**Proof.** For clarity, we give the proof of the theorem in case $n = 2$ and we successively show that (a) ⇒ (b) and (b) ⇒ (a) in this case. The proof of the general case follows a similar scheme.

• *Proof of (a) ⇒ (b).* Let $f \in w_1 w_2 \mathcal{D}'_{L^1}$. In view of Definition 15.2, we will obtain assertion (b) if we prove that $\mathcal{H}\varphi \in w_1^{-1} w_2^{-1} \mathcal{B}$ for every $\varphi \in \mathcal{S}$. Given $\varphi \in \mathcal{S}$, we write

$$\begin{aligned} \mathcal{H}\varphi(x) &= \lim_{\epsilon_1,\epsilon_2\to 0} \int_{|y_1|>\epsilon_1} \int_{|y_2|>\epsilon_2} \frac{\varphi(x_1-y_1, x_2-y_2)}{y_1 y_2} dy_1 dy_2 \\ &= \sum_{m=0}^{3} \lim_{\epsilon_1,\epsilon_2\to 0} \int_{E_m} \frac{\varphi(x_1-y_1, x_2-y_2)}{y_1 y_2} dy_1 dy_2 \\ &= \sum_{m=0}^{3} \lim_{\epsilon_1,\epsilon_2\to 0} (T_m\varphi)(x) \quad , \end{aligned}$$

where

$$\begin{aligned} E_0 &= E_0(\epsilon_1, \epsilon_2) = \{(y_1, y_2) : \epsilon_i < |y_i| \leq 1\}, \\ E_1 &= E_1(\epsilon_1) = \{(y_1, y_2) : \epsilon_1 < |y_1| \leq 1, |y_2| \geq 1\}, \\ E_2 &= E_2(\epsilon_2) = \{(y_1, y_2) : |y_1| \geq 1, \epsilon_2 < |y_2| \leq 1\}, \\ E_3 &= \{(y_1, y_2) : |y_i| \geq 1, i = 1, 2\} \quad . \end{aligned}$$

We estimate $(T_1\varphi)(x)$ as a typical term. Each of the other three terms can be dealt with using an appropriate combination of similar majorizations.

Fix $0 < \epsilon_1 < 1$. Using the Mean Value Theorem we can write

$$(T_1\varphi)(x_1, x_2) = -\int_{|y_2|\geq 1} \frac{1}{y_2} \int_{\epsilon_1<|y_1|<1} \int_0^1 (\partial_1\varphi)(x_1 - sy_1, x_2 - y_2) ds\, dy_1 dy_2.$$

Now we observe that the following inequalities hold :

$$w_1(x_1) \leq Cw_1(x_1 - sy_1)w_1(sy_1) \leq Cw_1(x_1 - sy_1)$$

for $0 \leq s \leq 1$, $|y_1| < 1$;

$$w_2(x_2) \leq Cw_2(x_2 - y_2)w_2(y_2) \leq Cw_2(x_2 - y_2)|y_2|$$

for $|y_2| > 1$.

Thus we have the estimate

$$\begin{aligned} &w_1(x_1)w_2(x_2)\,|T_1\varphi(x_1, x_2)| \\ &\leq C \int_{|y_2|>1} w_2^{-2}(x_2 - y_2) \int_{\epsilon_1<|y_1|\leq 1} \int_0^1 |\Phi(x_1 - sy_1, x_2 - y_2)|\, ds\, dy_1 dy_2 \end{aligned}$$

where $\Phi : (X_1, X_2) \mapsto w_1(X_1) \cdot w_2^3(X_2)(\partial_1\varphi)(X_1, X_2) \in \mathcal{S}(\mathbb{R}^2)$. This estimate shows that $|T_1\varphi|$ is bounded by $Cw_1^{-1}w_2^{-1}$, whenever $\varphi \in \mathcal{S}(\mathbb{R}^2)$.

Finally we observe that

$$\begin{aligned} \partial_1^{\alpha_1}\partial_2^{\alpha_2}(w_1w_2T_1\varphi) \quad &= \quad w_1w_2T_1(\partial_1^{\alpha_1}\partial_2^{\alpha_2}\varphi) \\ &+ \sum_{0<\beta_i\leq\alpha_i} b_{1,\beta_1}b_{2,\beta_2}w_1w_2T_1(\partial_1^{\alpha_1-\beta_1}\partial_2^{\alpha_2-\beta_2}\varphi) \end{aligned}$$

where the functions $b_{1,\beta_1}$, $b_{2,\beta_2}$ belong to $\mathcal{B}(\mathbb{R}^2)$.

This completes the proof of (a) $\Rightarrow$ (b).

- *Proof of (b) $\Rightarrow$ (a).* We choose the partition of unity

$$\begin{aligned} 1 \quad &= \quad (\xi_1 \otimes \xi_2) + (\xi_1 \otimes \eta_2) + (\eta_1 \otimes \xi_2) + (\eta_1 \otimes \eta_2) \\ &= \quad \alpha_0 \quad + \quad \alpha_1 \quad + \quad \alpha_2 \quad + \quad \alpha_3 \end{aligned}$$

and write $f = \sum_{\nu=0}^{3} \alpha_\nu f$. For $\nu = 1, 2, 3$, we denote by $(b_\nu)$, $(a_\nu)$ the following statements :

$$\begin{array}{ll} (b_\nu) & (\mathcal{H}\varphi)\alpha_\nu f \in \mathcal{D}'_{L^1} \text{ for every } \varphi \in \mathcal{S}; \\ (a_\nu) & \alpha_\nu f \in w_1 w_2 \mathcal{D}'_{L^1} \end{array}$$

Our purpose is to prove that $(b_\nu) \Rightarrow (a_\nu)$ for each $\nu = 1, 2, 3$. This will yield the global implication $(b) \Rightarrow (a)$ because the first term $\alpha_0 f$ in the decomposition of $f$ belongs to $\mathcal{E}'$. Furthermore, condition (b) implies all $(b_\nu)$ for $\nu = 1, 2, 3$.

We observe that for $\varphi = \varphi_1 \otimes \varphi_2$ with $\varphi_j \in \mathcal{D}(\mathbb{R})$, the conditions $(b_\nu)$ can be written as : $(\xi_1 H_1\varphi_1)(\eta_2 H_2 \varphi_2) f \in \mathcal{D}'_{L^1}$, $(\eta_1 H_1 \varphi_1)(\xi_2 H_2 \varphi_2) f \in \mathcal{D}'_{L^1}$, $(\eta_1 H_1 \varphi_1)(\eta_2 H_2 \varphi_2) f \in \mathcal{D}'_{L^1}$, respectively.

This remark leads us to seek a function $\mu \in \mathcal{D}(\mathbb{R})$ such that $\tilde{\xi}(H\mu)^{-1} \in \mathcal{D}(\mathbb{R})$ and a function $\psi \in \mathcal{D}(\mathbb{R})$ such that

$$\tilde{\eta}(H\psi)^{-1} \in (1+t^2)^{1/2}\mathcal{B}(\mathbb{R}).$$

Supposing that these functions exist and denoting

$$\mu_i(x_i) = \mu(x_i), \psi_i(x_i) = \psi(x_i), i = 1, 2,$$

we will then obtain $\alpha_1 f = \xi_1 (H\mu_1)^{-1} \cdot \eta_2 (H\psi_2)^{-1} \cdot g_{[\mu_1 \otimes \psi_2]}$ where $g_{[\cdot]} \in \mathcal{D}'_{L^1}(\mathbb{R}^2)$ and from there, $\alpha_1 f \in w_2 \mathcal{D}'_{L^1}(\mathbb{R}^2)$. Similarly, we will obtain $\alpha_2 f \in w_1 \mathcal{D}'_{L^1}(\mathbb{R}^2)$ and $\alpha_3 f \in w_1 w_2 \mathcal{D}'_{L^1}(\mathbb{R}^2)$. This means that all the assertions $(a_\nu)$, $\nu = 1, 2, 3$, will hold.

To end the proof, we consider a function $\mu \in \mathcal{D}(\mathbb{R})$ such that $\mu(t) > 0$ for $t \in (3, 4)$, $\mu(t) = 0$ elsewhere and $\int \mu(t)dt = 1$. Then, for $|x| \leq 2$, we have

$$|H\mu(x)| = \int_3^4 \frac{\mu(t)}{t-x} dt \geq C > 0 \quad .$$

Since $\partial^j (H\mu) = H(\partial^j \mu) \in C^\infty(\mathbb{R})$ for all $j \in \mathbb{N}$, this implies $\tilde{\xi}(H\mu)^{-1} \in \mathcal{D}(\mathbb{R})$, as needed.

Next, we consider $\psi \in \mathcal{D}(\mathbb{R})$ such that $\psi(t) > 0$ for $t \in \left(0, \frac{1}{2}\right)$, $\psi(t) = 0$ elsewhere, and $\int \psi dt = 1$. Then for $|x| \geq 1$,

$$|H\psi(x)| = \int_0^{1/2} \frac{\psi(t)}{|x-t|} dt \geq \frac{C}{|x|} > 0 \quad .$$

This shows that $\tilde{\eta}(H\psi)^{-1}$ is a $C^\infty$ function on $\mathbb{R}$ which behaves as $O(|x|)$ as $|x| \to \infty$. Furthermore, all the derivatives $\partial^\alpha[(H\psi)^{-1}]$ also satisfy this growth condition as $|x| \to \infty$. This yields $\tilde{\eta}(H\psi)^{-1} \in (1+t^2)^{1/2}\mathcal{B}(\mathbb{R})$ as required. ■

## 15.5 Necessary condition for the $\mathcal{S}'$-convolvability with a single Riesz kernel

In this section, we work in $\mathbb{R}^2$ and we consider again the partition of unity

$$\begin{aligned} 1 &= (\xi_1 \otimes \xi_2) + (\xi_1 \otimes \eta_2) + (\eta_1 \otimes \xi_2) + (\eta_1 \otimes \eta_2) \\ &= \quad \alpha_0 \quad + \quad \alpha_1 \quad + \quad \alpha_2 \quad + \quad \alpha_3 \end{aligned}$$

The support of $\alpha_\nu$ is denoted supp $\alpha_\nu$.
We also use the auxiliary one-variable functions $\mu$ and $\psi$ introduced in the proof of Theorem 15.4. Recall that

$$\mu \in \mathcal{D}(\mathbb{R});\ \mu(t) > 0 \text{ for } t \in (3,4);\ \mu(t) = 0 \text{ elsewhere and } \int \mu dt = 1 \quad . \tag{15.7}$$

$$\psi \in \mathcal{D}(\mathbb{R});\ \psi(t) > 0 \text{ for } t \in \left(0, \frac{1}{2}\right);\ \psi(t) = 0 \text{ elsewhere and } \int \psi dt = 1 \quad . \tag{15.8}$$

Also, we denote $\mu_i(x_i) = \mu(x_i)$, $\psi_i(x_i) = \psi(x_i)$, $i = 1, 2$.

In the following theorem, we give a necessary condition for the $\mathcal{S}'$-convolvability of $f \in \mathcal{S}'(\mathbb{R}^2)$ with the kernel $pv\frac{x_1}{|x|^3}$ in $\mathbb{R}^2$.

**Theorem 15.5** *Let $\{\alpha_\nu\}$ be defined as above and let $f \in \mathcal{S}'(\mathbb{R}^2)$. If $f$ is $\mathcal{S}'$-convolvable with $pv\frac{x_1}{|x|^3}$, then*

$$f = \alpha_0 f + |x_2|^3 \alpha_1 g_1 + |x_1|^2 \alpha_2 g_2 + \frac{|x|^3}{|x_1|} \alpha_3 g_3 \tag{15.9}$$

*where $g_\nu \in \mathcal{D}'_{L^1}(\mathbb{R}^2)$ for each $\nu = 1, 2, 3$.*
*Furthermore, the distribution $g = \alpha_0 f + \sum_{\nu=1}^{3} \alpha_\nu g_\nu$ can be written as*

$$g = \alpha_0 f + |x_2|^{-3} \alpha_1 f + |x_1|^{-2} \alpha_2 f + \frac{|x_1|}{|x|^3} \alpha_3 f \tag{15.10}$$

*where each term belongs to $\mathcal{D}'_{L^1}(\mathbb{R}^2)$.*

**Proof.** We first suppose $\nu = 1$. Using the notation (15.7), (15.8), we minorize $|R_1(\mu_1 \otimes \psi_2)(x)|$ for $x \in \text{supp}\ \alpha_1$ :

$$\begin{aligned} |R_1(\mu_1 \otimes \psi_2)(x)| &= \left| \int\limits_{(3,4)} \int\limits_{(0,\frac{1}{2})} \frac{\mu_1(y_1)\psi_2(y_2)\,|x_1-y_1|\text{sign}\,(x_1-y_1)}{[(x_1-y_1)^2+(x_2-y_2)^2]^{3/2}} dy_2 dy_1 \right| \\ &\geq C \int\limits_{(3,4)} \mu_1(y_1) \int\limits_{(0,\frac{1}{2})} \psi_2(y_2) \cdot \frac{1}{|x_2|^3} dy_2 dy_1 \\ &\geq \frac{C}{|x_2|^3} > 0 \quad . \end{aligned}$$

From this, we deduce that $\widetilde{\alpha_1}\,[R_1(\mu_1 \otimes \psi_2)]^{-1}$ is a $C^\infty$ function which behaves as $O\left(|x_2|^3\right)$ as $|x| \to \infty$, $x \in \text{supp}\ \alpha_1$. It can also be seen that all the derivatives $\partial^\beta \left\{ [R_1\left(\mu_1 \otimes \psi_2\right)]^{-1} \right\}$ have the same behaviour.
Thus, $\alpha_1 f = \alpha_1 \left|x_2\right|^3 g_1$ where $g_1 \in \mathcal{D}'_{L^1}$.

In the case $\nu = 2$, we minorize $|R_1\left(\psi_1 \otimes \mu_2\right)(x)|$ for $x \in \text{supp}\ \alpha_2$. We obtain $|R_1\left(\psi_1 \otimes \mu_2\right)(x)| \geq C\ x_1^2 > 0$, and an argument similar to that used in the preceding case shows that $\alpha_2 f = \alpha_2 x_1^2 g_2$ where $g_2 \in \mathcal{D}'_{L^1}$.

Let us now consider the case $\nu = 3$. Here we minorize $|R_1\left(\psi_1 \otimes \psi_2\right)(x)|$ for $x \in \text{supp}\ \alpha_3$. We note that for these $x$, one has $\text{sign}(x_1 - y_1) = \text{sign}\ x_1$ whenever $y_1 \in \left(0, \frac{1}{2}\right)$. Therefore,

$$\begin{aligned} |R_1\left(\psi_1 \otimes \psi_2\right)(x)| &= \int\limits_{(0,\frac{1}{2})} \int\limits_{(0,\frac{1}{2})} \frac{\psi_1(y_1)\psi_2(y_2)\,|x_1-y_1|}{[(x_1-y_1)^2+(x_2-y_2)^2]^{3/2}} dy_1 dy_2 \\ &\geq C\frac{|x_1|}{|x|^3} > 0 \end{aligned}$$

for $x \in \text{supp}\ \alpha_3$. From this, we deduce that $\widetilde{\alpha_3}\,[R_1\left(\psi_1 \otimes \psi_2\right)]^{-1}$ is a $C^\infty$ function which behaves like $\frac{|x|^3}{|x_1|}$ as $|x| \to \infty$, $x \in \text{supp}\ \alpha_3$. To study the behaviour of the derivatives of $[R_1(\psi_1 \otimes \psi_2)]^{-1}$, it is useful to observe that in supp $\alpha_3$,

$$O\left(\frac{|x|^3}{|x_1|}\right) = \begin{cases} O(|x|^2) & \text{for} \quad |x_2| \leq |x_1|, \\ O\left(\frac{|x_2|^3}{|x_1|}\right) & \text{for} \quad |x_1| \leq |x_2| \end{cases}$$

at infinity. It can be seen that all the derivatives under consideration satisfy this growth condition at infinity in supp $\alpha_3$, so that $\alpha_3 f = \alpha_3 |x|^3 |x_1|^{-1} g_3$ where $g_3 \in \mathcal{D}'_{L^1}$. Gathering the results obtained for $\nu = 1, 2, 3$, we deduce (15.9).

In order to prove (15.10), we remark that each $\alpha_\nu f$ ($\nu = 1, 2, 3$) is of the form

$$\alpha_\nu f = \alpha_\nu W_\nu g_\nu$$

where $g_\nu \in \mathcal{D}'_{L^1}$ and where both $W_\nu$ and $W_\nu^{-1}$ are $C^\infty$ and positive in a neighborhood of the support of $\alpha_\nu$. We may thus write

$$\begin{aligned} \alpha_\nu g_\nu &= \widetilde{\alpha_\nu}\alpha_\nu g_\nu = W_\nu^{-1}\widetilde{\alpha_\nu}\alpha_\nu W_\nu g_\nu = W_\nu^{-1}\widetilde{\alpha_\nu}\alpha_\nu f \\ &= W_\nu^{-1}\alpha_\nu f \quad , \end{aligned}$$

which yields (15.10). This completes the proof of the theorem. ■

The analogue of Theorem 29.5 for the second Riesz kernel in $\mathbb{R}^2$ is obtained by a symmetry argument. It reads :

**Theorem 15.6** *Let $\{\alpha_\nu\}$ be defined as above and let $f \in \mathcal{S}'(\mathbb{R}^2)$. If $f$ is $\mathcal{S}'$-convolvable with $pv\frac{x_2}{|x|^3}$, then*

$$f = \alpha_0 f + |x_2|^2\alpha_1 h_1 + |x_1|^3\alpha_2 h_2 + \frac{|x|^3}{|x_2|}\alpha_3 h_3 \tag{15.11}$$

*where $h_\nu \in \mathcal{D}'_{L^1}(\mathbb{R}^2)$ for each $\nu = 1, 2, 3$.*

*Furthermore, the distribution $h = \alpha_0 f + \sum_{\nu=1}^{3} \alpha_\nu h_\nu$ can be written as*

$$h = \alpha_0 f + |x_2|^{-2}\alpha_1 f + |x_1|^{-3}\alpha_2 f + \frac{|x_2|}{|x|^3}\alpha_3 f \tag{15.12}$$

*where each term belongs to $\mathcal{D}'_{L^1}(\mathbb{R}^2)$.*

As a corollary of the above theorems, we obtain a result that has been established by a different method in the proof of Theorem 15.1. It shows that a necessary condition for the $\mathcal{S}$'-convolution of $f \in \mathcal{S}'$ with the vector Riesz kernel $\frac{x}{|x|^3}$ in $\mathbb{R}^2$ is $f \in w^2\mathcal{D}'_{L^1}$. This corresponds to the implication (b) ⇒ (a) in Theorem 15.1.

**Corollary 15.7** *Suppose $f \in \mathcal{S}'(\mathbb{R}^2)$. If $f$ is $\mathcal{S}'$-convolvable with the vector Riesz kernel $pv\frac{x}{|x|^3}$ in $\mathbb{R}^2$, then $f \in w^2\mathcal{D}'_{L^1}(\mathbb{R}^2)$.*

**Proof.** From the hypothesis, it follows that for each $\nu = 1, 2, 3$, $\alpha_\nu f$ is $\mathcal{S}'$-convolvable with $\frac{x_1}{|x|^3}$ and $\frac{x_2}{|x|^3}$. For $\nu = 1$, this implies

$$\alpha_1 f = \alpha_1 |x_2|^3 g_1 = \alpha_1 |x_2|^2 h_1$$

where $g_1$, $h_1$ both belong to $\mathcal{D}'_{L^1}$. As a consequence, $\alpha_1 f \in w_2^2\mathcal{D}'_{L^1} \subset w^2\mathcal{D}'_{L^1}$. Similarly, we see that $\alpha_2 f \in w_1^2\mathcal{D}'_{L^1} \subset w^2\mathcal{D}'_{L^1}$.

Let us now suppose $\nu = 3$. Then,

$$\alpha_3 f = \frac{|x|^3}{|x_1|}\alpha_3 g_3 = \frac{|x|^3}{|x_2|}\alpha_3 h_3,$$

where $g_3$, $h_3$ both belong to $\mathcal{D}'_{L^1}$. We note that $\frac{|x|^3}{|x_1|} \sim |x|^2$ as $|x| \to \infty$ in the subset $\{(x_1, x_2) : |x_2| \leq |x_1|\}$ of the support of $\alpha_3$. On the other hand,

# Part III

# Integral Operators and Functional Analysis

# Chapter 16

# Asymptotic expansions and linear wavelet packets on certain hypergroups

**Khalifa Trimèche**

*ABSTRACT. We state harmonic analysis results on the Chébli-Trimèche hypergroups* $(R_+, *_A)$*, as asymptotic expansions, integral representations of Mehler and Schläfli type, and characterizations of maximal ideal spaces of some algebras. Next we study a continuous linear wavelet transform and a linear wavelet packet transform on the Chébli-Trimèche hypergroups* $(R_+, *_A)$*, and we prove for these transforms reconstruction formulas.*

## 16.1 Introduction

In the general case many results in analysis on locally compact groups suggest that there is a unifying structure within which the various convolution measure algebras can be studied without explicit reference to the groups themselves. The structure that has developed is that of a hypergroup, which is a locally compact space with a certain convolution structure.

In this paper we consider an important class of hypergroups called Chébli-Trimèche hypergroups and denoted $(\mathbb{R}_+, *_A)$. Many basis facts in classical harmonic analysis have been extended to the hypergroups $(\mathbb{R}_+, *_A)$. Among these results we can state the following

- Asymptotic expansions and integral representations of Mehler and Schläfli type.
- Maximal ideal spaces of some algebras.

Also we extend in this work the notion of linear wavelets to the case of the harmonic analysis on the Chébli-Trimèche hypergroups $(\mathbb{R}_+, *_A)$.

We study the continuous linear wavelet analysis on the hypergroups $(\mathbb{R}_+, *_A)$. Next according to [9] we present a construction of linear wavelet packets and of the corresponding linear wavelet packet transform.

It is believed that the linear wavelet analysis as discussed here will be a very useful tool in many areas of Mathematics.

The content of this paper is as follows. In the first section we give the definition of a hypergroup and in particular of the Chébli-Trimèche hypergroups $(\mathbb{R}_+, *_A)$. We characterize in the second section the dual of the Chébli-Trimèche hypergroups $(\mathbb{R}_+, *_A)$. In the third section results concerning asymptotics expansions and integral representations of Mehler and Schläfli type are developed on $(\mathbb{R}_+, *_A)$.In the fourth section we survey harmonic analysis on $(\mathbb{R}_+, *_A)$, and as applications we characterize the maximal ideal spaces of some algebras.

We define and study in the fifth section linear wavelets and the continuous linear wavelet transform on the Chébli-Trimèche hypergroups $(\mathbb{R}_+, *_A)$ and we prove for this transform reconstruction formulas. Next using these results we study linear wavelet packets and the corresponding linear wavelet packet transform on the hypergroups $(\mathbb{R}_+, *_A)$, and we prove also for this transform reconstruction formulas. We conclude this section by proving reconstruction formulas associated with the scale discrete linear scaling function corresponding to a linear wavelet packet on the Chébli-Trimèche hypergroups$(\mathbb{R}_+, *_A)$.

## 16.2 The Chébli-Trimèche hypergroups $(\mathbb{R}_+, *_A)$

Let K be a hypergroup in the sense of R.I. Jewett [10], this means that K is a nonvoid locally compact Hausdorff space with an associative convolution $(x, y) \to \delta_x * \delta_y \in M^1(K)$ (the space of probability measures on K), where $\delta_x$ is the unit point mass at $x$, and an involution $x \to x^\vee$ satisfying the following conditions

**C$_1$** - If we denote by $M_b(K)$ the space of bounded measures on $K$, then under the convolution $*$ the vector space $(M_b(K), +)$ is an algebra.

**C$_2$** - The mapping $(x, y) \to \delta_x * \delta_y$ of $K \times K$ into $M^1(K)$ is continuous.

**C$_3$** - There exists an element $e$ of K such that

$$\forall x \in K, \quad \delta_x * \delta_e = \delta_e * \delta_x = \delta_x.$$

**C$_4$** - The involution $x \to x^\vee$ is a homeomorphism of K onto itself with the property $(x^\vee)^\vee = x$, for all $x \in K$, such that

$$(\delta_x * \delta_y)^\vee = \delta_{y^\vee} * \delta_{x^\vee}, \text{ for all } x, y \in K.$$

Where $\mu^\vee$ denotes the image of $\mu$ under the involution.

**C$_5$** - If for all $x, y \in K$, we denote by $Supp\,(\delta_x * \delta_y)$ the support of the measure $\delta_x * \delta_y$, then the element $e$ belongs to $Supp\,(\delta_x * \delta_y)$ if and only if $x = y^\vee$.

**C$_6$** - For all $x, y \in K$, $Supp\,(\delta_x * \delta_y)$ is compact.

**C$_7$** - The mapping $(x, y) \to Supp\,(\delta_x * \delta_y)$ of $K \times K$ into $\mathcal{K}(K)$ (the space of nonvoid compact subsets of $K$) is continuous, where the latter space is

given the "Michael topology".
(See [1] [10] [18]).

A hypergroup is called commutative if $(M_b(K), +, *)$ is a commutative algebra.

A hypergroup $K$ for which the involution is the identity mapping is called hermitian. It is easy to see that every hermitian hypergroup is commutative.

By a theorem of R. Spector [18] a commutative hypergroup possesses a measure $m$ such that $Supp\,(m) = K$, and for every $x \in K$:

$$m * \delta_x = \delta_x * m = m$$

In this case, $m$ is called a Haar measure for $K$, and is unique up to a multiplicative constant.

The dual $\hat{K}$ of the hemitian hypergroup $K$ is the space of all real valued multiplicative functions $\varphi$ on $K$ with $\varphi(e) = \sup_{x\in K} |\varphi(x)| = 1$, i.e.

$$\forall x, y \in K, \quad \varphi(x)\varphi(y) = \int_K \varphi(z) d(\delta_x * \delta_y)(z). \tag{16.1}$$

The function $\varphi$ is called character of the hypergroup $K$.

In the sequel we consider the class of hypergroups on $K = \mathbb{R}_+ = [0, +\infty)$ defined as follows. Let $A$ be the function defined on $\mathbb{R}_+$ satisfying the following conditions

(i) $A(x) = x^{2\alpha+1} B(x)$, where $\alpha > -\frac{1}{2}$ and $B$ an even $\mathcal{C}^\infty-$ function on $\mathbb{R}$, such that $B(x) > 0$ for all $x \in \mathbb{R}_+$ and $B(0) = 1$.

(ii) $A$ is increasing and unbounded.

(iii) $\frac{A'}{A}$ is decreasing on $\mathbb{R}_+^* = (0, +\infty)$, and $\lim_{x\to+\infty} \frac{A'(x)}{A(x)} = 2\rho \geq 0$.

(iv) There exists a constant $\gamma > 0$ such that for all $x \in [x_0, +\infty)$, $\quad x_0 > 0$, we have

$$\frac{B'(x)}{B(x)} = \begin{cases} 2\rho - \frac{2\alpha+1}{x} + e^{-\gamma x} F(x) & \text{if } \rho > 0 \\ e^{-\gamma x} F(x) & \text{if } \rho = 0 \end{cases}$$

where F is a $\mathcal{C}^\infty-$ function on $[x_0, +\infty)$, bounded together with its derivatives.

Then there exists a unique hypergroup structure on $\mathbb{R}_+$ such that

$$\frac{\partial}{\partial x}\left[A(x)A(y)\frac{\partial}{\partial x}\left\{\int_{\mathbb{R}_+} f(z)d(\delta_x * \delta_y)(z)\right\}\right]$$
$$= \frac{\partial}{\partial y}\left[A(x)A(y)\frac{\partial}{\partial y}\left\{\int_{\mathbb{R}_+} f(z)d(\delta_x * \delta_y)(z)\right\}\right]$$

for every even $\mathcal{C}^\infty-$ function $f$ on $\mathbb{R}_+$ and $x, y \in \mathbb{R}_+$. The involution is the identity mapping and the neutral element is 0. Hypergroups of this type are called the Chébli-Trimèche hypergroups and denoted $(\mathbb{R}_+, *_A)$.

The Haar measure $m_A$ for the hypergroups $(\mathbb{R}_+, *_A)$ is absolutely continuous with respect to the Lebesgue measure, and can be choosen to have density $A(x)$. For all $x, y \in \mathbb{R}_+$, we have

$$\delta_x * \delta_y = \begin{cases} W(x,y,.)dm_A(.) & \text{if } x, y \in \mathbb{R}_+^* \\ \delta_x & \text{if } y = 0\ ,\ x \in \mathbb{R}_+ \end{cases} \tag{16.2}$$

where $W(x, y, .)$ is a continuous positive function on $(|x-y|, x+y)$ with support in $[|x-y|, x+y]$, such that

$$\int_{|x-y|}^{x+y} W(x,y,z)dm_A(z) = 1. \tag{16.3}$$

(See [23] chap. 6, [1, p. 201-248], [2] [26] [27] [28] [29]).
As particular cases, we have the following hypergroups.

**Example 16.1** *The Bessel-Kingmann hypergroups. These hypergroups correspond to the function*

$$A(x) = x^{2\alpha+1}, \quad \alpha > -\frac{1}{2}$$

*The function $W(x, y, .)$ defined by the relation (16.2) has the expression*

$$W(x,y,z) = \begin{cases} \frac{2^{1-2\alpha}\Gamma(\alpha+1)}{\sqrt{\pi}\Gamma(\alpha+1/2)} \frac{[(x+y)^2-z^2]^{\alpha-1/2}[z^2-(x-y)^2]^{\alpha-1/2}}{(xyz)^\alpha} \\ \qquad \textit{if } |x-y| < z < x+y \\ 0, \qquad \textit{elsewhere.} \end{cases} \tag{16.4}$$

*(See [20] p.93).*

**Example 16.2** *The Jacobi hypergroups. These hypergroups correspond to the function*

$$A(x) = 2^{2(\alpha+\beta+1)}(\sinh x)^{2\alpha+1}(\cosh x)^{2\beta+1},$$

*where* $\alpha \geq \beta \geq -\frac{1}{2}$ , $\alpha > -\frac{1}{2}$. *The function* $W(x, y, .)$ *defined by the relation (16.2) can be expressed as follows. Writing*

$$E = \frac{(\cosh x)^2 + (\cosh y)^2 + (\cosh z)^2 - 1}{2(\cosh x)(\cosh y)(\cosh z)}$$

*we have for* $|x - y| < z < x + y$ *(i.e.* $|E| < 1$*):*

$$W(x,y,z) = \frac{2^{-2(\alpha+\beta+1)}\Gamma(\alpha+1)(\cosh x \cosh y \cosh z)^{\alpha-\beta-1}}{\sqrt{\pi}\Gamma(\alpha+\frac{1}{2})(\sinh x \sinh y \sinh z)^{2\alpha}} \times (1-E^2)^{\alpha-\frac{1}{2}}{}_2F_1\left(\alpha+\beta, \alpha-\beta; \alpha+\frac{1}{2}; \frac{1-E}{2}\right) \tag{16.5}$$

*where* ${}_2F_1$ *is the Gauss hypergeometric function. For* $\alpha > \beta > -\frac{1}{2}$, *the function* $W(x, y, .)$ *has the following form*

$$W(x,y,z) = \begin{cases} b_{\alpha,\beta}(x,y,z) \int_0^\pi (W_{\alpha,\beta}(x,y,z,\theta))_+^{\alpha-\beta-1} (\sin\theta)^{2\beta} d\theta \\ \qquad \text{if } |x-y| < z < x+y \\ 0, \qquad \text{elsewhere} \end{cases}, \tag{16.6}$$

*where*

$$b_{\alpha,\beta}(x,y,z) = \frac{2^{1-2(\alpha+\beta+1)}\Gamma(\alpha+1)}{\sqrt{\pi}\Gamma(\alpha-\beta)\Gamma(\beta+\frac{1}{2})} (\sinh x \sinh y \sinh z)^{-2\alpha}$$

$$W_{\alpha,\beta}(x,y,z,\theta) = 1 - (\cosh x)^2 - (\cosh y)^2 - (\cosh z)^2 + 2\cosh x \cosh y \cosh z \cos\theta$$

*and*

$$(a)_+ = \begin{cases} a \text{ if } a > 0 \\ 0 \text{ if } a \leq 0 \end{cases}$$

*(See[8, p. 256]).*

## 16.3 The dual of the hypergroups $(\mathbb{R}, *_A)$

Let $L_A$ be the singular differential operator of second order on $\mathbb{R}_+^*$ :

$$L_A = \frac{d^2}{dx^2} + \frac{A'(x)}{A(x)}\frac{d}{dx}. \tag{16.7}$$

The equation

$$L_A u = -(\lambda^2 + \rho^2)u, \ \lambda \in \mathbb{C} \tag{16.8}$$

has a unique solution on $\mathbb{R}_+$ satisfying the conditions $u(0) = 1$ and $u'(0) = 0$. We extend this solution on $\mathbb{R}$ by parity and we denote it $\varphi_\lambda(x)$.

The function $\varphi_\lambda$ satisfies the following properties (see [2], [3, 4], [20, 21, 23, 26]).

i) For every $\lambda \in \mathbb{C}$, the function $x \to \varphi_\lambda(x)$ is an even $C^\infty$-function on $\mathbb{R}$.

ii) For all $x \in \mathbb{R}_+$, the function $\lambda \to \varphi_\lambda(x)$ is an even entire function on $\mathbb{C}$.

iii) For $\rho = 0 :\ \forall x \in \mathbb{R}_+,\ \varphi_0(x) = 1$;
For $\rho > 0$ : There exists a constant $k > 0$ such that

$$\forall x \in \mathbb{R}_+, \quad e^{-\rho x} \leq \varphi_0(x) \leq k(1+x)e^{-\rho x}.$$

iv) For all $\lambda \in \mathbb{C}$ and $x \in \mathbb{R}_+$, $|\varphi_\lambda(x)| \leq \varphi_{i\,\mathrm{Im}\,\lambda}(x) \leq \varphi_0(x)e^{|\mathrm{Im}\,\lambda|x}$.

v) For every $x \in \mathbb{R}_+$, we have $|\varphi_\lambda(x)| \leq 1$ if and only if $\lambda$ belongs to the set $\Sigma = \{\lambda \in \mathbb{C} \mid |\mathrm{Im}\,\lambda| \leq \rho\}$.

vi) The function $\varphi_\lambda$, $\lambda \in \mathbb{C}$, satisfies the product formula

$$\forall x, y \in \mathbb{R}_+^*, \quad \varphi_\lambda(x)\varphi_\lambda(y) = \int_{\mathbb{R}_+} \varphi_\lambda(z) W(x,y,z)\, dm_A(z) \tag{16.9}$$

where $W(x,y,z)$ is the function given by the relation (16.2).

Then the functions $\varphi_\lambda$ with $\lambda \in \mathbb{C}$, are multiplicative functions on the hypergroups $(\mathbb{R}_+, *_A)$. Using the properties of these functions we deduce that the dual $\widehat{\mathbb{R}}_+$ of these hypergroup is given by

$$\widehat{\mathbb{R}}_+ = \{\varphi_\lambda \mid \lambda \in \mathbb{R}_+ \cup i[0,\rho]\}. \tag{16.10}$$

It is convenient to identify $\widehat{\mathbb{R}}_+$ with the set $\mathbb{R}_+ \cup i[0,\rho]$. (see [1, 2]).

We remark that the equation (16.8) possesses also two solutions $\phi_{\pm\lambda}$, linearly independent having the following behavior at infinity:

$$\phi_{\pm\lambda}(x) \sim e^{(\pm i\lambda - \rho)x}. \tag{16.11}$$

Then there exists a function $c$ such that

$$\varphi_\lambda(x) = c(\lambda)\phi_\lambda(x) + c(-\lambda)\phi_{-\lambda}(x). \tag{16.12}$$

The function $c$ is called the Harish-Chandra function associated with the operator $L_A$. It possesses the following properties:

i) For all $\lambda \in \mathbb{R}_+$, $c(-\lambda) = \overline{c(\lambda)}$.

ii) The function $|c(\lambda)|^{-2}$ is continuous on $\mathbb{R}_+$, and there exist positive constants $k$, $k_1$, $k_2$ such that
-If $\rho \geq 0$, $\alpha > -1/2$, $\forall \lambda \in \mathbb{C}$ with $|\lambda| > k$,

$$k_1 |\lambda|^{2\alpha+1} \leq |c(\lambda)|^{-2} \leq k_2 |\lambda|^{2\alpha+1}. \tag{16.13}$$

-If $\rho > 0$, $\alpha > -1/2$, $\forall \lambda \in \mathbb{C}$ with $|\lambda| \leq k$,

$$k_1 |\lambda|^2 \leq |c(\lambda)|^{-2} \leq k_2 |\lambda|^2. \tag{16.14}$$

-If $\rho = 0$, $\alpha > 0$, $\forall \lambda \in \mathbb{C}$ with $|\lambda| \leq k$,

$$k_1 |\lambda|^{2\alpha+1} \leq |c(\lambda)|^{-2} \leq k_2 |\lambda|^{2\alpha+1}. \tag{16.15}$$

(see [22, 26])

We now give the expressions of the operator $L_A$ and the functions $\varphi_\lambda$, $\phi_\lambda$ and $c(\lambda)$ in the case of Bessel-Kingman and Jacobi hypergroups.

**Example 16.3** *(Bessel-Kingman) We have*

$$L_A = \frac{d^2}{dx^2} + \frac{2\alpha+1}{x}\frac{d}{dx}, \tag{16.16}$$

*the so called Bessel operator. In this case*

$$\varphi_\lambda(x) = j_\alpha(\lambda x) = \begin{cases} 2^\alpha \Gamma(\alpha+1) \frac{J_\alpha(\lambda x)}{(\lambda x)^\alpha}, & \text{if } \lambda x \neq 0 \\ 1 \quad \text{if } \lambda x = 0 \end{cases}, \tag{16.17}$$

*where $J_\alpha$ is the Bessel function of first kind and index $\alpha$. The function $j_\alpha(\lambda x)$ is called normalized Bessel function. Further we have*

$$\phi_\lambda(x) = \sqrt{\frac{\pi}{2}} e^{i(\alpha+1/2)\frac{\pi}{2}} \lambda^{1/2} x^{-\alpha} H_\alpha^{(1)}(\lambda x) \tag{16.18}$$

*where $H_\alpha^{(1)}$ is the Hankel function of first kind and index $\alpha$;*

$$c(\lambda) = 2^\alpha \Gamma(\alpha+1) e^{i(\alpha+1/2)\frac{\pi}{2}} \lambda^{-(\alpha+1/2)}, \ \lambda \in \mathbb{R}_+^*. \tag{16.19}$$

*Finally we deduce that*

$$|c(\lambda)|^{-2} = 2^{-2\alpha} (\Gamma(\alpha+1))^{-2} \lambda^{2\alpha+1}, \ \lambda \in \mathbb{R}_+. \tag{16.20}$$

**Example 16.4** *(Jacobi) Here we have*

$$L_A = \frac{d^2}{dx^2} + [(2\alpha+1)\coth x + (2\beta+1)\tanh x]\frac{d}{dx}, \tag{16.21}$$

*the so called Jacobi operator. In this case the following formulas hold:*

$$\varphi_\lambda(x) = \varphi_\lambda^{(\alpha,\beta)}(x) = F(\tfrac{\rho-i\lambda}{2}, \tfrac{\rho+i\lambda}{2}, \alpha+1; -\sinh^2 x), \tag{16.22}$$

*where* $\rho = \alpha+\beta+1$ *and* $F$ *the Gauss hypergeometric function;*

$$\phi_\lambda(x) = \phi_\lambda^{(\alpha,\beta)}(x) = (2\sinh x)^{i\lambda-\rho} F(\tfrac{\rho-2\alpha-i\lambda}{2}, \tfrac{\rho-i\lambda}{2}, 1-i\lambda; -\sinh^{-2} x) \tag{16.23}$$

*for* $x \in \mathbb{R}_+^*$ *and* $\lambda \in \mathbb{C} - i\mathbb{N}$;

$$c(\lambda) = \frac{2^{\rho-i\lambda}\Gamma(\alpha+1)\Gamma(i\lambda)}{\Gamma\left(\frac{\rho+i\lambda}{2}\right)\Gamma\left(\frac{\alpha-\beta+1+i\lambda}{2}\right)}. \tag{16.24}$$

*(see [8, 21, 23, 26])*

## 16.4 Asymptotic expansions and integral representations of Mehler and Schläfli type

### *16.4.1 The asymptotic expansions*

Using Olver's method (see [17] chap.12) we give in this subsection asymptotic expansions of the functions $\varphi_\lambda$ and $\phi_{-\lambda}$ with respect to $\lambda$ as $|\lambda| \to +\infty$, in terms respectively of Bessel functions of first kind and Mac-Donald functions, and we present estimates of the remainders which appear in these expansions.

In terms of Bessel functions the following results are known.

**Theorem 16.1** *For all* $m \in \mathbb{N}_+$, *we have*
*(i)* $\forall x \in \mathbb{R}_+\ ,\ \lambda \in \mathbb{R}\setminus\{0\}$,

$$\sqrt{A(x)}\varphi_\lambda(x) = \sum_{p=0}^{m} x^{p+\frac{1}{2}} B_p(x) \frac{J_{\alpha+p}(\lambda x)}{\lambda^{\alpha+p}} + \mathcal{R}_m(\lambda, x). \tag{16.25}$$

*The functions* $B_p\ ,\ p \in \mathbb{N}_+$, *are even* $C^\infty-$ *functions on* $\mathbb{R}$, *given by the recursive relations*

$$B_0(x) = 2^\alpha\Gamma(\alpha+1) \tag{16.26}$$

$$B_{p+1}(x) = \begin{cases} -\frac{1}{2x^{p+1}} \int_0^x t^p \left\{B_p''(t) + \frac{1-2\alpha}{t} B_p'(t) - \mathcal{X}(t)B_p(t)\right\} dt & \text{if } x \neq 0 \\ -\frac{1}{2(p+1)}\left[2(1-\alpha)B_p''(0) - \mathcal{X}(0)B_p(0)\right] & \text{if } x = 0 \end{cases} \tag{16.27}$$

*where*

$$\mathcal{X}(x) = (2\alpha+1)\frac{B'(x)}{2xB(x)} + \frac{1}{2}\left(\frac{B'(x)}{B(x)}\right)' + \frac{1}{4}\left(\frac{B'(x)}{B(x)}\right)^2. \qquad (16.28)$$

*(ii) The remainder $\mathcal{R}_m(\lambda, x)$ satisfies for all $x \in \mathbb{R}_+$ and $\lambda \in \mathbb{R}\setminus\{0\}$:*

$$|\mathcal{R}_m(\lambda, x)| \le \frac{d_m}{|\lambda|^{\alpha+m+\frac{3}{2}}}\left\{\int_0^x \left|\left(t^{m+1}B_{m+1}(t)\right)'\right| dt\right\} \exp\left(\frac{k}{|\lambda|}\int_0^x |\mathcal{X}(t)|dt\right) \qquad (16.29)$$

*where $d_m$ and $k$ are positive constants. (See [7]).*

**Remark 16.1** *The relation (16.25) can also be written in the form:* $\forall x \in \mathbb{R}_+\ ,\ \lambda \in \mathbb{R}\setminus\{0\}$,

$$\sqrt{A(x)}\varphi_\lambda(x) = \sum_{p=0}^{m} \frac{x^{\alpha+2p+\frac{1}{2}}B_p(x)}{2^{\alpha+p}\Gamma(\alpha+p+1)} j_{\alpha+p}(\lambda x) + \mathcal{R}_m(\lambda, x) \qquad (16.30)$$

*The function $\varphi_\lambda$ possesses also the following asymptotic expansion. (See [20]).*

**Theorem 16.2** *For all $m \in \mathbb{N}_+$, $m \ge 1$, $x \in \mathbb{R}_+$ , $\lambda \in \mathbb{R}\setminus\{0\}$, we have*

$$\sqrt{A(x)}\varphi_\lambda(x) = j_\alpha(\lambda x)x^{\alpha+\frac{1}{2}}\sum_{s=0}^{m}\frac{a_s(x)}{\lambda^{2s}} + j_{\alpha+1}(\lambda x)\frac{x^{\alpha+\frac{5}{2}}}{2(\alpha+1)}\sum_{s=0}^{m-1}\frac{b_s(x)}{\lambda^{2s}} + \theta_{\lambda,m}(x) \qquad (16.31)$$

*The functions $a_s$ and $b_s$, $s \in \mathbb{N}_+$, are even $\mathcal{C}^\infty-$ functions on $\mathbb{R}$, given for all $x \in \mathbb{R}_+$ by the recursive relations*

$$\begin{cases} a_0(x) = 1 \\ a_{s+1}(x) = \frac{1}{2}xb_s'(x) - \alpha b_s(x) - \frac{1}{2}\int_0^x \mathcal{X}(t)b_s(t)tdt + k_{s+1} \\ b_s(x) = -\frac{1}{2x}a_s'(x) + \frac{1}{2x}\int_0^x \left[t\mathcal{X}(t)a_s(t) - (2\alpha+1)a_s'(t)\right]\frac{dt}{t}. \end{cases} \qquad (16.32)$$

*the constants $k_{s+1}$ are choosen such that*

$$\forall s \ge 1\ ,\ a_s(0) = 0$$

*and the remainder $\theta_{\lambda,m}$ satisfies for all $x \in \mathbb{R}_+$ and $\lambda \in \mathbb{R}\setminus\{0\}$:*

$$|\theta_{\lambda,m}(x)| \le \frac{k_1}{|\lambda|^{\alpha+2m+\frac{3}{2}}}\left\{\int_0^x |(tb_m(t))'|\,dt\right\}\exp\left(\frac{k_0}{|\lambda|}\int_0^x |\mathcal{X}(t)|dt\right) \qquad (16.33)$$

*where $k_1$ and $k_0$ are positive constants.*

For the case $m = 0$, we have the following result. (See [20]).

**Theorem 16.3** *For all $x \in \mathbb{R}_+$ , $\lambda \in \mathbb{R} \setminus \{0\}$, we have*

$$\sqrt{A(x)}\varphi_\lambda(x) = x^{\alpha+\frac{1}{2}} j_\alpha(\lambda x) + \theta_\lambda(x). \tag{16.34}$$

*The function $\theta_\lambda$ satisfies for all $x \in \mathbb{R}_+$:*

$$|\theta_\lambda(x)| \leq \frac{k_1}{|\lambda|^{\alpha+3/2}} \left\{ \int_0^x |\mathcal{X}(t)|\, dt \right\} \exp\left( \frac{k_0}{|\lambda|} \int_0^x |\mathcal{X}(t)|\, dt \right) \tag{16.35}$$

*where $k_1$ and $k_0$ are positive constants.*

The asymptotic expansion given by Theorems 16.1 and 16.2 are related, because we can write the coefficients $a_s$ and $b_s$ by using the coefficients $B_p$. More precisely we have the following relations

$$\begin{cases} a_0(x) = 1 \\ a_s(x) = \Gamma(\alpha+s+1)\left(-\frac{x^2}{4}\right)^{-s} \sum_{p=s+1}^{2s} \binom{s-1}{p-s-1} \frac{(-\frac{1}{2})^p B_p(x)}{\Gamma(\alpha+p-s+1)}, \\ b_s(x) = \frac{1}{2}\Gamma(\alpha+s+1)\left(-\frac{x^2}{2}\right)^{-s-1} \sum_{p=s+1}^{2s+1} \binom{s}{p-s-1} \frac{(-\frac{1}{2})^p B_p(x)}{\Gamma(\alpha+p-s)}, \end{cases} \tag{16.36}$$

where $s = 1, 2, \ldots$.(See [7]).

**Remark 16.2** *(i) From Theorems 16.1, 16.2 and 16.3 we have the following estimates of the remainders $\mathcal{R}_m(\lambda, x)$ , $\theta_{\lambda,m}$ and $\theta_\lambda$:*

$$\forall x \in \mathbb{R}_+ ,\ \lambda \in \mathbb{R},\ |\lambda| \geq 1, \quad |\mathcal{R}_m(\lambda, x)| \leq \frac{k_m^{(1)}(x)}{|\lambda|^{\alpha+m+\frac{3}{2}}}. \tag{16.37}$$

$$\forall x \in \mathbb{R}_+ ,\ \lambda \in \mathbb{R},\ |\lambda| \geq 1, \quad |\theta_{\lambda,m}(x)| \leq \frac{k_m^{(2)}(x)}{|\lambda|^{\alpha+2m+\frac{3}{2}}}. \tag{16.38}$$

$$\forall x \in \mathbb{R}_+ ,\ \lambda \in \mathbb{R},\ |\lambda| \geq 1, \quad |\theta_\lambda(x)| \leq \frac{k_0^{(2)}(x)}{|\lambda|^{\alpha+\frac{3}{2}}}. \tag{16.39}$$

*where $k_m^{(1)}$ , $k_m^{(2)}$ and $k_0^{(2)}$ are continuous functions on $\mathbb{R}_+$. We see that for $|\lambda|$ large enough and $m \in \mathbb{N}_+$ , $m \geq 1$, the estimate (16.38) is better than the (16.37).*
*(ii) H. Chébli, A. Fitouhi and M.M. Hamza have studied in [6] the series development of the function $\varphi_\lambda$. They have showed that if the function $\mathcal{X}(x)$ given by the relation (16.28), is the restriction to $\mathbb{R}$ of a holomorphic*

*function in the disc* $D(0, 2R) = \{z \in \mathbb{C} \ / \ |z| < 2R\}$, *the function* $\varphi_\lambda$ , $\lambda \in \mathbb{R} \setminus \{0\}$, *possesses the series development*

$$\sqrt{A(x)}\varphi_\lambda(x) = \sum_{p=0}^{\infty} x^{p+\frac{1}{2}} B_p(x) \frac{J_{\alpha+p}(\lambda x)}{\lambda^{\alpha+p}}. \tag{16.40}$$

*which converge uniformly on every subinterval of* $(0 \ , \ 1+|1-2\alpha|^{-\frac{1}{2}} R \, e^{-1})$. *The functions* $B_p$ , $p \in \mathbb{N}_+$, *are given by the relations (16.26) and (16.27).*

The series development (16.40) can also be written in the form

$$\sqrt{A(x)}\varphi_\lambda(x) = \sum_{p=0}^{\infty} \frac{x^{\alpha+2p+\frac{1}{2}} B_p(x)}{2^{\alpha+p}\Gamma(\alpha+p+1)} j_{\alpha+p}(\lambda x). \tag{16.41}$$

In terms of Mac-Donald functions the following asymptotic expansions are known.

**Theorem 16.4** *For all* $m \in \mathbb{N}_+$, *we have*
*(i)* $\forall x \in \mathbb{R}_+^*$ , $\lambda \in \mathbb{C}$ , $\operatorname{Im} \lambda < 0$,

$$\sqrt{A(x)} \frac{\phi_{-\lambda}(x)}{(i\lambda)^{\alpha+\frac{1}{2}}} = \sqrt{\frac{2x}{\pi}} \sum_{p=0}^{m} a_p(x) \frac{K_{\alpha+p}(i\lambda x)}{(i\lambda)^{\alpha+p}} + \mathcal{R}_{\lambda,m}(x) \tag{16.42}$$

*where* $K_\nu$ *is the Mac-Donald function (see [25] p. 78), and* $a_p$ , $p \in \mathbb{N}_+$, *are* $C^\infty -$ *functions on* $\mathbb{R}_+^*$ *given by the recursive relations*

$$\begin{cases} a_0(x) = 1 \\ a_{p+1}(x) = -\frac{1}{2} \int_x^\infty \left[ a_p''(s) + \frac{1-2(\alpha+p)}{s} a_p'(s) + \frac{p(2\alpha+p)}{s^2} - \mathcal{X}(s) a_p(s) \right] ds \end{cases} \tag{16.43}$$

*The function* $\mathcal{X}$ *is given by the relation (16.28).*
*(ii) The remainder* $\mathcal{R}_{\lambda,m}(x)$ *satisfies for all* $x \in \mathbb{R}_+^*$ *and* $\lambda \in \mathbb{C}$ , $\operatorname{Im} \lambda < 0$:

$$|\mathcal{R}_{\lambda,m}(x)| \leq \frac{k_{\lambda,m}(x) e^{-|\operatorname{Im} \lambda| x}}{|\lambda|^{\alpha+m+\frac{3}{2}}} \left\{ \int_x^\infty \left| a_{m+1}'(s) \right| ds \right\} \times \exp\left( \frac{1}{|\lambda|} \int_x^\infty \left| \frac{\alpha^2 - \frac{1}{4}}{s^2} + \mathcal{X}(s) \right| ds \right) \tag{16.44}$$

*where*

$$k_{\lambda,m}(x)=\begin{cases} c(m,\alpha)\frac{1+(|\lambda|x)^{\alpha+m-\frac{1}{2}}}{(|\lambda|x)^{\alpha+m-\frac{1}{2}}} & \text{if } m\geq 1 \\ c(\alpha)\frac{1+(|\lambda|x)^{\alpha-\frac{1}{2}}}{(|\lambda|x)^{\alpha-\frac{1}{2}}} & \text{if } m=0 \text{ and } \alpha\geq\frac{1}{2} \\ c(\alpha) & \text{if } m=0 \text{ and } 0\leq\alpha<\frac{1}{2} \\ c(-\alpha) & \text{if } m=0 \text{ and } -\frac{1}{2}<\alpha<0 \end{cases} \tag{16.45}$$

*with* $c(m,\alpha)$ , $c(\alpha)$ *and* $c(-\alpha)$ *are positive constants. (See [13]).*

**Remark 16.3** *We have the following estimates for the derivatives of the remainder* $\mathcal{R}_{\lambda,m}$
*(i) For all* $x\in\mathbb{R}_+^*$ , $\lambda\in\mathbb{C}$ , $\operatorname{Im}\lambda<0$*:*

$$\left|\frac{d}{dx}\mathcal{R}_{\lambda,m}(x)\right|\leq\frac{k_{\lambda,m}(x)e^{-|\operatorname{Im}\lambda|x}}{|\lambda|^{\alpha+m+\frac{1}{2}}}\left\{\int_x^\infty\left|a'_{m+1}(s)\right|ds\right\}$$
$$\times\exp\left(\frac{1}{|\lambda|}\int_x^\infty\left|\frac{\alpha^2-\frac{1}{4}}{s^2}+\mathcal{X}(s)\right|ds\right). \tag{16.46}$$

*(ii) If* $\alpha=k+r$ , $k\in\mathbb{N}_+$ , $r\in(-\frac{1}{2},\frac{1}{2}]$*, then for all* $m$ , $p$ *in* $\mathbb{N}_+$ *with* $0\leq p\leq m+k$ *and* $\sigma>0$*, there exists a* $\mathcal{C}^\infty-$ *function* $C_{m,p}$ *on* $\mathbb{R}_+^*$*, bounded on every interval* $[x_0,+\infty)$*,* $x_0>0$*, such that for all* $x\in\mathbb{R}_+^*$ *and* $\lambda\in\mathbb{C}$ , $\operatorname{Im}\lambda<-\sigma$*, we have*

$$\left|\frac{d^p}{dx^p}\mathcal{R}_{\lambda,m}(x)\right|\leq C_{m,p}(x)\frac{e^{-|\operatorname{Im}\lambda|x}}{|\lambda|^{\alpha+m-p+\frac{3}{2}}}. \tag{16.47}$$

*(See [13]).*

### *16.4.2 Integral representations of Mehler and Schläfli type*

In this subsection we give integral representations of Mehler and Schläfli type respectively of the functions $\varphi_\lambda$ and $\phi_{-\lambda}$.

The normalized Bessel function $j_\alpha(\lambda x)$ possesses the Poisson integral representation

$$\forall x\in\mathbb{R}_+^*\ ,\ \lambda\in\mathbb{C},\quad j_\alpha(\lambda x)=\frac{2\Gamma(\alpha+1)x^{-2\alpha}}{\sqrt{\pi}\Gamma(\alpha+\frac{1}{2})}\int_0^x(x^2-y^2)^{\alpha-\frac{1}{2}}\cos\lambda y dy. \tag{16.48}$$

(See [21] p. 38).

F.G. Mehler has showed in [15] that the Legendre function $\varphi_\lambda^{(0,-\frac{1}{2})}(x)$ admits the following integral representation

$$\forall x \in \mathbb{R}_+^* \ , \ \lambda \in \mathbb{C}, \quad \varphi_\lambda^{(0,-\frac{1}{2})}(x) = \frac{\sqrt{2}}{\pi} \int_0^x (\cosh x - \cosh y)^{-\frac{1}{2}} \cos \lambda y dy. \tag{16.49}$$

Using Theorems 16.1 and 16.2, the classical Paley-Wiener theorem and the relation (16.48) we deduce that the function $\varphi_\lambda$ , $\lambda \in \mathbb{C}$, possesses the *integral representation of Mehler type*

$$\forall x \in \mathbb{R}_+^* \ , \ \lambda \in \mathbb{C}, \quad \varphi_\lambda(x) = \int_0^x K(x,y) \cos \lambda y \, dy \tag{16.50}$$

with $K(x,.)$ an even continuous function on $(-x,x)$ with support in $[-x,x]$, which has the following properties

(i) For all $x \in \mathbb{R}_+^*$ , $y \in (-x,x), K(x,y) = (x^2 - y^2)^{\alpha - \frac{1}{2}} P(x,y)$. The function $P$ satisfies: For every $x \in \mathbb{R}_+^*$, the function $y \to P(x,y)$ is continuous on $(-x,x)$, even and with support in $[-x,x]$. For every $y \in (-x,x)$, the function $x \to P(x,y)$ is continuous on $\mathbb{R}_+^*$.

(ii) For all $q \in \mathbb{N}_+$, there exist two functions $F(x,y)$ and $G(x,y)$ continuous with respect to $x \in \mathbb{R}_+^*$, of class $\mathcal{C}^\infty$ on $\mathbb{R}$ with respect to $y$, such that the function

$$K(x,y) - \left[ (x-y)_+^{\alpha-\frac{1}{2}} F(x,y) + (x+y)_+^{\alpha-\frac{1}{2}} G(x,y) \right]$$

is of class $\mathcal{C}^q$ on $[-x,x]$. (See [20]).

**Remark 16.4** *(i) The relation (16.50) corresponding to the Jacobi hypergroups case is given in [11] with an explicit expression of the kernel $K(x,y)$.* *(ii) If the function $\mathcal{X}(x)$ given by the relation (16.28) is the restriction to $\mathbb{R}$ of a holomorphic function in the disc $D(0,2R)$, then using the relations (16.41) and (16.48), we deduce that for all $x \in (0 \ , \ 1 + |1-2\alpha|^{-\frac{1}{2}} R \ e^{-1})$ the function $y \to K(x,y)$ defined on $(-x,x)$, possesses the following series development*

$$K(x,y) = (x^2 - y^2)^{\alpha - \frac{1}{2}} \sum_{p=0}^{\infty} \frac{x^{-\alpha+\frac{1}{2}} (x^2 - y^2)^p B_p(x)}{2^{\alpha+p-1} \sqrt{\pi} \Gamma(\alpha + p + \frac{1}{2}) \sqrt{A(x)}}. \tag{16.51}$$

*(See [6]).*
*(iii) As applications of the integral representation of Mehler type (16.50),*

*K. Trimèche has defined and studied in [20] (see also [21] [22]) the Riemann Liouville and Weyl integral transforms on the Chébli-Trimèche hypergroups* $(\mathbb{R}_+, *_A)$.

The Mac-Donald function $K_\alpha(i\lambda x)$ , $\alpha > -\frac{1}{2}$, admits the Schläfli integral representation (see [25, p. 172])

$$\sqrt{\frac{2x}{\pi}}\frac{K_\alpha(i\lambda x)}{(i\lambda)^\alpha} = \frac{x^{-\alpha+\frac{1}{2}}}{2^{\alpha-\frac{1}{2}}\Gamma(\alpha+\frac{1}{2})}\int_x^\infty (y^2-x^2)^{\alpha-\frac{1}{2}}e^{-i\lambda y}dy, \tag{16.52}$$

for all $x > 0$ , $\lambda \in \mathbb{C}$ , $\operatorname{Im}\lambda < 0$.

The function $\phi_{-\lambda}$ is not, in general the Laplace transform of a function. For example in the case of the Jacobi hypergroup corresponding to the parameters $\alpha = \frac{1}{2}$ , $\beta = -\frac{1}{2}$, we have

$$\phi_{-\lambda}^{(\frac{1}{2},-\frac{1}{2})}(x) = \frac{e^{-i\lambda x}}{2\sinh x}$$

which is the Laplace transform of the measure $\frac{1}{2\sinh x}\delta_x$, where $\delta_x$ is the Dirac measure at the point $x$.

So we look for the function $\frac{\phi_{-\lambda}(x)}{(i\lambda)^{\alpha+\frac{1}{2}}}$, and using Theorem 16.4 and the relation (16.52), one has an integral representation of Schläfli type of the form

$$\frac{\phi_{-\lambda}(x)}{(i\lambda)^{\alpha+\frac{1}{2}}} = \int_x^\infty \mathcal{K}(x,y)e^{-i\lambda y}dy \text{ , for all } x > 0 \text{ , } \lambda \in \mathbb{C} \text{ , } \operatorname{Im}\lambda < 0 \tag{16.53}$$

where $\mathcal{K}(x,\cdot)$ is a continuous function on $(x,+\infty)$ with support in $[x,+\infty)$. Moreover the kernel $\mathcal{K}(x,\cdot)$ can be written for all $y > x$, in the form

$$\mathcal{K}(x,y) = \frac{(y^2-x^2)^{\alpha-\frac{1}{2}}}{2^{\alpha-\frac{1}{2}}\Gamma(\alpha+\frac{1}{2})x^{\alpha-\frac{1}{2}}\sqrt{A(x)}}\sum_{p=0}^m \frac{a_p(x)}{2^p\left(\alpha+\frac{1}{2}\right)_p}\left(\frac{y^2-x^2}{x}\right)^p + \tilde{\mathcal{K}}_m(x,y) \tag{16.54}$$

where

$$\left(\alpha+\frac{1}{2}\right)_p = \frac{\Gamma\left(\alpha+p+\frac{1}{2}\right)}{\Gamma\left(\alpha+\frac{1}{2}\right)}$$

and $\tilde{\mathcal{K}}_m$ , $m \in \mathbb{N}_+$, satisfies

(i) For all $x \in \mathbb{R}_+^*$, the function $y \to \tilde{\mathcal{K}}_m(x,y)$ is of class $\mathcal{C}^m$ on $(x,+\infty)$ with support in $[x,+\infty)$.

(ii) For all $y \in \mathbb{R}_+^*$, the function $x \to \tilde{\mathcal{K}}_m(x,y)$ is of class $\mathcal{C}^m$ on $(0,y)$ with support in $[0,y]$.

(iii) For all $y \geq x > 0$,

$$\tilde{\mathcal{K}}_m(x,y) = \frac{a_{m+1}(x)}{2^{m+1}\left(\alpha+\frac{1}{2}\right)_{m+1}}\left(\frac{y^2-x^2}{x}\right)^{m+1} + \tilde{\mathcal{K}}_{m+1}(x,y)$$

(See [13]).

**Remark 16.5** *Using the integral representation of Schläfli type (16.53), M.N. Lazhari has done in [14] a similar work to that of K. Trimèche [20], in particular he has defined a Laplace transform on the Chébli-Trimèche hypergroups $(\mathbb{R}_+, *_A)$, and characterized spaces of functions on which this transform is bijective, and he has given an inversion formula. These results generalize analogous results of M. Mizony obtained for the Laplace transform on the Jacobi hypergroups (see [16]).*

## 16.5 Harmonic analysis and maximal ideal spaces of some algebras

### 16.5.1 Harmonic analysis

**Notation 16.5** *We denote by*
• $\mathcal{D}_*(\mathbb{R})$ *the space of even $\mathcal{C}^\infty$−functions on $\mathbb{R}$ with compact support. We have*

$$\mathcal{D}_*(\mathbb{R}) = \cup_{\alpha\geq 0}\mathcal{D}_a(\mathbb{R})$$

*where $\mathcal{D}_a(\mathbb{R})$ is the space of even $\mathcal{C}^\infty$-functions on $\mathbb{R}$ with support in the interval $[-a,a]$. The topology on $\mathcal{D}_a(\mathbb{R})$ is defined by the semi-normes $q_n(f) = \sup_{x\in[-a,a]}\left|\frac{d^n}{dx^n}f(x)\right|$. The space $\mathcal{D}_*(\mathbb{R})$ equiped with the limit inductive topology is a Fréchet space.*
• $\mathbb{H}_*(\mathbb{C})$ *the space of even entire functions on $\mathbb{C}$ which are of exponential type and rapidly decreasing. We have*

$$\mathbb{H}_*(\mathbb{C}) = \cup_{\alpha\geq 0}\mathbb{H}_a(\mathbb{C})$$

*where $\mathbb{H}_a(\mathbb{C})$ is the space of even entire functions on $\mathbb{C}$ satisfying*

$$\forall m \in \mathbb{N}_+ \ , \ P_m(\psi) = \sup_{\lambda\in\mathbb{C}}(1+|\lambda|^2)^m|\psi(\lambda)|e^{-a|\operatorname{Im}\lambda|} < +\infty.$$

*The topology on $\mathbb{H}_a(\mathbb{C})$ is defined by the semi-normes $P_m$ , $m \in \mathbb{N}_+$. The space $\mathbb{H}_*(\mathbb{C})$ equiped with the limit inductive topology is a Fréchet space.*

• $L^p(m_A)$ , $p \in [1,+\infty]$, *the space of measurable functions $f$ on $\mathbb{R}_+$ satisfying*

$$\|f\|_p = \left( \int_{\mathbb{R}_+} |f(x)|^p dm_A(x) \right)^{\frac{1}{p}} < +\infty \ , \quad p \in [1,+\infty[$$

$$\|f\|_\infty = ess \sup_{x\in\mathbb{R}_+} |f(x)| < +\infty$$

• $L^p\left(\frac{d\lambda}{|c(\lambda)|^2}\right)$ , $p \in [1,+\infty]$, *the space of measurable functions $f$ on $\mathbb{R}_+$ verifying*

$$\|f\|_{p,c} = \left( \int_{\mathbb{R}_+} |f(\lambda)|^p \frac{d\lambda}{|c(\lambda)|^2} \right)^{\frac{1}{p}} < +\infty \ , \ p \in [1,+\infty[$$

$$\|f\|_{\infty,c} = ess \sup_{\lambda\in\mathbb{R}_+} |f(\lambda)| < +\infty$$

We define the Fourier transform $\mathcal{F}$ on the Chébli-Trimèche hypergroups $(\mathbb{R}_+, *_A)$ by

$$\forall\lambda \in \mathbb{C} \ , \ \mathcal{F}(f)(\lambda) = \int_{\mathbb{R}_+} f(x)\varphi_\lambda(x)dm_A(x) \ , \ f \in \mathcal{D}_*(\mathbb{R}). \tag{16.55}$$

For the Bessel-Kingmann hypergroups the transform $\mathcal{F}$ is the Fourier-Bessel transform.

The transform $\mathcal{F}$ satisfies the following properties

1. For $f$ in $L^1(m_A)$ the function $\mathcal{F}(f)$ is continuous on $\Sigma$ and we have

$$\forall\lambda \in \Sigma \ , \ |\mathcal{F}(f)(\lambda)| \leq \|f\|_1. \tag{16.56}$$

$$\lim_{\lambda\to+\infty} \mathcal{F}(f)(\lambda) = 0.$$

2. We suppose that $\rho > 0$. We consider $p \in [1,2)$, $q \in [1,+\infty]$ such that $\frac{1}{p}+\frac{1}{q}=1$, and $D_p = \left\{\lambda \in \mathbb{C} \,|\, |\operatorname{Im}\lambda| < \left(\frac{2}{p}-1\right)\rho\right\}$. Then
(i) For all $\lambda \in D_p$ the function $\varphi_\lambda$ belongs to $L^q(m_A)$ .

(ii) For all $f \in L^p(m_A)$, the function $\mathcal{F}(f)$ is analytic in $\mathrm{D}_p$ and we have

$$\forall \lambda \in D_p, \ |\mathcal{F}(f)(\lambda)| \leq \|\varphi_\lambda\|_q \|f\|_p .$$

(See [3] [4] [20] [21] [23] [26] [27]).

**Theorem 16.6** *(i) The transform $\mathcal{F}$ is a topological isomorphism from $\mathcal{D}_*(\mathbb{R})$ onto $\mathbb{H}_*(\mathbb{C})$. The inverse transform $\mathcal{F}^{-1}$ is given by*

$$\forall x \in \mathbb{R}_+ , \ \mathcal{F}^{-1}(h)(x) = \int_{\mathbb{R}_+} h(\lambda)\varphi_\lambda(x)\frac{d\lambda}{|c(\lambda)|^2}. \tag{16.57}$$

*(ii) Let $f$ be in $L^1(m_A)$ such that the function $\mathcal{F}(f)$ belongs to $L^1\left(\frac{d\lambda}{|c(\lambda)|^2}\right)$. Then we have the following inversion formula for the transform $\mathcal{F}$:*

$$f(x) = \int_{\mathbb{R}_+} \mathcal{F}(f)(\lambda)\varphi_\lambda(x)\frac{d\lambda}{|c(\lambda)|^2} \quad a.e. \tag{16.58}$$

*(See [3] [4] [23] [26]).*

**Theorem 16.7** *(i) Plancheral formula. For all $f$ in $\mathcal{D}_*(\mathbb{R})$ we have*

$$\int_{\mathbb{R}_+} |f(x)|^2 dm_A(x) = \int_{\mathbb{R}_+} |\mathcal{F}(f)(\lambda)|^2 \frac{d\lambda}{|c(\lambda)|^2}. \tag{16.59}$$

*(ii)Plancherel theorem. The transform $\mathcal{F}$ extends uniquely to an unitary isomorphism of $L^2(m_A)$ onto $L^2\left(\frac{d\lambda}{|c(\lambda)|^2}\right)$. (See [3] [4] [23] [26]).*

For $x \in \mathbb{R}_+$ and $f$ in $\mathcal{D}_*(\mathbb{R})$ the generalized $x-$translate of $f$ is defined by

$$\forall y \in \mathbb{R}_+ , \ T_x f(y) = \int_{\mathbb{R}_+} f(z) d(\delta_x * \delta_y)(z). \tag{16.60}$$

Using the relation (16.2), we deduce that the relation (16.60) can also be written in the form

$$\forall x, y \in \mathbb{R}_+^* , \ T_x f(y) = \int_{|x-y|}^{x+y} f(z) W(x,y,z) dm_A(z) \tag{16.61}$$

$$\forall x \in \mathbb{R}_+ , \quad T_x f(0) = f(x).$$

The generalized translation operators $T_x$, $x \in \mathbb{R}_+$, satisfy the following properties

(i) For $f$ in $\mathcal{D}_*(\mathbb{R})$ we have

$$\forall x, y \in \mathbb{R}_+ \ , \ T_x f(y) = T_y f(x). \tag{16.62}$$

(ii) we have

$$\forall x, y \in \mathbb{R}_+ \ , \ \lambda \in \mathbb{C} \ , \ T_x \varphi_\lambda(y) = \varphi_\lambda(x)\varphi_\lambda(y). \tag{16.63}$$

(iii) Let $f$ be in $\mathcal{D}_*(\mathbb{R})$ such that supp$f \subset [-a, a]$, $a \geq 0$. Then for all $x \in \mathbb{R}_+$, the function $T_x f$ belongs to $\mathcal{D}_*(\mathbb{R})$ with supp$T_x f \subset [-a-x, a+x]$, and we have

$$\forall \lambda \in \mathbb{C} \ , \ \mathcal{F}(T_x f)(\lambda) = \varphi_\lambda(x)\mathcal{F}(f)(\lambda). \tag{16.64}$$

(iv) If $f$ is in $L^p(m_A)$ , $p \in [1, +\infty]$, then for all $x \in \mathbb{R}_+$ the function $T_x f$ is in $L^p(m_A)$ , $p \in [1, +\infty]$, and we have

$$\|T_x f\|_p \leq \|f\|_p. \tag{16.65}$$

(v) For $f$ in $L^1(m_A)$ and $x \in \mathbb{R}_+$, we have

$$\forall \lambda \in \Sigma \ , \ \mathcal{F}(T_x f)(\lambda) = \varphi_\lambda(x)\mathcal{F}(f)(\lambda). \tag{16.66}$$

(vi) For $f$ in $L^2(m_A)$ and $x \in \mathbb{R}_+$, we have

$$\mathcal{F}(T_x f)(\lambda) = \varphi_\lambda(x)\mathcal{F}(f)(\lambda) \quad a.e. \text{ on } \mathbb{R}_+. \tag{16.67}$$

(See [3] [4] [20] [21] [23] [26] [27]).

We define the *convolution* $*$ on the Chébli-Trimèche hypergroups $(\mathbb{R}_+, *_A)$ by

$$\forall x \in \mathbb{R}_+ \ , \ f * g(x) = \int_{\mathbb{R}_+} T_x f(y) g(y) dm_A(y) \ , \quad f, g \in \mathcal{D}_*(\mathbb{R}). \tag{16.68}$$

This convolution possesses the following properties

(i) The convolution $*$ is commutative and associative.

(ii) Let $f$ be in $\mathcal{D}_*(\mathbb{R})$ with support in $[-a, a]$, $a \geq 0$, and $g$ in $\mathcal{D}_*(\mathbb{R})$ with support in $[-b, b]$, $b \geq 0$. Then the function $f * g$ belongs to $\mathcal{D}_*(\mathbb{R})$ with support in $[-a-b, a+b]$, and we have

$$\forall \lambda \in \mathbb{C} \ , \ \mathcal{F}(f * g)(\lambda) = \mathcal{F}(f)(\lambda)\mathcal{F}(g)(\lambda). \tag{16.69}$$

(iii) Let $f$ be in $L^p(m_A)$, $p \in [1,+\infty]$,and $g$ in $L^q(m_A)$, $q \in [1,+\infty]$. Then the function $f * g$ belongs to $L^r(m_A)$, $r \in [1,+\infty]$, such that $\frac{1}{p}+\frac{1}{q}-1=\frac{1}{r}$, and we have

$$\|f * g\|_r \leq \|f\|_p \|g\|_q. \tag{16.70}$$

Moreover

- For $f$ and $g$ in $L^1(m_A)$ we have

$$\forall \lambda \in \Sigma \ , \ \mathcal{F}(f*g)(\lambda) = \mathcal{F}(f)(\lambda)\mathcal{F}(g)(\lambda). \tag{16.71}$$

- For $f$ in $L^1(m_A)$ and $g$ in $L^2(m_A)$ we have

$$\mathcal{F}(f*g)(\lambda) = \mathcal{F}(f)(\lambda)\mathcal{F}(g)(\lambda) \quad a.e. \text{ on } \mathbb{R}_+. \tag{16.72}$$

(iv) The space $L^1(m_A)$ provided with the convolution $*$ is a commutative Banach algebra.

**Theorem 16.8** *Let $f$ be a locally integrable function on $[0,+\infty)$ and $g$ a measurable function on $[0,+\infty)$ satisfying: There exists $r \in \mathbb{N}_+$ such that the function $(1+\lambda^2)^{-r}g$ belongs to $L^1\left(\frac{d\lambda}{|c(\lambda)|^2}\right)$. We suppose that*

$$\int_{\mathbb{R}_+} f(x)\mathcal{F}^{-1}(\psi)(x)dm_A(x) = \int_{\mathbb{R}_+} g(\lambda)\psi(\lambda)\frac{d\lambda}{|c(\lambda)|^2} \tag{16.73}$$

*for all $\psi \in \mathbb{H}_*(\mathbb{C})$. Then the function $f$ belongs to $L^2(m_A)$ if and only if the function $g$ belongs to $L^2\left(\frac{d\lambda}{|c(\lambda)|^2}\right)$ and we have*

$$\mathcal{F}(f) = g. \tag{16.74}$$

**Theorem 16.9** *For all $f, g$ in $L^2(m_A)$ and $\psi$ in $\mathbb{H}_*(\mathbb{C})$ we have the relation*

$$\int_{\mathbb{R}_+} f*g(x)\mathcal{F}^{-1}(\psi)(x)dm_A(x) = \int_{\mathbb{R}_+} \mathcal{F}(f)(\lambda)\mathcal{F}(g)(\lambda)\psi(\lambda)\frac{d\lambda}{|c(\lambda)|^2}. \tag{16.75}$$

**Corollary 16.10** *(i) Let $f, g$ be in $L^2(m_A)$. The function $f*g$ belongs to $L^2(m_A)$ if and only if the function $\mathcal{F}(f).\mathcal{F}(g)$ belongs to $L^2\left(\frac{d\lambda}{|c(\lambda)|^2}\right)$ and we have*

$$\mathcal{F}(f*g) = \mathcal{F}(f).\mathcal{F}(g). \tag{16.76}$$

*(ii) For all $f, g$ in $L^2(m_A)$ we have*

$$\int_{\mathbb{R}_+} |f * g(x)|^2 dm_A(x) = \int_{\mathbb{R}_+} |\mathcal{F}(f)(\lambda)|^2 |\mathcal{F}(g)(\lambda)|^2 \frac{d\lambda}{|c(\lambda)|^2}. \qquad (16.77)$$

*The two sides are finite or infinite.*

(See [3] [4] [20] [21] [23] [26] [27]).

### 16.5.2 The maximal ideal spaces of the algebras $L^1(m_A)$ and $M_b(\mathbb{R}_+)$

In this subsection we give the maximal ideal spaces $S$ and $S^*$ respectively of the algebras $L^1(m_A)$ and $M_b(\mathbb{R}_+)$, and we prove that $S$ and $S^*$ are homeomorphic respectively to the sets $\Sigma$ and $\overline{\Sigma} = \Sigma \cup \{+\infty\}$ equiped with the usual topology. (See [35]).

The maximal ideal space of $L^1(m_A)$ is depicted in the following result.

**Theorem 16.11** *To each homomorphism $\aleph$ from $L^1(m_A)$ into $\mathbb{C}$ corresponds a unique element $\lambda$ in $\Sigma$ such that*

$$\forall f \in L^1(m_A)\ , \quad \aleph(f) = \mathcal{F}(f)(\lambda) \qquad (16.78)$$

**Remark 16.6** *(i) Theorem 16.11 proves that the Fourier transform $\mathcal{F}$ coincides with the Gelfand transform defined on $L^1(m_A)$ by*

$$\tilde{f}(\aleph) = \aleph(f)\ , \quad \aleph \in S \qquad (16.79)$$

*where $S$ denotes the set of all homomorphisms $\aleph$ from $L^1(m_A)$ into $\mathbb{C}$.*

*(ii) Let $L^1(m_A)^\sim$ be the space of all $\tilde{f}$ for $f$ in $L^1(m_A)$. The Gelfand topology of $S$ is the weak topology induced by $L^1(m_A)^\sim$, that is the weakest topology that makes every $\tilde{f}$ continuous. Then we have*

$$L^1(m_A)^\sim \subset C(S) \qquad (16.80)$$

*where $C(S)$ is the space of all complex continuous functions on $S$.*

*(iii) The set $S$ equiped with the Gelfand topology is usually called the maximal ideal space of $L^1(m_A)$.*

**Theorem 16.12** *The maximal ideal space $S$ of $L^1(m_A)$ is homeomorphic to $\Sigma$ equiped with the usual topology.*

The Fourier transform $\mathcal{F}(\mu)$ of a measure $\mu$ in $M_b(\mathbb{R}_+)$ is defined by

$$\forall \lambda \in \Sigma\,, \quad \mathcal{F}(\mu)(\lambda) = \int_{\mathbb{R}_+} \varphi_\lambda(x) d\mu(x). \tag{16.81}$$

We have the following properties:

(i) The function $\mathcal{F}(\mu)$ is continuous on $\Sigma$ and we have

$$\forall \lambda \in \Sigma\,, \quad |\mathcal{F}(\mu)(\lambda)| \leq \|\mu\|. \tag{16.82}$$

where $\|\cdot\|$ is the norm on the space $M_b(\mathbb{R}_+)$.

(ii) The transform $\mathcal{F}$ is injective on $M_b(\mathbb{R}_+)$.

(iii) If $\mu = f\, m_A$ with $f$ in $L^1(m_A)$, then $\mathcal{F}(\mu)$ is the Fourier transform of $f$ given by the relation (16.55).

The convolution $*$ which defines the Chébli-Trimèche hypergroups $(\mathbb{R}_+, *_A)$ can be written for two measures $\mu, \nu$ in $M_b(\mathbb{R}_+)$ in the form

$$\mu * \nu(f) = \int_{\mathbb{R}_+}\int_{\mathbb{R}_+} T_x f(y) d\mu(x) d\nu(y)\,, \quad f \in \mathcal{D}_*(\mathbb{R}). \tag{16.83}$$

In particular if $\mu = g\, m_A$ and $\nu = h\, m_A$, with $g$ and $h$ in $L^1(m_A)$, then $\mu * \nu = (g * h)\, m_A$, where $g * h$ is the function in $L^1(m_A)$ given by the relation (16.68).

We have the relation

$$\forall \lambda \in \Sigma\,, \quad \mathcal{F}(\mu * \nu)(\lambda) = \mathcal{F}(\mu)(\lambda)\mathcal{F}(\nu)(\lambda). \tag{16.84}$$

We denote by $M_b'(\mathbb{R}_+)$ the subspace of $M_b(\mathbb{R}_+)$ consisting of measures concentrated in $\mathbb{R}_+^*$. We have the unique decomposition of each measure $\mu$ in $M_b(\mathbb{R}_+)$:

$$\mu = a\delta_0 + \mu'\,, \quad a \in \mathbb{C}\,,\ \mu' \in M_b'(\mathbb{R}_+). \tag{16.85}$$

**Remark 16.7** *(i) Let $\mu'$ be in $M_b'(\mathbb{R}_+)$. we have*

$$\lim_{\lambda \to +\infty} \mathcal{F}(\mu')(\lambda) = 0. \tag{16.86}$$

*(ii) Let $\mu$ be in $M_b(\mathbb{R}_+)$. We have*

$$\mathcal{F}(\mu)(+\infty) = \mu(\{0\}). \tag{16.87}$$

*Then the Fourier transform $\mathcal{F}(\mu)$ can be extended to $\overline{\Sigma} = \Sigma \cup \{+\infty\}$. Furthermore a measure $\mu$ is in $M_b'(\mathbb{R}_+)$ if and only if $\mathcal{F}(\mu)(+\infty) = 0$.*

The set $S^*$ of all complex homomorphisms of $M_b(\mathbb{R}_+)$ is characterized by the following theorem.

**Theorem 16.13** *To each homomorphism $\aleph$ from $M_b(\mathbb{R}_+)$ into $\mathbb{C}$ corresponds a unique element $\lambda$ in $\overline{\Sigma}$ such that*

$$\forall \mu \in M_b(\mathbb{R}_+) \ , \quad \aleph(\mu) = \mathcal{F}(\mu)(\lambda). \tag{16.88}$$

**Remark 16.8** *(i) Theorem 16.13 proves that the Fourier transform $\mathcal{F}$ given by the relation (16.81), coincides with the Gelfand transform defined on $M_b(\mathbb{R}_+)$ by*

$$\tilde{\mu}(\aleph) = \aleph(\mu) \ , \quad \aleph \in S^*. \tag{16.89}$$

*(ii) Let $M_b(\mathbb{R}_+)^{\tilde{}}$ be the space of all $\tilde{\mu}$ for $\mu$ in $M_b(\mathbb{R}_+)$. The Gelfand topology of $S^*$ is the weak topology induced by $M_b(\mathbb{R}_+)^{\tilde{}}$, that is the weakest topology that makes every $\tilde{\mu}$ continuous. Then we have*

$$M_b(\mathbb{R}_+)^{\tilde{}} \subset C(S^*) \tag{16.90}$$

*where $C(S^*)$ is the space of continuous functions on $S^*$.*

*(iii) The set $S^*$ equiped with the Gelfand topology is usually called the maximal ideal space of $M_b(\mathbb{R}_+)$.*

**Theorem 16.14** *The maximal ideal space $S^*$ of $M_b(\mathbb{R}_+)$ is homeomorphic to $\overline{\Sigma}$ equiped with the usual topology.*

**Remark 16.9** *A. Schwartz has obtained in [19] analogous results in the case of the Bessel-Kingman hypergroups.*

## 16.6 Continuous linear wavelet transform and its discretization

### 16.6.1 Linear wavelets on $(\mathbb{R}_+, *_A)$

We define in this subsection linear wavelets (also called L-wavelets), and the continuous L-wavelet transform on the Chébli-Trimèche hypergroups $(\mathbb{R}_+, *)$, and we prove for this transform reconstruction formulas.

**Proposition 16.15** *Let $a \in \mathbb{R}_+^*$ and $g$ be a function in $L^2(m_A)$. Then there exists a function $g_a$ in $L^2(m_A)$ such that*

$$\forall \lambda \in \mathbb{R}_+, \quad \mathcal{F}(g_a)(\lambda) = \mathcal{F}(g)(a\lambda). \tag{16.91}$$

*This function is given by the relation*

$$g_a = \mathcal{F}^{-1} \circ H_a \circ \mathcal{F}(g) \tag{16.92}$$

*and satisfies*

$$\|g_a\|_2 \leq \frac{k(a)}{\sqrt{a}} \|g\|_2 \tag{16.93}$$

*with $H_a$ the dilation operator given by*

$$H_a(f)(x) = \frac{1}{a} f(\frac{x}{a}) \tag{16.94}$$

*and $k(a)$ the function defined on $\mathbb{R}_+^*$ by*

$$k(a) = \sup_{\lambda \in \mathbb{R}_+^*} \frac{|c(\lambda)|}{\left|c\left(\frac{\lambda}{a}\right)\right|}. \tag{16.95}$$

**Proposition 16.16** *There exist two positive constants $M_1$ and $M_2$ such that*

- *If $\rho = 0$ and $\alpha > 0$:* $\quad \frac{M_1}{a^{\alpha+\frac{1}{2}}} \leq k(a) \leq \frac{M_2}{a^{\alpha+\frac{1}{2}}}$ *, for all $a \in \mathbb{R}_+^*$*
- *If $\rho > 0$ and $\alpha > -\frac{1}{2}$:*

$$\begin{cases} \frac{M_1}{a^{\gamma}} \leq k(a) \leq \frac{M_2}{a^{\gamma_0}} , & \text{if } a \geq 1 \\ \frac{M_1}{a^{\gamma_0}} \leq k(a) \leq \frac{M_2}{a^{\gamma}} , & \text{if } 0 < a < 1 \end{cases}$$

*where $\gamma = \max(1, \alpha + \frac{1}{2})$ and $\gamma_0 = \min(1, \alpha + \frac{1}{2})$.*

**Proposition 16.17** *(i) Let $g$ be a function in $\mathcal{D}_*(\mathbb{R})$ with support in $[-b, b]$, $b \geq 0$. Then for all $a \in \mathbb{R}_+^*$, the function $g_a$ belongs to $\mathcal{D}_*(\mathbb{R})$ with support in $[-ab, ab]$.*

*(ii) Let $g$ be a function in $L^2(m_A)$. Then the function $a \longrightarrow g_a$ is continuous from $(0, +\infty)$ into $L^2(m_A)$.*

**Example 16.5** *The Bessel Kingman hypergroups. Let $a \in \mathbb{R}_+^*$ and $g$ be a function in $L^2(m_{x^{2\alpha+1}})$. The function $g_a$ is in $L^2(m_{x^{2\alpha+1}})$ and is given by*

$$\forall x \in \mathbb{R}_+ , \quad g_a(x) = \frac{1}{a^{2\alpha+2}} g\left(\frac{x}{a}\right) = \frac{1}{a^{2\alpha+1}} H_a(g)(x)$$

*we have*

$$\|g_a\|_2 = \frac{1}{a^{\alpha+1}} \|g\|_2.$$

*Then for this case the inequality (16.93) is an equality and we have*

$$k_a = \frac{1}{a^{\alpha+\frac{1}{2}}}$$

**Example 16.6** *The Jacobi hypergroups. We consider the Jacobi hypergroup corresponding to the parameters $\alpha = \frac{1}{2}$, $\beta = -\frac{1}{2}$. Let $a \in \mathbb{R}_+^*$ and $g$ be a function in $L^2(m_{\sinh^2 x})$. The function $g_a$ is in $L^2(m_{\sinh^2 x})$ and is given for all $x \in \mathbb{R}_+^*$ by*

$$g_a(x) = \frac{\sinh \frac{x}{a}}{a^2 \sinh x} g\left(\frac{x}{a}\right) = \frac{1}{a \sinh x} H_a(\sinh(\cdot)g)(x)$$

*and we have*

$$\|g_a\|_2 = \frac{1}{a^{\frac{3}{2}}} \|g\|_2.$$

*Also for this case the inequality (16.93) is an equality and we have*

$$k(a) = \frac{1}{a}.$$

**Definition 16.1** *A linear wavelet (also called L-wavelet) is a measurable function $g^L$ on $\mathbb{R}_+$ satisfying the conditions*

$$\int_{\mathbb{R}_+} |\mathcal{F}(g^L)(\lambda)| \frac{d\lambda}{\lambda} < +\infty. \tag{16.96}$$

$$C_{g^L} = \int_{\mathbb{R}_+} \mathcal{F}(g^L)(\lambda) \frac{d\lambda}{\lambda} \neq 0. \tag{16.97}$$

**Example 16.7** *Let $t \in \mathbb{R}_+^*$. For all $p \in \mathbb{N}_+$, the function $\lambda^p e^{-t\lambda^2}$ belongs to $(L^1 \cap L^2)\left(\frac{d\lambda}{|c(\lambda)|^2}\right)$. We consider the function*

$$E_t(x) = \int_{\mathbb{R}_+} e^{-t\lambda^2} \varphi_\lambda(x) \frac{d\lambda}{|c(\lambda)|^2}, \quad x \in \mathbb{R}_+. \tag{16.98}$$

*From Theorems 16.6, 16.7 and Definition 16.1 we deduce that the function*

$$g^L(x) = -\frac{d}{dt} E_t(x), \quad x \in \mathbb{R}_+. \tag{16.99}$$

*which belongs to $L^2(m_A)$, is a L-wavelet and we have*

$$C_{g^L} = \frac{1}{2t}. \tag{16.100}$$

**Proposition 16.18** *Let $g^L$ be a L-wavelet in $L^2(m_A)$. Then the function $g^L_a$, $a \in \mathbb{R}^*_+$, defined by the relation (16.91) is a L-wavelet in $L^2(m_A)$ and we have*

$$C_{g^L_a} = C_{g^L}. \tag{16.101}$$

**Proof.** From Proposition 16.15 the function $g^L_a$, $a \in \mathbb{R}^*_+$, belongs to $L^2(m_A)$. Using the relation (16.91) and a change of variables we obtain

$$\int\limits_{\mathbb{R}_+} |\mathcal{F}(g^L_a)(\lambda)| \frac{d\lambda}{\lambda} = \int\limits_{\mathbb{R}_+} |\mathcal{F}(g^L)(a\lambda)| \frac{d\lambda}{\lambda} = \int\limits_{\mathbb{R}_+} |\mathcal{F}(g^L)(u)| \frac{du}{u} < +\infty \tag{16.102}$$

and

$$C_{g^L_a} = \int\limits_{\mathbb{R}_+} \mathcal{F}(g^L_a)(\lambda)\frac{d\lambda}{\lambda} = \int\limits_{\mathbb{R}_+} \mathcal{F}(g^L)(a\lambda)\frac{d\lambda}{\lambda} = \int\limits_{\mathbb{R}_+} \mathcal{F}(g^L)(u)\frac{du}{u} \neq 0$$

then the function $g^L_a$, $a \in \mathbb{R}^*_+$, is a L-wavelet and we have

$$C_{g^L_a} = C_{g^L}.$$

■

Let $g^L$ be a L-wavelet in $L^2(m_A)$ and $g^L_a$, $a \in \mathbb{R}^*_+$, the function defined by the relation (16.91). We consider for all $y \in \mathbb{R}_+$ the function $g^L_{a,y}$ given by

$$\forall x \in \mathbb{R}_+ \ , \quad g^L_{a,y}(x) = T_x g^L_a(y) \tag{16.103}$$

where $T_x$ , $x \in \mathbb{R}_+$, are the generalized translation operators given by the relation (16.61).

This function satisfies the following properties.

**Proposition 16.19** *(i) For all $y \in \mathbb{R}_+$, the function $g^L_{a,y}$ belongs to $L^2(m_A)$ and we have*

$$\|g^L_{a,y}\|_2 \leq \frac{k(a)}{\sqrt{a}}\|g^L\|_2 \tag{16.104}$$

*where $k(a)$ is the function given by the relation (16.95).*

*(ii) For all $y \in \mathbb{R}_+$, we have the relation*

$$\forall \lambda \in \mathbb{R}_+ \ , \quad \mathcal{F}(g^L_{a,y})(\lambda) = \varphi_\lambda(y)\mathcal{F}(g^L_a)(\lambda). \tag{16.105}$$

**Definition 16.2** *Let $g^L$ be a L-wavelet in $L^2(m_A)$. The continuous L-wavelet transform $\Phi^L_{g^L}$ is defined for regular functions $f$ on $\mathbb{R}_+$ by*

$$\Phi^L_{g^L}(f)(a,y) = \int_{\mathbb{R}_+} f(x) g^L_{a,y}(x) dm_A(x) \ , \ a \in \mathbb{R}^*_+ \ , \ y \in \mathbb{R}_+. \tag{16.106}$$

This transform can also be written in the form

$$\Phi^L_{g^L}(f)(a,y) = f * g^L_a(y) \tag{16.107}$$

where $*$ is the convolution defined by the relation (16.68).

**Proposition 16.20** *Let $g^L$ be a L-wavelet in $L^2(m_A)$. Then for all $a \in \mathbb{R}^*_+$ and $f$ in $L^q(m_A)$, $q \in [1,2]$, the function $y \to \Phi^L_{g^L}(f)(a,y)$ belongs to $L^r(m_A)$ with $r \in [1,+\infty]$, such that $\frac{1}{r} = \frac{1}{q} - \frac{1}{2}$, and we have*

$$\|\Phi^L_{g^L}(f)(a,\cdot)\|_r \leq \frac{k(a)}{\sqrt{a}} \|f\|_q \|g^L\|_2. \tag{16.108}$$

**Proof.** The relation (16.107), the properties of the convolution given earlier, and Proposition 16.19 give the results. ∎

The following result is the so-called *Calderón's reproducing formula.*

**Theorem 16.21** *Let $g^L$ be a L-wavelet in $L^2(m_A)$. Then for $f$ in $L^2(m_A)$ and $0 < \varepsilon < \delta < +\infty$, the function*

$$f^L_{\varepsilon,\delta}(y) = \frac{1}{C_{g^L}} \int_{\varepsilon}^{\delta} \Phi^L_{g^L}(f)(a,y) \frac{da}{a} \tag{16.109}$$

*belongs to $L^2(m_A)$ and we have*

$$\lim_{\varepsilon \to 0^+, \delta \to \infty} \|f^L_{\varepsilon,\delta} - f\|_2 = 0. \tag{16.110}$$

To prove this theorem we need the following two lemmas.

**Lemma 16.22** *Let $g^L$ be a L-wavelet in $L^2(m_A)$. For $0 < \varepsilon < \delta < +\infty$, we define the functions $G^L_{\varepsilon,\delta}$ and $K^L_{\varepsilon,\delta}$ by*

$$G^L_{\varepsilon,\delta}(\cdot) = \frac{1}{C_{g^L}} \int_{\varepsilon}^{\delta} g^L_a(\cdot) \frac{da}{a}. \tag{16.111}$$

$$K^L_{\varepsilon,\delta}(\lambda) = \frac{1}{C_{g^L}} \int_{\varepsilon}^{\delta} \mathcal{F}(g^L_a)(\lambda) \frac{da}{a}, \quad \lambda \in \mathbb{R}_+. \tag{16.112}$$

*Then*

*(i)* $G^L_{\varepsilon,\delta} \in L^2(m_A)$.

*(ii)* $K^L_{\varepsilon,\delta} \in L^\infty\left(\frac{d\lambda}{|c(\lambda)|^2}\right)$.

*(iii)* $\mathcal{F}(G^L_{\varepsilon,\delta}) = K_{\varepsilon,\delta}$.

*(iv)* $\forall \lambda \in \mathbb{R}^*_+$, $\lim_{\varepsilon\to 0^+,\delta\to\infty} K^L_{\varepsilon,\delta}(\lambda) = 1$.

**Proof.** From Proposition 16.17 ii) the function $G^L_{\varepsilon,\delta}$ belongs to $L^2(m_A)$. We have

$$\forall \lambda \in \mathbb{R}_+ \ , \ \left|K^L_{\varepsilon,\delta}(\lambda)\right| \le \frac{1}{C_{g^L}} \int\limits_{\mathbb{R}_+} |\mathcal{F}(g^L)(a\lambda)| \frac{da}{a}.$$

By change of variables and the relation (16.96) we obtain

$$\forall \lambda \in \mathbb{R}^*_+ \ , \ \left|K^L_{\varepsilon,\delta}(\lambda)\right| \le \frac{1}{C_{g^L}} \int\limits_{\mathbb{R}_+} |\mathcal{F}(g^L)(u)| \frac{du}{u} < +\infty.$$

Thus the function $K^L_{\varepsilon,\delta}$ belongs to $L^\infty\left(\frac{d\lambda}{|c(\lambda)|^2}\right)$. For all $\psi \in \mathbb{H}_*(\mathbb{C})$we have

$$\int\limits_{\mathbb{R}_+} G^L_{\varepsilon,\delta}(x)\mathcal{F}^{-1}(\psi)(x)dm_A(x)$$

$$= \frac{1}{C_{g^L}} \int\limits_{\mathbb{R}_+} \left( \int\limits_{\varepsilon}^{\delta} g^L_a(x) \frac{da}{a} \right) \mathcal{F}^{-1}(\psi)(x)dm_A(x).$$

As the function $\mathcal{F}^{-1}(\psi)$belongs to $\mathcal{D}_*(\mathbb{R})$and from Proposition 16.17(ii) the function $a \longrightarrow \|g^L_a\|_2$is continuous on $\mathbb{R}^*_+$, then from Fubini-Tonelli's theorem and Hölder's inequality we have

$$\int\limits_{\mathbb{R}_+} \int\limits_{\varepsilon}^{\delta} \left|g^L_a(x)\right| \left|\mathcal{F}^{-1}(\psi)(x)\right| dm_A(x) \frac{da}{a} \le \|\mathcal{F}^{-1}(\psi)\|_2 \int\limits_{\varepsilon}^{\delta} \|g^L_a\|_2 \frac{da}{a}$$

which is finite. Then from Fubini theorem we obtain

$$\int\limits_{\mathbb{R}_+} G^L_{\varepsilon,\delta}(x)\mathcal{F}^{-1}(\psi)(x)dm_A(x)$$

$$= \frac{1}{C_{g^L}} \int\limits_{\varepsilon}^{\delta} \left( \int\limits_{\mathbb{R}_+} g^L_a(x)\mathcal{F}^{-1}(\psi)(x)dm_A(x) \right) \frac{da}{a}.$$

By applying Theorem 16.7 we obtain

$$\int_{\mathbb{R}_+} G^L_{\varepsilon,\delta}(x)\mathcal{F}^{-1}(\psi)(x)dm_A(x)$$

$$= \frac{1}{C_{g^L}} \int_{\varepsilon}^{\delta} \left( \int_{\mathbb{R}_+} \mathcal{F}(g^L_a)(\lambda)\psi(\lambda)\frac{d\lambda}{|c(\lambda)|^2} \right) \frac{da}{a}.$$

From Fubini-Tonelli's theorem, a change of variables and the relation (16.96) we have

$$\int_{\varepsilon}^{\delta} \int_{\mathbb{R}_+} |\mathcal{F}(g^L_a)(\lambda)|\, |\psi(\lambda)| \frac{d\lambda}{|c(\lambda)|^2}\frac{da}{a} \leq \|\psi\|_{1,c} \int_{\mathbb{R}_+} |\mathcal{F}(g^L)(u)|\frac{du}{u} < +\infty.$$

Thus from Fubini theorem and the relation (16.112) we obtain

$$\int_{\mathbb{R}_+} G^L_{\varepsilon,\delta}(x)\mathcal{F}^{-1}(\psi)(x)dm_A(x) = \int_{\mathbb{R}_+} \left( \frac{1}{c_{g^L}} \int_{\varepsilon}^{\delta} \mathcal{F}(g^L_a)(\lambda)\frac{da}{a} \right) \psi(\lambda)\frac{d\lambda}{|c(\lambda)|^2}$$

$$= \int_{\mathbb{R}_+} K^L_{\varepsilon,\delta}(\lambda)\psi(\lambda)\frac{d\lambda}{|c(\lambda)|^2}.$$

Then Theorem 16.8 implies that the function $K^L_{\varepsilon,\delta}$belongs to $L^2\left(\frac{d\lambda}{|c(\lambda)|^2}\right)$and we have

$$\mathcal{F}(G^L_{\varepsilon,\delta}) = K^L_{\varepsilon,\delta}$$

We obtain the last result by change of variables and the relation (16.97). ∎

**Lemma 16.23** *The function $f^L_{\varepsilon,\delta}$ defined in Theorem 16.21 belongs to $L^2(m_A)$ and satisfies the relations*

$$f^L_{\varepsilon,\delta} = f * G^L_{\varepsilon,\delta}. \tag{16.113}$$

$$\mathcal{F}(f^L_{\varepsilon,\delta}) = \mathcal{F}(f)K^L_{\varepsilon,\delta}. \tag{16.114}$$

**Proof.** For all $\psi$ in $\mathbb{H}_*(\mathbb{C})$ we have from the relation (16.107):

$$\int_{\mathbb{R}_+} f^L_{\varepsilon,\delta}(x)\mathcal{F}^{-1}(\psi)(x)dm_A(x)$$

$$= \frac{1}{C_{g^L}} \int_{\mathbb{R}_+} \left( \int_{\varepsilon}^{\delta} f * g^L_a(x)\frac{da}{a} \right) \mathcal{F}^{-1}(\psi)(x)dm_A(x).$$

As the function $\mathcal{F}^{-1}(\psi)$ belongs to $\mathcal{D}_*(\mathbb{R})$, and from Proposition 16.17 ii) the function $a \longrightarrow \|g_a^L\|_2$ is continuous on $\mathbb{R}_+^*$, we obtain from the relation (16.70) and Fubini-Tonelli's theorem

$$\int_{\mathbb{R}_+}\int_{\varepsilon}^{\delta} \left|f * g_a^L(x)\right| \left|\mathcal{F}^{-1}(\psi)(x)\right| dm_A(x)\frac{da}{a} \leq \|f\|_2\|\mathcal{F}^{-1}(\psi)\|_1 \int_{\varepsilon}^{\delta} \|g_a^L\|_2\frac{da}{a}.$$

Thus from Fubini theorem we have

$$\int_{\mathbb{R}_+} f_{\varepsilon,\delta}^L(x)\mathcal{F}^{-1}(\psi)(x)dm_A(x)$$

$$= \frac{1}{C_{g^L}}\int_{\varepsilon}^{\delta}\left(\int_{\mathbb{R}_+} f * g_a^L(x)\mathcal{F}^{-1}(\psi)(x)dm_A(x)\right)\frac{da}{a}.$$

Using Theorem 16.9 we obtain

$$\int_{\mathbb{R}_+} f_{\varepsilon,\delta}^L(x)\mathcal{F}^{-1}(\psi)(x)dm_A(x)$$

$$= \frac{1}{C_{g^L}}\int_{\varepsilon}^{\delta}\left(\int_{\mathbb{R}_+} \mathcal{F}(f)(\lambda)\mathcal{F}(g_a^L)(\lambda)\psi(\lambda)\frac{d\lambda}{|c(\lambda)|^2}\right)\frac{da}{a}.$$

But from Fubini-Tonelli's theorem, Hölder's inequality, Theorem 16.7, a change of variables and the relation (16.96) we have

$$\int_{\varepsilon}^{\delta}\int_{\mathbb{R}_+} |\mathcal{F}(f)(\lambda)|\left|\mathcal{F}(g_a^L)(\lambda)\right| |\psi(\lambda)|\frac{d\lambda}{|c(\lambda)|^2}\frac{da}{a}$$

$$\leq \|f\|_2\|\psi\|_{2,c}\int_{\mathbb{R}_+} \left|\mathcal{F}(g^L)(u)\right|\frac{du}{u} < +\infty.$$

Then we deduce from Fubini theorem and the relation (16.112):

$$\int_{\mathbb{R}_+} f_{\varepsilon,\delta}^L(x)\mathcal{F}^{-1}(\psi)(x)dm_A(x)$$

$$= \int_{\mathbb{R}_+} \mathcal{F}(f)(\lambda)\left(\frac{1}{C_{g^L}}\int_{\varepsilon}^{\delta}\mathcal{F}(g_a^L)(\lambda)\frac{da}{a}\right)\psi(\lambda)\frac{d\lambda}{|c(\lambda)|^2}$$

$$= \int_{\mathbb{R}_+} \mathcal{F}(f)(\lambda)K_{\varepsilon,\delta}^L(\lambda)\psi(\lambda)\frac{d\lambda}{|c(\lambda)|^2}.$$

Thus Theorem 16.8 implies that the function $f^L_{\varepsilon,\delta}$ belongs to $L^2(m_A)$ and we have

$$\mathcal{F}(f^L_{\varepsilon,\delta}) = \mathcal{F}(f) K^L_{\varepsilon,\delta}.$$

Using Lemma 16.22 iii) we obtain

$$\mathcal{F}(f^L_{\varepsilon,\delta}) = \mathcal{F}(f)\mathcal{F}(G^L_{\varepsilon,\delta}).$$

Corollary 16.10 and Lemma 16.22 ii) imply

$$f^L_{\varepsilon,\delta} = f * G^L_{\varepsilon,\delta}.$$

■

**Proof.** (of Theorem 16.21) From Theorem 16.7 we have

$$\|f^L_{\varepsilon,\delta} - f\|_2 = \|\mathcal{F}(f^L_{\varepsilon,\delta}) - \mathcal{F}(f)\|_{2,c}.$$

Using the relation (16.114) we obtain

$$\|f^L_{\varepsilon,\delta} - f\|_2 = \|\mathcal{F}(f)\left(1 - K^L_{\varepsilon,\delta}\right)\|_{2,c}$$

we deduce the result from Lemma 16.22 iv) and dominated convergence theorem. ■

Next we formulate a pointwise reconstruction formula for the transform $\Phi^L_{g^L}$.

**Theorem 16.24** *Let $g^L$ be a L-wavelet in $L^2(m_A)$. For $f$ in $(L^1 \cap L^2)(m_A)$ such that $\mathcal{F}(f)$ belongs to $L^1\left(\frac{d\lambda}{|c(\lambda)|^2}\right)$, we have the following reconstruction formula:*

$$f(y) = \frac{1}{C_{g^L}} \int_{\mathbb{R}_+} \Phi^L_{g^L}(f)(a,y)\frac{da}{a}, \quad y \in \mathbb{R}_+. \tag{16.115}$$

*Where for all $y \in \mathbb{R}_+$, this integral converges absolutely.*

**Proof.** From Proposition 16.19 and Theorem 16.7 ii) we deduce that for all $y \in \mathbb{R}_+$:

$$\Phi^L_{g^L}(f)(a,y) = \int_{\mathbb{R}_+} \mathcal{F}(f)(\lambda)\varphi_\lambda(y)\mathcal{F}(g^L_a)(\lambda)\frac{d\lambda}{|c(\lambda)|^2}. \tag{16.116}$$

Using the fact that

$$\forall y \in \mathbb{R}_+ \ , \ \lambda \in \mathbb{R}_+ \ , \ |\varphi_\lambda(y)| \leq 1,$$

we obtain for all $y \in \mathbb{R}_+$:

$$\int_{\mathbb{R}_+} \left|\Phi^L_{g^L}(f)(a,y)\right| \frac{da}{a} \leq \int_{\mathbb{R}_+}\int_{\mathbb{R}_+} |\mathcal{F}(f)(\lambda)|\,|\mathcal{F}(g^L_a)(\lambda)|\,\frac{d\lambda}{|c(\lambda)|^2}\frac{da}{a}.$$

From Fubini-Tonelli's theorem we deduce that for all $y \in \mathbb{R}_+$:

$$\int_{\mathbb{R}_+} \left|\Phi^L_{g^L}(f)(a,y)\right| \frac{da}{a} \leq \int_{\mathbb{R}_+} |\mathcal{F}(f)(\lambda)| \left(\int_{\mathbb{R}_+} |\mathcal{F}(g^L_a)(\lambda)|\,\frac{da}{a}\right)\frac{d\lambda}{|c(\lambda)|^2}.$$

But from the relation (16.91) and a change of variables we have

$$\int_{\mathbb{R}_+} |\mathcal{F}(g^L_a)(\lambda)|\,\frac{da}{a} = \int_{\mathbb{R}_+} |\mathcal{F}(g^L)(a\lambda)|\,\frac{da}{a} = \int_{\mathbb{R}_+} |\mathcal{F}(g^L)(u)|\,\frac{du}{u}$$

then

$$\int_{\mathbb{R}_+} \left|\Phi^L_{g^L}(f)(a,y)\right| \frac{da}{a} \leq \left(\int_{\mathbb{R}_+} |\mathcal{F}(g^L)(u)|\,\frac{du}{u}\right)\|\mathcal{F}(f)\|_{1,c} < +\infty.$$

Thus the integral given in the relation (16.116) converges absolutely. On the other hand from the relation (16.110) and Fubini theorem we deduce that for all $y \in \mathbb{R}_+$:

$$\int_{\mathbb{R}_+} \Phi^L_{g^L}(f)(a,y)\frac{da}{a} = \int_{\mathbb{R}_+} \mathcal{F}(f)(\lambda)\varphi_\lambda(y)\left(\int_{\mathbb{R}_+} \mathcal{F}(g^L_a)(\lambda)\frac{da}{a}\right)\frac{d\lambda}{|c(\lambda)|^2}.$$

But from the relation (16.91) and a change of variables we have

$$\int_{\mathbb{R}_+} \mathcal{F}(g^L_a)(\lambda)\frac{da}{a} = \int_{\mathbb{R}_+} \mathcal{F}(g^L)(a\lambda)\frac{da}{a} = \int_{\mathbb{R}_+} \mathcal{F}(g^L)(u)\frac{du}{u} = C_{g^L}$$

then

$$\frac{1}{C_{g^L}}\int_{\mathbb{R}_+} \Phi^L_{g^L}(f)(a,y)\frac{da}{a} = \int_{\mathbb{R}_+} \mathcal{F}(f)(\lambda)\varphi_\lambda(y)\frac{d\lambda}{|c(\lambda)|^2}.$$

We obtain the relation (16.115) from Theorem 16.6 ii). ∎

**Remark 16.10** *The linear wavelet analysis on the Chébli-Trimèche hypergroups* $(\mathbb{R}_+, *_A)$ *discussed here may be formally interpreted as bilinear wavelet analysis studied in [22] [23] [24], in taking the Dirac distribution. Decomposition is done by the L-wavelet, reconstruction is formally performed by the Dirac distribution.*

### 16.6.2 Linear wavelet packet on $(\mathbb{R}_+, *_A)$

In this subsection we define and study the linear wavelet packet (also called L-wavelet packet), and the L-wavelet packet transform on the Chébli-Trimèche hypergroups $(\mathbb{R}_+, *_A)$, and we prove for this transform reconstruction formulas. Analogous results for the linear wavelet packet on the unit sphere in $\mathbb{R}^3$ was studied in [9].

Let $g^L$ be the L-wavalet in $L^2(m_A)$ studied in the subsection 16.6.1, $g_a^L$ , $a \in \mathbb{R}_+^*$, the function given by the relation (16.91), and $\{r_j\}_{j\in\mathbb{Z}}$ a scale sequence in $\mathbb{R}_+^*$ which is decreasing and such that

$$\lim_{j\to-\infty} r_j = +\infty \quad , \quad \lim_{j\to+\infty} r_j = 0.$$

We consider the function $g_j^{L,P}$, $j \in \mathbb{Z}$, defined by

$$g_j^{L,P}(.) = \frac{1}{C_{g^L}} \int_{r_{j+1}}^{r_j} g_a^L(.) \frac{da}{a}. \tag{16.117}$$

From Lemma 16.22 this function belongs to $L^2(m_A)$ and we have

$$\forall \lambda \in \mathbb{R}_+ \; , \; \mathcal{F}(g_j^{L,P})(\lambda) = \frac{1}{C_{g^L}} \int_{r_{j+1}}^{r_j} \mathcal{F}(g_a^L)(\lambda) \frac{da}{a}. \tag{16.118}$$

**Definition 16.3** *The sequence $\{g_j^{L,P}\}_{j\in\mathbb{Z}}$ defined by the relation (16.117) is called linear wavelet packet (also called L-wavelet packet).*

**Remark 16.11** *The function $g_j^{L,P}$, $j \in \mathbb{Z}$, is called L-wavelet packet member of step $j$.*

The L-wavelet packet $\{g_j^{L,P}\}_{j\in\mathbb{Z}}$ satisfies the following properties.

**Proposition 16.25** *(i) We have*

$$\forall \lambda \in \mathbb{R}_+ \; , \quad \sum_{j=-\infty}^{\infty} \left| \mathcal{F}\left(g_j^{L,P}\right)(\lambda) \right| \leq \frac{1}{|C_{g^L}|} \int_{\mathbb{R}_+} \left| \mathcal{F}(g^L)(u) \right| \frac{du}{u}. \tag{16.119}$$

*(ii) We have*

$$\forall \lambda \in \mathbb{R}_+ \; , \quad \sum_{j=-\infty}^{\infty} \mathcal{F}\left(g_j^{L,P}\right)(\lambda) = 1. \tag{16.120}$$

**Proof.** We obtain these results from the relations (16.118), (16.91) and Definition 16.1 ∎

**Example 16.8** *We consider the L-wavelet packet* $\left\{g_j^{L,P}\right\}_{j\in\mathbb{Z}}$ *associated with the L-wavelet defined by the relation (16.99). For all* $j \in \mathbb{Z}$ *we have*

$$\int_{r_{j+1}}^{r_j} \mathcal{F}(g_a^L)(\lambda)\,\tfrac{da}{a} = \lambda^2 \int_{r_{j+1}}^{r_j} a\,e^{-t\lambda^2 a^2}\,da$$

$$= \frac{1}{2t}\left[e^{-t\lambda^2 r_{j+1}^2} - e^{-t\lambda^2 r_j^2}\right].$$

*Then from the relations (16.100), (16.118) we have*

$$\forall \lambda \in \mathbb{R}_+\,, \quad \mathcal{F}\left(g_j^{L,P}\right)(\lambda) = e^{-t\lambda^2 r_{j+1}^2} - e^{-t\lambda^2 r_j^2}. \tag{16.121}$$

*Let* $\{g_j^{L,P}\}_{j\in\mathbb{Z}}$ *be a L-wavelet packet. We consider for all* $j \in \mathbb{Z}$ *and* $y \in \mathbb{R}_+$, *the function* $g_{j,y}^{L,P}$ *given by*

$$\forall x \in \mathbb{R}_+\,, \quad g_{j,y}^{L,P}(x) = T_x g_j^{L,P}(y) \tag{16.122}$$

*where* $T_x$ , $x \in \mathbb{R}_+$, *are the generalized translation operators given by the relation (16.61).*

This function satisfies the following properties.

**Proposition 16.26** *(i) For all* $j \in \mathbb{Z}$ *and* $y \in \mathbb{R}_+$, *the function* $g_{j,y}^{L,P}$ *belongs to* $L^2(m_A)$ *and we have*

$$\|g_{j,y}^{L,P}\|_2 \le \|g_j^{L,P}\|_2. \tag{16.123}$$

*(ii) For all* $j \in \mathbb{Z}$ *and* $y \in \mathbb{R}_+$, *we have the relations*

$$\forall \lambda \in \mathbb{R}_+\,, \quad \mathcal{F}\left(g_{j,y}^{L,P}\right)(\lambda) = \varphi_\lambda(y)\mathcal{F}\left(g_j^{L,P}\right)(\lambda) \tag{16.124}$$

$$\forall \lambda \in \mathbb{R}_+\,, \quad \sum_{j=-\infty}^{\infty} \left|\mathcal{F}\left(g_{j,y}^{L,P}\right)(\lambda)\right| \le \frac{1}{|C_{g^L}|}\int_{\mathbb{R}_+} \left|\mathcal{F}\left(g^L\right)(u)\right| \frac{du}{u}. \tag{16.125}$$

**Proof.** We deduce these results from the relations (16.122), (16.119), the properties of the generalized translation operators $T_x$ , $x \in \mathbb{R}_+$, and the fact that

$$\forall y \in \mathbb{R}_+\,,\ \lambda \in \mathbb{R}_+\,, \quad |\varphi_\lambda(y)| \le 1.$$

∎

**Definition 16.4** *Let* $\{g_j^{L,P}\}_{j\in\mathbb{Z}}$ *be a L-wavelet packet. The L-wavelet packet transform* $\Phi_{g^L}^{L,P}$ *is defined for regular functions* $f$ *on* $\mathbb{R}_+$ *by*

$$\Phi_{g^L}^{L,P}(f)(j,y) = \int_{\mathbb{R}_+} f(x) g_{j,y}^{L,P}(x) dm_A(x) \ , \quad j \in \mathbb{Z} \ , \ y \in \mathbb{R}_+. \qquad (16.126)$$

*This transform can also be written in the form*

$$\Phi_{g^L}^{L,P}(f)(j,y) = f * g_j^{L,P}(y) \qquad (16.127)$$

*where* $*$ *is the convolution defined by the relation (16.68).*

**Proposition 16.27** *Let* $\{g_j^{L,P}\}_{j\in\mathbb{Z}}$ *be a L-wavelet packet. Then for all* $j \in\mathbb{Z}$ *and for all* $f$ *in* $L^q(m_A)$, $q \in [1,2]$, *the function* $y \to \Phi_{g^L}^{L,P}(f)(j,y)$ *belongs to* $L^r(m_A)$ *with* $r \in [1,+\infty]$, *such that* $\frac{1}{r} = \frac{1}{q} - \frac{1}{2}$, *and we have*

$$\|\Phi_{g^L}^{L,P}(f)(j,.)\|_r \leq \|f\|_q \|g_j^{L,P}\|_2. \qquad (16.128)$$

**Proof.** The relations (16.127) and (16.70) give the results. ∎

**Theorem 16.28** *Let* $\left\{g_j^{L,P}\right\}_{j\in\mathbb{Z}}$ *be a L-wavelet packet. Then for* $f$ *in* $L^2(m_A)$ *and* $M, N \in \mathbb{Z}$ *with* $M < N$, *the function*

$$f_{M,N}^{L,P}(y) = \sum_{j=M}^{N-1} \Phi_{g^L}^{L,P}(f)(j,y) \qquad (16.129)$$

*belongs to* $L^2(m_A)$ *and we have*

$$\lim_{N\to+\infty,\, M\to-\infty} \|f_{M,N}^{L,P} - f\|_2 = 0. \qquad (16.130)$$

To prove this theorem we need the following Lemma.

**Lemma 16.29** *Let* $\left\{g_j^{L,P}\right\}_{j\in\mathbb{Z}}$ *be a L-wavelet packet. For* $M, N \in \mathbb{Z}$ *with* $M < N$, *we have*

$$\forall \lambda \in \mathbb{R}_+ \ , \ \mathcal{F}\left(\sum_{j=M}^{N-1} g_j^{L,P}\right)(\lambda) = K_{r_N, r_M}^L(\lambda) \qquad (16.131)$$

*where* $K_{r_N,r_M}^L$ *is the function defined in Lemma 16.22.*

**Proof.** We have from the relation (16.118): $\forall \lambda \in \mathbb{R}_+$,

$$\mathcal{F}\left(\sum_{j=M}^{N-1} g_j^{L,P}\right)(\lambda) = \sum_{j=M}^{N-1} \mathcal{F}(g_j^{L,P})(\lambda)$$
$$= \sum_{j=M}^{N-1}\left(\frac{1}{C_{g^L}}\int_{r_{j+1}}^{r_j} \mathcal{F}(g_a^L)(\lambda)\frac{da}{a}\right)$$
$$= \frac{1}{C_{g^L}}\int_{r_N}^{r_M} \mathcal{F}(g_a^L)(\lambda)\frac{da}{a}.$$

Thus

$$\forall \lambda \in \mathbb{R}_+,\ \mathcal{F}\left(\sum_{j=M}^{N-1} g_j^{L,P}\right)(\lambda) = K_{r_N,r_M}^L(\lambda).$$

■

**Proof.** (of Theorem 16.28) From the relation (16.127) we have

$$\forall y \in \mathbb{R}_+\ ,\ f_{M,N}^{L,P}(y) = \sum_{j=M}^{N-1} \Phi_{g^L}^{L,P}(f)(j,y) = \sum_{j=M}^{N-1} f * g_j^{L,P}(y).$$

Thus

$$\forall y \in \mathbb{R}_+\ ,\ f_{M,N}^{L,P}(y) = f * \left(\sum_{j=M}^{N-1} g_j^{L,P}\right)(y). \tag{16.132}$$

Using the relation (16.131) and Lemma 16.22 ii) we deduce that the function $\mathcal{F}(f)\mathcal{F}\left(\sum_{j=M}^{N-1} g_j^{L,P}\right)$ belongs to $L^2\left(\frac{d\lambda}{|c(\lambda)|^2}\right)$. Then from Corollary 16.10, the function $f_{M,N}^{L,P}$ belongs to $L^2(m_A)$ and we have

$$\forall \lambda \in \mathbb{R}_+\ ,\ \mathcal{F}\left(f_{M,N}^{L,P}\right)(\lambda) = \mathcal{F}(f)(\lambda)\mathcal{F}\left(\sum_{j=M}^{N-1} g_j^{L,P}\right)(\lambda).$$

Thus from the relation (16.131):

$$\forall \lambda \in \mathbb{R}_+\ ,\ \mathcal{F}\left(f_{M,N}^{L,P}\right)(\lambda) = \mathcal{F}(f)(\lambda)K_{r_N,r_M}^L(\lambda). \tag{16.133}$$

From Theorem 16.7 we have

$$\left\| f_{M,N}^{L,P} - f \right\|_2 = \left\| \mathcal{F}\left(f_{M,N}^{L,P}\right) - \mathcal{F}(f) \right\|_{2,c}.$$

By applying the relation (16.133) we obtain

$$\left\| f_{M,N}^{L,P} - f \right\|_2 = \left\| \mathcal{F}(f)\left(1 - K_{r_N, r_M}^L\right) \right\|_{2,c}.$$

We deduce the result from Lemma 16.22 iv) and dominated convergence theorem. ■

Next we give a pointwise reconstruction formula for the transform $\Phi_{g^L}^{L,P}$.

**Theorem 16.30** *Let $\{g_j^{L,P}\}_{j\in\mathbb{Z}}$ be a L-wavelet packet. For $f$ in $(L^1 \cap L^2)(m_A)$ such that $\mathcal{F}(f)$ belongs to $L^1\left(\frac{d\lambda}{|c(\lambda)|^2}\right)$, we have the following reconstruction formula*

$$f(y) = \sum_{j=-\infty}^{\infty} \Phi_{g^L}^{L,P}(f)(j,y) , \quad y \in \mathbb{R}_+ \tag{16.134}$$

*where for all $y \in \mathbb{R}_+$, the series converge absolutely.*

**Proof.** Using the relation (16.126), Proposition 16.26 and Theorem 16.7 ii), we deduce that for all $j \in \mathbb{Z}$ and $y \in \mathbb{R}_+$:

$$\Phi_{g^L}^{L,P}(f)(j,y) = \int_{\mathbb{R}_+} \mathcal{F}(f)(\lambda)\mathcal{F}\left(g_j^{L,P}\right)(\lambda)\varphi_\lambda(y)\frac{d\lambda}{|c(\lambda)|^2}. \tag{16.135}$$

But from the fact that

$$\forall y \in \mathbb{R}_+ , \ \lambda \in \mathbb{R}_+ , \quad |\varphi_\lambda(y)| \leq 1$$

and Fubini-Tonelli's theorem we obtain

$$\sum_{j=-\infty}^{\infty} \left|\Phi_{g^L}^{L,P}(f)(j,y)\right| \leq \int_{\mathbb{R}_+} |\mathcal{F}(f)(\lambda)| \left( \sum_{j=-\infty}^{\infty} \left| \mathcal{F}\left(g_j^{L,P}\right)(\lambda) \right| \right) \frac{d\lambda}{|c(\lambda)|^2}$$

From Proposition 16.25 we deduce

$$\sum_{j=-\infty}^{\infty} \left|\Phi_{g^L}^{L,P}(f)(j,y)\right| \leq \frac{\|\mathcal{F}(f)\|_{1,c}}{|C_{g^L}|} \int_{\mathbb{R}_+} \left|\mathcal{F}(g^L)(\lambda)\right| \frac{d\lambda}{\lambda} < \infty.$$

Then for all $y \in \mathbb{R}_+$, the series (16.134) converges absolutely. On the other hand for all $y \in \mathbb{R}_+$, the relation (16.135) implies for every $M_1, N_1 \in \mathbb{N}_+$:

$$\sum_{j=-N_1}^{M_1} \Phi_{g^L}^{L,P}(f)(j,y) = \int_{\mathbb{R}_+} \mathcal{F}(f)(\lambda)\varphi_\lambda(y) \left( \sum_{j=-N_1}^{M_1} \mathcal{F}\left(g_j^{L,P}\right)(\lambda) \right) \frac{d\lambda}{|c(\lambda)|^2}.$$

Using Proposition 16.25 and dominated convergence theorem we obtain

$$\lim_{M_1,N_1\to+\infty}\sum_{j=-N_1}^{M_1}\Phi_{g^L}^{L,P}(f)(j,y)=\int_{\mathbb{R}_+}\mathcal{F}(f)(\lambda)\varphi_\lambda(y)\frac{d\lambda}{|c(\lambda)|^2}.$$

We deduce the relation (16.134) from Theorem 16.6 ii). ∎

### *16.6.3 Scale discrete L-scaling function on* $(\mathbb{R}_+,*_A)$

In this subsection we define and study a scale discrete L-scaling function on the Chébli-Trimèche hypergroups $(\mathbb{R}_+,*_A)$, corresponding to the L-wavelet packet $\{g_j^{L,P}\}_{j\in\mathbb{Z}}$ studied in the previous subsection.

**Lemma 16.31** *Let $g^L$ be a L-wavelet in $L^2(m_A)$. For all $J\in\mathbb{Z}$, we define the function $G_J^{L,P}$ by*

$$G_J^{L,P}(.)=\frac{1}{C_{g^L}}\int_{r_J}^{\infty}g_a^L(\cdot)\frac{da}{a}.\tag{16.136}$$

*Then this function belongs to $L^2(m_A)$ and we have*

$$\forall\lambda\in\mathbb{R}_+\ ,\ \mathcal{F}\left(G_J^{L,P}\right)(\lambda)=\frac{1}{C_{g^L}}\int_{r_J}^{\infty}\mathcal{F}(g_a^L)(\lambda)\frac{da}{a}.\tag{16.137}$$

**Proof.** By applying Hölder's inequality we obtain

$$\left(\int_{r_J}^{\infty}\frac{1}{a}\left|\mathcal{F}(g_a^L)(\lambda)\right|da\right)^2\le\frac{1}{r_J}\int_{r_J}^{\infty}\left|\mathcal{F}(g_a^L)(\lambda)\right|^2da.$$

Thus Fubini-Tonelli's theorem, Theorem 16.7 and the relation (16.93) imply

$$\begin{aligned}\int_{\mathbb{R}_+}\left|\int_{r_J}^{\infty}\mathcal{F}(g_a^L)(\lambda)\frac{da}{a}\right|^2\frac{d\lambda}{|c(\lambda)|^2}&\\ \le\frac{1}{r_J}\int_{r_J}^{\infty}\left(\int_{\mathbb{R}_+}\left|\mathcal{F}(g_a^L)(\lambda)\right|^2\frac{d\lambda}{|c(\lambda)|^2}\right)da&\\ \le\frac{\|g^L\|_2^2}{r_J}\int_{r_J}^{\infty}\frac{k^2(a)}{a}da.&\end{aligned}$$

But Proposition 16.16 shows that the integral $\int_{r_J}^{\infty} \frac{k^2(a)}{a} da$ is finite. Then the function $\lambda \longrightarrow \int_{r_J}^{\infty} \mathcal{F}(g_a^L)(\lambda)\frac{da}{a}$ belongs to $L^2(m_A)$. From this result, Lemma 16.22 and Theorem 16.7, the function $G_J^{L,P}$ given by the relation (16.136) is well defined, belongs to $L^2(m_A)$ and satisfies the relation (16.137). ■

**Definition 16.5** *The sequence $\{G_J^{L,P}\}_{J\in\mathbb{Z}}$ defined by the relation (16.136) is called scale discrete L-scaling function.*

**Example 16.9** *We consider the L-wavelet packet $\{g_j^{L,P}\}_{j\in\mathbb{Z}}$ given by the relation (16.121). From the relation (16.137) we deduce that the scale discrete L-scaling function $\{G_J^{L,P}\}_{J\in\mathbb{Z}}$ corresponding to the previous $\{g_j^{L,P}\}_{j\in\mathbb{Z}}$ satisfies*

$$\forall \lambda \in \mathbb{R}_+ , \quad \mathcal{F}\left(G_J^{L,P}\right)(\lambda) = e^{-t\lambda^2 r_J^2}. \tag{16.138}$$

From Definition 16.3 and Proposition 16.25 we deduce that the functions $G_J^{L,P}$, $J \in \mathbb{Z}$, possess the following properties.

**Proposition 16.32** *(i) For all $J \in \mathbb{Z}$, we have*

$$\forall \lambda \in \mathbb{R}_+ , \quad \mathcal{F}\left(G_J^{L,P}\right)(\lambda) = \sum_{j=-\infty}^{J-1} \mathcal{F}\left(g_j^{L,P}\right)(\lambda). \tag{16.139}$$

*(ii) For all $J \in\mathbb{Z}$, we have*

$$\forall \lambda \in \mathbb{R}_+ , \quad \left|\mathcal{F}\left(G_J^{L,P}\right)(\lambda)\right| \leq \frac{1}{|C_{g^L}|}\int_{\mathbb{R}_+} \left|\mathcal{F}\left(g^L\right)(u)\right| \frac{du}{u}. \tag{16.140}$$

*(iii) we have*

$$\forall \lambda \in \mathbb{R}_+ , \quad \lim_{J\to+\infty} \mathcal{F}\left(G_J^{L,P}\right)(\lambda) = 1. \tag{16.141}$$

Let $\{G_J^{L,P}\}_{J\in\mathbb{Z}}$ be a scale discrete L-scaling function. We consider for all $J \in \mathbb{Z}$ and $y \in \mathbb{R}_+$, the function $G_{J,y}^{L,P}$ defined by

$$\forall x \in \mathbb{R}_+ , \quad G_{J,y}^{L,P}(x) = T_x G_J^{L,P}(y) \tag{16.142}$$

where $T_x$ , $x \in \mathbb{R}_+$, are the generalized translation operators given by the relation (16.61).

The function $G_{J,y}^{L,P}$ possesses the following properties.

**Proposition 16.33** *(i) The function $G_{J,y}^{L,P}$ belongs to $L^2(m_A)$ and we have*

$$\|G_{J,y}^{L,P}\|_2 \leq \|G_J^{L,P}\|_2. \tag{16.143}$$

*(ii) We have the relations*

$$\forall \lambda \in \mathbb{R}_+ , \quad \mathcal{F}\left(G_{J,y}^{L,P}\right)(\lambda) = \varphi_\lambda(y)\mathcal{F}\left(G_J^{L,P}\right)(\lambda). \tag{16.144}$$

$$\forall \lambda \in \mathbb{R}_+ , \quad \left|\mathcal{F}\left(G_{J,y}^{L,P}\right)(\lambda)\right| \leq \frac{1}{|C_{g^L}|}\int_{\mathbb{R}_+} |\mathcal{F}\left(g^L\right)(u)|\,\frac{du}{u}. \tag{16.145}$$

**Proof.** We deduce these results from the relations (16.65),(16.67) and (16.142). ■

**Proposition 16.34** *(i) For all $J \in \mathbb{Z}$, we have*

$$\forall \lambda \in \mathbb{R}_+ , \quad \mathcal{F}\left(G_J^{L,P}\right)(\lambda) + \sum_{j=J}^{\infty} \mathcal{F}\left(g_j^{L,P}\right)(\lambda) = 1. \tag{16.146}$$

*(ii) For all $j \in \mathbb{Z}$, we have*

$$\forall \lambda \in \mathbb{R}_+ , \quad \mathcal{F}\left(g_j^{L,P}\right)(\lambda) = \mathcal{F}\left(G_{j+1}^{L,P}\right)(\lambda) - \mathcal{F}\left(G_j^{L,P}\right)(\lambda). \tag{16.147}$$

*(iii) We have*

$$\forall \lambda \in \mathbb{R}_+ , \quad \sum_{j=-\infty}^{\infty} \left[\mathcal{F}\left(G_{j+1}^{L,P}\right)(\lambda) - \mathcal{F}\left(G_j^{L,P}\right)(\lambda)\right] = 1. \tag{16.148}$$

**Proof.** From the relations (16.118) and (16.120) we obtain

$$\begin{aligned}
\frac{1}{C_{g^L}}\int_{\mathbb{R}_+} \mathcal{F}(g_a^L)(\lambda)\frac{da}{a} &= \sum_{j=-\infty}^{\infty} \mathcal{F}\left(g_j^{L,P}\right)(\lambda) \\
&= \left(\sum_{j=-\infty}^{J-1} + \sum_{j=J}^{\infty}\right)\left(\mathcal{F}\left(g_j^{L,P}\right)(\lambda)\right) = 1.
\end{aligned}$$

The relation (16.139) gives (16.146) and (16.147). We deduce the relation (16.148) from the relations (16.147) and (16.120). ■

**Theorem 16.35** *Let $\{G_J^{L,P}\}_{J\in\mathbb{Z}}$ be a scale discrete L-scaling function which corresponds to the L-wavelet packet $\{g_j^{L,P}\}_{j\in\mathbb{Z}}$. For $f$ in $L^2(m_A)$ we have*

$$f(y) = \lim_{J\to+\infty} \int_{\mathbb{R}_+} f(x) G_{J,y}^{L,P}(x) dm_A(x). \tag{16.149}$$

*The convergence is in $L^2(m_A)$.*

$$f(y) = \int_{\mathbb{R}_+} f(x) G_{J,y}^{L,P}(x) dm_A(x) + \sum_{j=J}^{\infty} \Phi_{g^L}^{L,P}(f)(j,y). \tag{16.150}$$

*The convergence is in $L^2(m_A)$.*

To prove this theorem we need the following lemma.

**Lemma 16.36** *Let $g^L$ be a L-wavelet in $L^2(m_A)$. For all $\varepsilon > 0$, the function $K_{\varepsilon,\infty}^L$ defined by*

$$\forall \lambda \in \mathbb{R}_+ \ , \ K_{\varepsilon,\infty}^L(\lambda) = \frac{1}{C_{g^L}} \int_{\varepsilon}^{\infty} \mathcal{F}(g_a^L)(\lambda) \frac{da}{a} \tag{16.151}$$

*belongs to $L^\infty\left(\frac{d\lambda}{|c(\lambda)|^2}\right)$ and we have*

$$\forall \lambda \in \mathbb{R}_+^* \ , \ \lim_{\varepsilon\to 0} K_{\varepsilon,\infty}^L(\lambda) = 1. \tag{16.152}$$

**Proof.** By a change of variables we obtain

$$\forall \lambda \in \mathbb{R}_+, \ K_{\varepsilon,\infty}^L(\lambda) = \frac{1}{C_{g^L}} \int_{\lambda\varepsilon}^{\infty} \mathcal{F}(g^L)(u) \frac{du}{u}.$$

Thus from the relation (16.96):

$$\forall \lambda \in \mathbb{R}_+^* \ , \ \left|K_{\varepsilon,\infty}^L(\lambda)\right| \leq \frac{1}{|C_{g^L}|} \int_{\mathbb{R}_+} \left|\mathcal{F}(g^L)(u)\right| \frac{du}{u} < +\infty.$$

We deduce that the function $K_{\varepsilon,\infty}^L$ belongs to $L^\infty\left(\frac{d\lambda}{|c(\lambda)|^2}\right)$. On the other hand from the relation (16.97) we have

$$\forall \lambda \in \mathbb{R}_+^* \ , \ \lim_{\varepsilon\to 0} K_{\varepsilon,\infty}^L(\lambda) = \lim_{\varepsilon\to 0} \left[\frac{1}{C_{g^L}} \int_{\lambda\varepsilon}^{\infty} \mathcal{F}(g^L)(u) \frac{du}{u}\right] = 1.$$

**Proof.** (of Theorem 16.35) We have

$$\forall y \in \mathbb{R}_+ \ , \ \int_{\mathbb{R}_+} f(x) G_{J,y}^{L,P}(x) dm_A(x) = f * G_J^{L,P}(y)$$

As the functions $f$ and $G_J^{L,P}$ are in $L^2(m_A)$, and from Lemma 16.36 the function $\mathcal{F}(f) K_{r_J,\infty}^L$ is in $L^2\left(\frac{d\lambda}{|c(\lambda)|^2}\right)$, we deduce from Corollary 16.10 i) and (16.137) that the function $f * G_J^{L,P}$ belongs to $L^2(m_A)$ and we have

$$\mathcal{F}\left(f * G_J^{L,P}\right) = \mathcal{F}(f) K_{r_J,\infty}^L. \tag{16.153}$$

From Theorem 16.7 we have

$$\left\| f * G_J^{L,P} - f \right\|_2 = \left\| \mathcal{F}\left(f * G_J^{L,P}\right) - \mathcal{F}(f) \right\|_{2,c}.$$

Using the relation (16.153) we obtain

$$\begin{aligned} \left\| f * G_J^{L,P} - f \right\|_2 &= \left\| \mathcal{F}(f) K_{r_J,\infty}^L - \mathcal{F}(f) \right\|_{2,c} \\ &= \left\| \mathcal{F}(f) \left(1 - K_{r_J,\infty}^L\right) \right\|_{2,c}. \end{aligned}$$

We deduce formula (16.149) from Lemma 16.36 and dominated convergence theorem.
Using proof of Theorem 16.28 and the first relation of this proof we obtain for all $y \in \mathbb{R}_+$ ,

$$\int_{\mathbb{R}_+} f(x) G_{J,y}^{L,P}(x) dm_A(x) + \sum_{j=J}^{N-1} \Phi_{g^L}^{L,P}(f)(j,y) = f * G_J^{L,P}(y) + f_{J,N}^{L,P}(y).$$

But from the relations (16.133), (16.153) we have for all $\lambda \in \mathbb{R}_+$ ,

$$\mathcal{F}\left(f * G_J^{L,P} + f_{J,N}^{L,P}\right)(\lambda) = \mathcal{F}(f)(\lambda) \left(K_{r_N,r_J}^L(\lambda) + K_{r_J,\infty}^L(\lambda)\right).$$

Thus for all $\lambda \in \mathbb{R}_+$ :

$$\mathcal{F}\left(f * G_J^{L,P} + f_{J,N}^{L,P}\right)(\lambda) = \mathcal{F}(f)(\lambda). K_{r_N,\infty}^L(\lambda). \tag{16.154}$$

From Theorem 16.7 we have

$$\left\| \left(f * G_J^{L,P} + f_{J,N}^{L,P}\right) - f \right\|_2 = \left\| \mathcal{F}\left(f * G_J^{L,P} + f_{J,N}^{L,P}\right) - \mathcal{F}(f) \right\|_{2,c}.$$

Using the relation (16.154) we obtain

$$\left\| \left(f * G_J^{L,P} + f_{J,N}^{L,P}\right) - f \right\|_2 = \left\| \mathcal{F}(f) \left(1 - K_{r_N,\infty}^L\right) \right\|_{2,c}.$$

Lemma 16.36 and dominated convergence theorem give the relation (16.150). ■

We are now able to prove a pointwise reconstruction formulas associated with the function $\{G_J^{L,P}\}_{J\in\mathbb{Z}}$.

**Theorem 16.37** *Let $\{G_J^{L,P}\}_{J\in\mathbb{Z}}$ be a scale discrete L-scaling function which corresponds to the L-wavelet packet $\{g_j^{L,P}\}_{j\in\mathbb{Z}}$. For $f$ in $(L^1\cap L^2)(m_A)$ such that $\mathcal{F}(f)$ belongs to $L^1\left(\frac{d\lambda}{|c(\lambda)|^2}\right)$, we have the following reconstruction formulas*

$$f(y) = \lim_{J\to+\infty} \int_{\mathbb{R}_+} f(x) G_{J,y}^{L,P}(x) dm_A(x) \ , \ y \in \mathbb{R}_+. \tag{16.155}$$

$$f(y) = \int_{\mathbb{R}_+} f(x) G_{J,y}^{L,P}(x) dm_A(x) + \sum_{j=J}^{\infty} \Phi_{g^L}^{L,P}(f)(j,y) \ , \ y \in \mathbb{R}_+. \tag{16.156}$$

*where for all $y \in \mathbb{R}_+$, the series converge absolutely.*

**Proof.** From Theorem 16.7 ii) and the relation (16.144) we obtain for all $y \in \mathbb{R}_+$:

$$\int_{\mathbb{R}_+} f(x) G_{J,y}^{L,P}(x) dm_A(x) = \int_{\mathbb{R}_+} \mathcal{F}(f)(\lambda)\varphi_\lambda(y)\mathcal{F}\left(G_J^{L,P}\right)(\lambda)\frac{d\lambda}{|c(\lambda)|^2}. \tag{16.157}$$

Using the relations (16.140), (16.141), the fact that

$$\forall y \in \mathbb{R}_+ \ , \ \lambda \in \mathbb{R}_+ \ , \ |\varphi_\lambda(y)| \leq 1$$

and dominated convergence theorem, we obtain

$$\lim_{J\to+\infty} \int_{\mathbb{R}_+} f(x) G_{J,y}^{L,P}(x) dm_A(x) = \int_{\mathbb{R}_+} \mathcal{F}(f)(\lambda)\varphi_\lambda(y)\frac{d\lambda}{|c(\lambda)|^2}.$$

Then Theorem 16.6 ii) gives the relation (16.155). By applying Definition 16.4 and the proof of Theorem 16.30 we deduce that the series $\sum_{j=J}^{\infty} \Phi_{g^L}^{L,P}(f)(j,y)$ converges absolutely and we have for all $y \in \mathbb{R}_+$:

$$\sum_{j=J}^{\infty} \Phi_{g^L}^{L,P}(f)(j,y) = \int_{\mathbb{R}_+} \mathcal{F}(f)(\lambda)\varphi_\lambda(y)\left(\sum_{j=J}^{\infty} \mathcal{F}\left(g_j^{L,P}\right)(\lambda)\right)\frac{d\lambda}{|c(\lambda)|^2}. \tag{16.158}$$

On the other hand by using the relation (16.139), the equality (16.157) can be written in the form

$$\int_{\mathbb{R}_+} f(x) G_{J,y}^{L,P}(x) dm_A(x)$$

$$= \int_{\mathbb{R}_+} \mathcal{F}(f)(\lambda)\varphi_\lambda(y) \left( \sum_{j=-\infty}^{J-1} \mathcal{F}\left(g_j^{L,P}\right)(\lambda) \right) \frac{d\lambda}{|c(\lambda)|^2}. \tag{16.159}$$

Then for all $y \in \mathbb{R}_+$, we obtain from the relations (16.158), (16.159) and (16.120):

$$\int_{\mathbb{R}_+} f(x) G_{J,y}^{L,P}(x) dm_A(x) + \sum_{j=J}^{\infty} \Phi_{g^L}^{L,P}(f)(j,y) = \int_{\mathbb{R}_+} \mathcal{F}(f)(\lambda)\varphi_\lambda(y) \frac{d\lambda}{|c(\lambda)|^2}.$$

We obtain the relation (16.156) from Theorem 16.6 ii). ∎

*References*


[1] W.R. Bloom, H. Heyer, Harmonic analysis of probability measures on hypergroups. de Gruyter Studies in Mathematics, Vol 20. Editors: H. Bauer - J.L. Kazdan - E. Zehnder. de Gruyter. Berlin. New York 1994.

[2] W.R. Bloom, Z. Xu, The Hardy-Littlewood maximal function for Chébli-Trimèche hypergroups. Contemporary Math. Vol. 183, p. 45-69.

[3] H. Chébli, Opérateurs de translation généralisée et semi-groupes de convolution. Lecture notes in Math N° 404, Springer-Verlag, Berlin, 1974.

[4] H. Chébli, Sur un théorème de Paley-Wiener associé à la décomposition spectrale d'un opérateur de Sturn-Liouville sur $]0, +\infty[$. J. Func. Anal. 17, 1974, p. 447-461.

[5] H. Chébli, Sturn-Liouville hypergroups. Contemporary Math. Vol. 183, 1995, p. 71-88.

[6] H. Chébli - A. Fitouhi and M.M. Hamza, Expansion in series of Bessel functions and transmutations for perturbed Bessel operators. Math. Anal. Appl. Vol. 181. N° 3, 1994, p. 789-802.

[7] A. Fitouhi and M.M. Hamza, A uniform expansion for the eigenfunction of a singular second order differential operator. SIAM. J. Math. Anal. Vol. 21, N° 6, 1990, p. 1619-1632.

[8] M. Flensted-Jensen and T.H. Koornwinder, The convolution structure for Jacobi functions expansions. Ark. Math. 11, 1973, p. 245-262.

[9] W. Freeden and U. Windheuser, Spherical wavelet transform and its discretization. Adv. in Computational Math. 5, 1996, p. 51-94.

[10] R.I. Jewett, Spaces with an abstract convolution of measures. Adv. Math. 18, N° 1, 1973, p.1-101.

[11] T.H. Koornwinder, A new proof of a Paley-Wiener type theorem for Jacobi transform. Ark. Math. Vol. 13, N° 1, 1975, p. 145-159.

[12] M.N. Lazhari and K.Trimèche, Convolution algebras and factorization of measures on Chébli-Trimèche Hypergroups. C. R. Math. Rep. Acad. Sci. Canada, Vol. 17, N° 4, 1995, p. 165-169.

[13] M.N. Lazhari - L.T. Rachdi and K.Trimèche, Asymptotic expansion and generalized Schläfli integral representation for the eigenfunction of a singular second order differential operator, J. Math. Anal. Appl. 217, 1998, p263-292.

[14] M.N. Lazhari, Transformation de Radon exponentielle. Transformations intégrales et probabilité sur les hypergroupes. Thèse de Doctorat d'Etat es-Sciences Mathématiques. Faculté des Sciences de Tunis, 1997.

[15] F.G. Mehler, Uber eine mit den kugel und cylinder functionen verwandte function und ihre anwenung in der theorie der electricitätsvertheilung. Math. Ann. Vol. 18, 1881, p. 161-194.

[16] M. Mizony, Une transformation de Laplace-Jacobi. SIAM. J. Anal. Vol. 14, N° 5, 1983, p. 987-1003.

[17] F.W.J. Olver, *Asymptotics and special functions.* Academic Press, New-York, 1974.

[18] R. Spector, Aperçu de la théorie des hypergroupes, analyse harmonique sur les groupes de Lie. Séminaires Nancy-Strasbourg (1973-1975). p. 643-673. Lecture notes in Math. N° 497, Springer Verlag, Berlin, 1975.

[19] A.L. Schwartz, The structure of the algebra of Hankel transforms and the algebra of Hankel-Stieltjes transforms. Can. J. Math. Vol. 23, N° 2, 1971, p. 236-246.

[20] K.Trimèche, Transformation intégrale de Weyl et théorème de Paley-Wiener associés à un opérateur différentiel singulier sur $(0, +\infty)$. J. Math. Pures et Appl. (9), 60, 1981, p. 51-98.

[21] K.Trimèche, Transmutation operators and mean-periodic functions associated with differential operators. Mathematical reports Vol. 4,Part. I, 1988, p. 1-282. Harwood Academic Publishers, Chur - London - Paris - New-York - Melbourne.

[22] K.Trimèche, Inversion of Lions transmutation operators using generalized wavelets. Applied and Computational Harmonic Analysis 4, 1997, p. 1-16.

[23] K.Trimèche, *Generalized wavelets and hypergroups.* Gordon and Breach Science Publishers. 1997.

[24] K.Trimèche, Generalized wavelet packets associated with a singular differential operator on $]0, +\infty[$. Preprint. Faculty of Sciences of Tunis, 1997.

[25] G.N. Watson, *A treatise on the theory of Bessel functions.* 2nd ed. Cambridge University Press. London - New-York, 1966.

[26] Z. Xu, Harmonic analysis on Chébli-Trimèche hypergroups. Ph D Thesis, Murdoch University, Australia, 1994.

[27] Hm. Zeuner, One-dimensional hypergroups. Adv. Math. 76, 1989, N° 1, p. 1-18.

[28] Hm. Zeuner, The central limit theorem for the Chébli-Trimèche hypergroups. J. Theor. Prob. Vol. 2, N° 1, 1989, p.51-63.

[29] Hm. Zeuner, Limit theorems for one-dimensional hypergroups. Habilitationsschrift. Tübingen, 1990.

# Chapter 17

# Hardy-type inequalities for a new class of integral operators

**Gord Sinnamon**[1]

*ABSTRACT. Mapping properties between weighted Lebesgue spaces of the operator that integrates a function over a difference of two dilations of a starshaped set in $\mathbb{R}^n$ are completely characterized. This includes integrals over annuli whose inner and outer radii are arbitrary increasing functions. The general result is applied to give sufficient conditions for boundedness between weighted Lebegue spaces for operators with a large class of non-negative kernels.*

## 17.1 Introduction

The weight conditions which characterize the weighted Hardy inequality

$$\left(\int_0^\infty \left(\int_0^s f\right)^q v(s)\,ds\right)^{1/q} \le C\left(\int_0^\infty f^p u\right)^{1/p}$$

have set a standard for weight conditions because they are easy to estimate and to verify in particular cases, they relate well to the size of the constant $C$ in the inequality, and they are themselves essentially just weighted norms. See [2], [5] and [7].

The problem of finding comparably simple weight conditions for weighted norm inequalities involving other operators, for example, replacing $\int_0^s f$ in the above inequality by $\int_0^\infty k(s,t)f(t)\,dt$ for some kernel $k(s,t)$, is a difficult one even for non-negative $k$. Some results for particular kernels are available and certain classes of kernels have been investigated with success. One of the general results in this direction, begun by Martin-Reyes and Sawyer [4] and by Bloom and Kerman [1] and improved by Stepanov [8], solves the weight characterization problem for Generalized Hardy Operators: The operator $f \to \int_0^s k(s,t)f(t)\,dt$ is a GHO provided

[1]Support from the Natural Sciences and Engineering Research Council of Canada is gratefully acknowledged.

that $k(s,t)$ is non-negative, non-decreasing in $s$ or non-increasing in $t$, and for some $D > 0$ satisfies the growth condition

$$D^{-1}(k(s,\xi) + k(\zeta,t)) \leq k(s,t) \leq D(k(s,\xi) + k(\xi,t))$$

whenever $0 < t < \xi < s$. These are strong assumptions which, in particular, exclude kernels with sharp jumps or with zeros. The results are general enough, however, to include weighted norm inequalities for the Riemann-Liouville fractional integral operators.

Recently, in [3], weight characterizations were given for operators of the form $f \to \int_{a(s)}^{b(s)} f$ where $a$ and $b$ are increasing functions. In these results the kernel is the characteristic function of a set and does have sharp jumps and zeros. In this paper we extend the results of [3] to higher dimensions using arguments introduced in [6]. Necessary and sufficient weight conditions which live up to the standard of simplicity set for the Hardy operator are given for operators which integrate over differences of general star-shaped regions in $\mathbb{R}^n$. The weight conditions simplify futher in the case of integration over annuli.

The higher dimensional results are then applied to give sufficient conditions for one-dimensional inequalities for operator with a large class of kernels, namely those that can be expressed in the form

$$k(s,t) = \varphi(t/b(s)) - \varphi(t/a(s))$$

for some non-increasing function $\varphi$ and some increasing functions $a$ and $b$ with $a < b$. No growth condition is assumed.

## 17.2 Starshaped regions

We will call a region $S \in \mathbb{R}^n$ *smoothly starshaped* provided there exists a non-negative, piecewise-$C^1$ function $\psi$ defined on the unit sphere in $\mathbb{R}^n$ with $S = \{x \in^n \setminus \{0\} : |x| \leq \psi(x/|x|)\}$.

If $S$ is smoothly starshaped, let $B = \{x \in \mathbb{R}^n \setminus \{0\} : |x| = \psi(x/|x|)\}$ and note that $B$ is contained in the boundary of $S$. Since $\psi$ is not assumed to be continuous, $B$ may not be the whole boundary of $S$. The family of regions we integrate over is the collection of dilations of $S$.

Let $E$ be the union of all dilations of $S$, $E = \cup_{\alpha>0}\alpha S$, and note that $E = \mathbb{R}^n$ whenever 0 is in the interior of $S$. For non-zero $x \in E$, since $S$ is starshaped, there is a least positive dilation $\alpha_x S$ which contains $x$. We write $S_x = \alpha_x S$ and note that $x/\alpha_x \in B$ so that $x$ is on the boundary of $S$.

Throughout, $a$ and $b$ are taken to be increasing differentiable functions on $\mathbb{R}^+$, satisfying $a(0) = b(0) = 0$, $a(x) < b(x)$ for $x > 0$, and $a(\infty) = b(\infty) = \infty$.

The $n$-dimensional weighted inequality that we characterise in this section involves integrals over differences of the form $b(\alpha_x)S \setminus a(\alpha_x)S$.

Our first result reduces the problem to a one-dimensional one of the type studied in [3].

**Theorem 17.1** *Let $S$ be a smoothly starshaped region in $\mathbb{R}^n$ and let $B$, $E$ and $\alpha_x$ be defined as above. Suppose $0 < q < \infty$, $1 < p < \infty$, and $u$ and $v$ are non-negative weight functions on $E$. Then the inequality*

$$\left(\int_E \left|\int_{b(\alpha_x)S\setminus a(\alpha_x)S} f(y)\,dy\right|^q v(x)\,dx\right)^{1/q} \le C\left(\int_E |f(y)|^p u(y)\,dy\right)^{1/p} \tag{17.1}$$

*holds for all locally integrable functions $f$ on $E$ if and only if*

$$\left(\int_0^\infty \left|\int_{a(s)}^{b(s)} F(t)\,dt\right|^q V(s)\,ds\right)^{1/q} \le C\left(\int_0^\infty |F(t)|^p U(t)\,dt\right)^{1/p} \tag{17.2}$$

*holds for all locally integrable functions $F : (0,\infty) \to \mathbb{R}$. Here*

$$V(s) = \int_B v(s\eta)s^{n-1}\,d\eta, \quad \text{and} \quad U(t) = \left(\int_B u(t\tau)^{1-p'} t^{n-1}\,d\tau\right)^{1-p}. \tag{17.3}$$

*In particular, the best constants in inequalities (17.1) and (17.2) coincide.*

**Proof.** Suppose (17.2) holds and fix a locally integrable function $f : E \to \mathbb{R}$. Set

$$F(t) = \int_B f(t\tau)t^{n-1}\,d\tau. \tag{17.4}$$

Make the changes of variable $x = s\eta$ and $y = t\tau$ in the left hand side of

(17.1) and notice that for $\eta \in B$, $\alpha_{s\eta} = s$.

$$\left( \int_E \left| \int_{b(\alpha_x)S \setminus a(\alpha_x)S} f(y)\,dy \right|^q v(x)\,dx \right)^{1/q}$$

$$= \left( \int_E \left| \int_{a(\alpha_x)}^{b(\alpha_x)} \int_B f(t\tau) t^{n-1} d\tau\,dt \right|^q v(x)\,dx \right)^{1/q}$$

$$= \left( \int_0^\infty \int_B \left| \int_{a(\alpha_x)}^{b(\alpha_x)} \int_B f(t\tau) t^{n-1} d\tau\,dt \right|^q v(s\eta) s^{n-1} d\eta\,ds \right)^{1/q}$$

$$= \left( \int_0^\infty \left| \int_{a(\alpha_x)}^{b(\alpha_x)} F(t) dt \right|^q V(s)\,ds \right)^{1/q}$$

$$\leq C \left( \int_0^\infty |F(t)|^p\, U(t)\,dt \right)^{1/p}$$

The last inequality is the hypothesis (17.2). Use Hölder's inequality in the integral defining $F$ to estimate the last line above as follows.

$$C \left( \int_0^\infty |F(t)|^p U(t)\,dt \right)^{1/p} = C \left( \int_0^\infty \left| \int_B f(t\tau) t^{n-1}\,d\tau \right|^p U(t)\,dt \right)^{1/p}$$

$$\leq C \left( \int_0^\infty \left( \int_B |f(t\tau)|^p u(t\tau) t^{n-1}\,d\tau \right) \left( \int_B u(t\tau)^{1-p'} t^{n-1}\,d\tau \right)^{p/p'} U(t)\,dt \right)^{1/p}$$

$$= C \left( \int_0^\infty \int_B |f(t\tau)|^p u(t\tau) t^{n-1}\,d\tau\,dt \right)^{1/p} = C \left( \int_E |f(y)|^p u(y)\,dy \right)^{1/p}.$$

Thus (17.1) holds. To prove the converse, suppose that (17.1) holds and fix a locally integrable function $F : (0,\infty) \to \mathbb{R}$. Define $f : E \to \mathbb{R}$ by

$$f(t\tau) = F(t) U(t)^{p'-1} u(t\tau)^{1-p'}$$

and use the definition of $U$ to see that the relationship (17.4) is still valid.

As in the first part of the proof we have

$$\int_0^\infty \left| \int_{a(s)}^{b(s)} F(t)\, dt \right|^q V(s)\, ds = \int_E \left| \int_{b(\alpha_x)S \setminus a(\alpha_x)S} f(y)\, dy \right|^q v(x)\, dx.$$

Now the inequality (17.1) becomes

$$\left( \int_0^\infty \left| \int_{a(s)}^{b(s)} F(t)\, dt \right|^q V(s)\, ds \right)^{1/q} \leq C \left( \int_E |f(y)|^p u(y)\, dy \right)^{1/p}.$$

Using the definitions of $f$ and $U$ we recognize the right hand side above as the right hand side of (17.2).

$$\begin{aligned} C\left( \int_E |f(y)|^p u(y)\, dy \right)^{1/p} &= C\left( \int_0^\infty \int_B |f(t\tau)|^p u(t\tau) t^{n-1}\, d\tau\, dt \right)^{1/p} \\ &= C\left( \int_0^\infty |F(t)|^p U(t)^{p'} \int_B u(t\tau)^{(1-p')p+1} t^{n-1}\, d\tau\, dt \right)^{1/p} \\ &= C\left( \int_0^\infty |F(t)|^p U(t)\, dt \right)^{1/p}. \end{aligned}$$

This establishes (17.2) and completes the proof. ■

To complete the characterization we apply the results of [3] to give the following two theorems.

**Theorem 17.2** *Let $S$ be a smoothly starshaped region in $\mathbb{R}^n$ and let $B$, $E$ and $\alpha_x$ be defined as above. Suppose $1 < p \leq q < \infty$, and $u$ and $v$ are non-negative weight functions on $E$. Then (17.1) holds for all locally integrable $f$ on $E$ if and only if*

$$\sup_{\substack{t \leq s \\ a(s) \leq b(t)}} \left( \int_{b(t)S \setminus a(s)S} u^{1-p'} \right)^{1/p'} \left( \int_{sS \setminus tS} v \right)^{1/q} < \infty. \qquad (17.5)$$

**Proof.** Theorem 2.2 of [3] shows that (17.2) holds for all locally integrable $F$ if and only if

$$\sup_{\substack{t \le s \\ a(s) \le b(t)}} \left( \int_{a(s)}^{b(t)} U^{1-p'} \right)^{1/p'} \left( \int_t^s V \right)^{1/q} < \infty.$$

Here $U$ and $V$ are defined by (17.3) as before. Since Theorem 17.1 shows that (17.1) holds for all $f$ if and only if (17.2) holds for all $F$, it only remains to show that the last expression is equivalent to (17.5). Using the definitions of $U$ and $V$ the last expression shows that

$$\sup_{\substack{t \le s \\ a(s) \le b(t)}} \left( \int_{a(s)}^{b(t)} \int_B u(\xi\tau)^{1-p'} \xi^{n-1}\, d\tau\, d\xi \right)^{1/p'} \left( \int_t^s \int_B v(\xi\eta)\xi^{n-1}\, d\eta\, d\xi \right)^{1/q}$$

is finite. This reduces easily to (17.5). ■

To give an analogue of the above corollary in the case $q < p$ we need to define the normalizing function introduced in [3]. Since $a^{-1}$ and $b^{-1}$ exist and are increasing we may define the sequence $\{M_k\}_{k \in \mathbb{Z}}$ recursively as follows: Fix $M_0 = b^{-1}(1)$ and define

$$M_{k+1} = a^{-1}(b(M_k)), \text{ if } k \ge 0 \text{ and } M_k = b^{-1}(a(M_{k+1})), \text{ if } k < 0.$$

Clearly $a(M_{k+1}) = b(M_k)$ for all $k \in$. The *normalizing function* $\sigma$ is defined by

$$\sigma(t) = \sum_{k \in} \chi_{(M_k, M_{k+1})}(t) \frac{d}{dt}(b^{-1} \circ a)^k(t)$$

where $(b^{-1} \circ a)^k$ denotes $k$ times repeated composition.

**Theorem 17.3** *Let $S$ be a smoothly starshaped region in $\mathbb{R}^n$ and let $B$, $E$ and $\alpha_x$ be defined as above. Suppose $0 < q < p$, $1 < p < \infty$, and $u$ and $v$ are non-negative weight functions on $E$. Then (17.1) holds for all locally integrable $f$ on $E$ if and only if both*

$$\int_0^\infty \int_{tS \backslash b^{-1}(a(t))S} \left( \int_{b(\alpha_x)S \backslash a(t)S} u^{1-p'} \right)^{r/p'} \left( \int_{tS \backslash \alpha_x S} v \right)^{r/p} v(x)\, dx\, \sigma(t) dt \tag{17.6}$$

*and*

$$\int_0^\infty \int_{a^{-1}(b(t))S \backslash tS} \left( \int_{b(t)S \backslash a(\alpha_x)S} u^{1-p'} \right)^{r/p'} \left( \int_{\alpha_x S \backslash tS} v \right)^{r/p} v(x)\, dx\, \sigma(t) dt \tag{17.7}$$

*are finite*

**Proof.** Proceed as in the previous proof applying Theorem 2.5 of [3] instead of Theorem 2.2 of [3]. ∎

As mentioned, these results include operators which integrate over annuli. If we take the starshaped region $S$ to be the unit ball in $\mathbb{R}^n$ we see that $E = \mathbb{R}^n$ and $\alpha_x = |x|$. The operator becomes

$$f \to \int\limits_{b(|x|)\le|y|\le a(|x|)} f(y)\,dy,$$

an integral over annuli whose inner and outer radii are given by the increasing functions $a$ and $b$. In the next two corollaries we record this result in the special case that $a$ and $b$ are lines through the origin.

**Corollary 17.4** *Suppose* $1 < p \le q < \infty$ *and* $u$ *and* $v$ *are weights on* $\mathbb{R}^n$. *Fix real numbers* $A$ *and* $B$ *with* $0 < A < B$. *Then the inequality*

$$\left(\int\limits_{\mathbb{R}^n}\left(\int\limits_{A|x|\le|y|\le B|x|} f(y)\,dy\right)^q v(x)\,dx\right)^{1/q} \le C\left(\int\limits_{\mathbb{R}^n} f^p u\right)^{1/p} \tag{17.8}$$

*holds for all non-negative* $f$ *if and only if*

$$\sup_{|y|\le|x|\le(B/A)|y|}\left(\int\limits_{A|x|\le|z|\le B|y|} u(z)^{1-p'}\,dz\right)^{1/p'}\left(\int\limits_{|y|\le|z|\le|x|} v(z)\,dz\right)^{1/q}$$

*is finite.*

**Proof.** This is Theorem 17.2, taking $S$ to be the unit ball in $\mathbb{R}^n$, $a(t) = At$ and $b(t) = Bt$. ∎

**Corollary 17.5** *Suppose* $0 < q < p$, $1 < p < \infty$, $1/r = 1/q - 1/p$ *and* $u$ *and* $v$ *are weights on* $\mathbb{R}^n$. *Fix real numbers* $A$ *and* $B$ *with* $0 < A < B$. *Then the inequality (17.8) holds for all non-negative* $f$ *if and only if both*

$$\left(\int\limits_0^\infty \int\limits_{At/B\le|x|\le t}\left(\int\limits_{At\le|y|\le B|x|} u(y)^{1-p'}\,dy\right)^{r/p'} \times\left(\int\limits_{|x|\le|y|\le t} v(y)\,dy\right)^{r/p} v(x)\,dx\,\frac{dt}{t}\right)^{1/r}$$

*and*

$$\left(\int_0^\infty \int_{t\le|x|\le Bt/A} \left(\int_{A|x|\le|y|\le Bt} u(y)^{1-p'}\,dy\right)^{r/p'} \times \left(\int_{t\le|y|\le|x|} v(y)\,dy\right)^{r/p} v(x)\,dx\,\frac{dt}{t}\right)^{1/r}$$

*are finite.*

**Proof.** We apply Theorem 17.3, taking $S$ to be the unit ball in $\mathbb{R}^n$, $a(t) = At$ and $b(t) = Bt$. It was shown in the proof of Corollary 2.6 of [3] that the normalizing function $\sigma(t)$ satisfies $1/B \le t\sigma(t) \le 1/A$ in this case. This estimate completes the proof. ■

## 17.3 From regions to kernels

In this section we define particular starshaped regions and use the results of the previous section to prove one-dimensional inequalities for operators with more general kernels.

**Definition 17.1** *Suppose $\varphi : (0,\infty) \to (0,\infty)$ is a decreasing, continuously differentiable function with $\varphi(0) = 1$ and $\varphi(\infty) = 0$. Define*

$$S = \{(x_1, x_2) : 0 \le x_1, 0 \le x_2 \le x_1\varphi(x_1)\}$$

**Lemma 17.6** *$S$ is smoothly starshaped and the union of all dilations of $S$ is*

$$E = \{(x_1, x_2) : 0 \le x_1, 0 \le x_2 < x_1\}.$$

*For $x = (x_1, x_2) \in E$, $S_x$, the least dilation of $S$ that contains $x$, is $\alpha_x S$, where $\alpha_x$ satisfies $x_2 = x_1\varphi(x_1/\alpha_x)$.*

**Proof.** If $(x_1, x_2) \in S$ and $0 \le r \le 1$ then we have $0 \le rx_2 \le rx_1\varphi(x_1) \le rx_1\varphi(rx_1)$ so $(rx_1, rx_2) \in S$. Thus $S$ is starshaped. It is clear that $S$ is smoothly starshaped.

If $(x_1, x_2) \in \alpha S$ with $\alpha > 0$ then $0 \le x_1/\alpha$ and hence $0 \le x_1$. Also $0 \le x_2/\alpha \le (x_1/\alpha)\varphi(x_1/\alpha)$ so $0 \le x_2 \le x_1\varphi(x_1/\alpha) \le x_1\varphi(0)$. Thus $(x_1, x_2) \in E$. Conversely, if $(x_1, x_2) \in E$ then for sufficiently large $\alpha$ we have both $0 \le x_1/\alpha \le 1$ and $0 \le x_2/\alpha \le (x_1/\alpha)\varphi(x_1/\alpha)$ so $(x_1, x_2)$ is in some dilation of $S$.

The least dilation of $S$ that contains the point $(x_1, x_2)$ is the unique $\alpha$ for which $(x_1/\alpha, x_2/\alpha)$ is on the graph of $x_1\varphi(x_1)$. Thus $x_2 = x_1\varphi(x_1/\alpha_x)$. This completes the proof. ■

**Theorem 17.7** *Suppose $1 < p \le q < \infty$ and $u$ and $v$ are weights. Let $a$ and $b$ be as above and $\varphi$ and $S$ be as in Definition 17.1. Then the inequality*

$$\left(\int_0^\infty \left(\int_0^\infty [\varphi(t/b(s)) - \varphi(t/a(s))]g(t)\,dt\right)^q v(s)\,ds\right)^{1/q} \le C\left(\int_0^\infty g^p u\right)^{1/p} \tag{17.9}$$

*holds for all non-negative $g$ provided*

$$\sup_{\substack{t\le s\\ a(s)\le b(t)}} \left(\int_0^\infty [\varphi(\xi/b(t)) - \varphi(\xi/a(s))]u(\xi)^{1-p'}\,d\xi\right)^{1/p'} \left(\int_t^s v\right)^{1/q} < \infty. \tag{17.10}$$

**Proof.** We apply Theorem 17.2 with $S$ defined in terms of $\varphi$ as in Definition 17.1 and with $v(x_1, x_2)$ replaced by $\delta_1(x_1)v(1/\varphi^{-1}(x_2))\frac{d}{dx_2}(1/\varphi^{-1}(x_2))$ and $u(y_1, y_2)$ replaced by $y_1^{p-1}u(y_1)$. Here $\delta_1$ is the measure consisting of a single atom of weight 1 at 1. It is straightforward to extend Theorem 17.2 to such measures.

We begin by verifying the weight condition (17.5). Using the definition of $S$ in terms of $\varphi$ we have

$$\begin{aligned} b(t)S \setminus a(s)S &= \{(y_1, y_2) : 0 \le y_1, y_1\varphi(y_1/a(s)) \le y_2 \le y_1\varphi(y_1/b(t))\},\\ sS \setminus tS &= \{(x_1, x_2) : 0 \le x_1, x_1\varphi(x_1/t) \le x_2 \le x_1\varphi(x_1/s)\}, \text{ and}\\ (sS \setminus tS) \cap \{(x_1, x_2) : x_1 = 1\} &= \{(1, x_2) : \varphi(1/t) \le x_2 \le \varphi(1/s)\} \end{aligned}$$

The weight condition (17.5) becomes

$$\sup_{\substack{t\le s\\ a(s)\le b(t)}} \left(\int_0^\infty \int_{y_1\varphi(y_1/a(s))}^{y_1\varphi(y_1/b(t))} y_1^{-1}u(y_1)^{1-p'}\,dy_2\,dy_1\right)^{1/p'} \times \left(\int_{\varphi(1/t)}^{\varphi(1/s)} v(1/\varphi^{-1}(x_2))\,d(1/\varphi^{-1}(x_2))\right)^{1/q} < \infty$$

which, replacing the variable $y_1$ by $\xi$ in the first factor and making the substitution $\xi = 1/\varphi^{-1}(x_2)$ in the second factor, reduces to the hypothesis (17.10).

We have verified the weight condition of Theorem 17.2 so we may conclude that the inequality (17.1) holds for all $f(y_1, y_2)$. In particular, it holds with $f(y_1, y_2)$ replaced by $y_1^{-1} g(y_1)$ for any non-negative function $g$. To simplify the left hand side of (17.1) we observe that the choice of $v$ means that we may take $x_1 = 1$. Lemma 17.6 shows that $\alpha_{(1,x_2)} = 1/\varphi^{-1}(x_2)$ and applying Definition 17.1 we see that

$$b(\alpha_{(1,x_2)})S \setminus a(\alpha_{(1,x_2)})S = \{(y_1, y_2) : y_1\varphi(y_1/a(1/\varphi^{-1}(x_2))) \\ \leq y_2 \leq y_1\varphi(y_1/b(1/\varphi^{-1}(x_2)))\}.$$

The left hand side of (17.1) becomes

$$\left( \int_0^1 \left( \int_0^\infty \int_{y_1\phi(y_1/a(1/\varphi^{-1}(x_2)))}^{y_1\phi(y_1/b(1/\varphi^{-1}(x_2)))} y_1^{-1} g(y_1)\, dy_2\, dy_1 \right)^q \\ \times v(1/\varphi^{-1}(x_2))\, d(1/\varphi^{-1}(x_2)) \right)^{1/q}$$

which, replacing $y_1$ by $t$ and making the substitution $s = 1/\varphi^{-1}(x_2)$, becomes the left hand side of (17.9).

The description of $E$ in Lemma 17.6 shows that the right hand side of (17.1) is just

$$\left( \int_0^\infty \int_0^{y_1} (y_1^{-1} g(y_1))^p y_1^{p-1} u(y_1)\, dy_2\, dy_1 \right)^{1/p} = \left( \int_0^\infty g^p u \right)^{1/p}.$$

This completes the proof. ■

**Theorem 17.8** *Suppose* $0 < q < p$, $1 < p < \infty$, $1/r = 1/q - 1/p$ *and* $u$ *and* $v$ *are weights. Let* $a$ *and* $b$ *be as above and* $\varphi$ *and* $S$ *be as in Definition 17.1. Then the inequality (17.9) holds for all non-negative* $g$ *provided both*

$$\left( \int_0^\infty \int_{b^{-1}(a(t))}^t \left( \int_0^\infty [\varphi(\xi/b(s)) - \varphi(\xi/a(t))] u(\xi)^{1-p'}\, d\xi \right)^{r/p'} \\ \times \left( \int_s^t v \right)^{r/p} v(s)\, ds\sigma(t)\, dt \right)^{1/r}$$

*and*

$$\left(\int_0^\infty \int_t^{a^{-1}(b(t))} \left(\int_0^\infty [\varphi(\xi/b(t)) - \varphi(\xi/a(s))]u(\xi)^{1-p'}\, d\xi\right)^{r/p'} \times \left(\int_t^s v\right)^{r/p} v(s)\, ds\sigma(t)\, dt\right)^{1/r}$$

*are finite.*

**Proof.** Proceed as in the previous proof applying Theorem 17.3 instead of Theorem 17.2. ■

*References*


[1] S. Bloom and R. Kerman, *Weighted norm inequalities for operators of Hardy type*, Proc. Amer. Math. Soc., 113, 1991, 135–141.

[2] J. S. Bradley, *Hardy inequalities with mixed norms*, Canad. Math. Bull., 21, 1978, 405–408.

[3] H. P. Heinig and G. Sinnamon, *Mapping properties of integral averaging operators*, Studia Math. 129, 1998, 157-177.

[4] P. J. Martin-Reyes and E. T. Sawyer, *Weighted inequalities for Riemann-Liouville fractional integrals of order one and greater*, Proc. Amer. Math. Soc., 106, 1989, 727–733.

[5] B. Opic and A. Kufner, *Hardy-type Inequalities*, Longman Scientific & Technical, Longman House, Burnt Mill, Harlow, Essex, England, 1990.

[6] G. Sinnamon, *One-dimensional Hardy-type inequalities in many dimensions*, To appear in Royal Society of Edinburgh, Proceedings A.

[7] G. Sinnamon and V. Stepanov, *The weighted Hardy inequality: New proofs and the case $p = 1$*, J. London Math. Soc. (2), 54, 1996, 89–101.

[8] V. D. Stepanov, *Weighted norm inequalities of Hardy type for a class of integral operators*, J. London Math. Soc. (2) 50, 1994, 105–120.

# Chapter 18

# Regularly bounded functions and Hardy's inequality

**Tatyana Ostrogorski**

## 18.1 Introduction

We consider Hardy's inequality

$$\int_0^\infty \left(\frac{1}{t}\int_0^t F(s)\,ds\right)^p W(t)\,dt \leq C\int_0^\infty F^p(t)W(t)\,dt \tag{HI}$$

where $W$, a positive function, is the weight and $1 \leq p < \infty$. In [6, Th. 330] this inequality is given with the weight $W(t) = t^\alpha$, for $0 < \alpha < p-1$.

Muckenhoupt [7] found a necessary and sufficient condition for the weights in (HI)

$$\int_t^\infty s^{-p}W(s)\,ds \left(\int_0^t (W(s))^{-1/(p-1)}ds\right)^{p-1} < C \tag{H$_p$}$$

for every $t > 0$. Also in [3] a necessary and sufficient condition is given for the weights when (HI) is considered for nonincreasing functions only. This is the following condition

$$\int_t^\infty s^{-p}W(s)ds \leq Ct^{-p}\int_0^t W(s)\,ds \tag{Bp}$$

for every $t > 0$ (see also [2]). While the necessity of both these conditions is trivial to verify (by taking a particular choice for F), the sufficiency takes some work, in particular it makes use of integration by parts.

In this paper we consider an "easy" sufficient condition for (HI). Consider functions $W$ such that

$$\int_t^\infty s^{-p}W(s)ds \leq Ct^{1-p}W(t) \tag{*}$$

for every $t > 0$. Then $W$ is a weight in (HI) (see Corollary 18.23 below).

There is a similar class of functions which is well known. These are functions satisfying, for *large* values of $t$, a two-sides inequality like (*), that is, such that the integral is also $\geq C_1 t^{1-p} W(t)$ (with some other positive constant $C_1$). Such functions, called $O$-regularly varying functions, were defined by Karamata (see [1]). We introduce analogous class of functions, called *regularly bounded* for which (a two-sided inequality) (*) holds for *every* t.

Of course, since condition (*) is not necessary for (HI), the class of regularly bounded functions is larger than ($H_p$). Still it has the following two advantages: this same condition implies the weighted inequality not only for Hardy's operator, but for a whole class of integral operators (see section 18.7), and this condition generalizes easier to higher dimensions.

In the first four sections we develop the theory of regularly bounded functions, which is in many respect analogous to the theory of $O$-regularly varying function and to the better-known theory of regularly varying functions [9, 4, 5]. In particular we have the Uniform Boundedness Theorem, the global estimates and the Representation Theorem, and all these are characteristics properties of these functions. In section 18.6 we consider integral transforms of regularly bounded functions and in section 6 weighted inequalities generalizing Hardy's inequality. The relation of regularly bounded functions with the conditions ($H_p$) and (Bp) is discussed at the end of section 18.7.

## 18.2 Definition and Uniform Boundedness

**Definition 18.1** *A continuous function $F : \mathbb{R}_+ \to \mathbb{R}_+$ is called (multiplicative) regularly bounded if for every $s \epsilon \mathbb{R}_+$ there is an $M(s)$ such that*

$$\sup_{t>0} \frac{F(ts)}{F(t)} = M(s). \tag{18.1}$$

*The function $M : \mathbb{R}_+ \to \mathbb{R}_+$ is called the index function of $F$. We shall write $RB = RB(\mathbb{R}_+)$, for the class of all regularly bounded functions.* ◇

When the supremum in (18.1) is replaced by $\limsup_{(t\to\infty)}$ then we obtain the class of $O$-regularly varying functions ($O{-}RV$) and when it is replaced by $\lim_{(t\to\infty)}$ we are reduced to ordinary regularly varying functions ($O{-}RV$) [9, 4, 1].

Obviously, $RB$ is a subclass of $)O{-}RV$. On the other hand, an $O-$regularly varying function is in the fact defined for large values of the argument (in the sense that its form on any finite interval doesn't affect its regular variation). Then we can modify it so that it becomes bounded on any interval of the form $(0, N]$. This reasoning is applied every time we need to control the behaviour of the function on the whole $\mathbb{R}_+$, like when we consider integral transforms of $O-$regularly varying functions.

*Examples.*

(a) A function which is $O - RV$ and bounded away from 0 and $\infty$ on any interval $(O, N]$ is regularly bounded.

(b) A function which is $O-RV$ at $\infty$ and at zero and bounded away from 0 and $\infty$ on any interval $[a, b]$ is regularly bounded. (Here by definition a function is $O - RV$ at zero if the function $F^{\curlyvee}(x) = F(1/x)$ is $O - RV$ at infinity.))

Since a prototype of a weight in Hardy's inequality is $t^\alpha$, which has a singularity both at 0 and at $\infty$, we see the functions in (b) are more suitable as weights than functions in (a). (Lemma 18.20 below also shows that 0 and $\infty$ are in a similar position.)

The theory of regularly bound functions turns out to be completely analogous to Karamata's theory of regular variation (only the writing is simplified due to the fact that we need not worry about $t$ tending to infinity). In particular, there is a Uniform Boundness Theorem 18.3, from which everything else follows).

With a function $F : \mathbb{R}_+ \to \mathbb{R}_+$ we consider also a function $f : \mathbb{R} \to \mathbb{R}$ corresponding to it by the formula

$$f(x) = \log F(e^x). \tag{18.2}$$

**Definition 18.2** *A continuous function $f : \mathbb{R} \to \mathbb{R}$ is called (additively) regularly bounded if for every $y \epsilon \mathbb{R}$ there is an $m(y)$ such that*

$$\sup_{x \in \mathbb{R}} (f(x+y) - f(x)) = m(y).$$

*The function $m : \mathbb{R} \to \mathbb{R}$ is called the index function of $f$. We shall write $RB = RB(\mathbb{R})$ for the class of all regularly bounded functions.*

Obviously, for $F$ and $f$ related as in (18.2) we have $F \epsilon RB(\mathbb{R}_+)$ if and only if $f \epsilon RB(\mathbb{R})$, and in this case we also have $m(x) = \log M(e^x)$. We shall always let capital letters denote functions on $\mathbb{R}_+$ and script letters denote functions on $\mathbb{R}$. Then we can write simply $RB$, as it will be evident form the context in which of the classes we are. Until section 18.4 we shall work in $RB(\mathbb{R})$, since the additive notation is simpler. This choice is made only to simplify the writing, and for every additive result there is a corresponding multiplication result obtained via (18.2). In section 18.4 we have some "Abelian type" results for which there are no additive counterparts.

We start with some obvious observations about regularly bounded functions. Put for short $f_x(y) = f(x+y) - f(x)$; then we have

$$f_x(y_1 + y_2) = f_x(y_1) + f_{x+y_1}(y_2) \tag{18.3}$$

$$f_x(y) = -f_{x+y}(-y). \tag{18.4}$$

If $f$ is a regularly bounded function, then $\inf_{x\in\mathbb{R}}(f(x+y)-f(x)) = -m(-y)$ and thus we have for every $x\epsilon\mathbb{R}$

$$-m(-y) \leq f(x+y) - f(x) \leq m(y). \tag{18.5}$$

Indeed, we have by an application of (18.4)

$$\begin{aligned}\inf_x f(y) &= -\sup_x(-f_x(y)) = -\sup_x(f_{x+y}(-y)) \\ &= -\sup_u f_u(-y) = -m(-y).\end{aligned}$$

Now we proceed to prove the most important property of regularly bounded functions - the uniformity of the defining relation with respect to $y$ belonging to any compact set $K$. This is Theorem 18.3 below. The proof of this theorem goes by proving first this property for some open set $U$, then for an arbitrary neighborhood of 0, and finally for an arbitrary compact set.

**Proposition 18.1** *If $f$ is regularly bounded, that is if for every $y\epsilon\mathbb{R}$ there is an $m(y)$ such that*

$$\sup_x f_x(y) \leq m(y)$$

*then there is an open set $U$ and a constant $M = M_U$ such that*

$$\sup_x \sup_{y\in U} f_x(y) \leq M_U. \tag{18.6}$$

**Proof.** Put $E_{x,N} = \{y : f_x(y) \leq N\}$ and $E_N \cap E_{x,N} = \{y : \sup_x f_x(y) \leq N\}$,with $N$ a positive integer. Then by the continuity of $\mathrm{E}_{x,N}$ is closed and $E_N$ is closed and by the assumption of the proposition every $y\epsilon\mathbb{R}$ belongs to some $E_N$ (the one with $N \geq m(y)$). Thus $\mathbb{R} = \cup\mathbb{E}_N$ and by Baire's Theorem there is an $N_0$ and an open set $U \subseteq E_{N_0}$. This means that for every $\mathrm{y}\epsilon U$ we have $\mathrm{y}\epsilon E_{N_0}$, which by the definition of the set $E_{N_0}$ means that $\sup_x f_x(y) \leq M$ , for every $\mathrm{y}\epsilon U$ and this is (18.6). ∎

**Corollary 18.2** *If $f$ is regularly bounded, then there is a symmetric neighborhood of zero $V$ (i.e. $V = -V$) and a constant $M = M_V$ such that*

$$-M \leq \inf_{y\in V} f_x(y) \leq \sup_{y\in V} f_x(y) \leq M. \tag{18.7}$$

**Proof.** Let $U$ and $M_U$ be as in the proposition and for $u_0\epsilon U$ put $W = U - u_0$. Then

$$\sup_{y\in U} f_x(y) = \sup_{y\in U} f_x(u-u_0) = f_x(-u_0) + \sup_{u\in U} f_{x-u_0}(u)$$

by (18.3) and this is bounded by $m(-u_0) + M_U$, according to (18.6). Now every neighborhood of zero $W$ contains a symmetric neighborhood of zero

$V$, i.e. $V \subseteq W$ and $V = -V$. Then finally $\sup_{y \in W} f_x(y) \leq \sup_{y \in W} \mathrm{f}_x(y) \leq m(-u_0) + M_U$, which proves the right-hand side inequality in (18.7). Next we have by (18.4)

$$\inf_{y \in V} f_x(y) = -\sup_{y \in V}(-f_x(y)) = -\sup_{y \in V} f_{x+y}(-y) = -\sup_{u \in V} f_{x-u}(u)$$

since $V = -V$. Now the last expression is $\geq -M$, by the above.
■

**Theorem 18.3** *(Uniform Boundedness Theorem). If $f$ is regularly bounded, then for every compact set $K$ there is a constant $M_K$ such that for every $x \in \mathbb{R}$ we have*

$$-M_K \leq \inf_{y \in K} f_x(y) \leq \sup_{y \in K} f_x(y) \leq M_K. \tag{18.8}$$

**Proof.** By Corollary 18.2 we find $V$ and $M$ so that (18.7) holds. For $K$ we find points $y_i, i = 1, ..., m$,such that $K \subseteq i = 1\overset{m}{\cup}(y_i + V) = i = 1\overset{m}{\cup}V_i$. Then we have

$$\sup_{y \in K} f_x(y) \leq \max_i \sup_{y \in V_i} f_x(y). \tag{18.9}$$

Now compute

$$\sup_{y \in K} f_x(y) = \sup_{u \in V} f_x(y_i + u) = f_x(y_i) \ + \sup_{u \in V} f_{x+y_i} \leq m(y_i) + M$$

by (18.7). Now substitute this into (18.9), and obtain $\sup_{y \in K} f_x(y) \leq \max_i(m(y_i) + M) = M_K$. The proof of the left-hand side inequality is

similar. ■

Now we consider the properties of the index function m.

**Lemma 18.4** *If $f$ is regularly bounded, then its index function $m$ is subadditive, i.e. $m(y_1 + y_2) \leq m(y_1) + m(y_2)$, and locally bounded. The function $f$ is also locally bounded.*

**Proof.** The subadditivity follows by taking the supremum in (18.3). The local boundedness is a direct consequence of the Uniform Boundedness Theorem. Indeed, let $I$be any compact interval, the by (18.8)

$$\sup_{y \in I} m(y) = \sup_x \sup_{y \in I} f_x(y) \leq M_I$$

and similarly for the lower bound. Finally, by the preceding inequality we have $f(x + y) - f(x) \leq M_I$ for every $y \in I$ and every $y \in I$ and every $x$. Now put $x = 0$ and obtain $f(y) \leq M_I + f(0)$, for every $y \in I$. ■

**Proposition 18.5** *([1]) (a) Let $m : [0, \infty) \to \mathbb{R}$ be subadditive and locally bounded. Then the following limit exists*

$$\lim_{t \to \infty} m(t)/t = \inf_{t > 0} m(t)/t.$$

*(b) Let $m$: $(-\infty, 0] \to \mathbb{R}$ be subadditive and locally bounded. Then the following limit exists*

$$\lim_{t \to -\infty} m(t)/t = \sup_{t > 0} m(t)/t.$$

**Definition 18.3** *Let $f$ be a regularly bounded function. Put*

$$\sigma = \lim_{t \to -\infty} m(t)/t, \qquad \tau = \lim_{t \to \infty} m(t)/t \tag{18.10}$$

*and call $\sigma$ the lower index and $\tau$ the upper index of $f$. We shall write $RB_{\sigma}^{\tau}$ for the class of all regularly bounded functions with indices $\sigma$ and $\tau$.*

**Lemma 18.6** *Let $\widetilde{f}$ be defined by $\widetilde{f}\,(x) = f(-x)$. Then $f \in RB_{\sigma}^{\tau}$ if and only if $f \in RB_{-\tau}^{-\sigma}$.*

**Corollary 18.7** *([1)] Let $m : \mathbb{R} \to \mathbb{R}$ be subadditive and locally bounded and let $\sigma$ and $\tau$ be defined by (18.10). Then - $\infty < \sigma \leq \tau < \infty$. Moreover, for every $y$*

$$m(y) \geq \max(\sigma y, \tau y)$$

*and for every $\alpha_1$ and $\alpha_2$ such that $\alpha_1 < \sigma$ and $\tau < \alpha_2$ there is a constant $C$ such that for every $y$*

$$m(y) \leq \max(\alpha_1 y, \alpha_2 y) + C. \tag{18.11}$$

## 18.3 The global bounds

Introduce the notation, for $\alpha, \beta \in \mathbb{R}$

$$q^{\alpha,\beta}(y) = \max(\alpha y, \beta y) \qquad p^{\alpha,\beta}(y) = \min(\alpha y, \beta y) \tag{18.12}$$

**Lemma 18.8** *We have*

$$\begin{aligned} p^{\alpha,\beta}(y) &\leq q^{\alpha,\beta}(x+y) - q^{\alpha,\beta}(x) \leq q^{\alpha,\beta}(y) \\ p^{\alpha,\beta}(y) &\leq p^{\alpha,\beta}(x+y) - p^{\alpha,\beta}(x) \leq q^{\alpha,\beta}(y) \end{aligned} \tag{18.13}$$

*and consequently, $q^{\alpha,\beta}$ and $p^{\alpha,\beta}$belong to $RB_{\alpha}^{\beta}$.*

**Proof.** Indeed, (18.13) are elementary properties of the functions *max* and *min*, thus the functions $q^{\alpha,\beta}$ and $p^{\alpha,\beta}$ are regularly bounded with index function $q^{\alpha,\beta}$. Now it is easy to verify that the indices are $\alpha$ and $\beta$. ■

**Theorem 18.9** *Let $f \in RB_\sigma^\tau$. Then for every $\alpha_1$, $\alpha_2$ such that $\alpha_1 < \sigma$ and $\tau < \sigma_2$ there is a $C$ such that for every $y$ and for every $x$*

$$p^{\alpha_1,\alpha_2}(y) - C \leq f(x+y) - f(x) \leq q^{\alpha_1,\alpha_2}(y) + C. \tag{18.14}$$

**Proof.** The right-hand side inequality follows directly from (18.11), since $f(x+y) - f(x) \leq m(y)$. For the left-hand side inequality we use (18.5) $f(x+y) - f(x) \geq -m(-y)$. Now again by (18.11) we have $-m(y) \geq -\max(-y) \geq -\max(-\alpha_1 y, -\alpha_2 y) - C$. ■

We shall see that (18.14) is a characteristic property of regularly bounded functions. Other such properties will be give in sections 3 and 4. Note that property (18.14) for $f$ defines new indices for $f$ as in the following definition.

**Definition 18.4** *Let $f$ be a function such that there are $\alpha_1$, $\alpha_2$ and $C$ such that for every $y > 0$ we have*

$$\alpha_1 y - C \leq f(x+y) - f(x) \leq \alpha_2 y + C. \tag{18.15}$$

*Put $a_1$ equal to the supremum of all $\alpha_1$ satisfying (18.15) and pit $a_2$ equal to the infimum of all $\alpha_2$ satisfying (18.15).*

**Corollary 18.10** *Let $f : \mathbb{R} \to \mathbb{R}$ be continuous function. Then $f \in RB_\sigma^\tau$ if and only if there are $\alpha_1$, $\alpha_2$ such that $f$ satisfies (18.15). In case this is true we have*

$$\sigma = a_1 \qquad \tau = a_2 \tag{18.16}$$

**Proof.** Suppose $f \in RB_\sigma^\tau$. Then by Theorem 18.9 $\alpha_2 > \tau$ implies (18.15) and by Definition 18.4 this means that $\alpha_2 > a_2$. Thus $a_2 \leq \tau$. In a similar way we find $a_1 \geq \sigma$. On the other hand, suppose that (18.15) and by Definition 18.4 this means that $\alpha_2 > a_2$. Then we have $m(y) \leq \alpha_2 y + C$, and by definition (18.10) of $\tau$ we have $\tau \leq \alpha_2$, that is $\tau \leq a_2$. The inequality for the lower indices is analogous. ■

## 18.4 The representation theorem

Let us introduce the following notation for the equivalence relation

$$f \asymp g \iff (\exists m)(\exists M)(\forall x \in \mathbb{R}) \quad m \leq f(x) - g(x) \leq M. \tag{18.17}$$

Let us write $T_y f(x) = f(x+y)$. Then by the definition (see also (18.5)) a function is regularly bounded if and only if $T_y f \asymp f$. We deduce that every function equivalent with a regularly bounded function is regularly bounded.

**Lemma 18.11** *(a) Every bounded function is regularly bounded.*
*(b) Every function with bounded derivative is regularly bounded.*

**Proof.** (a) Indeed, $f$ bounded means $f \asymp 0$; but then also $T_y f \asymp 0$.
(b) Let $f$ be bounded by constants $m \leq f\ (x) \leq M$. Then since $f(x+y)$ - $f(x) = f(x+\theta y)y$ (with some $\theta$), we gave for $y > 0$ that $f(x+y)$ - $f(x) \leq My$ and for $y < 0$ that $f(x+y)$ - $f(x) \leq my$. This proves that $f$ is regularly bounded. ■

In Lemma 18.11 we saw that bounded functions and functions with bounded derivatives are regularly bounded. The Representation Theorem says that in some way the converse is also true. Let us denote by $C^k$ the class of functions having continuous k-th derivative.

**Lemma 18.12** *Let $f$ be regularly bounded. The there exists an equivalent function $f_k$, such that $f_k \epsilon C^k$.*

**Proof.** Indeed, the function $f_1(x) = \int_0^1 f(x+y)dy$ is equivalent with $f$ since

$$f_1(x) - f(x) = \int_0^1 (f(x+y) - f(x))dy \leq \sup_{y \in I}(f(x+y) - f(x)) \leq M_I \tag{18.18}$$

by the Uniform Boundedness Theorem for $I = [0,1]$. A similar estimate holds from below and the rest of the lemma follows by induction. ■

**Theorem 18.13** *(Representation Theorem). A continuous function $f : \mathbb{R} \longrightarrow \mathbb{R}$ is regularly bounded if and only if there is a continuous function $\delta : \mathbb{R} \longrightarrow \mathbb{R}$ and a differentiable function $\eta : \mathbb{R} \longrightarrow \mathbb{R}$ such that*

$$f(x) = \delta(x) + \eta(x) \tag{18.19}$$

*with*

$$\delta \asymp 0 \qquad \eta' \asymp 0. \tag{18.20}$$

**Proof.** The sufficiency follows from Lemma 18.11. Now for the converse put

$$\eta(x) = \int_0^1 f(x+y)dy \qquad \textit{and} \qquad \delta(x) = f(x) - \eta(x).$$

Thus $\eta$ is the function $f_1$ from Lemma 18.12 and by (18.18) it follows that $\delta \asymp 0$. On the other hand $\eta'(x) = f(x+1) - f(x)$ and thus

$$-m(-1) \leq \eta'(x) \leq m(1).$$

■

**Remark 18.1** *The representation (18.19) is not unique, since if $\gamma$ is any function such that $\gamma \asymp 0$ and $\gamma' \asymp 0$, then $f(x) = (\delta(x) + \gamma(x)) + (\eta(x) - \gamma(x))$ is also a representation.*

## 18.5 The multiplicative class

It is obvious that for the function $F : \mathbb{R}_+ \longrightarrow \mathbb{R}_+$ there are statements analogous to all the statements of the previous sections.

The indices $\sigma$ and $\tau$, defined in (18.10), will be also called the indices of $F$.

The equivalence relation (18.17) is transformed by (18.2) into

$$F \asymp G \iff \qquad (\exists c, C > 0)(\forall t > 0) \qquad c < F(t)/G(t) < C.$$

In the following theorem part (a) gives a characterization of regular boundedness analogous to that of Theorem 18.9, part (c) analogous to Theorem 18.9 and parts ($b_1$) and ($b_2$) have no additive analogues.

**Theorem 18.14** *The following conditions are equivalent:*
*(a) there are $\alpha_1$, $\alpha_2$ and there is a $C > 0$ such that for every $t > 0$ and for every $s > 0$*

$$1/C\, P^{\alpha_1,\alpha_2}(s) \leq \frac{F(ts)}{F(t)} \leq C\, Q^{\alpha_1,\alpha_2}(s). \tag{18.21}$$

*($b_1$) there a $\beta_1$ is such that*

$$\int_0^t s^{-\beta_1} F(s) ds/s \approx t^{-\beta_1} F(t). \tag{18.22}$$

*($b_2$) there is a $\beta_2$ such that*

$$\int_t^\infty s^{-\beta_2} F(s)\, ds/s \approx t^{-\beta_2} F(t). \tag{18.23}$$

*(c) there are $\gamma_1$, $\gamma_2$ and there is a continuous function $\Delta : \mathbb{R}_+ \longrightarrow \mathbb{R}_+$ and a differentiable function $H : \mathbb{R}_+ \longrightarrow \mathbb{R}_+$ such that $F(t) = \Delta(t)H(t)$ and*

$$\Delta(t) \asymp 1 \qquad \gamma_1 < \frac{tH'(t)}{H(t)} < \gamma_2 \tag{18.24}$$

*If any of the conditions holds, then $F$ is regularly bounded. Let $A_i$ be the set of all $\alpha_1$ such that (a) holds, $B_1$ be the set of all $\beta_1$ be the set of all $\beta_1$ such that ($b_1$) holds and $C_i$ be the set of all $\gamma_i$ such that (c) holds. Put*

$$\begin{array}{lll} a_1 = \sup A_1 & b_1 = \sup B_1 & c_1 = \sup C_1 \\ a_2 = \inf A_2 & b_2 = \inf B_2 & c_2 = \inf C_2 \end{array}$$

*then we have*

$$a_1 = b_1 = c_1 = \sigma \qquad a_2 = b_2 = c_2 = \tau. \tag{18.25}$$

**Proof.** (a)$\Longrightarrow$ $(b_1)$. Let $0 < s < 1$, then (18.21) reads

$$1/C\, s^{\alpha_2} \le \frac{F(ts)}{F(t)} \le C\, s^{\alpha_1}.$$

Then

$$\frac{\int_0^t s^{-\beta_1} F(s)\, ds/s}{t^{-\beta_1} F(t)} = \int_0^1 s^{-\beta_1} \frac{F(ts)}{F(t)}\, ds/s \le \int_0^1 s^{-\beta_1} s^{\alpha_1} ds/s \tag{18.26}$$

and this is finite for every $\beta_1 < \alpha_1$. Likewise, the quotient in (18.26) is bounded from below by $\int_0^1 s^{-\beta_1} s^{\alpha_2} ds/s$ which is obviously $> 0$. Thus we obtain (18.22) for every $\beta_1 < \alpha_1$. This shows that

$$a_1 \le b_1 \tag{18.27}$$

(Indeed, let $\lambda <$a$_1$. The we can dine an $\alpha_1$ such that $\lambda < \alpha_1 < a_1$. But then by the above $\lambda$ satisfies (18.22), thus $\lambda < b_1$.

(a)$\Longrightarrow$ $(b_2)$. Let 1<s<$\infty$. Then (18.17) reads

$$1/C\, s^{\alpha_1} \le \frac{F(ts)}{F(t)} \le C\, s^{\alpha_2}.$$

and the rest of the proof is similar to (a) $\Longrightarrow$ (b$_1$), and we obtain that (18.23) is true under the condition $\beta_2 > \alpha_2$. Then we obtain

$$a_2 \ge b_2. \tag{18.28}$$

(b$_1$) $\Longrightarrow$ (c). Take $\beta_1$ such that (18.21) holds and put $G(s) = s^{-\beta_1 F(s)}$ and $H_1(t) = \int_0^t G(s)\, ds/s$. Then by the assumption (18.21)

$$G(t) = H_1(t)\Delta(t) \tag{18.29}$$

with $1/C_1 \le \Delta(t) \le C_1$. Then $\frac{tH_1(t)}{H_1(t)} = \frac{G(t)}{H_1(t)} = \Delta(t)$ and thus

$$1/C_1 \le \frac{tH_1(t)}{H_1(t)} \le C_1 \tag{18.30}$$

Finally we have for $F(t) = t^{\beta_1} G(t)$, by (18.29)

$$F(t) = t^{\beta_1} H_1(t)\Delta(t) = H(t)\Delta(t)$$

and since $\frac{tH'(t)}{H(t)} = \beta_1 + \frac{tH_1(t)}{H_1(t)}$, we have by (18.30)

$$\beta_1 + 1/c_1 \leq \frac{tH'(t)}{H(t)} \leq \beta_1 + C_1.$$

Thus if we put $\gamma_1 = \beta_1 + 1/C_1$. we have proved that (c) holds with some $\gamma_1 > \beta_1$. This proves that

$$b_1 \leq c_1. \tag{18.31}$$

($b_2$) $\Longrightarrow$ (c). The proof is similar to the one for ($b_1$) $\Longrightarrow$ (c) and we obtain

$$b_2 \leq c_2. \tag{18.32}$$

(c)$\Longrightarrow$ ($a$). Suppose $F(t) = H(t)\Delta(t)$ with (18.24). The for $t > 0$ we have

$$\gamma_1/t \leq (\log H(t))' \leq \gamma_2/t \tag{18.33}$$

and the for $s > 1$ we have

$$\begin{aligned} \log\ H(ts) - \log H(t) &= \int_t^{ts} (\log H(u))' du \\ &\leq \gamma_2 \int_t^{ts} du/u = \gamma_2(\log\ ts - \log\ t) = \gamma_2 \log\ s \end{aligned}$$

and finally $H(ts)/H(t) \leq s^{\gamma_2}$. From the other inequality in (18.33) we obtain $s^{\gamma_1} \leq H(ts)/H(t)$. Then

$$1/C\ s^{\gamma_1} \leq \frac{F(ts)}{F(t)} \leq C\ s^{\gamma_2}$$

for every $s > 1$, and then (18.21) follows with $\alpha_1 = \gamma_1$ and $\alpha_2 = \gamma_2$. Thus

$$a_1 = c_1, \quad a_2 = c_2. \tag{18.34}$$

Finally, (18.25) follows from (18.27), (18.31), and (18.33). ■

## 18.6 Abelian Theorems

Now consider integral transforms of regularly bounded functions. Let $k : \mathbb{R}_+ \longrightarrow \mathbb{R}$ be the *kernel* of the integral transform

$$KF(t) = \int_0^\infty F(s)k(s/t)\, ds/s = \int_0^\infty F(ts)k(s)\, ds/s. \tag{18.35}$$

An example of such an integral transform is Hardy's operator

$$HF(t) = \int_0^t F(s)\,ds/s$$

which we had already in (18.21) of Theorem 18.14.

**Lemma 18.15** *Let $\phi^\gamma(s) = s^\gamma$ be the power of function.*
*(a) If $C_\alpha = \int_0^\infty s^\alpha k(s)\,ds/s < \infty$, then $K\phi^\alpha(t) = C_\alpha \phi^\alpha(t)$, for every $\dot{t} > 0$.*
*(b) If $C_{\alpha,\beta} = \int_0^\infty Q^{\alpha,\beta}(s)k(s)\,ds/s < \infty$, then $KQ^{\alpha,\beta}(t) \leq C_{\alpha,\beta}Q^{\alpha,\beta}(t)$, for every $\dot{t} > 0$.*
*(c) If $\alpha \leq \gamma \leq \beta$ and $C_{\alpha,\beta} < \infty$, then $C_\gamma < \infty$.*

The following proposition is a generalization of the (a)$\Longrightarrow$ ($b_1$) part of this Theorem 18.14. (Also Corollary 18.18 is a generalization of (a)$\Longrightarrow$ ($b_2$) in Theorem 18.14)

**Proposition 18.16** *Let $F \epsilon RB_\sigma^\tau$ and let $\alpha_1$ and $\alpha_2$, with $\alpha_1 < \sigma$ and $\tau < \alpha_2$, be such that*

$$\int_0^\infty Q^{\alpha_1,\alpha_2}(s)k(s)\,ds/s < \infty.$$

*Then $KF \asymp F$.*

**Proof.** Let $\alpha_1$ and $\alpha_2$ satisfy the assumptions. By Theorem 18.14 (a) there is a C such that

$$\frac{KF(t)}{F(t)} = \int_0^\infty \frac{F(ts)}{F(t)} k(s)\,ds/s \leq C \int_0^\infty Q^{\alpha_1,\alpha_2}(s)k(s)\,ds/s$$

and this is finite by assumption. Also we have an estimate from below

$$\frac{KF(t)}{F(t)} \geq 1/c \int_0^\infty P^{\alpha_1,\alpha_2}(s)k(s)\,ds/s > 0.$$

■

The *dual operator* for the operator (18.35) is defined by taking the kernel $k^*(t) = k(1/t)$ and then putting

$$K^*F(t) = \int_0^\infty F(s)k^*(s/t)ds/s = \int_0^\infty F(s)k(t/s)\,ds/s.$$

**Lemma 18.17** *We have*

$$K^*F^*(t) = KF(1/t) \tag{18.36}$$

*and consequently,* $KF \asymp F$ *if and only if* $K^*F^* \asymp F^*$.

**Corollary 18.18** *Let* $F \epsilon RB_\sigma^\tau$ *and let* $\alpha_1$ *and* $\alpha_2$, *with* $\alpha_1 < \sigma$ *and* $\tau < \alpha_2$, *be such that*

$$\int_0^\infty Q^{-\alpha_2, -\alpha_1}(s)k(s)\, ds/s < \infty. \tag{18.37}$$

*Then* $K^*F \asymp F$.

**Proof.** Put $G = F^*$. Then $G \epsilon RB^{-\sigma}{}_{-\tau}$ and by assumption we have for $-\alpha_2 < -\tau$ and $-\sigma < -\alpha_1$ that (18.37) holds. Thus by Proposition 18.16 we have $KG \asymp G$. But then by the preceding lemma we have $K^*G^* \asymp G$ and this is $K^*F \asymp F$. ∎

## 18.7 Hardy's Inequality

In this section we show that regularly bounded functions can be taken as weights in the generalized Hardy's inequality. In the next theorem we first see how Hardy's inequality follows easily from condition (18.39). Now by Corollary 18.18 we know that (18.39) holds, in particular, for regularly bounded functions.

**Theorem 18.19** *(a) Let* $1 \le p < \infty$. *Let* $K$ *be an integral operator such that for some* $\gamma$ *we have*

$$K\phi^\gamma(1) < \infty \tag{18.38}$$

*and let* $U$ *be a function such that*

$$K^*(\phi^{\gamma(p-1)}U) \le C\phi^{\gamma(p-1)}U. \tag{18.39}$$

*Then the weighted inequality holds for* $K$.

$$\int_0^\infty (KF(t)^p U(t)dt/t \le C \int_0^\infty F^p(t)U(t)dt/t. \tag{18.40}$$

*(b) In equality, let* $\alpha_1$ *and* $\alpha_2$ *be such that*

$$\int_0^\infty Q^{-\alpha_2, -\alpha_1}(s)k(s)\, ds/s < \infty \tag{18.41}$$

*and let* $U \epsilon RB_\sigma^\tau$ *where* $\alpha_1 < \sigma/p < \alpha_2$. *Then (18.40) holds.*

**Proof.** Let $1 < p < \infty$. By Hölder's inequality we have

$$KF(t) = \int_0^\infty k(s/t)F(s)s^{-\gamma/p'}s^{\gamma/p'}\,ds/s$$
$$\leq \left(\int_0^\infty k(s/t)F^p(s)s^{-\gamma p/p'}ds/s\right)^{1/p}\left(\int_0^\infty d(s/t)s^\gamma ds/s\right)^{1/p'} \tag{18.42}$$

Now the second integral in (18.42) is equal to $K\phi^\gamma(t)$, and by Lemma 18.15(a) and (18.38) it is equal to the $Ct^\gamma$. Now substitute this into (18.42), and then (18.42) into the left-hand side of (18.40)

$$\int_0^\infty (KF(t))^p U(t)dt/t$$
$$\leq C\int_0^\infty t^{\gamma(p-1)}U(t)\left(\int_0^\infty k(s/t)F^p(s)s^{-\gamma(p-1)}ds/s\right)dt/t$$
$$= C\int_0^\infty F^p(s)s^{-\gamma(p-1)}\left(\int_0^\infty k(s/t)^{\gamma(p-1)}U(t)dt/t\right)\,ds/s, \tag{18.43}$$

by interchanging the order of integration. Now the inner integral in (18.43) is $K^*(\phi^{\gamma(p-1)}U)(s)$ and by (18.39) it is $\leq Cs^{\gamma(p-1)}U(s)$ and then (18.40) follows. Let $p = 1$. Then by interchanging the order of integration

$$\int_0^\infty\left(\int_0^\infty k(s/t)F(s)\,ds/s\right)U(t)dt/t = \int_0^\infty F(s)\left(\int_0^\infty k(s/t)U(t)dt/t\right)ds/s$$

Now the inner integral is equal to $K^*U(s)$ and is $\leq CU(s)$ by (18.39). This proves the theorem in this case.
(b) We only have to prove that we can find a $\gamma$ satisfying both (18.38) and (18.39). By Lemma 18.15(c) and (18.41), if we chose $\gamma$ such that

$$-\alpha_2 < \gamma < -\alpha_1 \tag{18.44}$$

then (18.38) holds. Moreover, since $U$ is regularly bounded, the $U_1 = \phi^{\gamma(p-1)}U$ is regularly bounded with indices $\sigma + \gamma(p-1)$ and $\tau + \gamma(p-1)$. By Corollary 18.18 if $\gamma$ is such that

$$\alpha_1 < \sigma + \gamma(p-1), \qquad \tau + \gamma(p-1) < \alpha_2 \tag{18.45}$$

we have $K^*U_1 \asymp U_1$,i.e. $K^*(\phi^{\gamma(p-1)}U) \asymp \phi^{\gamma(p-1)}U$ which proves (18.39). The only thing is left to check that it is possible to have $\gamma$ satisfying both (18.44) and (18.45), which reduces to $(\alpha_1 - \sigma)/(p-1) < -\alpha_1$ and $-\alpha_2 < (\alpha_2 - \tau)/(p-1)$, and this is true by assumption. ■

**Lemma 18.20** *Let $K$ be an integral operator. Then the weighted inequality holds for $K$ with some weight $U$*

$$\int_0^\infty (KF(t))^p\, U(t)\, dt/t \leq C \int_0^\infty F^p(t)U(t)\, dt/t \tag{18.46}$$

*if and only if weighted inequality holds for $K^*$ with the weight $U^*$*

$$\int_0^\infty (K^*F(t))^p\, U^*(t)\, dt/t \leq C \int_0^\infty F^p(t)U^*(t)\, dt/t. \tag{18.47}$$

**Proof.** By changing the variable $t = 1/s$ in (18.46), since $KF(1/t) = K^*F^*(s)$, (by (18.36)) we have

$$\int_0^\infty (K^*F^*(t))^p U^*(t)\, dt/t \leq C \int_0^\infty (F^*(t))^p\, U^*(t)\, dt/t$$

and since this is true for every function $F$, then (18.47) follows. ■

**Corollary 18.21** *Let $1 \leq p < \infty$. Suppose $U$ is a function which satisfies any of the following four conditions: (a) there is a $\gamma > 0$ and $C > 0$ such that*

$$\int_0^\infty t^{\gamma(p-1)} U(t) dt/t \leq C\, s^{\gamma(p-1)} U(s) \tag{18.48}$$

*(b) there is a $C_1 > 0$ such that*

$$\int_s^\infty U(t) dt/t \leq C_1 U(s) \tag{18.49}$$

*(c) there is an $\alpha < 0$ and there is a $C_2 > 0$ such that for every $s > 0$ and every $t > 1$ we have*

$$\frac{U(st)}{U(s)} \leq C_2 t^\alpha \tag{18.50}$$

*(d) $U \in RB_\sigma^\tau$ with $\tau < 0$.*
*Then Hardy's inequality*

$$\int_0^\infty \left( \int_0^\tau F(s) ds/s \right)^p U(t) dt/t \leq C \int_0^\infty F^p(t) U(t) dt/t.$$

*holds.*

**Proof.** (a) We apply Theorem 18.19 to Hardy's operator. Condition (18.38) is

$$\int_0^t s^\gamma ds/s < \infty$$

which holds since $\gamma > 0$. Also (18.48) is exactly condition (18.39) for Hardy's operator.
(b) Indeed, it is easy to verify that condition (18.49) implies (18.48). If the estimates were two-sided, this would follow from Theorem 18.14, but even for one-sided estimates this is obtained simply by integration by parts. Suppose (18.49) holds and write $\varepsilon = \gamma(p-1)$, then

$$\begin{aligned}\int_s^R t^\varepsilon U(t)/t\,dt &= s^\varepsilon \int_s^R U(t)/t\,dt + \varepsilon \int_s^R x^{\varepsilon-1}\left(\int_x^R U(t)/t\,dt\right)dx \\ &\leq C_1 s^\varepsilon U(s) + \varepsilon C_1 \int_s^R x^{\varepsilon-1}U(x)dx.\end{aligned}$$

Now let $R$ tend to infinity, we obtain (18.48) with $C = C_1/(1-C_1\varepsilon)$.
(c) The proof that (18.50) implies (18.49) is contained in the proof of (a)$\Longrightarrow$($b_2$) in theorem 18.14 (and $\geq$ inequality in (a) implies the $\geq$ inequality in ($b_2$)).
(d) Indeed, $U \in RB_\sigma^\tau$ with $\tau < 0$ obviously implies (18.50) since (18.50) is half of regular boundness. ■

**Remark 18.2** *We see that "half" of the condition of regular boundedness (18.50) is sufficient for Hardy's equality. (it is not difficult to check directly that (18.49) implies ($A_p$).) Note however that integration by parts was used to deduce (18.48) from (18.49) and this makes this argument unusable in higher dimension; also condition (18.49) is not applicable to operators different from Hardy's.*

**Corollary 18.22** *Let $U \in RB_\sigma^\tau$ and $\sigma > 0$. Then*

$$\int_0^\infty (H^*F(t))^p U(t)dt/t \leq \int_0^\infty F^p(t)U(t)dt/t. \tag{18.51}$$

**Proof.** By Lemma 18.15 we have that (18.51) holds if and only if

$$\int_0^\infty (HF(t))^p U^*(t)dt/t \leq C \int_0^\infty F^p(t)U^*(t)dt/t \tag{18.52}$$

Now by assumptions of Corollary 18.21 (d) and then (18.52) follows. This proves the Corollary. ■

In the introduction we considered Hardy's operator in the form $\int_0^t F(s)ds$ (with $ds$ rather than the invariant measure $ds/s$); indeed, it is then applicable to a larger class of functions. It is easy to rewrite the conditions for the weights when the inequality is in the form (HI) from the introduction. Put

$$U(t)/t = W(t)/t^p \tag{18.53}$$

in Corollary 18.21 (and replace $F(t)/t$ by $F(t)$).

**Corollary 18.23** *Let $1 \leq p < \infty$. Then (HI) holds if any of the following four conditions is satisfied.*
*(a) there exists an $\varepsilon > 0$ such that*

$$\int_s^\infty W(t)/t^{p-\varepsilon} dt \leq C\, sW(s)/s^{p-\varepsilon} \tag{18.54}$$

*(b) $W$ is such that*

$$\int_s^\infty W(t)/t^p\, dt \leq C\, sW(s)/s^p \tag{18.55}$$

*(c) there is an $\alpha < p-1$ and there is a $C_2 > 0$ such that for every $s > 0$ and every $t > 1$ we have*

$$\frac{W(st)}{W(s)} \leq C_2 t^\alpha$$

*(d) $W \in RB_\sigma^\tau$ with $\tau < p-1$.*

**Proof.** Indeed, with the notation (18.53) condition (18.54) is condition (18.48) (with $\varepsilon = \gamma(p-1)$) and condition (18.55) is condition (18.49); for (d) note that if $\tau(U) < 0$ as in Corollary 18.21, then $\tau(W) = p-1+\tau(U) < p-1$. ■

Note that for $p = 1$ condition (18.54) is also necessary for (HI) since it is equivalent with ($H_p$) is interpreted as *ess sup* $1/W(x)$).

Now compare (18.54) with condition (Bp), which is necessary and sufficient for (HI) for nonincreasing functions $F$. Write $W_1(t) = \int_0^t W(s)ds$; then it is easy to show that (Bp) is equivalent with

$$\int_0^\infty W_1(t)/t^{p+1} dt \asymp C\, W_1(s)/s^p$$

Indeed, the inequality $\leq$ is obtained from (Bp) by integration by parts. Now $W_1$ is monotone increasing, so that we have trivially the converse inequality.

**Corollary 18.24** ***Lemma 18.25*** *(a) Condition (Bp) is equivalent with the fact that* $W_1 \in RB_\sigma^\tau$ *with* $\tau < p$.
*(b) If* $W \in RB_\sigma^\tau$ *then* $W_1 \in RB_{\sigma+1}^{\tau+1}$, *and the converse is true only in the case* $W$ *is monotone.*

*References*

[1] Aljancic, A. and Arandjelovic, D. *O-regularly varying functions*, *Publ.Inst.Math.* (Beograd), 22 (36) 1977, 5-22.

[2] Andersen, K. *Weighted generalized Hardy inequalities for non-increasing functions*, *Can. J. Math.* 43 1991, 1121-1135.

[3] Arino, M.M. and Muckenhoupt, B. *Maximal functions on classical Lorentz spaces and Hardy's inequality with weights for nonincreasing functions*, *Trans. Amer. Math. Soc.* 320 1990, 727-735.

[4] Bingham, N.H., Goldie, C.M. and J.L. Teugels, *Regular Variation*, Cambridge University Press, Cambridge, 1987.

[5] Geluk, J.L. and De Haan, L., *Regular Variation, Extensions and Tauberian Theorems,* CWI Tract, 40, Stichting Mathematish Centrum, Centrum voor Wiskunde en Informatica, Amsterdam, 1987.

[6] Hardy, G., Littlewood, J. and Polya, G., *Inequalities*, Cambridge University Press, 1952.

[7] Muckenhoupt, B., *Hardy's inequality with weights*, *Studia Math.* 44 1972, 31-38.

[8] Sawyer, E. *Weighted inequalities for the two-dimensional Hardy operator*, *Studia Math.* 82 1985, 1-16.

[9] Seneta, E., *Regularly Varying Functions*, Lecture Notes in Mathematics 508, Springer, 1976.

# Chapter 19

# Extremal problems in generalized Sobolev classes

**Sergey K. Bagdasarov**

## 19.1 Introduction

### *19.1.1 General problem of sharp inequalities for intermediate derivatives*

Let $\mathbb{I}$ be either the entire line $\mathbb{R}$ or the positive half-line $\mathbb{R}_+$. Let also $p, s, q \in [1, +\infty)$, and $r,\ m \in \mathbb{N} : m < r$.

**Definition 19.1** *A function $f$ belongs to the class $W^r_{p,s}(\mathbb{I})$, if $f^{(r-1)}$ is absolutely continuous on any interval $[\sigma, \xi] \in \mathbb{I}$, and both norms $\|f\|_{L_p(\mathbb{I})}$ and $\|f^{(r)}\|_{L_s(\mathbb{I})}$ are finite.*

The first results concerning *inequalities for intermediate derivatives* of functions $f \in W^r_{p,s}(\mathbb{I})$ in the multiplicative form

$$\|f^{(m)}\|_{L_q(\mathbb{I})} \leq K\|f\|^{\alpha}_{L_p(\mathbb{I})}\|f^{(r)}\|^{\beta}_{L_s(\mathbb{I})}, \qquad (0 < m < r), \tag{19.1}$$

are due to E. Landau [30] and J. Hadamard [19], who constructed extremal functions in the sharp inequalities (19.1) in the case $m = 1$, $r = 2$, $p = q = s = \infty$ for $\mathbb{I} = \mathbb{R}_+$ and $\mathbb{I} = \mathbb{R}$, respectively.

V. N. Gabushin [17] described the exponents $\alpha$ and $\beta$ in the inequalities (19.1): if $\dfrac{r-m}{p} + \dfrac{m}{s} \geq \dfrac{r}{q}$, then $\alpha$ and $\beta$ can be determined uniquely:

$$\alpha = \frac{r - m - s^{-1} + q^{-1}}{r - s^{-1} + p^{-1}}, \qquad \beta = 1 - \alpha.$$

In the late 1930's G. E. Shilov [9] found the form of sharp inequalities (19.1) for $p = q = s = \infty$, $I = \mathbb{R}$, $2 \leq r \leq 5$, and formulated the hypothesis that proved to be true: *the set of extremal functions in the inequality (19.1) for $p = q = s = \infty$ and $=$ coincides with the set of functions of the form $f(t) = \gamma\phi_{\lambda,r}(t+\rho)$, $\gamma$, $\rho \in \mathbb{R}$, $\lambda \in \mathbb{R}_+$, where $\phi_{\lambda,r}$ is a $2\pi/\lambda$-periodic function such that $\phi^{(r)}_{\lambda,r}(t) = $* $\operatorname{sign}\sin(\lambda t))$. These functions had appeared in the works of J. Favard [14], N. I. Akhieser and M. G. Krein [1] and even earlier in Euler's investigations – sometimes they are referred to as the *Euler splines.*

In 1939 A. N. Kolmogorov [23] confirmed the Shilov's hypothesis and characterized the sharp constants in (19.1) in the case $p = q = s = \infty$, $\mathbb{I} = \mathbb{R}$ ::

$$\|f^{(m)}\|_{L_\infty(\mathbb{R})} \leq c_{rm} \|f\|_{L_\infty(\mathbb{R})}^{1-\frac{m}{r}} \|f^{(r)}\|_{L_\infty(\mathbb{R})}^{\frac{m}{r}}, \tag{19.2}$$

where $c_{rm} := K_{r-m}/K_m^{1-\frac{k}{r}}$, and $K_l := \frac{4}{\pi}\sum_{\nu=0}^{\infty} \dfrac{(-1)^{\nu(l+1)}}{(2\nu+1)^{l+1}}$, $l \in \mathbb{Z}_+$ are known as *the Favard constants.* A. S. Cavaretta [11], [13] suggested an elementary proof and refinements of the Kolmogorov inequalities (19.2).

The Kolmogorov's result led to the development of *the theory of sharp inequalities for intermediate derivatives on* $\mathbb{R}$ and $\mathbb{R}_+$. Since 1939, the complete solution of *the problem of sharp constants*

$$K = K_{p,q,s}^{r,m} = \sup_{f \in W_{p,s}^r(\mathbb{I}),\, f \neq 0} \left\| f^{(m)} \right\|_{L_q(\mathbb{I})} \|f\|_{L_p(\mathbb{I})}^{-\alpha} \left\| f^{(r)} \right\|_{L_s(\mathbb{I})}^{-\beta}$$

in (19.1) for all $m, r : 0 < m < r$, and some fixed constants $p, q, s$, has been obtained in three cases for the entire line $\mathbb{R}$: G. H. Hardy, J. Littlewood, G. Polya [20], $p = q = s = 2$; E. M. Stein [37], $p = q = s = 1$; L. V. Taikov [39], $q = \infty$, $p = s = 2$; and in two cases for the half-line $\mathbb{R}_+$: N. P. Kupcov [29], $p = q = s = 2$; V. N. Gabushin [18], $q = \infty$, $p = s = 2$. Besides, a series of results by B. Szek\"efalvi-Nagy[38], H. Cartan [10], V. V. Arestov [2], V. N. Gabushin [16] and G. G. Magaril-Il'yaev [31] deals with the problem of exact constants in the inequality (19.1) in special *partial* cases (i.e. for some fixed $m, r$ and $p, q, s$). The qualitative characterization of extremal function in (19.1) in various settings ($\mathbb{I} = \mathbb{R}_+$, $[0, 1]$) of the case $p = q = s = \infty$ is due to V. M. Tihomirov [40], A. Pinkus [35], S. Karlin [22], and A. S. Cavaretta, I. J. Schoenberg [12].

A comprehensive survey of Kolmogorov inequalities for various choices of $p, q, s$ and $r, m$ in (19.1) can be found in the commentary by V. M. Tihomirov and G. G. Magaril-Il'yaev in [24].

Due to the homogeneity of classes $W_{p,s}^r(\mathbb{I})$, it suffices to pose the problem of sharp inequalities (19.1) in the Sobolev classes of functions $f$ whose norms $\|f^{(r)}\|_{L_s(\mathbb{I})}$ are bounded by the unity.

**Definition 19.2** *Let* $\mathbb{I} = [a, b]$, $-\infty \leq a < b \leq +\infty$. *The Sobolev class* $W_\infty^n(\mathbb{I})$ *is defined as follows:*

$$W_\infty^n(\mathbb{I}) := \{ f \in W_{\infty,\infty}^n(\mathbb{I}) \mid \|f^{(n)}\|_{L_\infty(\mathbb{I})} \leq 1 \}, \qquad n \in \mathbb{N}. \tag{19.3}$$

Note that functions from the Sobolev class $f \in W_\infty^{r+1}(\mathbb{I})$ obey the constraint $\|f^{(r+1)}\|_{L_\infty(\mathbb{I})} \leq 1$. This inequality is equivalent to the restraint

$\omega\left(f^{(r)};t\right) \le t$, where $\omega(g;t) := \inf_{|x-y|\le t} |g(x)-g(y)|$ stands for the *modulus of continuity* of the continuous function $g(t)$.

In our generalizations, we consider constraints of the form $\omega(f^{(r)};t) \le \omega(t)$ for some fixed *concave modulus of continuity* $\omega$. This discussion leads us to the following setting of new problems and formulation of results in the theory of functional classes defined by a common majorizing modulus of continuity.

### *19.1.2 Functional classes $W^r H^\omega(\mathbb{I})$*

In Jackson inequalities the errors of approximation of an individual function $f \in C^r[a,b]$ by a finite dimensional subspace are expressed in terms of the modulus of continuity of the $r$-th derivative of the function $f$ (consult, e.g. [25],[27]) . Instead, S. M. Nikol'skii suggested to consider classes of functions with a common bounding concave modulus of continuity $\omega$.

**Definition 19.3** *A concave increasing function* $\omega : \mathbb{R}_+ \to \mathbb{R}_+$, *originating at zero* ($\omega(0)=0$), *is called a concave modulus of continuity.*

**Definition 19.4** *Let* $\omega$ *be a concave modulus of continuity. The functional class* $W^r H^\omega[a,b]$ *is introduced as follows:*

$$W^r H^\omega[a,b] := \{x \in C^r[a,b] \mid \omega(x^{(r)};t) \le \omega(t),\ t \in [0,b-a]\}. \tag{19.4}$$

*In the case* $r=0$ *we use the notation* $H^\omega[a,b] := W^0 H^\omega[a,b]$.

The standard Sobolev class can be viewed as a particular case of the class $W^r H^{\widetilde{\omega}}[a,b]$ with $\widetilde{\omega}(t)=t$. Another example is provided by the *Hölder modulii of continuity* $\omega_\alpha(t) = t^\alpha,\ 0<\alpha\le 1$. In this case, we sometimes use the notation

$$W^r H^\alpha[a,b] := W^r H^{\omega_\alpha}[a,b].$$

We will also use the subset of functions in $H^\omega[a,b]$ vanishing at a point $\tau \in [a,b]$:

$$H^\omega_\tau[a,b] := \{f \in H^\omega[a,b] \mid f(\tau)=0\}. \tag{19.5}$$

The classes $W^r H^\omega[a,b]$ were introduced in 1946 by S. M. Nikol'skii [34] in connection with approximation of functions by Fourier sums. Naturally, a number of problems arose concerning best characteristics of approximation of these classes by algebraic and trigonometric polynomials and other finite-dimensional subspaces.

An excellent summary of various results in the theory of classes $W^r H^\omega(\mathbb{I})$ up to 1985 and a rich bibliography are given by N. P. Korneichuk in his survey article [28] dedicated to the contribution of S. M. Nikol'skii to the Approximation Theory.

### 19.1.3 The Kolmogorov problem in $W^r H^\omega(\mathbb{I})$

The main objective of this paper is to describe extremal functions of the problem

$$\|f^{(m)}\|_{\infty()} \to \sup, \qquad f \in W^r H^\omega(\mathbb{I}), \quad \|f\|_{L_p(\mathbb{I})} \le B \tag{19.6}$$

where $0 \le m \le r$, if $1 \le p < \infty$, and $0 < m \le r$, if $p = \infty$.

The solution of all variations of (19.6) for $m = r$ and $\mathbb{I} = \mathbb{R}_+$ requires the description of the form and properties of extremal functions in the problem

$$\int_a^b h(t)\psi(t)\,dt \to \sup, \qquad h \in H_a^\omega[a,b], \tag{19.7}$$

for integrable kernels $\psi$ with a finite number of sign changes on $[a,b]$. The corresponding formulae for extremal functions in (19.7) are given in Theorem 19.6 in Section 19.2.

Relying on these results, in Section 19.3 we give a complete characterization of the structural properties of extremal function of the general problem (19.6). Finally, the special case $r = 1$ of the problem (19.6) is treated in Section 19.4. In particular, we expose a remarkable property of extremal functions $X_B$ of (19.6) in the case $r = m = 1$, $B > 0$, $\mathbb{I} = \mathbb{R}_+$, and Hölder modulii $\omega(t) = t^\alpha$, $0 < \alpha \le 1$. The collection of knots $\{\xi_i\}_{i=1}^\infty$ (points of sign changes of $X_B''$) constitutes a *geometric mesh*:

$$\frac{\xi_{i+1} - \xi_i}{\xi_i - \xi_{i-1}} = \gamma, \quad \gamma = \gamma(\alpha) \in (0,1), \qquad i \in \mathbb{N},$$

where $\xi_0 := 0$. In particular, the support of $X_B$ is a *finite* interval $[0,d]$. Moreover, the extremal function $X_B$ enjoys the fractal property of *self-similarity*: for $i \in$,

$$X_B(t + \xi_i) = (-1)^i \left(\frac{\xi_{i+1} - \xi_i}{\xi_1}\right)^{1+\alpha} X_B\left(\frac{\xi_1}{\xi_{i+1} - \xi_i} t\right), \qquad t \in [0, \xi_{i+1} - \xi_i].$$

This phenomenon for $\omega(t) = t$ and $p = 2$ was first discovered by A. T. Fuller [15], then by V. N. Gabushin [16], and later by G. G. Magaril-Il'yaev [31] for all $p \in [1, +\infty)$.

## 19.2 Maximization of integral functionals over $H^\omega[a,b]$

Our goal in this section is to introduce the reader to the notion of *perfect $\omega$-splines* as extremal functions of *linear integral functionals*. We also give a comprehensive list of various properties of $\omega$-splines used in our arguments.

### 19.2.1 *Simple kernels $\Psi(\cdot)$ and their rearrangements $\Re(\Psi;\cdot)$*

The Korneichuk lemma describes extremal functions of the functional

$$h \mapsto \int_a^b h(t)\psi(t)\,dt, \qquad h \in H^\omega[a,b], \tag{19.8}$$

where $\psi$ is the derivative of *a simple kernel* on $[a,b]$.

**Definition 19.5** *Let a kernel $\psi(\cdot) \in L_1[a,b]$ be such that $\int_a^b \psi(x)\,dx = 0$ and for some points $a', b' : a < a' \le b' < b$,*

$$\begin{array}{lll} (i) & \psi(x) < 0, & \text{for a.e. } x \in [a,a']; \\ (ii) & \psi(x) = 0, & \text{for a.e. } x \in [a',b']; \\ (iii) & \psi(x) > 0, & \text{for a.e. } x \in [b',b]. \end{array} \tag{19.9}$$

*Then $\Psi(x) = \xi \int_a^x \psi(t)\,dt$, $a \le x \le b$, $\xi \in \{1,-1\}$, is called a simple kernel.*

If $\Psi$ is a simple kernel, the equation $|\Psi(t)| = y$, for $0 < y < \|\Psi\|_{C[a,b]}$, has precisely two solutions: $\alpha_y \in (a,a')$ and $\beta_y \in (b',b)$ (see Figure 19.1). The value of the maximum of the functional (19.8) is expressed in terms of *the rearrangement of the simple kernel* $\Psi$.

**Definition 19.6** *Let $\Psi(x)$ be a simple kernel on $[a,b]$ introduced in Definition 19.5, and $c := \frac{1}{2}(a'+b')$. Let the function $\rho : [a,c] \longrightarrow [c,b]$ be derived from*

$$\begin{cases} \Psi(t) = \Psi(\rho(t)), & t \in [a,a'], \quad \rho(t) \in [b',b]; \\ \rho(t) = a' + b' - t, & t \in [a',c]. \end{cases} \tag{19.10}$$

*The decreasing rearrangement $\Re(\Psi;t)$ of the simple kernel $\Psi(t)$ is defined on the interval $[0, b-a]$ as follows:*

$$\Re(\Psi;t) := \begin{cases} \|\Psi\|_{C[a,b]}, & t \in [0, b'-a'], \\ |\Psi(y_t)|, & \begin{cases} t & \in (b'-a', b-a], \\ y_t & \in [a,a'] : \rho(y_t) - y_t = t. \end{cases} \end{cases} \tag{19.11}$$

Figure 19.1 also illustrates the graph of the decreasing rearrangement $\Re_\omega(\Psi;\cdot)$ of a simple kernel $\Psi$. Systematic exposition of various properties of rearrangements is given in the G. G. Hardy, G. E. Littlewood, G. Polya monograph [20] and in the A. Zygmund's book [41].

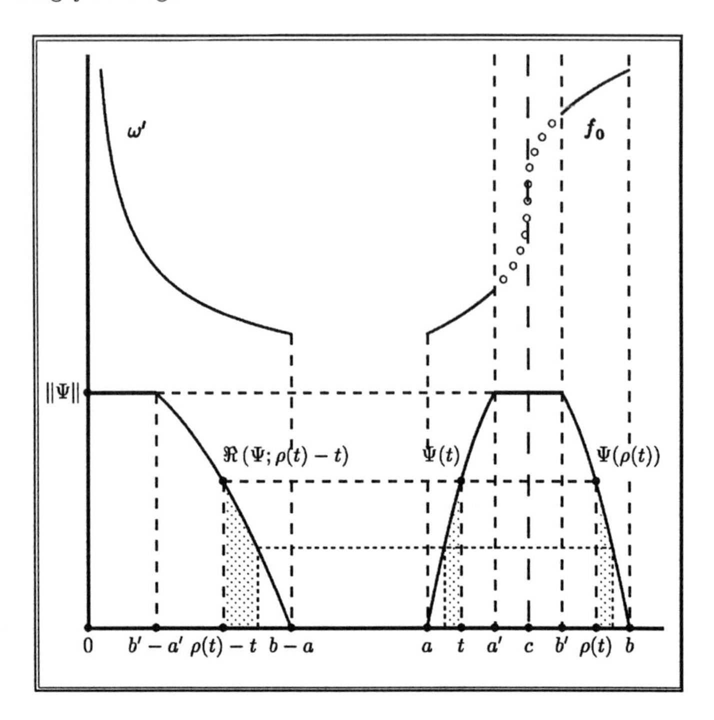


**FIGURE 19.1. Simple kernel $\Psi$ and its rearrangement $\mathcal{R}\Psi$.**

### *19.2.2 The Korneichuk lemma*

The following result, due to N. P. Korneichuk (cf. [27]), is crucial in the characterization of extremal functions of the functional (19.8) for kernels $\psi$ with a finite or countable set of points of sign change.

**Lemma 19.1** *Let $\Psi(t) := \int_a^t \psi(y)\,dy$, $a \le t \le b$, be a simple kernel with the derivative $\psi$ satisfying (19.9). Let $\omega(t)$ be a concave modulus of continuity. Then,*

$$\sup_{f\in H^\omega[a,b]} \int_a^b f(t)\psi(t)\,dt = \int_0^{b-a} \Re(\Psi;t)\,\omega'(t)\,dt, \tag{19.12}$$

*where $\rho(\cdot)$ and $\Re(\Psi;\cdot)$ are introduced by (19.10) and (19.11), respectively. The upper bound in (19.12) is attained on the functions with the derivative given by the formula*

$$\frac{d}{dx} f_0(x) = \begin{cases} \omega'(\rho(x)-x), & a \le x \le c, \\ \omega'(x-\rho^{-1}(x)), & c \le x \le b, \end{cases} \tag{19.13}$$

*where $c = (a'+b')/2$.*

By Definition 19.5, the kernel $\psi$ has the zero mean on $[a,b]$, so extremal functions of the functional (19.8) are determined up to an additive constant. Therefore,

$$\sup_{h\in H^\omega[a,b]} \int_a^b h(t)\psi(t)\,dt = \sup_{h\in H_a^\omega[a,b]} \int_a^b h(t)\psi(t)\,dt. \tag{19.14}$$

Moreover, it can be observed from (19.10) and (19.13) that the derivative $\frac{d}{dt}f_0(t)$ of the extremal function of the functional (19.8) is determined *uniquely* by (19.13) only on *the support* $[a,a'] \cup [b',b]$ of the kernel $\psi$. We illustrated this phenomenon on Figure 19.1 by graphing $f_0(t)$ with solid lines on the support of $\psi$ and by putting circles along the graph of $f_0$ on the *zero-interval* $[a',b']$ of $\psi$.

We mention one corollary from Lemma 19.1 clarifying the structure of the extremal function $f_0$.

**Theorem 19.2** *Let the function* $\frac{d}{dx}f_0(x)$ *be defined by* (19.13). *Then,* $f_0$ *has the full modulus of continuity on* $[0,b-a]: \omega(f_0;t)=\omega(t),\ 0\le t\le b-a$, *or, more precisely,*

$$f_0(\rho(t)) - f_0(t) = \omega(\rho(t)-t), \qquad 0\le t\le c. \tag{19.15}$$

**Proof.** By (19.13), for all $x$, $0\le x\le c := \frac{1}{2}(a'+b')$, we have

$$\begin{aligned} f_0(\rho(x)) - f_0(x) &= \int_c^{\rho(x)} \omega'(u-\rho^{-1}(u))\,du - \int_c^x \omega'(\rho(u)-u)\,du \\ &= \int_c^x \omega'(\rho(u)-u)\,d(\rho(u)-u) = \omega(\rho(x)-x). \end{aligned}$$

It remains to notice that the function $\rho(t)-t$ increases from $0$ to $b-a$, as $t$ decreases from $c$ to $0$. ■

### *19.2.3 Extremal functions of functionals over $H^\omega[a,b]$*

Throughout this section we fix an interval $[a,b]$, $-\infty < a < b \le +\infty$. Our objective in the section is to characterize extremal functions of the problem

$$\int_a^b h(t)\psi(t)\,dt \to \sup, \qquad h\in H_a^\omega[a,b], \tag{19.16}$$

for integrable kernels $\psi$ with a finite number or a countable set of points of sign change on $[a,b]$, accumulating to the endpoint $b$. In partucular, we give the solution of the problem (19.16) on the half-line $\mathbb{R}_+$.

We adopt the following notations.

**Notation 19.3** *Let $N_1 \in \mathbb{N}$, $N_2 \in \mathbb{N} \cup \{\infty\}$, and $\xi = \{\xi_i = \xi(i)\}_{i=N_1}^{N_2}$ be a collection of points. Then, $\alpha \triangleright \xi \blacktriangleright \beta$, if and only if*

$$\alpha = \xi(N_1) \le \xi(N_1+1) \le \cdots \le \xi(N_2) = \beta, \ \textit{for } N_2 \in \mathbb{N}$$
$$\alpha = \xi(N_1); \quad \xi(i) \le \xi(i+1), \ i \ge N_1; \ \lim_{i\to\infty} \xi(i) = \beta; \ \textit{for } N_2 = +\infty.$$

*Analogously, $\alpha \blacktriangleleft \xi \triangleleft \beta$, if and only if*

$$\beta = \xi(N_1) \ge \xi(N_1-1) \ge \cdots \ge \xi_{N_2} = \alpha, \qquad \textit{for } N_2 \in \mathbb{N};$$
$$\beta = \xi(N_1); \quad \xi(i) \ge \xi(i+1), \ i \ge N_1; \ \lim_{i\to\infty} \xi(i) = \alpha; \ \textit{for } N_2 = +\infty.$$

**Definition 19.7** *Let $[\alpha,\beta]$ be a finite interval, and $\psi \in L_1[\alpha,\beta]$. By the definition,*

$$\begin{aligned}
\operatorname{sign}\psi &= 1 \quad \textit{on} \quad [\alpha,\beta] \iff \operatorname{meas}\{t \in [\alpha,\beta] \,:\, \psi(t) > 0\} = \beta - \alpha;\\
\operatorname{sign}\psi &= -1 \quad \textit{on} \quad [\alpha,\beta] \iff \operatorname{sign}(-\psi) = 1 \quad \textit{on} \quad [\alpha,\beta].
\end{aligned}$$

**Notation 19.4** *The following notation will be used for intervals $I = [\gamma,\gamma]$ with coincident endpoints: $I = \boxdot$.*

**Notation 19.5** *Let $E \subset \mathbb{R}$. The function*

$$\chi(E;t) = \begin{cases} 1, & t \in E; \\ 0, & t \notin E; \end{cases} \tag{19.17}$$

*is called the indicator of the set $E$..*

**Definition 19.8** *Let $j \in \{-1,0,+1\}$, $n \in \mathbb{N} \cup \{\infty\}$, and $\psi \in L_1[a,b]$. Then, $\psi$ belongs to the class $\mathcal{M}_n^j[a,b]$ for $n \ge 2$, if and only if $\operatorname{sign} \int_a^b \psi(x)\,dx = j$, and there exist such $\alpha = \{\alpha_i\}_{i=0}^n$ that $\alpha_{i-1} < \alpha_i$, $i = 1,\dots,n$, $a \triangleright \alpha \blacktriangleright b$, and*

$$\operatorname{sign}\psi = (-1)^i \ \textit{on } [\alpha_{i-1},\alpha_i], \qquad i = 1,\dots,n.$$

*By the definition,*

$\mathcal{M}_1^l[a,b] := \varnothing$, $l = 0, 1$, $\mathcal{M}_1^{-1}[a,b] := \{\psi \in_1 [a,b] \mid \operatorname{sign}\psi = -1 \ \textit{on } [a,b]\}$.

*The class of kernels with $n-1$ sign changes is then*

$$\mathcal{M}_n[a,b] := \bigcup_{j=-1}^{1} \mathcal{M}_n^j[a,b]. \tag{19.18}$$

**Definition 19.9** *Let $N \in \mathbb{N} \cup \{\infty\}$.*

*The sets of indices $\{J_i(N)\}_{i \in \mathbb{N}}$, $\{L_i\}_{i \in \mathbb{N}}$ and $\mathcal{P}(N)$ are defined as follows.*

(1) *For $N = 1,\ 2,\ 3,\quad J_i(N) = L_i = \varnothing,\qquad i = 1, \dots, N.$*

(2) *For $N \geq 4$,*

$$J_i(N) = \varnothing,\quad i = N-2,\ N-1,\ N;\quad L_i = \varnothing,\quad l = 1,\ 2,\ 3;$$

$$J_i(N) = \{j = i + 1 + 2k,\ k \in \ \big|\ j \leq N\},\quad 1 \leq i \leq N-3;$$

$$L_i = \{l = i - 1 - 2k,\ k \in \ \big|\ l \geq 1\},\quad 4 \leq i \leq N;$$

(3) $\mathcal{P}(N) = \{(i, j) \in \times \ \big|\ 1 \leq i \leq N-3,\ j \in J_i(N)\ \}.$

The structure of extremal functions of the problem (19.16) will be characterized in terms of special partitions of the interval $[a, b]$.

**Definition 19.10** *Let $n \in \mathbb{N} \cup \{\infty\}$.*

*A partition $= \big(\{A_i,\ B_i,\ C_i,\ D_i\}_{i=1}^n;\ \{B_{ij},\ C_{ji}\}_{(i,j) \in \mathcal{P}(\mathbb{N})}\big)$ of the interval $[a, b]$ into subintervals is called a $V_n^j$-partition, $j \in \{-1, 0, 1\}$, if the following conditions are satisfied.*

(A) $C_i = [\gamma_{4i-4}, \gamma_{4i-3}];\ D_i = [\gamma_{4i-3}, \gamma_{4i-2}];$

$B_i = [\gamma_{4i-2}, \gamma_{4i-1}];\quad A_i = [\gamma_{4i-1}, \gamma_{4i}];$

*for $i = 1, \dots, n$ and such $\gamma = \{\gamma_i\}_{i=0}^{4n}$ that $a \rhd \gamma \blacktriangleright b$;*

(B) $C_i = \boxdot,\qquad i = 1,\ 2,\ 3;\quad B_i = \boxdot,\qquad i = n-2,\ n-1,\ n;$

($C_1$) $j = 0 \implies D_i = \boxdot,\qquad i = 1, \dots, n;$

($C_2$) $j = -1 \implies D_{2k} = \boxdot,\qquad k = 1, \dots, [n/2];$

($C_3$) $j = +1 \implies D_{2k-1} = \boxdot,\qquad k = 1, \dots, \lceil n/2 \rceil;$

(D) $B_i = \bigcup\limits_{j \in J_i(n)} B_{ij},\quad 1 \leq i \leq n-3,$ *where*

$B_{ij} = [\xi_i(\frac{j-i+1}{2}), \xi_i(\frac{j-i-1}{2})],\qquad j \in J_i(n),$

*for such $\xi_i = \{\xi_i(k)\}_{k=1}^{|J_i(n)|}$ that $\gamma_{4i-2} \blacktriangleleft \xi_i \lhd \gamma_{4i-1}$;*

(E) $C_i = \bigcup\limits_{l \in L_i} C_{il},\quad 4 \leq i \leq n,$ *where*

$C_{il} = [\varkappa_i(\frac{i-l-1}{2}), \varkappa_i(\frac{i-l+1}{2})],\qquad l \in L_i,$

*for such $\varkappa_i = \{\varkappa_i(k)\}_{k=1}^{|L_i|}$ that $\gamma_{4i-4} \rhd \varkappa_i \blacktriangleright \gamma_{4i-3}$.*

**Remark 19.1** *We list the atoms of $V_n^j$-partitions of the interval $[a, b]$ into the intervals $\{A_i,\ B_i,\ C_i,\ D_i\}_{i=1}^n$ in their natural order excluding the degenerated intervals $A_N$, $\{B_i\}_{i=N-2}^N$, $\{C_i\}_{i=1}^3$:*

$N = 2:\ D_1 A_1 D_2;$

$N = 3:\ D_1 A_1 D_2 A_2 D_3;$

$N = 4:\ D_1 B_1 A_1 D_2 A_2 D_3 A_3 C_4 D_4;$

$N = 5:\ D_1 B_1 A_1 D_2 B_2 A_2 D_3 A_3 C_4 D_4 A_4 C_5 D_5;$

$N = 6:\ D_1 B_1 A_1 D_2 B_2 A_2 D_3 B_3 A_3 C_4 D_4 A_4 C_5 D_5 A_5 C_6 D_6,$

$N \geq 7:\ D_1 B_1 A_1 D_2 B_2 A_2 D_3 B_3,\quad A_k C_{k+1} D_{k+1} B_{k+1},\quad 3 \leq k \leq N-4,$

$A_{N-3} C_{N-2} D_{N-2} A_{N-2} C_{N-1} D_{N-1} A_{N-1} C_N D_N.$

Figure 19.2 clarifies the order of atoms in a $V_8^0$-partition.

Given a $\psi \in \mathcal{M}_n^j[a, b]$, $j \in \{-1,\ 0 + 1\}$, we describe extremal functions of the problem (19.16). In particular, in Theorem 19.6 below we show that

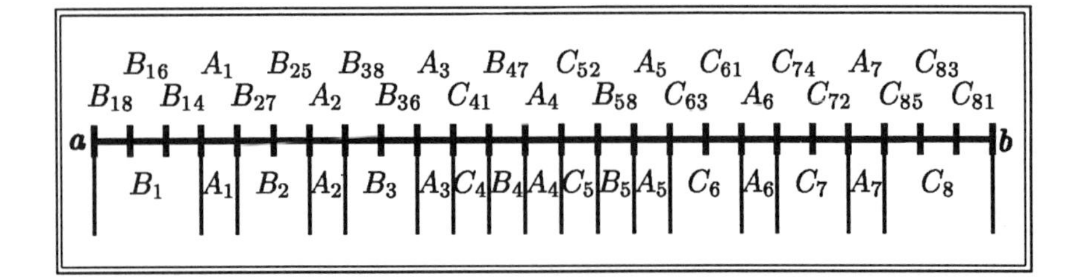


**FIGURE 19.2. The $V_8^0$ partition.**

*if $\psi \in \mathcal{M}_n^0[a,b]$ and $n \geq 2$, then the kernel $\Psi(t) = \int_a^t \psi(x)\,dx$ can be decomposed into the sum of simple kernels $\{\Phi_i(\cdot) = \Phi_i(\omega;\cdot)\}_{i=1}^{I_n}$ such that*

$$(i) \quad \Psi(t) = \sum_{i=1}^{I_n} \Phi_i(t), \qquad a \leq t \leq b;$$

$$(ii) \quad \sup_{h\in H^\omega[a,b]} \int_a^b h(t)\psi(t)\,dt = \sum_{i=1}^{I_n} \sup_{h\in H^\omega[a,b]} \int_a^b h(t)\Phi_i'(t)\,dt.$$

In the following theorem we describe the structure of extremal functions of the problem (19.16) called *perfect $\omega$-splines.*

**Theorem 19.6** *Let $\psi \in \mathcal{M}_n^j[a,b]$, $j \in \{-1,0,+1\}$, and $\{\alpha_i\}_{i=1}^{n-1}$ be the points of sign change of $\psi$ as in Definition 19.5. Then, there exist a solution $x_{\omega,\psi}$ of the problem (19.16) and a $V_n^j$-partition $\mathcal{V}$ of the interval $[a,b]$ with the following properties:*

$(A)$ $\alpha_i \in A_i, \quad i = 1,\dots,n-1;$

$(B)$ $\int_{B_{ij}\cup C_{ji}} \psi(t)\,dt = 0, \quad (i,j) \in \mathcal{P}(n);$

$(C)$ $\int_{A_i} \psi(t)\,dt = 0, \quad i = 1,\dots,n-1;$

$(D_1)$ $j = -1 \Rightarrow x_{\omega,\psi}(t) = -\omega(t-a),\ t \in D_{2k-1} \neq \Box,\ k = 1,\dots,\lceil n/2\rceil;$

$(D_2)$ $j = 1 \implies x_{\omega,\psi}(t) = \omega(t-a),\ t \in D_{2k} \neq \Box,\ k = 1,\dots,[n/2];$

$(E)$ *for $(i,j) \in \mathcal{P}(n)$, the function $x_{\omega,\psi}$ is extremal in the problem*

$$\int_a^b h(t)\psi_{ij}(t)\,dt \to \sup, \qquad h \in H^\omega[a,b], \tag{19.19}$$

$$\psi_{ij}(t) := \psi(t)\cdot\chi(B_{ij}\cup C_{ji};t), \qquad t \in [a,b]; \tag{19.20}$$

$(F)$ *for $i = 1,\dots,n-1$, the function $x_{\omega,\psi}$ is extremal in the problem*

$$\int_a^b h(t)\psi_i(t)\,dt \to \sup, \qquad h \in H^\omega[a,b], \tag{19.21}$$

$$\psi_i(t) := \psi(t) \cdot \chi(A_i; t), \qquad t \in [a, b]. \tag{19.22}$$

## *19.2.4 Structural properties of extremal functions* $x_{\omega,\psi}$

First of all, note that all kernels

$$\Psi_{ij}(t) = \int_a^t \psi_{ij}(y)\, dy, \quad (i, j) \in (n), \tag{19.23}$$

$$\Psi_i(t) = \int_a^t \psi_i(y)\, dy, \quad i = 1, \ldots, n-1, \tag{19.24}$$

are simple on their respective supports in the sense of Definition 19.5. Indeed, from the inclusions $\alpha_i \in A_i = [\gamma_{4i-1}, \gamma_{4i}]$, $i = 1, \ldots, n-1$, and the order of atoms in the $V_n^j$-partition, shown in Remark 19.1, it follows that

$$\operatorname{sign}\psi = \begin{cases} (-1)^i & \text{on } \begin{cases} B_{ij}, & (i,j) \in (n); \\ [\gamma_{4i-1}, \alpha_i], & i = 1, \ldots, n-1; \end{cases} \\ (-1)^{i+1} & \text{on } \begin{cases} C_{ji}, & (i,j) \in (n); \\ [\alpha_i, \gamma_{4i}], & i = 1, \ldots, n-1; \end{cases} \end{cases} \tag{19.25}$$

where $[\gamma_{4i-1}, \gamma_{4i}] := A_i$, $i = 1, \ldots, n-1$, and

$$\operatorname{sign}\psi(t) = \begin{cases} -1 & \text{on } D_{2i-1},\ i = 1, \ldots, \lceil n/2 \rceil; \\ 1 & \text{on } D_{2i},\ i = 1, \ldots, [n/2]. \end{cases} \tag{19.26}$$

Therefore, by (19.25) and the statements (B) and (C) of Theorem 19.6, each of the kernels $\{\Psi_{ij}\}_{(i,j)\in(n)}$ in (19.23) and $\{\Psi_i\}_{i=1}^{n-1}$ in (19.24) is simple. Then, the Korneichuk Lemma 19.1 provides us with the following formulas for the derivative $\frac{d}{dt}x_{\omega,\psi}(t)$:

$$\frac{d}{dt}x_{\omega,\psi}(t) = \begin{cases} (-1)^{i+1}\omega'(\rho_{ij}(t) - t), & t \in B_{ij}; \\ (-1)^{i+1}\omega'(t - \rho_{ij}^{-1}(t)), & t \in C_{ji}; \end{cases} \tag{19.27}$$

for all $(i, j) \in \mathcal{P}(n)$, where $\rho_{ij} : B_{ij} \to C_{ji}$ is determined from the equation

$$\Psi_{ij}(t) = \Psi_{ij}(\rho_{ij}(t)), \quad t \in B_{ij}, \quad \rho_{ij}(t) \in C_{ji}, \tag{19.28}$$

and

$$\frac{d}{dt}x_{\omega,\psi}(t) = \begin{cases} (-1)^{i+1}\omega'(\rho_i(t) - t), & t \in [\gamma_{4i-1}, \alpha_i]; \\ (-1)^{i+1}\omega'(t - \rho_i^{-1}(t)), & t \in [\alpha_i, \gamma_{4i}]; \end{cases} \tag{19.29}$$

where $\rho_i : [\gamma_{4i-1}, \alpha_i] \to [\alpha_i, \gamma_{4i}]$ is determined from the equation

$$\Psi_i(t) = \Psi_i(\rho_i(t)), \quad t \in [\gamma_{4i-1}, \alpha_i], \quad \rho_i(t) \in [\alpha_i, \gamma_{4i}]. \tag{19.30}$$

Schematic graphs of the extremal function $x_{\omega,\psi}$, $\psi \in \mathcal{M}_n^j[a, b]$ for various values of $n$ and $j$ are illustrated on Figures 19.3–19.5.

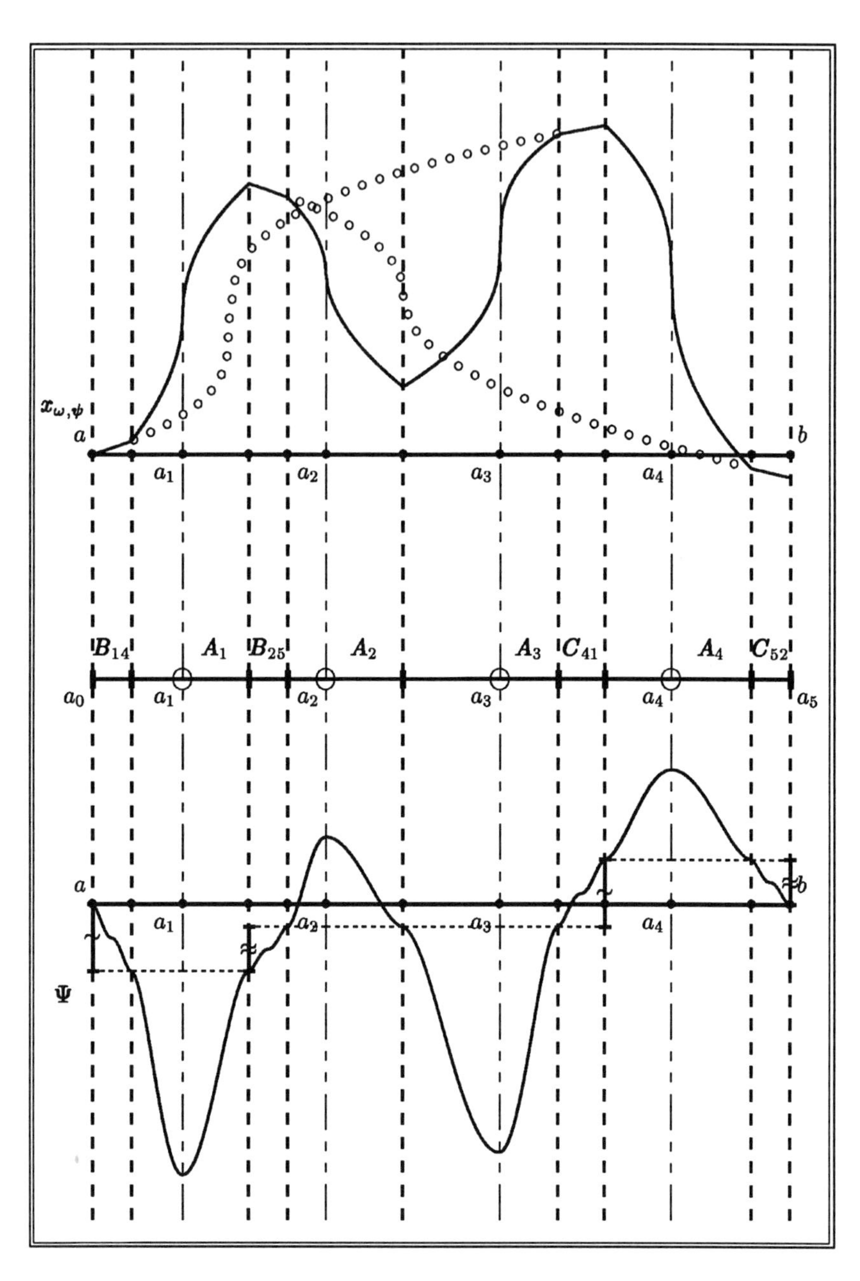


**FIGURE 19.3.** $V_5^0$ **partition, graphs of** $x_{\omega,\psi}$, $\Psi$ **for** $\psi \in M_5^0[a,b]$.

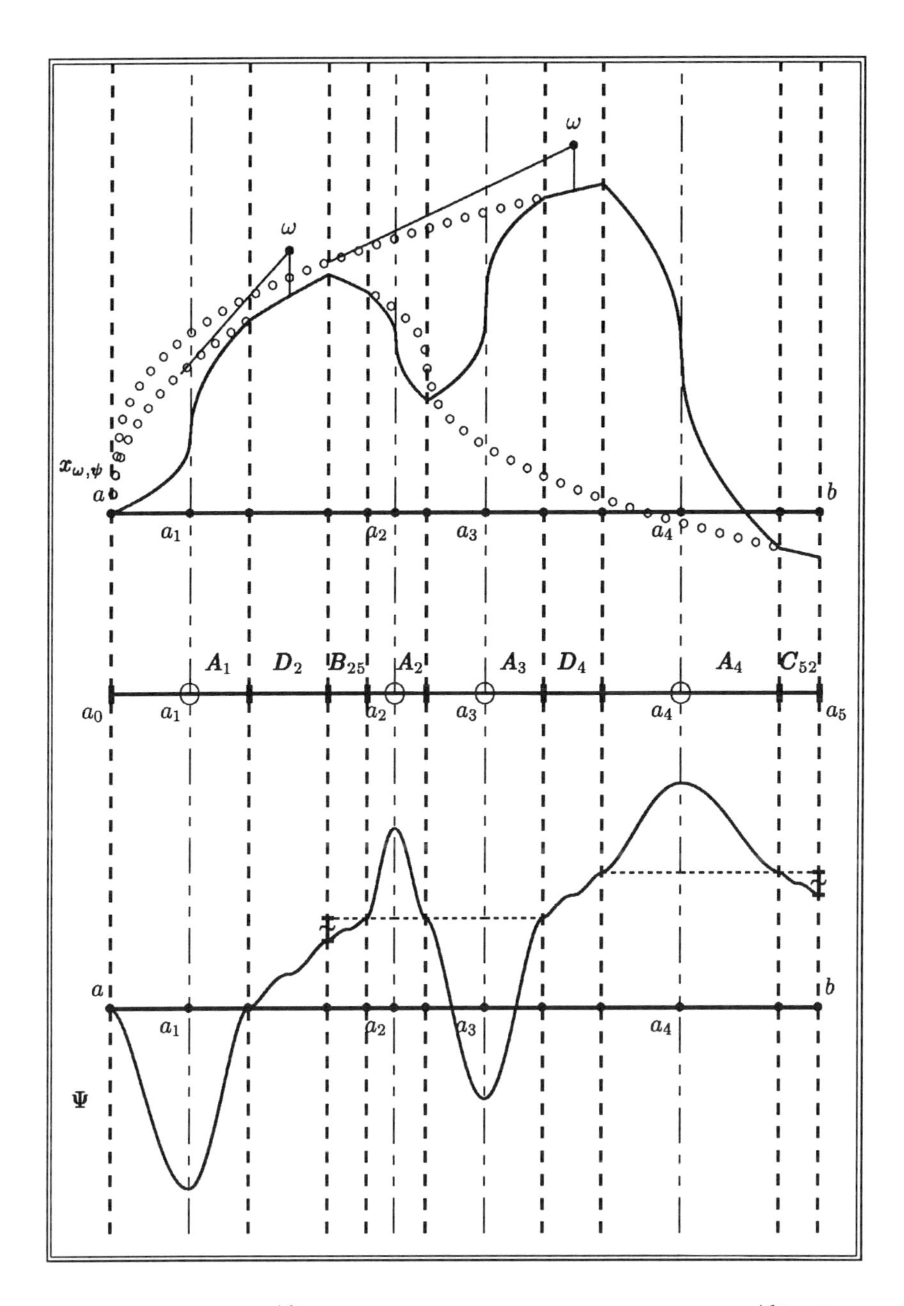


**FIGURE 19.4.** $V_5^{+1}$ **partition, graphs of** $x_{\omega,\psi}$, $\Psi$ **for** $\psi \in M_5^{+1}[a,b]$.

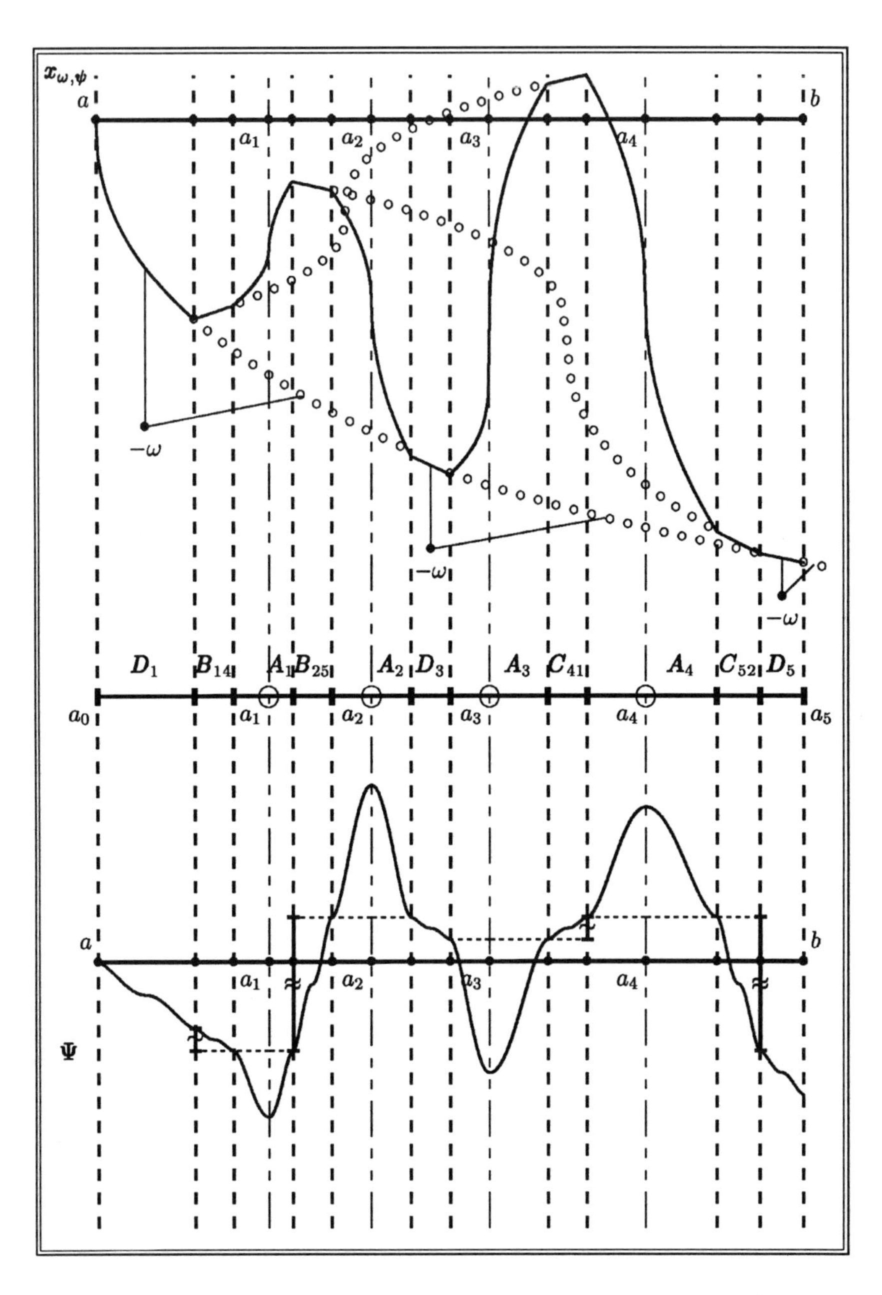


**FIGURE 19.5.** $V_5^{-1}$ **partition, graphs of** $x_{\omega,\psi}$, $\Psi$ **for** $\psi \in M_5^{-1}[a,b]$.

**Corollary 19.7** *The derivative of the extremal function $x_{\omega,\psi}$ of the problem* (19.16) *and the extremal $V_n^j$-partition $= (\omega, \psi)$ are unique.*

The $V_n^j$-partition $(\omega, \psi)$ from Theorem 19.6 is called *the extremal $V_n^j$-partition of the problem* (19.16).

**Remark 19.2** *By the formulas* (19.27)–(19.30),

$$\operatorname{sign} \frac{d}{dt} x_{\omega,\psi}(t) = \begin{cases} (-1)^{i+1}, & t \in A_i, \quad i = 1, \ldots, n-1; \\ (-1)^{i+1}, & t \in B_{ij},\ C_{ji}, \quad (i,j) \in P(n); \end{cases} \tag{19.31}$$

*while according to statements $(D_1)$, $(D_2)$ of Theorem X,*

$$\operatorname{sign} \frac{d}{dt} x_{\omega,\psi}(t) = \begin{cases} -1, & t \in D_{2i-1}, \quad i = 1, \ldots, \lceil n/2 \rceil; \\ 1, & t \in D_{2i}, \quad i = 1, \ldots, [n/2]. \end{cases} \tag{19.32}$$

Therefore, by (19.31) and (19.32), the function $\frac{d}{dt} x_{\omega,\psi}(t)$ can have at most $n-2$ sign changes on the interval $[a, b]$, if $\psi$ belongs to $\mathcal{M}_n^0[a, b]$ or $\mathcal{M}_n^{+1}[a, b]$, and $n-1$ sign changes on $[a, b]$, if $\psi \in \mathcal{M}_n^{-1}[a, b]$. These values are the upper bounds for the number of sign changes of $\frac{d}{dt} x_{\omega,\psi}(t)$, because some of the intervals of the extremal $V_n^j$-partition may degenerate into points. However, the following result shows that in the case of *a strictly concave* modulus of continuity $\omega$, the function $\frac{d}{dt} x_{\omega,\psi}(t)$ has the maximum possible number of sign changes.

**Corollary 19.8** *Let $\omega(\cdot)$ be a strictly concave modulus of continuity, and $\psi \in \mathcal{M}_n^j[a, b]$, $j \in \{-1, 0, +1\}$ and $\mathcal{V}$ be the extremal $V_n^j$-partition for the problem (19.16). Then,*

*(I)* $\quad A_i \neq \Box, \qquad i = 1, \ldots, n-1;$

*(II)* $\quad$ *if* $\psi \in \mathcal{M}_n^{-1}[a, b]$, *then* $D_1 \neq \Box$.

Corollary 19.8 in combination with relations (19.31) and (19.32) for $j = -1$ imply that the derivative $\frac{d}{dt} x_{\omega,\psi}(t)$ has precisely $n-2$ sign changes, if $\psi \in \mathcal{M}_n^j[a, b]$, $j = 0, 1$, and exactly $n-1$ sign changes, if $\psi \in \mathcal{M}_n^{-1}[a, b]$.

**Remark 19.3** *Remark* 19.2 *and Remark* 19.1 *on the order of atoms in $V_n^j$-partitions enable us to find the following relations between the points $\{z_i\}_{i=1}^{n-2}$ of sign change of $\frac{d}{dt} x_{\omega,\psi}(t)$ and the points $\{\alpha_i\}_{i=1}^{n-1}$ of sign change of the kernel $\psi \in \mathcal{M}_n^j[a, b]$, $j = 0,\ 1$:*

$$\alpha_i < z_i < \alpha_{i+1}, \qquad i = 1, \ldots, n-2. \tag{19.33}$$

*If* $\psi \in \mathcal{M}_n^{-1}[a,b]$ *and* $\{z_i\}_{i=1}^{n-1}$ *is a collection of sign changes of* $\frac{d}{dt}x_{\omega,\psi}(t)$, *then*

$$\alpha_{i-1} < z_i < a_i, \qquad i = 1, \dots, n-1. \tag{19.34}$$

The quantitative solution of the problem (19.16) is given in terms of *the extremal rearrangement* $\Re_\omega(\Psi;\cdot)$.

**Definition 19.11** *Let* $\psi \in \mathcal{M}_n[a,b]$ *and* $\Psi(t) = \int_b^t \psi(y)\,dy$, $t \in [a,b]$. *The extremal rearrangement* $\Re_\omega(\Psi;\cdot)$ *of the kernel* $\Psi$ *is defined as follows:*

$$\Re_\omega(\Psi;t) = \sum_{(i,j)\in\mathcal{P}(n)} \Re(\Psi_{ij};t) + \sum_{i=1}^{n-1} \Re(\Psi_i;t) + |\Psi(t+a)| \sum_{i=0}^{n} \chi(D_i;t), \tag{19.35}$$

*for* $t \in [0, b-a]$, *where the indicator function* $\chi(E;\cdot)$ *is introduced in* (19.17).

**Corollary 19.9** *Let the assumptions and notations be as in Theorem* 19.6. *Then,*

$$\sup_{h\in H_a^\omega[a,b]} \int_a^b h(t)\psi(t)\,dt = \int_0^{b-a} \Re_\omega(\Psi;x)\,\omega'(x)\,dx. \tag{19.36}$$

## 19.3 Kolmogorov problem for intermediate derivatives

All results of this section on the structure of extremal functions of the problem

$$\|f^{(m)}\|_{L_\infty(\mathbb{R}_+)} \to \sup, \qquad f \in W^r H^\omega(\mathbb{R}_+), \quad \|f\|_{L_p(\mathbb{R}_+)} \le B, \tag{19.37}$$

for $1 \le p < \infty$ are borrowed from [8]. The solution of the problem (19.37) in the case of $\omega(t) = t$ is due to G. G. Magaril-Il'yaev [32]. Various settings of the problem (19.37) for $p = \infty$ are treated in [5]–[7].

In the first two subsections of the section we establish the differentiation formula for functions from $f \in W^r H^\omega[0,d]$ with a finite $L_p$-norm. Then, we list sufficient conditions of extremality of a function $f \in W^r H^\omega[0,d]$ in the corresponding Kolmogorov–Landau problem

$$f^{(m)}(0) \to \sup, \qquad f \in W^r H^\omega[0,d], \quad \|f\|_{L_p[0,d]} \le B. \tag{19.38}$$

Relying on these conditions, expressed in the form of *operator equations,* we describe extremal functions and sharp additive inequalities of the problem (19.37). In conclusion, we give the corresponding exact *multiplicative* inequalities in Hölder classes $W^r H^\alpha(\mathbb{R}_+)$.

### *19.3.1 Differentiation formulae for $f^{(m)}(0)$, $0 \le m < r$*

Fix an interval $[0,d]$ and integers $r,\, m \in \mathbb{Z}_+ : 0 \le m \le r$.

Let $1 \le p < \infty$, and

$$p' = p'(p) = \begin{cases} \dfrac{p}{p-1}, & 1 < p < \infty, \\ \infty, & p = 1. \end{cases}$$

Fix a positive constant $B > 0$, a concave modulus of continuity $\omega$, and a function $f \in W^r H^\omega[0,d]$ such that $\|f\|_{L_p[0,d]} \le B$.

Let a function $Z \in W^r[0,d]$ be endowed with the properties

$$\begin{array}{lll} (i) & Z^{(i)}(d) = 0, & i = 0,\dots,r-1; \\ (ii) & Z^{(i)}(0) = 0, & i = 0,\dots,r-2-m, r-m,\dots,r-1; \\ (iii) & (-1)^{m+1} Z^{(r-1-m)}(0) > 0; & \\ (iv) & \int\limits_0^d Z(t)\,dt = 0. & \end{array} \tag{19.39}$$

Integrating by parts and using properties (19.39) of the function $Z$, we arrive at the formula for $f^{(m)}(0)$:

$$f^{(m)}(0) = \frac{(-1)^{m+1}}{Z^{(r-1-m)}(0)} \left[ \int\limits_0^d f(t) Z^{(r)}(t)\,dt + (-1)^{r+1} \int\limits_0^d f^{(r)}(t) Z(t)\,dt \right]. \tag{19.40}$$

Therefore, employing the Hölder inequality, we obtain the following estimate for the value of the $m$-th derivative of the function $f$ at the origin:

$$f^{(m)}(0) \le \frac{1}{|Z^{(r-1-m)}(0)|} \left[ \|Z^{(r)}\|_{L_{p'}[0,d]} B + \sup_{h \in H^\omega[0,d]} \int\limits_0^d h(t) Z(t)\,dt \right] \tag{19.41}$$

**Remark 19.4** *The property* (19.39) – $(iv)$ *of the function* $Z$ *assures that*

$$\sup_{h \in H^\omega[0,d]} \int\limits_0^d h(t) Z(t)\,dt = \sup_{h \in H_0^\omega[0,d]} \int\limits_0^d h(t) Z(t)\,dt. \tag{19.42}$$

### *19.3.2 Differentiation formula for $f^{(r)}(0)$*

Let a function $Z \in C^r[0,d]$ enjoy the properties

$$\begin{array}{lll} (i) & Z^{(i)}(0) = Z^{(i)}(d) = 0, & i = 0,\dots,r-1; \\ (ii) & \int\limits_0^d Z(t)\,dt > 0. & \end{array} \tag{19.43}$$

Integrating by parts and using properties (19.43) of the function $Z$, we derive the following formula for $f^{(r)}(0)$:

$$f^{(r)}(0) = \left( \int_0^d Z(t)\,dt \right)^{-1} \times$$
$$\times \left[ (-1)^r \int_0^d f(t) Z^{(r)}(t)\,dt - \int_0^d [f^{(r)}(t) - f^{(r)}(0)] Z(t)\,dt \right]$$

Therefore, employing the Hölder inequality, we obtain the estimate

$$|f^{(r)}(0)| \leq \left| \int_0^d Z(t)\,dt \right| \left[ \|Z^{(r)}\|_{L_{p'}[0,d]} B + \sup_{h \in H_0^\omega[0,d]} \int_0^d h(t) Z(t)\,dt \right] \tag{19.44}$$

### 19.3.3 *Sufficient conditions of extremality*

From the identities (19.40) for $m \in \{0, \dots, r-1\}$ and (??) for $m = r$ we derive the following *sufficient conditions of extremality* of the functions $X \in W^r H^\omega[0,d]$ and $Z \in C^r[0,d]$ in inequalities (19.41) for $0 \leq m < r$ and (19.44) for $m = r$ (transforming these inequalities into the equalities), respectively:

$$\begin{aligned}
&(i) && Z^{(r)}(t) = |X(t)|^{p-1} \operatorname{sign} X(t), \qquad t \in [0,d]; \\
&(ii) && \int_0^d [X^{(r)}(t) - X^{(r)}(0)] Z(t)\,dt = \sup_{h \in H_0^\omega[0,d]} \int_0^d h(t) Z(t)\,dt; \\
&(iii) && \|X\|_{L_p[0,d]} = B.
\end{aligned} \tag{19.45}$$

**Remark 19.5** *The equation* (19.45)-($i$) *represents the case of the equality in the Hölder inequality*

$$\int_0^d f(t) Z^{(r)}(t)\,dt \leq \|f\|_{p[0,d]} \|Z^{(r)}\|_{p'[0,d]}. \tag{19.46}$$

### 19.3.4 *Extremality conditions in the form of an operator equation*

We introduce the *differentiation operator* $D_r : C^{\,r}[0,d] \to C[0,d]$,

$$(D_r\, x)(t) := x^{(r)}(t) - x^{(r)}(0), \qquad x \in C^r[0,d], \tag{19.47}$$

the *Hölder operator* $\mathcal{H}\colon [0,d] \to [0,d]$,

$$\mathcal{H}(x)(t) := |x(t)|^{p-1} \operatorname{sign} x(t), \qquad x \in C[0,d], \tag{19.48}$$

the *integration operator* $I_r : C[0,d] \to C^r[0,d]$,

$$(I_r\, x)(t) := \frac{(-1)^r}{r!} \int_0^d (y-t)_+^{r-1} x(y)\, dy, \qquad x \in C[0,d], \tag{19.49}$$

and the operator $M$ from the set $\bigcup_{n \in \mathbb{N} \cup \{\infty\}} \mathcal{M}_n[0,d]$ (consult 19.18) of integrable functions on $[0,d]$ with a finite or countable ordered set of points of sign change to the functional class $H_0^\omega[0,d]$:

$$(M\, x)(t) := z(t), \qquad x \in \mathcal{M}_n[0,d], \tag{19.50}$$

where the function $z \in H_0^\omega[0,d]$ is uniquely defined by the property

$$\int_0^d z(t)x(t)\, dt = \sup_{h \in H_0^\omega[0,d]} \int_0^d h(t)x(t)\, dt. \tag{19.51}$$

The structure of the extremal function $z(t)$ in (19.51) is described in Theorem 19.6. We also introduce the mapping $\Gamma : C^r[0,d] \to C^{r-1}$: for $x \in C^r[0,d]$,

$$\Gamma\, x = \left( x'(0), x''(0), \ldots, x^{(r-m-2)}(0), x^{(r-m)}(0), \ldots, x^{(r)}(0) \right). \tag{19.52}$$

Then, the conditions (19.45) of extremality of a function $X(\cdot)$ in the Kolmogorov-Landau problem (19.38) can be reformulated as follows.

**Theorem 19.10** *The function $X \in W^r H^\omega[0,d]$ is extremal in the problem (19.38), if $\|X\|_{L_p[0,d]} = B$, and $X(\cdot)$ satisfies the operator equation*

$$[D_r - M \circ I_r \circ]\, X = 0 \tag{19.53}$$

*with the zero boundary conditions*

$$\Gamma\left((I_r \circ) X\right) = 0. \tag{19.54}$$

### *19.3.5 Sharp additive inequalities for intermediate derivatives*

The following result describes functions $X_n \in W^r H^\omega[0,d_n]$ with $n$ simple zeroes on $[0,d_n]$, extremal in the problem (19.38) for $d = d_n$.

**Theorem 19.11** *Let* $r, m \in \mathbb{N} : 0 \le m \le r$; $n \in \mathbb{N} : n \ge r$, $B > 0$, $1 \le p < +\infty$, *and* $\omega$ *be a modulus of continuity. There exists a positive number* $d_n = d_n(\omega, B, r, m, p)$ *and a function* $X_n = X_{n,\omega,B,r,m,p} \in W^r H^\omega[0, d_n]$ *endowed with the following properties:*

(1) $X_n$ *has precisely* $n$ *simple zeroes* $\{t_i = t_i(\omega, B, r, m, n)\}_{i=1}^{n}$ *on* $[0, d_n]$;

(2) $X_n^{(r)}$ *exhibits exactly* $n - r$ *simple zeroes on* $[0, d_n]$, *if* $0 \le m < r$, *and* $n - r + 1$ *simple zeroes on* $[0, d_n]$, *if* $m = r$;

(3) $\|X_n\|_{L_p[0,d_n]} = B$;

(4) $X_n$ *is a solution of the equation* (19.53) *satisfying the boundary conditions* (19.54).

In [8] we apply the limiting procedure to the sequence $\{X_n\}_{n\in\mathbb{N}}$ to construct the extremal function in the problem (19.37).

**Theorem 19.12** *There exists a function* $Z(t) = Z_{\omega,r,m,B,p}(t)$, $t \in \mathbb{R}_+$, *satisfying conditions* (19.39) *in the case* $0 \le m < r$ *and* (19.43) *in the case of* $m = r$ *for* $d = +\infty$, *and the function* $X = X_{\omega,r,m,B,p} \in W^r H^\omega(\mathbb{R}_+)$ *endowed with the properties* (19.45) *for* $d = +\infty$.

**Remark 19.6** *If* $d = +\infty$ *the equalities* $Z^{(i)}(d) = 0$, $i = 0, \dots, r-1$, *are understood in the sense that* $\lim_{t\to+\infty} Z^{(i)}(t) = 0$, $i = 0, \dots, r-1$.

In [10] we also established the differentiation formulae

$$f^{(m)}(0) = \alpha(Z) \left[ \int_{\mathbb{R}_+} f(t) Z^{(r)}(t)\, dt + (-1)^{r+1} \int_{\mathbb{R}_+} f^{(r)}(t) Z(t)\, dt \right]. \quad (19.55)$$

for $0 \le m < r$, where $\alpha(Z) := \left(Z^{(r-1-m)}(0)\right)^{-1}$, and

$$f^{(r)}(0) = \beta(Z) \left[ (-1)^r \int_{\mathbb{R}_+} f(t) Z^{(r)}(t)\, dt - \int_{\mathbb{R}_+} [f^{(r)}(t) - f^{(r)}(0)] Z(t)\, dt \right], \quad (19.56)$$

where $\beta(Z) := \left( \int_0^d Z(t)\, dt \right)^{-1}$.

Therefore, we have the following estimates for the norm $\|f^{(m)}\|_{L_\infty(\mathbb{R}_+)}$ of the intermediate derivative:

$$\|f^{(m)}\|_{L_\infty(\mathbb{R}_+)} \le |\alpha(Z)| \left[ \|Z^{(r)}\|_{L_{p'}(+)} B + \sup_{h\in H^\omega(\mathbb{R}_+)} \int_{\mathbb{R}_+} h(t) Z(t)\, dt \right], \quad (19.57)$$

if $0 \le m < r$ and

$$\|f^{(r)}\|_{L_\infty(\mathbb{R}_+)} \le |\beta(Z)| \left[ \|Z^{(r)}\|_{L_{p'}(\mathbb{R}_+)} B + \sup_{h \in H_0^\omega(\mathbb{R}_+)} \int_{\mathbb{R}_+} h(t) Z(t)\, dt \right] \tag{19.58}$$

Properties of functions $X = X_{\omega,r,m,B,p}$ and $Z = Z_{\omega,r,m,B,p}$ guarantee that inequalities (19.57) and (19.58) are sharp.

### *19.3.6 Kolmogorov problem in Hölder classes*

Fix $r \in$, $m \in \mathbb{Z}_+ : 0 \le m < r$, and $\alpha \in (0,1]$, $p \in [1, +\infty)$.

Let $\widehat{X} := X_{\omega_\alpha,r,m,1,p}$ be the extremal function of the problem (K) for $\omega(t) = \omega_\alpha(t) = t^\alpha$ and $B = 1$. Let also $\widehat{Z}$ be the corresponding function $Z_{\omega_\alpha,r,m,1}$. We introduce the constants $P = P_{\alpha,r,m,p}$, $Q = Q_{\alpha,r,m,p}$ and $S = S_{\alpha,r,m,p}$:

$$P = \begin{cases} \|Z^{(r)}\|_{L_{p'}(+)} \cdot |\widehat{Z}^{r-1-m}(0)|^{-1}; \\ \|Z^{(r)}\|_{L_{p'}(+)} \cdot \left| \int_+ \widehat{Z}(t)\, dt \right|; \end{cases} \qquad Q = \int_{\mathbb{R}_+} \Re_{\omega_\alpha}(\widehat{Z}; t) \omega_\alpha'(t)\, dt;$$

$$S := (r + \alpha + p^{-1}) \left( \frac{P}{r + \alpha - m} \right)^{\frac{r+\alpha-m}{r+\alpha+p^{-1}}} \left( \frac{Q}{m + p^{-1}} \right)^{\frac{m+p^{-1}}{r+\alpha+p^{-1}}}. \tag{19.59}$$

The sharp *multiplicative* inequalities are characterized in the following corollary of Theorem 19.12.

**Corollary 19.13** *If $f \in W^r H^\alpha(\mathbb{R}_+)$, then*

$$\|f^{(m)}\|_{L_\infty(\mathbb{R}_+)} \le S \|f\|_{L_p(\mathbb{R}_+)}^{\frac{r+\alpha-m}{r+\alpha+p^{-1}}}, \tag{19.60}$$

*where the constant $S = S_{\alpha,r,m,p}$ is introduced in* (??), (19.59).

## 19.4 Kolmogorov problem in $W^1 H^\omega(\mathbb{R}_+)$ and $W^1 H^\omega(\mathbb{R})$

### *19.4.1 Preliminary remarks*

In this section we describe the structure of extremal functions of the Kolmogorov problem in $W^1 H^\omega(\mathbb{R})$ and $W^1 H^\omega(\mathbb{R}_+)$:

$$\|f^{(m)}\|_{L_\infty(\mathbb{I})} \to \sup, \qquad f \in W^1 H^\omega(\mathbb{I}), \quad \|f\|_{L_p(\mathbb{I})} \le B, \tag{19.61}$$

where $m = 0, 1,\ 1 \le p < +\infty$, and $\mathbb{I} = \mathbb{R} \vee \mathbb{R}_+$. The problem (19.61) in the case of $\omega(t) = t$ was solved by V. N. Gabushin [16] and G. G. Magaril-Il'yaev [31].

Let a function $\mathcal{Z}(t) = \mathcal{Z}_{\omega,m,B,p,}$ be endowed with the properties

$$\begin{array}{ll} (i) & \mathcal{Z}^{(m)}(0) = 0, \qquad (-1)^{m+1}\mathcal{Z}^{1-m}(0) > 0; \\ (ii) & \lim_{t\to\infty} \mathcal{Z}(t) = \lim_{t\to\infty} \mathcal{Z}'(t) = 0. \end{array} \tag{19.62}$$

In [8] we established the existence of functions $X \in W^1H^\omega(\mathcal{I})$ and $\mathcal{Z}$ satisfing conditions (19.62) and exhibiting the following properties:

$$\begin{array}{ll} (i) & \mathcal{Z}''(t) = |X(t)|^{p-1} \operatorname{sign} X(t), \qquad t \in \mathbb{I}; \\ (ii) & \int\limits_{\mathbb{I}} [X'(t) - X'(0)]'(t)\, dt = \sup\limits_{h\in H_0^\omega(\mathbb{I})} \int\limits_{\mathbb{I}} h(t)'(t)\, dt; \\ (iii) & \|X\|_{L_p(\mathbb{I})} = B. \end{array} \tag{19.63}$$

**Remark 19.7** *In the case* $\mathbb{I} = \mathbb{R}_+$ *the function* $\mathcal{Z}(x)$ *is the indefinite integral* $\int\limits_{+\infty}^{x} Z(t)\, dt$ *of the function* $Z(t)$ *from Theorem* 19.12 *for* $r = 1$.

In the following four subsections of this section we examine specific features of the functions $\mathcal{Z}$ and $X$ in each of the possible cases $m \in \{0, 1\}$ and $\mathbb{I} = \{\mathbb{R}, \mathbb{R}_+\}$.

### *19.4.2 Maximization of the norm $\|f\|_{L_\infty(\mathbb{R}_+)}$*

First, we describe the structure of extremal functions of the problem

$$\|f\|_{L_\infty(\mathbb{R}_+)} \to \sup, \qquad f \in W^1H^\omega(\mathbb{R}_+), \quad \|f\|_{L_p(\mathbb{R}_+)} \le B. \tag{19.64}$$

In addition to enjoying the properties (19.62), (19.63), the function $\mathcal{Z}(t)$ satisfies the following conditions:

$$\begin{array}{ll} \text{(i)} & \mathcal{Z} \text{ vanishes at } \{\xi_i\}_{i=0}^\infty : 0 = \xi_0 < \xi_1 < \cdots; \\ \text{(ii)} & \mathcal{Z}' \text{ has one zero } \eta_i \text{ on } [\xi_i, \xi_{i+1}] \text{ for each } i \in \mathbb{Z}_+. \end{array} \tag{19.65}$$

By (19.65)-(ii), each of the restrictions $\mathcal{Z}\big|_{[\xi_i,\xi_{i+1}]}$ is *a simple kernel* on $[\xi_i, \xi_{i+1}]$ for $i \in \mathbb{Z}_+$.

The following differentiation formula holds for all $f \in W^1H^\omega(\mathbb{R}_+)$ with $\|f\|_{L_p(\mathbb{R}_+)} \le B$:

$$f(0) = -\mathcal{Z}'(0)^{-1} \left( \int\limits_0^\infty f(x)\mathcal{Z}''(x)\, dx + \int\limits_0^\infty f'(x)\mathcal{Z}'(x)\, dx \right). \tag{19.66}$$

Consequently,

$$|f(0)| \leq |\mathcal{Z}'(0)|^{-1} \left( \|\mathcal{Z}''\|_{L_{p'}(\mathbb{R}_+)} B + \int_+ \Re_\omega(\mathcal{Z};t)\omega'(t)\,dt \right). \tag{19.67}$$

We showed in [8] that the lengths of the intervals $\{[\xi_{i-1}, \xi_i]\}_{i\in\mathbb{N}}$ decrease:

$$\xi_i - \xi_{i-1} > \xi_{i+1} - \xi_i, \qquad i \in \mathbb{N}. \tag{19.68}$$

The inequalities (19.68) imply (cf. [7]) that for all $i \in \mathbb{N}$ the restriction $X'\big|_{[\xi_{i-1},\xi_i]}$ is extremal in the problem

$$\int_{\xi_{i-1}}^{\xi_i} h(t)\mathcal{Z}'(t)\,dt \to \sup, \qquad h \in H^\omega[\xi_{i-1}, \xi_i], \tag{19.69}$$

and the extremal rearrangement $\Re_\omega(\mathcal{Z};t)$ is the sum of rearrangements of simple kernels $\mathcal{Z}\big|_{[\xi_{i-1},\xi_i]}$:

$$\Re_\omega(\mathcal{Z};t) := \sum_{i=1}^{\infty} \Re\left(\mathcal{Z}\big|_{[\xi_{i-1},\xi_i]};t\right). \tag{19.70}$$

The Korneichuk's Lemma 19.1 gives the formulas for the derivative of the extremal function $X$:

$$X'(t) = \begin{cases} (-1)^{j+1}\omega'(\rho_i(t) - t), & \xi_j \leq t < \eta_j, \\ (-1)^{j+1}\omega'(t - \rho_i^{-1}(t)), & \eta_j < t \leq \xi_{j+1}, \end{cases} \quad i \in \mathbb{Z}_+, \tag{19.71}$$

where $\rho_j : [\xi_j, \eta_j] \to [\eta_j, \xi_{j+1}]$ is defined by the equations

$$\mathcal{Z}(x) = \mathcal{Z}(\rho_j(x)), \qquad x \in [\xi_j, \eta_j], \quad \rho_j \in [\eta_j, \xi_{j+1}]. \tag{19.72}$$

For any $j \in \mathbb{N}$ we introduce functions $X_j(t)$ and $\mathcal{Z}_j(t)$ by the formulae

$$X_j(t) = X(t + \xi_j), \qquad \mathcal{Z}_j(t) = \mathcal{Z}(t + \xi_j), \qquad t \in \mathbb{R}_+. \tag{19.73}$$

For $j \in \mathbb{N}$, set also

$$B_j := \|X_j\|_{L_p(\mathbb{R}_+)}; \qquad \eta_i(j) = \eta_{i+j}, \quad \xi_i(j) = \xi_{i+j}, \quad i \in \mathbb{Z}_+, \tag{19.74}$$

where $\{\eta_i\}_{i\in\mathbb{N}}$ and $\{\xi_i\}_{i\in\mathbb{N}}$ are the points of sign change of the functions $\mathcal{Z}'$ and $\mathcal{Z}$, respectively.

The reference to the properties (19.62), (19.63) and (19.65), (19.66) of $X$ and $\mathcal{Z}$ reveals that functions $X_j$ and $\mathcal{Z}_j$ enjoy all properties of the functions $X$ and $\mathcal{Z}$ *with respect to the collections of points* $\{\eta_i\}_{i\in\mathbb{N}}$ and $\{\xi_i\}_{i\in\mathbb{N}}$. Therefore, for each $j \in \mathbb{N}$, the function $X_j$ solves the problem

$$\|f\|_{L_\infty(\mathbb{R}_+)} \to \sup, \qquad f \in W^r H^\omega(\mathbb{R}_+), \quad \|f\|_{L_p(\mathbb{R}_+)} \leq B_j. \tag{19.75}$$

### 19.4.3 Extremal functions in Hölder classes $W^1H^\alpha(\mathbb{R}_+)$

Let us consider the variant of the problem (19.64) in the particular case of the Hölder modulus of continuity $\omega_\alpha(t) = t^\alpha$ for $\alpha \in (0,1]$:

$$\|f\|_{L_\infty(\mathbb{R}_+)} \to \sup, \qquad f \in W^1H^\alpha(\mathbb{R}_+), \quad \|f\|_{L_p(\mathbb{R}_+)} \le B. \tag{19.76}$$

Let $\mathcal{R}$ be the extremal function in (19.76) for $B = 1$, and $\mathcal{R}_j$ be the shifts $\mathcal{R}(\xi_j + \cdot)$. then, the solution $R_B$ of the problem (19.76) is a dilated and rescaled version of $\mathcal{R}$ :

$$R_B(t) = B^{\frac{1+\alpha}{2+\alpha}} \mathcal{R}(B^{-\frac{1}{2+\alpha}} t), \ t \in \mathbb{R}_+. \tag{19.77}$$

Moreover, we showed in [8] that

$$\mathcal{R}(t) = (-1)^j B^{\frac{1+\alpha}{2+\alpha}} \mathcal{R}(B^{-\frac{1}{2+\alpha}} (t + \xi_j), \ t \in \mathbb{R}_+. \tag{19.78}$$

This property of the function $\mathcal{R}$ immediately implies that the points $\{\xi_i\}_{i\in\mathbb{N}}$ constitute a *geometric mesh:*

$$\frac{\xi_{j+1} - \xi_j}{\xi_j - \xi_{j-1}} = \gamma(\alpha), \quad j \in \mathbb{N}, \tag{19.79}$$

for some constant $\gamma = \gamma(\alpha) \in [0,1]$. In particular, (19.79) implies that $\mathcal{R}$ has compact support.

Consequently, in the Hölder class $W^1H^\alpha(\mathbb{R}_+)$ we have the familiar variant of *the self-similarity property*: for all $i \in$ and $0 \le t \le \xi_{i+1} - \xi_i$,

$$\mathcal{R}\big|_{[\xi_i,\xi_{i+1}]}(t + \xi_i) = (-1)^i \left(\frac{\xi_{i+1} - \xi_i}{\xi_1}\right)^{1+\alpha} \mathcal{R}\big|_{[0,\xi_1]} \left(\frac{\xi_1}{\xi_{i+1} - \xi_i} t\right). \tag{19.80}$$

### 19.4.4 Maximization of the norm $\|f'\|_{L_\infty(\mathbb{R}_+)}$

In this section we examine the structure of extremal functions of the problem

$$\|f'\|_{L_\infty(\mathbb{R}_+)} \to \sup, \qquad f \in W^1H^\omega(\mathbb{R}_+), \quad \|f\|_{L_p(\mathbb{R}_+)} \le B. \tag{19.81}$$

In addition to (19.62) and (19.63), the function $\mathcal{Z}(t)$ exhibits the following properties:

$$\begin{array}{ll} \text{(i)} & \mathcal{Z} \text{ vanishes at } \{\xi_i\}_{i=0}^\infty : 0 = \xi_0 < \xi_1 < \cdots ; \\ \text{(ii)} & \mathcal{Z}' \text{ has one zero } \eta_i \text{ on } [\xi_i, \xi_{i+1}] \text{ for each } i \in \mathbb{Z}_+. \end{array} \tag{19.82}$$

By (19.82), each of the restrictions $\mathcal{Z}\big|_{[\xi_i,\xi_{i+1}]}$ is a *simple kernel* on $[\xi_i, \xi_{i+1}]$ for $i \in \mathbb{N}$, while $\mathcal{Z}$ is monotone and positive on $[0, \xi_1)$.

The following differentiation formula holds for any function $f \in W^1H^\omega(\mathbb{R}_+)$ : $\|f\|_{L_p(\mathbb{R}_+)} \le B$:

$$f'(0) = \mathcal{Z}(0)^{-1}\left(\int\limits_{+} f(x)\mathcal{Z}''(x)\,dx + \int\limits_{+}[f'(x) - f'(0)]\mathcal{Z}'(x)\,dx\right). \quad (19.83)$$

Consequently,

$$\|f'\|_{l_\infty(\mathbb{R}_+)} \le |\mathcal{Z}(0)|^{-1}\left(\|\mathcal{Z}''\|_{L_{p'}(\mathbb{R}_+)}B + \int\limits_{\mathbb{R}_+}\Re_\omega(\mathcal{Z};t)\omega'(t)\,dt\right). \quad (19.84)$$

We also showed that the lengths of the intervals $\{[\xi_{i-1},\xi_i]\}_{i\in\mathbb{N}}$ decrease:

$$\xi_i - \xi_{i-1} > \xi_{i+1} - \xi_i, \qquad i \in \mathbb{N}. \quad (19.85)$$

Therefore, for $i \in$ the restriction $X'\big|_{[\xi_i,\xi_{i+1}]}$ is extremal in the problem

$$\int\limits_{\xi_i}^{\xi_{i+1}} h(t)\mathcal{Z}'(t)\,dt \to \sup, \qquad h \in H^\omega[\xi_i,\xi_{i+1}], \quad (19.86)$$

while the restriction $X'\big|_{[0,\xi_1]}$ is given by the formula

$$X'(t) = X'(0) - \omega(t), \qquad t \in [0,\xi_1]. \quad (19.87)$$

In accordance with (19.85), the extremal rearrangement $\Re_\omega(\mathcal{Z};t)$ of the kernel $\mathcal{Z}$ has support $[0,\xi_1]$ and is given by the formula

$$\Re_\omega(\mathcal{Z};t) = \mathcal{Z}(t)\chi|_{[0,\xi_1]} + \sum_{i=1}^{\infty}\Re\left(\mathcal{Z}|_{[\xi_{i-1},\xi_i]};t\right), \quad (19.88)$$

where $\chi|_{[0,\xi_1]}$ is the indicator of the interval $[0,\xi_1]$.

Let us introduce the functions $\hat{X}(t)$ and $\hat{\mathcal{Z}}(t)$ on the half-line $\mathbb{R}_+$ by

$$\hat{X}(t) = X(t+\xi_1), \qquad \hat{\mathcal{Z}}(t) = \mathcal{Z}(t+\xi_1), \qquad t \in \mathbb{R}_+. \quad (19.89)$$

and set

$$\hat{B} := \|\hat{X}\|_{L_p(\mathbb{R}_+)}; \qquad \hat{\xi}_i = \xi_{i+1}, \quad \hat{\eta}_i = \eta_{i+1}, \qquad i \in \mathbb{Z}_+, \quad (19.90)$$

where $\{\xi_i\}_{i\in\mathbb{N}}$ and $\{\eta_i\}_{i\in\mathbb{N}}$ are the points of sign change of $\mathcal{Z}$ and $\mathcal{Z}'$, respectively.

The reference to the properties (19.62), (19.63) and (19.82), (19.83) of the extremal functions $X$ and $\mathcal{Z}$ demonstrates that the functions $\hat{X}(t)$ and

$\hat{\mathcal{Z}}(t)$ enjoy all of the properties of extremal functions $X$ and $\mathcal{Z}$ from section 19.4.2 with respect to the collections $\{\xi_i\}_{i\in\mathbb{N}}$ and $\{\eta_i\}_{i\in\mathbb{N}}$. Therefore, the function $\hat{X}$ is extremal in the problem

$$\|f'\|_{L_\infty(\mathbb{R}_+)} \to \sup, \qquad f \in W^r H^\omega(\mathbb{R}_+), \quad \|f\|_{L_p(\mathbb{R}_+)} \le \widehat{B}. \tag{19.91}$$

In particular, in the Hölder class $W^1 H^\alpha(\mathbb{R}_+)$ the lengths $\Delta_i = \xi_{i+1} - \xi_i$ of the interval $[\xi_i, \xi_{i+1}]$ constitute a geometric mesh:

$$\frac{\Delta_{i+1}}{\Delta_i} = \gamma(\alpha), \quad i \in \mathbb{Z}_+, \tag{19.92}$$

with the *same constant* as in (19.79). In addition, the function $X|_{[\xi_1,\infty)}$ is self-similar : for $i \in \mathbb{N}$ and $0 \le t \le \xi_{i+1} - \xi_i$,

$$X|_{[\xi_i,\xi_{i+1}]}(t+\xi_i) = (-1)^{i+1}\left(\frac{\xi_{i+1}-\xi_i}{\xi_i-\xi_{i-1}}\right)^{1+\alpha} X|_{[\xi_1,\xi_2]}\left(\frac{\xi_2-\xi_1}{\xi_{i+1}-\xi_i}t\right). \tag{19.93}$$

### 19.4.5 *Maximization of the norm* $\|f\|_{L_\infty(\mathbb{R})}$

The following problem is under our consideration:

$$\|f\|_{L_\infty(\mathbb{R})} \to \sup, \qquad f \in W^1 H^\omega(\mathbb{R}), \quad \|f\|_{L_p(\mathbb{R})} \le B. \tag{19.94}$$

In this case the function $\mathcal{Z}(t)$ is *even,* and its restriction to the positive half-line enjoys the following properties:

$$\begin{array}{l} (i)\ \mathcal{Z}\big|_+ \text{ vanishes at zeroes } \{\xi_i\}_{i=1}^\infty : 0 < \xi_1 < \xi_2 < \dots; \\ (ii)\ \mathcal{Z}' \text{ has one zero } \eta_i \text{ on } [\xi_i, \xi_{i+1}] \text{ for each } i \in \mathbb{N}. \end{array} \tag{19.95}$$

Let $f$ be a function from $W^1 H^\omega(\mathbb{R})$ such that $\|f\|_{L_p(\mathbb{R})} \le B$. Integration by parts gives us the following formula for the value of the function $f$ at zero:

$$f(0) = -\frac{1}{2}\mathcal{Z}'(0)^{-1}\left(\int_{\mathbb{R}} f(x)\mathcal{Z}''(x)\,dx + \int_{\mathbb{R}} f'(x)\mathcal{Z}'(x)\,dx\right). \tag{19.96}$$

Let $\xi_1$ be the first zero of $\mathcal{Z}$ on $(0,\infty)$. Then,

$$\int_{\mathbb{R}} f'(x)\mathcal{Z}'(x)\,dx = \int_{-\xi_1}^{\xi_1} f'(x)\mathcal{Z}'(x)\,dx + \int_{\mathbb{R}\setminus[-\xi_1,\xi_1]} f'(x)\mathcal{Z}'(x)\,dx. \tag{19.97}$$

We showed that the lengths of the intervals $\{[\xi_i, \xi_{i+1}]\}$ decrease. Thus, taking into account the fact that $\mathcal{Z}'$ is odd and applying the Korneichuk

Lemma 19.1 on each of the intervals $[\xi_i, \xi_{i+1}]$, we estimate the second summand in (19.97) as follows:

$$\sup_{h \in H^\omega(\mathbb{R})} \int\limits_{\mathbb{R}\setminus[-\xi_1,\xi_1]} h(x)\mathcal{Z}'(x)\,dx \leq \sum_{i=1}^{\infty} \int\limits_0^{\xi_{i+1}-\xi_i} \left[2\sum_{i=1}^{\infty} \mathcal{R}(\mathcal{Z}|_{[\xi_{i-1},\xi_i]};t)\right] \omega'(t)dt. \tag{19.98}$$

The function $\mathcal{Z}'|_{[-\xi_1,\xi_1]}$ is odd and has 0 as its only zero. Therefore, $\mathcal{Z}'|_{[-\xi_1,\xi_1]}$ is a simple kernel, and the function $\mathcal{Z}'$ is given on the interval by the formula

$$\mathcal{Z}'(x) = \begin{cases} -1/2\omega(2x), & x \in [0, \xi_1] \\ 1/2\omega(-2x), & x \in [-\xi_1, 0]. \end{cases} \tag{19.99}$$

Setting

$$\mathcal{R}_\omega(\mathcal{Z};t) = \mathcal{R}\left(\mathcal{Z}|_{[-\xi_1,\xi_1]};t\right) + 2\sum_{i=1}^{\infty} \mathcal{R}(\mathcal{Z}|_{[\xi_{i-1},\xi_i]};t), \quad t \in \mathbb{R}_+, \tag{19.100}$$

we arrive at the estimate

$$\|f\|_{L_\infty(\mathbb{R})} \leq \frac{1}{2}\left|\mathcal{Z}'(0)\right|^{-1} \left( \|\mathcal{Z}'\|_{L_{p'}(\mathbb{R})} B + \int\limits_{\mathbb{R}_+} \omega'(t)\mathcal{R}_\omega(\mathcal{Z};t)\,dt \right). \tag{19.101}$$

As in previous cases, for each $j \in \mathbb{N}$ the shift $\widehat{X}(t) = X(t+t_j)$ is extremal in the problem

$$\|f\|_{L_\infty(\mathbb{R}_+)} \to \sup, \qquad f \in W^r H^\omega(\mathbb{R}_+), \quad \|f\|_{L_p(\mathbb{R}_+)} \leq B_j,$$

where $B_j = \|X\|_{L_p[\xi_j,\infty)}$.

### 19.4.6 Maximization of the norm $\|f'\|_{\infty(\mathbb{R})}$

Finally, we consider the problem

$$\|f'\|_{L_\infty(\mathbb{R})} \to \sup, \qquad f \in W^1 H^\omega(\mathbb{R}), \quad \|f\|_{L_p(\mathbb{R})} \leq B. \tag{19.102}$$

In this case the function $\mathcal{Z}(t)$ is *odd*, and its restriction to the positive half-line enjoys the following properties:

$$\begin{aligned} &(i) \quad \mathcal{Z}\big|_+ \text{ vanishes at zeroes } \{\xi_i\}_{i=0}^{\infty} : 0 = \xi_0 < \xi_1 < \dots; \\ &(ii) \quad \mathcal{Z}'\big|_+ \text{ has one zero } \eta_i \text{ on } [\xi_i, \xi_{i+1}] \text{ for all } i \in \mathbb{N}. \end{aligned} \tag{19.103}$$

Let $f$ be a function from $W^1H^\omega(\mathbb{R})$ such that $\|f\|_{L_p(\mathbb{R})} \le B$. The differentiation formula for the value of $f'$ at zero is as follows:

$$f'(0) = \frac{1}{2}\mathcal{Z}(0)^{-1}\left(\int_{\mathbb{R}} f(x)\mathcal{Z}''(x)\,dx + \int_{\mathbb{R}} [f'(x) - f'(0)]\mathcal{Z}'(x)\,dx\right). \tag{19.104}$$

For $t \in \mathbb{R}_+$ set

$$\mathfrak{R}_\omega(\mathcal{Z};t) := 2|\mathcal{Z}(t)|\chi_{[0,\xi_1]}(t) + 2\sum_{i=1}^{\infty}\mathfrak{R}\left(\mathcal{Z}\big|_{[\xi_i,\xi_{i+1}]};t\right). \tag{19.105}$$

where, as usual, $\chi_{[a,b]}(\cdot)$ is the indicator function of the interval $[a,b]$. Thus, we arrive at the estimate

$$\|f'\| \le \frac{1}{2}|\mathcal{Z}(0)|^{-1}\left(\|f\|_{L_p(\mathbb{R})}\,B + \int_{\mathbb{R}_+} \omega'(t)\mathfrak{R}_\omega(\mathcal{Z};t)\,dt\right). \tag{19.106}$$

Each of the kernels $\mathcal{Z}\big|_{[\xi_i,\xi_{i+1}]}$ is simple for $i \in \mathbb{N}$, and $\{\Delta_i := \xi_{i+1} - \xi_i\}_{i\in\mathbb{N}}$ constitute a decreasing sequence, so one can use the Korneichuk's formulae (19.71), (19.72) for the characterization of the derivative $X'$ on \$[-\xi_1, \xi_1]$. The reasoning of §§ 19.4.4 enables us to conclude that

$$X'(t) = X'(0) - \omega(|t|), \qquad t \in [-\xi_1, \xi_1]. \tag{19.107}$$

Finally, the shift $\widetilde{X}_j(t) := X(t + t_j)$ is extremal in the problem

$$\|f\|_{L_\infty(\mathbb{R}_+)} \to \sup, \qquad f \in W^rH^\omega(\mathbb{R}_+), \quad \|f\|_{L_p(\mathbb{R}_+)} \le B_j, \tag{19.108}$$

where $B_j := \|X\|_{L_p[\xi_j,+\infty)}$, $j \in \mathbb{N}$.

*References*

[1] N. I. Akhiezer, M. G. Krein, *On the best approximation of differentiable periodic functions by trigonometric sums*, DAN SSSR 15 (1937), p.p. 107–112. (Russian)

[2] V. V. Arestov, *On sharp inequalities between the norms of functions and their derivatives*, Acta Sci. Math. 33 (1972), p.p. 243–267

[3] S. K. Bagdasarov, *Maximization of functionals in $H^\omega[a,b]$*, Matem. Sbornik 189, no. 2 (1998)

[4] S. K. Bagdasarov, *Extremal functions of integral functionals in $H^\omega[a,b]$*, Izvestiya RAN (1998), to appear

[5] S. K. Bagdasarov, *Zolotarev $\omega$-polynomials in $W^rH^\omega[0,1]$*, J. Approx. Theory 90, no. 3 (1997), p.p. 340–378

[6] S. K. Bagdasarov, *The general construction of Chebyshev $\omega$-splines of the given norm*, Algebra and Analysis (St.-Petersburg Math. Journal) (1998), to appear

[7] S. K. Bagdasarov, Chebyshev Splines and Kolmogorov Inequalities, Operator Series: Advances and Applications, Birkhäuser, Basel (1998)

[8] S. K. Bagdasarov, *Kolmogorov problem for intermediate derivatives and optimal control* (1998), (in preparation)

[9] Yu. G. Bosse (G. E. Shilov), *Inequalities between derivatives*, Sbornik rabot Nauch. Stud. Kruzhkov Mosk. Un-ta (1937), p.p. 17–27

[10] H. Cartan, *Sur les classes de fonctions définies par des inegalites portant sur leurs dérivées successives*, Act. Sci. Ind. 867 (1940), Hermann, Paris

[11] A. S. Cavaretta, *An elementary proof of Kolmogorov's theorem*, Amer. Math. Monthly 81 (1974), p.p. 480–486

[12] A. S. Cavaretta, I. J. Schoenberg *Solution of Landau's problem concerning higher derivatives on the half-line*, MRC T.S.R. 1050 (1970), Madison, Wisconsin; Also in Proc. of the Intern. Conf. on Constructive Function Theory, Golden Sands (Varna), May 19–25, 1970, Publ. House Bulg. Acad. Sci., Sofia 1972, p.p. 297–308

[13] A. S. Cavaretta, *A refinement of Kolmogorov's inequality*, Journal of Approximation Theory 27 (1979), p.p. 45–60

[14] J. Favard, *Sur les meilleures procédes d'approximation de certaines classes des fonctions far des polynomes trigonometriques*, Bull. Sci. Math. 61 (1937), p.p. 209–224, 243–256

[15] A. T. Fuller, *Optimization of nonlinear control systems with transient inputs*, J. of Electron. and Control 8, no. 6 (1960), p.p. 465–479

[16] V. N. Gabushin, *Inequalities for norms of functions and their derivatives in the $L_p$ metrics*, Matem. Zametki 1, no. 3 (1967), p.p. 291–298; transl. in Math. Notes 1, no. 3 (1967), p.p. 194–198

[17] V. N. Gabushin, *Exact constants in inequalities between norms of the derivatives of a function*, Math. Notes 4, no. 2, (1968), p.p. 630–634

[18] V. N. Gabushin, *Best approximation of a differentiation operator on the half-line*, Matem. Zametki 6 (1969), p.p. 573–582; transl. in Math. Notes 6, (1969), p.p. 573–582

[19] J. Hadamard, *Sur le module maximum d'une fonction et de ses dérivées*, Soc. math. France, Comptes rendus, des Séances 41 (1914)

[20] G. H. Hardy, J. Littlewood, G. Pölya, Inequalities, Cambridge Univ. Press, New York (1934)

[21] D. Jackson, *Über die Genauigkett des Annäherung stetigen Funktionen durch ganze rationale Funktionen gegebenen Grades und trigonometrischen Summen gegebener*, Ordnung Diss., Göttingen (1911)

[22] S. Karlin, *Oscillatory perfect splines and related extremal problems, in* "Studies in Spline Functions and Approximation Theory" (1976), p.p. 371–460, (S. Karlin, C. A. Micchelli, A. Pinkus, and I. J. Schoenberg, eds.), Academic Press, New York, N. Y.

[23] A. N. Kolmogorov, *On inequalities between upper bounds of successive derivatives of an arbitrary function defined on an infinite interval*, Uch. zap. MGU, Matematika 3 (1939), p.p. 3–16, transl. in Amer. Math. Soc. Transl. Ser. 1, 2 (1962), p.p. 233–243

[24] A. N. Kolmogorov, Izbrannye trudy. Matematika i mehanika (Selected works. Mathematics and mechanics), Nauka, Moscow (1985)

[25] N. P. Korneichuk, Ekstremal'nye Zadachi Teorii Priblizheniya (Extremal Problems of Approximation Theory), Nauka, Moscow (1976)

[26] N. P. Korneichuk, Splainy v Teorii Priblizheniya (Splines in Approximation Theory), Nauka, Moscow (1984)

[27] N. P. Korneichuk, Exact Constants in Approximation Theory, Cambridge University Press, Cambridge, New York (1990), Series: Encyclopedia of Mathematics and Its Applications, vol. 38

[28] N. P. Korneichuk, *S. M. Nikol'skii and the development of research on approximation theory in the USSR,* Russian mathematical surveys 40 (1985), p.p. 83–156

[29] N. P. Kupcov, *Kolmogorov estimates for derivatives in* $L_2[0,\infty)$, Proc. Steklov Institute of Mathematics 138 (1975), p.p. 101–125

[30] E. Landau, *Einige Ungleichingen fur zweimal differentierbare Funktionen*, Proc. London Math. Soc. 13 (1913), p.p. 43–49

[31] G. G. Magaril-Il'yaev, *On Kolmogorov inequalities on a half-line*, Vestnik Mosk. Univ. (Moscow Univ. Bulletin), Matematika 31, no. 5 (1976), p.p. 33–41

[32] G. G. Magaril-Il'yaev, *Inequalities for the derivatives and the duality*, Trudy Mat. Inst. Steklov 161 (1983), p.p. 183–194

[33] A. P. Matorin, *On inequalities between the maxima of absolute values of a function and its derivatives on a half-line*, Amer. Math. Soc. Transl., Series 2, 8 (1958), p.p. 13–17

[34] S. M. Nikol'skii, *La série de Fourier d'une fonction dont be module de continuité est donné*, Dokl. Akad. Nauk SSSR 52 (1946), p.p. 191–194

[35] A. Pinkus, *Some extremal properties of perfect splines and the pointwise Landau problem on the finite interval*, J. Approx. Theory 23, no. 2 (1978), p.p. 37–64

[36] S. B. Stechkin, *Inequalities between norms of derivatives of an arbitrary function*, Acta Sci. Math. 26 (1965), p.p. 225–230

[37] E. M. Stein, *Functions of exponential type*, Ann. of Math. 65, no. 3 (1957), p.p. 582–592

[38] B. Szekëfalvi-Nagy, *Über Integralungleichungen zwischen einer Funktion und ihrer Ableitung*, Acta Sci. Math. 10 (1941), p.p. 64–74

[39] L. V. Taikov, *Inequalities of Kolmogorov type and best formulas for numerical differentiation*, Math. Zametki 4 (1968), p.p. 233–238, transl. in Math. Notes 4 (1968), p.p. 631–634

[40] V. M. Tihomirov, *Best methods of approximation and interpolation of differentiable functions in the space* $C[-1,1]$, Math USSR Sbornik 9 (1969), p.p. 277–289

[41] A.Zygmund, Trigonometric Series, Cambridge, Univ. Press, 1959.

# Chapter 20

# On angularly perturbed Laplace equations in the unit ball and their distributional boundary values

*Dedicated to the Memory of John Kelingos*

**Peter R. Massopust**[1]

*ABSTRACT. All solutions of a continuously perturbed Laplace-Beltrami equation in its angular coordinates in the open unit ball $\mathbb{B}^{n+2} \subset \mathbb{R}^{n+2}$, $n \geq 1$, are characterized. Moreover, it is shown that such pertubations yield distributional boundary values which are different from, but algebraically and topologically equivalent to, the hyperfunctions of Lions & Magenes. This is different from the case of radially perturbed Laplace-Beltrami operators (cf. [7]) where one has stability of distributional boundary values under such pertubations.*

## 20.1 Introduction

The present paper is a continuation of work that commenced in [11, 6, 7], namely the investigation of solutions of perturbed Laplace equations. More precisely, given a pertubation function $\sigma \in C^1(\mathbb{R}^{n+2}, \mathbb{R})$, obtain a characterization of all solutions of

$$Lu := \operatorname{div}(\sigma \operatorname{grad} u) = 0, \quad \text{on } \mathbb{B}^{n+2}. \tag{20.1}$$

where $\mathbb{B}^{n+2} := \{x \in \mathbb{R}^{n+2} \colon |x| < 1\}$ denotes the unit open ball and $|\cdot| : \mathbb{R}^{n+2} \to \mathbb{R}_0^+$ the canonical Euclidean norm on $\mathbb{R}^{n+2}$, and describe its distributional boundary values. The main issue is that $\sigma$ is *not* assumed to be real-analytic in its variables.

For a particular choice of $\sigma$, namely $\sigma$ describing a pertubation of the polar and azimuthal angles, and under rather mild regularity and growth

[1] This work was supported by Sandia National Laboratories. Sandia is a multiprogram laboratory operated by Sandia Corporation, a Lockheed Martin Company, for the United States Department of Energy (DOE) under contract DE-ACo4-94AL85000.

conditions it is shown that Eqn. (20.1) has bounded solutions in the open unit ball of $\mathbb{R}^{n+1}$, and that these solutions are completely characterized in terms of certain constant vector sequences satisfying essentially the same lim sup condition as in [5, 11, 6, 7]. The angular pertubation problem, as studied in this paper, involves solving generalizations of coupled Legendre differential equations yielding coupled eigenvalue problems for the polar angles. The solutions of these coupled equations may be interpreted as *generalized spherical harmonics* on $\mathbb{S}^{n+1}$, the unit sphere in $\mathbb{R}^{n+2}$. Once the characterization of these generalized spherical harmonics in terms of certain vector sequences is obtained, it is shown that the same vector sequences may be employed to describe the distributional boundary values of these solutions. Unlike in the case of an only radially perturbed Laplace equation [7], it is found that the distributional boundary values are different from, but algebraically and topologically equivalent to, the hyperfunctions or analytical functionals studied by among others Lions & Magenes (cf. for instance, [9]). The description of these distributional boundary values is done entirely in terms of vector sequences and avoids the usual methods involving the inductive limit topology and intermediate normed linear spaces.

The structure of this paper is as follows: In Section 2 preliminaries are presented and the relevant notation is introduced. Section 3 concentrates on the bounded solutions of Eqn. (20.1) and the main theorems characterizing these solutions are stated. In Section 4 the distributional boundary values are described and a representation theorem of the solution of Eqn. (20.1) is presented which may be interpreted as a generalized Poisson integral. The last section combines the results of this paper and those of [7] to give a structure theorem involving a fully – radially as well as angularly – perturbed Laplace equation on $\mathbb{B}^{n+2}$.

## 20.2 Notation and Preliminaries

Let $x \in \mathbb{R}^{n+2}$ and let $x' \in \mathbb{S}^{n+1} := \{x \in \mathbb{R}^{n+2} : |x| = 1\}$. Then, setting $r := |x|$, one can express $x := (x_1, \ldots, x_{n+2})^T$ ($^T$ denotes the transpose), in spherical coordinates as

$$\begin{aligned} x_1 &= r\cos\theta_1 \\ x_2 &= r\sin\theta_1\cos\theta_2 \\ &\vdots \\ x_n &= r\sin\theta_1\sin\theta_2\cdots\sin\theta_{n-1}\cos\theta_n \\ x_{n+1} &= r\sin\theta_1\sin\theta_2\cdots\sin\theta_{n-1}\sin\theta_n\cos\phi \\ x_{n+2} &= r\sin\theta_1\sin\theta_2\cdots\sin\theta_{n-1}\sin\theta_n\sin\phi, \end{aligned} \tag{20.2}$$

where $\boldsymbol{\theta} := (\theta_1, \ldots, \theta_n)^T$ and $0 \le \phi \le 2\pi$ represents the vector of polar angles $0 \le \theta_j \le \pi$, $j = 1, \ldots, n$, and the azimuth angle, respectively.

To shorten notation, let

$$s := \sin\phi, \quad c := \cos\phi, \quad s_j := \sin\theta_j, \quad c_j := \cos\theta_j, \quad j = 1, \ldots, n. \tag{20.3}$$

The gradient $\nabla$ and the Laplace-Beltrami operator $\Delta$ have the following representations in spherical coordinates:

$$\nabla = \frac{\partial}{\partial r} e_r + \frac{1}{r} \sum_{i=1}^{n} \frac{1}{\prod_{j=1}^{i=1} s_j} \frac{\partial}{\partial \theta_j} e_{\theta_j} + \frac{1}{r \prod_{j=1}^{n} s_j} \frac{\partial}{\partial \phi} e_\phi, \tag{20.4}$$

and

$$\begin{aligned} \Delta &= \frac{1}{r^{n+1}} \frac{\partial}{\partial r}\left(r^{n+1} \frac{\partial}{\partial r}\right) + \frac{1}{r^2} \sum_{i=1}^{n} \frac{1}{\prod_{j=1}^{i-1} s_j^2} \frac{1}{s_i^{n+1-i}} \frac{\partial}{\partial \theta_i}\left(s_i^{n+1-i} \frac{\partial}{\partial \theta_i}\right) \\ &\quad + \frac{1}{r^2} \frac{1}{\prod_{j=1}^{n} s_j^2} \frac{\partial^2}{\partial \phi^2}. \end{aligned} \tag{20.5}$$

Here, $(e_r, e_{\theta_1}, \ldots, e_{\theta_n}, e_\phi)^T$ is the canonical spherical basis of $\mathbb{R}^{n+2}$. Also, the usual notation concerning the empty product is assumed.

In spherical coordinates, the differential equation (20.1) then reads

$$\begin{aligned} Lu &= \operatorname{div}(\sigma \operatorname{grad} u) = \sigma\Delta u + \langle \nabla\sigma, \nabla u \rangle \\ &= \sigma\Delta u + \sigma_r u_r + \frac{1}{r^2} \sum_{i=1}^{n} \frac{1}{\prod_{j=1}^{i-1} s_j^2} \sigma_{\theta_i} u_{\theta_i} + \frac{1}{r^2 \prod_{j=1}^{n} s_j^2} \sigma_\phi u_\phi \\ &= 0. \end{aligned} \tag{20.6}$$

As usual, $u_x := \partial u / \partial x$, and $\langle \cdot, \cdot \rangle$ denotes the canonical inner product on $\mathbb{R}^{n+2}$.

Now assume that the function $\sigma \in C^1(\mathbb{R}^{n+2}, \mathbb{R})$ is of the form

$$\sigma(x) := \left( \prod_{j=1}^{n} \sigma_j(\theta_j) \right) \widetilde{\sigma}(\phi), \tag{20.7}$$

where $\sigma_j \in C^1([0, \pi], \mathbb{R})$, $j = 1, \ldots, n$, and $\widetilde{\sigma} \in C^1([0, 2\pi], \mathbb{R})$. Using this particular form of $\sigma$ in Eqn. (20.6) yields

$$\sigma\Delta u + \frac{1}{r^2} \sum_{i=1}^{n} \frac{1}{\prod_{j=1}^{i-1} s_j^2} \frac{\sigma_i'(\theta_i)}{\sigma_i(\theta_i)} u_{\theta_i} + \frac{1}{r^2 \prod_{j=1}^{n} s_j^2} \frac{\widetilde{\sigma}_i'(\phi)}{\widetilde{\sigma}_i(\phi)} u_\phi = 0. \tag{20.8}$$

**Notation 20.1** *In what follows, $'$ will always denote the derivative with respect to the variable a function depends upon.*

Comparing this last equation with the expression for $\Delta u = 0$, leads to the following definitions

$$\varepsilon_i(\theta_i) := s_i^{n+1-i} : \frac{\sigma_i'(\theta_i)}{\sigma_i(\theta_i)}, \quad i = 1, \dots, n, \quad \widetilde{\varepsilon}(\phi) := \frac{\widetilde{\sigma}_i'(\phi)}{\widetilde{\sigma}_i(\phi)}.$$

The perturbed equation (26.1) now can be expressed as

$$\frac{1}{r^{n+1}}\frac{\partial}{\partial r}\left(r^{n+1}\frac{\partial u}{\partial r}\right) + \frac{1}{r^2}\sum_{i=1}^{n}\frac{1}{\prod_{j=1}^{i-1}s_j^2}\frac{1}{s_i^{n+1-i}}\left[\frac{\partial}{\partial\theta_i}\left(s_i^{n+1-i}\frac{\partial u}{\partial\theta_i}\right)\right.$$
$$\left. +\varepsilon_i(\theta_i)\frac{\partial u}{\partial\theta_i}\right] + \frac{1}{r^2}\frac{1}{\prod_{j=1}^{n}s_j^2}\left[\frac{\partial^2 u}{\partial\phi^2} + \widetilde{\varepsilon}(\phi)\frac{\partial u}{\partial\phi}\right] = 0. \tag{20.9}$$

In other words, the pertubations are in the lower order derivative terms. Note that it is *not* assumed that $\widetilde{\varepsilon}$ or $\varepsilon_j$, $j = 1, \dots, n$, are real-analytic. The objective of this paper is to characterize all bounded solutions of Eqn. (20.9) in the open unit ball $\mathbb{B}^{n+2}$ subject to these *continuous* pertubations, and to investigate the distributional boundary values.

## 20.3 Bounded Solutions on $\mathbb{B}^{n+2}$

To obtain solutions to Eqn. (20.9), the method of separation of variables is employed. To this end, let

$$u = u(r, \boldsymbol{\theta}, \phi) := R(r)\left(\prod_{j=1}^{n}\Theta_j(\theta_j)\right)\Phi(\phi).$$

Substitution of this ansatz into Eqn. (20.9) and division by $R(\prod_{j=1}^n \Theta_j)\Phi$ yields

$$\prod_{j=1}^{n}\frac{s_j^2}{Rr^{n-1}}\frac{\partial}{\partial r}\left(r^{n+1}R'(r)\right) \quad + \quad \prod_{i=1}^{n}s_i^2\sum_{i=1}^{n}\frac{1}{\left(\prod_{j=1}^{i-1}s_j^2\right)\Theta_i\, s_i^{n+1-i}} \times$$
$$\times \quad \left[\frac{\partial}{\partial\theta_i}\left(s_i^{n+1-i}\Theta_i'\right) + \varepsilon_i\Theta_i'\right]$$
$$+ \quad \frac{1}{\Phi}\left(\Phi'' + \widetilde{\varepsilon}\Phi'\right) = 0.$$

The usual separtion of variables arguments give the existence of a real constant $\alpha$ such that

$$\Phi'' + \widetilde{\varepsilon}\Phi' + \alpha\Phi = 0. \tag{20.10}$$

As $\phi$ is the azimuthal angle one needs to impose periodicity conditions:

$$\Phi(0) = \Phi(2\pi), \qquad \Phi'(0) = \Phi'(2\pi). \tag{20.11}$$

Eqns. (20.10) and (20.11) constitute a regular Sturm-Liouville system. Such a system yields a complete set of eigenfunctions together with asymptotic estimates for the eigenvalues. More precisely, if

$$\widetilde{\varepsilon} \in C^1(\mathbb{R}), \quad \text{and} \quad \int_0^{2\pi} \widetilde{\varepsilon}(\phi)d\phi = 0,$$

there exists a countably infinite set of nonnegative eigenvalues $\{\lambda_k\}_{k\in\mathbb{Z}_0^+}$ such that $\lambda_0 = 0$ is simple with eigenfunction $\Phi_0 \equiv 1$, whereas the remaining eigenvalues occur in pairs:

$$\{0 = \lambda_0 < \lambda_1' \leq \lambda_1'' < \lambda_2' \leq \lambda_2'' < \cdots \to \infty\}. \tag{20.12}$$

Furthermore, these eigenvalues satisfy the following asymptotics

$$\sqrt{\lambda_k} = k + o(1) \quad \text{as } k \to \infty.$$

Here $\lambda_k$ refers to $\lambda_k'$ or $\lambda_k''$. The complete system of eigenfunctions $\{\Phi_\lambda\}$ may be expressed – due to the pairing of the eigenvalues – in the form

$$\{\Phi_\lambda\} = \{C_0(\phi), C_1(\phi), S_1(\phi), C_2(\phi), S_2(\phi), \dots\}. \tag{20.13}$$

These eigenfunctions are orthonormal with respect to the weight function $\eta(\phi) := \exp\left(\int_0^\phi \widetilde{\varepsilon}\, dt\right)$ and exhibit the following asymymptotic behaviour for $k \geq 1$

$$\begin{aligned} C_k(\phi) &= \frac{1}{\sqrt{\pi\eta(\phi)}} \cos k\phi + O(1) \\ S_k(\phi) &= \frac{1}{\sqrt{\pi\eta(\phi)}} \sin k\phi + O(1). \end{aligned} \tag{20.14}$$

This particular regular Sturm-Liouville problem was considered in [6] and the interested reader may refer to this reference for more details.

As this part of the separation process is well understood, and for the sake of simplicity, it is assumed for the remainder of this section that $\widetilde{\varepsilon} \equiv 0$. (The general case is considered later and involves only minor changes, mostly in notation and terminology.) In this case one has $\alpha = \alpha_k = k^2$, $C_k(\phi) = (1/\sqrt{\pi}) \cos k\phi$, and $S_k(\phi) = (1/\sqrt{\pi}) \sin k\phi$. To shorten notation, let $e_k(\phi) := (1/\sqrt{2\pi})e^{ik\phi} = ((1/\sqrt{\pi}) \cos k\phi, (1/\sqrt{\pi}) \sin k\phi)$. Returning to

Eqn. (20.10) with $\alpha = \alpha_k = k^2$ yields

$$\begin{aligned} &\frac{1}{Rr^{n-1}}\frac{\partial}{\partial r}\left(r^{n+1}R'(r)\right) \\ &+\sum_{i=1}^{n}\frac{1}{\left(\prod_{j=1}^{i-1}s_j^2\right)\Theta_i s_i^{n+1-i}}\left[\frac{\partial}{\partial\theta_i}\left(s_i^{n+1-i}\Theta_i'\right)+\varepsilon_i\Theta_i'\right] \\ &-\frac{k^2}{\prod_{j=1}^{n}s_j^2}=0. \end{aligned} \tag{20.15}$$

Thus, there exists a real constant $\gamma_1$, independent of $k$, such that

$$\frac{1}{Rr^{n-1}}\frac{\partial}{\partial r}\left(r^{n+1}R'(r)\right)=\gamma_1,$$

or, after simplification

$$r^2R''+(n+1)rR'-\gamma_1R=0. \tag{20.16}$$

This Cauchy-Euler equation is solved by the ansatz $R(r) := r^\ell$, $\ell \in \mathbb{R}$. This yields

$$\gamma_1 = \ell(\ell+n).$$

As only bounded solutions are of interest, $\ell \geq 0$ and therefore $\gamma_1 \geq 0$. In fact,

$$\ell = \frac{-n+\sqrt{n^2+4n\gamma_1}}{2}.$$

Eqn. (20.15) now gives the following system of equations for the polar angles $\theta_i$, $i = 1, \ldots, n$.

$$\sum_{i=1}^{n}\frac{1}{\left(\prod_{j=1}^{i-1}s_j^2\right)\Theta_i s_i^{n+1-i}}\left[\frac{\partial}{\partial\theta_i}\left(s_i^{n+1-i}\Theta_i'\right)+\varepsilon_i\Theta_i'\right]-\frac{k^2}{\prod_{j=1}^{n}s_j^2}=-\gamma_1. \tag{20.17}$$

Now the variables $\Theta_i$ have to be separated one at a time. One begins by splitting off the $i = 1$ term. After some algebra this yields

$$\begin{aligned} &\sum_{i=2}^{n}\frac{1}{\left(\prod_{j=2}^{i-1}s_j^2\right)\Theta_i s_i^{n+1-i}}\left[\frac{\partial}{\partial\theta_i}\left(s_i^{n+1-i}\Theta_i'\right)+\varepsilon_i\Theta_i'\right]-\frac{k^2}{\prod_{j=2}^{n}s_j^2} \\ &\qquad=\frac{-s_1^{2-n}}{\Theta_1}\left[\frac{\partial}{\partial\theta_1}\left(s_1^n\Theta_1'\right)+\varepsilon_1\Theta_1'\right]-s_1^2\gamma_1. \end{aligned} \tag{20.18}$$

Again, there exists a real constant $-\gamma_2$ such that

$$\frac{1}{s_1^{n-2}}\left[\frac{\partial}{\partial\theta_1}\left(s_1^n\Theta_1'\right)+\varepsilon_1\Theta_1'\right]+\left(s_1^2\gamma_1-\gamma_2\right)\Theta_1=0.$$

To simplify this last equation, a new variable $y_1 := \cos\theta_1$ is introduced. Substitution into the above equation and subsequent simplification gives

$$(1-y_1^2)\Theta_1''-\left[(n+1)y_1+\frac{\varepsilon_1}{(1-y_1^2)^{(n-1)/2}}\right]\Theta_1'=\left[\frac{\gamma_2}{1-y_1^2}-\gamma_1\right]\Theta.$$

Returning to Eqn. (20.18), one obtains

$$\sum_{i=2}^{n}\frac{1}{\left(\prod_{j=2}^{i-1}s_j^2\right)\Theta_i s_i^{n+1-i}}\left[\frac{\partial}{\partial\theta_i}\left(s_i^{n+1-i}\Theta_i'\right)+\varepsilon_i\Theta_i'\right]-\frac{k^2}{\prod_{j=2}^{n}s_j^2}=-\gamma_2.$$

Notice that this is just Eqn. (20.17) with the $i=j=1$ terms missing and $\gamma_1$ replaced by $\gamma_2$. Hence, repetition of the above process gives for all $j=1,\dots,n$

$$(1-y_j^2)\Theta_j''-\left[(n+2-j)y_j+\frac{\varepsilon_j}{(1-y_j^2)^{(n-j)/2}}\right]\Theta_j'=\left[\frac{\gamma_{j+1}}{1-y_j^2}-\gamma_j\right]\Theta_j, \tag{20.19}$$

with real constants $\gamma_j$ and $\gamma_{n+1} := k^2$. (Here, $y_j := \cos\theta_j$.)

**Remark 20.1** *For $\varepsilon_j \equiv 0$, $j=1,\dots,n$, the above equations are solved by the* associated Legendre functions $\mathcal{P}_{m_{j-1}}^{m_j}(n+2-j;y_j)$ *of degree $m_{j-1}$ and order $m_j$ in dimension $n+2-j$, where $\gamma_j = m_j(m_j+n+1-j)$ and $m_j = 0,1,\dots,m_{j-1}$. Moreover, these associated Legendre functions are expressible in terms of a Gauß ${}_2F_1$ function as*

$$(1-y_j^2)^{p_j}\,{}_2F_1\left(-2p_j-m_j, m_j-2p_j+n-j; 2p_j+\frac{n+2-j}{2}; \frac{1-y_j}{2}\right),$$

*for $-1<y_j<1$, and with $p_j := (1/4)\left(1-n+\sqrt{(1-n)^2+4j(j+n-1)}\right)$, $j=1,\dots,n$.*

To obtain bounded solutions on $\mathbb{B}^{n+2}$, one has to require that

$$\forall j=1,\dots,n : |\Theta_j(y_j)| \le M < \infty \text{ as } |y_j|\to 1.$$

For this purpose, fix $j\in\{1,\dots,n\}$ and let $\Theta_+ := \begin{cases}\Theta_j & 0\le y_j<1\\ 0 & \text{otherwise}\end{cases}$ and $\Theta_- := \begin{cases}\Theta_j & -1<y_j\le 0\\ 0 & \text{otherwise}\end{cases}$. Clearly, $\Theta_j = \Theta_+ + \Theta_-$ and $\Theta_j(0) =$

$\Theta_\pm(0)$. Let $t := y_j/(1-y_j)$. Then $\Theta_+(t) : [0,\infty) \to \mathbb{R}$. In this new variable, Eqn. (20.19) now reads (For notational simplicity, the subscript $j$ is dropped from all variables and $A := \gamma_{j+1}$ and $B := \gamma_j$.)

$$\begin{aligned}\ddot{\Theta} &+ \left[\frac{2}{1+2t} - \frac{(n-j)t}{(1+2t)(1+t)} - \frac{\varepsilon(1+t)^{n-j}}{(1+2t)^{(n-j-2)/2}}\right]\dot{\Theta} \\ &+ \left[\frac{B}{(1+2t)(1+t)^2} - \frac{A}{(1+2t)^2}\right]\Theta = 0.\end{aligned} \tag{20.20}$$

Notice that $\varepsilon = \varepsilon(t/(1+t))$ and that $\dot{} := d/dt$. Let

$$p(t) := \frac{2}{1+2t} - \frac{(n-j)t}{(1+2t)(1+t)} - \frac{\varepsilon(1+t)^{n-j}}{(1+2t)^{(n-j-2)/2}}$$

and

$$q(t) := \frac{B}{(1+2t)(1+t)^2} - \frac{A}{(1+2t)^2}.$$

Applying a Liouville transform $v : [0,\infty) \to \mathbb{R}$, $v := \Theta \exp\left((1/2)\int_0^t p(s)ds\right)$, to Eqn. (20.20) yields

$$\ddot{v} + \left[q(t) - \frac{\dot{p}(t)}{2} - \frac{p(t)^2}{4}\right] v = 0. \tag{20.21}$$

The expression in brackets is explicitly given by

$$\frac{\left[\left(\frac{n-j}{2}\right)^2 + (1-A)\right]t^2 + [(n-j+2) + 2(B-A)]\,t + \left[\frac{n-j+2}{2} + (B-A)\right]}{(1+2t)^2(1+t)^2}$$

$$+ \frac{(1+t)^{n-j-1}\left[(n-j+2)(t+1) - 4\right]}{2(1+2t)^{(n-j)/2}}\,\varepsilon$$

$$- \frac{(1+t)^{2(n-j)}}{4(1+2t)^{n-j-2}}\,\varepsilon^2 + \frac{(1+t)^{n-j-2}}{2(1+2t)^{(n-j-2)/2}}\,\dot{\varepsilon}.$$

Let $B' = B - A$ and define

$$a(t) := \frac{\left[\left(\frac{n-j}{2}\right)^2 + (1-A)\right]t^2 + [(n-j+2) + 2B']\,t + \left[\frac{n-j+2}{2} + B'\right]}{(1+2t)^2(1+t)^2},$$

and

$$b(t) := \frac{(1+t)^{n-j-1}\left[(n-j+2)(t+1)-4\right]}{2(1+2t)^{(n-j)/2}}\,\varepsilon - \frac{(1+t)^{2(n-j)}}{4(1+2t)^{n-j-2}}\,\varepsilon^2 + \frac{(1+t)^{n-j-2}}{2(1+2t)^{(n-j-2)/2}}\,\dot{\varepsilon}.$$

Theorem 2, p. 112 of [1], states that *if* there are bounded solutions of

$$\ddot{v} + a(t)v = 0,$$

then there exist bounded solutions of

$$\ddot{v} + \left[a(t) + b(t)\right]v = 0,$$

provided that $b \in L^1[0,\infty)$. Now, if

**E1.** $\ell \in \mathbb{Z}_0^+$;

**E2.** $\gamma_j := m_j(m_j + n + 1 - j)$, $j = 1, \ldots, n+1$, where $m_1 := \ell$, $m_{n+1} := k$, and $m_j = 0, 1, \ldots, m_{j-1}$; $j = 2, \ldots, n$;

then there exist bounded solutions of $\ddot{v} + a(t)v = 0$; namely, the associated Legendre functions of degree $m_{j-1}$ and order $m_j$ in dimension $n+2-j$. Hence, it suffices to impose conditions on $\varepsilon_j$ that guarantee that $b = b_j \in L^1[0,\infty)$.

**Proposition 20.2** *If* $\varepsilon_j \in C^1$, $(1+t)^{(n-j+1)/2}\varepsilon_j \in L^1[0,\infty) \cap L^2[0,\infty)$, *and* $(1+t)^{(n-j-2)/2}\,\dot{\varepsilon}_j \in L^1[0,\infty)$, *then* $b_j \in L^1[0,\infty)$, $j = 1, \ldots, n$.

**Proof.** These conditions follow easily from estimates involving the three terms in the expression for $b(t)$. ■

To show that $\Theta_-$ is bounded, the substitution $y := -t/(1+t)$ is employed. It is not difficult to verify that this implies the substitution $\varepsilon \mapsto -\varepsilon$ giving the boundedness of $\Theta_-$ under the same conditions as above. Since $\Theta_j = \Theta_- + \Theta_+$ and $\Theta_j(0) = \Theta_\pm(0)$, the boundedness of $\Theta_j$ is established for all $j = 1, \ldots, n$. Hence, for each $\ell \in \mathbb{Z}_0^+$, there exist bounded solutions of Eqn. (20.19) if the constants $\gamma_j$, $j = 1, \ldots, n+1$, are chosen according to **E1** and **E2**. Moreover, $\gamma_j = m_j(m_j + n + 1 - j)$ is an eigenvalue of $\Theta_j$ with $m_j = 0, 1, \ldots, m_{j-1}$, $j = 1, \ldots, n$. The corresponding system of eigenfunctions is denoted by $\{\Theta_{m_j}\}_0^{m_{j-1}}$. Furthermore, these eigenfunctions are bounded independent of the eigenvalue, complete and orthonormal with respect to the weight functions (cf. [2])

$$w_j(y_j) := (1-y_j^2)^{(n-j)/2} \exp\left(-\int_0^{y_j} \frac{\varepsilon_j\, d\xi_j}{(1-\xi_j^2)^{(n+2-j)/2}}\right).$$

This follows immediately from transforming Eqn. (20.19) into self-adjoint form:

$$\left(\left[1-y_j^2\right]^{(n+2-j)/2}\bar{\varepsilon}_j\,\Theta_j'\right)' \;-\; \left[1-y_j^2\right]^{(n-2-j)/2}\bar{\varepsilon}_j\,\gamma_{j+1}\,\Theta_j$$
$$+\;\left[1-y_j^2\right]^{(n-j)/2}\bar{\varepsilon}_j\,\gamma_j\,\Theta_j = 0.$$

Here, $\bar{\varepsilon}_j := \exp\left(-\int_0^{y_j}\frac{\varepsilon_j\,d\xi_j}{(1-\xi_j^2)^{(n+2-j)/2}}\right)$. In the following it will become necessary to order products of the eigenfunctions $\{\boldsymbol{\theta}_{m_j}\}$ and $e_k$. For this purpose, let $\mathsf{W}$ denote the set of all words $\mathsf{w} = \mathsf{l}_1\mathsf{l}_2\cdots\mathsf{l}_n\mathsf{l}_{n+1}$ of length $n+1$ subject to the conditions

$$\mathsf{l}_1 \in \mathbb{Z}_0^+,\quad \mathsf{l}_\nu \in \{0,1,\ldots,m_1\},\ \nu = 2,\ldots,n+1, \tag{20.22}$$

$$\mathsf{l}_\nu \geq \mathsf{l}_{\nu+1},\quad \nu = 1,\ldots,n. \tag{20.23}$$

Note that these requirements are motivated by conditions **E1** and **E2**. An ordering $\preceq$ can be introduced into $\mathsf{W}$ via

$$\preceq(\mathsf{w},\mathsf{w}') \iff \mathsf{l}_\nu \leq \mathsf{l}_\nu',\ \forall\nu = 1,\ldots,n+1.$$

A simple combinatorial argument shows that for a fixed $\mathsf{l}_1 = m_1 = \ell$ there are exactly $d(\ell,n+1) := \binom{\ell+n+1}{n+1}$ words of length $n+1$ ordered according to $\preceq$.[2]

An *($\preceq$-)ordered vector* is a sequence of real constants of length $d(\ell,n+1)$ whose elements are indexed by words $\mathsf{w}\in\mathsf{W}$ and ordered according to $\preceq$:

$$\mathbf{c}_{\mathsf{l}_1} := (c_{\mathsf{w}})_{\mathsf{w}\in\mathsf{W}}.$$

The letter $\mathsf{l}_1$ is singled-out since it is the only independent letter. For notational simplicity, we let $\ell := \mathsf{l}_1$.

The *norm of an ordered vector* $\mathbf{c}_\ell$ is defined by

$$\|\mathbf{c}_\ell\| := \sqrt{\sum_{\mathsf{w}\in\mathsf{W}}|c_{\mathsf{w}}|^2},$$

and the *inner product between two ordered vectors* $\mathbf{a}_\ell$ and $\mathbf{b}_\ell$ by

$$\mathbf{a}_\ell \bullet \mathbf{b}_\ell := \sum_{\mathsf{w}\in\mathsf{W}} a_{\mathsf{w}}\cdot b_{\mathsf{w}}.$$

---

[2]This is not too surprising a result since in the unperturbed case the $\Theta_{m_j}$ are the associated Legendre functions, or the spherical harmonics, and $d(\ell,n+1)$ is then the dimension of the space of harmonic polynomials on the sphere $\mathbb{S}^{n+1}$. As the eigenvalues are the same in the perturbed case, the number of eigenfunctions remains the same.

For a word $\mathsf{w} = \ell \, \mathsf{l}_2 \mathsf{l}_3 \cdots \mathsf{l}_n \mathsf{l}_{n+1} \in W$ an *ordered vector function* $\mathbf{X}_\ell(\boldsymbol{\theta}, \phi)$ is defined by

$$X_{\mathsf{w}}(\boldsymbol{\theta}, \phi) := (\mathbf{X}_\ell)_{\ell \mathsf{l}_2 \mathsf{l}_3 \cdots \mathsf{l}_n \mathsf{l}_{n+1}}(\boldsymbol{\theta}, \phi) := \Theta_\ell(\theta_1) \Theta_{\mathsf{l}_2}(\theta_2) \cdots \Theta_{\mathsf{l}_n}(\theta_n) e_{\mathsf{l}_{n+1}}(\phi). \tag{20.24}$$

The next theorem summarizes the results obtained sofar.

**Theorem 20.3** *Assume that conditions **E1** and **E2** are satisfied. Then every real solution $u(r, \boldsymbol{\theta}, \phi)$ of Eqn. (20.9) with $\widetilde{\varepsilon} \equiv 0$ in the punctured open unit ball $\mathbb{B}^{n+2} \setminus \{0\}$ which is bounded in a neighborhood of the origin is continuous at the origin, and may be written in the form*

$$u(r, \boldsymbol{\theta}, \phi) = \sum_{\ell \in \mathbb{Z}_0^+} r_\ell^\ell \mathbf{c} \bullet \mathbf{X}_\ell(\boldsymbol{\theta}, \phi), \tag{20.25}$$

*for a unique set constants $\mathbf{c}_\ell$ satisfying*

$$\limsup_{\ell \to \infty} \|\mathbf{c}_\ell\|^{1/\ell} \leq 1. \tag{20.26}$$

*Conversely, every such set of constants with property (20.26) determines a solution through (20.25).*

**Proof.** By assumption, each $\varepsilon_j \in C^1$ and thus any solution $u(r, \boldsymbol{\theta}, \phi)$ of Eqn. (20.9) has Hölder continuous second derivatives in the punctured open unit ball. Now define the Fourier-Laplace coefficients of $u(r, \boldsymbol{\theta}, \phi) = u_r(\boldsymbol{\theta}, \phi)$ ($r$ here is regarded as a parameter) by

$$\widehat{\mathbf{u}}_r(\ell) := \mathbf{c}_\ell = (c_{\mathsf{w}})_{\mathsf{w} \in \mathsf{W}} \tag{20.27}$$

with

$$c_{\mathsf{w}} := \int_{[0,\pi]^n} \int_{[0,2\pi]} u_r(\boldsymbol{\theta}, \phi) \prod_{j=1}^n w_j(\theta_j) X_{\mathsf{w}}(\boldsymbol{\theta}, \phi) \, d\boldsymbol{\theta} \, d\phi. \tag{20.28}$$

Note that these Fourier-Laplace coefficients are uniquely determined for any function $f \in L^2(\mathbb{S}^{n+1}; dx')$ and that each such function has an expansion of the form

$$f = \sum_{\ell \in \mathbb{Z}_0^+} \widehat{\boldsymbol{f}}(\ell) \bullet \mathbf{X}_\ell,$$

with absolutely and uniformly convergent series. Thus,

$$u_r(\boldsymbol{\theta}, \phi) = \sum_{\ell \in \mathbb{Z}_0^+} \widehat{\boldsymbol{u}}_r(\ell) \bullet \mathbf{X}_\ell(\boldsymbol{\theta}, \phi), \quad 0 < r < 1,$$

and this series converges absolutely and uniformly on compact subsets of $\overset{\circ}{\mathbb{B}}^{n+2} \setminus \{0\}$. Applying the differential operator $L$ to the above equation and using its linearity yields

$$\sum_{\ell \in \mathbb{Z}_0^+} L\left(\widehat{\boldsymbol{u}}_r(\ell) \bullet \boldsymbol{X}_\ell\right) = \left(L_r \widehat{\boldsymbol{u}}_r\right) \bullet \boldsymbol{X}_\ell + \frac{1}{r^2}\, \widehat{\boldsymbol{u}}_r \bullet L_{\boldsymbol{\theta},\phi} \boldsymbol{X}_\ell \equiv 0.$$

(Here, $L_r$, respectively, $L_{\boldsymbol{\theta},\phi}$ denote the radial, respectively, angular parts of $L$.)

But $L_{\boldsymbol{\theta},\phi} \boldsymbol{X}_\ell = -\ell(\ell+n)\boldsymbol{X}_\ell$, and thus

$$Lu = \sum_{\ell \in \mathbb{Z}_0^+} \left[\left(\frac{1}{r^{n+1}} \frac{\partial}{\partial r}\left(r^{n+1} \frac{\partial \widehat{\boldsymbol{u}}_r(\ell)}{\partial r}\right)\right) - \frac{\ell(\ell+n)}{r^2} \widehat{\boldsymbol{u}}_r(\ell)\right] \bullet \boldsymbol{X}_\ell \equiv 0.$$

By the uniqueness of the Fourier-Laplace coefficients, the expression inside the brackets is equal to zero, i. e., $\widehat{\boldsymbol{u}}_r(\ell)$ is a solution of the Cauchy-Euler equation (20.16). As $u(r, \boldsymbol{\theta}, \phi)$ is bounded near the origin, so is $\widehat{\boldsymbol{u}}_r(\ell)$. Therefore, $\widehat{\boldsymbol{u}}_r(\ell)$ is the unique constant multiple of $r^\ell$.

The condition on the coefficient vector remains to be shown. However, as the functions $\Theta_{m_j}$ are bounded independently of $\ell$ (see above), (20.26) follows from application of the Cauchy-Schwartz inequality to the inner product $\bullet$, the root test, and the fact that

$$\limsup_{\ell \to \infty} \|\boldsymbol{X}_\ell\|^{1/\ell} \le \limsup_{\ell \to \infty} (C\sqrt{d_\ell}\,)^{1/\ell} = 1,$$

some $C > 0$.

The last statement of the theorem follows from (20.26), the fact that the series in Eqn. (20.25) therefore converges absolutely and subuniformly, and that $\mathbf{1} \bullet r^\ell \boldsymbol{X}_\ell$ is a solution of $Lu = 0$. ∎

**Remark 20.2** *The ordered vector functions $\boldsymbol{X}_\ell$ may be considered as* generalized *spherical harmonics. For, in the case that $\varepsilon_j \equiv 0$, for all $j = 1, \ldots, n$, Theorem (20.3) gives the following well-known characterization of harmonic functions in $\mathbb{R}^{n+2}$ (cf. also [7]):*
*A function $u(x) = u(rx')$ is harmonic in $\mathbb{B}^{n+2}$ iff it is of the form*

$$u(x) = \sum_{\ell \in \mathbb{Z}_0^+} r^\ell \boldsymbol{c}_\ell \bullet \mathbf{Y}_\ell(x'), \tag{20.29}$$

*for some unique sequence of constant vectors $\boldsymbol{c}_\ell$ with the property*

$$\limsup_{\ell \to \infty} \|\boldsymbol{c}_\ell\|^{1/\ell} \le 1. \tag{20.30}$$

*Here $\mathbf{Y}_\ell$ denotes the ordered vector of spherical harmonics on $\mathbb{S}^{n+1}$.*

The *Poisson kernel* $P(x, y')$ is – in the unperturbed case – defined in terms of the spherical harmonics as

$$P(x, y') = \sum_{\ell \in \mathbb{Z}_0^+} r^\ell \mathbf{Y}_\ell(y') \bullet \mathbf{Y}_\ell(x'). \tag{20.31}$$

For each fixed $y' \in \mathbb{S}^{n+2}$ the convergence is absolute and subcompact in $\mathbb{B}^{n+2}$, and moreover, $P(x, y')$ is harmonic in $\mathbb{B}^{n+2}$ as a function of $x$. The above remark suggests the definition of a *generalized Poisson kernel* $Q(x, y')$:

$$Q(x, y') := \sum_{\ell \in \mathbb{Z}_0^+} r^\ell \boldsymbol{X}_\ell(y') \bullet \boldsymbol{X}_\ell(x'). \tag{20.32}$$

As $\limsup_{\ell \to \infty} \|\boldsymbol{X}_\ell\|^{1/\ell} \leq 1$ the above series converges absolutely and subcompactly on $\mathbb{B}^{n+2}$. Since $\boldsymbol{X}_\ell$ is a solution of Eqn. (20.9), the following corollary holds.

**Corollary 20.4** *The generalized Poisson kernel $Q(x, y')$ defined by (20.32) is, for each fixed $y' \in \mathbb{S}^{n+1}$, a solution to the differential equation (20.9) with $\widetilde{\varepsilon} \equiv 0$.*

Finally, a remark about $\widetilde{\varepsilon} \not\equiv 0$ is in order. In the case of a perturbed azimuthal angle $\phi$, the eigenfunctions are given by Eqns. (20.13) and (20.14), and the eigenvalues by Eqn. (20.12). Denoting the complete orthonormal – with respect to the weight function $\eta(\phi)$ – set of eigenfunctions by $\{\Phi_k(\phi)\}$, and introducing the ordered vector function $\Xi_\ell(\boldsymbol{\theta}, \phi)$,

$$\Xi_{\mathsf{w}}(\boldsymbol{\theta}, \phi) := (\Xi_\ell)_{\ell \mathsf{l}_2 \mathsf{l}_3 \cdots \mathsf{l}_n \mathsf{l}_{n+1}}(\boldsymbol{\theta}, \phi) := \Theta_\ell(\theta_1)\Theta_{\mathsf{l}_2}(\theta_2) \cdots \Theta_{\mathsf{l}_n}(\theta_n)\Phi_{\mathsf{l}_{n+1}}(\phi), \tag{20.33}$$

a theorem similar to 20.3 can be proven.

**Theorem 20.5** *Under the assumptions of Theorem 20.3, every real solution $u(r, \boldsymbol{\theta}, \phi)$ of Eqn. (20.9) in the punctured open unit ball $\mathbb{B}^{n+2} \setminus \{0\}$ which is bounded in a neighborhood of the origin is continuous at the origin, and may be expressed in the form*

$$u(r, \boldsymbol{\theta}, \phi) = \sum_{\ell \in \mathbb{Z}_0^+} r^\ell \boldsymbol{c}_\ell \bullet \Xi_\ell(\boldsymbol{\theta}, \phi) \tag{20.34}$$

*for a unique set constants $\{\boldsymbol{c}_\ell\}$ satisfying*

$$\limsup_{\ell \to \infty} \|\boldsymbol{c}_\ell\|^{1/\ell} \leq 1. \tag{20.35}$$

*Conversely, every such set of constants with property (20.35) determines a solution through (20.34).*

**Proof.** The proof follows immediately from the proof of Theorem 20.3 with the obvious changes and replacements. ■

Analogously to Eqn. (20.32) one defines a generalized Poisson kernel by

$$\mathcal{Q}(x,y') := \mathcal{Q}_r(\boldsymbol{\theta},\phi) = \sum_{\ell\in\mathbb{Z}_0^+} r^\ell \Xi_\ell(y') \bullet \Xi_\ell(x'). \tag{20.36}$$

**Corollary 20.6** *The generalized Poisoon kernel $\mathcal{Q}(x,y')$ defined by (20.36) is, for each fixed $y' \in \mathbb{S}^{n+1}$, a solution to the differential equation (20.9).*

## 20.4 Distributional Boundary Values

In this section, the distributional boundary values of a solution $u(r,\boldsymbol{\theta},\phi) = u_r(\boldsymbol{\theta},\phi)$ are investigated as $r \to 1-$. In this context, $u(r,\boldsymbol{\theta},\phi)$ is viewed as a family of functions with trace values $u_r(\boldsymbol{\theta},\phi)$, $0 < r < 1$. The approach undertaken in [6, 7] which is based on results obtained in [5] is followed here as well. The main idea is to characterize the distributional boundary values in terms of ordered vector sequences satisfying condition (20.35). To this end, define a test space $\mathcal{J}$ on $\mathbb{S}^{n+1}$ as follows:

$$\mathcal{J} := \left\{ \varphi\colon \varphi(x') = \sum_{\ell\in\mathbb{Z}_0^+} \widehat{\varphi}(\ell) \bullet \Xi_\ell(x'),\ \limsup_{\ell\to\infty} \|\widehat{\varphi}(\ell)\|^{1/\ell} < 1 \right\}, \tag{20.37}$$

where $\widehat{\varphi}(\ell)$ denotes the $\ell$th ordered vector-valued Laplace-Fourier coefficient with respect to the ordered vector function $\Xi_\ell$ (see also (20.28)). Clearly, the linear space $\mathcal{J}$ depends on the pertubations $\widetilde{\varepsilon}$ and $\{\varepsilon_j\}_{j=1}^n$ and therefore on the weight functions $\eta(\phi)$ and $w_j(\theta_j)$, $j = 1, \ldots, n$. If all these pertubations are identically equal to zero, then $\mathcal{J}$ is identical to the space $\mathcal{H}$ of hyperfunctions or analytical functionals as, for instance, investigated by Lions & Magenes [9].

The completeness of the eigenfunctions $\{\Xi_\ell\}$ implies the existence of a injective mapping between elements of $\mathcal{J}$ and the set $\mathcal{S}$ of all ordered vector sequences $\boldsymbol{c}_\ell$. This mapping is explicitly given by $\varphi \mapsto \widehat{\varphi}(\ell)$.

To topologize $\mathcal{J}$, or equivalently, $\mathcal{S}$, one employs the approach developed by Johnson [5]. For this purpose, let $\mathsf{A}$ denote all real sequences $\mathsf{a} = \{\mathsf{a}_\ell\}_{\ell\in\mathbb{Z}_0^+}$ with the properties

$$\mathsf{a}_\ell \geq \mathsf{a}_{\ell+1} > 0, \qquad \ell \in \mathbb{Z}_0^+,$$

$$\mathsf{a}_{\ell+1}/\mathsf{a}_\ell \to 1 \qquad \text{as } \ell \to \infty.$$

The collection of sets

$$\mathsf{V}(\mathsf{a}) := \left\{\varphi \in \mathcal{J}\colon \|\widehat{\varphi}(\ell)\| \leq \mathsf{a}_\ell,\ \ell \in \mathbb{Z}_0^+\right\},$$

for all $\mathsf{a} \in \mathsf{A}$, forms a base for the neighborhood system of the origin for the desired topology on $\mathcal{J}$. Following the arguments in [5], only replacing moduli $|\cdot|$ with norms $\|\cdot\|$, one shows that the thus-defined topology is equivalent to the inductive limit topology and renders $\mathcal{J}$ into a locally convex topological vector space that is a nonmetrizable, complete Montel space ([5], Propositions 3, 4, and 6). Johnson's arguments also show – with no change – the following.

**Proposition 20.7** *The Laplace-Fourier series for each $\varphi \in \mathcal{J}$ converges to $\varphi$ in $\mathcal{J}$.*

Moreover, the collection

$$E(a, \varrho) := \left\{\varphi \in \mathcal{J} \colon \|\widehat{\varphi}(\ell)\| \le a\varrho^{\ell},\ \ell \in \mathbb{Z}_0^{+}\right\},$$

for all $a > 0$ and $0 < \varrho < 1$, is a fundamental system of bounded sets for $\mathcal{J}$, and each $E(a, \varrho)$ is compact (cf. [5], Proposition 5). Let $\mathcal{J}'$ be the linear space consisting of all continuous linear functionals on $\mathcal{J}$. The Fourier-Laplace coefficients of $f \in \mathcal{J}'$ are defined by

$$\widehat{\boldsymbol{f}}(\ell) := \boldsymbol{c}_\ell = (c_{\mathsf{w}})_{\mathsf{w}\in\mathsf{W}},$$

where the $d(\ell, n+1)$ components of the ordered vector $\boldsymbol{c}_\ell$ are given by

$$c_{\mathsf{w}} := \left\langle f, \prod_{j=1}^{n} \omega_j(\cdot)\, \eta(\cdot)\, \Xi_{\mathsf{w}} \right\rangle, \qquad \mathsf{w} \in \mathsf{W}. \tag{20.38}$$

Here $\langle\cdot,\cdot\rangle$ denotes the canonical pairing between the space and its dual. Elements of $\mathcal{J}'$ are identified with their Fourier-Laplace coefficients.

The next result is again a direct consequence of Theorem 2 in [5]. The proof goes through without any change. (NB. The uniqueness is a result of the completeness of the eigenfunctions $\{\Theta_{m_j}(\cdot)\Phi_k(\cdot)\}$.)

**Theorem 20.8** *If $f \in \mathcal{J}'$, then*

$$\limsup_{\ell\to\infty} \|\widehat{\boldsymbol{f}}(\ell)\|^{1/\ell} \le 1, \tag{20.39}$$

*and for each $\varphi \in \mathcal{J}$*

$$\langle f, \varphi\rangle = \sum_{\ell\in\mathbb{Z}_0^+} \widehat{\boldsymbol{f}}(\ell) \bullet \widehat{\varphi}(\ell) \tag{20.40}$$

*The above series converges absolutely. Conversely, any sequence of vectors $\boldsymbol{c}_\ell$ of length $d(\ell, n+1)$ satisfying $\limsup_{\ell\to\infty} \|\boldsymbol{c}_\ell\|^{1/\ell} \le 1$ determines a unique $f \in \mathcal{J}'$ with $\widehat{\boldsymbol{f}}(\ell) = \boldsymbol{c}_\ell$.*

Now suppose that $f \in \mathcal{J}$. Define a continuous linear functional $j_f : \mathcal{J} \to \mathbb{R}$ by $\langle j_f, \varphi \rangle := \int_{[0,\pi]^n} \int_{[0,2\pi]} f(\boldsymbol{\theta}, \phi)\, \varphi(\boldsymbol{\theta}, \phi) \prod_{j=1}^n w_j(\theta_j)\, \eta(\phi)\, d\boldsymbol{\theta}\, d\phi$. Then

$$\begin{aligned}
\widehat{j}_f(\ell) &= \left( \left\langle j_f, \prod_{j=1}^n \omega_j(\cdot)\, \eta(\cdot)\, \Xi_{\mathsf{w}} \right\rangle \right)_{\mathsf{w} \in \mathsf{W}} \\
&= \left( \int_{[0,\pi]^n} \int_{[0,2\pi]} f(\boldsymbol{\theta}, \phi)\, \Xi_{\mathsf{w}}(\boldsymbol{\theta}, \phi) \prod_{j=1}^n w_j(\theta_j)\, \eta(\phi)\, d\boldsymbol{\theta}\, d\phi \right)_{\mathsf{w} \in \mathsf{W}} \\
&= \widehat{\boldsymbol{f}}(\ell).
\end{aligned}$$

Thus, $\widehat{\boldsymbol{f}}(\ell)$ and therefore $\widehat{j}_f(\ell)$ satisfies the lim sup condition in (20.37) and hence (20.39). By Theorem (20.8), $j_f \in \mathcal{J}'$. Thus, the following corollary holds.

**Corollary 20.9** $\mathcal{J} \subseteq \mathcal{J}'$.

The linear space $\mathcal{J}'$ will be endowed with the strong topology, i. e., the topology of uniform convergence on bounded subsets of $\mathcal{J}$. For this purpose, define a class $\mathsf{B}$ of sequences $\mathsf{b} = \{\mathsf{b}_\ell\}_{\ell \in \mathbb{Z}_0^+}$ satisfying

$$\begin{aligned}
0 < \mathsf{b}_\ell \le \mathsf{b}_{\ell+1}, \qquad & \ell \in \mathbb{Z}_0^+ \\
\mathsf{b}_{\ell+1}/\mathsf{b}_\ell \to 1, \qquad & \text{as } \ell \to \infty.
\end{aligned} \tag{20.41}$$

The next proposition summarizes Propositions 9 and 10, and Theorem 3 of [5].

**Proposition 20.10** *The sets*

$$\mathsf{V}'(a, \varrho) := \{ f \in \mathcal{J}' \colon \|\widehat{\boldsymbol{f}}(\ell)\| \le a \varrho^\ell,\ \ell \in \mathbb{Z}_0^+ \},$$

*for all $a > 0$ and all $0 < \varrho < 1$, form a base for the neighborhood system at the origin for the strong topology of $\mathcal{J}'$. Furthermore, the sets*

$$\mathsf{F}'(\mathsf{b}) := \{ f \in \mathcal{J}' \colon \|\widehat{\boldsymbol{f}}(\ell)\| \le \mathsf{b}_\ell,\ \ell \in \mathbb{Z}_0^+ \},$$

*for all $\mathsf{b} \in \mathsf{B}$, form a fundamental system of bounded sets for $\mathcal{J}'$. The topological space $\mathcal{J}'$ is a Montel space whose strong dual is $\mathcal{J}$.*

As the eigenfunctions $\{\Xi_{\mathsf{w}}\}_{\mathsf{w} \in \mathsf{W}}$ are complete, they are total in $\mathcal{J}$. Therefore, boundedness in $\mathcal{J}'$ and convergence of the Fourier-Laplace coefficients imply strong convergence. In particular, for each $f \in \mathcal{J}'$

$$\sum_{\ell \in \mathbb{Z}_0^+} \widehat{\boldsymbol{f}}(\ell) \bullet \Xi_\ell(x') \to f \quad \text{in } \mathcal{J}'.$$

Also, weak convergence in $\mathcal{J}$ or $\mathcal{J}'$ implies strong convergence to the same limit (cf. [8]).

For $f$ and $g$ in $\mathcal{J}$ or $\mathcal{J}'$ a convolution is defined in the usual pointwise sense:

$$f * g := \sum_{\ell \in \mathbb{Z}_0^+} (\widehat{f}\,\widehat{g})(\ell) \bullet \Xi_\ell, \tag{20.42}$$

where $(\widehat{f}\,\widehat{g})(\ell) := (c_{\mathsf{w}}^f c_{\mathsf{w}}^g)_{\mathsf{w} \in \mathsf{W}}$ with $c_{\mathsf{w}}^f$ and $c_{\mathsf{w}}^g$ given by Eqn. (20.38). For each $g \in \mathcal{J}'$, the mapping $* \, g$ is continuous. Note that $(\mathcal{J}', *)$ is a convolution algebra whose unique identity is given by

$$\delta = \sum_{\ell \in \mathbb{Z}_0^+} \mathbf{1}_\ell \bullet \Xi_\ell, \tag{20.43}$$

where $\mathbf{1}_\ell$ is the ordered constant vector $(1)_{\mathsf{w} \in \mathsf{W}}$.

Employing the distributional setting developed sofar in this section, Theorem 20.5 may be recast in the following form to obtain a generalized Poisson integral representation of all solutions to Eqn. (20.9).

**Theorem 20.11** *A function $u(r, \boldsymbol{\theta}, \phi)$ in $\mathbb{B}^{n+2}$ is a solution of Eqn. (20.9) iff there exists a generalized function $f \in \mathcal{J}'$ such that*

$$u(r, \boldsymbol{\theta}, \phi) = u_r(\boldsymbol{\theta}, \phi) = (\mathcal{Q}_r * f)(\boldsymbol{\theta}, \phi) \tag{20.44}$$

*for each $0 < r < 1$. Furthermore, $u_r \to f$ in $\mathcal{J}'$ and consequently $f$ is uniquely determined.*

**Proof.** Suppose that $u(r, \boldsymbol{\theta}, \phi)$ is a solution of Eqn. (20.9). Then $u$ may be expressed in the form (20.34) with uniquely determined ordered vector coefficients $\boldsymbol{c}_\ell$ satisfying (20.35). Defining

$$f := \sum_{\ell \in \mathbb{Z}_0^+} \boldsymbol{c}_\ell \bullet \Xi_\ell,$$

and appealing to Theorem (20.8), gives immediately $f \in \mathcal{J}'$ and the equivalence of (20.34) and (20.44). Moreover, choosing $\mathsf{b} := (1, 1, \ldots) \in \mathsf{B}$, Proposition (21.5) implies that the family of functions $\mathcal{Q}_r \in \mathcal{J} \subseteq \mathcal{J}'$ is bounded in $\mathcal{J}'$, since $\widehat{\mathcal{Q}}_{r_{\mathsf{w}}}(\ell) = r^\ell \leq 1$. Thus, $\mathcal{Q}_r \to \delta$ in $\mathcal{J}'$ as $r \to 1-$. Therefore, $u_r = \mathcal{Q}_r * f \to \delta * f = f$ in $\mathcal{J}'$, by the continuity of $*$. This proves the necessity part of the theorem.

The sufficiency part is immediate. ■

It should be remarked that the above result is consistent with the classical solution of the Dirichlet problem for Eqn. (20.9). Suppose that $f \in C(\mathbb{S}^{n+1})$ and $v(r, \boldsymbol{\theta}, \phi) = v_r(\boldsymbol{\theta}, \phi)$ is the unique solution to (20.9) on $\mathbb{B}^{n+2}$, with $v \in C(\mathbb{B}^{n+2} \cup \mathbb{S}^{n+1})$ and $v \equiv f$ on $\mathbb{S}^{n+1}$. The $v_r \to f$ uniformly as $r \to 1-$ and therefore $v_r \to f$ in $\mathcal{J}'$. By the uniqueness part of Theorem (20.11), $v_r \equiv u_r = \mathcal{Q}_r * f$.

## 20.5 Generalities

In this section a slight generalization is being presented. The pertubation function in Eqn. (20.1) is assumed to also contain a radial pertubation. More precisely, the radial part of the Laplace-Beltrami operator $\Delta$ on $\mathbb{B}^{n+2}$ is assumed to have the form

$$\breve{L}_r := \frac{1}{r^{n+1}} \frac{\partial}{\partial r} \left( r^{n+1} \frac{\partial}{\partial r} \right) + \frac{\breve{\varepsilon}(r)}{r} \frac{\partial}{\partial r}, \tag{20.45}$$

where the radial pertubation function $\breve{\varepsilon}$ satisfies the following conditions:

- $\breve{\varepsilon} \in C[0,1]$ and $\int_0^1 \frac{|\breve{\varepsilon}|}{r} dr$ converges;
- $\breve{\varepsilon}'$ is Hölder continuous for $r_0 \leq r \leq 1$, for all $0 < r_0 < 1$;
- $\breve{\varepsilon}'(r) = o(1/r)$ as $r \to 0+$ and $\int_0^1 |\breve{\varepsilon}'| dr$ converges.

Under these conditions it was shown in [7] that there exist bounded solutions to an *only radially* perturbed Laplace equation; the radial functions $r^\ell$ are replaced by bounded functions $R_\ell(r)$ with controlled growth rates on the derivatives. Taking these results into account, one arrives at the following structure theorem whose proof is a straighforward extension of that presented in [7] (Theorem 3.1).

**Theorem 20.12** *A function $u(x)$ defined in $\mathbb{B}^{n+2}$ is a solution of the fully perturbed Laplace equation*

$$(\breve{L}_r + L_{\boldsymbol{\theta},\phi})u = 0 \tag{20.46}$$

*iff there exists a sequence of Borel measures $\{\mu_q\}$ on $\mathbb{S}^{n+1}$ satisfying*

$$\sum_{q \in \mathbb{Z}_0^+} (2q)! M^q \operatorname{var} \mu_q < \infty, \quad \textit{for all } M > 0, \tag{20.47}$$

*such that*

$$u(x) = \sum_{q \in \mathbb{Z}_0^+} \int_{\mathbb{S}^{n+1}} \Delta_{y'}^q \breve{\mathfrak{Q}}(x, y') \, d\mu_q(y'), \tag{20.48}$$

*where $\Delta_{y'}^q$ denotes the $q$th iterate of the Laplace-Beltrami operator on $\mathbb{S}^{n+1}$ and $\breve{\mathfrak{Q}}(x, y') := \sum_{\ell \in \mathbb{Z}_0^+} R_\ell(r) \, \Xi_\ell(y') \bullet \Xi_\ell(x')$.*

The distributional boundary values remain $\mathcal{J}'$ in this case of a fully perturbed Laplace equation, and Theorem 20.11 is still valid where, of course, the new generalized Poisson kernel is given as above.

*References*

[1] R. E. Bellman, *Stability Theory of Differential Equations*, McGraw - Hill, New York, 1953.

[2] R. Courant and D. Hilbert, *Methoden der Mathematischen Physik I.* Springer Verlag, Heidelberg, 1968.

[3] M. S. P. Eastham, *The spectral theory of periodic differential equations*, McGraw - Hill, New York, 1973.

[4] E. W. Hobson, *The Theory of Spherical and Ellipsoidal Harmonics*, Chelsea Publishing Company, New York, 1955.

[5] G. Johnson, *Harmonic functions on the unit disk I*, Illinois J. Math. **12** (1968), 143 – 154.

[6] J. Kelingos and P. R. Massopust, *A characterization of solutions to a perturbed Laplace equation II,* Rocky Mountain J. Math. **24**, No. 2 (1994), 549 – 562.

[7] J. Kelingos and P. R. Massopust, *A characterization of solutions to a radially perturbed Laplace equation in the unit n-ball*, J. Math. Anal. and Appl. **199** (1996), 728 – 747.

[8] G. Köthe, *Topological Vector Spaces*, Springer Verlag, New York, 1969.

[9] J. L. Lions and E. Magenes, *Problèmes aux limites non homogènes, VII*, Ann. Mat. Pura Appl. **63** (1963), 201 – 224.

[10] M. Sato, *Theory of Hyperfunctions I and II*, J. Fac. Sci. Univ. Tokyo **8** (1959 - 60), 139 – 193, 387 – 437.

[11] P. O. Staples and J. Kelingos, *A characterization of solutions to a perturbed Laplace equation*, Illinois J. Math. **22** (1978), 208 – 216.

# Chapter 21

# Nonresonant semilinear equations and applications to boundary value problems

**P. S. Milojević**

*ABSTRACT. In this paper we present some new solvability results for nonresonant semi-abstract semilinear equations involving nonlinear perturbations of Fredholm maps of index zero. Applications are given to semilinear elliptic boundary value problems and to semilinear parabolic and hyperbolic periodic-boundary value problems.*

## 21.1 Introduction

Let $A : D(A) \subset X \to Y$ be a linear map and $N : X \to Y$ a nonlinear map. We say that the equation $Ax + Nx = f$ is not in resonance if $A$ and $N$ are such that it is solvable for each $f \in Y$. Such equations have been studied extensively using various topological and variational methods. In [Mi-1-4], we have studied such equations using the (pseudo) $A$-proper mapping approach.

In Section 21.2, using the abstract theorem stated below, we study semi-abstract nonresonant semilinear equations. Sections 21.3 and 21.4 are devoted to applications of the results of Section 21.2 to BV problems for semilinear elliptic equations and to periodic-BVP's for semilinear parabolic and hyperbolic equations assuming nonuniform nonresonance conditions. Such problems have been studied earlier in [Ma-1-4], [MW], [NW], etc.

Recall the definition of (pseudo) $A$-proper maps.

**Definition 21.1** *A map $T : D \subset X \to Y$ is (pseudo) $A$-proper w.r.t. a scheme $\Gamma = \{X_n, Y_n, Q_n\}$ with $dim X_n = dim Y_n$ on $D$ if whenever $\{x_{n_k} \in D \cap X_{n_k}\}$ is bounded and such that $Q_{n_k} T x_{n_k} - Q_{n_k} f \to 0$ for some $f \in Y$, then $\{x_{n_k}\}$ has a subsequence converging to $x \in D$ (there is $x \in D$) with $Tx = f$.*

The classes of $A$-proper and pseudo $A$-proper maps are very general and we refer to [Mi-1-4] for many examples of such maps.

We need the following abstact result. For a map $M$, define its quasinorm by $|M| = \limsup_{||x|| \to \infty} ||Mx||/||x||$.

**Theorem 21.1** *(cf. [Mi-4]) Let $A : D(A) \subset X \to Y$ be a linear densely defined map and $N : X \to Y$ be bounded and of the form $Nx = B(x)x + Mx$ for some linear maps $B(x) : X \to X$. Assume that there is a $c > |M|$ and a positively homogeneous map $C : X \to Y$ such that*

$$||Ax - (1-t)Cx - tB(x)x|| \geq c||x||, \; x \in D(A) \setminus B(0,R).$$

*(i) $H_t = A - (1-t)C - tN$ is A-proper w.r.t. $\Gamma = \{X_n, Y_n, Q_n\}$ for $t \in [0,1)$ and $A - N$ is pseudo A-proper*

*(ii) For all $r > R$, $deg(Q_n(A-C), B(0,r) \cap X_n, 0) \neq 0$ for each large $n$. Then the equation $Ax + Nx = f$ is solvable for each $f \in Y$.*

## 21.2 Semi-abstract nonresonance problems

Let $Q \subset R^n$ be a bounded domain, $V$ be a closed subspace of $W_2^{2m}(Q)$ containing the test functions and $L : V \to L_2$ be a linear map with closed range in $H = L_2(Q)$. Let $V_1$ be a closed subspace of $V$ and $L_1$ be the restriction of $L$ to $V_1$. Assume

(L1) Each eigenvalue $\lambda_j$ of $L_1$ has a finite multiplicity and the corresponding eigenfunctions $\{..., w_{-1}, w_0, w_1, ..\}$ form a complete set in $V_1$.

Let $A = A_1 + L$ for some linear map $A_1 : V \to H$. For a fixed integer $j$, define $B : V \to H$ by $Bu = -Au + \lambda_j u$.

(B1) There is $\lambda \neq \lambda_j$, $j = 1, 2, ...$, such that the map $B - \lambda I = -A_1 - L - (\lambda - \lambda_j)I : V \to L_2$ is bijective.

Let $\lambda \neq \lambda_j$ for each $j = 1, 2, ...$ be fixed, $\Gamma = \{Y_n, Q_n\}$ be a projectionally complete scheme for $L_2$ and $X_n = (B - \lambda I)^{-1}(Y_n) \subset V$ for each $n$. Then $\Gamma_B = \{X_n, Y_n, Q_n\}$ is an admissible or a projectionally complete scheme for $(V, L_2)$. Since $B - \lambda I : V \to L_2$ is linear, one-to-one and $A$-proper w.r.t. $\Gamma_B$, there is a constant $c > 0$ (depending) only on $\lambda$) such that

$$||(B - \lambda I)u|| \geq c||u||_V, \;\; u \in V. \tag{21.1}$$

Consider the following semilinear equation in $V$

$$Au + g(x, u, Du, ..., D^{2m-1}u)u + f(x, u, Du, ..., D^{2m}u) = h(x) \tag{21.2}$$

For $u \in H$, set $u^{\pm}$ =max$(\pm u, 0)$. Let $r = \lambda_{j+1} - \lambda_j$. We require that $B$ has the following properties:

**Property I** $B$ is a closed densely defined map in $H$ with closed range $R(B)$, $(Bu, u) \geq -r^{-1}||Bu||^2$ and $R(B) = N(B)^{\perp}$ in $H$, $N(-L_1 + \lambda_j I) \subset N(B)$ and $(Bu, u) = (-Lu + \lambda_j u, u)$ on $V$.

**Property II** If $(Bu, u) = -r^{-1}||Bu||^2$ for some $u \in V$, then $u \in N(-L_1 + \lambda_j I) \oplus N(-L_1 + \lambda_{j+1} I)$.

Let us note that if $B^{-1}$ is a partial inverse of $B$ and $B^{-1} + r^{-1}I$ is strongly monotone on $R(B)$, i.e. it is a bounded linear map on $R(B)$ and

$((B^{-1}+r^{-1}I)u,u) \geq c_0||(B^{-1}+r^{-1}I)u||^2$ on $R(B)$ for some $c_0 > 0$, then ([BF]) Property II holds in the sense that if $(Bu,u) = -r^{-1}||Bu||^2$ for some $u \in V$, then $u \in N(B) \oplus N(B+rI)$. If $B$ is selfadjoint or angle bounded in the sense of H. Amann, it is known that $B^{-1}+r^{-1}I$ is strongly monotone. If $B \neq B^*$ and $B$ is a normal map, the strong monotonicity of $B^{-1}+r^{-1}I$ has been discussed in Hetzer [H].

Some properties of $B$ are given next.

**Lemma 21.2** *Let $B$ have Properties I and II. Suppose that $p_\pm \in L_\infty(Q)$ are such that $0 \leq p_\pm(x) \leq r$ for a.e. $x \in Q$ and*

$$\int_Q [p_+(v^+)^2 + p_-(v^-)^2] > 0 \textit{ for all } v \in N(-L_1+\lambda_j I) \setminus \{0\}$$

*and*

$$\int_Q [(r-p_+)(w^+)^2 + (r-p_-)(w^-)^2] > 0 \textit{ for all } w \in N(-L_1+\lambda_{j+1} I) \setminus \{0\}.$$

*Then the equation*

$$Bu + p_+u^+ - p_-u^- = 0 \tag{21.3}$$

*has only the trivial solution.*

**Proof.** Define $p : Q \times R \to R$ by

$$p(x,u) = p_+(x) \text{ if } u \geq 0,$$

$$p(x,u) = p_-(x) \text{ if } u \leq 0.$$

Then

$$0 \leq p(x,u) \leq r \text{ for } (x,u) \in Q \times R \tag{21.4}$$

and, for $u \in H$ and a.e. $x \in Q$,

$$p(x,u(x))u(x) = p(x,u(x))u^+(x) - p(x,u(x))u^-(x) =$$

$$p_+(x)u^+(x) - p_-(x)u^-(x).$$

Define $P : V \subset H \to H$ by $(Pu)(x) = p(x,u(x))u(x)$ for a.e. $x \in Q$. Then Eq. (21.3) is equivalent to

$$Bu + Pu = 0, \quad u \in V. \tag{21.5}$$

By (21.4), we have that $||Pu||^2 \leq r(Pu, u)$ on $V$. Moreover, for each solution $u \in V$ of (21.5), we get by Property I that

$$-r^{-1}||Pu||^2 = -r^{-1}||Bu||^2 \leq (Bu, u) = (-Pu, u)$$

and so $||Pu||^2 \geq r(Pu, u)$. Hence, $||Pu||^2 = r(Pu, u)$ and $(Bu, u) = -r^{-1}||Bu||^2$. By Property II, we get that

$$u \in N(-L_1 + \lambda_j I) \oplus N(-L_1 + \lambda_{j+1} I).$$

Hence, $u = v + w$ with $v \in N(-L_1 + \lambda_j I)$ and $w \in N(-L_1 + \lambda_{j+1} I)$. Since $u$ is a solution of (21.3), we get that

$$(Bu, u) = (-Lu + \lambda_j u, u) = (-Lv + \lambda_j v, v) + (-Lv + \lambda_j v, w)$$

$$+(-Lw + \lambda_j w, v + w) = (-Lw + \lambda_{j+1} w - rw, v + w) = (-rw, w)$$

and so $(-rw, w) + (p(.,u(.))(v + w), v + w) = 0$. Then

$$(v - w, -rw + p(.,u(.))(v + w)) = (v + w, -rw + p(.,u(.))(v + w))$$

$$-2(w, -rw + p(.,u(.))(v + w)) = -2(w, -rw + p(.,u(.))(v + w))$$

$$= -2(v + w, -rw - B(v + w)) + 2(v, -rw - B(v + w)) =$$

$$-2(v, rw + B(v + w)) = -2(v, B(v + w)) = 0$$

since $v \in N(-L_1 + \lambda_j I) \subset N(B)$ and $R(B) = N(B)^\perp$. Since

$$\begin{aligned} (p(.,u(.))(v + w), v - w) &= (p(.,u(.))v, v) + ([r - p(.,u(.))]w, w) \\ &\quad +([r - p(.,u(.))]w, -v) + (p(.,u(.))v, -w) \\ &= (p(.,u(.))v, v) + ([r - p(.,u(.))]w, w) \end{aligned}$$

we get that

$$\begin{aligned} (p(.,u(.))v, v) + ([r - p(.,u(.))]w, w) &= (p(.,u(.))(v + w), v - w) \\ &\quad +(rw, -v + w) \\ &= 0. \end{aligned} \tag{21.6}$$

Since each term in (21.6) is nonnegative by (21.4), we get that each term is zero, i.e.,

$$\int_Q p(x, v(x) + w(x))v^2(x)dx = 0 \tag{21.7}$$

$$\int_Q [(r - p(x, v(x) + w(x))]w^2(x)dx = 0. \tag{21.8}$$

Set $Q_v = \{x \in Q \,|v(x) \neq 0\}$ and $Q_w = \{x \in Q \,|w(x) \neq 0\}$.

By (21.7)-(21.8), we get $p(x, v(x) + w(x)) = 0$ for a.e. $x \in Q_v$ and $p(x, v(x) + w(x)) = r$ for a.e. $x \in Q_w$ and so $Q_v \cap Q_w = \emptyset$. If $Q_v = \emptyset$, then $u = w$ and the equation (21.8) becomes

$$0 = \int_Q [(r - p(x, w(x))]w^2(x)dx = \int_Q (r - p_+)(w^+)^2 + (r - p_-)(w^-)^2$$

so that by our hypothesis, $w = 0$ and therefore $u = 0$.

Next, suppose that $Q_v \neq \emptyset$. Then we have that $p(x, v(x) + w(x)) = 0$ on $Q_v$ and, by (21.8), $\int_{Q_v} rw^2(x) = 0$, i.e., $w(x) = 0$ for a.e. $x \in Q_v$. Then by (21.7)

$$\begin{aligned} 0 &= \int_{Q_v} p(x, v(x))v^2(x) = \int_{Q_v} (p_+(v^+)^2 + p_-(v^-)^2) \\ &= \int_Q (p_+(v^+)^2 + p_-(v^-)^2). \end{aligned}$$

By our assumption, this implies that $v = 0$, in contradiction to $Q_v \neq \emptyset$. Hence, $Q_v = \emptyset$ and $u = 0$. ∎

**Lemma 21.3** *Let (L1) and (B1) hold and $B$ have Properties I and II. Suppose that $a_\pm$, $b_\pm \in L_\infty(Q)$ are such that $0 \leq a_\pm(x) \leq b_\pm \leq r$ for a.e. $x \in Q$ and*

$$\int_Q [a_+(v^+)^2 + a_-(v^-)^2] > 0 \textit{ for all } v \in N(-L_1 + \lambda_j I) \setminus \{0\} \tag{21.9}$$

*and*

$$\int_Q [(r - b_+)(w^+)^2 + (r - b_-)(w^-)^2] > 0 \textit{ for all } w \in N(-L_1 + \lambda_{j+1} I) \setminus \{0\}. \tag{21.10}$$

*Then there exists $\epsilon = \epsilon(a_\pm, b_\pm) > 0$ and $\delta = \delta(a_\pm, b_\pm) > 0$ such that for all $p_\pm \in L_\infty(Q)$ with*

$$a_+(x) - \epsilon \leq p_+(x) \leq b_+(x) + \epsilon \tag{21.11}$$

$$a_-(x) - \epsilon \leq p_-(x) \leq b_-(x) + \epsilon \tag{21.12}$$

*for a.e. $x \in Q$ and for all $u \in V$, one has*

$$||Bu + p_+u^+ - p_-u^-|| \geq \delta ||u||_V. \tag{21.13}$$

**Proof.** If this is not the case, then we can find the sequences $\{u_k\} \subset V$, with $||u_k||_V = 1$ for each $k$ and $\{p^k_\pm\} \subset L_\infty(Q)$ such that

$$a_\pm(x) - k^{-1} \leq p_\pm(x) \leq b_\pm(x) + k^{-1} \text{ a.e. on } Q \tag{21.14}$$

and

$$Bu_k + p^k_+ u^-_k - p^k_- u^-_k = v_k \to 0 \text{ as } k \to \infty. \tag{21.15}$$

Then $p^k_\pm \to p_\pm$ weakly in $H$ with $a_\pm(x) \leq p_\pm(x) \leq b_\pm(x)$ a.e. on $Q$. Let $\mu \neq \lambda_j$ and consider the identity

$$u_k + (B - \mu I)^{-1}[(p^k_+ - p_+)u^+_k - (p^k_- - p_-)u^-_k] \tag{21.16}$$

$$= (B - \mu I)^{-1}(-p_+ u^+_k + p_- u^-_k - \mu u_k + v_k).$$

By the compactness of the embedding of $V$ into $L_2$, we have that $u_k \to u$ in $L_2$ as well as $u^\pm_k \to u^\pm$ in $L_2$. Since $(B - \mu I)^{-1}$ is continuous both as a map from $L_2$ to $V$ and from $L_2$ to $L_2$, we get that

$$\begin{aligned} &(B - \mu I)^{-1}(-p_+ u^+_k + p_- u^-_k - \mu u_k + v_k) \\ &\to (B - \mu I)^{-1}(-p_+ u^+ + p_- u^- - \mu u) \end{aligned} \tag{21.17}$$

in $L_2$ and $V$.

Next, we shall show that $p^k_\pm \to p_\pm u^\pm$ weakly in $H$. For $\phi \in C^\infty_0(Q)$, we have that

$$(p^k_+ u^+_k - p_+ u^+, \phi) = (p^k_+(u^+_k - u^+), \phi) + ((p^k_+ - p_+)u^+, \phi)$$

$$\leq c||u^+_k - u^+|| + ((p^k_+ - p_+)u^+, \phi) \to 0$$

as $k \to \infty$. Hence, $p^k_+ u^+_k \to p_+ u^+$ weakly in $L_2$ by the density of $C^\infty_0(Q)$ in $L_2$, and similarly, $p^k_- u^-_k \to p_- u^-$ weakly in $L_2$. Hence, (21.16)-(21.17) imply that $u = (B - \mu I)^{-1}(-p_+ u^+ + p_- u^- - \mu u)$, i.e., $Bu + p_+ u^+ - p_- u^- = 0$. Moreover, for each $v \in N(-L_1 + \lambda_j I) \setminus \{0\}$, we have that

$$\int_Q p_+(v^+)^2 + p_-(v^-)^2 \geq \int_Q a_+(v^+)^2 + a_-(v^-)^2 > 0$$

and, for each $w \in N(-L_1 + \lambda_{j+1} I) \setminus \{0\}$, we have that

$$\int_Q [(r - p_+)(w^+)^2 + (r - p_-)(w^-)^2] \geq \int_Q [(r - b_+)(w^+)^2 + (r - b_-)(w^-)^2] > 0.$$

Hence, by Lemma 21.2, $u = 0$ a.e.on $Q$. Thus, $u_k \to 0$ in $L_2$, $||u_k||_V = 1$ and $||p_+^k u_k^+ - p_-^k u_k^- - \mu u_k|| \to 0$. By (21.1), we get that

$$||Bu_k + p_+^k - p_-^k u_k^-|| \geq ||Bu_k - \mu u_k|| - ||\mu u_k - p_+^k u_k^+ + p_-^k u_k^-||$$

$$\geq c - ||p_+^k u_k^+ - p_-^k u_k^- - \mu u_k||.$$

By (21.15), passing to the limit as $k \to \infty$, we get that $0 \geq c > 0$, a contradiction. Hence, the lemma is valid. ∎

**Remark 21.1** *Modifying suitably the proof of Lemma 21.3, condition (B1) can be replaced by*
*(B2) $dimN(B) < \infty$ and the partial inverse of $B$ is compact.*

Let $R^{s_k}$ be the vector space whose elements are

$$\xi = \{\xi_\alpha \mid |\alpha = (\alpha_1, ..., \alpha_n)| \leq k\}.$$

Each $\xi \in R^k$ may be written as a pair $\xi = (\eta, \zeta)$ with $\eta \in R^{s_{k-1}}$, $\zeta = \{\xi_\alpha \mid |\alpha| = k\} \in R^{s_k - s_{k-1}} = R^{s_k'}$ and $|\xi| = (\sum_{|\alpha| \leq k} |\xi_\alpha|^2)^{1/2}$. Set $\eta(u) = (Du, ..., D^{2m-1}u)$ and $\xi(u) = (u, Du, ..., D^{2m}u)$. Define $Nu = g(x, u, \eta(u))u + Fu$, where $Fu = f(x, \xi(u))$. Set $k = s_{2m-1} - 1$.

**Theorem 21.4** *Let (L1) and (B1) hold, $B$ have properties I and II and there are functions $\gamma_\pm, \Gamma_\pm \in L_\infty(Q)$ such that for some $j \in J \subset Z$, one has*

$$\lambda_j \leq \gamma_\pm(x) \leq \Gamma_\pm(x) \leq \lambda_{j+1} \text{ for a.e. } x \in Q$$

*and*

$$\int_Q [(\gamma_+ - \lambda_j)(v^+)^2 + (\gamma_- - \lambda_j)(v^-)^2] > 0 \text{ for all } v \in N(L_1 - \lambda_j I) \setminus \{0\} \tag{21.18}$$

*and*

$$\int_Q [(\lambda_{j+1} - \Gamma_+)(w^+)^2 + (\lambda_{j+1} - \Gamma_-)(w^-)^2] > 0 \tag{21.19}$$

*for all $w \in N(L_1 - \lambda_{j+1} I) \setminus \{0\}$. Suppose that for $\epsilon > 0$ and $\delta > 0$ given in Lemma 21.3,*
*(G1) there is $\rho > 0$ such that for a.e. $x \in Q$*

$$\gamma_+(x) - \epsilon \leq g(x, u, \eta(u)) \leq \Gamma_+(x) + \epsilon \text{ if } u > \rho, \eta(u) \in R^k$$

$$\gamma_-(x) - \epsilon \leq g(x, u, \eta(u)) \leq \Gamma_-(x) + \epsilon \text{ if } u < -\rho, \eta(u) \in R^k$$

*(G2) There are functions $b(x) \in L_\infty(Q)$ and $k_s(x) \in L_2(Q)$ for each $s > 0$ such that*

$$|g(x, u, \eta(u))| \leq sb(x)(\sum_{|\alpha| \leq 2m-1} |D^\alpha u|^2)^{1/2} + k_s(x), u \in V.$$

*(F) $||Fu|| = ||f(x, u, Du, ..., D^{2m}u)|| \leq \beta||u||_V + \gamma$ for $\beta \in (0, \delta), \gamma > 0$.*
*(H) $H_t = A - \lambda_j I - tF : V \to H$ is A-proper w.r.t. $\Gamma_B$ for $t \in [0, 1)$ and $H_1 = B - F$ is pseudo A-proper.*
*Then the equation (21.2) has at least one solution in $V$ for each $h \in H$.*

**Proof.** Let $g_1 : Q \times R \times R^k \to R$ be given by $g_1(x, u, \eta(u)) = g(x, u, \eta(u)) - \lambda_j$. Define functions

$$g_+(x, u, \eta(u)) = g_1(x, u, \eta(u)) \quad (x, \eta(u)) \in Q \times R^k, u \geq \rho$$

$$g_+(x, u, \eta(u)) = g_1(x, \rho, \eta(u)) \quad (x, \eta(u)) \in Q \times R^{s_{2m-1}}, 0 \leq u \leq \rho$$

$$g_-(x, u, \eta(u)) = g_1(x, u, \eta(u)) \quad (x, \eta(u)) \in Q \times R^k, u \leq -\rho$$

$$g_-(x, u, \eta(u)) = -g_1(x, -\rho, \eta(u)) \quad (x, \eta(u)) \in Q \times R^{s_{2m-1}}, -\rho \leq u \leq 0.$$

$$q(x, 0, \eta(u)) = g_1(x, 0, \eta(u)) \quad (x, \eta(u)) \in Q \times R^k$$

$$q(x, u, \eta(u)) = g_1(x, u, \eta(u))u - g_+(x, u, \eta(u))u \quad (x, \eta(u)) \in Q \times R^k$$

and $u > 0$

$$q(x, u, \eta(u)) = g_1(x, u, \eta(u))u - g_-(x, u, \eta(u))u \text{ for all } (x, \eta(u)) \in Q \times R^k$$

and $u < 0$. Then $q$ satisfies Caratheodory conditions. Set $a_\pm(x) = \gamma_\pm(x) - \lambda_j$ and $b_\pm(x) = \Gamma_\pm(x) - \lambda_j$. Then

$$a_+(x) - \epsilon \leq g_+(x, u, \eta(u)) \leq b_+(x) + \epsilon \text{ on } Q \times R_+ \times R^k$$

$$a_-(x) - \epsilon \leq g_-(x, u, \eta(u)) \leq b_-(x) + \epsilon \text{ on } Q \times R_- \times R^k.$$

Then problem (21.2) is equivalent in $V$ to

$$Bu + g_+(x, u^+, \eta(u))u^+ - g_-(x, -u^-, \eta(u))u^- \\ + f(x, \xi(u)) + q(x, u, \eta(u)) = -h.$$

Then for $u \in H$, set

$$Q_+(u) = \{x \in Q \mid u(x) > 0\}, Q_-(u) = \{x \in Q \mid u(x) < 0\}$$

and let $\chi_{Q_\pm}$ be the corresponding characteristic functions. Define the maps $E : H \to L_\infty(Q)$, $F, G, H : V \to H$, respectively, by

$$E(u)(x) = g_+(x, u^+(x), \eta(u))\chi_{Q_+(u)} + g_-(x, -u^-(x), \eta(u))\chi_{Q_-(u)}$$

$G(u)(x) = [E(u)(x)]u(x) = (E(u)u)(x)$ so that

$$G(u)(x) = g_+(x, u^+(x), \eta(u))u^+(x) - g_-(x, -u^-(x), \eta(u))u^-(x),$$

$F(u)(x) = f(x, \xi(u))$ and $H(u)(x) = q(x, u(x), \eta(u))$. Hence (21.2) can be written in the operator form

$$Bu + Gu + Fu + Hu = -h, \quad u \in V. \tag{21.20}$$

We know that $G, F$ and $H$ are well defined, continuous and bounded in $H$.

Let $C : H \to H$ be defined by $C(u)(x) = b_+(x)u^+(x) - b_-(x)u^-(x)$. Clearly, $C$ is a positively homogeneous map and $C, G, H : V \to H$ are completely continuos maps, i.e. they map weakly convergent sequences in $V$ into strongly convergent sequences in $H$. Indeed, let us show this, for example, for $G$. Since $V$ is compactly embedded in $H$, it follows from the construction of $G$ and (G2) that if $\{u_k\} \subset V$ converges weakly to $u_0$ in $V$, then ([K])

$$\begin{aligned} ||g_+(x, u_k^+, \eta(u_k)) &- g_-(x, -u_k^-, \eta(u_k)) \\ &- g_+(x, u_0^+, \eta(u_0)) + g_-(x, -u_0^-, \eta(u_0))|| \to 0. \end{aligned}$$

Hence, the map $G : V \to L_p$ is completely continuous since

$$\begin{aligned} ||Gu_k - Gu_0|| &= ||E(u_k)u_k - E(u_0)u_0|| \\ &\le ||E(u_k)(u_k - u_0)|| + ||E(u_k) - E(u_0)||||u_0|| \end{aligned}$$

$$\begin{aligned} \le \max\{||a_+||_\infty &+ ||a_-||_\infty + 2\epsilon, ||b_+||_\infty + ||b_-||_\infty + 2\epsilon\}||u_k - u_0|| \\ &+ ||b_-||_\infty + 2\epsilon\}||u_k - u_0|| + ||E(u_k) - E(u_0)||\, ||u_0|| \to 0. \end{aligned}$$

Thus, we have that $H_t = B + (1-t)C + t(F + G + H)$ is $A$-proper for each $t \in [0, 1)$ from $V \to H$ and $H_1 : V \to H$ is pseudo $A$-proper.

Next, by the construction

$$(1-t)Cu + tGu = [(1-t)b_+(x) + tg_+(x, u^+, \eta(u))]u^+(x) -$$

$$[(1-t)b_-(x) + tg_-(x, -u^-, \eta(u))]u^-(x)$$

and, for a.e. $x \in Q$, $\eta(u) \in R^k$

$$a_+(x) - \epsilon \le (1-t)b_+(x) + tg_+(x, u^+(x), \eta(u)) \le b_+(x) + \epsilon$$

$$a_-(x) - \epsilon \leq (1-t)b_-(x) + tg_-(x, -u^-(x), \eta(u)) \leq b_(x) + \epsilon$$

and $|q(x,u,\eta(u))| \leq d_\rho(x)$ for a.e. $x \in Q$ and all $(u, \eta(u)) \in R \times R^k$, where $d_\rho \in L_2(Q)$ is independent of $u$ since $g$ (and hence $g_1$) grows at most linearly. Hence, by Lemma 21.3 with $p_+(x) = (1-t)b_+(x) + tg_+(x, u^+(x), \eta(u))$ and $p_-(x) = (1-t)b_-(x) + tg_-(x, -u^-(x), \eta(u))$, we get for some $c > 0$

$$||Bu + (1-t)Cu + tGu|| \geq \delta ||u||^2 \text{ for all } u \in V.$$

It is left to show that $\deg(Q_n(B+C), B_R \cap V_n, 0) \neq 0$ for all $n$. Let $\eta \in (0,r)$ be fixed. Then, for each $t \in [0,1]$, and a.e. $x \in Q$, we have that $0 \leq (1-t)\eta + tb_\pm \leq r$. It is easy to show that $p_+ = (1-t)\eta + tb_+$ and $p_- = (1-t)\eta + tb_-$ satisfy $0 \leq p_\pm \leq r$ for a.e. $x \in Q$, and

$$\int_Q [p_+(v^+)^2 + p_-(v^-)^2] > 0 \text{ for all } v \in N(-L_1 + \lambda_j I) \setminus \{0\}$$

and

$$\int_Q [r - p_+)(w^+)^2 + (r - p_-)(w^-)^2] > 0 \text{ for all } w \in N(-L_1 + \lambda_{j+1} I) \setminus \{0\}.$$

Hence, one gets that the equation

$$Bu + [(1-t)\eta + tb_+]u^+ + [(1-t)\eta + tb_-]u^- = 0 \tag{21.21}$$

has only the trivial solution for each $t \in [0,1]$. Since the homotopy given by (21.21) is $A$-proper, there is an $n \geq n_0$ such that for each $R > 0$

$$\deg\,(P_n(B + b_+(.)^+ - b_-(.)^-, B(0,R) \cap H_n, 0) =$$

$$\deg\,(P_n(B + \eta I), B(0,R) \cap H_n, 0) = \pm 1 \text{ for all } n \geq n_0.$$

Hence, Eq. (21.2) is solvable in $V$ by Theorem 21.1. ■

**Remark 21.2** *Conditions (21.18)-(21.19) hold for a wide class of nonlinearities $g$. For example, they are implied by $\lambda_j < \lambda_j + \epsilon \leq \gamma_+(x) \leq \Gamma_+(x) \leq \lambda_{j+1}$ and $\lambda_j \leq \gamma_-(x) \leq \Gamma_-(x) \leq \lambda_{j+1} - \epsilon < \lambda_{j+1}$, or $\lambda_j \leq \gamma_+(x) \leq \Gamma_+(x) \leq \lambda_{j+1} - \epsilon < \lambda_{j+1}$ and $\lambda_j < \lambda_j + \epsilon \leq \gamma_-(x) \leq \Gamma_-(x) \leq \lambda_{j+1}$, in the case when the eigenfunctions associated to $\lambda_j$ and $\lambda_{j+1}$ change sign in $Q$.*

Next, we shall give some concrete assumptions on $f$ and $g$ that imply (F)-(H) in Theorem 21.4.

**Theorem 21.5** *Let (L1) and (B1) hold, $B$ have properties I and II and there be functions $\gamma_\pm, \Gamma_\pm \in L_\infty(Q)$ such that for some $j \in J \subset Z$, one has*

$$\lambda_j \leq \gamma_\pm(x) \leq \Gamma_\pm(x) \leq \lambda_{j+1} \quad \textit{for a.e.} \ \ x \in Q$$

*and*

$$\int_Q [(\gamma_+ - \lambda_j)(v^+)^2 + (\gamma_- - \lambda_j)(v^-)^2] > 0 \; for\ all\, v \in N(L_1 - \lambda_j I) \setminus \{0\}$$

*and*

$$\int_Q [(\lambda_{j+1} - \Gamma_+)(w^+)^2 + (\lambda_{j+1} - \Gamma_-)(w^-)^2] > 0$$

*for all* $w \in N(L_1 - \lambda_{j+1} I) \setminus \{0\}$. *Suppose that for* $\epsilon > 0$ *and* $\delta > 0$ *given in Lemma 21.3,*

*(G1) there is* $\rho > 0$ *such that for a.e.* $x \in Q$, $\eta(u) \in R^k$

$$\gamma_+(x) - \epsilon \le g(x, u, \eta(u)) \le \Gamma_+(x) + \epsilon \;\; if \;\; u > \rho$$

$$\gamma_-(x) - \epsilon \le g(x, u, \eta(u)) \le \Gamma_-(x) + \epsilon \;\; if \;\; u < -\rho$$

*(G2) There are functions* $b(x) \in L_\infty(Q)$ *and* $k_s(x) \in L_2(Q)$ *for each* $s > 0$ *such that*

$$|g(x, u, \eta(u))| \le sb(x)( \sum_{|\alpha| \le 2m-1} |D^\alpha u|^2)^{1/2} + k_s(x), u \in V.$$

*(F1) There are functions* $a(x) \in L_\infty(Q)$ *and* $d_r(x) \in L_2(Q)$ *for each* $r > 0$ *such that*

$$|f(x, \xi(u))| \le ra(x)( \sum_{|\alpha| \le 2m} |D^\alpha u|^2)^{1/2} + d_r(x), \;\; for\ all \;\; u \in V.$$

*(F2) There is a constant* $k > 0$ *such that* $k \le c$ *and*

$$|f(x, \eta, \zeta) - f(x, \eta, \zeta')| \le k \sum_{|\alpha| = 2m} |\zeta_\alpha - \zeta'_\alpha|$$

*for a.e.* $x \in Q$, *all* $\eta \in R^k$ *and* $\zeta, \zeta' \in R^{s'_{2m}} = R^{s_{2m}} - R^{s_{2m-1}}$, *where* $c$ *is a constant in (21.1).*
*Then there is a* $u \in V$ *that satisfies Eq. (21.2) for a.e.* $x \in Q$.

**Proof.** It is easy to see that (F) of Theorem 21.4 holds. Hence, it remains to verify (H) of that theorem, i.e. that $H_t = B - tF$ is $A$-proper w.r.t. $\Gamma_B$ for each $t \in [0, 1)$ and $H_1$ is pseudo $A$-proper. Since the embedding of $V$ into $H$ is compact, it sufficies to show these facts for $F_t = L - tF$. Set $B_\mu = B - \mu I$ for some $\mu \neq \lambda_j$ for each $j$. Then, for each $t \in [0, 1]$, it follows from (F2), the Holder inequality, and an easy calculation that

$$(F_t u - F_t v, B_\mu(u - v)) \ge (1 - k/c)||B_\mu(u - v)||^2 + \phi(u - v) \qquad (21.22)$$

where the functional $\phi : V \to R$ is given by

$$\phi(u-v) = t(M(u,v) - M(v,v), B_\mu(u-v)) + \mu(u-v, B_\mu(u-v)),$$

with $M : V \times V \to H$ being a continuous form given by $M(u,v) = f(x, \eta(u), \zeta(v))$. The functional $\phi$ is weakly continuous. Indeed, let $u_k \to u$ weakly in $V$. Then $u_k \to u$ in the $W_2^{2m-1}$-norm by the Sobolev imbedding theorem and by the results from [K], it is not hard to show that $\phi(u_k - u) \to 0$ as $k \to \infty$. If $k < c$, then (F2) implies that $F_t$ is $A$-proper w.r.t. $\Gamma_B$ (see, e.g., in [Mi-2-4]). If $k = c$, then $F_t$ is again $A$-proper for each $t \in [0,1)$ and it is easy to see that $F_1$ is pseudo $L_\mu$-monotone. Hence, $F_1$ is pseudo $A$-proper w.r.t. $\Gamma_B$ ([Mi-3]) and (H) of Theorem 21.4 holds. ∎

**Corollary 21.6** *Let all conditions of Theorem 21.5 hold with (G1) replaced by*
*(G1') $\gamma_\pm(x) \le \liminf_{u\to\pm\infty} g(x,u,\eta(u)) \le \limsup_{u\to\pm\infty} g(x,u,\eta(u)) \le \Gamma_\pm(t,x)$ uniformly for a.e. $(x,\eta(u)) \in Q \times R^k$.*
*Then there is a $u \in V$ that satisfies Eq. (21.2) for a.e. $x \in Q$.*

**Proof.** It is easy to see that (G1') implies (G1). ∎

## 21.3 Strong solvability of elliptic BVP's

Now, we shall apply the results of Section 2 to strong solvability of elliptic boundary value problems in $V$ of the form

$$\sum_{|\alpha|\le 2m} A_\alpha(x) D^\alpha u(x) + g(x,u,Du,\dots,D^{2m-1}u)u + f(x,u,Du,\dots,D^{2m}) = 0, \tag{21.23}$$

under nonuniform nonresonance conditions. Here $Q \subset R^n$ is a bounded smooth domain, $V$ is a closed subspace of $W_2^{2m}(Q)$ containing the test functions and the linear part is elliptic. Assume the linear map $L : V \to L_2(Q)$, induced by the linear elliptic operator in (21.23), has closed range in $H = L_2(Q)$ and satisfies conditions (L1), (B1) in Section I with $B = -L + \lambda_j I$. Here, $L_1 = L$ and $A_1 = 0$.

Let $\lambda \ne \lambda_j$ for each $j = 1, 2, \dots$ be fixed, $\Gamma = \{Y_n, Q_n\}$ be a projectionally complete scheme for $L_2$ and $X_n = (B - \lambda I)^{-1}(Y_n) \subset V$ for each $n$. Then $\Gamma_L = \{X_n, Y_n, Q_n\}$ is an admissible or a projectionally complete scheme for $(V, L_2)$. Since $B - \lambda I : V \to L_2$ is linear, one-to-one and $A$-proper w.r.t. $\Gamma_L$, there is a constant $c > 0$ (depending) only on $\lambda$) such that

$$||(B-\lambda I)u|| \ge c||u||_V, \quad u \in V. \tag{21.24}$$

**Theorem 21.7** *Let $B = -L + \lambda I$ be a closed densely defined map in $H$ such that $R(B) = N(B)^\perp$, $(Bu,u) \ge -r^{-1}||Bu||^2$ on $V$ and if $(Bu,u) =$*

$-r^{-1}||Bu||^2$ *for some* $u \in V$, *then* $u \in N(-L+\lambda I) \oplus N(-L+\lambda_{j+1}I)$. *Suppose that there are functions* $\gamma_\pm, \Gamma_\pm \in L_\infty(Q)$ *such that for some* $j \in J \subset Z$, *one has*

$$\lambda_j \le \gamma_\pm(x) \le \Gamma_\pm(x) \le \lambda_{j+1} \text{ for a.e. } x \in Q$$

*and*

$$\int_Q [(\gamma_+ - \lambda_j)(v^+)^2 + (\gamma_- - \lambda_j)(v^-)^2] > 0 \text{ for all } v \in N(L-\lambda_j I) \setminus \{0\}$$

*and*

$$\int_Q [(\lambda_{j+1} - \Gamma_+)(w^+)^2 + (\lambda_{j+1} - \Gamma_-)(w^-)^2] > 0$$

$w \in N(L-\lambda_{j+1}I) \setminus \{0\}$. *Suppose that for* $\epsilon > 0$ *and* $\delta > 0$ *given in Lemma 21.3,*

*(G1) there is* $\rho > 0$ *such that for a.e.* $x \in Q$

$$\gamma_+(x) - \epsilon \le g(x,u,\eta(u)) \le \Gamma_+(x) + \epsilon \text{ if } u > \rho, \eta(u) \in R^k$$

$$\gamma_-(x) - \epsilon \le g(x,u,\eta(u)) \le \Gamma_-(x) + \epsilon \text{ if } u < -\rho, \eta(u) \in R^k$$

*(G2) There are functions* $b(x) \in L_\infty(Q)$ *and* $k_s(x) \in L_2(Q)$ *for each* $s > 0$ *such that*

$$|g(x,u,\eta(u))| \le sb(x)(\sum_{|\alpha|\le 2m-1} |D^\alpha u|^2)^{1/2} + k_s(x), u \in V.$$

*(F)* $||Fu|| = ||f(x,u,...,D^{2m}u)|| \le \beta||u||_V + \gamma$ *for some* $\beta \in (0,\delta)$, $\gamma > 0$.

*(H)* $H_t = L - tF : V \to H$ *is A-proper w.r.t.* $\Gamma_L$ *for* $t \in [0,1)$ *and* $L - F$ *is pseudo A-proper.*

*Then BVP (21.23) has a solution* $u \in V$.

**Proof.** It follows from Theorem 21.4 with $L_1 = L$ and $A_1 = 0$. ■

As before, we give now some concrete conditions on $f, g$ so that (H) holds.

**Theorem 21.8** *Let* $B = -L + \lambda I$ *be a closed densely defined map in* $H$ *such that* $R(B) = N(B)^\perp$, $(Bu,u) \ge -r^{-1}||Bu||^2$ *on* $V$ *and if* $(Bu,u) = -r^{-1}||Bu||^2$ *for some* $u \in V$, *then* $u \in N(-L+\lambda I) \oplus N(-L+\lambda_{j+1}I)$. *Suppose that there are functions* $\gamma_\pm, \Gamma_\pm \in L_\infty(Q)$ *such that for some* $j \in J \subset Z$, *one has*

$$\lambda_j \le \gamma_\pm(x) \le \Gamma_\pm(x) \le \lambda_{j+1} \text{ for a.e. } x \in Q$$

*and*

$$\int_Q [(\gamma_+ - \lambda_j)(v^+)^2 + (\gamma_- - \lambda_j)(v^-)^2] > 0 \text{ for all } v \in N(L - \lambda_j I) \setminus \{0\}$$

*and*

$$\int_Q [(\lambda_{j+1} - \Gamma_+)(w^+)^2 + (\lambda_{j+1} - \Gamma_-)(w^-)^2] > 0$$

$w \in N(L - \lambda_{j+1} I) \setminus \{0\}$. *Suppose that for* $\epsilon > 0$ *and* $\delta > 0$ *given in Lemma 21.3,*

*(G1) there is* $\rho > 0$ *such that for a.e.* $x \in Q$

$$\gamma_+(x) - \epsilon \leq g(x, u, \eta(u)) \leq \Gamma_+(x) + \epsilon \text{ if } u > \rho, \eta(u) \in R^k$$

$$\gamma_-(x) - \epsilon \leq g(x, u, \eta(u)) \leq \Gamma_-(x) + \epsilon \text{ if } u < -\rho, \eta(u) \in R^k$$

*(G2) There are functions* $b(x) \in L_\infty(Q)$ *and* $k_s(x) \in L_2(Q)$ *for each* $s > 0$ *such that*

$$|g(x, u, \eta(u))| \leq sb(x)(\sum_{|\alpha| \leq 2m-1} |D^\alpha u|^2)^{1/2} + k_s(x), u \in V.$$

*(F1) There are functions* $a(x) \in L_\infty(Q)$ *and* $d_r(x) \in L_2(Q)$ *for each* $r > 0$ *such that*

$$|f(x, \xi(u))| \leq ra(x)(\sum_{|\alpha| \leq 2m} |D^\alpha u|^2)^{1/2} + d_r(x), \text{ for all } u \in V.$$

*(F2) There is a constant* $k > 0$ *such that* $k \leq c$ *and*

$$|f(x, \eta, \zeta) - f(x, \eta, \zeta')| \leq k \sum_{|\alpha| = 2m} |\zeta_\alpha - \zeta'_\alpha|$$

*for a.e.* $x \in Q$, *all* $\eta \in R^k$ *and* $\zeta, \zeta' \in R^{s'_{2m}} = R^{s_{2m}} - R^{s_{2m-1}}$, *where* $c$ *is a constant in (21.24). Then there is a* $u \in V$ *that satisfies (21.23) for a.e.* $x \in Q$.

**Proof.** It follows from Theorem 21.5 with $L_1 = L$ and $A_1 = 0$. ■

For our next result, we assume also

(L2) There is an integer $j \geq 1$ such that $\lambda_j < \lambda_{j+1}$ and $Lw = \lambda_k w$ for $k = j$ and $k = j + 1$, has the continuation property, that is if $w(x) = 0$ on a set of positive measure, then $w(x) = 0$ a.e. on $Q$.

**Theorem 21.9** *Let $L$ satisfy (L1)-(L2) and (B1) with $B = -L + \lambda I$ and $\gamma(x), \Gamma(x) \in L_\infty(Q)$ be such that*

*(H1) $\lambda_j \leq \gamma(x) \leq \Gamma(x) \leq \lambda_{j+1}$ with $meas\{x \in Q | \lambda_j \neq \alpha(x)\} > 0$ and $meas\{x \in Q | \lambda_{j+1} \neq \beta(x)\} > 0$.*

*Suppose that (G1) of Theorem 8 holds and for $\epsilon > 0$ and $\delta > 0$ given by Lemma 3*

*(H2) $\gamma(x) - \epsilon \leq g(x, \xi) \leq \Gamma(x) + \epsilon$ for all $(x, \xi) \in Q \times R^{s_{2m-1}}$*

*(H3) $||Fu|| = ||f(x, u, ..., D^{2m}u)|| \leq \beta ||u||_V + \gamma$ for some $\beta \in (0, \delta)$, $\gamma > 0$.*

*(H4) $H_t = L - tF$ is A-proper w.r.t. $\Gamma_L$ for $t \in [0, 1)$ and $L - F$ is pseudo A-proper.*

*Then BVP(23) has a solution $u \in V$.*

**Proof.** Clearly, (L2) and (H1) imply the integral inequalities in Theorem 21.8. Hence, the conclusion follows from this theorem. ■

**Theorem 21.10** *Let $L$ and $\gamma(x)$, $\Gamma(x)$ be as in Theorem 9. Let $f : Q \times R^{s_{2m}} \to R$ and $g : Q \times R^{s_{2m-1}} \to R$ be Caratheodory functions such that*

*(F1) There are functions $a(x) \in L_\infty(Q)$ and $d_r(x) \in L_p(Q)$ for each $r > 0$ such that*

$$|f(x, \xi(u))| \leq ra(x)(\sum_{|\alpha| \leq 2m} |D^\alpha u|^2)^{1/2} + d_r(x), \quad \text{for all } u \in V.$$

*(F2) There is a constant $k > 0$ such that $k \leq c$ and*

$$|f(x, \eta, \zeta) - f(x, \eta, \zeta')| \leq k \sum_{|\alpha| = 2m} |\zeta_\alpha - \zeta'_\alpha|$$

*for a.e. $x \in Q$, all $\eta \in R^k$ and $\zeta, \zeta' \in R^{s'_{2m}} = R^{s_{2m}} - R^{s_{2m-1}}$.*

*(G1) $\lambda_j \leq \gamma(x) \leq liminf_{|u| \to \infty} g(x, u, \eta(u)) \leq limsup_{|u| \to \infty} g(x, u, \eta(u)) \leq \Gamma(x) \leq \lambda_{j+1}$ uniformly for $x \in Q$ and the non-$u$ components $\eta(u)$.*

*(G2) There are functions $b(x) \in L_\infty(Q)$ and $k_s(x) \in L_p(Q)$ for each $s > 0$ such that*

$$|g(x, u, \eta(u))| \leq sb(x)(\sum_{|\alpha| \leq 2m-1} |D^\alpha u|^2)^{1/2} + k_s(x), u \in V.$$

*Then there is a $u \in V$ that satisfies Eq. (21.23) for a.e. $x \in Q$.*

**Proof.** It follows from Theorem 21.9, as in the case of Theorem 21.5. ■

Theorem 21.8 extends a result of Beresticki-de Figueiredo [BF] who assumed $f = 0$ and $g$ to depend only on $u$. A simplified proof of their results has been given by Mawhin [Ma-2]. If $f$ does not depend on derivatives of order $2m$, Theorem 21.10 reduces to a result of Mawhin-Ward [MW]. Their proofs are based on the Leray-Schauder and the coincidence degree theories respectively.

## 21.4 Time periodic solutions of BVP's for nonlinear parabolic and hyperbolic equations

We shall prove the existence of generalized periodic solutions (GPS), under nonuniform nonresonance conditions, for the nonlinear parabolic equation

$$u_t + A_0 u + g(t, x, u, u_t, D_x u, ..., D_x^{2m-1} u)u$$
$$+ f(t, x, u, u_t, D_x u, ..., D_x^{2m} u) = h \tag{21.25}$$

in $H = L_2(\Omega)$, where $\Omega = [0, 2\pi] \times Q$ with $Q \subset R^n$, $A_0$ is a uniformly strongly elliptic operator of order $2m$ in $x \in Q$ for each $t \in [0, 2\pi]$, and the nonlinear hyperbolic equations with damping

$$\sigma u_t + u_{tt} + A_0 u = g(t, x, u, u_t, u_{tt}, D_x u, ..., D_x^{2m-1} u)u$$

$$+ f(t, x, u, u_t, u_{tt}, D_x u, ..., D_x^{2m} u) = h \tag{21.26}$$

h in H, $\sigma \neq 0$ with boundary conditions

$$u(t, .) \in H_0^m(Q) \ \text{ for all } \ t \in (0, 2\pi) \tag{21.27}$$

and periodicity conditions

$$u(0, x) = u(2\pi, x) \ \text{ for all } \ x \in Q. \tag{21.28}$$

The results of this section extend the corresponding results in Nkashama-Willem [NW], who assumed only the u dependence in g and $f = 0$ and used the coincidence degree theory.

### *21.4.1 Nonlinear parabolic equations*

We assume that $L_1 : D(L_1) = H_0^{2m}(Q) = H^{2m}(Q) \cap H_0^m(Q) \subset L_2(Q) \to L_2(Q)$ is a linear selfadjoint uniformly strongly elliptic operator induced by $A_0$.

Define a linear map $A : D(A) \subset H \to H$ by

$$V = D(A) = \{u \in L_2((0, 2\pi), H_0^{2m}(Q)) = H_0^{2m}(\Omega) | u_t \in H,$$

$$u(0, x) = u(2\pi, x), x \in Q\}$$

$Au = u_t + Lu$, where $(Lu)(t, x) = (A_0 u)(t, x) = L_1 u(t, x)$ on $Q$ for each $t \in [0, 2\pi]$ fixed. Set $Bu = -Au + \lambda_j u$, $j \in J \subset Z$, $j$-fixed, with $D(B) = D(A)$. Define the norm on $V$ by

$$||u||_V = \int_\Omega [u^2 + (u_t)^2 + \sum_{|\alpha| \leq 2m} (D_x^\alpha u)^2].$$

Since a $2\pi$- periodic function on $Q$ can be expended as

$$u(t,x) = \sum_{k\in Z, l\in J} u_{kl} e^{ikt} w_l(x), \quad u_{k,l} = \bar{u}_{-k,l}$$

then

$$Bu = \sum_{k\in Z, l\in J} u_{kl}(-ik - \lambda_l + \lambda_j) e^{ikt} w_l(x)$$

Facts about $B$ ([NW]): 1) The null space $N(B) = N(L_1 - \lambda_j I) = N(B^*)$ since $L_1 - \lambda_j I$ is selfadjoint. Hence, each function $u \in N(B)$ is independent of $t$.

2) dim $N(L_1 - \lambda_j I) < \infty$ implies that dim$N(B)$ is finite.

3) $B$ is closed densely defined linear map with range $R(B) = (N(B))^{\perp}$ and so $H = N(B) \oplus R(B)$. Hence, $B$ is Fredholm of index 0.

4) The generalized right inverse $K = (B|_{D(B)\cap R(B)})^{-1} : R(B) \to R(B)$ is compact.

5) $(Bu, u) \geq -1/r||Bu||^2$ for all $u \in D(B)$, where $r = \lambda_{j+1} - \lambda_j$.

6) Property II holds on $D(B)$.

**Definition 21.2** *A GPS of the problem (21.25)-(21.27)-(21.28) is a function $u \in V$ that satisfies (21.25) for a.e. $(t,x) \in \Omega$.*

For $u \in V$, define $\eta(u) = (u_t, D_x u, ..., D_x^{2m-1} u)$, $\xi(u) = (u, u_t, D_x u, ..., D_x^{2m} u)$, $\zeta(u) = (D_x^{\alpha} u \mid |\alpha| = 2m)$. Let $k$ be the number of components in $\eta(u)$ and $R^{k'} = \{\zeta = (\xi_\alpha) : |\alpha = (\alpha_1, ..., \alpha_n)| = 2m\}$.

**Theorem 21.11** *Let $L_1$ be as above and (L1) and (B1) hold. Suppose that $\gamma_\pm, \Gamma_\pm \in L_\infty(\Omega)$ are such that for some $j \in J \subset Z$, we have*

$$\lambda_j \leq \gamma_\pm(t,x) \leq \Gamma_\pm(t,x) \leq \lambda_{j+1} \text{ for a.e. } (t,x) \in \Omega.$$

*Assume also*

$$\int_\Omega [(\gamma_+ - \lambda_j)(v^+)^2 + (\gamma_- - \lambda_j)(v^-)^2] > 0$$

*for all $v \in N(L_1 - \lambda_j I) = N(B), v \neq 0$ and*

$$\int_\Omega [(\lambda_{j+1} - \Gamma_+)(w^+)^2 + (\lambda_{j+1} - \Gamma_-)(w^-)^2] > 0 \text{ for all } w \in N(L_1 - \lambda_{j+1} I)$$

$$= N(B + rI), w \neq 0.$$

*(G1) there is $\rho > 0$ such that for a.e. $(t,x) \in \Omega$, $(u, \eta(u)) \in R \times R^k$*

$$\gamma_+(t,x) - \epsilon \leq g(t,x,u,\eta(u)) \leq \Gamma_+(t,x) + \epsilon \text{ if } u > \rho$$

$$\gamma_-(t,x) - \epsilon \le g(t,x,u,\eta(u)) \le \Gamma_-(t,x) + \epsilon \text{ if } u < -\rho$$

*(G2) There are functions $b(t,x) \in L_\infty(\Omega)$ and $k_s(t,x) \in L_2(\Omega)$ for each $s > 0$ such that*

$$|g(t,x,u,\eta(u))| \le sb(t,x)(u_t^2 + \sum_{|\alpha|\le 2m-1} |D_x^\alpha u|^2)^{1/2} + k_s(t,x), u \in V.$$

*(F1) There are functions $a(t,x) \in L_\infty(\Omega)$ and $d_r(t,x) \in L_2(\Omega)$ for each $r > 0$ such that*

$$|f(t,x,\xi(u))| \le ra(t,x)(u_t^2 + \sum_{|\alpha|\le 2m} |D_x^\alpha u|^2)^{1/2} + d_r(t,x), \quad \text{for all } u \in V.$$

*(F2) There is a constant $k > 0$ such that $k \le c$ and*

$$|f(t,x,u,\eta,\zeta) - f(t,x,u,\eta,\zeta')| \le k \sum_{|\alpha|=2m} |\zeta_\alpha - \zeta'_\alpha|$$

*for a.e. $(t,x) \in \Omega$, all $(u,\eta) \in R \times R^k$ and $\zeta,\zeta' \in R^{k'}$, where c is a constant in (1).*
*Then problem (21.25)-(21.27)-(21.28) has at least one GPS for each $h \in H$.*

**Proof.** Set $A_1 = u_t$, $L = A_0$ and note that $(Bu,u) = (-Lu + \lambda_j u, u)$ on $V = D(B)$ by using Fourier series. By the above remarks, we have that the theorem follows from Theorem 21.5 with Q replaced by $\Omega$. ∎

**Remark 21.3** *The integral conditions above hold for a wide class of nonlinearities g. For example, they are implied by $\lambda_j < \lambda_j + \epsilon \le \gamma_+(t,x) \le \Gamma_+(t,x) \le \lambda_{j+1}$ and $\lambda_j \le \gamma_-(t,x) \le \Gamma_-(t,x) \le \lambda_{j+1} - \epsilon < \lambda_{j+1}$, or $\lambda_j \le \gamma_+(t,x) \le \Gamma_+(t,x) \le \lambda_{j+1} - \epsilon < \lambda_{j+1}$ and $\lambda_j < \lambda_j + \epsilon \le \gamma_-(t,x) \le \Gamma_-(t,x) \le \lambda_{j+1}$, in the case when the eigenfunctions associated to $\lambda_j$ and $\lambda_{j+1}$ change sign in $\Omega$.*

**Remark 21.4** *Functions $v \in N(L_1 - \lambda_j I)$ and $w \in N(L_1 - \lambda_{j+1} I)$ in the theorem are understood as functions of two variables $(t,x)$ which are independent of t.*

### *21.4.2 Applications to the heat equation*

We apply the above results to the nonlinear heat equation

$$u_t(t,x) - u_{xx}(t,x) + g(t,x,u(t,x)) = h(t,x) \quad (t,x) \in (0,2\pi) \times (0,\pi) \tag{21.29}$$

$$u(t,0) = u(t,\pi) = 0, \quad t \in (0,2\pi) \tag{21.30}$$

$$u(0,x) = u(2\pi, x), \quad x \in (0,\pi) \tag{21.31}$$

under nonuniform nonresonnce conditions. Now, $\lambda_j = j^2, j \in N^*$ and the corresponding eigenfunctions are $w_j = sinjx, j \in N^*$. Hence, the above theorem implies the following result.

**Corollary 21.12** *( [NW] ) Let*

$$\gamma(t,x) \leq \liminf_{|u|\to\infty} u^{-1}g(t,x,u) \leq \limsup_{|u|\to\infty} u^{-1}g(t,x,u) \leq \Gamma(t,x) \tag{21.32}$$

*hold uniformly for a.e.* $(t,x) \in \Omega = (0,2\pi)\times(0,\pi)$*, where* $\gamma, \Gamma \in L_\infty(\Omega)$ *satisfy, for some* $j \in N^*$*,*

$$j^2 \leq \gamma(t,x) \leq \Gamma(t,x) \leq (j+1)^2 \; for \; a.e.(t,x) \in \Omega \tag{21.33}$$

*with*

$$\int_\Omega (\gamma(t,x) - j^2) sin^2 jx dx dt > 0 \tag{21.34}$$

$$\int_\Omega [(j+1)^2 - \Gamma(t,x)] sin^2 (j+1) dx dt > 0. \tag{21.35}$$

*Then Eq. (21.29) has at least one GPS for each* $h \in H$.

**Remark 21.5** *Assumptions (21.34)-(21.35) are equivalent to* $j^2 < \gamma(t,x)$ *and* $\Gamma(t,x) < (j+1)^2$ *respectively on a subset of* $\Omega$ *of positive measure.*

**Remark 21.6** *The proof of the theorem implies that it is still valid if (21.32) is replaced by*

$$\gamma_\pm(t,x) - \eta \leq \liminf_{u\to\pm\infty} u^{-1}g(t,x,u) \leq \limsup_{u\to\pm\infty} u^{-1}g(t,x,u) \leq \Gamma_\pm(t,x) + \eta$$

*for some* $0 \leq \eta < \epsilon$*. Hence, some "jumping" over two consequtive eigenvalues is allowed. Indeed, let* $\gamma_+ = \gamma_- = j^2, \Gamma_+ = \Gamma_- = (j+1)^2, j \geq 2$*, and* $g(t,x,u) = g(u)$ *continuous with*

$$lim_{u\to\infty} u^{-1} g(u) = \mu \; and \; lim_{u\to-\infty} u^{-1} g(u) = \nu.$$

*Then there is* $\epsilon = \epsilon(j^2, (j+1)^2) > 0$ *such that either* $j^2 - \epsilon < \mu \leq j^2 \leq \nu < (j+1)^2 + \epsilon$ *or* $j^2 - \epsilon < \nu \leq j^2 < (j+1)^2 \leq \mu < (j+1)^2 + \epsilon$.

**Remark 21.7** *When* $\gamma_\pm$ *and* $\Gamma_\pm$ *are constants, conditions (21.34)-(21.35) become: if* $j \geq 2$*, since all eigenfunctions associated with* $j^2$ *and* $(j+1)^2$ *change sign,* $[\gamma_+, \Gamma_+] \subset [j^2, (j+1)^2], [\gamma_-, \Gamma_-] \subset [j^2, (j+1)^2]$ *with* $(\gamma_+, \gamma_-) \neq (j^2, j^2), (\Gamma_+, \Gamma_-) \neq ((j+1)^2, (j+1)^2)$*. If* $j = 1$*, the corresponding eigenspace is spanned by* $\{sinx\}$ *which does not change sign in* $\Omega$*, so that we have* $1 < \gamma_+ \leq \Gamma_+ \leq 4, 1 < \gamma_- \leq \Gamma_- \leq 4$ *and* $(\Gamma_+, \Gamma_-) \neq (4,4)$.

### *21.4.3 Nonlinear hyperbolic equtions*

Let $A_0$ be as in part A. Now, we shall study (21.26) under the nonuniform nonresonance conditions of the form given above.

Define a linear map $A : D(A) \subset H \to H$ by

$$V = D(A) = \{u \in L_2((0, 2\pi), H_0^{2m}(Q)) = H_0^{2m}(\Omega) | u_t \in H, u_{tt} \in H,$$

$$u(0, x) = u(2\pi, x), x \in Q\}$$

$Au = \sigma u_t + Lu$, where $(Lu)(t, x) = u_{tt} + A_0 u(t, x)$ on $\Omega = Q \times (0, 2\pi)$, $\sigma \neq 0$ fixed. Set $Bu = -Au + \lambda_j u$, $j \in J \subset Z$, $j$-fixed with $D(B) = D(A)$. Assume that $\Omega$ is such that $R(L)$ is closed in $H = L_2(\Omega)$. For some explicit assumptions on $\Omega$ ensuring the closedness of $R(L)$ in $H$, see [V]. Let $...\lambda_{-1} < 0 \leq \lambda_1 \leq \lambda_2 ...$ be the eigenvalues of $L$ in $V$ and $...w_{-1}, w_0, w_1, ...$ be the eigenfunctions associated with the eigenvalues of the linear elliptic problem. Then $\{w_j e^{ikt} : j \in J, k \in Z\}$ form an orthogonal system in $H$ of eigenfunctions of $Lu = \lambda u$ in $V$.

Since a $2\pi$- periodic function on $Q$ can be expended as

$$u(t, x) = \sum_{k \in Z, l \in J} u_{kl} e^{ikt} w_l(x), \quad u_{k,l} = \bar{u}_{-k,l}$$

then

$$Bu = \sum_{k \in Z, l \in J} u_{kl}(-ik\sigma + k^2 - \lambda_l + \lambda_j) e^{ikt} w_l(x)$$

Equip $V = D(B)$ with the norm

$$||u||_V = \int_\Omega [u^2 + (u_t)^2 + (u_{tt})^2 + \sum_{|\alpha| \leq 2m} (D_x^\alpha u)^2].$$

Using Fourier series method, Parseval equality and integration by parts, we get that

$$-(Bu, u_t) = \sigma |u_t|^2 \text{ for all } u \in V.$$

Hence, $N(B) = N(A_0 - \lambda_j I) = N(B^*)$ is finite dimensional and $R(B) = (N(B))^\perp$, so that $H = N(B) \oplus R(B)$. Thus, $B$ is Fredholm of index 0, and each function $u \in N(B)$ is independent of $t$. Moreover, as in [V], one gets that the partial inverse of $B$ is compact in $H$. It is well-known ([BN]) that there exists $r = \lambda_{j+1} - \lambda_j$ such that

$$(-Lu + \lambda_j u, u) \geq -r^{-1} || - Lu + \lambda_j u||^2. \tag{21.36}$$

Using Fourier series, we get that $(Bu, u) = (-Lu + \lambda_j u, u)$ so that, by (21.36), we get

$$(Bu, u) \geq -r^{-1}||Bu||^2 \tag{21.37}$$

since $||Bu|^2 = || - Lu + \lambda_j u||^2 + \sigma^2 ||u_t||^2$. Moreover, if equality holds in (21.37), then $(-Lu + \lambda_j u, u) = -r^{-1}|| - Lu + \lambda_j u||^2$ which is equivalent to $u \in N(-Lu + \lambda_j I) \oplus N((-Lu + \lambda_j) + rI)$ by Remark 1 and Lemma 1 in [BF, p. 99] since $-L + \lambda_j I$ is selfadjoint. Hence, $B$ has Properties I and II. with $L_1 = L$.

**Definition 21.3** *A GPS of the problem (21.26)-(21.27)-(21.28) is a function $u \in V$ that satisfies (21.26) for a.e. $(t, x) \in \Omega$.*

Define $\eta(u) = (u_t, u_{tt}, D_x u, ..., D_x^{2m-1} u)$, $\xi(u) = (u, u_t, u_{tt}, D_x u, ..., D_x^{2m} u)$ and $\zeta(u) = (D_x^\alpha u \mid |\alpha = (\alpha_1, ..., \alpha_n)| = 2m)$. Let $R^k$ and $R^{k'}$ be be as above for these $\eta(u)$, $\xi(u)$ and $\zeta(u)$.

**Theorem 21.13** *Let $A_0$ be as above and $\gamma_\pm, \Gamma_\pm \in L_\infty(\Omega)$ be such that for some $j \in J \subset Z$, we have*

$$\lambda_j \leq \gamma_\pm(t, x) \leq \Gamma_\pm(t, x) \leq \lambda_{j+1} \textit{ for a.e. } (t, x) \in \Omega.$$

*Assume also*

$$\int_\Omega [(\gamma_+ - \lambda_j)(v^+)^2 + (\gamma_- - \lambda_j)(v^-)^2] > 0 \textit{ for all} \tag{21.38}$$

*for all $v \in N(L - \lambda_j I) = N(B), v \neq 0$ and*

$$\int_\Omega [(\lambda_{j+1} - \Gamma_+)(w^+)^2 + (\lambda_{j+1} - \Gamma_-)(w^-)^2] > 0 \textit{ for all } w \in N(L - \lambda_{j+1} I)$$

$$= N(B + rI), w \neq 0. \tag{21.39}$$

*(G1) there is $\rho > 0$ such that for a.e. $(t, x) \in \Omega$, $(u, \eta(u)) \in R \times R^k$*

$$\gamma_+(t, x) - \epsilon \leq g(t, x, u, \eta(u)) \leq \Gamma_+(t, x) + \epsilon \textit{ if } u > \rho$$

$$\gamma_-(t, x) - \epsilon \leq g(t, x, u, \eta(u)) \leq \Gamma_-(t, x) + \epsilon \textit{ if } u < -\rho$$

*(G2) There are functions $b(t, x) \in L_\infty(\Omega)$ and $k_s(t, x) \in L_2(\Omega)$ for each $s > 0$ such that*

$$|g(t, x, u, \eta(u))| \leq sb(t, x)(u_t^2 + u_{tt}^2 + \sum_{|\alpha| \leq 2m-1} |D_x^\alpha u|^2)^{1/2} + k_s(t, x), u \in V.$$

*(F1) There are functions $a(t,x) \in L_\infty(\Omega)$ and $d_r(t,x) \in L_2(\Omega)$ for each $r > 0$ such that*

$$|f(t,x,\xi(u))| \leq ra(t,x)(u_t^2 + u_{tt}^2 + \sum_{|\alpha| \leq 2m} |D_x^\alpha u|^2)^{1/2} + d_r(t,x),$$

*for all $u \in V$.*

*(F2) There is a constant $k > 0$ such that $k \leq c$ and*

$$|f(t,x,\eta,\zeta) - f(t,x,\eta,\zeta')| \leq k \sum_{|\alpha|=2m} |\zeta_\alpha - \zeta'_\alpha|$$

*for a.e. $(t,x) \in Q$, all $\eta \in R^{k+1}$ and $\zeta, \zeta' \in R^{k'}$, where $c$ is a constant in (1).*
*Then problem (21.26)-(21.27)-(21.28) has at least one GPS for each $h \in H$.*

**Proof.** Set $A_1 = u_t$, $Lu = u_{tt} + A_0$ and $L_1 = L$, and note that that $(Bu,u) = (-Lu + \lambda_j u, u)$ on $V$ by using Fourier series. Then the theorem follows from Theorem 21.5 with $Q$ replaced by $\Omega$ and the above remarks. ■

### *21.4.4 Applications to the telegraph equation*

Here we shall apply the above results to the telegraph equation

$$\begin{aligned} \sigma u_t(t,x) + u_{tt}(t,x) - u_{xx}(t,x) &= g(t,x,u(t,x)) + h(t,x), \quad (21.40) \\ (t,x) &\in (0,2\pi) \times (0,\pi) \end{aligned}$$

$$u(t,0) = u(t,\pi) = 0, \quad t \in (0,2\pi) \tag{21.41}$$

$$u(0,x) = u(2\pi,x), \quad x \in (0,\pi), \tag{21.42}$$

with $\sigma \neq 0$.

Clearly, the spectrum of $L$ is given by

$$\sum = \{j^2 - k^2 : k \in Z, j \in N^*\} = \{\lambda_j : j \in J = Z\} \tag{21.43}$$

since each $u \in H$ can be expanded in the Fourier series

$$u(t,x) = \sum_{k \in Z, j \in N^*} u_{jk} e^{ikt} sin(jx).$$

**Corollary 21.14** *( [NW] ) Assume that (21.32) hold uniformly for a.e. $(t,x) \in \Omega = (0,2\pi) \times (0,\pi)$, where $\gamma, \Gamma \in L_\infty(\Omega)$ are such that $\lambda_j \leq \gamma(t,x) \leq \Gamma(t,x) \leq \lambda_{j+1}$ for a.e. $(t,x) \in \Omega$ with $\lambda_j < \gamma(t,x)$ and $\Gamma(t,x) < \lambda_{j+1}$ on subsets of $\Omega$ of positive measure. Then problem (21.40)-(21.42) has at least one GPS for each $h \in H$.*

**Proof.** The eigenfunctions associated to $\lambda_j$ and $\lambda_{j+1}$ satisfy the unique continuation property, i.e. if it vanishes on a subset of $\Omega$ of positive measure, then it vanishes everywhere on $\Omega$. Hence, conditions (21.38)-(21.39) hold and the result follows from Theorem 21.13. ■

**Remark 21.8** *For more general results on telegraph equations with nonlinerities depending also on derivatives, we refer to [Mi-3].*

### *21.4.5 Application to the beam equation with damping*

Corollary 21.14 holds also for the nonlinear beam equation with damping

$$\begin{aligned} \sigma u_t(t,x) + u_{tt}(t,x) + u_{xxxx}(t,x) &= g(t,x,u(t,x)) + h(t,x), \\ (t,x) &\in (0,2\pi)\times(0,\pi) \end{aligned}$$

$$u(0,x) = u(2\pi,x), \quad x \in (0,\pi),$$

$$u(t,0) = u(t,\pi) = u_x(t,0) = u_x(t,\pi) = 0, \quad t \in (0,2\pi)$$

with $\sigma \neq 0$, if $\sum$ is given by

$$\sum = \{j^4 - k^2 : k \in Z, j \in N^*\} = \{\lambda_j : j \in J \subset Z\}.$$

**Remark 21.9** *Dirichlet boundary conditions may be replaced by $2\pi$-periodic boundary conditions in the above subsections and generalize in that way some of the earlier results in [Ma-1] and others (see [V]).*

*References*


[BF] H. Berestycki and D.G. De Figueiredo, Double resonance in semilinear elliptic problems, Comm. Partial Differential Equations, 6 (1981), 91-120.

[BN] H. Brezis and L. Nirenberg, Characterizations of the ranges of some nonlinear operators and applications to boundary value problems, Ann. Scuola Norm. Sup. Pisa Cl. Sci. (4) 5 (1978), 225-326.

[H] H. Hetzer, A spectral characterization of monotonicity properties of normal linear operators and applications to nonlinear telegraph equations, J. Operator Theory, 12 (1984), 333-341.

[K] M. A. Krasnoselskii, Topological methods in the theory of nonlinear integral equations, GITTL, Moscow, 1956.

[Ma-1] J Mawhin, Periodic solutions of nonlinear telegraph equations, Dynamical systems (Bednarek and Cesari eds), Academic Press, NY 1977, 193-210.

[Ma-2] ......, Nonresonnce conditions of nonuniform type in nonlinear boundary value problems, Dynamical systems II (Bednarek and Cesari eds), Academic Press, NY, 1982, 255-276.

[MW] J. Mawhin and J.R. Ward, Jr., Nonresonance and existence for nonlinear elliptic boundary value problems, Nonlinear Analysis, TMA, 5(1981), 677-684.

[Mi-1] P. S. Milojević, Some generalizations of the firs Fredholm theorem to multivalued $A$-proper mappings with applications to nonlinear elliptic equations, J. Math. Anal. Appl. 65(1978), 468-502.

[Mi-2] ....., Theory of $A$-proper and $A$-closed mappings, Habilitation Memoir, UFMG, Belo Horizonte, Brasil, 1980, pp. 1-207

[Mi-3] ......, Solvability of semilinear equations and applications to semilinear hyperbolic equations, in Nonlinear Functional Analysis ( P. S. Milojević ed ), Lecture Notes in Pure and Applied Math., vol. 121, 1989, pp. 95-178, M. Dekker, NY.

[Mi-4] ......, Approximation solvability of semilinear equations and applications, Theory and Applications of Nonlinear Operators of Accretive and Monotone Type, ( A.G. Kartsatos ed. ) Lecture Notes in Pure and Applied Math. , vol. 178, 1996, 149-208, M. Dekker, NY.

[NW] M.N. Nkashama- M. Willem, Time periodic solutions of boundary value problems for nonlinear heat, telegraph and beam equations, Colloquia Mathematica Societatis Janos Bolyai 47. Differential Equations: qualitative theory, Szeged, 1984, 809-846.

[V] Vejvoda et al., Partial Differential Equations: Time - periodic Solutions, 1981, Noordhoff, Groningen.

# Chapter 22

# A topological and functional analytic approach to statistical convergence

**Jeff Connor**

*ABSTRACT. This paper gives an overview of the theory of statistical convergence and extends a result of Fridy and Orhan. A sequence $x$ is said to be statistically convergent to $L$ with respect to the finitely additive measure $\mu$ provided that 'almost all' of the values $x$ are arbitarily close to $L$ with respect to $\mu$. One can also define what is meant by a $\mu$-statistical cluster point of a sequence, $\mu$-statistical limit superior of a sequence and so forth and thus create a theory of convergence that includes ordinary convergence. In this note we review some of the basic results of $\mu$-statistical convergence, indicate either topological or functional analytic proofs of some basic results and provide a means of isolating the invariants of statistical convergence.*

## 22.1 Introduction

A sequence is $\mu$-statistically convergent to a number $L$ if almost all of its values are close to $L$ where 'almost all' is defined using a finitely additive set function $\mu$. This notion has been in the literature, under different guises, since at least 1913. Since that time it has been used in the theory of Fourier analysis, ergodic theory, and number theory, usually in the guise of bounded strong summability or convergence in density. This note gives a short overview of the basic theory of statistical convergence and, using a new proof, generalizes a known result (Theorem 22.7).

Fast introduced the definition of statistical convergence in 1951 using the asymptotic density of a set to give meaning to the phrase 'almost all' ([14]). In particular, if $A \subseteq \mathbb{N}$ then the *asymptotic density* of $A$, denoted $\delta(A)$, is given by

$$\delta(A) = \lim_{n\to\infty} \frac{1}{n} |\{k \leq n | k \in A\}|$$

where $|B|$ denotes the cardinality of the set $B$. A real (or complex) valued sequence $x = \langle x_k \rangle$ is *statistically convergent to* $L$ if

$$\delta\left(\{k : |x_k - L| > \varepsilon\}\right) = 0$$

for every $\varepsilon > 0$. In this case we write $st_\delta - \lim x = L$ and call $L$ the statistical limit of $x$. The statistically convergent sequences form a linear subspace of the space of all scalar valued sequences and the statistical limit acts as a linear functional on this space. Note that convergent sequences are statistically convergent.

Statistical convergence, as defined above, has been investigated in a number of recent papers ([8], [16], [18], [36], [37]). The above definition has also been extended in a variety of ways. For instance, Maddox has introduced the notion of the statistical convergence for sequences taking values in a locally convex space ([30], [31]). Other authors have replaced the asymptotic density with one generated by a nonnegative matrix summability method ([29], [28], [19], [32]), a lacunary sequence ([19], [20], [35]) or, more generally, statistical convergence determined by a finitely additive set function satisfying some elementary properties ([5], [9], [26]). Statistical convergence is also an example of an R-type summability method as introduced by Freedman and Sember in [15] and further developed by Chun and Freedman in [6] and [7].

This note uses an extension of Fast's definition of statistical convergence where the asymptotic density is replaced by a finitely additive set function, as presented in [9]. Throughout this note, let $\mu$ be a finitely additive set function taking values in $[0,1]$ defined on a field of subsets of $\mathbb{N}$ such that if $|A| < \infty$, then $\mu(A) = 0$; if $A \subset B$ and $\mu(B) = 0$, then $\mu(A) = 0$; and $\mu(\mathbb{N}) = 1$. Throughout this note we will call a set function satisfying the above criteria a *measure*. Let $x = \langle x_k \rangle$ be a real valued sequence and $L \in \mathbb{R}$. We say that $x$ is *$\mu$-statistically convergent* to $L$, and write $st_\mu - \lim x = L$, provided $\mu\left(\{k : |x_k - L| > \varepsilon\}\right) = 0$ for every $\varepsilon > 0$.

It is easy to generate finitely additive measures meeting the above requirements. If $T = (t_{n,k})$ is a nonnegative regular summability method then $T$ can be used to generate a measure as follows: for each $n \in \mathbb{N}$ , set $\mu_n(A) = \sum_{k=1}^{\infty} t_{n,k}\chi_A(k)$ for each $A \subseteq \mathbb{N}$. Define $\mu_T$ by

$$\mu_T(A) = \lim_{n\to\infty} \mu_n(A) = \lim_{n\to\infty} \sum_{k=1}^{\infty} t_{n,k}\chi_A(k).$$

Observe that if $T$ is the identity matrix, then $\mu_T$-statistical convergence is just the ordinary definition of convergence and that if $T$ is the Cesàro matrix, then $\mu_T$-statistical convergence is statistical convergence as introduced by Fast.

Fridy and Orhan have introduced the notions of statistical cluster point, statistical limit superior and statistical limit inferior in the context $\delta$–statistical convergence ([17], [21]). These concepts can be easily extended to the setting of $\mu$-statistical convergence for an arbitrary measure $\mu$. Given a real valued sequence $x$, a real number $L$ and a measure $\mu$, $L$ is a *$\mu$-statistical cluster point* of $x$ provided $\mu(\{k : |x_k - L| < \varepsilon\}) \neq 0$ for every $\varepsilon > 0$ (note should be taken that $\mu(A) \neq 0$ denotes that either $\mu(A) > 0$ or that $\mu(A)$ is not defined). Set

$$A_x = \{a \in \mathbb{R} : \mu(\{k : x_k < a\}) \neq 0\}.$$

and

$$B_x = \{b \in \mathbb{R} : \mu(\{k : x_k > b\}) \neq 0\}$$

Define the *$\mu$-statistical limit superior* and the *$\mu$-statistical limit inferior* as follows:

$$st_\mu - \limsup x = \begin{cases} \sup B_x & B_x \neq \phi \\ -\infty & B_x = \phi \end{cases}$$

and

$$st_\mu - \liminf x = \begin{cases} \inf A_x & A_x \neq \phi \\ \infty & A_x = \phi \end{cases}.$$

It is worth noting that

$$\liminf x \leq st_\mu - \liminf x \leq st_\mu - \limsup x \leq \limsup x.$$

A sequence $x$ is *$\mu$-statistically bounded* if there is a number $B$ such that $\mu(\{k : |x_k| > B\}) = 0$.

Many of the standard results of the usual theory of convergence pass through to the theory $\mu$-statistical convergence. For instance:

1. (Cauchy criteria) Let $x$ be a sequence. The following are equivalent:

   - The sequence $x$ is $\mu$-statistically convergent.
   - The sequence $x$ is *$\mu$-statistically Cauchy:* for every $\varepsilon > 0$ there is an $n(\varepsilon)$ such that $\mu(\{k : |x_k - x_{n(\varepsilon)}| < \varepsilon\}) = 1$.

2. If $x$ is bounded then the $st_\mu - \limsup x$ is the greatest $\mu$-statistical cluster point of $x$ .

3. If $\beta = st_\mu - \limsup x$ is finite, then for all $\varepsilon > 0$

   $$\mu(\{k : x_k > \beta - \varepsilon\}) \neq 0 \text{ and } \mu(\{k : x_k > \beta + \varepsilon\}) = 0$$

   Conversely, if the above holds for all $\varepsilon > 0$ then $\beta = st_\mu - \limsup x$.

4. Suppose that $x$ is $\mu$-statistically bounded. Then $st_\mu - \limsup x = st_\mu - \liminf x = L$ if and only if $st_\mu - \lim x = L$.

The above definitions and results were introduced by Fridy ([16], [17]) or Fridy and Orhan [21] for the case where $\mu$ is the asymptotic density; the extension of these results to $\mu$-statistical convergence for an arbitrary measure $\mu$ is straightforward (for instance, see [10], [12], [26]). Note that if $\upsilon$ is the measure defined by

$$\upsilon(A) = \begin{cases} 0 & |A| < \infty \\ 1 & |\mathbb{N} \backslash A| < \infty \end{cases}$$

then $\upsilon$ generates the ordinary definitions of limit, limit superior and limit inferior.

The theory of $\delta$-statistical convergence differs from the ordinary theory of convergence in at least one important way: a $\delta$-statistically convergent sequence need not be bounded and there is no control as to how quickly a $\delta$-statistically convergent sequence becomes unbounded, i.e. for any sequence $y$ there is a sequence $x$ such that $st_\delta - \lim x = 0$ and $x \neq O(y)$. This, in effect, yields that the $\delta$-statistically convergent sequences cannot be given a locally convex Fréchet topology with the property that the functionals $x \to x_n$ are continuous (i.e. an FK topology) ([8]). Clearly the first remark extends to any measure that has an infinite null set; the second has been extended to measures of the form $\mu_T$ where $T$ is a nonnegative regular matrix summability method with the property $\lim_n \max_k t_{n,k} = 0$ ([25]).

It was the absence of an FK topology for the statistically convergent sequences that motivated the development of the approach used in this note.

## 22.2 The support set of a measure

Before we can apply functional analysis or topology to the study of statistical convergence, it is necessary to describe the statistically convergent sequences in an amenable fashion. The key tool is the support set of a measure; this is a generalization of the support set of a matrix, which was introduced by Henrikson ([23]) and further discussed by Atalla and Rainwater in connection with bounded strong matrix summability ([3], [4], [34]) and discussed in relationship to $\mu$-statistical convergence in ([10], [12], [13], [26]).

The support set of $\mu$ is defined to be the set

$$K_\mu = \bigcap \{ cl_{\beta\mathbb{N}}(A) | \quad A \subseteq \mathbb{N} \text{ and } \mu(A) = 1 \}$$

where $\beta\mathbb{N}$ denotes the Stone-Čech compactification of $\mathbb{N}$. The concepts introduced in the previous section can be described using the support set of the matrix.

**Proposition 22.1** *Let $\mu$ be a measure, $x$ a bounded real valued sequence and $x^\beta$ its extension to $\beta\mathbb{N}$. Then:*

1. *$x$ is $\mu$-statistically convergent to $\alpha$ if and only if $x^{\beta}(p) = \alpha$ for all $p \in K_{\mu}$.*

2. *$\alpha$ is $\mu$-statistical cluster point of a bounded sequence $x$ if and only if there is a point $p \in K_{\mu}$ such that $x^{\beta}(p) = \alpha$.*

3. *$st_{\mu} - \limsup x = \sup\left\{x^{\beta}(p) : p \in K_{\mu}\right\}$ and $st_{\mu} - \liminf x = \inf\left\{x^{\beta}(p) : p \in K_{\mu}\right\}$*

The first and second parts of Proposition 22.1 essentially appear in [12]; the third is an immediate consequence of earlier remarks made in this note.

## 22.3 Invariants of statistical convergence

A property $\mathfrak{P}$ of a measure is an invariant of statistical convergence if any other measure that determines the same statistically convergent sequences also has property $\mathfrak{P}$. It is clear from the preceding proposition that two measures generate the same support set if and only if they determine the same statistically convergent sequences; hence the invariants of the measures are those that can be described in terms of their support sets. The following result gives an invariant of statistical convergence with respect to a nonnegative regular summability matrix.

**Proposition 22.2** *Let $S = [s_{nk}]$ and $T = [t_{nk}]$ be two nonnegative regular summability methods such that $\mu_S$-statistical convergence and $\mu_T$-statistical convergence are equivalent. If $\lim_n \max\{t_{nk} | k \in \mathbb{N}\} = 0$, then $\lim_n \max\{s_{nk} | k \in \mathbb{N}\} = 0$.*

**Proof.** In [34] it is shown that, for a nonnegative regular matrix $T$, $\lim_n \max_k t_{nk} = 0$ if and only if $K_{\mu_T}$ is nowhere dense in $\beta\mathbb{N}\backslash\mathbb{N}$. ■

The measures considered in this note are never countably additive; this follows from noting that $0 = \sum_{k\in\mathbb{N}} \mu(\{k\}) \neq \mu(\mathbb{N}) = 1$ for any measure $\mu$. However, a measure may possess the following weaker property: $\mu$ has the *APO (additive property for null sets)* provided that whenever $< A_n >$ is a disjoint sequence of $\mu$-null sets, then there is a sequence of sets $< B_n >$ such that $\bigcup_{n=1}^{\infty} B_n = B$ is a $\mu$-null set and $B_n \vartriangle A_n$ is finite for each $n$. The APO was defined in [15] during a discussion of convergence in density (defined below). In particular, Freedman and Sember established the following key result:

**Theorem 22.3** *If $T$ is a nonnegative regular matrix summability method, then $\mu_T$ has the APO.*

This useful result had been established and generalized repeatedly in the literature both before and after Freedman and Sember's paper; for instance, in one form or another, this result has appeared in [3], [5], [9], [16], [24], and [26].

A sequence $x$ is *convergent in $\mu$-density* to $L$ provided there is a set $A \subset \mathbb{N}$ such that $\mu(A) = 1$ and for every $\varepsilon > 0$ there is a $K_\varepsilon$ such that $k > K_\varepsilon$ and $k \in A$ implies that $|x_k - L| < \varepsilon$. This note discusses the role the APO plays in relating the connection between convergence in $\mu$-density and $\mu$-statistical convergence. Let

$$\begin{aligned} D_\mu &= \{x \in \ell_\infty : x \text{ is convergent in } \mu\text{-density to } 0\} \\ S_\mu &= \{x \in \ell_\infty : x \text{ is } \mu\text{-statistically convergent to } 0\} \end{aligned}$$

One can show that $D_\mu$ and $S_\mu$ are ideals in $\ell_\infty$; $c_0 \subset D_\mu \subset S_\mu$ and $\overline{D_\mu} = S_{\mu.}$(where the closure is taken in the sup topology of $\ell_\infty$) ([9]). It is also clear that $\mu$-statistical convergence is equivalent to convergence in $\mu$-density if and only if $D_\mu = S_\mu$.

**Theorem 22.4** *([9])$D_\mu = S_\mu$ if and only if $\mu$ has the APO.*

The next result shows that the APO is an invariant of statistical convergence.

**Theorem 22.5** *([10]) The measure $\mu$ has the APO if and only if $K_\mu$ is interior to any $G_\delta$ subset of $\beta\mathbb{N}\backslash\mathbb{N}$ that contains it (i.e. $K_\mu$ is a* P-set *in $\beta\mathbb{N}\backslash\mathbb{N}$).*

Observe that if $T$ is a nonnegative regular summability matrix, then $\{x \in \ell_\infty : \lim_n \sum_k t_{n,k} |x_k| = 0\}$ is an ideal of $\ell_\infty$ that contains $D_{\mu_T}$ and is contained in $S_{\mu_T}$; since $\mu_T$ has the APO, the next result is immediate:

**Corollary 22.6** *Let $T$ be a nonnegative regular summability method and $x$ be a bounded sequence. The following are equivalent:*

1. *$x$ is $\mu_T$-statistically convergent to $L$.*
2. *$x$ is strongly $T$-summable to $L$.*
3. *$x$ is convergent in $\mu_T$-density to $L$.*

## 22.4 Summability theorems

In this section we use the above tools to obtain some typical summability theorems for the $\mu$-statistical convergence of real valued sequences. In particular we give a consistency theorem, a Tauberian theorem, an inclusion theorem and a core theorem. All of these results apply to bounded strong summability with respect to a nonnegative matrix.

Proposition 22.1 gives enough information to characterize when $\nu$-statistical and $\mu$- statistical convergence are consistent for bounded sequences; this happens precisely when $K_\mu \bigcap K_\nu \neq \phi$. Observe that if $x$ is $\mu$-statistically convergent to $L$ and $\nu$-statistically convergent to $M$ and $p \in K_\mu \bigcap K_\upsilon$, then $x^\beta(p) = L = M$. If $K_\mu \bigcap K_\upsilon = \phi$, then, as $K_\mu$ and $K_\nu$ are disjoint closed subsets of $\beta\mathbb{N}$, then there is function $f \in C(\beta\mathbb{N})$ such that $f \restriction K_\mu = 0$ and $f \restriction K_\upsilon = 1$. Letting $x = f \restriction \mathbb{N}$ yields a sequence such that $\mu$-statistically convergent to 0 and $\nu$-statistically convergent to 1. Also note that $K_\mu \bigcap K_\nu \neq \phi$ if and only if, for a subset $A \subset \mathbb{N}$, $\mu(A) = 1$ implies that $A$ is not a $\nu$-null set. (See [12] for a similar discussion.)

Next we give a gap Tauberian condition for $\mu$-statistical convergence. Let be an $\gamma : \mathbb{N} \to \mathbb{N}$ be an increasing function with $\gamma(0) = 0$. Then

$$(\Delta x)_k \neq 0 \Rightarrow \exists\, r \in \mathbb{N} \text{ such that } k = \gamma(r) \tag{*}$$

is a Tauberian condition for $\mu$-statistical convergence if and only if

$$cl_{\beta\mathbb{N}\setminus\mathbb{N}}\left(\bigcup_{r\in\mathbb{N}}(\gamma(n_r), \gamma(n_r+1)]\right)\bigcap K_\mu \neq \phi \tag{**}$$

for any increasing sequence of integers $\langle n_r \rangle$. In order to establish the sufficiency of (**), suppose that $x$ is statistically convergent to $L$ and satisfies the gap condition (*). As the intersection given in (**) is nonempty, every limit point of $x$ is a statistical cluster point of $x$. Now, since the sequence is $\mu$-statistically convergent to $L$, it follows that $\lim x = L$ since $L$ is the only statistical cluster point - and hence only limit point - of $x$. In order to see the necessity of (**), note that if there is a sequence $\langle p_r \rangle$ such that the intersection given in (**) is empty, then the sequence $x = \langle x_k \rangle$ given by

$$x_k = \begin{cases} 1 & k \in \bigcup_{r\in\mathbb{N}}(\gamma(p_r), \gamma(p_r+1)] \\ 0 & \text{otherwise} \end{cases}$$

is divergent, satisfies the gap condition given in (*), and $\mu$-statistically convergent to 0. (This is a small extension of Theorem 4.2 of [11]).

The preceding results, when translated into the language of analysis, indicate precisely which inequalities need to be established to show two nonnegative methods are consistent for bounded strong summability and when $\gamma$ generates a gap Tauberian condition for strong summability. In particular:

- Strong $T$-summability and strong $S$-summability are consistent if and only if, for all $A \subset \mathbb{N}$,

$$\lim_{n\to\infty} \sum_{k\in A} t_{n,k} = 1 \Rightarrow \limsup_{n\to\infty} \sum_{k\in A} s_{n,k} > 0.$$

- $\gamma : \mathbb{N} \to \mathbb{N}$ generates a gap Tauberian condition for strong $T$-summability if and only if

$$\limsup_{n\to\infty} \sum_r \sum_{k>\gamma(n_r)}^{\gamma(n_r+1)} t_{n,k} > 0$$

for any increasing sequence of integers $\langle n_r \rangle$.

The above results essentially appear in [12] and [11], respectively.

Following an outline suggested by Atalla and Bustoz ([2]), we can also give a generalization and alternative proof of a statistical core theorem due to Fridy and Orhan. Recall that for a bounded real valued sequence $x$ the *Knopp core* of $x$, denoted $K$-core$\{x\}$, is the interval $[\liminf x,\ \limsup x]$. Following Fridy and Orhan ([21]), the *$\mu$-statistical core* of a real valued statistically bounded sequence $x$ is defined to be the interval $[st_\mu - \liminf x,\ st_\mu - \limsup x]$. The Knopp core and $\mu$-statistical core of a sequence are denoted by $K$-core$\{x\}$ and $st_\mu$-core$\{x\}$ respectively. Knopp showed, for complex valued sequences, that if $T$ is a nonnegative regular matrix summability method, then $K$-core$\{Tx\} \subseteq K$-core$\{x\}$ ([27]); Agnew later characterized the matrix maps which map the bounded sequences into themselves and have the property that the Knopp core of the transform of a sequence is contained in the Knopp core of the sequence for every bounded sequence ([1]). In a similar vein, Fridy and Orhan gave necessary and sufficient conditions for a matrix $T$ which maps the bounded sequences into themselves to have the property that $K$-core$\{Tx\} \subseteq st_\delta$-core$\{x\}$ for every bounded real valued sequence $x$ in [21]. The next result extends Fridy and Orhan's result to $\mu$-statistical convergence and hence includes Agnew's result for real valued sequences.

**Theorem 22.7** *Let $T$ be a matrix operator which maps $\ell_\infty$ into itself and $\mu$ be a measure. The following are equivalent:*

1. *$K$-core$\{Tx\} \subseteq st_\mu$-core$\{x\}$ for all $x \in \ell_\infty$.*

2. *$T$ is regular, $\lim_n \sum_{k \in A} |t_{n,k}| = 0$ whenever $\mu(A) = 0$, and $\lim_n \sum_k |t_{n,k}| = 1$.*

Before starting the proof of the theorem, it will be helpful to list a few observations and a theorem:

- Observe that $\lim_n \sum_{k \in A} |t_{n,k}| = 0$ whenever $\mu(A) = 0$ and $T$ maps null sequences into null sequences if and only if $T$ maps bounded $\mu$-statistically null sequences to null sequences. The sufficiency follows from a standard sliding hump argument and the necessity can be established by first showing that $T(D_\mu) \subseteq c_0$ and then noting that, since $T$ extends to a continuous map from $\ell_\infty$ to itself and $c_0$ is closed in $\ell_\infty$, $S_\mu = \overline{D_\mu} \subseteq T^{\leftarrow}(c_0)$. (Kolk establishes similar results for measures generated by nonnegative matrices in [28].)

- Observe that $\ell_\infty / S_\mu$ is isometrically isomorphic to $C(K_\mu)$: the proof is similar to the proof that $\ell_\infty / c_0$ is isometrically isomorphic to $C(\beta\mathbb{N} \backslash \mathbb{N})$.

- Proposition 22.1 yields that $st_\mu$-core$\{x\}$ is the closed convex hull of $\{x^\beta(p) : p \in K_\mu\}$ and $K$-core$\{x\}$ is the closed convex hull of $\{x^\beta(p) : p \in \beta\mathbb{N}\backslash\mathbb{N}\}$.

- Let $X$ and $Y$ be sets, $C(X)$ and $C(Y)$ normed with the sup norm, and $T : C(X) \to C(Y)$ a linear operator. Then $||T|| = 1$ and $T(1) = 1$ if and only if $(Tf)(Y)$ is contained in the closed convex hull of the range of $f$ for every $f \in C(X)$ (Phelps, [33]).

**Proof.** ($2 \Rightarrow 1$): Note that since $T$ maps statistically null sequences to null sequences we have that $T$ induces an operator $\widetilde{T} : \ell_\infty/S_\mu \to \ell_\infty/c_0$ which in turn induces an operator $\widehat{T} : C(K_\mu) \to C(\beta\mathbb{N}\backslash\mathbb{N})$ where $\widehat{T}f = (Tx)^\beta \restriction (\beta\mathbb{N}\backslash\mathbb{N})$ when $x$ is a bounded sequence such that $x^\beta \restriction K_\mu = f$. Note that

$$\left\|\widehat{T}\right\| = \left\|\widetilde{T}\right\| = \limsup_n \sum_k |t_{n,k}| = 1,$$

and, since $T$ is regular, $T(1) = 1$. Now Phelp's theorem yields that, for each bounded sequence $x$, $\widehat{T}x^\beta(\beta\mathbb{N}\backslash\mathbb{N})$ is contained in the closed convex hull of $x^\beta(K_\mu)$ and hence $K$-core$\{Tx\} \subseteq st_\mu$-core$\{x\}$ for all $x \in \ell_\infty$.

($1 \Rightarrow 2$). First note that the hypothesis forces $T$ to be a regular matrix summability method and that statistically null sequences are mapped to null sequences, thus $T$ can be extended to an operator from $C(K_\mu)$ to $C(\beta\mathbb{N}\backslash\mathbb{N})$ as before. Now Phelp's theorem, in conjunction with the regularity of $T$ yields

$$1 = \lim_n \sum_k t_{n,k} \leq \limsup_n \sum_k |t_{n,k}| = 1$$

and hence $\lim_n \sum_k |t_{n,k}| = 1$ ■

Fridy and Orhan present a core theorem regarding the $\delta$-statistical convergence of complex valued sequences in [22].

*References*

[1] Agnew, R.P., Cores of complex sequences and of their transforms, Amer. J. Math. 61 (1939), 178-186

[2] Atalla, R. and Bustoz, J., On sequential cores and a theorem of R.R. Phelps, Proc. Amer. Math. Soc. 21 (1969), 36-42.

[3] Atalla, R. On the multiplicative behavior of regular matrices, Proc. Amer. Math. Soc. 26 (1970), 437-446.

[4] Atalla, R., On the consistency theorem in matrix summability, Proc. Amer. Math. Soc. 35 (1972) no. 2, 416-422.

[5] Buck, R.C., Generalized Asymptotic Densities, American J. Math. 75 (1953), 335-46.

[6] Chun, C.S. and Freedman, A.R., Theorems and examples for R-type summability methods, J. Korean Math. Soc. 25 (1988), 315-324.

[7] Chun, C.S. and Freedman, A.R., A bounded consistency theorem for strong summabilities, Inter. J. Math. Sci. 12 (1989), 39-46.

[8] Connor, J., The statistical and strong p-Cesàro convergence of sequences, Analysis 8 (1988), 47-63.

[9] Connor, J., Two valued measures and summability, Analysis 10 (1990), 373-385.

[10] Connor, J., R-type summability methods, Cauchy criteria, P-sets and statistical convergence, Proc. Amer. Math. Soc. 115 (1992), no. 2, 319-327.

[11] Connor, J., Gap tauberian theorems, Bull. Austral. Math. Soc. 47 (1993), no. 3, 385-393.

[12] Connor, J. and Kline, J., On Statistical limit points and the consistency of statistical convergence, J. Math. Anal. Appl. 197 (1996), no. 2 392-399.

[13] Connor, J. and Swardson, M.A., Measures and ideals of $C^*(X)$, Papers on general topology and applications, 80-91, Ann. New York Acad. Sci., 704, New York Acad. Sci., New York, 1993.

[14] Fast, H., Sur la convergence statistique. Colloq. Math 2 (1951), 241-244.

[15] Freedman, A. R. and Sember, J.J., Densities and summability, Pacific J. Math 95 (1981), 293-305.

[16] Fridy, J., On statistical convergence, Analysis 5 (1985), 301-313.

[17] Fridy, J., Statistical limit points, Proc. Amer. Math. Soc. 118 (1993), no. 4, 1187-1192.

[18] Fridy, J. and Miller, H.I., A matrix characterization of statistical convergence, Analysis 11 (1991), 59-66.

[19] Fridy, J. and Orhan, C., Lacunary Statistical Convergence, Pacific Journal of Math., 160 (1993), 43-51.

[20] Fridy, J. and Orhan, C., Lacunary Statistical Summability, Journal of Mathematical Analysis and Applications, Vol. 173 (1991), no. 2, 497-504.

[21] Fridy, J. and Orhan, C., Statistical Limit Superior and Limit Inferior, Proc. Amer. Math. Soc., to appear.

[22] Fridy, J.;Orhan, C., Statistical core theorems, J. Math. Anal. Appl. 208 (1997), no. 2, 520-527.

[23] Henriksen, M., Multiplicative summability methods and the Stone-Čech compactification, Math. Z. 71 (1959), 427-435.

[24] Hill, J.D. and Sledd, W.T., Approximation in bounded summability fields, Canad. J. Math. 20 (1968), 410-415.

[25] Kline, J., The $T$-statistically convergent sequences are not an FK space, Internat. J. Math. Math. Sci. 18 (1995), no. 4, 825-827.

[26] Kline, J. Statistical convergence and densities generated by sequences of measures, Ph.D. dissertation, Ohio University, 1995.

[27] Knopp, K., Zur Theorie de Limitierungsverfahren (Erste Mitteiiung), M. Zeit. 31 (1930), 115-127.

[28] Kolk, E., Matrix summability of statistically convergent sequences. Analysis 13 (1993), no. 2, 497-504.

[29] Kolk, E., The statistical convergence in Banach spaces, Tartu Ul. Toimetised No. 928 (1991), 41-52.

[30] Maddox, I.J., Statistical convergence in a locally convex space, Math. Proc. Cambridge Philos. Soc. 104 (1988), 141-145.

[31] Maddox, I.J., A tauberian theorem for statistical convergence, Math. Proc. Camb. Phil. Soc. 106 (1989), 277-280.

[32] Miller, H., A measure theoretical subsequence characterization of statistical convergence. Trans. Amer. Math. Soc. 347 (1995) no. 5, 1811-1819.

[33] Phelps, R.R., The range of $Tf$ for certain linear operators $T$, Proc. Amer. Math. Soc. 16 (1965), 381-382.

[34] Rainwater, J., Regular matrices with nowhere dense support, Proc. Amer. Math. Soc. 29 (1971), 90-91.

[35] Savac, E.; Faith, N, On $\sigma$-statistically convergence and lacunary $\sigma$ statistically convergence. Math. Slovaca 43 (1993), no. 3, 309-315.

[36] Salat, T., On statistically convergent sequences of real numbers, Math. Slovaca 30 (1980), No. 2, 139-150.

[37] Schoenberg, I. J., The integrability of certain functions and related summability methods, Amer. Math. Monthly 66 (1959), 361-375.

# Part IV

# Asymptotics and Applications

# Chapter 23

# Optimal control of divergent control systems

**Dean A. Carlson**[1]

## 23.1 Introduction and History

The study of Lagrange problems defined on an unbounded interval begins in the arena of mathematical economics when Ramsey [36] formulated a model of economic growth and, through the development of the Euler-Lagrange equations, stated the first "golden rule of accumulation" now known as Ramsey's rule. The study of these problems remained dormant until the 1940's when the students of Tonelli initiated a systematic treatment for free problems in the calculus of variations defined on $[0, \infty)$. These works; Cinquini [18], [19], Faedo [22], [23], and Darbo [21]; provide a rather complete theory providing existence theorems, necessary conditions, and most importantly (at least for the purposes here) the asymptotic convergence of optimal trajectories to a steady state. In all of these works, the objective or cost function was assumed to be a convergent improper integral. That this may not be the case was realized by Ramsey in his seminal paper. To circumvent this he introduced what we now call the "optimal steady-state problem," which in Ramsey's model was referred to as the "maximal sustainable rate of enjoyment," or more simply as "Bliss." In his approach, Ramsey introduces a new problem in which the optimal value is finite whenever it is possible to "control" the system to reach Bliss in a finite time. This approach leads to what we now know as the *asymptotic turnpike theory.* In this theory it is demonstrated that there exists an "optimal steady state" which asymptotically attracts the optimal trajectory. This idea of Ramsey's is related to the finite interval Turnpike Theorems of mathematical economics explored by von Neumann [41], R. Radner [35], H. Atsumi [1], the nobel laureate P. A. Samuelson [39], L. McKenzie [33], and many others. For a survey of this literature we refer the reader to the survey paper of McKenzie [34]. In that theory, the solution of the optimal steady state is shown to be an attractor of the optimal trajectories so that the trajectories bend toward the "turnpike" leaving it only when the optimal trajectory is required to meet a terminal target.

[1] Research supported by NSF (INT-9500782)

Another approach for assuring convergence of the objective functional for at least some of admissible trajectories is through the use of discounting. This is done by introducing an exponential weight $t \to e^{-rt}$ in which $r > 0$ is a discount rate. This approach, while considered in his paper, was rejected by Ramsey as unimaginative and morally incorrect since it weights preferences to the present at the expense of the future. Nevertheless, problems with discounting are frequently discussed in economic growth and therefore have an important role to play. It is surprising how different problems with discounting are when compared to undiscounted problems. This is most easily seen in the corresponding finite horizon theorem. For these problems, the optimal steady state is now defined by an *implicit mathematical programming problem* (see Feinstein and Luenberger [26]) and is not an attractor. Instead the best that can be said is that the optimal trajectories are attracted toward an exponential funnel whose axis is the solution to the implicit steady state (see Feinstein and Oren [25]). On the other hand it is now known that if the discount rate $r > 0$ is small enough relative to the convexity/concavity of the Hamiltonian, then the optimal implicit steady state is an asymptotic attractor for the optimal trajectories. This however introduces a tradeoff since to insure convergence of the improper integral it may be necessary to choose a larger discount rate. Thus, again we see that one encounters divergent improper integrals.

Under the assumption of Bliss, it was commonly believed that, because of the strong convexity conditions imposed in economic models, a solution of the Euler-Lagrange equations led to the desired solution with a finite objective. However, in 1960 Chakravarty [17] presented a simple example in which a solution to these necessary conditions produced a feasible trajectory for which the cost was unbounded! This fact led to several new definitions of optimality introduced by C.C. von Weizäcker [42] in which the convergence of the improper integral was not required. The most relevant of these is now referred to as *overtaking optimality.* Related to the "turnpike theory" was the notion of transversality conditions at infinity. A typical problem considered in these early investigations required that the admissible trajectories only satisfy a fixed initial condition and not a boundary condition at infinity. As is well known in the finite interval case, this leads to a boundary condition at the terminal point known as a transversality condition. In these models, these are most easily stated as requiring the "shadow price" (in classical mechanics this is the momentum) to be zero at the end time. For the infinite horizon case considered here this condition was typically assumed by early researchers (as an asymptotic boundary condition) and it was not until 1974 that an example, given in Halkin [27], demonstrated that this assumption was incorrect. The correct first order necessary conditions for these problems was given in [27] in the framework of optimal control and Pontryagin's Maximum principle. These revelations led for the search of appropriate transversality conditions for these problems and resulted, in particular, with an entire volume of the

Journal of Economic Theory devoted to a series of articles on Hamiltonian dynamics arising in economic growth models in 1976 (see also Cass and Shell [15]). Also in 1976 the first general existence result for "overtaking optimality" was given by Brock and Haurie [5] for convex problems. Additionally, the spirit of Tonelli appeared in the works of Baum [4], Bates [3], and Balder [2] in which the existence results of Tonelli's students mentioned above were generalized to ordinary control systems. Throughout the 1980's a large number of existence results for overtaking optimality under increasingly weaker and more general dynamics have been discovered. This work, as well as additional references relating to the above ideas is nicely collected in the monograph of Carlson, Haurie, and Leizarowitz [13] and for brevity we refer the reader there for these results. More recently Zaslavski [43] has produced a series of articles in which the general question is to determine how large is the class of integrands for which an overtaking optimal solutions exists is.

With this lengthy introduction to the history of these interesting variational problems in hand we now consider problems in which the admissible trajectories are required to to satisfy **explicit pointwise state constraints**. From the point of view of existence of overtaking optimal solutions constraints of this type pose no particular difficulties since typically it is assumed that admissible trajectories exist. On the other hand, pointwise state constraints create difficulties when Pontryagin's maximum principle is applied since in these situations the Lagrange multiplier corresponding to the state constraint is known only to be a function of bounded variation (see e.g., Clarke, F. H. [20]) which will often have discontinuities when the trajectory meets the boundary of the state constraint. The ideas of the asymptotic turnpike theory enable us circumvent this difficulty since we will be able to look at an associated uncoupled optimal control problem which does not include the explicit state constraint as a constraint but as part of the objective function. In addition, it is possible to devise a "tax scheme" by solving the steady state problem and using the constant Lagrange multiplier as the tax rate.

In the work presented below, we present a scheme for investigating the existence of an optimal solution for infinite horizon optimal control in which an explicit pointwise inequality constraint is imposed, perhaps by a regulatory agency. By using the Lagrange multipliers for an associated steady state optimization problem we define an associated infinite horizon control problem without the explicit constraint. The theory presented below establishes a relationship between the overtaking optimal solutions of this associated problem with the original problem. Specificly we show that, as a consequence of the turnpike property, the overtaking solutions of the associated problem is averaging optimal for the original problem. Additionally we introduce the notion of satisfying the explicit pointwise inequality constraint "most of the time." The motivation for this new notion arises from two ideas. The first of these is the clasical turnpike theory in dynamic optimization and the second from a the practical view that in a long term model it is unlikely that a given firm can meet such a constraint all of the time. This notion determines the class of admissible trajectories that we consider. These results, in a single player setting, extend the calculus of variations results given in Carlson and Haurie [11] for dynamic games to an optimal control setting.

The potential application of these results in a practical setting can be envisioned by the following scenario. A regulatory agency, such as the Environmental Protection Agency, wishes to impose an output emissions constraint on a firm. This firm takes the form of a pointwise inequality constraint. To penalize a violation of the constraint the agency imposes a constanat tax rate on the firm which is determined by solving the *optimal steady state problem*. This tax rate is the resulting Lagrange multiplier. The firm however is interested in solving a dynamic optimization problem and therefore solves the associated dynamic optimization problem which embeds the tax constraint as a penalty term. The theory presented below demonstrates that the overtaking optimal solution obtained by the firm is averaging optimal for the problem with the agency emission restriction over the class of all admissible trajectories that asymptotically satisfy the pointwise inequality constraint.

With these remarks, the plan of our paper is as follows. In Section 2 we introduce the basic hypotheses and definitions as well as describe the general model with and without discounting that we wish to consider. Section 3 is devoted to reviewing the optimal steady state problem and presenting the basic existence results for convex infinite horizon optimal control problems with and without discounting which we will utilize in our discussion. These problems will not include the explicit pointwise inequality constraint. In Section 4 we introduce the associated uncoupled infinite horizon optimal control problem, again both with and without discounting. Finally, in Section 5 we present our existence results for the state constrained models.

## 23.2 Basic models and hypotheses

We begin by considering a control system whose dynamics are described by the ordinary differential equation

$$\begin{aligned} \dot{x}(t) &= f(x(t), u(t)) \quad \text{a.e.} \quad t \geq 0 \\ x(0) &= x_0 \\ x(t) &\in X \subset \mathbb{R}^n \quad \text{for} \quad t \geq 0 \\ u(t) &\in U \subset \mathbb{R}^m \quad \text{a.e.} \quad t \geq 0, \end{aligned} \tag{23.1}$$

in which $f(\cdot,\cdot) : \mathbb{R}^n \times \mathbb{R}^m \to \mathbb{R}^n$ is assumed to be a continuous mapping, $X \subset \mathbb{R}^n$ and $U \subset \mathbb{R}^m$ are closed, convex sets, and $x_0 \in \mathbb{R}^n$ is a fixed constant initial condition.

**Definition 23.1** *A pair of functions* $\{x(\cdot), u(\cdot)\} : [0, +\infty) \to \mathbb{R}^n \times \mathbb{R}^m$ *is called a* trajectory-control pair *if* $x(\cdot)$ *is a bounded, locally absolutely continuous function on* $[0, +\infty)$, $u(\cdot)$ *is Lebesgue measurable, and if the control system (23.1) is satisfied.*

In addition to satisfying the above control system we wish to impose (or single out) an explicit pointwise state inequality constraint which takes the form

$$h_l(x(t)) \leq 0 \quad \text{for} \quad t \geq 0 \quad l = 1, 2, \dots k, \tag{23.2}$$

where $h_l : \mathbb{R}^n \to \mathbb{R}$ is a continuous convex function, $l = 1, 2, \dots k$. One interpretation of this constraint in applications is to view it as a constraint that is imposed on a firm by a regulatory agency. Obeying such a constraint at each instance of time, while desirable, is often neither practical nor expected by the regulatory agency. These ideas motivate the following definitions.

**Definition 23.2** *A trajectory-control pair* $\{x(\cdot), u(\cdot)\}$ *asymptotically satisfies the state constraint (23.2) if for each* $\epsilon > 0$ *there exists a constant* $B \doteq B(\epsilon, x(\cdot), u(\cdot)) > 0$ *such that*

$$meas\left[\{t \geq 0 : \quad h_l(x(t)) > \epsilon \quad l = 1, 2, \dots k\}\right] \leq B.$$

The above definition is inspired from the *turnpike property* that arises in the context of optimal economic growth models (see Carlson, Haurie, and Leizarowitz [13] for a detailed discussion of this idea). In the present context it is clear that by satisfying the constraint in this manner any asymptotically trajectory-control pair will spend "most of its time" being close to satisfying the pointwise inequality constraint.

To consider discounted problems we introduce the following definition in which $r > 0$ is a fixed discount rate.

**Definition 23.3** *Let* $\{\hat{x}(\cdot), \hat{u}(\cdot)\}$ *be a fixed trajectory-control pair and let* $(\hat{x}_\infty, \hat{u}_\infty) \in X \times U$ *be such that*

$$\begin{aligned} \lim_{t \to +\infty} \hat{x}(t) &= \hat{x}_\infty \\ f(\hat{x}_\infty, \hat{u}_\infty) &= 0 \\ h(\hat{x}_\infty) &\leq 0. \end{aligned}$$

*We say* $\{x(\cdot), u(\cdot)\} \in \Omega$ *asymptotically satisfies the coupled state constraint (23.2) for the discounted optimal control problem relative to* $\{\hat{x}(\cdot), \hat{u}(\cdot)\}$ *if*

$$\int_0^{+\infty} e^{-rt} h_l(x(t))\, dt \leq \int_0^{+\infty} e^{-rt} h_l(\hat{x}(t))\, dt$$

*for* $l = 1, 2, \dots k$.

The above definition views the coupled state constraint (23.2) more as an isoperimetric-type constraint rather than a pointwise constraint. Nevertheless, the above definition says that, relative to the trajectory-control pair $\{\hat{x}(\cdot), \hat{u}(\cdot)\}$, the average discounted value of the state constraint for any asymptotically trajectory-control pair will not exceed the average discounted value of the constraint for the pair $\{\hat{x}(\cdot), \hat{u}(\cdot)\}$.

To measure the performance of a trajectory-control pair, say $\{x(\cdot), u(\cdot)\}$, up to time $T > 0$ we use a cost functional described by the following integral

$$J_T(x(\cdot), u(\cdot)) = \int_0^T e^{-rt} g(x(t), u(t))\, dt \tag{23.3}$$

where $g : \mathbb{R}^n \times \mathbb{R}^m \to \mathbb{R}$ is a lower semicontinuous function. When $r = 0$ we will refer to (23.3) as the undiscounted cost and when $r > 0$ we call (23.3) the discounted cost. We now define the set of feasible points for our optimal control problem as follows.

**Definition 23.4** *A trajectory-control pair $\{x(\cdot), u(\cdot)\}$ will be called an admissible pair for*

1. *the unconstrained problem if the map $t \to e^{-rt} g(x(t), u(t))$ is Lebesgue integrable on $[0, T]$ for each $T > 0$.*

2. *the constrained problem without discounting if $t \to g(x(t), u(t))$ is Lebesgue integrable on $[0, T]$ for each $T > 0$ and asymptotically satisfies the state constraint (23.2).*

3. *the constrained problem with discounting relative to the trajectory-control pair $\{\hat{x}(\cdot), \hat{u}(\cdot)\}$, if $t \to e^{-rt} g(x(t), u(t))$ is Lebesgue integrable on $[0, T]$ for each $T > 0$ and if it asymptotically satisfies the coupled state constraint (23.2) for the discounted optimal control problem relative to $\{\hat{x}(\cdot), \hat{u}(\cdot)\}$.*

Following standard terminology we will refer to $x(\cdot)$ as an *admissible trajectory* and to $u(\cdot)$ as an *admissible control* where admissible refers to any one of the above concepts.

As we require only local integrability of the cost functional we must specify how to decide the best or optimal admissible pair. For our purposes we will use the following concepts.

**Definition 23.5** *An admissible pair $\{x^*(\cdot), u^*(\cdot)\}$ will be called*

1. strongly optimal *if for any other admissible pair $\{x(\cdot), u(\cdot)\}$ we have*

$$J_\infty(x^*(\cdot), u^*(\cdot)) = \lim_{T \to +\infty} J_T(x^*(\cdot), u^*(\cdot))$$

*is finite and if*

$$J_{\infty}(x^*(\cdot), u^*(\cdot)) \leq \lim_{T \to +\infty} J_T(x(\cdot), u(\cdot)). \tag{23.4}$$

2. overtaking optimal *if for every* $\epsilon > 0$ *and admissible pair* $\{x(\cdot), u(\cdot)\}$ *there exists* $\hat{T} = \hat{T}(\epsilon, x(\cdot), u(\cdot)) > 0$ *such that for all* $T \geq \hat{T}$ *we have*

$$J_T(x^*(\cdot), u^*(\cdot)) \leq J_T(x(\cdot), u(\cdot)) + \epsilon \tag{23.5}$$

*(or equivalently*

$$\liminf_{T \to +\infty} [J_T(x(\cdot), u(\cdot)) - J_T(x^*(\cdot), u^*(\cdot))] \geq 0) \tag{23.6}$$

3. average cost optimal *if for any admissible pair* $\{x(\cdot), u(\cdot)\}$ *we have*

$$\liminf_{T \to +\infty} \frac{1}{T} [J_T(x(\cdot), u(\cdot)) - J_T(x^*(\cdot), u^*(\cdot))] \geq 0. \tag{23.7}$$

In the above "admissible" means in any of the senses described above and of course we only compare pairs that are admissible in the same sense. The above concepts of optimality are listed in descending order with strong optimality, which is the standard concept of a minimizer, being the strongest and average optimality being the weakest. The notion of overtaking optimality was introduced into 1965 by von Weizäcker [42] to address precisely those cases when a strong minimum does not exist as a result of the divergence of the improper integrals. This notion is one of a hierarchy of similar types of optimality and we refer the interested reader to the monograph of Carlson, Haurie, and Leizarowitz [13] for more information (see also Stern [40] or Carlson [8] as well).

## 23.3 Existence of optimal solutions

In this section we review the existence theory of infinite horizon optimal control. As was indicated in the introduction, the inclusion of a pointwise inequality constraint (23.2) poses no particular difficulty since it is typically assumed that admissible pairs exist and consequently this explicit constraint can be embedded in the constraint $x(t) \in X$ for all $t \geq 0$.

### *23.3.1 Existence of overtaking optimal solutions without discounting*

For this problem we assume that the following assumptions are satisfied

1. The set
$$\Omega \doteq \{(z_0, z, x) : \; z_0 \geq g(x,u), \; z = f(x,u), \; x \in X, \text{ for some } u \in U\}, \tag{23.8}$$
is assumed to be convex and closed.

2. For each $\epsilon > 0$ there exists $c_\epsilon > 0$ such that
$$|f(x,u)| \leq c_\epsilon + \epsilon \cdot g(x,u) \tag{23.9}$$
holds for $(x,u) \in X \times U$.

3. There exists a vector $\bar{p} \in \mathbb{R}^n$ and a unique $\bar{x} \in X$ such that for at least one $\bar{u} \in U$ the inequality
$$g(\bar{x},\bar{u}) \leq g(x,u) + \bar{p} \cdot f(x,u) \tag{23.10}$$
holds for all $(x,u) \in X \times U$.

4. There exists an admissible pair $\{\tilde{x}(\cdot), \tilde{u}(\cdot)\}$ for the unconstrained problem such that for some finite time $\tau > 0$ we have
$$\tilde{x}(t) = \bar{x} \quad \text{and} \quad \tilde{u}(t) = \bar{u}$$
for all $t \geq \tau$.

The above three assumptions have become proto-typical in the theory of infinite horizon optimal control. Although they are not the weakest possible hypotheses, they are the simplest to state and are sufficient for our purposes here. The first of these assumptions above is clearly a convexity condition. Perhaps its role can be more clearly understood by informing the reader that this condition is equivalent to assuming that the extended real-valued function $\mathfrak{L} : \mathbb{R}^n \times \mathbb{R}^n \to \mathbb{R} \cup \{+\infty\}$ given by the formula
$$\mathfrak{L}(x,z) = \begin{cases} \inf\{z^0 : \quad (z^0, z, x) \in \Omega\} & \text{if} \quad \Omega \neq \emptyset \\ +\infty \quad \text{if} \quad \Omega = \emptyset, \end{cases}$$
is lower semicontinuous and convex and that our optimal control problem is equivalent to the nonsmooth calculus of variations problem of finding an overtaking optimal trajectory corresponding to the objective of minimizing.
$$\mathfrak{J}_T(x(\cdot)) = \int_0^T \mathfrak{L}(x(t), \dot{x}(t))\, dt.$$

The growth condition defined by (23.9), referred to as the growth condition $(\gamma)$, was introduced by Cesari [16] and is utilized to establish the required sequential weak compactness of the set of admissible trajectories when considered as a subset of the locally absolutely continuous functions. This topology is easily understood when we identify $x \in AC_{loc}([0, +\infty); \mathbb{R}^n)$ with the space $\mathbb{R}^n \times L^1_{loc}([0, +\infty); \mathbb{R}^n)$ through the mapping

$$x \rightarrow (x(0), \dot{x}(\cdot)).$$

Further, it is known that for the autonomous case considered here this growth condition is equivalent to the "seminormality" conditions (in the sense of Tonelli and McShane) used in almost all existences theorems for optimal controls when applied to the problem considered here. For a discussion of these ideas we refer the reader to Cesari [16].

The existence of the vectors $\bar{x} \in X$ and $\bar{p} \in \mathbb{R}^n$ is simply a consequence of solving the optimal steady state problem which is described by the finite dimensional mathematical programming problem

$$\text{minimize} : \{g(x, u) : \quad 0 = f(x, u), : x \in X, : u \in U\}.$$

Under the convexity hypotheses considered here the existence of the requisite $\bar{x}$ and $\bar{p}$ and are easily seen to be satisfied through a standard application of the Kuhn-Tucker theorem (see e.g, [20, Chapter 6] or Rockafellar [37, Chapter VI; Corollary 28.3.1]). Finally, the last assumption is a controllability result that permits an associated problem, known as the Associated Problem of Lagrange to have a finite infimum.

With these hypotheses we now state the following theorem.

**Theorem 23.1** *Under the above assumptions, there exists an admissible trajectory-control pair for the unconstrained problem, say $\{x^*(\cdot), u^*(\cdot)\}$, which is overtaking optimal and satisfies the asymptotic stability condition*

$$\lim_{t \rightarrow +\infty} x^*(t) = \bar{x}.$$

**Proof.** See Carlson, Haurie, and Leizarowitz [13], Theorem 4.10. ∎

The earliest version of the above result was given in Brock and Haurie [5] under somewhat stronger conditions. In the present form considered here it is due to Leizarowitz [28] with considerably weaker conditions than those found in [5]. Other results with weaker convexity conditions were given in Carlson [7] for nonautonomous systems and also in Leizarowitz [30]. In addition to these results we mention that versions of this result have also been proven for the case when the dynamics are described by an abstract evolution equation on a Hilbert space in Carlson, Haurie, and Jabrane [12], for a certain classes of integro-differential equations in Carlson [14] and most recently to more general delay systems in Carlson [10], as well as to stochastic control problems in Leizarowitz [29], [31].

The approach to all of these results follows the main theme of what is now referred to as a "Reduction to finite costs" argument. That is, an associated problem is defined through the use of the optimal steady state problem and a controllability assumption is used to demonstrate that this associated problem has a finite infimum. Using known results for strongly optimal solutions (such as those given in Bates [3], Baum [4], Balder [2], as well as the much earlier results given by Cinquini [19], [18], Darbo [21], and Faedo [24], [23], [22]).

An interesting strengthening of these existence results concerns the relaxing, or in some cases replacing, the support property given by (23.10). In Carlson [7] most of these ideas were shown to be related to the classical Hamilton-Jacobi equation through the notion of an equivalent variational problem originally due to Carathéodory in the late 1920's. Carathéodory's work is described in great detail in his book [6]. All of this work is described in the monograph of Carlson, Haurie, and Leizarowitz [13].

Before concluding this section we point out that the asymptotic convergence property can also be established by studying the corresponding Hamiltonian system. The corresponding Hamiltonian for this optimal control problem is defined as a maximized Hamiltonian, $\mathfrak{H} : \mathbb{R}^n \times \mathbb{R}^n \to \mathbb{R} \cup \{-\infty\}$, through the formula

$$\mathfrak{H}(x,p) = \sup_{z \in \mathbb{R}^n} \{p \cdot z - \mathfrak{L}(x,z)\} . \tag{23.11}$$

Under the joint convexity conditions listed above it is an easy matter to see that the Hamiltonian is concave in $x$ and convex in $p$. With this notation, the dynamical system that is satisfied by an optimal trajectory $x^*(\cdot)$ takes the form

$$\begin{aligned} \dot{x}^*(t) &\in \partial_p \mathfrak{H}(x^*(t), p(t)) \\ -\dot{p}(t) &\in \partial_x \mathfrak{H}(x^*(t), p(t)) \end{aligned}$$

in which $\partial_p$ denotes the partial subgradient of $\mathfrak{H}$ with respect to $p$ in the sense of convex analysis and and $\partial_x$ the partial subgradient of $\mathfrak{H}$ with respect to $x$ in the sense of concave analysis. In addition, it can be established that the solution of the optimal steady state problem, $\bar{x}$, and the multiplier, $\bar{p}$, are a steady state for the above dynamical system and moreover under strict concavity/convexity hypotheses in $x$ and $p$ it is possible to show that we have

$$\lim_{t \to +\infty} (x^*(t), p(t)) = (\bar{x}, \bar{p}) .$$

This result is referred to as an asymptotic turnpike property. The variable $p(\cdot)$, of course, corresponds to the momentum of the system. For our economic problem, if we interpret the state $x^*(t)$ as a capital stock (e.g., goods that are to be sold), then the momentum, $p(t)$, can be interpreted as a price rate per unit of stock and so the Hamiltonian represents the maximal rate of growth of profit (i.e. total revenue $p(t) \cdot x^*(t)$ minus the cost $\mathfrak{L}(x^*(t), \dot{x}^*(t))$). In this way it is possible to interpret all of the basic physical laws as corresponding economic principles. For example, since we are dealing with an autonomous system (i.e., a conservative system) the total energy is conserved. That is, along an optimal "trajectory-price" path we have

$$\mathfrak{H}(x^*(t), p(t)) = \mathfrak{H}(\bar{x}, \bar{p}) \quad \text{for all} \quad t \geq 0.$$

For our economic problem this states that the maximal rate of growth of profit is constant for all time so that accumulated profit is essentially linear in time. For further economic interpretations of the classical physical laws we refer the interested reader to MaGill [32].

### 23.3.2 *Existence of overtaking optimal solutions with discounting*

The introduction of a nonzero discount rate, while an aid in insuring that the objective functional described by letting $T \to +\infty$ in equation (23.3) is well defined, opposes the asymptotic turnpike property in that if the discount rate ($r > 0$) is too large, then it can not be assured that this property holds. Here we present the following results regarding this problem. It can be shown that under the conditions that we have imposed on our model that a necessary condition for an optimal trajectory $x^*(\cdot)$ to be overtaking optimal is that there exists $p(\cdot) : [0, +\infty) \to \mathbb{R}^n$ such the modified Hamiltonian system

$$\begin{aligned} \dot{x}^*(t) &\in \partial_p \mathfrak{H}(x^*(t), p(t)) \\ -\dot{p}(t) &\in \partial_x \mathfrak{H}(x^*(t), p(t)) + rp(t) \end{aligned} \tag{23.12}$$

is satisfied. This system of differential inclusions differs from the Hamiltonian system described above for the undiscounted case. In view of the turnpike property for the undiscounted case it seems reasonable to expect some sort of asymptotic stability property for small discount rates. However, the steady state for the modified Hamiltonian system is not an optimal steady state. Instead, it has been shown by Feinstein and Luenberger [26] that this steady state pair $(\bar{x}, \bar{p})$ is related to the so called *implicit programming problem.* This problem is most easily described by considering the following family of mathematical programming problems

$$\begin{aligned} &\text{minimize} && g(x, u) \\ &\text{subject to} && \\ &\qquad 0 = && f(x, u) - r(x - c) \\ &x \in X \quad \text{and} && u \in U \end{aligned}$$

in which $c \in X$ is viewed as a parameter. If we let $(\bar{x}(c), \bar{u}(c))$ denote a solution of the above problem corresponding to the parameter $c$, then the appropriate steady state we seek corresponds to a fixed point of the map $c \to \bar{x}(c)$ (i.e., $c = \bar{x}(c)$). Specificly, the following theorem due to Feinstein and Luenberger [26], holds.

**Theorem 23.2** *Suppose that the Hamiltonian, $\mathfrak{H}$ as defined above is concave in $x$ and convex in $p$ . Further assume that the pair $(\bar{x}, \bar{p}) \in X \times \mathbb{R}^n$ is a stationary point of the modified Hamiltonian system (23.12). Then the pair $(\bar{x}, \bar{u})$, where $\bar{u} \in \mathbb{R}^m$ satisfies*

$$g(\bar{x}, \bar{u}) = \min \{ g(\bar{x}, u) : \quad f(\bar{x}, u) = 0, \quad u \in U(\bar{x}) \},$$

*is a solution to the implicit programming problem. Moreover, if we additionally assume that $\mathfrak{H}$ is continuously differentiable, then the converse is true. That is, given any pair $(\bar{x}, \bar{u})$ which solves the implicit programming problem there exists $\bar{p} \in \mathbb{R}^n$ so that the pair $(\bar{x}, \bar{p})$ is a steady state for the modified Hamiltonian system.*

With regards to the turnpike property we have the following result due to Rockafellar [38].

**Theorem 23.3** *Suppose that $(\bar{x}, \bar{p})$ is a steady state for the modified Hamiltonian system (23.12). Further assume that there exists $\alpha > 0$ and $\eta > 0$ such that the Hamiltonian $\mathfrak{H}$ is $\alpha$-concave in $x$ and $\eta$-convex in $p$ (see below). If the discount rate $r > 0$ satisfies*

$$r^2 < \alpha \eta, \tag{23.13}$$

*then the asymptotic turnpike property holds. That is, for each solution $t \to (x(t), p(t))$ of the modified Hamiltonian system (23.12) we have*

$$\lim_{t \to +\infty} x(t) = \bar{x}$$

**Proof.** See Rockafellar [38]. ■

A survey of this result and other results with discounting are discussed in Carlson [9].

**Remark 23.1** *The strengthened convexity/concavity conditions given above simply mean that the function*

$$(x, p) \to \mathfrak{H}(x, p) + \frac{1}{2} \left( \eta \|p\|^2 - \alpha \|x\|^2 \right)$$

*is concave in $x$ and convex in $p$.*

We close this section by combining the above turnpike results with the undiscounted existence theorem to give the following existence result for the discounted optimal control problem.

**Theorem 23.4** *Assume that the set $\Omega$, defined by (23.8), is closed, convex, the growth condition (23.9) holds, there exists a solution unique steady state pair to the modified Hamiltonian system, and there exists an admissible trajectory $\{\hat{x}, \hat{u}\}$ such that the trajectory reaches $\bar{x}$ in some finite time. Then if the Hamiltonian is $\alpha$-concave in $x$ and $\eta$-convex in $p$, there exists a strongly optimal solution for the discounted optimal control problem whenever the discount rate $r > 0$ satisfies the condition (23.13). Moreover, the asymptotic turnpike property holds as well.*

**Proof.** See Carlson, Haurie, and Leizarowitz [13]. ∎

With this summary of the relevant existence and turnpike theorems for overtaking optimality we are now ready to investigate an explicit state constraint.

## 23.4 The associated uncoupled optimal control problems

In this section we introduce two uncoupled optimal control problems which are related to the coupled undiscounted and discounted problems. Briefly, we utilize the steady state optimization problems to embed the state constraint into the objective integrand and consider the resulting uncoupled problem. We begin with the undiscounted problem.

### *23.4.1 The undiscounted case*

We now explicitly single out (or impose) an explicit state constraint of the form given in equation (23.2) in which we assume that the functions $h_l : \mathbb{R}^n \to \mathbb{R}$ are continuous and convex. With this constraint we consider the question of determining an overtaking optimal control for the problem in which the performance up to time $T > 0$ is given by

$$J_T(x(\cdot), u(\cdot)) = \int_0^T g(x(t), u(t))\, dt$$

over all admissible trajectory-control pairs satisfying the control system

$$\begin{aligned} \dot{x}(t) &= f(x(t), u(t)) \quad \text{a.e. } t \geq 0 \\ x(0) &= x_0 \\ x(t) &\in \tilde{X} = \{x \in X : \quad h_l(x) \leq 0 \quad l = 1, 2, \ldots, k\} \quad \text{for } t \geq 0 \\ u(t) &\in U(x(t)) \quad \text{a.e. } t \geq 0. \end{aligned}$$

It is an easy matter to see that the optimal steady state problem associated with this optimal control problem is given by

$$\begin{array}{rcl} \text{minimize} & & g(x,u) \\ 0 & = & f(x,u) \\ 0 & \geq & h_l(x), \quad l=1,2,\dots,k \\ x & \in & X \\ u & \in & U. \end{array}$$

As a result of the convexity conditions placed on the model the Kuhn-Tucker conditions for this problem state that if $(\bar{x},\bar{u})$ is an optimal steady state then there exists a vector $\bar{p}\in\mathbb{R}^n$ and a vector $\beta=(\beta_1,\beta_2,\dots,\beta_k)\in\mathbb{R}^k$ such that

$$\begin{array}{rcl} g(\bar{x},\bar{u}) & \leq & g(x,u)+\bar{p}\cdot f(x,u)+\beta\cdot h(x) \quad \text{for all} \quad (x,u)\in X\times U \\ \beta_l & \geq & 0 \quad \text{for} \quad l=1,2,\dots k \\ \beta_l h_l(\bar{x}) & = & 0 \quad \text{for} \quad l=1,2,\dots k. \end{array} \tag{23.14}$$

These conditions motivate the consideration of the following related "unconstrained " optimal control problem in which the objective functional is replaced by

$$\tilde{J}_T(x,u)=\int_0^T g(x(t),u(t))+\beta\cdot h(x(t))\,dt \tag{23.15}$$

and an admissible trajectory-control pair satisfies the control system (23.1). We observe that this problem has the form of those discussed in the previous section and consequently we can apply those results to the problem considered here. We further observe that the interpretation of this new objective is that it is a "penalized" cost functional in which a tax is paid for violating the constraint. The constant vector $\beta$ can be viewed as a vector of constant tax rates based on the optimal steady state.

An natural question to be addressed in the above framework is the relationship between an overtaking optimal solution of this "unconstrained" optimization problem and the original problem exhibiting the explicit state constraints (23.2). We explore this question in the next section.

### *23.4.2 The discounted case*

For this case we must replace the optimal steady state problem with the implicit programming problem. That is, we must consider the family of mathematical programs

$$\begin{array}{rcl} \text{minimize} & & g(x,u) \\ 0 & = & f(x,u) - r(x-c) \\ 0 & \geq & h_l(x), \quad l = 1, 2, \ldots, k \\ x & \in & X \\ u & \in & U. \end{array}$$

Once again, the Kuhn-Tucker theorem provides us with an appropriate auxiliary problem. Specificly, if $(\bar{x}, \bar{u})$ is a solution to the above implicit programming problem then there exists a vector $\bar{p} \in \mathbb{R}^n$ and a vector $\beta = (\beta_1, \beta_2, \ldots, \beta_k)$ satisfying

$$g(\bar{x},\bar{u}) \leq g(x,u) + \bar{p} \cdot [f(x,u) - r(x-\bar{x})] + \beta \cdot h(x) \quad (23.16)$$

$$\beta_l \geq 0 \quad k = 1, 2, \ldots, k \quad (23.17)$$

$$\beta_k h_k(\bar{x}) = 0 \quad k = 1, 2, \ldots, k$$

for $(x, u) \in X \times U$.

From the above we are motivated to consider the auxiliary unconstrained discounted optimal control problem whose performance objective is described by the integral functional

$$\tilde{J}_T(x,u) = \int_0^T e^{-rt}[g(x(t),u(t)) + \beta \cdot h(x(t))]\, dt. \quad (23.18)$$

Here again, an admissible trajectory-control pair is assumed to satisfy (23.1) and we observe that the framework for discounted problems described in the previous can be used to discuss the existence of optimal solution for this problem for this problem as well.

## 23.5 Optimal solutions of the explicitly state constrained optimal control problem

In this section we demonstrate that, as a result of the turnpike property, we can attack the question of existence of the explicit state constrained problem by examining the question for the associated unconstrained problems. Of course, these results are not without a price. To do this we must set our sights somewhat lower and replace the notion of overtaking optimality with averaging optimality. In addition, we also must weaken the concept of admissibility to that of asymptotic admissibility as described in Definition 23.3. We now present our results.

**Theorem 23.5** *Let us assume that*

1. *There exists a vector* $\bar{p} \in \mathbb{R}^n$, *a vector* $\beta \in \mathbb{R}^k$, *and a unique* $\bar{x} \in X$ *such that there exists at least one* $\bar{u} \in U$ *for which the inequality*

$$\begin{aligned} g(\bar{x},\bar{u}) &\leq g(x,u) + \bar{p}\cdot f(x,u) + \beta \cdot h(x) \\ h_l(\bar{x}) &\leq 0 \quad \text{for} \quad l = 1,2,\ldots k \\ \beta_l &\geq 0 \quad \text{for} \quad l = 1,2,\ldots k \\ \beta_l h(\bar{x}) &= 0 \quad \text{for} \quad l = 1,2,\ldots k \end{aligned} \tag{23.19}$$

*holds for all* $(x,u) \in X \times U$.

2. *The set*

$$\Omega \doteq \{(z_0, z, x) :\ z_0 \geq g(x,u),\ z = f(x,u),\ x \in X,\ \text{for some } u \in U\}, \tag{23.20}$$

*is a convex, closed set.*

3. *For each* $\epsilon > 0$ *there exists* $c_\epsilon > 0$ *such that*

$$|f(x,u)| \leq c_\epsilon + \epsilon\{g(x,u) + \beta \cdot h(x)\} \tag{23.21}$$

*holds for* $(x,u) \in X \times U$.

4. *There exists an admissible pair* $\{\tilde{x}(\cdot), \tilde{u}(\cdot)\}$ *for the unconstrained problem such that for some finite time* $\tau > 0$ *we have*

$$\tilde{x}(t) = \bar{x} \quad \text{and} \quad \tilde{u}(t) = \bar{u}$$

*for all* $t \geq \tau$.

*Then there exists a trajectory-control pair* $\{x^*, u^*\}$ *that is average cost optimal over all trajectory-control pairs that are admissible for the unconstrained problem without discounting.*

**Proof.** We begin by observing that the associated uncoupled undiscounted optimal control problem satisfies the conditions needed to insure the existence of an overtaking optimal solution. Indeed the requisite convexity condition is a consequence of our assumptions and the convexity of the functions $h_l(\cdot)$. The support property follows trivially from our assumptions and the fact that $\beta \cdot h(x) \leq 0$ for all $x \in X$. All other conditions are simple translations of those discussed in section 3. Thus, there exists a trajectory-control pair $\{x^*, u^*\}$ such that for each $\epsilon > 0$ and trajectory-control pair $\{x, u\}$ that is admissible for the unconstrained, undiscounted optimal control problem we have there exists $\hat{T} \geq 0$ for which

$$\int_0^T [g(x^*(t), u^*(t)) + \beta \cdot h(x^*(t))]\, dt \leq \int_0^T [g(x(t), u(t)) + \beta \cdot h(x(t))]\, dt + \epsilon$$

holds for all $T > \hat{T}$ and for which we have

$$\lim_{t \to +\infty} x^*(t) = \bar{x}.$$

That is, $x^*$ enjoys the asymptotic turnpike property.

We now fix $\epsilon > 0$ and assume that the pair $\{x, u\}$ is admissible for the constrained problem without discounting. As a consequence of the turnpike property we can further assume that $\hat{T} > 0$ has been chosen sufficiently large so that for all $t \geq \hat{T}$ we have

$$h_l(\bar{x}) - \epsilon < h_l(x^*(t)) \leq h_l(\bar{x}) + \epsilon \leq \epsilon,$$

for $l = 1, 2, \ldots k$. From this we see that $\{x^*, u^*\}$ is asymptotically admissible for the constrained problem (i.e., choose $B = \hat{T}$). Now, for $T > \hat{T}$ we observe that from the above we have

$$\begin{aligned}
\frac{1}{T}\int_0^T [g(x(t), u(t)) \quad &- \quad g(x^*(t), u^*(t))]\, dt > -\frac{1}{T}\epsilon + \frac{1}{T}\int_0^{\hat{T}} \beta \cdot h(x^*(t))\, dt \\
&+\frac{1}{T}(\beta \cdot (h(\bar{x}) - \epsilon)(T - \hat{T}) \\
&-\frac{1}{T}\int_{S_\epsilon^T} \beta \cdot h(x(t))\, dt - \frac{1}{T}\int_{[0,T]\setminus S_\epsilon^T} \beta \cdot h(x(t))\, dt \\
&> \left[-\frac{1}{T} - \frac{T - \hat{T}}{T}\right]\epsilon - \frac{\epsilon}{T}\text{meas}\left[[0, T] \setminus S_\epsilon^T\right] \\
&+\frac{1}{T}\int_0^{\hat{T}} \beta \cdot h(x^*(t))\, dt - \frac{1}{T}\int_{S_\epsilon^T} \beta \cdot h(x(t))\, dt \\
&= \left[\frac{\hat{T} - 1}{T} - 2\right]\epsilon + \frac{1}{T}\int_0^{\hat{T}} \beta \cdot h(x^*(t))\, dt \\
&-\frac{1}{T}\int_{S_\epsilon^T} \beta \cdot h(x(t))\, dt,
\end{aligned}$$

where we have used the complementary slackness condition $\beta_l h_l(\bar{x}) = 0$ for $l = 1, 2, \ldots, k$ and

$$S_\epsilon^T = \{t \geq [0, T] : h_l(x(t)) > \epsilon \quad \text{for} \quad l = 1, 2, \ldots k\}.$$

By hypothesis, the Lebesgue measure of the set $S_\epsilon^T$ is uniformly bounded, independent of $T > 0$, for all $T > 0$. From the above string of inequalities, and since all admissible trajectories are bounded we have that

$$\liminf_{T\to+\infty}\frac{1}{T}\int_0^T [g(x(t),u(t)) - g(x^*(t),u^*(t))]\,dt \geq -2\epsilon,$$

giving us that

$$\liminf_{T\to+\infty}\frac{1}{T}\int_0^T [g(x(t),u(t)) - g(x^*(t),u^*(t))]\,dt \geq 0,$$

since $\epsilon > 0$ was arbitrary. Further since $\{x,u\}$ was an arbitrary asymptotically admissible pair, the above inequality holds for all asymptotically admissible pairs. Thus, $\{x^*,u^*\}$ is an averaging optimal solution for the constrained optimal control problem without discounting. ∎

With regards to the optimal value the next result show that the long-term average optimal cost is the value of the optimal steady state.

**Theorem 23.6** *For the averaging optimal solution given in Theorem 23.5 we have the*

$$\lim_{T\to+\infty}\frac{1}{T}\int_0^T g(x^*(t),u^*(t))\,dt = g(\bar{x},\bar{u}). \tag{23.22}$$

**Proof.** Let $\{\tilde{x},\tilde{u}\}$ be the trajectory-control pair satisfying the controllability assumption given in Theorem 23.5. Then clearly this pair is admissible for the constrained problem without discounting (see Definition 2.4). Thus, for each $\epsilon > 0$ there exists $\tilde{T} > 0$ such that for all $T > \tilde{T}$ we have

$$\frac{1}{T}\int_0^T g(x^*(t),u^*(t))\,dt \leq \frac{1}{T}\int_0^T g(\tilde{x}(t),\tilde{u}(t))\,dt + \frac{\epsilon}{3}.$$

From this we easily see for all $T > \max\{\tilde{T},\tau\}$ that

$$\frac{1}{T}\int_0^T g(x^*(t),u^*(t))\,dt - g(\bar{x},\bar{u}) \leq \frac{1}{T}\int_0^\tau g(\tilde{x}(t),\tilde{u}(t))\,dt - \frac{\tau g(\bar{x},\bar{u})}{T} + \frac{\epsilon}{3}.$$

Therefore we can find $\hat{T} > 0$ such that for all $T > \hat{T}$ we have

$$\frac{1}{T}\int_0^T g(x^*(t),u^*(t))\,dt - g(\bar{x},\bar{u}) \leq \epsilon.$$

On the other hand, the fact that $(\bar{x}, \bar{u})$ is the optimal steady state, we have from equation (23.19) that for almost all $t \geq 0$ we have

$$\bar{p} \cdot f(x^*(t), u^*(t)) - \beta \cdot h(x^*(t)) \leq g(x^*(t), u^*(t)) - g(\bar{x}, \bar{u}),$$

from which we immediately obtain

$$\frac{1}{T}\{\bar{p} \cdot (x^*(T) - x_0) \quad + \quad \int_0^T \beta \cdot h(x^*(t))\, dt\} \leq \frac{1}{T} \int_0^T g(x^*(t), u^*(t))\, dt - g(\bar{x}, \bar{u}).$$

Now since an admissible trajectory is bounded by hypothesis we immediately see that the left-hand side of the above inequality tends to zero as $T \to +\infty$. Therefore we can find a $T^* > \hat{T}$ such for all $T > T^*$ we have

$$-\epsilon \leq \frac{1}{T} \int_0^T g(x^*(t), u^*(t))\, dt - g(\bar{x}, \bar{u}) \leq \epsilon,$$

from which the desired conclusion follows. ■

### *23.5.1 The undiscounted case*

We now direct our attention to the discounted case. In this case we need not concern ourselves with the overtaking optimality concept since the boundedness hypotheses and the positive discount rate insure that the optimal value of each admissible trajectory-control pair is finite. We do however have to strengthen the convexity hypotheses to insure that the asymptotic turnpike property holds. With these remarks we have the following theorem.

**Theorem 23.7** *Assume that the following hypotheses hold:*

1. *There exists a steady-state pair $(\bar{x}, \bar{u})$, a vector $\bar{p} \in \mathbb{R}^n$ and a vector $\beta \in \mathbb{R}^k$ such that the conditions given in equation (4.3) hold.*

2. *For each $t > 0$ the set $\Omega(t)$ defined by*

$$\Omega(t) \doteq \left\{(z_0, z, x) : \; z_0 \geq e^{-rt} g(x, u), \; z = f(x, u), \; x \in X, \; \text{for some } u \in U\right\}, \quad (23.23)$$

*is a convex, closed set.*

3. *For each* $\epsilon > 0$ *there exists* $c_\epsilon : [0, +\infty) \to [0, +\infty)$, *locally integrable such that*

$$|f(x,u)| \leq c_\epsilon(t) + \epsilon\{e^{-rt}g(x,u) + \beta \cdot h(x)\} \tag{23.24}$$

*holds for* $(x,u) \in X \times U$.

4. *The Hamiltonian* $\mathcal{H} : \mathbb{R}^n \times \mathbb{R}^n \to \mathbf{R} \bigcup \{-\infty\}$ *defined by*

$$\mathcal{H}(x,p) = \sup_{u \in U}\{\bar{p} \cdot f(x,u) - g(x,u) - \beta \cdot h(x)\},$$

*where we assume that the function* $\mathcal{H}$ *is defined to be* $-\infty$ *whenever* $x \notin X$, *is* $\alpha$*-convex in* $x$ *and* $\eta$*-concave in* $p$.

*Then there exists a strongly optimal trajectory-control pair* $\{x^*, u^*\}$ *over the class of trajectory-control pairs that are admissible relative to the pair* $\{x^*, u^*\}$ *for the constrained problem with discounting (see Definitions 2.3 and 2.4).*

**Proof.** Under the above hypotheses there exist a trajectory-control pair that is optimal for the associated discounted uncoupled optimal control problem, say $\{x^*, u^*\}$. Now let $\{x, u\}$ be a trajectory-control pair that asymptotically satisfies the coupled state constraint for the discounted control problem relative to the pair $\{x^*, u^*\}$. In this case we have, by the optimality for the associated uncoupled control problem we have

$$\int_0^{+\infty} e^{-rt}[g(x^*(t), u^*(t)) + \beta \cdot h(x^*(t))]\, dt \leq \int_0^{+\infty} e^{-rt}[g(x(t), u(t)) + \beta \cdot h(x(t))]\, dt.$$

From this we immediately obtain

$$\begin{aligned}\int_0^{+\infty} e^{-rt}g(x^*(t), u^*(t))\, dt &\leq \int_0^{+\infty} e^{-rt}g(x(t), u(t))\, dt \\ &\quad + \int_0^{+\infty} e^{-rt}[\beta \cdot h(x(t)) - \beta \cdot h(x^*(t))]\, dt \\ &\leq \int_0^{+\infty} e^{-}rtg(x(t), u(t))\, dt,\end{aligned}$$

since the pair asymptotic satisfies the coupled state constraint relative to the pair $\{x^*, u^*\}$. This, of course, gives the desired result. ∎

## 23.6 Conclusions

In the previous work we indicated a scheme in which the existence of an optimal control for an implicit state constrained infinite horizon optimal control problem could be treated by embedding the state constraint into the objective to create an associated optimal control problem without this explicit state constraint. The novel feature in these results is that through the use of the asymptotic turnpike theory we are able to effect this result with a constant Lagrange multiplier rather than a continuous function, as is done in the classical theory, or more generally a function of bounded variation as is found in the literature for similar finite interval problems. From a practical point of view, if the state constraint is viewed as a constraint that is imposed by a regulatory board, then the constant multiplier may be viewed as a constant tax rate that penalizes the firm for violating the constraint. In addition, the results we establish do not require the firms to satisfy the explicit constraint in a pointwise manner, but rather in either a "most of the time" framework (as described in Definition 2.2) for the undiscounted case or in some average senses for the discounted problem (see Definition 2.3). In closing we remark that the results reported here were first established in Carlson and Haurie [11] in the setting of dynamic games that are described in a variational manner. The problem considered here was formulated as a single player game (i.e., an optimization problem) with the dynamics described by a control system. The results here are obtained under weaker hypotheses than those of [11] since we do not need to concern ourselves with the other players.

*References*


[1] H. Atsumi, Neoclassical growth and the efficient program of capital accumulation, Review of Economic Studies **30** (1963), 127–136.

[2] E. J. Balder, An existence result for optimal economic growth, Journal of Mathematical Analysis and its Applications **95** (1983), 195–213.

[3] G. R. Bates, Lower closure and existence theorems for optimal control problems with infinite horizon, Journal of Optimization Theory and Applications **24** (1978), 639–649.

[4] R. F. Baum, Existence theorems for lagrange control problems with unbounded time domain, Journal of Optimization Theory and Applications **19** (1976), 89–116.

[5] W. A. Brock and A. Haurie, On existence of overtaking optimal trajectories over an infinite time horizon, Mathematics of Operations Research **1** (1976), 337–346.

[6] C. Carathéodory, Calculus of variations and partial differential equations, Chelsea, New York, New York, 1982.

[7] D. A. Carlson, On the existence of catching up optimal solutions for lagrange problems defined on unbounded intervals, Journal of Optimization Theory and Applications **49** (1986), 207–225.

[8] ______, The existence of catching-up optimal solutions for a class of infinite horizon optimal control problems with time delay, SIAM Journal on Control and Optimization **28** (1990), 402–422.

[9] ______, Asymptotic stability for optimal trajectories of infinite horizon optimal control models with state and control dependent discounting, Proceedings of the International Workshop in Analysis and its Applications (Novi Sad, Yugoslavia) (C. Stanojević and O. Hadžić, eds.), Institute of Mathematics, 1991, pp. 281–300.

[10] ______, Overtaking optimal solutions for convex lagrange problems with time delay, Journal of Mathematical Analysis and Applications **208** (1997), 31–48.

[11] D. A. Carlson and Haurie A., Overtaking equilibria for infinite horizon dynamic games with coupled state constraints,to appear in the Annals of International Society of Dynamic Games.

[12] D. A. Carlson, A. Haurie, and A. Jabrane, Existence of overtaking solutions to infinite dimensional control problems on unbounded intervals, SIAM Journal on Control and Optimization **25** (1987), 1517–1541.

[13] D. A. Carlson, A. Haurie, and A. Leizarowitz, Infinite horizon optimal control: Deterministic and stochastic systems, 2nd ed., Springer-Verlag, New York, 1991.

[14] D.A. Carlson, Infinite-horizon optimal controls for problems governed by a volterra integral equation with state-and-control-dependent discount factor, Journal of Optimization Theory and Applications **66** (1990), no. 2, 311–336.

[15] D. Cass and K. Shell, The hamiltonian approach to dynamic economies, Academic Press, New York, New York, 1976.

[16] L. Cesari, Optimization-theory and applications: Problems with ordinary differential equations, Applications of Applied Mathematics, vol. 17, Springer-Verlag, New York, 1983.

[17] S. Chakravarty, The existence of an optimum savings program, Econometrica **30** (1962), 178–187.

[18] S. Cinquini, Una nuova estensione dei moderni metodi del calcolo delle variazioni, Annali della Scuola Normale Superiore di Pisa, Serie 2 **9** (1940), 258–261.

[19] ———, Sopra l'esistenza dell'estremo assoluto per integrali estesi a intervalli infiniti, Rendiconti della Accademia Nazionale dei Lincei, Serie 8 **32** (1962), 320–325 and 845–851.

[20] F. H. Clarke, Optimization and nonsmooth analysis, John Wiley and Sons, Inc., New York, New York, 1983.

[21] G. Darbo, L'estremo assoluto per gli integrali su intervallo infinito, Rendiconti del Seminario Matematico dell'Universitá Padova **22** (1953), 319–416.

[22] S. Faedo, Il calcolo dell variazioni per gli integrali su intervalli infiniti, Commentationes Pontificia Academia Scientarium **8** (1944), 319–421.

[23] ———, Il calcolo delle variazoni per gli integrali su intervalli infiniti, Rendiconti di Matematica e delle Sue Applicazioni **8** (1949), 94–125.

[24] ———, Il calcolo delle variazioni per gli integrali estesi a intervalli infiniti, Annali della Suola Normale Superiore di Pisa **7** (1968), 91–132.

[25] C. D. Feinstein and S. S. Oren, A "funnel" turnpike theorem for optimal growth problems with discounting, Journal of Economic Dynamnics and Control **9** (1985), 25–39.

[26] C.D. Feinstein and D.G. Luenberger, Analysis of the asymptotic behavior of optimal control trajectories, SIAM Journal on Control and Optimization **19** (1981), 561–585.

[27] H. Halkin, Necessary conditions for optimal control problems with infinite horizon, Econometrica **42** (1974), 267–273.

[28] A. Leizarowitz, Existence of overtaking optimal trajectories for problems with convex integrands, Mathematics of Operations Research **10** (1985), 450–461.

[29] ———, Infinite-horizon stochastic regulation and tracking with the overtaking criterion, Stochastics **22** (1987), 117–150.

[30] ———, Optimal trajectories of infinite horizon deterministic control systems, Applied Mathematics and Optimization **19** (1989), 11–39.

[31] ———, Optimal control for diffusions in $\mathbb{R}^d$–a min-max formula for the minimal cost-growth rate, Journal of Mathematical Analysis and its Applications (1990).

[32] M. Magill, On a general economic theory of motion, Springer-Verlag, 1970.

[33] L. W. McKenzie, Turnpike theorems for a generalized leontief model, Econometrica **31** (1963), 165–180.

[34] Lionel W. McKenzie, Turnpike theory, Econometrica **44** (1976), no. 5, 841–865.

[35] R. Radner, Paths of economic growth that are optimal with regard only to final state: A turnpike theorem, Review of Economic Studies **28** (1961), 98–104.

[36] F. Ramsey, A mathematical theory of saving, Economic Journal **38** (1928), 543–549.

[37] R. T. Rockafellar , Convex Analysis, Princeton University Press, Princeton, New Jersey, 1970.

[38] R. T. Rockafellar, Saddle points of hamiltonian systems in convex lagrange problems having nonzero discount rate, Journal of Economic Theory **12** (1976), 71–113.

[39] P.A. Samuelson, A caternary turnpike theory involving consumption and the golden rule, American Economic Review **55** (1965), 486–496.

[40] L. J. Stern, Criteria for optimality in the infinite-time optimal control problem, Journal of Optimization Theory and Applications **44** (1984), 497–508.

[41] J. Von Neumann, A model of general economic equilibrium, Review of Economic Studies **13** (1945), 1–9.

[42] C.C. Von Weizäcker, Existence of optimal programs of accumulation for an infinite time horizon, Review of Economic Studies **32** (1965), 85–104.

[43] A. Zaslavski, Optimal programs on infinite horizons 1 and 2, SIAM Journal on Control and Optimization **33** (1995), 1643–1660,1661–1686.

# Chapter 24

# Surfaces minimizing integrals of divergent integrands

**Harold R. Parks**

*ABSTRACT. We consider parametric integrands which are unbounded above as a function of position. When one seeks a surface $T$ minimizing the integral of such an integrand, subject to a boundary condition $\partial T = R$, the type of minimizing surface that can be found—even what is meant by "minimizing"—is dependent on various characteristics of the boundary $R$ and the divergent behavior of the integrand. A notion of overtaking minimization is defined, and the existence of overtaking minimizers is proved under fairly weak hypotheses. As a class of examples which is hoped to be prototypical, we consider the area integrand in $\mathbb{R}^3$ multiplied by a function that diverges as the $z$-axis is approached. For such integrands, satisfying some additional technical conditions, overtaking minimizers are shown to be well-behaved near the $z$-axis.*

## 24.1 Introduction

In this paper, we will consider divergent parametric integrands (that is, integrands that are unbounded as a function of position), what it means for a surface to minimize the integral of such an integrand, and how such a minimizing surface must behave. The study of surfaces that minimize divergent integrands has a surprisingly long history dating back at least to the 1939 University of California at Berkeley Ph.D. thesis of A. T. Lonseth (published in [10]) in which he considered the Plateau problem in hyperbolic space. We can consider this a problem involving a divergent integrand because the upper half-space model of $(n+1)$-dimensional hyperbolic space is the set

$$\mathbb{H}^{n+1} = \{\ (x, y) \in \mathbb{R}^n \times \mathbb{R}\ :\ y > 0\}$$

equipped with the metric

$$ds^2 = y^{-2}(dx^2 + dy^2),$$

so as the sphere at infinity, $\mathbb{S}^n(\infty) = \mathbb{R}^n \times \{0\}$, is approached, the area integrand grows without bound. Forty years later in another University of California at Berkeley Ph.D. thesis (published in [2]), M. T. Anderson proved the existence of complete minimal varieties in hyperbolic space asymptotic to a given manifold in the sphere at infinity. In later papers, R. M. Hardt and F. H. Lin (see [8] and [9]) investigated the boundary behavior of such surfaces.

In this paper, we explore the problem of minimizing integrals of integrands with even stronger and more concentrated singularities than those provided by the hyperbolic metric. We will restrict our attention to $n$-dimensional surfaces in $\mathbb{R}^{n+1}$ and consider integrands which are infinite on a set having $n$-dimensional Hausdorff measure zero. Later we will restrict to $n = 2$. The set on which the integrand is infinite will be called the "singular set" of the integrand. A notion of minimization is introduced that applies even when all admissible surfaces produce infinite integrals. This type of minimization involves approximating the divergent integrand $\Phi$ by a sequence of integrands $\Phi_\rho$, $\rho > 0$, which converge to $\Phi$ as $\rho \downarrow 0$. We then say that $T$ is an overtaking minimizer in a class of admissible surfaces if

$$\liminf_{\rho \downarrow 0} \left[ \int_S \Phi_\rho - \int_T \Phi_\rho \right] \geq 0$$

holds for any admissible $S$. We use the terminology "overtaking minimizer" because of the similarity to the overtaking optimum used in studying infinite horizon control problems (see [3]).

In Theorem 24.2, we give a sufficient condition for the existence of an overtaking minimizer. This existence result is obtained by subtracting off the expected singularity and then dealing with the remaining finite problem. The technique of subtracting off the expected singularity is a method also used in the study of physical problems such as vorticity (*e.g.* see [14]).

By considering rotationally symmetric examples in $\mathbb{R}^3$ in Section 24.4, we are able to gain some insight into the behavior that might be expected near the singular set of the integrand. While regularity of overtaking minimizers away from the singular set of the integrand is essentially automatic from standard results in geometric measure theory, the study of the asymptotic behavior of an overtaking minimizer near the singular set of the integrand requires additional effort. Motivated by the examples of Section 24.4, we consider the regularity question for an integrand that becomes infinite as a singular line is approached. For simplicity of notation, the singular line is assumed to be the $z$-axis. We prove that under certain conditions, an overtaking minimizer can be expressed in the form of the graph of a function $z = u(x, y)$ near the singular line (Theorem 24.14) and that the overtaking minimizer meets the singular line in a disc that is orthogonal to the singular line and that decays at a rate of $r^{\beta_0}$ as the distance $r$ to the singular line decreases to zero, where $\beta_0 \geq 2$ is a constant (Theorem 24.15). The technique of proof for the regularity results is to use barriers obtained from the study of quasilinear partial differential equations and to apply curvature estimates known to hold for surfaces that minimize parametric integrals. Both of these tools require that we impose some technical conditions on our integrands. These conditions are set forth in Section 24.5.

## 24.2 Surfaces and Integrands

Our category of surfaces will be the rectifiable and integral currents. The comprehensive reference is [5]. A more accessible introduction is [11]. An $m$-dimensional **rectifiable current** $T$ in $\mathbb{R}^{n+1}$ is comprised of three things:

**(i)** A bounded, $\mathcal{H}^m$ measurable and $(\mathcal{H}^m, m)$ rectifiable set with finite $\mathcal{H}^m$ measure which we will also denote by $T$. The support of $T$, denoted $\operatorname{spt}(T)$, is the complement of the largest open set for which $U \cap T$ has $\mathcal{H}^m$ measure zero. We may and shall assume $T \subset \operatorname{spt}(T)$.

**(ii)** An orientation function $\overrightarrow{T}(x)$ taking values in the Grassmann manifold of oriented $m$-dimensional subspaces of $\mathbb{R}^{n+1}$. At $\mathcal{H}^m$-almost every point of $T$, $\overrightarrow{T}$ must equal one or the other of the oriented approximate tangent planes to $T$, and the function $x \mapsto \overrightarrow{T}(x)$ must be a measurable function with respect to $\mathcal{H}^m$ restricted to $T$. In case $m = n$, the orientation function can be thought of as representing the unit normal vector and thus can be assumed to take its values in the unit sphere.

**(iii)** An integer valued multiplicity function $\mu_T(x)$ which also must be measurable with respect to Hausdorff measure restricted to $T$.

In general, currents are elements of the dual space to smooth differential forms. Supposing $\omega$ to be a differential $m$-form, we define the pairing of the rectifiable current $T$ against $\omega$ by

$$T(\omega) = \int_T \langle \omega(x), i(\overrightarrow{T}(x)) \rangle \, \mu_T(x) \, d\mathcal{H}^m x,$$

where $i$ embeds the Grassmann manifold of oriented subspaces into the space $\bigwedge_m(\mathbb{R}^{n+1})$ of $m$-vectors in $\mathbb{R}^{n+1}$.

The **boundary operator** is defined on currents by postulating Stokes's Theorem. Thus $\partial T$ is characterized by

$$\partial T(\omega) = T(d\omega),$$

where $d\omega$ is the exterior derivative. If $T$ is a rectifiable current, then the resulting boundary current $\partial T$ might or might not agree with the $(m-1)$-dimensional current corresponding to a rectifiable set, orientation function, and multiplicity function. If there does exist an $(m-1)$-dimensional rectifiable current that agrees with $\partial T$, then we say that $T$ is an **integral current.**

The set of $m$-dimensional integral currents is denoted $\mathbb{I}_m$. We will use the notation $\mathbb{B}(p,r)$ and $\overline{\mathbb{B}}(p,r)$ for the open and closed balls in $\mathbb{R}^{n+1}$ of radius $r$ centered at $p$, respectively.

Two important norms are the **mass** and the **flat norm.**[1] For a rectifiable current $T$ the mass of $T$ is

$$\mathbb{M}(T) = \int_T \mu_T \, d\mathcal{H}^m x.$$

The mass is the Hausdorff measure of the set, but counting multiplicity. The flat norm of the rectifiable current $T$ is

$$\mathcal{F}(T) = \inf \{\mathbb{M}(A) + \mathbb{M}(B) : \ T = A + \partial B, \ A \in \mathcal{R}_m, \ B \in \mathcal{R}_{m+1}\}.$$

The flat norm $\mathcal{F}(S - T)$ gives a good measure of how geometrically close $S$ and $T$ are.

One of the most significant results in geometric measure theory is the **Compactness Theorem** (first proved in [6]) which says that, for any choice of $0 < c < \infty$,

$$\{\, T \in \mathbb{I}_m \ : \ \operatorname{spt}(T) \subset \overline{\mathbb{B}}(0,c), \ \mathbb{M}(T) \le c, \ \mathbb{M}(\partial T) \le c \,\}$$

is totally bounded under the flat norm.

By a **parametric integrand** of degree $n$ on $\mathbb{R}^{n+1}$ is meant a function

$$\Phi : \mathbb{R}^{n+1} \times \mathbb{R}^{n+1} \to \mathbb{R}$$

that is positively homogeneous of degree 1 in the second argument. The integral of the integrand $\Phi$ over the $n$-dimensional rectifiable current $T$ is defined by

$$\int_T \Phi = \int_T \Phi\left(p, \vec{T}(p)\right) \mu_T(p) \, d\mathcal{H}^n p$$

(remember that in the case of $n$-dimensional currents in $\mathbb{R}^{n+1}$ we are identifying the orientation with the unit normal vector). Conventional practice is to assume that $\Phi$ is a smooth, bounded function, so that $\int_T \Phi$ will be defined. As an example, the **area integrand** is given by

$$\Phi(p, v) = |v|.$$

---

[1] The term flat norm was introduced by H. Whitney who borrowed the word from musical terminology; he also introduced a sharp norm.

By a **Φ-minimizing surface** is meant an $n$-dimensional rectifiable current $T$ in some admissible family of currents—a typical admissible family would be those rectifiable currents $S$ satisfying the boundary condition $\partial S = R$— such that

$$\int_T \Phi \leq \int_S \Phi \tag{24.1}$$

holds for all admissible $S$. We also say that a surface $T$ is a **locally Φ-minimizing surface** in some open set $U$ when, for each $x \in U$, there is a neighborhood $V$ with $x \in V \subset U$ for which

$$\int_{T \cap V} \Phi \leq \int_{S \cap V} \Phi \tag{24.2}$$

holds for every admissible $S$ that equals $T$ outside of $V$. Area-minimization is a special case, though arguably the most important special case, of Φ-minimization.

An integrand $\Phi(p, v)$, $p \in \mathbb{R}^{n+1}$, $v \in \mathbb{R}^{n+1}$, that is independent of $p$ is called a **constant coefficient integrand.** Let an integrand $\Phi$ be given. Then for any choice of $a \in \mathbb{R}^{n+1}$, we define the constant coefficient integrand $\Phi_{\langle a \rangle}$ by setting $\Phi_{\langle a \rangle}(p, v) = \Phi(a, v)$.

The **direct method in the calculus of variations** is used to prove the existence of a Φ-minimizer by selecting a sequence of admissible surfaces for which the values of the integrals converge to the infimum of the values of integrals over all admissible surfaces and then proving that a subsequence of the sequence of surfaces converges to a minimizer. In the setting of rectifiable currents, one uses the compactness theorem to select a subsequence that is convergent in the flat norm. If the integral is a lower-semi-continuous functional in the flat norm topology, then a minimizing surface will have been found. The most effective hypothesis for guaranteeing this lower-semi-continuity is to assume that, for each $a \in \mathbb{R}^{n+1}$, the associated constant coefficient integrand $\Phi_{\langle a \rangle}(p, v) = \Phi(a, v)$ is a convex function of $v$.

The ultimate goal in the program of the direct method is to prove that the minimizers obtained are, in fact, classical surfaces. In the setting of rectifiable currents that is the most difficult part of the theory. The standard assumption used in this context is that there exists a positive $c$ such that

$$\int_T \Phi_{\langle a \rangle} - \int_R \Phi_{\langle a \rangle} \geq c[\mathbb{M}(T) - \mathbb{M}(R)]$$

holds whenever $R$ is a piece of an $n$-dimensional plane and $\partial T = \partial R$. Such integrands are said to be **elliptic.**[2]

[2] The provenance of the term is not at all obvious: It is used because the associated Euler-Langrange equations, which will be discussed later, are elliptic.

The known **Regularity Theorems**[3] tell us that, *away from its boundary, an $n$-dimensional area minimizing current in $\mathbb{R}^{n+1}$ is real analytic except possibly on a set of dimension $n-7$. Also, if the boundary is a $C^2$ submanifold, then at every boundary point the area minimizing current is a $C^1$ embedded manifold with boundary. Moreover, the interior regularity results apply in a smooth Riemannian manifold, though real analyticity must be weakened if the manifold is not real analytic.*

In this paper, we will consider integrands that may be unbounded as a function of position. In the next section, we will define a notion of "overtaking minimizer" that applies even if the integrals in question are infinite. We also provide a sufficient condition to conclude that such an overtaking minimizer exists. In later sections, we will consider integrands that are multiples of the area integrand and which have the specific property of diverging as the $z$-axis is approached. By the regularity theorem above, we see that any notion of minimizing surface that implies a surface is minimizing in the usual sense away from the $z$-axis, would imply that minimizing surfaces are also regular away from the $z$-axis. Thus the interesting regularity question is how does the surface behave as it approaches the $z$-axis? Examples indicate that it is reasonable to conjecture that as the singular line is approached, an overtaking minimizer will decay to a disc orthogonal to the singular line. We will give a set of sufficient conditions under which that conjecture is valid (see Section 24.5).

## 24.3 Overtaking Minimizers

Here we propose a notion of minimization in terms of finite approximating integrands.

**Definition 24.1** *Suppose $Z \subset \mathbb{R}^{n+1}$ satisfies*

$$\mathcal{H}^n(Z) = 0.$$

*By a* ***divergent integrand of degree $n$ with singular set $Z$*** *we mean a non-negative continuous function*

$$\Phi : (\mathbb{R}^{n+1} \setminus Z) \times \mathbb{R}^{n+1} \to \mathbb{R}$$

*that is positively homogeneous of degree 1 in the second argument and satisfies*

$$\lim_{p \to p_0} \Phi(p, v) = \infty$$

*for each $p_0 \in Z$ and $0 \neq v \in \mathbb{R}^{n+1}$.*

---

[3] Due to Fleming, Almgren, Simons, Federer, and Hardt and Simon.

**Condition A.** We assume there is a sequence of lower-semi-continuous integrands $\Phi_\rho$ that converge pointwise to the divergent integrand $\Phi$ as $\rho \downarrow 0$, *i.e.*

$$\lim_{\rho\downarrow 0} \Phi_\rho(p,v) = \Phi(p,v),$$

for $(p,v) \in (\mathbb{R}^{n+1} \setminus Z) \times \mathbb{R}^{n+1}$, and

$$\lim_{\rho\downarrow 0} \Phi_\rho(p,v) = \infty$$

for $(p,v) \in Z \times \left(\mathbb{R}^{n+1} \setminus \{0\}\right)$. We impose the following conditions on this convergence:

**(i)** For each $(p,v) \in \mathbb{R}^{n+1} \times \mathbb{R}^{n+1}$, $\Phi_\rho(p,v)$ is a non-increasing function of $\rho \in (0,\infty)$.

**(ii)** For each compact $K$ with $K \cap Z = \emptyset$, $\Phi_\rho = \Phi$ on $K \times \mathbb{R}^{n+1}$ for all sufficiently small $\rho > 0$.

**(iii)** For each compact $K$, there exists $0 < c$ such that

$$c \leq \Phi_\rho(p,v)$$

holds for all $p \in K$, all $v \in \mathbb{R}^{n+1}$ with $|v| = 1$, and all sufficiently small $\rho > 0$.

**Remark 24.1** *While Condition A is somewhat technical, the purpose is to insure that $\Phi_\rho$ does not have any unnecessary pathologies and that $\Phi_\rho$ converges to $\Phi$ in a natural way as $\rho \downarrow 0$.*

**Definition 24.2** *A surface $T$ is an* ***overtaking minimizer*** *if*

$$\liminf_{\rho\downarrow 0} \left[ \int_S \Phi_\rho - \int_T \Phi_\rho \right] \geq 0 \tag{24.3}$$

*holds for all surfaces $S$ with $\partial S = \partial T$.*

**Proposition 24.1** *If $T$ is an overtaking minimizer, then $T$ is locally $\Phi$-minimizing on the complement of the singular set of the integrand.*

**Proof.** Suppose $x \notin Z$. Let $V$ be a bounded open set containing $x$ such that $\overline{V} \cap Z = \emptyset$.

Arguing by contradiction, suppose $S$ and $T$ agree outside of $V$, but there is an $\epsilon > 0$ such that

$$\int_{S\cap V} \Phi + \epsilon \leq \int_{T\cap V} \Phi.$$

By Condition A(ii) applied with $K = \overline{V}$, we see that there is $\rho_0 > 0$ such that, for $0 < \rho < \rho_0$, we have $\Phi = \Phi_\rho$ on $V \times \mathbb{R}^{n+1}$. Since $S$ and $T$ agree outside of $V$, we have

$$\int_S \Phi_\rho + \epsilon \leq \int_T \Phi_\rho$$

for all $0 < \rho < \rho_0$, contradicting Eqn. (24.3). ■

The next theorem gives a sufficient condition under which we can prove the existence of an overtaking minimizer.

**Theorem 24.2** *Let an $(n-1)$-dimensional rectifiable current $R$ with $\partial R = 0$ be given. If there exists a surface $Q$ such that for any surface $S$ satisfying the boundary condition $\partial S = R$*

$$\int_S \Phi_\rho - \int_Q \Phi_\rho$$

*is non-decreasing as $\rho$ decreases and if there exists some surface $S_0$ with $\partial S_0 = R$ so that*

$$\lim_{\rho \downarrow 0} \left[ \int_{S_0} \Phi_\rho - \int_Q \Phi_\rho \right] < \infty,$$

*then an overtaking minimizer $T$ with $\partial T = R$ exists.*

**Proof.** Clearly, we can select a sequence of surfaces $T_i$ with

$$\lim_{i \to \infty} \left( \lim_{\rho \downarrow 0} \left[ \int_{T_i} \Phi_\rho - \int_Q \Phi_\rho \right] \right) = \inf \left\{ \lim_{\rho \downarrow 0} \left[ \int_S \Phi_\rho - \int_Q \Phi_\rho \right] \; : \; \partial S = R \right\}$$

and we may pass to a subsequence, but without changing notation, so that the $T_i$ are flat convergent to an integral current $T$ with $\partial T = R$.

Let $S$ with $\partial S = R$ be an arbitrary surface satisfying the boundary condition. We now want to prove Eqn. (24.3). Set

$$L_1 = \lim_{i \to \infty} \left( \lim_{\rho \downarrow 0} \left[ \int_{T_i} \Phi_\rho - \int_Q \Phi_\rho \right] \right)$$

and

$$L_2 = \lim_{\rho \downarrow 0} \left[ \int_S \Phi_\rho - \int_Q \Phi_\rho \right],$$

so we have

$$L_1 \le L_2.$$

Let $\epsilon > 0$ be given. We can choose $\rho_0 > 0$ so that

$$L_2 - \epsilon \le \int_S \Phi_\rho - \int_Q \Phi_\rho \le L_2 \text{ holds for all } 0 < \rho \le \rho_0.$$

We can also choose $i_0$ so that

$$L_1 \le \lim_{\rho \downarrow 0} \left[ \int_{T_i} \Phi_\rho - \int_Q \Phi_\rho \right] \le L_1 + \epsilon \text{ holds for all } i_0 \le i.$$

Now, for a fixed $\rho$ we have

$$\int_{T_i} \Phi_\rho - \int_Q \Phi_\rho \le \lim_{t \downarrow 0} \left[ \int_{T_i} \Phi_t - \int_Q \Phi_t \right] \le L_1 + \epsilon \text{ for all } i_0 \le i,$$

so

$$\int_T \Phi_\rho - \int_Q \Phi_\rho \le \liminf_{i \to \infty} \left[ \int_{T_i} \Phi_\rho - \int_Q \Phi_\rho \right] \le L_1 + \epsilon.$$

In particular, for $0 < \rho \le \rho_0$, we have

$$\int_T \Phi_\rho - \int_Q \Phi_\rho \le L_1 + \epsilon \le L_2 + \epsilon \le \int_S \Phi_\rho - \int_Q \Phi_\rho + 2\epsilon.$$

Thus

$$-2\epsilon \le \int_S \Phi_\rho - \int_T \Phi_\rho \text{ holds for all } 0 < \rho \le \rho_0.$$

Thus we have

$$-2\epsilon \le \liminf_{\rho \downarrow 0} \left[ \int_S \Phi_\rho - \int_T \Phi_\rho \right],$$

and since $\epsilon > 0$ was arbitrary, Eqn. (24.3) holds. ■

**Example 24.1** *Fix $\alpha \leq -2$. Let $Z$ be the $z$-axis in $\mathbb{R}^3$. Define a degree 2 divergent integrand on $(\mathbb{R}^3 \setminus Z) \times \mathbb{R}^3$ by setting*

$$\Phi\Big((x,y,z),v\Big) = \Big(\sqrt{x^2+y^2}\Big)^{\alpha}\,|v| \quad \text{if } \sqrt{x^2+y^2} > 0.$$

*The approximating integrands satisfying Condition A are*

$$\Phi_\rho\Big((x,y,z),v\Big) = \begin{cases} \Big(\sqrt{x^2+y^2}\Big)^{\alpha}\,|v| & \text{if } \sqrt{x^2+y^2} \geq \rho \\ \rho^{\alpha}\,|v| & \text{if } \sqrt{x^2+y^2} < \rho \end{cases}$$

*Let $Q$ be the unit disc considered as a current. Then for any current $S$ that projects onto a neighborhood of the origin in $\mathbb{R}^2$, the function*

$$\int_S \Phi_\rho - \int_Q \Phi_\rho$$

*will be non-decreasing for all small enough $\rho$, namely all $\rho$ such that $S$ projects onto the ball of radius $\rho$. Thus the hypotheses of Theorem 24.2 will be satisfied if we take $R$ to be any 1-dimensional current that forms a non-trivial link with the $z$-axis.*

## 24.4 A Radially Symmetric Example

In this section we motivate the regularity results that will be proved later. We continue to let $Z$ denote the $z$-axis in $\mathbb{R}^3$, and we refer to it as the singular line of the integrand. In $(\mathbb{R}^3 \setminus Z) \times \mathbb{R}^3$ we define the integrand

$$\Phi\Big((x,y,z),v\Big) = \Big(\sqrt{x^2+y^2}\Big)^{\alpha}\,|v|.$$

If the specific surface under consideration happens to be the graph of a function, a so-called **non-parametric surface,**

$$z = u(x,y)$$

over a domain $\Omega \subset \mathbb{R}^2$, then the integral of $\Phi$ over the surface is given by

$$\int_\Omega \Big(\sqrt{x^2+y^2}\Big)^{\alpha} \sqrt{1+|Du|^2}\, d\mathcal{L}^2$$

or in polar coordinates

$$\iint_\Omega r^{\alpha+1}\sqrt{1+|Du|^2}\, dr\, d\theta. \tag{24.4}$$

We see that if

$$\alpha \le -2$$

the integral in (24.4) will be seriously divergent in that any $R$ that links with the $z$-axis will bound only surfaces $T$ with infinite integral.

If the surface is radially symmetric and defined over the unit disc in $\mathbb{R}^2$, then it can be described by a function of one variable, say $u(r,\theta) = z(r)$, $0 \le r \le 1$, which simplifies all calculations. For convenience, we will write

$$\gamma = 1 + \alpha.$$

The integral (24.4) then equals

$$\int_0^1 r^\gamma \sqrt{1 + (z')^2}\, dr. \tag{24.5}$$

The integral in (24.5) is convergent if $-1 < \gamma$, but we will consider

$$\gamma \le -1.$$

Admissible functions will be absolutely continuous and satisfy $z(0) = 0$ and $z(1) = z^*$, where $z^*$ is some constant. Since the integral in (24.5) is divergent when $\gamma \le -1$, a reasonable approach would be to make a change of variables $t = 1/r$ and consider an infinite horizon optimal control problem. To keep the geometry fixed, we instead introduce a small parameter $\epsilon > 0$ and consider minimizing

$$\int_\epsilon^1 r^\gamma \sqrt{1 + (z')^2}\, dr$$

subject to $z(\epsilon) = 0$ and $z(1) = z^*$.

Using the Euler equation (see [4], page 30), we have

$$\frac{d}{dr}\left[\frac{\partial}{\partial z'} r^\gamma \sqrt{1 + (z')^2}\right] = 0$$

or

$$\frac{(z')^2}{1 + (z')^2} = C r^{-2\gamma}.$$

If $z'$ vanishes anywhere we arrive at the constant solution $z \equiv 0$, which applies when $z^* = 0$.

So suppose $z^* > 0$ and $z' > 0$. Then setting $a = \sqrt{C}$, we have

$$z'(r) = \frac{a\, r^{-\gamma}}{\sqrt{1 - a^2 r^{-2\gamma}}}. \tag{24.6}$$

Note that the sign of $\gamma$ matters. If $\gamma$ were positive, then $z'$ would be undefined near $r = 0$. With $\gamma < 0$, as we are assuming, the right-hand-side of Eqn. (24.6) is actually $O(r^{-\gamma})$ as $r \downarrow 0$. (Here we have the interesting situation that making the integrand sufficiently divergent has broadened the range of solutions to the Euler equation.)

If the surface under consideration is an overtaking minimizer, then the principle of optimality ([4], page 27) tells us that in fact the Euler equation will be satisfied for $r > 0$. So we may dispense with the parameter $\epsilon$ and conclude that

$$z(r) = \int_0^r \frac{a\,t^{-\gamma}}{\sqrt{1 - a^2 t^{-2\gamma}}}\,dt.$$

Notice that $z(r)$ is $O(r^{-\alpha})$ as $r \downarrow 0$.

This example shows the sort of result we might hope to prove. Namely that any overtaking minimizer rapidly approaches a disc as the singular line is approached.

## 24.5 Hypotheses for Regularity

Continue to let $Z$ denote the singular line which will be the $z$-axis in $\mathbb{R}^3$, and let $Z_\rho$ denote the set of points that are within distance less than or equal to $\rho$ from the $z$-axis. Throughout the remainder of this paper, we will consider integrands of degree 2 on $(\mathbb{R}^3 \setminus Z) \times \mathbb{R}^3$ of the special form

$$\Phi\Big((x, y, z), v\Big) = \phi(x, y)\,|v|, \tag{24.7}$$

with $\phi$ diverging as the origin is approached. We assume the following:

**Condition B.** There exist $2 \le \beta_0$ and $0 < R_0 \le 1$, such that

$$\frac{D\phi \cdot p}{\phi} \le -\beta_0 \tag{24.8}$$

holds for all $p = (x, y)$ with $|p| \le R_0$.

**Condition C.** There exist $0 < \delta_1 < 1$, $0 < R_1$, and $C_1 < \infty$ such that for $p_0 = (x_0, y_0)$ with $|p_0| < R_1$

$$C_1^{-1} \le \frac{\phi(p)}{\phi(p_0)} \le C_1 \tag{24.9}$$

$$\frac{|p_0|\,|D\phi(p)|}{|\phi(p_0)|} \le C_1 \tag{24.10}$$

$$\frac{|p_0|^2 \, |D^2\phi(p)|}{|\phi(p_0)|} \leq C_1 \tag{24.11}$$

hold for $p \in \mathbb{B}(p_0, \delta|p_0|)$.

**Condition D.**

$$\lim_{p \to (0,0)} \left( \int\limits_{\mathbb{B}(p,\delta|p|)} \phi \, d\mathcal{L}^2 \right) = \infty \tag{24.12}$$

for all $0 < \delta < 1$.

**Remark 24.2** *Conditions B and C are imposed as technical devices to allow the proof of regularity to proceed. In particular, Condition B will be used to insure that we can construct appropriate barriers for the quasilinear partial differential equation considered in [12], §21. Condition C allows us to control the effect of blowing-up pieces of the overtaking minimizer in Lemma 24.12. The type of bound given in Condition C is determined by the measure of the size of an integrand used in [1]. While Condition D also plays a technical role, it is natural to impose a condition on how fast the integrand diverges as the singular line is approached and that is what Condition D expresses.*

We will also be considering a 2-dimensional rectifiable current $T$ in $\mathbb{R}^3$ with $\operatorname{spt}(\partial T) \cap Z = \emptyset$ and $\operatorname{spt}(T) \cap Z \neq \emptyset$ that is locally $\Phi$-minimizing away from the singular line and that satisfies

**Condition E.**

$$\lim_{\rho \downarrow 0} \left( \int\limits_{T \setminus Z_\rho} \Phi(p, \overrightarrow{T}) \, \mu_T \, d\mathcal{H}^2 p - \int\limits_{r > \rho} \phi \, d\mathcal{L}^2 \right) < \infty. \tag{24.13}$$

**Remark 24.3** *Condition E will be satisfied by overtaking minimizers such as those whose existence is guaranteed by Theorem 24.2 applied with $Q$ the unit disc (or with $Q$ any open subset of the plane that contains the origin).*

## 24.6 Barriers

**Lemma 24.3** *Any local $\Phi$-minimizer given by $z = u(x, y)$ must satisfy the following **Euler-Lagrange equation**:*

$$\mathcal{L}u := [(1 + |Du|^2)I - Du \, Du] D^2 u + \frac{D\phi \cdot Du}{\phi} (1 + |Du^2|) = 0.$$

**Proof.** See [12], page 479. ■

**Lemma 24.4** *Suppose $2 \le \beta$. Then there exist $0 < \alpha < \infty$ and $0 < R' \le 1$ such that, for each $0 < R \le R'$, the maximum of the function*

$$f(r) = \frac{\beta + \alpha^2\,\beta^2\,R^{2-2\beta}\,r^{-2+2\beta}}{1 + \alpha^2\,\beta^2\,R^{2-2\beta}\,r^{\beta}}$$

*on the interval $[0, R]$ is $\beta$.*

**Proof.** We see that for $f'$ to vanish we must have

$$2(\beta - 1)r^{\beta-2} + \alpha^2\,\beta^2\,(\beta - 2)\,R^{2-2\beta}\,r^{2\beta-2} = \beta^2.$$

When $\beta = 2$ this trivially cannot happen. When $2 < \beta$, choose $R'$ so that $0 < R' < \frac{1}{2}\beta^2$ and $\alpha$ so that $\alpha^2\,(\beta - 2) < \frac{1}{2}$. These choices insure that $f'(r)$ does not vanish in $(0, R)$ for $0 < R \le R'$. One then verifies that $f(0) = \beta$ is at least as large as $f(R) = (\beta + \alpha^2\,\beta^2)/(1 + \alpha^2\,\beta^2\,R^{2-\beta})$. ■

**Theorem 24.5** *If $\phi$ satisfies Condition B, then there exist $0 < \alpha < \infty$ and $0 < R_2 \le R_0$ such that, for each $0 < R \le R_2$,*

$$w(x, y) = \alpha\,R^{1-\beta_0}\,\left(x^2 + y^2\right)^{\beta_0/2}$$

*satisfies*

$$\mathcal{L}w \le 0 \quad \text{for} \quad x^2 + y^2 \le R^2.$$

*Here $\beta_0$ is as in Condition B, Eqn. (24.8).*

**Proof.** Direct computation shows that if $w(x, y) = C\,\left(x^2 + y^2\right)^{\beta_0/2}$ with $C = \alpha\,R^{1-\beta_0}$, then

$$\begin{aligned} \mathcal{L}w &= C\,\beta_0^2\,\left(x^2 + y^2\right)^{-1+\beta_0/2} + C^3\,\beta_0^3\,\left(x^2 + y^2\right)^{-2+3\beta_0/2} \\ &+ \frac{D\phi \cdot \overrightarrow{P}}{\phi}\,C\,\beta_0\,\left(x^2 + y^2\right)^{-1+\beta_0/2}\,[1 + C^2\,\beta_0^2\,\left(x^2 + y^2\right)^{\beta_0/2}]. \end{aligned}$$

Removing a common factor of $C\,\beta_0\,\left(x^2 + y^2\right)^{\beta_0/2}$, we see that the condition $\mathcal{L}w \le 0$ is equivalent to

$$\frac{D\phi \cdot p}{\phi} \le -\frac{\beta_0 + C^2\,\beta_0^2\,\left(x^2 + y^2\right)^{-1+\beta_0}}{1 + C^2\,\beta_0^2\,(x^2 + y^2)^{\beta_0/2}}. \tag{24.14}$$

Now, choose $R_2$ to be the minimum of $R_0$ and the $R'$ of Lemma 24.4 and let $\alpha$ also be as in Lemma 24.4. Then Lemma 24.4 tells us that the right-hand-side of the Eqn. (24.14) is greater than or equal to $-\beta_0$ and the theorem follows. ■

## 24.7 A Result in Differential Geometry

In this section, we give what seems to be the optimal interpretation for a bound on the principal curvatures of a graph. We will use standard notation from differential geometry (see [7]): A surface in $\mathbb{R}^3$ will be parameterized by $r(x_1, x_2)$. The first fundamental form will have coefficients $g_{ij}$, the second fundamental form will have coefficients $b_{ij}$, the unit normal vector will be $m$, and the Christoffel symbols will be $\Gamma_{ij}^k$.

**Lemma 24.6** *For a graph $z = u(x, y)$ considered as parametrized by*

$$(x_1, x_2) \mapsto (x_1, x_2, u(x_1, x_2)) = r(x_1, x_2),$$

*the following hold:*

$$\begin{aligned}
g &= \sqrt{1 + |Du|^2}, \\
\Gamma_{11}^1 &= \frac{b_{11}\, D_1 u}{\sqrt{g}}, \qquad \Gamma_{11}^2 = \frac{b_{11}\, D_2 u}{\sqrt{g}}, \\
\Gamma_{12}^1 &= \frac{b_{12}\, D_1 u}{\sqrt{g}}, \qquad \Gamma_{12}^2 = \frac{b_{12}\, D_2 u}{\sqrt{g}}, \\
\Gamma_{22}^1 &= \frac{b_{22}\, D_1 u}{\sqrt{g}}, \qquad \Gamma_{22}^2 = \frac{b_{22}\, D_2 u}{\sqrt{g}}.
\end{aligned}$$

**Proof.** We compute

$$r_1 = (1, 0, D_1 u), \qquad r_2 = (0, 1, D_2 u).$$

So

$$g_{11} = 1 + (D_1 u)^2, \quad g_{12} = g_{21} = (D_1 u)(D_2 u), \quad g_{22} = 1 + (D_2 u)^2.$$

Thus

$$\begin{aligned}
g &= g_{11} g_{22} - g_{12} g_{21} \\
&= 1 + (D_1 u)^2 + (D_2 u)^2_{(} D_1 u)^2 (D_2 u)^2 - ((D_1 u)(D_2 u))^2 \\
&= 1 + |Du|^2.
\end{aligned}$$

Also, we compute

$$r_{11} = (0, 0, D_{11} u), \quad r_{12} = r_{21} = (0, 0, D_{12} u), \quad r_{22} = (0, 0, D_{22} u).$$

We have

$$\begin{aligned}
r_1 \times r_2 &= (-D_1 u, -D_2 u, 1), \\
m &= \frac{r_1 \times r_2}{|r_1 \times r_2|} \\
&= (-g^{-1/2}\, D_1 u, -g^{-1/2}\, D_2 u, g^{-1/2}).
\end{aligned}$$

By Equation (18.1), page 180, of [7], we have

$$\begin{aligned}
&(0,0,D_{11}u) = \\
&\quad \Gamma_{11}^{1}\,(1,0,D_1u) + \Gamma_{11}^{2}\,(0,1,D_2u) \\
&\qquad +b_{11}\,(-g^{-1/2}\,D_1u, -g^{-1/2}\,D_2u, g^{-1/2}), \\
&(0,0,D_{12}u) = \\
&\quad \Gamma_{12}^{1}\,(1,0,D_1u) + \Gamma_{12}^{2}\,(0,1,D_2u) \\
&\qquad +b_{12}\,(-g^{-1/2}\,D_1u, -g^{-1/2}\,D_2u, g^{-1/2}), \\
&(0,0,D_{22}u) = \\
&\quad \Gamma_{22}^{1}\,(1,0,D_1u) + \Gamma_{22}^{2}\,(0,1,D_2u) \\
&\qquad +b_{22}\,(-g^{-1/2}\,D_1u, -g^{-1/2}\,D_2u, g^{-1/2}).
\end{aligned}$$

All the formulae for the Christoffel symbols follow from considering the first and second components. ■

**Lemma 24.7** *For general parametric surfaces in* $\mathbb{R}^3$, *we have*

$$\begin{aligned}
\frac{1}{2}\frac{\partial g}{\partial x_1} &= -g\,(\Gamma_{11}^{1} + \Gamma_{12}^{2}), \\
\frac{1}{2}\frac{\partial g}{\partial x_2} &= -g\,(\Gamma_{12}^{1} + \Gamma_{22}^{2}).
\end{aligned}$$

**Proof.** Since

$$g = (r_1 \cdot r_1)\,(r_2 \cdot r_2) - (r_1 \cdot r_2)^2,$$

we simply differentiate to find

$$\begin{aligned}
\frac{\partial g}{\partial x_1} = {} & 2\,(r_{11} \cdot r_1)\,(r_2 \cdot r_2) + 2\,(r_1 \cdot r_1)\,(r_{12} \cdot r_2) \\
& -2\,(r_{11} \cdot r_2)\,(r_1 \cdot r_2) - 2\,(r_1 \cdot r_{12})\,(r_1 \cdot r_2).
\end{aligned}$$

Using Equation (18.6), page 180, of [7], we have

$$\begin{aligned}
\frac{1}{2}\frac{\partial g}{\partial x_1} &= \Gamma_{111}\,g_{22} + \Gamma_{122}\,g_{11} - \Gamma_{112}\,g_{12} - \Gamma_{121}\,g_{12} \\
&= (\Gamma_{11}^{j}\,g_{j1}\,g_{22} - \Gamma_{11}^{j}\,g_{j2}\,g_{12}) + (\Gamma_{12}^{j}\,g_{j2}\,g_{11} - \Gamma_{12}^{j}\,g_{j1}\,g_{12}) \\
&= \Gamma_{11}^{j}\,(g_{j1}\,g_{22} - g_{j2}\,g_{12}) + \Gamma_{12}^{j}\,(g_{j2}\,g_{11} - g_{j1}\,g_{12}).
\end{aligned}$$

Notice that $g_{j1}\,g_{22} - g_{j2}\,g_{12}$ equals $g$ when $j = 1$ and equals 0 when $j = 2$. Likewise, $g_{j2}\,g_{11} - g_{j1}\,g_{12}$ equals 0 when $j = 1$ and equals $g$ when $j = 2$. Thus we have the formula for $(1/2)\,(\partial g/\partial x_1)$.

Similarly, we compute

$$\begin{aligned}
\frac{\partial g}{\partial x_2} = {} & 2\,(r_{12} \cdot r_1)\,(r_2 \cdot r_2) + 2\,(r_1 \cdot r_1)\,(r_{22} \cdot r_2) \\
& -2\,(r_{12} \cdot r_2)\,(r_1 \cdot r_2) - 2\,(r_1 \cdot r_{22})\,(r_1 \cdot r_2),
\end{aligned}$$

so

$$\begin{aligned}\frac{1}{2}\frac{\partial g}{\partial x_2} &= \Gamma_{121}\,g_{22}+\Gamma_{222}\,g_{11}-\Gamma_{122}\,g_{12}-\Gamma_{221}\,g_{12}\\ &= (\Gamma^j_{21}\,g_{j1}\,g_{22}-\Gamma^j_{21}\,g_{j2}\,g_{12})+(\Gamma^j_{22}\,g_{j2}\,g_{11}-\Gamma^j_{22}\,g_{j1}\,g_{12})\\ &= \Gamma^j_{21}\,(g_{j1}\,g_{22}-g_{j2}\,g_{12})+\Gamma^j_{22}\,(g_{j2}\,g_{11}-g_{j1}\,g_{12})\\ &= \Gamma^1_{21}\,g+\Gamma^2_{22}\,g.\end{aligned}$$

■

**Corollary 24.8** *For non-parametric surfaces in $\mathbb{R}^3$, we have*

$$\begin{aligned}\begin{pmatrix}\frac{\partial g}{\partial x_1}\\ \frac{\partial g}{\partial x_2}\end{pmatrix} &= -\frac{2}{\sqrt{g}}\begin{pmatrix} b_{11} & b_{12}\\ b_{21} & b_{22}\end{pmatrix}\begin{pmatrix} D_1 u\\ D_2 u\end{pmatrix}\\ &= -\frac{2}{\sqrt{g}}\begin{pmatrix} g_{11} & g_{12}\\ g_{21} & g_{22}\end{pmatrix}\begin{pmatrix} b^1_1 & b^1_2\\ b^2_1 & b^2_2\end{pmatrix}\begin{pmatrix} D_1 u\\ D_2 u\end{pmatrix},\end{aligned}$$

*and if the principal curvatures are bounded in absolute value by $\kappa$, then*

$$|Dg| \le 2\,g^{3/2}\,\kappa$$

*holds.*

**Proof.** For the non-parametric surface, we can use Lemma 24.6 to replace the Christoffel symbols in Lemma 24.7 by expressions of the form $(b_{ij}\,D_k u)/\sqrt{g}$. Then notice that the eigenvalues of

$$\begin{pmatrix} g_{11} & g_{12}\\ g_{21} & g_{22}\end{pmatrix}$$

are 1 and $g$, and that the principal curvatures are the eigenvalues of

$$\begin{pmatrix} b^1_1 & b^1_2\\ b^2_1 & b^2_2\end{pmatrix}.$$

■

**Theorem 24.9** *Suppose the graph $z = u(x, y)$ has all its principal curvatures at $(x_0, y_0, z_0)$ bounded in absolute value by $\kappa$. Then*

$$\left| D\left(\frac{1}{\sqrt{1+|Du|^2}}\right)\right| \le \kappa$$

*holds.*

**Proof.** We have

$$\left| D\left(\frac{1}{\sqrt{g}}\right)\right| = \frac{1}{2}\,g^{-3/2}\,|Dg| \le \kappa.$$

■

## 24.8 Bounding the Curvature

The following result generalizes Theorem 1 of [13]. It follows from the results of [1], in particular Theorem 1.2 of Part I.

**Theorem 24.10** *For each compact $K \subset \mathbb{R}^3$ and each set of positive multiples of the area integrand that is compact in the $C^2$ topology, there is a constant $C_2 < \infty$ such that if a 2-dimensional oriented surface in $K$ minimizes the integral of one of the multiples of the area integrand and if $P$ is a point of the support of the minimizer that is at distance at least $R \leq 1$ from the support of the boundary of the minimizer, then all the principal curvatures of the surface at $P$ are bounded in absolute value by $C_2/R$.*

**Proof.** As on page 547 of [13], we see that it suffices to be able to express the surface in non-parametric form over a disc of radius proportional to the distance to the boundary, and to have a Hölder 1/2 bound on the rate of change of the tangent direction. One then appeals to the Schauder interior estimates.

Recall that Theorem 1.2 of [1] tells us that there are constants $\epsilon > 0$, $\beta \in (0,1)$, depending only on $n$ and $\lambda$, such that if $T \in \mathcal{M}(\lambda, \rho_0)$, if $x_0 \in \operatorname{spt}(T)$, if $\rho \in (0, \rho_0 - |x_0|)$ and if

$$\operatorname{spt}(T) \cap \mathbb{U}^{n+1}(x_0, \rho) \subset \{x : \operatorname{dist}(x, H) < \epsilon\rho\}$$

for some hyperplane $H$ containing $x_0$, then $\operatorname{spt}(T) \cap \mathbb{U}^{n+1}(x_0, \beta\rho)$ is a connected $C^2$ hypersurface $M$ with $\overline{M} \setminus M \subset \partial\mathbb{U}^{n+1}(x_0, \beta\rho)$ and with unit normal $\nu = \nu^T$ satisfying

$$|\nu(x) - \nu(\bar{x})| \leq c\frac{|x - \bar{x}|}{\rho}, \qquad x, \bar{x} \in M.$$

Here $c$ is a constant depending only on $n$ and $\lambda$, and $\mathcal{M}(\lambda, \rho_0)$ is the notation of [1] for the minimizing surfaces in a certain $C^2$ neighborhood.

By Theorem 1.2 of [1], we need only show that a minimizer with the origin in its support and with boundary outside the unit ball has the part of its support within the ball of radius $\rho$ lying within $\epsilon\rho$ of some hyperplane. Arguing by contradiction, note that if not, then we can take a sequence of surfaces each of which fails for every $\rho$, and a subsequence thereof that converges. In a compact family of integrands the final minimizer does have a tangent plane. So the limit surface is as closely approximated as we wish by its tangent plane. There is a uniform estimate from below for the density of the surfaces in the sequence, which forces the existence of positive mass of the limit minimizer at a significant distance from the tangent plane. Thus, we have a contradiction. ∎

**Notation 24.11** *We will use $\mathbb{B}^2(p, r)$ to denote the open ball in $\mathbb{R}^2$ of radius $r$ centered at $p$.*

**Lemma 24.12** *For $p_0 = (x_0, y_0) \neq (0,0)$, set $R = |p_0|$. If $\phi$ satisfies Condition C, then the mapping*

$$(x,y,z) \mapsto (\xi,\eta,\zeta) = (\frac{x-x_0}{R}, \frac{y-y_0}{R}, \frac{z}{R}) \tag{24.15}$$

*sends a surface minimzing the integrand $\Phi$ on the region $\mathbb{B}^2(p_0, \delta_1 R) \times \mathbb{R}$ to a surface minimizing an integrand in a $C^2$-bounded family of integrands on $\mathbb{B}^2(0,\delta_1) \times \mathbb{R}$ as in Theorem 24.10. Here $\delta_1$ is as in Condition C.*

**Proof.** In order to avoid the notational complexity of the formula for mapping integrands that occurs in [5], §5.1.1, we will consider what happens when the surface is a graph.

We are mapping the graph $z = u(x,y)$ to the graph $\zeta = w(\xi,\eta)$ with the variables connected by Eqn. (24.15). Let us also set $\xi_0 = x_0/R$ and $\eta_0 = y_0/R$. Thus

$$w(\xi,\eta) = \frac{1}{R} u(x,y) = \frac{1}{R} u(R\xi + R\xi_0, R\eta + R\eta_0).$$

It follows that $|D_{(\xi,\eta)} w| = |D_{(x,y)} u|$ holds at corresponding points. Note that $\partial(x,y)/\partial(\xi,\eta) = R^2$. So we have

$$\int\int_{\mathbb{B}^2(p_0,\delta_1 R)} \phi(x,y) \sqrt{1 + |D_{(x,y)}u|^2}\, dx\, dy =$$

$$R^2 \int\int_{\mathbb{B}^2(0,\delta_1)} \phi(R\xi + R\xi_0, R\eta + R\eta_0) \sqrt{1 + |D_{(\xi,\eta)}w|^2}\, d\xi\, d\eta.$$

Thus the image surface minimizes the integral of the area integrand multiplied by the function

$$R^2 \phi(R\xi + R\xi_0, R\eta + R\eta_0). \tag{24.16}$$

If we multiply the function in Eqn. (24.16) by any positive number that does not depend on position or orientation, such as $R^{-2}/\phi(R\xi_0, R\eta_0)$, the image surface is still minimizing. Thus the image surface minimizes area multiplied by

$$\frac{\phi(R\xi + R\xi_0, R\eta + R\eta_0)}{\phi(R\xi_0, R\eta_0)}. \tag{24.17}$$

Condition C is precisely the condition needed to bound the multipliers in Eqn. (24.17) in the $C^2$ topology. ∎

**Corollary 24.13** *For any surface that is locally $\Phi$-minimizing away from the singular line, the curvature at distance $R$ from the singular line is bounded by $C_2/R$.*

**Theorem 24.14** *Near the singular line, a locally $\Phi$-minimizing surface satisfying Condition E of Section 24.5 can be represented as a graph $z = u(x,y)$ in a sufficiently small neighborhood of the singular line and it holds that $|Du| \to 0$ as $(x,y) \to (0,0)$.*

**Proof.** If not we can fix $\epsilon > 0$ such that there are points arbitrarily close to $(0,0)$ with $|Du| \geq \epsilon$. Say we have such a point $p$ at distance $R$ from the singular line. Define $\sigma > 0$ by requiring

$$\frac{1}{\sqrt{1+\epsilon^2}} = 1 - \sigma.$$

Then we have

$$\frac{1}{\sqrt{1+|Du|^2}} \leq 1 - \sigma$$

at that point, so in a ball of radius

$$\delta\, R = \frac{\sigma\, R}{4\, C_2} \tag{24.18}$$

(or smaller, so we can assume $\delta < 1$), we have

$$\frac{1}{\sqrt{1+|Du|^2}} \leq 1 - \frac{\sigma}{2}.$$

For $\rho > 0$ set

$$E(\rho) = \int_{T \setminus Z_\rho} \Phi(p, \overrightarrow{T})\, \mu_T\, d\mathcal{H}^2 p - \int_{r>\rho} \phi\, d\mathcal{L}^2.$$

As long as the radius in Eqn. (24.18) is less than $R/2$ we can estimate

$$\sqrt{1+|Du|^2} - 1 \geq \frac{1}{1-\sigma/2} - 1 = \frac{\sigma}{2-\sigma}$$

so

$$E[(1-\delta)R] \geq \frac{\sigma}{2-\sigma} \int_{\mathbb{B}(p,\delta\, R)} \phi\, d\mathcal{L}^2 \to \infty \quad \text{as } R \downarrow 0,$$

contradicting Condition E, Eqn. (24.13). ■

**Theorem 24.15** *If Conditions B–E of Section 24.5 hold, then any overtaking minimizer meets the singular line in a disc perpendicular to the singular line and decays to that disc at the rate $r^{\beta_0}$ as $r \downarrow 0$, i.e. there exists $C < \infty$ such that*

$$|u(x,y) - u(0,0)| \leq C|(x,y)|^{\beta_0}$$

*holds for all $(x,y)$ sufficiently close to $(0,0)$. Here $\beta_0$ is as in Condition B, Eqn. (24.8).*

**Proof.** Note that the function $w$ in Theorem 24.5 has the value $\alpha R$ at distance $R$ from the origin, and $\alpha$ is a fixed positive number. By Theorem 24.5 and Theorem 24.14, for small enough $R > 0$, we can use $w(x, y) + u(0, 0)$ as a barrier above the graph of $u$, and likewise we can use $-w(x, y) + u(0, 0)$ as a barrier below the graph of $u$. ■

**Acknowledgement.** I would like to thank the program committee of the Seventh Meeting of the International Workshop in Analysis and its Applications for encouraging this work by inviting me to participate in the workshop. I would like to thank Časlav Stanojević for his suggestions for improving the significance and generality of the results of this paper. I would also like to thank the referee for the useful references he provided and for his suggestions for improving the clarity of exposition of this paper.

*References*

[1] Frederick J. Almgren, Richard M. Schoen & Leon M. Simon, *Regularity and singularity estimates on hypersurfaces minimizing parametric elliptic variational integrals*, Acta Mathematica **139** (1977), 217–265.

[2] Michael T. Anderson, *Complete minimal varieties in hyperbolic space*, Inventiones Mathematicae **69** (1982), 477–494.

[3] Dean A. Carlson, Alain B. Haurie & Arie Leizarowitz, *Infinite Horizon Optimal Control*, 2*nd* edition, Springer-Verlag, New York, 1991.

[4] Lamberto Cesari, *Optimization—Theory and Applications*, Springer-Verlag, New York, 1983.

[5] Herbert Federer, *Geometric measure theory*, Springer-Verlag, New York, 1969.

[6] Herbert Federer & Wendell H. Fleming, *Normal and integral currents*, Annals of Mathematics, Second Series, **72** (1960), 458–520.

[7] Abraham Goetz, *Introduction to differential geometry*, Addison Wesley, Reading, Massachusetts, 1970.

[8] Robert M. Hardt & Fang Hua Lin, *Regularity at infinity for area-minimizing hypersurfaces in hyperbolic space*, Inventiones Mathematicae **88** (1987), 217–224.

[9] Fang Hua Lin, *Asymptotic behavior of area-minimizing currents in hyperbolic space*, Communications on Pure and Applied Mathematics **42** (1989), 229–242.

[10] Arvid T. Lonseth, *The problem of Plateau in hyperbolic space*, American Journal of Mathematics **64** (1942), 229–259.

[11] Frank Morgan, *Geometric measure theory : a beginner's guide*, 2nd edition, Academic Press, New York, 1995.

[12] James B. Serrin, *The problem of Dirichlet for quasilinear elliptic differential equations with many independent variables*, Philosophical Transactions of the Royal Society of London, Series A, **264** (1969), 413–496.

[13] Leon M. Simon, *Remarks on curvature estimates for minimal hypersurfaces*, Duke Mathematical Journal **43** (1976), 545–553.

[14] Joel Spruck & Yisong Yang, *On multivortices in the electroweak theory, II*, Communications in Mathematical Physics **144** (1992), 215–234.

# Chapter 25

# Sparse exponential sums with low sidelobes

**George Benke**

*ABSTRACT. Let $k_1, \cdots, k_N$ be integers satisfying $0 \le k_j < Np$ for positive interges $N$ and $p$. For infinitely many integers $N$ a construction is given for sets of integers $\{k_1, \cdots, k_N\}$ such that*

$$\sup_{-1/2<x\le 1/2} \left( \left| \sum_{j=1}^{N} e^{2\pi i k_j x} \right| - \left| \frac{\sin \pi N p x}{p \sin \pi x} \right| \right) = O\left(N^{1/2}\right).$$

*The $N$ are of the form $p^{M+g(M)}$ where $g(M) = O\left(p^{M/2}\right)$.*

## 25.1 Introduction

In this paper we consider trigonometric polynomials of the form

$$f(x) = \sum_{j=1}^{N} e^{2\pi i k_j x}$$

where $k_1, \cdots, k_N$ are integers taken from the interval $[0, Np)$ for some positive integer $p$. Clearly $f(0) = N$ regardless of the particular integers $k_j$. Since $f$ is periodic with period 1, we restrict attention to $x \in (-1/2, 1/2]$ which we denote by $I$. We ask: "How close can we come to the situation where $f(0) = N$ and $f(x) = 0$ for all other $x \in I$?". To gain some insight into this question consider the case $k_j = (j-1)p$ for $j = 1, \cdots, N$. This case gives

$$|f(x)| = \left| \frac{\sin \pi N p x}{\sin \pi p x} \right|$$

and since $f(m/p) = N$ for any integer $m$, $f(x)$ has $p-1$ "extra lobes" in $I$. On the other hand, suppose we randomize the $k_j$. Thus let $K_j$ be independent random variables uniformly distributed in the set of integers $0, \cdots, Np-1$. Then

$$\begin{aligned} Ef(x) &= E\sum_{j=1}^{N} e^{2\pi i K_j x} = \sum_{j=1}^{N} E\left(e^{2\pi i K_j x}\right) \\ &= N\sum_{k=0}^{Np-1} e^{2\pi i k x}\left(\frac{1}{Np}\right) \\ &= e^{\pi i (Np-1)x}\left(\frac{\sin \pi N p x}{p \sin \pi x}\right). \end{aligned}$$

Therefore a randomized $f(x)$ is expected to show peaks of size $N$ only at $x = 0$, and to fall off like $N/x$. Thus the "extra lobes" in the first example are the result of the regular arithmetic structure of the $k_j$. Note also that

$$\int_I |f(x)|^2 dx = N$$

while

$$\int_I |E(f(x))|^2 dx = \frac{N}{p}.$$

It can be shown that the structure of the central lobe of $|f(x)|$ is relatively insensitive to the location of the integers $k_j$, and therefore resembles the central lobe of $|Ef(x)|$. Energy considerations, as expressed by the previous integrals, therefore show that there must be a considerable discrepancy between $|f(x)|$ and $|Ef(x)|$ for $x$ bounded away from 0. We therefore reformulate our earlier question more precisely.

Our problem is to find $k_1, \cdots, k_N$ in $[0, Np]$ which minimize

$$F(k_1, \cdots, k_N) = \sup_{-1/2 < x \le 1/2} \left( \left| \sum_{j=1}^{N} e^{2\pi i k_j x} \right| - \left| \frac{\sin \pi N p x}{p \sin \pi x} \right| \right).$$

The case of $k_j$ chosen randomly has been treated in [2] where it was shown that

$$F(k_1, \cdots, k_N) = O\left((N \log N)^{1/2}\right).$$

This is the best result that can be obtained by simply choosing $k_j$ at random. On the other hand, energy considerations show that we cannot hope to do better than $O\left(N^{1/2}\right)$.

In this paper we utilize the theory of Generalized Rudin-Shapiro systems to construct $k_1, \cdots, k_N$ such that

$$F(k_1, \cdots, k_N) = O\left(N^{1/2}\right).$$

Our construction lacks full generality for two reasons. First, we require the density factor $p$ to be prime. Second, $N$ is a number which is approximately $p^M$ for some $M$. We cannot pre-determine the exact value of $N$, although a simple additional step in the construction allows us to have $N = p^M$.

This problem is related to an important engineering problem in the following way. The polynomials $f(x)$ describe the sensitivity of a phased array antenna of omnidirectional sensors. The variable $x$ is (essentially) the direction of the incoming signal, and the $k_j$ specify the positions of the sensors on a line. When $p > 1$ such arrays are said to be sparse and the "extra lobes" are called grating lobes. The determination of sensor locations in large sparse phased array antennas such that peak sidelobes are minimized is an important consideration in antenna design. For a more detailed discussion concerning phased array antennas see [2, 7, 8, 12, 13].

## 25.2 Generalized Rudin-Shapiro Polynomials

The Rudin-Shapiro polynomials were first introduced by M.Golay in 1949 in a paper entitled "Multislit Spectrometry" [4, 5]. In the 1950's H.Shapiro [10] and W. Rudin [9] independently rediscovered the same polynomials. Since then a large literature has developed in both the mathematics and engineering communities, where the coefficients of these polynomials are referred to as "complementary series" [6, 11, 14, 15]. Generalizations have been considered in [1] and [3]. In this section we recall the definition of Generalized Rudin-Shapiro polynomials, as treated in [1], and state and prove the results we need later.

Let $A$ is a $p{\times}p$ matrix with rows and columns enumerated by the elements of $\mathbf{Z}_p = \{0, 1, \ldots, p-1\}$. Denote by $\mathbf{p}$ the n-tuple $(1, p, p^2, \ldots, p^{n-1})$ and by $\langle\, ,\, \rangle$ the standard inner product on $\mathbf{R}^n$. Thus for $\boldsymbol{\omega} \in \mathbf{Z}_p^n$,

$$\langle \mathbf{p}, \boldsymbol{\omega} \rangle = (\omega_0 + \omega_1 p + \cdots + \omega_{n-1} p^{n-1})$$

is the integer whose base-$p$ digits are given by the components of $\boldsymbol{\omega}$. For $\boldsymbol{\nu} = (\nu_0, \ldots, \nu_{n-1}) \in \mathbf{Z}_p^n$ and $\lambda \in \mathbf{Z}_p$ denote by $S(\boldsymbol{\nu}, \lambda)$ the $(n-1)$-tuple $(\nu_1 + \lambda, \nu_2, \ldots, \nu_{n-1})$ where the addition is computed modulo $p$. The Generalized Rudin-Shapiro polynomials are given in the next definition.

**Definition 25.1** *For $n \geq 1$, $p \geq 2$ and $\boldsymbol{\nu} = (\nu_0, \ldots, \nu_{n-1}) \in \mathbf{Z}_p^n$ define polynomial $A_{n,\boldsymbol{\nu}}$ by*

$$A_{n,\boldsymbol{\nu}}(x) = \sum_{\boldsymbol{\omega} \in \mathbf{Z}_p^n} a_{\nu_0, \omega_{n-1}} \left( \prod_{k=1}^{n-1} a_{v_{n-k}+\omega_k, \omega_{k-1}} \right) e^{2\pi i \langle \mathbf{p}, \boldsymbol{\omega} \rangle x}$$

*where the subscript additions are computed modulo $p$. When $n = 1$, the product is taken to be empty.*

The next theorem gives the fundamental recursion relation for the $A_{n,\nu}$.

**Theorem 25.1** . *Let $p \geq 2$, $n \geq 2$ and $\boldsymbol{\nu} \in \mathbf{Z}_p^n$. Then*

$$A_{n,\nu}(x) = \sum_{\lambda \in \mathbf{Z}_p} a_{\nu_0,\lambda} e^{2\pi i \lambda p^{n-1} x} A_{n-1,S(\nu,\lambda)}(x).$$

**Proof.** The definition of $A_{n,\nu}$ involves a $p$-fold sum. Breaking this sum into a single sum and a $(p-1)$-fold sum, we may write

$$A_{n,\nu}(x) = \sum_{\omega_{n-1} \in \mathbf{Z}_p} a_{\nu_0,\omega_{n-1}} e^{2\pi i \omega_{n-1} p^{n-1} x} \sum_{\omega \in \mathbf{Z}_{p^{n-1}}} \left( \prod_{k=1}^{n-1} a_{v_{n-k}+\omega_k,\omega_{k-1}} \right) e^{2\pi i \langle \mathbf{p},\omega \rangle x}$$

where $\omega = (\omega_1, \ldots, \omega_{n-1})$ and $\mathbf{p} = (1, p, p^2, \ldots p^{n-2})$. A comparison of the inner sum in the last expression with the definition of $A_{n,\nu}$ shows that the

$$\text{inner sum} = A_{n-1,S(\nu,\lambda)}(x).$$

Changing the outer summation index from $\omega_{n-1}$ to $\lambda$ establishes the recursion relation. ■

The classical Rudin-Shapiro polynomials are defined as a sequence pair $P_n(x)$ and $Q_n(x)$ where

$$\begin{aligned} P_0(x) &= 1 \\ Q_0(x) &= 1 \\ P_{n+1}(x) &= P_n(x) + e^{2\pi i 2^n x} Q_n(x) \\ Q_{n+1}(x) &= P_n(x) - e^{2\pi i 2^n x} Q_n(x). \end{aligned}$$

If in the previous definition we take

$$A = \begin{pmatrix} 1 & 1 \\ 1 & -1 \end{pmatrix}$$

then

$$\begin{aligned} A_{n,(0,0,\ldots,0)}(x) &= P_n(x) \\ A_{n,(1,0,\ldots,0)}(x) &= Q_n(x). \end{aligned}$$

We now suppose that $A$ is a multiple of a unitary matrix, and prove a set of theorems which establish the flatness properties of these generalized Rudin-Shapiro systems of polynomials.

**Theorem 25.2** . *Let $n \geq 2$ and $p \geq 2$. Suppose $A = \sqrt{\beta}U$ for some unitary matrix $U$ and constant $\beta$. Then for all $(\nu_1, \dots, \nu_{n-1}) \in \mathbf{Z}_p^{n-1}$*

$$\sum_{\nu_0 \in \mathbf{Z}_p} |A_{n,\boldsymbol{\nu}}(x)|^2 = \beta \sum_{\nu' \in \mathbf{Z}_p} |A_{n-1,\boldsymbol{\nu}'}(x)|^2$$

*where $\boldsymbol{\nu} = (\nu_1, \dots, \nu_{n-1})$ and $\boldsymbol{\nu}' = (\nu_0, \dots, \nu_{n-1})$.*

**Proof.** According to the basic recursion formula in Theorem 25.1

$$\begin{aligned}\sum_{\nu_0 \in \mathbf{Z}_p} |A_{n,\boldsymbol{\nu}}(x)|^2 &= \sum_{\nu_0 \in \mathbf{Z}_p} \sum_{\lambda \in \mathbf{Z}_p} \sum_{\lambda' \in \mathbf{Z}_p} \left(a_{\nu_0,\lambda} e^{2\pi i \lambda p^{n-1} x} A_{n-1,S(\boldsymbol{\nu},\lambda)}(x)\right) \\ &\quad \cdot \overline{a}_{\nu_0,\lambda'} e^{2\pi i \lambda' p^{n-1} x} \overline{A}_{n-1,S(\boldsymbol{\nu},\lambda')}(x) \\ &= \sum_{\lambda \in \mathbf{Z}_p} \sum_{\lambda' \in \mathbf{Z}_p} \left( \sum_{\nu_0 \in \mathbf{Z}_p} a_{\nu_0,\lambda} \overline{a}_{\nu_0,\lambda'} \right) e^{2\pi i (\lambda - \lambda') p^{n-1} x} \\ &\quad \cdot A_{n-1,S(\boldsymbol{\nu},\lambda)}(x) \overline{A}_{n-1,S(\boldsymbol{\nu},\lambda')}(x).\end{aligned}$$

Since $A = \sqrt{\beta}U$, with $U$ unitary, the sum over $\nu_0$ vanishes unless $\lambda = \lambda'$ in which case it equals $\beta$. Therefore the previous expression gives

$$\sum_{\nu_0 \in \mathbf{Z}_p} |A_{n,\boldsymbol{\nu}}(x)|^2 = \beta \sum_{\lambda \in \mathbf{Z}_p} \left|A_{n-1,S(\boldsymbol{\nu},\lambda)}(x)\right|^2.$$

But as $\lambda$ runs over $\mathbf{Z}_p$, $\nu_1 + \lambda$ also runs over $\mathbf{Z}_p$. Hence we may express the right hand side of the previous equation as given in the statement of the theorem. ∎

**Corollary 25.3** *Let $n \geq 1$ and $p \geq 2$. Suppose $A = \sqrt{\beta}U$ for some constant $\beta$ and unitary matrix $U$. Then for all $\boldsymbol{\nu} = (\nu_0, \dots, \nu_{n-1}) \in \mathbf{Z}_p^n$ and all real $x$*

$$\sum_{\nu_0 \in \mathbf{Z}_p} |A_{n,\boldsymbol{\nu}}(x)|^2 = p\beta^n.$$

**Proof.** Suppose first that $n = 1$. By definition

$$A_{1,\nu} = \sum_{\omega \in \mathbf{Z}_p} a_{\nu\omega} e^{2\pi i \omega x}.$$

Therefore

$$\begin{aligned}\sum_{\nu \in \mathbf{Z}_p} |A_{1,\nu}(x)|^2 &= \sum_{\nu \in \mathbf{Z}_p} \sum_{\omega \in \mathbf{Z}_p} \sum_{\omega' \in \mathbf{Z}_p} a_{\nu\omega} \overline{a}_{\nu\omega'} e^{2\pi i (\omega - \omega') x} \\ &= \sum_{\omega \in \mathbf{Z}_p} \sum_{\omega' \in \mathbf{Z}_p} \left( \sum_{\nu \in \mathbf{Z}_p} a_{\nu\omega} \overline{a}_{\nu\omega'} \right) e^{2\pi i (\omega - \omega') x} \\ &= \beta \sum_{\omega \in \mathbf{Z}_p} 1 = \beta p. \qquad (25.1)\end{aligned}$$

Now suppose that $n \geq 2$. Repeated application of Theorem 25.2 gives

$$\sum_{\nu_0 \in \mathbf{Z}_p} |A_{n,\nu}(x)|^2 = \beta^{n-1} \sum_{\nu \in \mathbf{Z}_p} |A_{1,\nu}(x)|^2 = \beta^n p.$$

■

**Corollary 25.4** *Let $n \geq 1$ and $p \geq 2$. Suppose $A = \sqrt{\beta} U$ for some unitary matrix $U$ and constant $\beta$. Then for all $\boldsymbol{\nu} \in \mathbf{Z}_p^n$*

$$|A_{n,\boldsymbol{\nu}}(x)| \leq \sqrt{p}\, \beta^{n/2}.$$

**Theorem 25.5** *Let $n \geq 1$ and $p \geq 2$. Suppose $A = \sqrt{\beta} U$ for some unitary matrix $U$ and constant $\beta$. Then for all $\boldsymbol{\nu} \in \mathbf{Z}_p^n$*

$$\int_0^1 |A_{n,\boldsymbol{\nu}}(x)|^2 \, dx = \beta^n.$$

**Proof.** The proof is by induction. Suppose $n = 1$. Then

$$A_{1,\nu} = \sum_{\omega \in \mathbf{Z}_p} a_{\nu\omega} e^{2\pi i \omega x}$$

so that by Parseval's theorem

$$\int_0^1 |A_{1,\nu}(x)|^2 \, dx = \sum_{\omega \in \mathbf{Z}_p} |a_{\nu\omega}|^2 = \beta.$$

Now since $n \geq 2$, then again by Parseval's theorem

$$\begin{aligned} \int_0^1 |A_{n,\nu}(x)|^2 \, dx &= \sum_{\omega \in \mathbf{Z}_p^n} |a_{\nu_0,\omega_{n-1}}|^2 \left| \prod_{k=1}^{n-1} a_{\nu_{n-k}+\omega_k,\omega_{k-1}} \right|^2 \\ &= \sum_{\omega_{n-1} \in \mathbf{Z}_p} |a_{\nu_0,\omega_{n-1}}|^2 \\ &\quad \cdot \sum_{\omega \in \mathbf{Z}_p^{n-1}} \left| \prod_{k=1}^{n-1} a_{\nu_{n-k}+\omega_k,\omega_{k-1}} \right|^2 . \end{aligned} \tag{25.2}$$

Applying Parseval's theorem to the inner sum gives

$$\int_0^1 |A_{n,\nu}(x)|^2 \, dx = \sum_{\lambda \in \mathbf{Z}_p} |a_{\nu_0,\lambda}|^2 \int_0^1 |A_{n-1,S(\nu,\lambda)}(x)|^2 \, dx.$$

By the induction hypothesis and the unitary nature of $U$ the last expression becomes

$$\int_0^1 |A_{n,\nu}(x)|^2 \, dx = \sum_{\lambda \in \mathbf{Z}_p} |a_{\nu_0,\lambda}|^2 \beta^{n-1} = \beta^n$$

■

The next theorem expresses the flatness property of $A_{n,\nu}$. It is an immediate consequence of Corollary 25.4 and Theorem 25.5.

**Theorem 25.6** *Let $n \geq 1$ and $p \geq 2$. Suppose $A = \sqrt{\beta}U$ for some unitary matrix $U$ and constant $\beta$. Then for $\boldsymbol{\nu} \in \mathbf{Z}_p^n$,*

$$\|A_{n,\nu}\|_\infty \leq \sqrt{p}\, \|A_{n,\nu}\|_2 \, .$$

## 25.3 Exponential Sums with Low Sidelobes

In this section we show how to use the $A_{n,\nu}$ polynomials to construct sparse exponential sums such that $F(k_1, \cdots, k_N) = O\left(N^{1/2}\right)$. Given a positive integer $p$, let the matrix $A_r$ be defined by having

$$a_{kj} = \exp\left(\frac{2\pi i r k j}{p}\right)$$

as the entry in row $k \in \{0, \cdots, p-1\}$ and column $j \in \{0, \cdots, p-1\}$. For $r = 1, \ldots, p-1$, $A_r$ is $\sqrt{p}$ times a unitary matrix and $A_0$ is a matrix of all 1's. Substitution into the definition of $A_{n,\nu}$ gives

$$A_{n,\nu}(x) = \sum_{\omega \in \mathbf{Z}_p^n} \exp\left(2\pi i r/p\right)(\langle R\boldsymbol{\nu}, \boldsymbol{\omega}\rangle + \langle Q\boldsymbol{\omega}, \boldsymbol{\omega}\rangle) \exp 2\pi i \langle \mathbf{p}, \boldsymbol{\omega}\rangle x \quad (25.3)$$

where $R\boldsymbol{\nu} = (\nu_{n-1}, \cdots, \nu_0)$ and $Q\boldsymbol{\omega} = (0, \omega_0, \cdots, \omega_{n-2})$. Let $A_{r,n,\nu}$ denote the polynomial $A_{n,\nu}$ determined by the matrix $A_r$. Now define for each $q = 0, \ldots, p-1$

$$H_{n,q,\nu}(x) = \frac{1}{p} \sum_{r=0}^{p-1} e^{-2\pi i r q/p} A_{r,n,\nu}(x).$$

**Theorem 25.7** . *Let $n \geq 1$ and $p$ be prime. Then each coefficient in $H_{n,q,\nu}$ is either 0 or 1. Moreover writing*

$$H_{n,q,\nu}(x) = \sum_{k=0}^{p^n - 1} h_{n,q,\nu}(k) e^{2\pi i k x}$$

*gives*

$$h_{n,q,\nu}(k) = \begin{cases} 1 \text{ if } \langle R\nu, \omega\rangle + \langle Q\omega, \omega\rangle = q \bmod p \\ 0 \text{ if } \langle R\nu, \omega\rangle + \langle Q\omega, \omega\rangle \neq q \bmod p \end{cases}$$

*where*

$$k = \omega_0 + \omega_1 p + \cdots + \omega_{n-1} p^{n-1}.$$

**Proof.** Let

$$\phi(\nu, k) = \langle R\nu, \omega\rangle + \langle Q\omega, \omega\rangle$$

where

$$k = \omega_0 + \omega_1 p + \cdots + \omega_{n-1} p^{n-1}.$$

Substituting (25.3) into the definition of $H_{n,q,\nu}$ gives

$$\begin{aligned} H_{n,q,\nu}(x) &= \frac{1}{p} \sum_{r=0}^{p-1} \sum_{k=0}^{p^n-1} \exp\left(\frac{2\pi i r \phi(\nu, k)}{p}\right) \\ &\quad \cdot \exp(2\pi i k x) \exp\left(\frac{-2\pi i r q}{p}\right) \end{aligned}$$

and interchanging the order of summation gives

$$H_{n,q,\nu}(x) = \sum_{k=0}^{p^n-1} \frac{1}{p} \sum_{r=0}^{p-1} \exp\left(\frac{2\pi i r \left(\phi(\nu, k) - q\right)}{p}\right) \exp(2\pi i k x).$$

Since $p$ is prime

$$\sum_{r=0}^{p-1} \exp\left(\frac{2\pi i r m}{p}\right) = \begin{cases} p \text{ if } m = 0 \bmod p \\ 0 \text{ if } m \neq 0 \bmod p \end{cases}.$$

This allows us to write

$$H_{n,q,\nu}(x) = \sum_{k=0}^{p^n-1} h_{n,q,\nu}(k) e^{2\pi i k x}$$

where

$$h_{n,q,\nu}(k) = \begin{cases} 1 \text{ if } \phi(\nu, k) = q \bmod p \\ 0 \text{ if } \phi(\nu, k) \neq q \bmod p \end{cases}.$$

■

Since $h_{n,q,\nu}(k)$ is 0 or 1, we can rewrite $H_{n,q,\nu}$ as

$$H_{n,q,\nu}(x) = \sum_{j=1}^{N} e^{2\pi i \xi_j (n,q,\nu) x}$$

where the $\xi_j$ are those integers $k$ for which $h_{n,q,\nu}(k) = 1$.

**Theorem 25.8** *Given $n \geq 1$, $p$ prime, and $\boldsymbol{\nu} \in \mathbf{Z}_p^n$, let*

$$H_{n,q,\nu}(x) = \sum_{j=1}^{N} e^{2\pi i \xi_j(n,q,\nu)x}$$

*be as defined above. Then*

$$|H_{n,q,\nu}(x)| \leq \left| \frac{\sin(\pi p^n x)}{p \sin(\pi x)} \right| + (p-1)\sqrt{p^{n-1}}$$

**Proof.**

$$\begin{aligned} H_{n,q,\nu}(x) &= \frac{1}{p} \sum_{r=0}^{p-1} e^{-2\pi i r q/p} A_{r,n,\nu}(x) \\ &= \frac{1}{p} A_{0,n,\nu}(x) + \frac{1}{p} \sum_{r=1}^{p-1} e^{-2\pi i r q/p} A_{r,n,\nu}(x) \end{aligned}$$

Since $A_0$ is a matrix consisting entirely of 1's

$$A_{0,n,\nu}(x) = \sum_{k=0}^{p^n-1} e^{2\pi i k x} = \frac{e^{2\pi i p^n x} - 1}{e^{2\pi i x} - 1}.$$

Hence

$$|A_{0,n,\nu}(x)| = \left| \frac{\sin \pi p^n x}{\sin \pi x} \right|.$$

For $1 \leq r \leq p-1$ Corollary 25.4 shows that

$$|A_{r,n,\nu}(x)| \leq p^{(n+1)/2}.$$

Therefore

$$\left| \frac{1}{p} \sum_{r=1}^{p-1} A_{r,n,\nu}(x) e^{2\pi i r q/p} \right| \leq \left( \frac{p-1}{p} \right) p^{(n+1)/2} = (p-1)p^{(n-1)/2}$$

and

$$|H_{n,q,\nu}(x)| \leq \left| \frac{\sin \pi p^n x}{p \sin \pi x} \right| + (p-1)p^{(n-1)/2}.$$

■

Expressing the polynomials $H_{n,q,\nu}(x)$ in form

$$H_{n,q,\nu}(x) = \sum_{j=1}^{N} e^{2\pi i k_j x}$$

it remains to determine $N$ or, equivalently, it remains to determine $|S_{n,q,\nu}|$ where $H_{n,q,\nu}(x)$ is expressed as

$$H_{n,q,\nu}(x) = \sum_{k \in S_{n,q,\nu}} c^{2\pi i k x}.$$

The next theorem shows that $N$ is asymptotic to $p^{n-1}$. Since $S_{n,q,\nu} \subset \{0, \cdots, p^n - 1\}$, the density of $S_{n,q,\nu}$ is therefore asymptotic to $1/p$.

**Theorem 25.9** *Let $n \geq 1$, $p$ be prime and $\boldsymbol{\nu} \in \mathbf{Z}_p^n$. For each $q = 0, \ldots, p-1$ define a set of integers $S_{n,q,\nu}$ by*

$$S_{n,q,\nu} = \{\, k \,|\, 0 \leq k \leq p^n - 1,\ \phi(k, \boldsymbol{\nu}) \equiv q \bmod p\}$$

*where*

$$\phi(k, \boldsymbol{\nu}) = \langle R\boldsymbol{\nu}, \boldsymbol{\omega} \rangle + \langle Q\boldsymbol{\omega}, \boldsymbol{\omega} \rangle$$

*and*

$$k = \omega_0 + \omega_1 p + \cdots + \omega_{n-1} p^{n-1}.$$

*Letting $|S_{n,q,\nu}|$ denote the number of elements in $S_{n,q,\nu}$, we have*

$$p^{n-1} - (p-1)^2 p^{(n-1)/2} \leq |S_{n,q,\nu}| \leq p^{n-1} + (p-1)p^{(n-1)/2}$$

*for all $0 \leq q \leq p-1$, and*

$$p^{n-1} \leq |S_{n,q,\nu}| \leq p^{n-1} + (p-1)p^{(n-1)/2}$$

*for at least one $q$.*

**Proof.** First we establish the upper bound in the assertion. Since we can write

$$H_{n,q,\nu}(x) = \sum_{k \in S_{n,q,\nu}} e^{2\pi i k x}$$

it follows that

$$\begin{aligned} |S_{n,q,\nu}| &= H_{n,q,\nu}(0) \\ &\leq p^{n-1} + (p-1)p^{(n-1)/2} \end{aligned} \tag{25.4}$$

where the last inequality is obtained by applying theorem (25.5).

It is clear from the definition of $S_{n,q,\nu}$ that for fixed $n$ and $\boldsymbol{\nu}$ these sets are disjoint as $q$ ranges over $0, \ldots, p-1$. It is also clear that

$$\{0, \ldots, p^n - 1\} = \bigcup_{j=0}^{p-1} S_{n,j,\nu}$$

and therefore that

$$p^n = \sum_{j=0}^{p-1} |S_{n,j,\nu}| . \tag{25.5}$$

Choosing any $q$, we can rearrange (25.5) and write

$$|S_{n,q,\nu}| = p^n - \sum_{j \neq q} |S_{n,j,\nu}| .$$

Using the bound in (25.4) gives

$$\begin{aligned} |S_{n,q,\nu}| &\geq p^n - (p-1) \\ &= p^{n-1} - (p-1)^2 p^{(n-1)/2} \end{aligned}$$

which establishes the first assertion of the theorem. As for the second assertion, note that if every $|S_{n,j,\nu}|$ in the right hand side of (25.5) satisfied $|S_{n,j,\nu}| < p^{n-1}$, then (25.5) would give

$$p^n = \sum_{j=0}^{p-1} |S_{n,j,\nu}| < p$$

which is a contradiction. ■

## *REFERENCES*


[1] G. Benke, *Generalized Rudin-Shapiro systems,* J. of Fourier Anal. and Appl. **1** (1994), 87–101.

[2] G. Benke and W.J. Hendricks, *Estimates for large deviations in random trigonometric polynomials,* Siam. J. Math. Anal. **24** (1993), 1067–1085.

[3] J. S. Byrnes, *Quadrature mirror filters, low crest factor arrays, functions achieving optimal uncertainty principle bounds, and complete orthonormal sequences–a unified approach,* J. of Appl. and Comp. Harmonic Anal. **1** (1994), 261–266.

[4] M. J. E. Golay, *Multislit spectrometry,* J. Optical Soc. Amer. **39** (1949), 437.

[5] *Complementary series,* IRE Trans. Inform. Theory **IT-7** (1961), 82–87.

[6] I. Habib, L. Turner *New class of M-ary communication systems using complementary sequences,* IEE Proceedings **3 Pt. F** (1986), 293–300.

[7] W. J. Hendricks, *The Totally Random Versus the Bin Approach for Random Arrays,* IEEE Trans. Ant. and Prop. **39** (1991), 1757–1762.

[8] Y.T. Lo and V.D. Agrawal, *Distribution of sidelobe level in random arrays,* Proc. IEEE **57** (1969), 1764–1765.

[9] W.G. Rudin, *Some theorems on Fourier Coefficients,* Proc. Amer. Math. Soc. **36** (1959) 855–859.

[10] H.S. Shapiro, *Extremal problems for polynomials and power series,* Thesis, Massachussetts Institute of Technology, (1957).

[11] R. Sivaswamy, *Multiphase complementary codes,* IEEE Trans. on Inform. Theory **IT-24** (1978) 564–552.

[12] B.D.Steinberg, *The peak sidelobe of the phased array having randomly located elements,* IEEE Trans. Antennas and Propagation **AP-20** (1972). 129–136.

[13] –, sl Principles of aperture and array system design, New York: Wiley, 1976.

[14] Y. Taki, H. Miyakawa, M. Hatori, S. Namba, *Even-shift orthogonal sequences,* IEEE Trans. on Inform. Theory **IT-15** (1969), 295–300.

[15] C. C. Tseng, C. L. Lui, *Complementary sets of sequences,* IEEE Trans. on Inform. Theory **IT-18** (1972) 664–651.

# Chapter 26

# Spline type summability for multivariate sampling

**W. R. Madych**

*ABSTRACT. The introductory material briefly describes certain mathematical aspects of signal processing which are centered around the celebrated Whittaker-Kotelnikov-Shannon sampling theorem and the spline type summability method for cardinal series developed by Schoenberg. The bulk of this article is devoted to an exposition of a version of the spline summability method for recovering multivariate band limited functions from discrete lattice samples which uses polyharmonic splines. The pertinent theory of such splines is outlined and the details of the reconstruction method, whose range of application includes data which may grow at infinity, are developed. Some of the proofs and results are new, even when reduced to the univariate case. Certain related material, including generalizations and numerical aspects, is also briefly discussed.*

## 26.1 Introduction

### *26.1.1 Sampling theory*

Sampling theory is a fairly well established component of the mathematical aspect of signal processing, for example see [2, 3, 5, 6, 9, 22, 28, 46, 47, 17, 50, 52, 53, 58, 60, 65]. For the most part, basic sampling theory is concerned with questions involving the sampling and reconstruction of members of the class of entire functions of exponential type. For an interesting essay on the relationship between such functions and the signals found in nature see [58].

A complex valued function $F$ of the complex variable $z$ is called an *entire function of exponential type* if it is holomorphic in the whole complex plane and for some $p$, $p < \infty$, and positive constants $A$ and $\sigma$ satisfies the bound

$$|F(z)| \leq A(1 + |z|)^p \exp(\sigma|\text{Im}z|) \qquad (26.1)$$

for all $z$ where Im$z$ is the imaginary part of $z$. This class of functions[1] is often parametrized by the parameter $\sigma$. Thus when the bound (26.1) holds

[1] The usual definition of exponential type is somewhat more general, see [4, 30, 59].

for some specific value of $\sigma$ the function $F$ is said to be of *type* $\sigma$ and $E_\sigma$ denotes the class of all such functions $F$.

When restricted to the real line, in view of (26.1), an entire function of exponential type $F$ may be regarded as a tempered distribution and thus has a Fourier transform $f$ which is also a tempered distribution. By virtue of a celebrated theorem due to Paley and Wiener the distribution $f$ is supported in the interval $[-\sigma, \sigma]$. In the language of engineering, $F$ is frequency *band limited* to the band $[-\sigma, \sigma]$. Since the inverse Fourier transform $F$ of any distribution $f$ whose support is in the interval $[-\sigma, \sigma]$ can be extended to the whole complex plane as an entire function which satisfies a bound like (26.1), it should be clear that the class of entire functions of exponential type can also be characterized as the class of frequency band limited functions.

A subclass of exponential type functions often encountered in sampling theory is the so-called Paley-Wiener class, parametrized by $\sigma$ and denoted by $PW_\sigma$, which may be defined as follows: The Paley-Wiener class $PW_\sigma$ consists of those elements $F$ in $E_\sigma$ whose restriction to the real line is square integrable; in other words, $PW_\sigma = E_\sigma \bigcap L^2(\mathbb{R})$.

In what follows, to avoid unnecessary notational complexity, we restrict our attention to type $\pi$. All the conclusions recorded below can be easily reformulated for general type $\sigma$ by using a simple scaling argument.

A fundamental result in sampling theory concerns the class $PW_\pi$ and can be expressed as follows:

**(F1)** If $F$ is in $PW_\pi$ then the sequence of values $\{F(m)\}_{m\in\mathbb{Z}}$ is in $l^2(\mathbb{Z})$.

**(F2)** If the sequence $\{y_m\}_{m\in\mathbb{Z}}$ is in $l^2(\mathbb{Z})$ then there is a unique $F$ in $PW_\pi$ such that $F(m) = y_m$ for all $m$.

**(F3)** If $F$ is in $PW_\pi$ it may be recovered from its sequence of values $\{F(m)\}_{m\in\mathbb{Z}}$ via the formula

$$F(z) = \sum_{m=-\infty}^{\infty} F(m) \frac{\sin \pi(z-m)}{\pi(z-m)} . \tag{26.2}$$

For real $x$ the partial sums of the right hand side of (26.2) with $z = x$ converge to $F(x)$ uniformly in $x$ and in $L^2(\mathbb{R})$.

Classically, formula (26.2) is known as the Whittaker cardinal series and the function

$$\frac{\sin \pi z}{\pi z}$$

The class of functions defined to be of exponential type here is exactly the class for which the general Paley-Wiener Theorem is valid, see [23].

is known as the cardinal sine, *sinus cardinalis*, or sinc. In contemporary literature this formula is often referred to as the Whittaker-Kotelnikov-Shannon [62, 52, 28] sampling theorem.

In view of the Paley-Wiener theorem alluded to above, the $L^2$ mapping properties of the Fourier transform, and properties of Fourier series, this result may seem quite superficial. To wit, every $F$ in $PW_\pi$ enjoys the representation

$$F(z) = \frac{1}{2\pi} \int_{-\pi}^{\pi} f(\xi) e^{iz\xi} d\xi \tag{26.3}$$

where $f$ is in $L^2([-\pi, \pi])$. Such an $f$ uniquely determines and, in turn, is uniquely determined[2] by the sequence of values

$$y_m = \frac{1}{2\pi} \int_{-\pi}^{\pi} f(\xi) e^{im\xi} d\xi = F(m) , \quad m \in \mathbb{Z}.$$

This implies (F1) and (F2). Item (F3) follows from the fact that

$$f(\xi) = \left\{ \sum_{m=-\infty}^{\infty} F(m) e^{-im\xi} \right\} \chi(\xi)$$

where $\chi$ is the characteristic (or indicator) function of the interval $[-\pi, \pi]$ and the partial sums converge to $f$ in the $L^2$ sense, the formula

$$\frac{\sin \pi(z-m)}{\pi(z-m)} = \frac{1}{2\pi} \int_{-\pi}^{\pi} e^{i(z-m)\xi} d\xi ,$$

and identity (26.3).

On the other hand, without the benefit of representation (26.3), the rationale for using the interpolation formula (26.2) is not quite so transparent. Thus Whittaker's original argument for using (26.2) found in [62] and other early work on the subject[3] can be quite enlightening.

In spite of its importance, formula (26.2) has several annoying peculiarities which restrict its applicability. For example:

- It does not reproduce polynomials. Indeed, the right hand side of (26.2) does not appear to make much sense when the values $F(m)$ of such functions are inserted. On the other hand, since polynomials

[2] In the $L^2$ sense of course; that is, almost everywhere.

[3] References to some of the early work can be found in the interesting historical article [5]. Also see [22, 65].

are the archetypal examples of band limited functions and of entire functions of exponential type, it seems not unreasonable to expect that a procedure purporting to reconstruct a meaningful subclass of entire functions should, at the very least, reconstruct polynomials.

- Because of the slow decay of the cardinal sine at infinity, values $F(m)$ corresponding to $m$'s far from $x$ can significantly effect the value of $F(x)$. Of course the basic hypothesis of analyticity forces this sort of behavior. Nevertheless, in view of the fact that in various applications the data $\{F(m)\}$ can be noisy and/or truncated, the problem of reducing this effect is of some importance.

In what follows we will outline an alternate reconstruction method.

### *26.1.2 Splines and sampling theory*

In 1946 Schoenberg [54] published an extensive paper concerning approximations of formula (26.2) which included piecewise polynomial interpolatory splines[4]. Such splines provide significant relief from the annoyances associated with representation (26.2) mentioned above.

A piecewise polynomial spline of order $2k$, $k = 1, 2, \ldots$, with knots on the integer lattice $\mathbb{Z}$ is a function $S$ which is $2k-2$ times continuously differentiable and on the complement of $\mathbb{Z}$ satisfies

$$\frac{d^{2k}S(x)}{dx^{2k}} = 0\,,$$

in other words, $S$ coincides with a polynomial of degree no greater than $2k-1$ when restricted to any interval of the form $[m, m+1]$, $m \in \mathbb{Z}$. A convenient expository account of such splines may be found in [55]. Here we briefly outline some of the basic facts which are germane to our development.

If the continuous function $F(x)$, $x \in \mathbb{R}$, grows no faster than a power of $|x|$ as $x \to \pm\infty$ then there is a piecewise polynomial spline of order $2k$ with knots on the integer lattice $\mathbb{Z}$, call it $S_k F$, which interpolates $F$ on $\mathbb{Z}$, that is $S_k F(m) = F(m)$ for all $m$ in $\mathbb{Z}$, and which grows no faster than a power of $|x|$ as $x \to \pm\infty$. The spline $S_k F$ is unique and enjoys the representation

$$S_k F(x) = \sum_{m=-\infty}^{\infty} F(m) L_k(x-m)\,. \tag{26.4}$$

---

[4]It should be mentioned that most of Schoenberg's publications which are cited here, together with many other interesting articles, have been reproduced in [13].

Here $L_k$ is the unique piecewise polynomial spline of order $2k$ with knots on the integer lattice $\mathbb{Z}$ which satisfies

$$L_k(m) = \begin{cases} 1 & \text{if } m = 0 \\ 0 & \text{if } m \in \mathbb{Z} \setminus \{0\} \end{cases} \tag{26.5}$$

and grows no faster than a power of $|x|$ as $x \to \pm\infty$. This "Lagrange like" or *fundamental spline* $L_k$ enjoys the property that for positive constants $A$ and $a$, which depend on $k$,

$$|L_k(x)| \leq Ae^{-a|x|} \tag{26.6}$$

so that (26.4) is well defined whenever $F(m)$ grows no faster than a power of $|m|$ as $m \to \pm\infty$.

In [54] Schoenberg detailed the use of (26.4) as a substitute or approximation of (26.2) and subsequently continued to refine various aspects of the theory, for example see [55, 13]. Among the results he developed along the way was a spline summability method for the recovery of certain classes of band limited functions from their values on the integer lattice. A significant portion of these results were recorded in [56] and concern those functions $F$ which have a derivative of some order in $PW_\pi$. This class[5] of functions is considerably larger than $PW_\pi$ and, in fact, includes all the polynomials. We briefly summarize some of this material, while avoiding technical details, as follows:

**(S)** Suppose that the derivative of order $m$ of $F$,

$$F^{(m)}(x) = \frac{d^m F(x)}{dx^m},$$

is in $PW_\pi$. Then the data sequence $y_j = F(j)$, $j \in \mathbb{Z}$, can be completely characterized. In particular such data can grow no faster than a power of $|j|$ as $j \to \pm\infty$ so that $S_k F(x)$ is well defined for each $k$ via formula (26.4). These splines converge to $F$, that is,

$$F(x) = \lim_{k\to\infty} S_k F(x) \tag{26.7}$$

uniformly in $x$. Futhermore, for any positive integer $p$ and sufficiently large $k$, the derivative of order $p$ of $S_k F$ exists and

$$F^{(p)}(x) = \lim_{k\to\infty} (S_k F)^{(p)}(x) \tag{26.8}$$

uniformly in $x$.

Note that the summability method summarized by (26.4) and (26.7) is sufficiently general to recover any polynomial $F$ and, for fixed $k$, the exponential decay of $L_k(x)$ precludes the instability associated with the evaluation of the cardinal series (26.2) caused by the slow decay of the sinc.

[5] Other classes have also been considered, for example see [51, 57].

### *26.1.3 Contents, notation, and acknowledgements*

There are many results and theories which may be considered to be generalizations, logical extensions, or natural continuations of the material outlined above. In this article we restrict our attention to following:

- The central topic, taken up in Section 2, concerns the recovery of multivariate band limited functions $F$ from data consisting of samples $F(j)$ where $j$ ranges over a lattice. The method of recovery is a spline summability procedure analogous to that described above in the univariate case.

  Some of the results reduce to those summarized by item (S) and recorded by Schoenberg [56] in the univariate case. On the other hand various important tools used in the univariate case are not available in the general multivariate senario. For example, in the case of more than one variable the polyharmonic splines which we use do not have compactly supported basis functions, an important feature in Schoenberg's work. Thus significant details of the development are completely different from those found in [56]. Indeed, when reduced to the univariate case, our development provides new alternate proofs for some of the theorems in [56].

  On the other hand some of results, such as the statement concerning the $L^2$ convergence of the splines and their derivatives, are new even in the univariate case.

- In Section 3 we briefly deal with several "loose ends". The specific topics can be easily identified by the sub and subsubsection headings. Some of this matter is of a tutorial nature but there is no attempt to achieve any sort of completeness.

We use standard mathematical notation as found in, for example, [23, 37, 59] and simply advise the reader of the normalization of the Fourier transform used here. If $\phi$ is in the Schwartz space of rapidly decreasing functions on $\mathbb{R}^n$, $\mathcal{S}(\mathbb{R}^n)$, then its Fourier transform $\mathcal{F}\phi = \hat{\phi}$ is defined by

$$\hat{\phi}(\xi) = \int_{\mathbb{R}^n} e^{-i\langle \xi, x\rangle} \phi(x)dx$$

so that the Fourier inversion formula reads

$$\phi(x) = \frac{1}{(2\pi)^n} \int_{\mathbb{R}^n} e^{i\langle x, \xi\rangle} \hat{\phi}(\xi)d\xi \ .$$

This article contains details of some of the material which I lectured on at the IWAA conference in Orono in June. I thank Bill Bray for inviting me to give that lecture and allowing me to publish this article. Karlheinz Gröchenig alerted me to the recently published book [22] on sampling theory.

# 26.2 Regular sampling of multivariate functions and their recovery via splines

### *26.2.1 Band limited functions and polyharmonic splines*

In the $n$-variate case there are many ways of defining classes of entire functions of exponential type. We are interested in those classes whose members have Fourier transforms which are supported in the parallelpiped (or hypercube) in $I\!R^n$ defined by

$$I_\sigma = \{\xi = (\xi_1, \dots \xi_n) : |\xi_m| \leq \sigma, m = 1, \dots, n\} \tag{26.9}$$

and parametrized by $\sigma, \sigma > 0$. Now, if $f$ is a distribution with support in $I_\sigma$ then its inverse Fourier transform $F$ is analytic with a holomorphic extension to all of complex Euclidean $n$-space $\mathbb{C}^n$ as an entire function which satisfies the bound

$$|F(z)| \leq A(1 + |z|)^p \exp(\sigma\{\sum_{m=1}^{n} |\text{Im} z_m|\}) \tag{26.10}$$

for some $p, p < \infty$, and positive constant $A$.

Conversely, if $F$ is an entire function of the variable $z = (z_1 \dots, z_n)$ in $\mathbb{C}^n$ which satisfies (26.10) then for all $x$ in $I\!R^n$

$$|F(x)| \leq A(1 + |x|)^p \tag{26.11}$$

where $A$ and $p$ are constants inherited from (26.10). The bound (26.11) implies that the restriction of $F$ to $I\!R^n$ is a tempered distribution and thus has a Fourier transform. It is a consequence of an extension of the Paley-Wiener theorem alluded to earlier, that the Fourier transform of $F$ is a distribution $f$ with support in $I_\sigma$.

The class of $n$-variate entire functions $F$ which satisfy the bound (26.10) for a given value of $\sigma$ is denoted by $E_\sigma(\mathbb{C}^n)$. Equivalently, $E_\sigma(\mathbb{C}^n)$ is the class of those tempered distributions $F$ whose Fourier transforms are supported in $I_\sigma$. In other words it is the class of those functions $F$ which are frequency band limited to $I_\sigma$.

If $F$ is an entire function then the sequence of values $\{F(j)\}_{j \in \mathbb{Z}^n}$ is well defined. Furthermore if $F$ is in the Paley-Wiener class $PW_\pi(I\!R^n) = E_\pi(\mathbb{C}^n) \cap L^2(I\!R^n)$ then a result analogous to that described by $(F1)$, $(F2)$, and $(F3)$ in the introduction is valid. In particular, the $n$-variate version of the Whittaker-Kotelnikov-Shannon sampling theorem holds in this case, for example see [19, 20, 22, 47, 50, 65]; it reads

$$F(z) = \sum_{j \in \mathbb{Z}^n} F(j) \text{sinc}(z - j) \tag{26.12}$$

where

$$\operatorname{sinc}(z) = \operatorname{sinc}(z_1, \dots, z_n) = \prod_{m=1}^{n} \frac{\sin(\pi z_m)}{\pi z_m}.$$

Formula (26.12) suffers from the same inadequacies as its univariate version. In what follows we outline the details of an $n$-variate counterpart of the spline summability method developed by Schoenberg in the univariate case. We begin by defining an $n$-variate analogue of the piecewise polynomial splines of even order with knots on the integer lattice.

If $k$ is a positive integer an $n$-variate *$k$-harmonic spline* with knots on $\mathbb{Z}^n$ is a tempered distribution $u$ such that $\Delta^k u$ is a Radon measure supported on the integer lattice $\mathbb{Z}^n$ in $\mathbb{R}^n$. Symbolically

$$\Delta^k u(x) = \sum_{j \in \mathbb{Z}^n} a_j \delta(x - j) \tag{26.13}$$

where $\Delta$ is the $n$-variate Laplacian

$$\Delta u = \sum_{m=1}^{n} D_m^2 u = \sum_{m=1}^{n} \frac{\partial^2 u}{\partial x_m^2},$$

$\Delta^k u = \Delta \Delta^{k-1}$ for $k > 1$, and $\delta(x)$ is the unit Dirac measure supported at the origin of $\mathbb{R}^n$. The sequence $\{a\} = \{a_j\}_{j \in \mathbb{Z}^n}$ is a sequence of scalars of course. The class of $k$-harmonic splines with knots on $\mathbb{Z}^n$ is denoted by $SH_k(\mathbb{R}^n)$. A *polyharmonic spline* is one which is $k$-harmonic for some $k$.

The basic properties of these splines are recorded in [37, 38]. Here we merely remind the reader of the following.

**Proposition 26.1** *If $2k \geq n + 1$ and $u$ is in $SH_k(\mathbb{R}^n)$ then*

**(i)** *$u$ is in the continuity class $C^{2k-n-1}(\mathbb{R}^n)$,*

**(ii)** *$\Delta^k u = 0$ on $\mathbb{R}^n \setminus \mathbb{Z}^n$*

**(iii)** *There is a number $p, p < \infty$, such that*

$$M_p(u) = \sup_{x \in \mathbb{R}^n} \frac{|u(x)|}{(1 + |x|)^p}$$

*is finite.*

In view of this proposition it should be clear that in the case $n = 1$ the class $SH_k(\mathbb{R}^n)$ is simply the class of piecewise polynomial splines of order $2k$ with knots on $\mathbb{Z}^n$ which grow no faster than a power of $|x|$ as $x \to \pm\infty$.

$\mathcal{Y}^p$ is the class of those scalar sequences $\{y\} = \{y_j\}_{j \in \mathbb{Z}^n}$ such that

$$N_p(\{y\}) = \sup_{j \in \mathbb{Z}^n} \frac{|y_j|}{(1 + |j|)^p}$$

is finite.

$$\mathcal{Y}^{\infty} = \bigcup_p \mathcal{Y}^p \quad \text{and} \quad \mathcal{Y}^{-\infty} = \bigcap_p \mathcal{Y}^p$$

where the union and intersection respectively are taken over all $p$, $-\infty < p < \infty$. The connection between polyharmonic splines and sequences in $\mathcal{Y}^{\infty}$ is related to the interpolation problem, that is, given a sequence $\{y\}$ in $\mathcal{Y}^{\infty}$, the problem of finding an element $u$ in $SH_k(\mathbb{R}^n)$ such that

$$u(j) = y_j \quad \text{for all } j \text{ in } \mathbb{Z}^n.$$

To this end we recall that the function $\Lambda_k$ defined via the formula for its Fourier transform[6] by

$$\hat{\Lambda}_k(\xi) = \frac{|\xi|^{-2k}}{\sum_{j\in\mathbb{Z}^n} |\xi - 2\pi j|^{-2k}} \tag{26.14}$$

is well-defined when $2k \geq n+1$, is in $SH_k(\mathbb{Z}^n)$, and satisfies

$$\Lambda_k(j) = \begin{cases} 1 & \text{if } j = 0 \\ 0 & \text{if } j \in \mathbb{Z}^n \setminus \{0\}. \end{cases} \tag{26.15}$$

Furthermore, there are positive constants $A$ and $a$ such that for all $x$ in $\mathbb{R}^n$

$$|\Lambda_k(x)| \leq Ae^{-a|x|}. \tag{26.16}$$

**Proposition 26.2** *Suppose $2k \geq n+1$ and $\{y\} = \{y_j\}_{j\in\mathbb{Z}^n}$ is in $\mathcal{Y}^p$ for some $p$, $p < \infty$. Then there is a unique $k$-harmonic spline, call it $S_k\{y\}$, which interpolates $\{y\}$ on $\mathbb{Z}^n$. This spline enjoys the representation*

$$S_k\{y\}(x) = \sum_{j\in\mathbb{Z}^n} y_j \Lambda_k(x-j). \tag{26.17}$$

*The series converges absolutely and uniformly in every compact subset of $\mathbb{R}^n$. Furthermore, $M_p(S_k\{y\}) < \infty$.*

In view of (26.11) and Proposition 26.2 it should be clear that, given the values $\{F(j)\}_{j\in\mathbb{Z}^n}$, of an entire function of exponential type $F$ then there is a unique $k$-harmonic spline, call it $S_kF$, which interpolates these values and enjoys the representation

$$S_kF(x) = \sum_{j\in\mathbb{Z}^n} F(j)\Lambda_k(x-j) \tag{26.18}$$

[6] Note that a different normalization was used for the Fourier transform in [35, 36, 37, 38] and other related articles.

for $x$ in $\mathbb{R}^n$. Furthermore, since in view of (26.14) it is not difficult to verify that

$$\lim_{k\to\infty} \Lambda_k(x) = \operatorname{sinc}(x) \tag{26.19}$$

uniformly in $x$ on $\mathbb{R}^n$, it is not unreasonable to suspect that

$$\lim_{k\to\infty} S_k F(x) = F(x) \tag{26.20}$$

in some sense for some subclass of functions $F$ in $E_\pi(\mathbb{C}^n)$. In [36] it was shown that this is indeed the case. To describe the result recorded there we have to consider a technical point.

If $f$ is a distribution with support in $I_\pi$ then its $2\pi\mathbb{Z}^n$ periodization $\wp f$ is well defined via the formula

$$\langle \wp f, \phi \rangle = \langle f, \wp\phi \rangle$$

where $\phi$ is any test function in the Schwartz class $\mathcal{S}(\mathbb{R}^n)$,

$$\wp\phi(\xi) = \sum_{j\in\mathbb{Z}^n} \phi(\xi - 2\pi j),$$

and $\langle f, \phi \rangle$ denotes the distribution $f$ evaluated at the test function $\phi$.

Next, observe that if the distribution $f$ has the property that its restriction to some neighborhood of the boundary of $I_\pi$ is a distribution of order 0 then, if $\chi$ is the characteristic or indicator function of $I_\pi$, the product $\chi(\xi)\wp f(\xi)$ is a well defined distribution with support in $I_\pi$.

**Proposition 26.3** *Suppose $F$ is in $E_\pi(\mathbb{C}^n)$, the restriction of its Fourier transform $f$ to some neighborhood of the boundary of $I_\sigma$ is a distribution of order 0, and $\chi(\xi)\wp f(\xi) = f(\xi)$. Then*

$$\lim_{k\to\infty} S_k F(x) = F(x)$$

*uniformly in $x$ on compact subsets of $\mathbb{R}^n$.*

Examples show that the technical condition concerning the behavior of $f$ near the boundary of $I_\sigma$ cannot be significantly relaxed. On the other hand there are relatively simple conditions one may impose on $F$ which imply the technical condition concerning the boundary of $I_\sigma$. For example, we may state the following.

**Corollary 26.4** The conclusions of Proposition 26.3 are valid whenever one of the following holds:
*(i)*$F$ is a polynomial.
*(ii)*$F$ is in $E_\sigma(\mathbb{C}^n)$ for $\sigma < \pi$.
*(iii)*$\Delta^k F$ is in $E_\pi(\mathbb{C}^n) \cap L^2(\mathbb{R}^n)$ for some integer $k$.

In the case that $F$ is a polynomial it should be clear that for $k$ sufficiently large $S_k F = F$. Also note that any polynomial enjoys properties (ii) and (iii).

Although Proposition 26.3 is quite general its usefulness is limited by the fact that the data $\{F(j)\}_{j\in\mathbb{Z}^n}$ corresponding to functions $F$ which satisfy its hypothesis has not yet been adequately characterized. On the other hand, the data of functions $F$ which satisfy condition (iii) of the Corollary can be characterized and even significantly stronger convergence results can be obtained. These are detailed in Subsections 2.3 and 2.4. First however, in Subsection 2.2, we review the variational and related properties of polyharmonic splines necessary for this development.

## *26.2.2 The spaces $L^{2,k}(\mathbb{R}^n)$ and $l^{2,k}(\mathbb{Z}^n)$ and the variational properties of polyharmonic splines*

The linear space $L^{2,k}(\mathbb{R}^n)$ is the class of those tempered distributions $u$ all of whose $k$th order derivatives are square integrable; in other words

$$L^{2,k}(\mathbb{R}^n) = \{u \in \mathcal{S}'(\mathbb{R}^n) : D^\nu u \text{ is in } L^2(\mathbb{R}^n) \text{ for all } |\nu| = k\}.$$

For this space a semi-inner product is given by

$$\langle u, v\rangle_{L^{2,k}} = \sum_{|\nu|=k} c_\nu \int_{\mathbb{R}^n} D^\nu u(x)\overline{D^\nu v(x)}dx, \tag{26.21}$$

where the positive constants $c_\nu$ are specified by

$$|\xi|^{2k} = \sum_{|\nu|=k} c_\nu \xi^{2\nu}. \tag{26.22}$$

The semi-norm corresponding to (26.21) is denoted by $||u||_{L^{2,k}}$, thus

$$||u||^2_{L^{2,k}} = \langle u, u\rangle_{L^{2,k}}.$$

The null space of this semi-norm is $\mathcal{P}_{k-1}(\mathbb{R}^n)$, the class of polynomials of degree less than or equal to $k-1$. The space $\mathcal{P}_{k-1}(\mathbb{R}^n)$ is finite dimensional: we denote its dimension by $\dim\mathcal{P}_{k-1}(\mathbb{R}^n)$.

We regard the space $L^{2,k}(\mathbb{R}^n)$ as a subspace of tempered distributions determined by the finiteness of a certain semi-norm, not as a collection of equivalence classes. In particular, if $2k > n$ then the elements of $L^{2,k}(\mathbb{R}^n)$ are continuous functions. Most of the properties of $L^{2,k}(\mathbb{R}^n)$ specific to our development can be found in [38]. For the reader's convenience, in the next proposition, we record several properties which are particularly important in what follows.

First we recall that a subset $\Omega$ of $\mathbb{R}^n$ is said to be *unisolvent* for $\mathcal{P}_{k-1}(\mathbb{R}^n)$ if for any scalar valued function $\phi(x)$ on $\Omega$ there is a unique polynomial $P$ in $\mathcal{P}_{k-1}(\mathbb{R}^n)$ such that $P(x) = \phi(x)$ for all $x$ in $\Omega$. Such a set is necessarily finite; it contains $\dim\mathcal{P}_{k-1}(\mathbb{R}^n)$ elements.

**Proposition 26.5** *If $2k \geq n+1$ then the elements of $L^{2,k}(\mathbb{R}^n)$ are continuous functions and there is a linear map* $\mathbf{P}$ *on $L^{2,k}(\mathbb{R}^n)$ with the following properties:*

**(i)** $\mathbf{P}u$ *is in* $\mathcal{P}_{k-1}(\mathbb{R}^n)$

**(ii)** $\mathbf{P}^2 u = \mathbf{P}u$

**(iii)** *If $\Omega$ is a unisolvent set for $\mathcal{P}_{k-1}(\mathbb{R}^n)$ then*

$$|\mathbf{P}u(x)| \leq C(1+|x|^{k-1})(||u||_{L^{2,k}} + ||u||_\Omega)$$

*where $||u||_\Omega$ denotes the maximum of $u$ on $\Omega$ and $C$ is a constant which depends on $\Omega$ but is independent of $u$.*

**(iv)** *If $\mathbf{Q}u = u - \mathbf{P}u$ then $\mathbf{Q}u$ is continuous and satisfies*

$$||\mathbf{Q}u||_{L^{2,k}} = ||u||_{L^{2,k}}$$

*and*

$$|\mathbf{Q}u(x)| \leq C(1+|x|^k)||u||_{L^{2,k}}$$

*where $C$ is a constant independent of $u$.*

Another property of $L^{2,k}(\mathbb{R}^n)$ which plays a significant role in our development is a routine consequence of Theorem 1 in [34]. Here we record the following convenient formulation.

**Proposition 26.6** *Suppose $2k \geq n+1$, $u$ is an element of $L^{2,k}(\mathbb{R}^n)$, and $u(j) = 0$ for all $j$ in $\mathbb{Z}^n$. Then there is a constant $c$, independent of $k$ and $u$, such that*

$$||u||_{L^2} \leq c^k ||u||_{L^{2,k}}.$$

A discrete analogue of $L^{2,k}(\mathbb{R}^n)$ is the class of sequences $l^{2,k}(\mathbb{Z}^k)$ which may be described as follows.

Given a sequence $\{y\} = \{y_j\}_{j\in\mathbb{Z}^n}$ indexed by $\mathbb{Z}^n$, for each $m$, $m = 1, \ldots n$, $\{T_m y\}$ is the sequence whose components are

$$(T_m y)_j = y_{j+e_m} - y_j, \quad \text{for all } j \in \mathbb{Z}^n$$

where $e_m$ is the $n$-tuple with 1 in the $m$th slot and 0 elsewhere. In other words $T_m$ is a difference operator in the direction $e_m$ and, in this context, $T = (T_1, \ldots, T_n)$ is a discrete version of the gradient $D$. For any multi-index $\nu = (\nu_1, \ldots, \nu_n)$, $T^\nu$ is the usual composition (product) of $T_1, \ldots, T_n$, namely $T^\nu = T_1^{\nu_1} \cdots T_n^{\nu_n}$. In particular if $\{y\}$ is in $\mathcal{Y}^{-\infty}$ and

$$\hat{y}(\xi) = \sum_{j\in\mathbb{Z}^n} y_j e^{-i\langle j,\xi\rangle}$$

then $\{T^\nu y\}$ is also in $\mathcal{Y}^{-\infty}$ and

$$\widehat{T^\nu y}(\xi) = \{\prod_{m=1}^{n}(e^{i\xi_m} - 1)^{\nu_m}\}\hat{y}(\xi)$$

where $\xi = (\xi_1, \dots, \xi_n)$.

The space $l^{2,k}(\mathbb{Z}^n)$ consists of those sequences $\{y\} = \{y_j\}_{j\in\mathbb{Z}^n}$ for which

$$||\{y\}||^2_{l^{2,k}} = \sum_{|\nu|=k}\sum_{j\in\mathbb{Z}^n} |(T^\nu y)_j|^2$$

is finite. This space and its connection with $L^{2,k}(\mathbb{R}^n)$ and $k$-harmonic splines is described in [38]. Some of this material is outlined below.

First we remark that an analogue of Proposition 26.5 is valid for $l^{2,k}(\mathbb{Z}^n)$. Also, it should be clear that $l^2(\mathbb{Z}^n) = l^{2,0}(\mathbb{Z}^n)$ and

$$l^{2,0}(\mathbb{Z}^n) \subset l^{2,1}(\mathbb{Z}^n) \subset \cdots \subset l^{2,k}(\mathbb{Z}^n) \subset l^{2,k+1}(\mathbb{Z}^n) \subset \cdots . \tag{26.23}$$

It is convenient to consider the union of these space as the space $l^{2,\infty}(\mathbb{Z}^n)$. Thus, by definition

$$l^{2,\infty}(\mathbb{Z}^n) = \bigcup_{k=0}^{\infty} l^{2,k}(\mathbb{Z}^n) .$$

Next note that if $u$ is in $L^{2,k}(\mathbb{R}^n)$ and $2k \geq n+1$ then by virtue of Proposition 26.5 for each $j$ in $\mathbb{Z}^n$ the value $u(j)$ is well defined and the sequence of values $\{u\} = \{u(j)\}_{j\in\mathbb{Z}^n}$ is in $\mathcal{Y}^k$. Thus the unique $k$-harmonic spline with knots on $\mathbb{Z}^n$ which interpolates $u$ on $\mathbb{Z}^n$ is well defined via $S_k u = S_k\{u\}$. Indeed, it turns out that $SH_k(\mathbb{R}^n)\bigcap L^{2,k}(\mathbb{R}^n)$ is a closed subspace of $L^{2,k}(\mathbb{R}^n)$.

**Proposition 26.7** *If $2k \geq n+1$, the mapping $u \to S_k u$ is an orthogonal projection of $L^{2,k}(\mathbb{R}^n)$ onto $L^{2,k}(\mathbb{R}^n)\bigcap SH_k(\mathbb{R}^n)$. In particular, if $u$ is in $L^{2,k}(\mathbb{R}^n)$ then the sequence of values $\{u\} = \{u(j)\}_{j\in\mathbb{Z}^n}$ is in $l^{2,k}(\mathbb{Z}^n)$ and*

$$||S_k u||_{L^{2,k}} \leq ||u||_{L^{2,k}} .$$

**Proposition 26.8** *If $2k \geq n+1$ and the sequence $\{y\} = \{y_j\}_{j\in\mathbb{Z}^n}$ is in $l^{2,k}(\mathbb{Z}^n)$ then $\{y\}$ is in $\mathcal{Y}^k$ and there is a unique $k$-harmonic spline $S_k\{y\}$ which interpolates $\{y\}$ on $\mathbb{Z}^n$. Furthermore:*

**(i)** *The mapping $\{y\} \to S_k\{y\}$ is a continuous isomorphism between $l^{2,k}(\mathbb{Z}^n)$ and $L^{2,k}(\mathbb{R}^n)\bigcap SH_k(\mathbb{R}^n)$.*

**(ii)** *$S_k\{y\}$ is the unique element of minimal $L^{2,k}(\mathbb{R}^n)$ norm which interpolates $\{y\}$ on $\mathbb{Z}^n$. That is,*

$$||S_k\{y\}||_{L^{2,k}} < ||u||_{L^{2,k}}$$

*for any $u$ in $L^{2,k}(\mathbb{R}^n)$ which interpolates $\{y\}$ on $\mathbb{Z}^n$ and is not $S_k\{y\}$.*

Statement (ii) of Proposition 26.8 summarizes the variational properties of polyharmonic splines alluded to earlier.

### *26.2.3 The Paley-Wiener space $PW_\pi^k$*

The space $PW_\pi(I\!R^n)$ is an $n$-variate analogue of the classical Paley-Wiener space. It consists of those functions $F$ in $E_\pi(\mathbb{C}^n)$ whose restriction to $I\!R^n$ is square integrable. In other words:

$$PW_\pi(I\!R^n) = E_\pi(\mathbb{C}^n) \bigcap L^2(I\!R^n).$$

For any positive integer $k$, $PW_\pi^k(I\!R^n)$ is the collection of those elements $F$ in $E_\pi(\mathbb{C}^n)$ all of whose $k$-th order derivatives are in $PW_\pi(I\!R^n)$. This class may also be described via the identity

$$PW_\pi^k(I\!R^n) = E_\pi(\mathbb{C}^n) \bigcap L^{2,k}(I\!R^n).$$

For $k = 0$, $PW_\pi^0(I\!R^n) = PW_\pi(I\!R^n)$. Note that

$$PW_\pi(I\!R^n) \subset PW_\pi^1(I\!R^n) \subset \cdots \subset PW_\pi^k(I\!R^n) \subset PW_\pi^{k+1}(I\!R^n) \subset \cdots . \tag{26.24}$$

Relations (26.24) are an easy consequence of the following.

**Proposition 26.9** *Suppose $F$ is in $L^{2,k}(I\!R^n)$ and its Fourier transform $f$ is supported in $\{\xi : |\xi| \le \epsilon\}$ for some $\epsilon, 0 \le \epsilon < \infty$. Then $F$ is in $L^{2,m}(I\!R^n)$ for every $m \ge k$ and*

$$\|F\|_{L^{2,m}} \le \epsilon^{m-k} \|F\|_{L^{2,k}}. \tag{26.25}$$

**Proof.**

$$\|F\|_{L^{2,m}}^2 = \frac{1}{(2\pi)^n} \int\limits_{|\xi|\le\epsilon} |\xi|^{2m} |f(\xi)|^2 d\xi$$

$$\le \frac{\epsilon^{2m-2k}}{(2\pi)^n} \int\limits_{|\xi|\le\epsilon} |\xi|^{2k} |f(\xi)|^2 d\xi \le \epsilon^{2m-2k} \|F\|_{L^{2,k}}^2 .$$

■

In view of relation (26.24) it is often convenient to consider the union of these spaces. Thus we set

$$PW_\pi^\infty(I\!R^n) = \bigcup_{k=0}^{\infty} PW_\pi^k(I\!R^n).$$

The relationships (26.23) and (26.24), taken together with the evident association of these spaces with the spaces $L^{2,k}(I\!R^n)$, seem to suggest a more intimate connection between the spaces $l^{2,k}(Z\!\!Z^n)$ and $PW_\pi^k(I\!R^n)$. Indeed, if the sequence $\{y\} = \{y\}_{j\in Z\!\!Z^n}$ is in $l^{2,k}(Z\!\!Z^n)$, consider the mapping

$$\{y_j\}_{j\in Z\!\!Z^n} \to \sum_{j\in Z\!\!Z^n} y_j \text{sinc}(x-j). \tag{26.26}$$

It is essentially the content of the so-called Whittaker-Kotelnikov-Shannon sampling theorem mentioned earlier that, in the case $k=0$, this mapping is a continuous isomorphism between $l^{2,k}(Z\!\!Z^n)$ and $PW_\pi^k(I\!R^n)$. Unfortunately in the cases $k \geq 1$ the series in (26.26) may fail to converge in any reasonable sense[7]. Nevertheless, in the case $n=1$, Schoenberg [56] (i) recorded the fact that, no matter the value of $k$, there is a unique function $F$ in $PW_\pi^k(I\!R^n)$ which interpolates $\{y\}$ on $Z\!\!Z^n$ and (ii) showed that the mapping of $\{y\}$ into its corresponding interpolating function $F$ in $PW_\pi^k(I\!R^n)$ gives a continuous isomorphism between $l^{2,k}(Z\!\!Z^n)$ and $PW_\pi^k(I\!R^n)$.

In what follows we show that this is also the case in the general $n$-variate situation.

**Proposition 26.10** *If $F$ is in $PW_\pi^k$ then the sequence of values $\{F\} = \{F(j)\}_{j\in Z\!\!Z^n}$ is in $l^{2,k}(Z\!\!Z^n)$ and*

$$\|\{F\}\|_{l^{2,k}} \leq C\|F\|_{PW_\pi^k}$$

*where $C$ is a constant independent of $F$.*

**Proof.** In view of Proposition 26.8 we may write

$$\|\{F\}\|_{l^{2,k}} \leq C\|S_k\{F\}\|_{L^{2,k}} < C\|F\|_{PW_\pi^k}$$

which is the desired result. ■

**Proposition 26.11** *If the sequence $\{y\} = \{y_j\}_{j\in Z\!\!Z^n}$ is in $l^{2,k}(Z\!\!Z^n)$ then there is a unique $F$ in $PW_\pi^k(I\!R^n)$ such that $F(y) = y_j$ for all $j$ in $Z\!\!Z^n$. This $F$ satisfies*

$$\|F\|_{PW_\pi^k} \leq C\|\{y\}\|_{l^{2,k}}$$

*where $C$ is a constant independent of $\{y\}$.*

**Proof.** First assume that $\{y\}$ is in $\mathcal{Y}^{-\infty}$ and let

$$F(z) = \sum_{j\in Z\!\!Z^n} y_j \text{sinc}(z-j).$$

[7] This may be readily seen by letting the sequence $\{y\}$ be the values of an appropriate polynomial $F$, that is $y_j = F(j)$ for all $j$ in $Z\!\!Z^n$.

In other words, the Fourier transform $f$ of $F$ is

$$f(\xi) = \sum_{j\in\mathbb{Z}^n} y_j e^{-i\langle j,\xi\rangle}\chi(\xi)$$

where $\chi(\xi)$ is the characteristic (indicator) function of $I_\pi$. To estimate the $L^{2,k}(\mathbb{R}^n)$ norm of $F$ observe that

$$(i\xi)^\nu f(\xi) = \frac{(i\xi)^\nu}{(\hat{T}(\xi))^\nu} \sum_{j\in\mathbb{Z}^n} (T^\nu y)_j e^{-i\langle j,\xi\rangle}\chi(\xi).$$

Since $(i\xi)^\nu/(\hat{T}(\xi))^\nu$ is bounded for $\xi$ in $I_\pi$, it follows that

$$\int_{\mathbb{R}^n} |(i\xi)^\nu f(\xi)|^2 d\xi \leq C\|\{T^\nu y\}\|_{l^2}^2. \tag{26.27}$$

Since

$$\|F\|_{PW_\pi^k} = \sum_{|\nu|=k} c_\nu \|D^\nu F\|_{L^2}^2,$$

$$(2\pi)^n \|D^\nu F\|_{L^2}^2 = \int_{\mathbb{R}^n} |(i\xi)^\nu f(\xi)|^2 d\xi,$$

and

$$\|\{y\}\|_{l^{2,k}} = \sum_{|\nu|=k} \|\{T^\nu y\}\|_{l^2},$$

inequality (26.27) implies that

$$\|F\|_{PW_\pi^k} \leq C\|\{y\}\|_{l^{2,k}} \tag{26.28}$$

where $C$ is a constant independent of $\{y\}$.

It follows that, except for uniqueness, the conclusions of the proposition are valid whenever $\{y\}$ is in $\mathcal{Y}^{-\infty}$ which is a dense subspace of $l^{2,k}(\mathbb{Z}^n)$.

Now let $\{y\}$ be any element in $l^{2,k}(\mathbb{Z}^n)$. Since $\mathcal{Y}^{-\infty}$ is dense in $l^{2,k}(\mathbb{Z}^n)$ there is a sequence $\{y_m\} = \{y_{m,j}\}_{j\in\mathbb{Z}^n}$, $m = 1,2,3\ldots$, of elements in $\mathcal{Y}^{-\infty}$ which converges to $\{y\}$ in $l^{2,k}(\mathbb{Z}^n)$. Let $\Omega$ be a subset of $\mathbb{Z}^n$ which is unisolvent for $\mathcal{P}_{k-1}(\mathbb{R}^n)$ and let $P_m$ be a sequence of polynomials in $\mathcal{P}_{k-1}(\mathbb{R}^n)$ such that

$$P_m(j) = y_j - y_{m,j} \text{ for all } j \text{ in } \Omega$$

In view of the $l^{2,k}(\mathbb{R}^n)$ variant of Proposition 26.5 it follows that

$$\lim_{m\to\infty} (P_m(j) + y_{m,j}) = y_j \text{ for all } j \text{ in } \mathbb{Z}^n. \tag{26.29}$$

Finally, let

$$F_m(z) = P_m(z) + \sum_{j\in\mathbb{Z}^n} y_{m,j}\operatorname{sinc}(z-j).$$

The sequence $F_m$, $m = 1, 2, 3, \dots$, is a Cauchy sequence in $PW_\pi^k$ by virtue of the fact that $\{y_m\}$, $m = 1, 2, 3, \dots$, is a Cauchy sequence in $l^{2,k}(\mathbb{Z}^n)$ and inequality (26.28). Also, since each $F_m$ interpolates $\{y\}$ on $\Omega$, that is,

$$F_m(j) = y_j, \text{ for all } j \text{ in } \Omega,$$

it follows that $F_m(x)$, $m = 1, 2, 3, \dots$, is a Cauchy sequence uniformly in $x$ whenever $x$ is restricted to a compact subset of $\mathbb{R}^n$. Let $F$ be the limit of this sequence. Then

$$F \text{ is in } PW_\pi^k(\mathbb{R}^n),$$

$$\|F\|_{PW_\pi^k} \le C\|\{y\}\|_{l^{2,k}}$$

where $C$ is a constant independent of $\{y\}$, and

$$F(j) = y_j \text{ for all } j \text{ in } \mathbb{Z}^n.$$

The first two of the above claims concerning $F$ follow from the fact that $F = \lim_{m\to\infty} F_m$ in $PW_\pi^k$, the definition of $F_m$, and inequality (26.28). The third follows from the fact that $F = \lim_{m\to\infty} F_m$ pointwise in $\mathbb{R}^n$, the fact that $F_m(j) = P_m(j) + y_{m,j}$, and identity (26.29).

It remains to show that the function $F$ obtained by the above procedure is unique. For this it suffices to show that the assumptions

$$F \text{ is in } PW_\pi^k(\mathbb{R}^n)$$

and

$$F(j) = 0 \text{ for all } j \text{ in } \mathbb{Z}^n$$

imply that $F$ is identically 0.

To this end, assume that $F$ enjoys the last two displayed properties. For any multi-index $\nu$ let $T^\nu$ be the finite difference operator defined by

$$\widehat{T^\nu F}(\xi) = \widehat{T^\nu}(\xi) f(\xi)$$

where $f$ is the Fourier transform of $F$ and

$$\widehat{T^\nu}(\xi) = \prod_{m=1}^{n} (e^{i\xi_m} - 1)^{\nu_m}.$$

Because of the assumptions on $F$ we have

$$T^{\nu}F(j) = \frac{1}{(2\pi)^n} \int\limits_{I_\pi} \frac{\widehat{T^{\nu}}(\xi)}{\xi^{\nu}} \xi^{\nu} f(\xi) e^{i\langle j,\xi\rangle} d\xi = 0$$

for all $j$ in $\mathbb{Z}^n$ and all multi-indexes $\nu$ such that $|\nu| = k$. This implies that the square integrable functions $\xi^{\nu} f(\xi)$ are 0 for all such multi-indexes which, in turn, implies that $F$ is in $\mathcal{P}_{k-1}(\mathbb{R}^n)$. However, the only polynomial which vanishes on $\mathbb{Z}^n$ is 0 so $F$ must be 0. ■

**Corollary 26.12** *Given a sequence* $\{y\} = \{y_j\}_{j\in\mathbb{Z}^n}$ *in* $l^{2,k}(\mathbb{Z}^n)$ *let* $F$ *be the unique element in* $PW_\pi^k(\mathbb{R}^n)$ *which interpolates* $\{y\}$ *on* $\mathbb{Z}^n$ *whose existence is guaranteed by Proposition 26.11. The mapping*

$$\{y\} \to F$$

*is a continuous isomorphism between* $l^{2,k}(\mathbb{Z}^n)$ *and* $PW_\pi^k(\mathbb{R}^n)$.

### *26.2.4 Convergence of m-harmonic splines as* $m \to \infty$

Suppose $\{y\} = \{y_j\}_{j\in\mathbb{Z}^n}$ is in $l^{2,k}(\mathbb{Z}^n)$. Then if $2m \geq n+1$, the $m$-harmonic spline interpolant of $\{y\}$ on $\mathbb{Z}^n$, the function $S_m\{y\}$, is uniquely defined. Since there is a unique $F$ in $PW_\pi^k$ which interpolates $\{y\}$ on $\mathbb{Z}^n$, Proposition 26.3 implies that

$$\lim_{m\to\infty} S_m\{y\}(x) = F(x)$$

uniformly on compact subsets of $\mathbb{R}^n$.

In what follows we will show that this convergence is, in fact, much stronger.

Before stating the main result, observe that $S_m\{y\}$ is in $L^{2,m}(\mathbb{R}^n)$ whenever $m \geq k$. This suggests that the derivatives of $S_m\{y\}$ also behave nicely.

**Theorem 26.13** *Suppose* $\{y\} = \{y_j\}_{j\in\mathbb{Z}^n}$ *is in* $l^{2,k}(\mathbb{Z}^n)$ *for some integer* $k$, $k \geq 0$, *and, assuming that* $2m \geq n+1$, *let* $S_m\{y\}$ *be the unique* $m$-*harmonic spline interpolant of* $\{y\}$ *on* $\mathbb{Z}^n$. *If* $F$ *is the unique element in* $PW_\pi^k(\mathbb{R}^n)$ *which interpolates* $\{y\}$ *on* $\mathbb{Z}^n$ *then*

$$\lim_{m\to\infty} S_m\{y\}(x) = F(x) \tag{26.30}$$

*in the* $L^2(\mathbb{R}^n)$ *sense and uniformly in* $x$ *on* $\mathbb{R}^n$. *Furthermore, given any multi-index* $\nu$, $D^{\nu}S_m\{y\}$ *is well defined and continuous for sufficiently large* $m$ *and*

$$\lim_{m\to\infty} D^{\nu}S_m\{y\}(x) = D^{\nu}F(x) \tag{26.31}$$

*in the* $L^2(\mathbb{R}^n)$ *sense and uniformly in* $x$ *on* $\mathbb{R}^n$.

**Proof.** Let $F$ be the unique element in $PW_\pi^k(\mathbb{R}^n)$ which interpolates $\{y\}$ on $\mathbb{Z}^n$. Then $S_m\{y\} = S_m F$ and it suffices to show that $\lim_{m\to\infty} S_m F = F$ in the appropriate senses.

To this end, for positive $\epsilon$ let $F_0$ be defined by the following formula for its Fourier transform

$$\hat{F}_0(\xi) = \hat{F}(\xi)\chi_\epsilon(\xi)\,,$$

where $\chi_\epsilon$ is the characteristic (or indicator) function of the ball $\{\xi : |\xi| < \epsilon\}$. A more precise value of $\epsilon$ will be chosen later.

Now write

$$F = F_0 + F_1 \tag{26.32}$$

so that by linearity of the mapping $F \to S_m F$

$$S_m F = S_m F_0 + S_m F_1. \tag{26.33}$$

Note that

$$\|F_0\|_{L^{2,k}} \leq \|F\|_{L^{2,k}}\,.$$

Since $F_0(j) - S_m F_0(j) = 0$ for all $j$ in $\mathbb{Z}^n$, Proposition 26.6 implies that $F_0 - S_m F_0$ is in $L^2(\mathbb{R}^n)$ and

$$\|F_0 - S_m F_0\|_{L^2} \leq c^m \|F_0 - S_m F_0\|_{L^{2,m}}$$

where $c$ is a constant independent of $m$. Proposition 26.7 implies that

$$\|F_0 - S_m F_0\|_{L^{2,m}} \leq \|F_0\|_{L^{2,m}}$$

and Proposition 26.9 gives the following estimate whenever $m \geq k$,

$$\|F_0\|_{L^{2,m}} \leq \epsilon^{m-k}\|F_0\|_{L^{2,k}}\,.$$

The last three displayed inequalities imply that

$$\|F_0 - S_m F_0\|_{L^2} \leq (c\epsilon)^m \epsilon^{-k} \|F_0\|_{L^{2,k}}\,. \tag{26.34}$$

Since the constant $c$ in the last inequality is independent of $m$ and $F$, choosing $\epsilon$ such that $c\epsilon < 1$ implies that

$$\lim_{m\to\infty} \|F_0 - S_m F_0\|_{L^2} = 0\,. \tag{26.35}$$

Now, if $m$ is also $\geq |\nu|$ then

$$\|D^\nu F_0 - D^\nu S_m F_0\|_{L^2} \leq C\{\|F_0 - S_m F_0\|_{L^2} + \|F_0 - S_m F_0\|_{L^{2,m}}\}$$

where $C$ is independent of $F$ and $m$ so that the same reasoning which led to (26.34) results in

$$\|D^\nu F_0 - D^\nu S_m F_0\|_{L^2} \le C\{(c\epsilon)^m \epsilon^{-k} + \epsilon^{m-k}\}\|F_0\|_{L^{2,k}}$$

and choosing $\epsilon = \min(c/2, 1/2)$ implies that

$$\lim_{m\to\infty} \|D^\nu F_0 - D^\nu S_m F_0\|_{L^2} = 0. \tag{26.36}$$

Next observe that $F_1$ is in $L^2(\mathbb{R}^n)$. This follows from

$$\|F_1\|^2_{L^2} = \frac{1}{(2\pi)^n} \int\limits_{|\xi|>\epsilon} |\hat{F}(\xi)|^2 d\xi$$

$$\le \frac{1}{(2\pi)^n \epsilon^{2k}} \int\limits_{|\xi|>\epsilon} |\xi|^{2k} |\hat{F}(\xi)|^2 d\xi \le \epsilon^{-2k}\|F\|^2_{L^{2,k}} .$$

Take the Fourier transform of $D^\nu F_1 - D^\nu S_m F_1$ and observe that

$$\mathcal{F}(D^\nu F_1 - D^\nu S_m F_1)(\xi) = (i\xi)^\nu\{\hat{F}_1(\xi) - \hat{\Lambda}_m(\xi)\wp\hat{F}_1(\xi)\}$$

$$= (i\xi)^\nu\{1 - \hat{\Lambda}_m(\xi)\}\hat{F}_1(\xi) + (i\xi)^\nu \hat{\Lambda}_m(\xi) \sum_{j\in Z - Z^n\{0\}} \hat{F}_1(\xi - 2\pi j).$$

Use (i) the fact that $\hat{F}_1(\xi)$ is square integrable and supported in $I_\pi$, (ii) the explicit formula (26.14) for $\hat{\Lambda}_k(\xi)$, and (iii) the dominated convergence theorem to conclude that

$$\lim_{m\to\infty} \mathcal{F}(D^\nu F_1 - D^\nu S_m F_1)(\xi) = 0$$

in $L^2(\mathbb{R}^n)$. Plancherel's formula now implies that

$$\lim_{m\to\infty} \|D^\nu F_1 - D^\nu S_m F_1\|_{L^2} = 0 \tag{26.37}$$

for all multi-indexes $\nu$.

Since $S_m\{y\} = S_m F$, relations (26.32) through (26.37) prove that identities (26.30) and (26.31) are valid in the $L^2(\mathbb{R}^n)$ sense.

The fact that (26.30) and (26.31) are also valid uniformly in $x$ on all of $\mathbb{R}^n$ follows from the $L^2$ convergence and the fact that for any $\varphi$ in the set $L^2(\mathbb{R}^n) \bigcap L^{2,p}(\mathbb{R}^n)$ there is a constant $C$ independent of $\varphi$ such that

$$|\varphi(x)| \le C\{\|\varphi\|_{L^2} + \|\varphi\|_{L^{2,p}}\}$$

whenever $2p > n$. ■

**Corollary 26.14** *For every sequence $\{y\} = \{y_j\}_{j\in\mathbb{Z}^n}$ in $l^{2,\infty}(\mathbb{Z}^n)$ there is a unique entire function $F$ in $PW_\pi^\infty(\mathbb{R}^n)$ which interpolates $\{y\}$ on $\mathbb{Z}^n$. Furthermore*

$$F(x) = \lim_{m\to\infty} S_m\{y\}(x)$$

*where $S_m\{y\}$ is the unique $m$-harmonic spline interpolant of $\{y\}$ on $\mathbb{Z}^n$ and the convergence is both in the $L^2(\mathbb{R}^n)$ sense and uniformly in $x$ on $\mathbb{R}^n$.*

## 26.3 Generalizations, related methods, and computational issues

### *26.3.1 Generalizations*

Sampling on a general lattice $\mathcal{L}$

In the preceding development, in order to avoid notational obfuscation, we restricted our attention to the case of sampling on the integer lattice $\mathbb{Z}^n$. However, the results can be readily reformulated for the case of sampling on a general non-degenerate lattice $\mathcal{L}$. We outline the significant modifications below.

The key ingredients which underly the development in the general case are (i) the fact that every non-degenerate lattice $\mathcal{L}$ is related to the integer lattice $\mathbb{Z}^n$ via an invertible linear transformation and (ii) the corresponding analogues of formula (26.18), the periodization operator $\wp$, and the corresponding Poisson summation formula.

Item (i) is more precisely expressed as follows: If $\mathcal{L}$ is a non-degenerate lattice then there is an invertible linear transformation $A$ such that

$$\mathcal{L} = A\mathbb{Z}^n = \{l \in \mathbb{R}^n : l = Aj, j \in \mathbb{Z}^n\} \tag{26.38}$$

In the succeeding development it is always assumed that $\mathcal{L}$ is related to $\mathbb{Z}^n$ via (26.38).

To see the formulation of the ideas mentioned in item (ii) in terms of the lattice $\mathcal{L}$, suppose $F$ is in $S(\mathbb{R}^n)$ and write

$$\sum_{l\in\mathcal{L}} F(l)\Lambda_k(x-l) = \sum_{j\in\mathbb{Z}^n} F(Aj)\Lambda_k(x-Aj).$$

The Fourier transform of the above expression is

$$\{\sum_{j\in\mathbb{Z}^n} F(Aj)e^{-i\langle Aj,\xi\rangle}\}\hat{\Lambda}_k(\xi). \tag{26.39}$$

Now, for any $m$ in $\mathbb{Z}^n$ we may write

$$\sum_{j\in\mathbb{Z}^n} F(A_j)e^{-i\langle Aj,\xi\rangle} = \sum_{j\in\mathbb{Z}^n} F(Aj)e^{-i\langle j,A^*\xi-2\pi m\rangle}$$

$$= \sum_{j\in\mathbb{Z}^n} F(Aj)e^{-i\langle Aj,\xi-2\pi Bm\rangle}$$

where $A^*$ is the adjoint of $A$ and $B = (A^*)^{-1}$. The last identity implies that $\sum_{j\in\mathbb{Z}^n} F(Aj)e^{-i\langle Aj,\xi\rangle}$ is $2\pi B\mathbb{Z}^n$ periodic so that, if $f$ is the Fourier transform of $F$, the appropriate variant of the Poisson summation formula is

$$\sum_{j\in\mathbb{Z}^n} F(Aj)e^{-i\langle Aj,\xi\rangle} = \sum_{j\in\mathbb{Z}^n} f(\xi - 2\pi Bj). \tag{26.40}$$

From (26.39) and (26.40) it should be clear that in this case the formula for $\hat{\Lambda}_k$ should be

$$\hat{\Lambda}_k(\xi) = \frac{|\xi|^{-2k}}{\sum_{j\in\mathbb{Z}^n} |\xi - 2\pi Bj|^{-2k}}$$

and the role played by the parallelpiped $I_\pi$ is now played by the polygonal region

$$\Omega = \{\xi \in \mathbb{R}^n : |\xi| \le |\xi - 2\pi Bj| \text{ for all } j \in \mathbb{Z}^n \setminus \{0\}\}. \tag{26.41}$$

Note that $\Omega$ is not necessarily $BI_\pi$.

The lattice $\mathcal{L}^* = 2\pi B\mathbb{Z}^n$ is sometimes referred to as the lattice dual to $\mathcal{L}$ and the region $\Omega$ is sometimes referred to as a tile or fundamental region for the lattice $\mathcal{L}^*$. See Subsection 3.2.

## Generalized splines

Note that the $k$-harmonic splines in Section 2 are simply linear combinations of lattice translates of $\Lambda_k(x)$ or, from a more primitive point of view, they are essentially linear combinations of lattice translates of the fundamental solution of the $k$th iterate of the Laplacian $\Delta^k$. It should be clear that the material in Section 2 can be easily extended to include interpolants which are solutions to more general variational problems. Indeed, it should not be surprising that certain portions of the summability theory described in Section 2 can be even further generalized by replacing the polyharmonic splines by linear combinations of translates of some appropriately prescribed function $H(x)$. What we have in mind is interpolators or approximations $S(x)$ of the form

$$S(x) = \sum_{v\in\Upsilon} c(v)H(x - v) \tag{26.42}$$

where $\Upsilon$ is a fixed discrete subset of $\mathbb{R}^n$ and $\{c(v)\}_{v\in\Upsilon}$ are scalar coefficients which are the free parameters to be chosen appropriately. The definition of $S$ may also include a term from some appropriate finite dimensional subspace such as $\mathcal{P}_k(\mathbb{R}^n)$. We find it convenient to refer to such functions $S$ as *generalized splines.* As in the case of polyharmonic splines the function $H$, and the resulting generalized spline, can depend on a parameter; such a parameter need not be discrete.

Interpolators or approximations of the form (26.42) arise in the theory and application of multiresolution analyses and wavelets, [3, 7, 8, 9, 26, 36, 43, 61], the theory of interpolation associated with variational problems, [1, 15, 29, 39, 40, 41], and other contexts, [10, 12, 18, 27, 45]. The choice of the function $H$ which results in reasonable approximators and/or interpolators can be quite wide.

A particularly simple example is the Poisson kernel

$$H_\sigma(x) = \frac{\sigma}{(\sigma^2 + |x|^2)^{(n+1)/2}}$$

or the Gaussian

$$H_\sigma(x) = \sigma^{-n/2} \exp(-\frac{|x|^2}{\sigma})$$

with $\sigma > 0$. In either case, if $h_\sigma(\xi)$ is the Fourier transform of $H_\sigma$ then the function $\Lambda_\sigma(x)$ defined by its Fourier transform via

$$\hat{\Lambda}_\sigma(x) = \frac{h_\sigma(\xi)}{\sum_{j\in\mathbb{Z}^n} h_\sigma(\xi - 2\pi j)}$$

enjoys the property that

$$\Lambda_\sigma(j) = \begin{cases} 1 & \text{if } j = 0 \\ 0 & \text{if } j \in \mathbb{Z}^n \setminus \{0\}, \end{cases}$$

Note that $\Lambda_\sigma$ may also be expressed as

$$\Lambda_\sigma(x) = \sum_{j\in\mathbb{Z}^n} a_j H_\sigma(x - j) \tag{26.43}$$

where the coefficients $a_j$ decay exponentially as $|j|$ tends to infinity. It is not difficult to verify that if $F$ is in the Paley-Wiener class $PW_\pi(\mathbb{R}^n)$ described in Subsection 2.3 then

$$S_\sigma(x) = \sum_{j\in\mathbb{Z}^n} F(j)\Lambda_\sigma(x - j)$$

is a well defined continuous function in $L^2(\mathbb{R}^n)$, which is of form (26.42) with $\Upsilon = \mathbb{Z}^n$, $H = H_\sigma$, and $c_j = \sum_{m\in\mathbb{Z}^n} a_{j-m}F(m)$. Furthermore $S_\sigma$

interpolates $F$ on $\mathbb{Z}^n$, that is $S_\sigma(j) = F(j)$ for all $j \in \mathbb{Z}^n$. Routine calculations show that that as $\sigma \to \infty$ then $\hat{\Lambda}_\sigma(\xi)$ converges to the characteristic (indicator) function of the parallepiped $I_\pi$ in the $L^2$ sense and in the $L^1$ sense. Consequently, as $\sigma$ tends to infinity, $S_\sigma(x)$ converges to $F(x)$ in $L^2(\mathbb{R}^n)$ and uniformly in $x$ on $\mathbb{R}^n$. However, unlike polyharmonic splines, these interpolators do not reproduce polynomials. These and a related family of examples are considered in [42].

We also mention that in the univariate case

$$H_k(x) = \phi_k * \tilde{\phi}_k(x)\,,$$

where $\tilde{\phi}_k(x) = \overline{\phi_k(-x)}$ and $\phi_k$ is the orthogonal scaling function introduced by Daubechies [9] whose support is in the interval $[0, 2k-1]$, enjoys the following properties:

- $H_k$ is continuous and has compact support. Indeed, its support is in the interval $[-(2k-1), 2k-1]$.

-
$$H_k(j) = \begin{cases} 1 & \text{if } j = 0 \\ 0 & \text{if } j \in \mathbb{Z}^n \setminus \{0\}\,. \end{cases}$$

- For any continous $F$ function on $\mathbb{R}$ the interpolator
$$S_k(x) = \sum_{j \in \mathbb{Z}} F(j) H_k(x-j)$$
is well defined, continuous, and satisfies $S_k(j) = F(j)$ for all $j$ in $\mathbb{Z}$.

- $S_k(x)$ reproduces polynomials of degree $\leq 2k-1$. Indeed, it coincides with the interpolant produced by the iterative interpolation process described in [14], also see [9, 26].

- If $F$ is in the Paley-Wiener class $PW_\pi(\mathbb{R})$ then $S_k(x)$ converges to $F(x)$ in $L^2(\mathbb{R})$ and uniformly in $x$ for all $x$ in $\mathbb{R}$.

Analogues of the above are also valid in the multivariate case and for more general classes of interpolatory scaling functions[8]. In Subsection 3.3 we outline the basic aspects of the exhaustively studied example of box splines. However, to limit the length of this article, we do not give details of any other explicit examples here.

---

[8] Indeed, in view of (26.50) the polyharmonic splines $\Lambda_k(x)$ are examples of such scaling functions.

Irregular sampling

The case of irregular sampling, namely, the case when the sampled values $F(x)$ do not correspond to points $x$ which lie on a lattice $\mathcal{L}$, is a very important, interesting, and broad mathematical topic in signal and image processing. It is far too vast and complicated to be seriously considered here in any detail, for example see [2, 3, 9, 16, 22, 17, 60, 64, 65] and the references cited there.

Here we very briefly describe several results in the univariate case when the sampling set $X$ is such that the corresponding exponentials $\{e^{ix\xi}\}_{x\in X}$ are a Riesz basis of $L^2([-\pi,\pi])$. This is a natural extension of the case when $X$ is the integer lattice $\mathbb{Z}$. Such sampling sets were studied by Paley-Wiener [48], Levin [30, Lecture 22], and other authors[9] and have been completely characterized in terms of zeros of certain entire functions by Pavlov [49], see also [24, 33]. Examples of such sampling sets include the following:

- $X = \{x_m\}_{m\in\mathbb{Z}}$ where the sequence $x_m$ satisfies the constraints $|x_m - m| \le r$ for some parameter $r < 1/4$.
- $X = \bigcup_{m=1}^{N}\{N\mathbb{Z} + a_m\} = \bigcup_{j\in\mathbb{Z}}\bigcup_{m=1}^{N}\{Nj + a_m\}$ where $0 \le a_1 < a_2 \ldots < a_N < N$ and $N$ is any positive integer.

For such sampling sets a version of Schoenberg's spline summability method is valid. That is, if $X$ is such a sampling set and $F$ is in $PW_\pi^k$ then $F$ can be recovered from its values $\{F(x)\}_{x\in X}$ as the limit of piecewise polynomial splines of order $2k$ as $k$ tends to infinity, see [31, 32].

### *26.3.2 Multivariate analogues of the Paley-Wiener Theorem and the sampling theorem*

The inverse Fourier transforms of those distributions $f$ which are supported in the set $\Omega$ can be characterized as a class of entire functions of exponential type if $\Omega$ is (E1) compact, (E2) convex, (E3) symmetric about the origin, and (E4) has a non-empty interior. To describe the situation precisely we first need to consider several technical points.

Given a norm $||\xi||$ on $\mathbb{R}^n$, the set $\Omega = \{\xi : ||\xi|| \le 1\}$ satisfies properties (E1)-(E4). Conversely, given a closed set $\Omega$ which enjoys properties (E1)-(E4), there is a norm $||\xi||$ such that $\Omega = \{\xi : ||\xi|| \le 1\}$.

In what follows we use the notation $|\xi|_\Omega$ to denote the norm $||\xi||$ associated with set $\Omega$.

If $\Omega$ satisfies (E1)-(E4) then the set

$$\Omega^* = \{x : \langle x, \xi\rangle \le 1 \text{ for all } \xi \text{ in } \Omega\}$$

[9] See [24] and the historical comments and appropriate references in [30, 31].

also satisfies (E1)-(E4). The associated norm $|x|_{\Omega^*}$ is called the norm dual to $|\xi|_\Omega$. For example, if $\Omega = I_\sigma$ is the parallelpiped defined in Subsection 2.1, then

$$|\xi|_\Omega = \max_m \left\{ \frac{|\xi_m|}{\sigma} : m = 1, \dots, n \text{ and } \xi = (\xi_1, \dots, \xi_n) \right\}$$

and

$$|x|_{\Omega^*} = \sigma \sum_{m=1}^{n} |x_m|$$

where $x = (x_1, \dots, x_n)$.

If $\Omega$ is a set which satisfies (E1)-(E4) then an entire function $F(z)$, $z \in \mathbb{C}^n$, is said to be of exponential type $\Omega$ or $F$ is in the class $E_\Omega$ whenever there are positive constants $A$ and $p$ so that $F$ satisfies the bound

$$|F(z)| \leq A(1 + |z|)^p \exp(|\mathrm{Im} z|_{\Omega^*})$$

for all $z$ in $\mathbb{C}^n$. An $n$-variate version of the Paley-Wiener Theorem[10] is the following statement:

*An entire function $F(z), z \in \mathbb{C}^n$, is the Fourier transform of a distribution with support in $\Omega$ if and only if $F$ is in $E_\Omega$.*

If $\mathcal{L}$ is a non-degenerate lattice and $\Omega$ is the polygonal region described by (26.41) then it should be clear that $\Omega$ satisfies (E1)-(E4) and that elements of $E_\Omega$ can be recovered from their samples on $\mathcal{L}$. In other words, variants of the Whittaker-Kotelnikov-Shannon sampling theorem, Proposition 3, Theorem 1, and related supporting material are valid in this context.

To see a general variant of the sampling theorem, view the dual lattice $\mathcal{L}^*$ as a discrete subgroup of $\mathbb{R}^n$ and let $\Omega$ be a complete set of coset representatives of the quotient group $\mathbb{R}^n/\mathcal{L}^*$. Such a set $\Omega$ is often referred to as a tile or fundamental region for $\mathcal{L}^*$. Note that

**(T1)** $\Omega \cap (\Omega + l^*) = \phi$ whenever $l^*$ is in $\mathcal{L}^* \setminus \{0\}$ and $\bigcup_{l^* \in \mathcal{L}^*} (\Omega + l^*) = \mathbb{R}^n$.

**(T2)** The measure of $\Omega$ is finite. In fact $|\Omega| = 2\pi \det B = 2\pi / \det A$. Recall that $\mathcal{L}$ and $\mathbb{Z}^n$ are related via (26.38) and $B = (A^*)^{-1}$.

**(T3)** The set $\Omega$ need not be convex or symmetric about the origin.

**(T4)** The set $\Omega$ need not be bounded.

Note that the set $\Omega$ defined by (26.41) is the closure of such a tile.

If $\Omega$ is a tile for $\mathcal{L}^*$ let $\chi_\Omega$ be its characteristic function. In view of (T1) and (T2) the following facts should be readily transparent:

---

[10] The statement given here is based on the development in [59, p. 111-112] and the proof in [23, p. 21-22].

- The collection of functions $\{ce^{-i\langle l^*,\xi\rangle}\chi_\Omega(\xi)\}_{l^*\in\mathcal{L}^*}$ is a complete orthonormal set for $L^2(\Omega)$. Here $c = (\det A/(2\pi)^n)^{1/2}$.

- If $\mathrm{sinc}_\Omega(x)$ is the inverse Fourier transform of $c\chi_\Omega(\xi)$, where $c = \det A$, then

  **(a)** $\mathrm{sinc}_\Omega(x)$ is continuous on $\mathbb{R}^n$.

  **(b)** $\mathrm{sin}_\Omega(l) = \begin{cases} 1 & \text{if } l = 0 \\ 0 & \text{if } l \in \mathcal{L}\setminus\{0\}, \end{cases}$

  **(c)** the collection of functions $\{\mathrm{sinc}_\Omega(x-l)\}_{l\in\mathcal{L}}$ is orthonormal in $L^2(\mathbb{R}^n)$ and spans $\mathcal{F}^{-1}L^2(\Omega)$.

- Any function $F$ in $L^2(\mathbb{R}^n)$ whose Fourier transform vanishes outside the set $\Omega$ is continuous, the sequence of values $\{F(l)\}_{l\in\mathcal{L}}$ is square summable, and $F$ enjoys the representation

$$F(x) = \sum_{l\in\mathcal{L}} F(l)\mathrm{sinc}_\Omega(x-l) \tag{26.44}$$

- Conversely, if the sequence of values $\{y_l\}_{l\in\mathcal{L}}$ is in square summable then there is a unique $F$ in $L^2(\mathbb{R}^n)$ whose Fourier transform has support in $\Omega$ such that $F(l) = y_l$ for all $l$ in $\mathcal{L}$. This $F$ is given by formula (26.44).

In general, functions $F$ which enjoy representation (26.44) cannot be neatly characterized as a subclass of entire functions of exponential type. Nevertheless, if $\Omega$ is bounded then such $F$ belong to a subclass of entire functions of exponential type.

Formula (26.44) may be viewed as a general multivariate analogue of the Whittaker-Kotelnikov-Shannon sampling theorem. More details and references may be found in [22, 47, 50, 65].

### *26.3.3 Box splines*

In the univariate case, $n = 1$, the fundamental cardinal spline $\Lambda_1(x)$ can be explicitly described via a simple algebraic formula, namely

$$\Lambda_1(x) = \begin{cases} 1-|x| & \text{if } |x| < 1 \\ 0 & \text{if } |x| \geq 1\,. \end{cases}$$

Another convenient representation of $\Lambda_1(x)$ in this case is also given by

$$\Lambda_1(x) = \chi * \chi(x)$$

where $\chi$ is the characteristic function of the interval $[-1/2, 1/2]$ and $*$ denotes convolution. Both formulas are consequences of representation (26.14).

However they can be, and usually are, easily derived independently of (26.14).

Remaining for the moment in the univariate case, it is clear that if $B_1(x) = \Lambda_1(x)$ and for $k = 2, 3, \ldots$

$$B_k(x) = \underbrace{\Lambda_1 * \ldots * \Lambda_1}_{k \text{ terms}}(x)$$

then $B_k$ is in $SH_k(I\!R)$ and has support in the interval $[-k, k]$. Note that $B_k$ can also be expressed as

$$B_k(x) = \underbrace{\chi * \ldots * \chi}_{2k \text{ terms}}(x) \tag{26.45}$$

or, via its Fourier transform, as

$$\hat{B}_k(\xi) = \left(\frac{\sin \xi/2}{\xi/2}\right)^{2k}. \tag{26.46}$$

It turns out, for example see [55], that every element $S$ in $SH_k(I\!R)$ can be uniquely represented as

$$S(x) = \sum_{j \in \mathbb{Z}} c_j B_k(x - j) \tag{26.47}$$

so that the functions $\{B_k(x - j)\}_{j \in \mathbb{Z}}$ can be regarded as a basis of compactly supported basic splines. The functions $B_k(x - j)$ are often referred to as $B$ splines. Representation (26.47) played an important role in the theoretical development and numerical applications of univariate piecewise polynomial splines, for example see [10, 55] and the references cited there.

Formulas (26.45) and (26.46) and the representation (26.47) are essentially the basis of another generalization of the piecewise polynomial spline theory to the multivariate scenario. This is the theory of so-called *box splines* championed by deBoor and thoroughly developed by him and many other authors. The book [12] contains a detailed description of this material and an extensive list of references. The basic building blocks $B$ of this theory are the compactly supported functions whose Fourier transforms are natural, but somewhat technically complicated, generalizations of (26.46). To be somewhat more specific, the functions $B$ are parametrized by certain discrete sets[11] $\Upsilon$ and defined via the formula for their Fourier transform as

$$\hat{B}(\xi) = \prod_{\upsilon \in \Upsilon} \frac{\sin\langle \upsilon, \xi \rangle}{\langle \upsilon, \xi \rangle}$$

[11] In [12] the symbols $\xi$ and $\Xi$ are used in this context.

where $\Upsilon$ is a finite collection of elements in $\mathbb{R}^n$ which must satisfy certain conditions so that $B$ is compactly supported, continuous, and enjoys other desirable properties; the nature of $B$ is determined by the choice of the set $\Upsilon$.

The connection with sampling theory is this: Consider the fundamental cardinal function of interpolation $\Gamma_k(x)$ for the linear space spanned by $\mathbb{Z}^n$ translates of

$$B_k(x) = \underbrace{B * \ldots * B}_{k \text{ terms}}(x) .$$

That is,

$$\Gamma_k(x) = \sum_{j \in \mathbb{Z}^n} c_j B_k(x - j)$$

and

$$\Gamma_k(j) = \begin{cases} 1 & \text{if } j = 0 \\ 0 & \text{if } j \in \mathbb{Z}^n \setminus \{0\} . \end{cases}$$

Then, under appropriate conditions on the set $\Upsilon$,

$$\lim_{k \to \infty} \Gamma_k(x) = \Phi(x)$$

where the Fourier transform of $\Phi$ is the characteristic function of a tile $\Omega$ for $2\pi\mathbb{Z}^n$. The exact nature of the tile is, of course, determined by $\Upsilon$. Thus, in this case, elements in the linear span of $\{B_k(x - j)\}_{j \in \mathbb{Z}^n}$ can be used to approximate the class of functions in $L^2(\mathbb{R}^n)$ whose Fourier transforms vanish almost everywhere off the tile $\Omega$.

The tiles $\Omega$ which arise in this way can be quite intricate. Details of this and the corresponding $L^2(\mathbb{R}^n)$ convergence theory may be found in [12] and the pertinent references cited there. It should be noted that the basic result concerning this convergence recorded in [11] has applications which go beyond box splines.

### *26.3.4 Computing polyharmonic splines*

Polyharmonic splines whose knots are on the lattice $\mathbb{Z}^n$ can be fairly efficiently computed via the fast Fourier transform algorithm, an observation which has been recorded by several authors [25, 35]. One rather transparent cause for this is the pair of formulas

$$S(x) = \sum_{j \in \mathbb{Z}^n} S(j) \Lambda_k(x - j) \tag{26.48}$$

$$\hat{\Lambda}_k(\xi) = \frac{|\xi|^{-2k}}{\sum_{j \in \mathbb{Z}^n} |\xi - 2\pi j|^{-2k}} \tag{26.49}$$

which are valid for any $k$-harmonic spline $S$ with knots on $\mathbb{Z}^n$, so that any such spline can be evaluated in terms of its values on $\mathbb{Z}^n$ by computing the inverse Fourier transform of

$$\hat{S}(\xi) = \{\sum_{j\in\mathbb{Z}^n} S(j)e^{-i\langle j,\xi\rangle}\}\hat{\Lambda}_k(\xi).$$

Another reasonably efficient algorithm for evaluating such splines can be based on (26.48) and the scaling equation

$$\Lambda_k(x) = \sum_{j\in\mathbb{Z}^n} \lambda_j \Lambda_k(2x - j), \tag{26.50}$$

which is an immediate consequence of (26.49). To wit, write

$$\hat{\Lambda}_k(\xi) = \left\{\frac{\sum_{j\in\mathbb{Z}^n} |(\xi/2) - 2\pi j|^{-2k}}{\sum_{j\in\mathbb{Z}^n} |\xi - 2\pi j|^{-2k}}\right\} \frac{|\xi|^{-2k}}{\sum_{j\in\mathbb{Z}^n} |(\xi/2) - 2\pi j|^{-2k}}$$

$$= \left\{\frac{\sum_{j\in\mathbb{Z}^n} |\xi - 4\pi j|^{-2k}}{\sum_{j\in\mathbb{Z}^n} |\xi - 2\pi j|^{-2k}}\right\} \frac{|\xi/2|^{-2k}}{\sum_{j\in\mathbb{Z}^n} |(\xi/2) - 2\pi j|^{-2k}}$$

and set

$$\frac{1}{2^n}\sum_{j\xi\mathbb{Z}^n} \lambda_j e^{-i\langle j,\xi/2\rangle} = \frac{\sum_{j\in\mathbb{Z}^n} |\xi - 4\pi j|^{-2k}}{\sum_{j\in\mathbb{Z}^n} |\xi - 2\pi j|^{-2k}} \tag{26.51}$$

to get

$$\hat{\Lambda}_k(\xi) = \left\{\frac{1}{2^n}\sum_{j\xi\mathbb{Z}^n} \lambda_j e^{-i\langle j,\xi/2\rangle}\right\} \hat{\Lambda}_k(\xi/2) .$$

Taking the inverse Fourier transform of the last identity results in formula (26.50). Note that in addition to providing a formula for the coefficients $\lambda_j$ identity (26.51) implies that the coefficients $\lambda_j$ enjoy exponential decay as $|j|$ tends to infinity.

To see how the coefficients $\lambda_j$ can be employed to evaluate $S$ from its values on $\mathbb{Z}^n$, use (26.50) in the right hand side of (26.48) to write

$$S(x) = \sum_{j\in\mathbb{Z}^n}\sum_{m\in\mathbb{Z}^n} S(j)\lambda_m\Lambda_k(2x - (2j + m)). \tag{26.52}$$

Now using the fact that

$$\Lambda_k(j) = \begin{cases} 1 & \text{if } j = 0 \\ 0 & \text{if } j \in \mathbb{Z}^n \setminus \{0\}, \end{cases}$$

it follows that

$$S(j/2) = \sum_{m \in \mathbb{Z}^n} S(m)\lambda_{j-2m} \,. \tag{26.53}$$

Formula (26.53) allows us to evaluate $S$ on the lattice $\frac{1}{2}\mathbb{Z}^n$ from its values on $\mathbb{Z}^n$. By using (26.50) in (26.52) and continuing in this way it should be clear that formula (26.53) can be iterated to obtain a recurrence relation for evaluating $S$ on the lattice $\frac{1}{2^{p+1}}\mathbb{Z}^n$ from its values on $\frac{1}{2^p}\mathbb{Z}^n$, $p = 0, 1, 2, \ldots$, namely

$$S(j/2^{p+1}) = \sum_{m \in \mathbb{Z}^n} S(m/2^p)\lambda_{j-2m}. \tag{26.54}$$

The recursive subdivision scheme suggested by (26.54) is the algorithm we alluded to.

The scaling equation (26.50) is, of course, the basis for the multiresolution analysis properties of polyharmonic splines. See [35] and, for the general theory, [9, 26]. Note that in this case

$$\lambda_{2j} = \begin{cases} 1 & \text{if } j = 0 \\ 0 & \text{if } j \in \mathbb{Z}^n \setminus \{0\}, \end{cases}$$

which is typical of interpolatory schemes based on scaling equations such as (26.50).

For illustration we exhibit the following explicit examples.

In the case $n = k = 2$ the coefficients $\lambda_j$, $j = (j_1, j_2)$ with $0 \le j_2 \le j_1 \le 7$, rounded to three decimal places are listed below; the remaining coefficients are either effectively 0 when rounded to three decimal places or, because of symmetry considerations, are already listed.

$$\begin{array}{cccccccc}
 & & & & 0 & 0.001 & 0 & -0.000 \\
 & & & -0.010 & 0.000 & 0.003 & 0.000 & -0.001 \\
 & & 0 & -0.021 & 0 & 0.006 & 0 & -0.002 \\
 & 0.337 & 0.037 & -0.049 & -0.009 & 0.009 & 0.002 & -0.002 \\
1.0 & 0.536 & 0 & -0.075 & 0 & 0.013 & 0 & -0.003
\end{array} \tag{26.55}$$

These coefficients were evaluated using identity (26.51) and the fast Fourier transform.

To illustrate the use of identity (26.53) we subsampled the 300 by 300 image in Figure 26.1 using one value in each 8 by 8 sub-block to get Figure 26.2 and then used (26.53) recurcively three times with the coefficients (26.55) to obtain the image in Figure 26.3. Note that (26.53) can be efficiently implemented by (i) subdividing the coefficients $\{\lambda_j\}_{j \in \mathbb{Z}^2}$ into four disjoint groups $\{\lambda^t\} = \{\lambda_j^t\}_{j \in \mathbb{Z}^2} = \{\lambda_{t+2j}\}_{j \in \mathbb{Z}^2}$, $t = (0,0)$, $(1,0)$, $(0,1)$, or $(1,1)$, (ii) computing the usual discrete convolution of each $\{\lambda^t\}$ with $\{S(j)\}_{j \in \mathbb{Z}^2}$, (iii) and combining the results appropriately to get (26.53).

**FIGURE 26.1. The original image.**

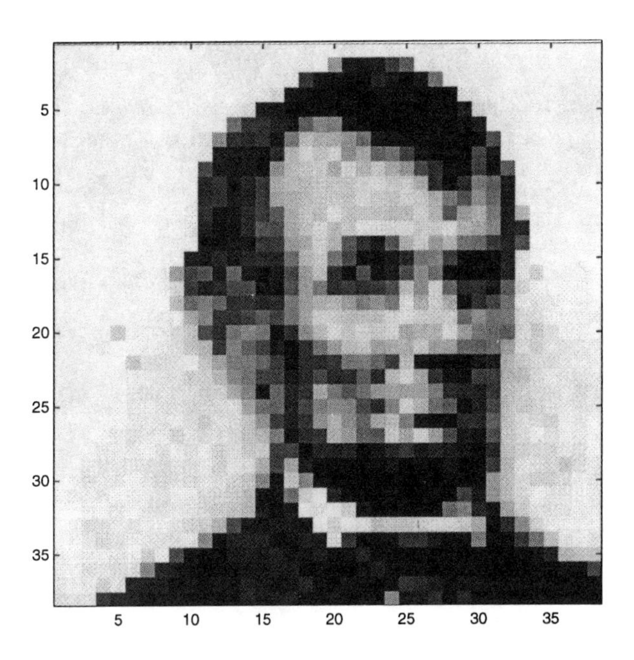


**FIGURE 26.2. The image interpolated from the subsampled image.**

FIGURE 26.3. The subsampled image.

FIGURE 26.4. The image interpolated from the original image.

Finally we used one iteration of the above procedure on the original image in Figure 26.1 and truncated the left, right, and bottom edges to obtain the 500 by 500 image in Figure 26.4.

MATLAB software was used to perform the computations and display the images. The original image file was obtained for me by K. Marinelli off the internet.

*References*


[1] M Atteia, *Hilbertian kernels and spline functions*, Studies in Computational Mathematics, 4, North-Holland Publishing Co., Amsterdam, 1992.

[2] J. J. Benedetto, Irregular sampling and frames, in [8], 445-507.

[3] J. J. Benedetto and M. W. Frazier, eds., *Wavelets: Mathematics and Applications*, CRC Press, 1993.

[4] R. P. Boas, *Entire Functions*, Academic Press, New York, 1954.

[5] P. L. Butzer and R. L. Stens, Sampling theory for not necessarily band-limited functions: a historical overview, *SIAM Rev.* **34**, (1992), 40-53.

[6] P. L. Butzer, W. Splettstösser, and R. L. Stens, The sampling theorem and linear prediction in signal analysis, *Jahresber. Deutsch, Math-Verein.*, **90**, (1988), 1-70.

[7] C. K. Chui, *An Introduction to Wavelets*, Academic Press, Boston, 1992.

[8] C. K. Chui, ed., *Wavelets: A Tutorial in Theory and Applications*, Academic Press, Boston, 1992.

[9] I. Daubechies, *Ten Lecture on Wavelets*, CBMS-NSF Regional Conference Series in Applied Mathematics, Vol. 61, SIAM, Philadelphia, 1992.

[10] C. deBoor, *A Practical Guide to Splines*, Springer-Verlag, New York, 1978.

[11] C. deBoor, K. Höllig, and S. Riemenschneider, Convergence of cardinal series, *Proc. Amer. Math. Soc.* **98**, (1986), 457-460.

[12] C. deBoor, K. Höllig, and S. Riemenschneider, *Box Splines*, Springer-Verlag, New York, 1993

[13] C. deBoor, ed., *I. J. Schoenberg: Selected Papers*, Contemporary Mathematicians Series, Birkhäuser-Verlag, Basel, 1988.

[14] G. Deslauriers and S. Dubuc, Symmetric iterative interpolation processes, *Constr. Approx.*, **5**, (1989), 49-58.

[15] J. Duchon, Splines minimizing rotation-invariant seminorms in Sobolev spaces, in *Constructive Theory of Functions of Several Variables*, W. Schempp and K. Zeller, eds., Springer-Verlag, New York, 1977, 85-100.

[16] H. G. Feichtinger and K. Gröchenig, Theory and practice of irregular sampling, in [3], 305-363.

[17] P. J. S. G. Ferreira, *Proceedings of the 1997 International Workshop on Sampling and Applications*, Aveiro, Portugal, 16-19 June, 1997.

[18] F. Franke, Scattered data interpolation: tests of some methods, *Math. Comp.*, **38**, (1982), 181-199.

[19] R. P. Gosselin, On the $L^p$ theory of cardinal series, *Ann. of Math.*, 78, (1963), 567-581.

[20] R. P. Gosselin, Singular integrals and cardinal series, *Studia Math.*, 44, (1972), 39-45.

[21] R. P. Gosselin and J. H. Neuwirth, On Paley-Wiener bases, *J. Math. Mech.*, 18 (1968/69), 871-879.

[22] J. R. Higgins, *Sampling Theory in Fourier and Signal Analysis*, Oxford University Press, Oxford, 1996.

[23] L. Hörmander, *Linear Partial Differential Operators*, Springer-Verlag, New York, 1969.

[24] S. V. Hrushchev, N. K. Nikol'skii, and P. S. Pavlov, Unconditional bases of exponentials and of reproducing kernels, in *Complex Analysis and Spectral Theory*, V. P. Havin and N. K. Nikol'skii (eds.), Lecture Notes in Mathematics **864**, Springer-Verlag, Berlin, 1981, 214–335.

[25] K. Jetter and J. Stöckler, Algorithms for cardinal interpolation using box splines and radial basis functions, *Numer. Math*, **60**, (1991), 97-114.

[26] J.-P. Kahane and P.-G. Lemarié-Rieusset, *Fourier Series and Wavelets*, Gordon and Breach, Amsterdam, 1995.

[27] E. J. Kansa, ed., *Advances in the Theory and Applications of Radial Basic Functions*, issue of *Computers and Mathematics* Vol. 24, No. 12, (1992).

[28] V. A. Kotelnikov, On the transmission capacity of "ether" and wire in electrocommunications, All-Union Energetics Committee, Izd. Red. Ups. Svyazi RKKA, Moscow, 1933.

[29] P.-G. Lemarié, Bases d'ondelettes sur les groupes de Lie stratifiés, *Bull. Soc. Math. France*, **117**, (1989), 211-232.

[30] B. Ya. Levin, *Lectures on Entire Functions*, Translation of mathematical monographs, Vol. 150, American Math. Society, Providence, 1996.

[31] Yu. Lyubarskii and W. R. Madych, The recovery of irregularly sampled band limited functions via tempered splines, *J. Functional Analysis*, **125** (1994), 201-222.

[32] Yu. Lyubarskii and W. R. Madych, The characterization and recovery of a class of band limited signals from irregular samples, in [17], 317-320.

[33] Yu. Lyubarskii and K. Seip, Complete interpolating sequences for Paley-Wiener spaces and Muckenhoupt $(A_p)$ condition, *Revista Math. Iber.*, to appear

[34] W. R. Madych and E. H. Potter, Error estimates for multivariate interpolation, *J. Approximation Theory*, **43**, (1985), 132-139.

[35] W. R. Madych, Cardinal interpolation with polyharmonic splines, in *Multivariate Aproximation IV*, C. K. Chui, W. Schempp, and K. Zeller, eds., Birkhäuser-Verlag, Basel, 1989, 241-248.

[36] W. R. Madych, Polyharmonic splines, multiscale analysis, and entire functions, in *Multivariate Approximation and Interpolation*, W. Haußmann and K. Jetter, eds., Birkhäuser Verlag, Basel, 1990, 205-216.

[37] W. R. Madych and S. A. Nelson, Polyharmonic cardinal splines, *J. Approx. Theory* **60**, (1990), 141-156.

[38] W. R. Madych and S. A. Nelson, Polyharmonic cardinal splines: a minimization property, *J. Approx. Theory* **63**, (1990), 303-320.

[39] W. R. Madych and S. A. Nelson, Multivariate interpolation and conditionally positive definite functions II, *Math. Comp.* **54**, (1990), 211-230.

[40] W. R. Madych, Error estimates for interpolation by generalized splines, in *Curves and Surfaces*, P. J. Laurent, A. LeMehaute, and L. L. Schumaker, eds. Academic Press, Boston, 1991, 297-306.

[41] W. R. Madych and S. A. Nelson, Bounds on multivariate polynomials and exponential error estimates for multiquadric interpolation, *J. Approx. Theory*, **70**, (1992), pp. 94-114

[42] W. R. Madych, Miscellaneous error bounds for multiquadic and related interpolators, in [27], 121-138.

[43] W. R. Madych, Orthogonal wavelet bases for $L^2(R^n)$, *Fourier analysis: analytic and geometric aspects*, W.O. Bray et al, eds, Marcel Dekker, New York, 1994, 243-303.

[44] Y. Meyer, *Wavelets and Operators*, translated from 1990 French original by D. H. Salinger, Cambridge Studies in Advanced Mathematics 37, Cambridge Univ. Press, Cambridge, 1992.

[45] C. A. Micchelli, Interpolation of scattered data: distance matrices and conditionally positive definite functions, *Cons. Approx.* **2**, (1986), 11-22.

[46] M. Z. Nashed and G. G. Walter, General sampling theorems for functions in reproducing kernel Hilbert spaces, *Math. Con. Sig. Sys.*, **4**, (1991), 363-390.

[47] F. Natterer, *The Mathematics of Computerized Tomography*, John Wiley & Sons Ltd and B. G. Teubner, Stuttgart, 1986.

[48] R.E.A.C. Paley and N. Wiener, *Fourier transforms in the complex domain*, Colloq. Pub. 19, Amer. Math. Soc., Providence, 1934.

[49] B. S. Pavlov, The basis property of a system of exponentials and the condition of Muckenhoupt, *Dokl. Acad. Nauk SSSR*, **247**, no. 1, (1979), 37-40.

[50] E. M. Petersen and D. Middleton, Sampling and reconstruction of wave-number-limited functions in N-dimensional Euclidean spaces, *Inform. Control* **5**, (1962), 279-223.

[51] S. D. Riemenschneider, Convergence of interpolating cardinal splines: power growth, *Israel J. Math.* **23**, (1976), 339-346.

[52] C. E. Shannon, Communication in the presence of noise, *Proc. IRE*, **37**, (1949), 20-21.

[53] C. E. Shannon and W. Weaver, *The Mathematical Theory of Communication*, Univ. of Illinois Press, Urbana, 1949.

[54] I. J. Schoenberg, Contributions to the problem of approximation of equidistant data by analytic functions, Part A and B, *Quart. Appl. Math.* **4**, (1946), 45-99 and 112-141.

[55] I. J. Schoenberg, *Cardinal Spline Interpolation*, CBMS Vol. 12, SIAM, Philadelphia, 1973.

[56] I. J. Schoenberg, Cardinal interpolation and spline function VII: The behavior of cardinal spline interpolation as their degree tends to infinity, *J. Analyse. Math.* **27**, (1974), 205-229.

[57] I. J. Schoenberg, On the remainders and the convergence of cardinal spline interpolation for almost peroidic functions, in *Studies in Spline Functions and Approximation Theory,* S. Karlin *et al.*, eds., Academic Press, Boston, 1976, 277-303.

[58] D. Slepian, On bandwidth, *Proc. IEEE,* **64**, (1976), 292-300.

[59] E. M. Stein and G. Weiss, *Introduction to Fourier Analysis on Euclidean Spaces*, Princeton University Press, Princeton, 1971.

[60] D. Walnut, Nonperiodic sampling of band limited functions on unions of rectangular lattices, *J. Fourier Anal. Appl.* **2**, (1996), 435-452.

[61] G. G. Walter, *Wavelets and Other Orthogonal Systems with applications,* CRC Press, Boca Raton, 1994.

[62] E. T. Whittaker, On functions which are represented by the expansions of the interpolatory theory, *Proc. Roy. Soc. Edinburgh* **35**, (1915), 181-194.

[63] J. M. Whittaker, On the cardinal function of interpolation theory, *Proc. Edinburgh Math. Soc.* **1** (1929), 41-46.

[64] R. M. Young, *An Introduction to Nonharmonic Fourier Series,* Academic Press, New York, 1980.

[65] A. I. Zayed, *Advances in Shannon's Sampling Theory,* CRC Press, Boca Raton, 1993.

# Chapter 27

# $B$-Splines and orthonormal sets in Paley-Wiener space

**Ahmed I. Zayed**

*ABSTRACT. It is known that the cardinal B-splines can be used to generate a multiresolution analysis, and hence an orthonormal wavelet basis for $L^2(\mathbb{R})$. We show that they can also be used to generate orthonormal sets, as well as, frames in the Paley–Wiener space $B^2_\sigma$, which is also known as the space of bandlimited functions with bandwidth $\sigma$. A new orthonormal set of functions in $B^2_\sigma$, obtained from one single function by translating it by integer multiples of $2\pi/\sigma$, is given explicitly in terms of Young's function that was introduced by W. H. Young in 1912.*

## 27.1 Introduction

Sampling theory is the theory that deals with the reconstruction of certain classes of entire functions from discrete data associated with these functions. Chief among those classes of entire functions is the class $B^2_\sigma$ of bandlimited functions (signals), which is also known as the Paley-Wiener class of functions. Provided with the $L^2$-norm, $B^2_\sigma$ is a reproducing-kernel Hilbert subspace of $L^2(\mathbb{R})$.

It is known that the cardinal $B$-spline of order $n, (n = 0, 1, 2, ...), \phi_n(x)$, with knots at the integers, generates a Riesz basis of a subspace, $V$, of $L^2(\mathbb{R})$. The space $V$ is then used to construct a multiresolution analysis in which $V$ is the ground subspace $V_0$. Having constructed a multiresolution analysis, a mother wavelet ${}_n\psi$ can then be constructed, which in turn can be used to generate an orthonormal wavelet basis of $L^2(\mathbb{R})$ of the form

$$ {}_n\psi_{m,k}(x) = 2^{-m/2}\psi(2^{-m}x - k); \quad k, m \in \mathbb{Z}. $$

In this article we borrow some ideas from wavelet analysis and sampling theory to show that the Fourier transform of the square root of the cardinal $B$-splines can be used to generate orthonormal sets and frames in various Paley-Wiener spaces consisting of functions with different bandwidth.

Although our results are of a general nature and provide only existence of such orthonormal sets of functions, we have been able to calculate one such set in closed form using an old and forgotten special function, known as Young's function. Young's function, which was introduced by W. H. Young in 1912 [6], is regarded as an extension of the sine and cosine functions, but unlike the sine and cosine, Young's function is not in general periodic.

## 27.2 Preliminaries

Throughout this paper we adopt the following conventional notation. The set of real numbers is denoted by $\mathbb{R}$ and the set of integers by $\mathbb{Z}$. The Fourier transform of a function $f(t)$ is defined as

$$\hat{f}(\omega) = \mathcal{F}(f)(\omega) = \frac{1}{\sqrt{2\pi}} \int_{-\infty}^{\infty} f(t) e^{it\omega} dt,$$

and the convolution operation, $*$, is defined so that $\left(\widehat{f * g}\right)(\omega) = \hat{f}(\omega)\hat{g}(\omega)$. To conform with electric engineering terminology, the variables $\omega$ and $t$ will stand for frequency and time, respectively.

By a frame in a Hilbert space $\mathcal{H}$, we mean a sequence $\{g_n\}_{n=-\infty}^{\infty}$ of elements of $\mathcal{H}$ that satisfies the inequality

$$A\|f\|^2 \leq \sum_{n\in\mathbb{Z}} |\langle f, g_n\rangle|^2 \leq B\|f\|^2, \tag{27.1}$$

for any $f \in \mathcal{H}$, where $A$ and $B$ are positive numbers that depend on the sequence $\{g_n\}_{n=-\infty}^{\infty}$ but not on $f$. The constants $A$ and $B$ are called the frame bounds. If $A = B$, the frame is said to be tight and if the sequence $\{g_n\}_{n=-\infty}^{\infty}$ ceases to be a frame whenever any single element is deleted from it, the frame is said to be exact.

A sequence $\{\lambda_n\}_{n=-\infty}^{\infty}$ of real numbers is said to have uniform density $d, d > 0$, if there are positive constants $L$ and $\delta$ such that

$$\left|\lambda_n - \frac{n}{d}\right| \leq L\,, \qquad n = 0, \pm 1, \pm 2, \dots\,,$$

and

$$|\lambda_n - \lambda_m| \geq \delta > 0\,, \quad n \neq m\,.$$

Let $\{\gamma_n\}_{n=-\infty}^{\infty}$ be a sequence of real numbers such that $\gamma_0 = 0$. We shall say that a function $S(t)$ is a sampling function with respect to the sequence $\{\gamma_n\}_{n=-\infty}^{\infty}$ if

$$S(\gamma_n) = \begin{cases} 0 & \text{if } n \neq 0 \\ 1 & \text{if } n = 0\,. \end{cases}$$

Throughout this article, we assume that $S$ is either in $L^1(\mathbb{R})$ or $L^2(\mathbb{R})$.

The Paley-Wiener space of bandlimited functions, denoted by $B_\sigma^2$, is defined as the space of all functions $f(t) \in L^2(\mathbb{R})$ whose Fourier transforms, $\hat{f}$, have support in $[-\sigma, \sigma]$, or equivalently it is the space of all functions $f(t)$ that can be written in the form

$$f(t) = \frac{1}{\sqrt{2\pi}} \int_{-\sigma}^{\sigma} F(\omega) e^{-it\omega} d\omega,$$

for some $F \in L^2(-\sigma, \sigma)$.

It is evident that the Fourier transformation is an isometric isomorphism from $L^2(-\sigma, \sigma)$ onto $B_\sigma^2$. Therefore, it is easy to show that if $\{\gamma_n\}_{n=-\infty}^{\infty}$ is a sequence of real numbers such that $\{e^{i\gamma_n \omega}\}_{n=-\infty}^{\infty}$ is a frame in $L^2(-\sigma, \sigma)$, then $\{\sin \sigma(t - \gamma_n)/[\sigma(t - \gamma_n)]\}_{n=-\infty}^{\infty}$ is also a frame in $B_\sigma^2$. One can easily verify that the space $B_\sigma^2$ has an orthogonal basis that is generated from one single function, $\phi$, by translating it by integer multiples of $\pi/\sigma$. More precisely, $\{\phi(t - t_n)\}_{n=-\infty}^{\infty}$, is an orthogonal basis of $B_\sigma^2$, where

$$\phi(t) = \sin(\sigma t)/(\sigma t),$$

and $t_n = n\pi/\sigma$. Moreover, the function $\phi(t)$ is readily seen to be a sampling function with respect to the sequence $\{t_n\}_{n=-\infty}^{\infty}$. As a consequence of that, we have for any $f \in B_\sigma^2$

$$f(t) = \sum_{n \in \mathbb{Z}} f(t_n) \frac{\sin \sigma(t - t_n)}{\sigma(t - t_n)},$$

which is known as the Whittaker-Shannon-Kotel'nikov sampling theorem for bandlimited functions [8].

Young's functions were introduced by W. H. Young in 1912 [6] in his investigation of non-converging Fourier series. They also appeared in connection with the Hardy transform, which generalizes the Hankel and the $Y$ transforms; for more details, see [7]. Young's function of order $\nu (\nu \geq 0)$, which will be denoted by $\mathbf{Y}_\nu$ to distinguish it from the Bessel function of the second kind $Y_\nu$, is defined by

$$\mathbf{Y}_\nu(z) = z^\nu \sum_{k=0}^{\infty} \frac{(-1)^k z^{2k}}{\Gamma(\nu + 2k + 1)}. \tag{27.2}$$

Clearly, $\mathbf{Y}_0(z) = \cos z$ and $\mathbf{Y}_1(z) = \sin z$. For $\nu \geq 0, \mathbf{Y}_\nu$ is entire if $\nu$ is an integer and is analytic in the plane cut along $[-\infty, 0]$ if $\nu$ is arbitrary. Moreover, $\mathbf{Y}_\nu(z)/z^\nu$ is an entire function of exponential type as can be seen from [5, p. 5]. For $0 < \nu < 1$, $\mathbf{Y}_\nu$ is no longer periodic, but fills up the analytical gap between the sine and cosine functions. For $0 \leq \nu \leq 2$,

$\mathbf{Y}_\nu$ is bounded in $x$ and $\nu$, where $x = \operatorname{Re} z$, and for $1 < \nu, x^{-\nu}\mathbf{Y}_\nu(x)$ is of bounded variation on $(0, \infty)$. The following relations are easy to establish:

$$\mathbf{Y}_\nu(x) = \frac{d}{dx}\mathbf{Y}_{\nu+1}(x)\,, \text{ and } \mathbf{Y}_\nu(x) = \frac{x^{\nu-2}}{\Gamma(\nu-1)} - \mathbf{Y}_{\nu-2}(x).$$

When combined together, they yield the differential equation for $\mathbf{Y}_\nu$

$$y'' + y = \frac{x^{\nu-2}}{\Gamma(\nu-1)}\,, \qquad \nu > 1$$

which, in turn, for $\nu = 0$ and 1, reduces to $y'' + y = 0$.

## 27.3 Sampling and Orthonormal Functions

In this section we show how sampling and orthogonal functions are related.

**Lemma 27.1** *(a) Let $\psi \in L^2(\mathbb{R})$ and $t_n = \pi n/\sigma, \sigma > 0$. Then $\{\psi(t - t_n)\}_{n=-\infty}^{\infty}$ is an orthonormal sequence in $L^2(\mathbb{R})$ if and only if*

$$\sum_{k=-\infty}^{\infty} |\hat{\psi}(\omega + 2k\sigma)|^2 = \frac{1}{2\sigma},$$

*and $\{\psi(t - t_n)\}_{n=-\infty}^{\infty}$ is orthogonal if and only if*

$$\sum_{k=-\infty}^{\infty} |\hat{\psi}(\omega + 2k\sigma)|^2 = \textit{constant} = c.$$

*(b) If $\{\gamma_n\}_{n=-\infty}^{\infty}$ is a sequence of real numbers such that*

$$|\gamma_n - n\pi/\sigma| \le \alpha < \pi/(4\sigma),$$

*then $\{\psi(t - \gamma_n)\}_{n=-\infty}^{\infty}$ is complete in $B_\sigma^2$ if $\hat{\psi}(\omega) \neq 0$ a.e in $[-\sigma, \sigma]$.*

*(c) If $\{\gamma_n\}$ has uniform density $d$ and $0 < A \le |\hat{\psi}(\omega)| \le B < \infty$, then $\{\psi(t - \gamma_n)\}_{n=-\infty}^{\infty}$ is a frame in $B_\sigma^2$, where $0 < \sigma < \pi d$.*

**Proof.** (a) The proof is standard and can be found in [2, p. 123].
(b) Let $f \in B_\sigma^2$ and assume that

$$\int_{-\infty}^{\infty} f(t)\overline{\psi}(t - \gamma_n)dt = 0 \quad \text{for all } n.$$

Then by Parseval's relation, we have

$$\int_{-\sigma}^{\sigma} \hat{f}(\omega)\overline{\hat{\psi}}(\omega)e^{-i\gamma_n\omega}d\omega = 0 \quad \text{for all } n.$$

But since $\{e^{i\gamma_n\omega}\}_{n=-\infty}^{\infty}$ is complete in $L^1(-\sigma,\sigma)$ [3], it follows that $\hat{f}(\omega)\overline{\hat{\psi}}(\omega) = 0$ a.e in $[-\sigma,\sigma]$, and in view of the fact that $\hat{\psi}(\omega) \neq 0$ a.e in $[-\sigma,\sigma]$, we have $\hat{f}(\omega) = 0$ a.e, and hence $f = 0$ a.e.
(c) The proof is similar to that of part (b), except we use Theorem I in [3].
■

**Lemma 27.2** *Let $S$ be a function in $L^1(\mathbb{R})$ or $L^2(\mathbb{R})$ such that*

$$\sum_{k\in\mathbb{Z}} \left|\widehat{S}(\omega + 2k\sigma)\right| \leq B < \infty.$$

*Then $S$ is a sampling function with respect to the sequence $\{t_n = \pi n/\sigma\}_{n=-\infty}^{\infty}$ if and only if*

$$\sum_{k=-\infty}^{\infty} \hat{S}(\omega + 2k\sigma) = \frac{\sqrt{\pi}}{\sqrt{2}\sigma}\ .$$

**Proof.** Since

$$\begin{aligned}
\int_{-\infty}^{\infty} |\hat{S}(\omega)|d\omega &= \sum_{k=-\infty}^{\infty} \int_{\sigma(2k-1)}^{\sigma(2k+1)} |\hat{S}(\omega)|d\omega \\
&= \int_{-\sigma}^{\sigma} \sum_{k=-\infty}^{\infty} |\hat{S}(\omega + 2k\sigma)|d\omega \leq 2\sigma B < \infty,
\end{aligned}$$

it follows that $\hat{S}(\omega) \in L^1(\mathbb{R})$, and therefore we can write

$$\begin{aligned}
S(t) &= \frac{1}{\sqrt{2\pi}} \int_{-\infty}^{\infty} \hat{S}(\omega)e^{-it\omega}d\omega = \frac{1}{\sqrt{2\pi}} \sum_{k=-\infty}^{\infty} \int_{\sigma(2k-1)}^{\sigma(2k+1)} \hat{S}(\omega)e^{-it\omega}d\omega \\
&= \frac{1}{\sqrt{2\pi}} \sum_{k=-\infty}^{\infty} \int_{-\sigma}^{\sigma} \hat{S}(\omega + 2\sigma k)e^{-it(\omega+2k\sigma)}d\omega \\
&= \frac{1}{\sqrt{2\pi}} \int_{-\sigma}^{\sigma} e^{-it\omega} \left( \sum_{k=-\infty}^{\infty} \hat{S}(\omega + 2\sigma k)e^{-2ik\sigma t} \right) d\omega.
\end{aligned}$$

Interchanging the summation and integration signs is permissible by the Lebesgue dominated convergence theorem. Thus,

$$S(t_n) = \frac{1}{\sqrt{2\pi}} \int_{-\sigma}^{\sigma} e^{-it_n\omega} \left( \sum_{k=-\infty}^{\infty} \hat{S}(\omega + 2k\sigma) \right) d\omega.$$

But since $S(t_n) = \delta_{n,0}$ for all $n$, it follows that

$$\sum_{k=-\infty}^{\infty} \hat{S}(\omega + 2k\sigma) = \frac{\sqrt{\pi}}{\sqrt{2}\sigma}.$$

■

**Theorem 27.3** *Let $S$ be a sampling function with respect to the sequence $\{t_n = n\pi/\sigma\}_{n\in\mathbb{Z}}$ such that $\hat{S}(w) > 0$ a.e. Then $S$ generates an orthonormal family of functions $\{\psi_n(t)\}_{n\in\mathbb{Z}}$ in $L^2(\mathbb{R})$, each of which is obtained from one single function $\psi$ by a translation by an integer multiple of $\pi/\sigma$, namely, $\psi_n(t) = \psi(t - t_n)$ for all $n \in \mathbb{Z}$.*

*Conversely, any orthonormal family in $L^2(\mathbb{R})$ that is generated from one single function by translations by $t_n$ can be obtained from a sampling function (with respect to the sequence $\{t_n\}_{n\in\mathbb{Z}}$ ) with positive Fourier transform.*

**Proof.** Define $\hat{\psi}(\omega)$ as

$$\hat{\psi}(\omega) = \frac{1}{\sqrt[4]{2\pi}} \sqrt{\hat{S}(\omega)} h(\omega),$$

where $h(\omega)$ is a complex-valued function with $|h(\omega)| = 1$. Then in view of Lemmas 27.1 and 27.2, we have

$$\sum_{k=-\infty}^{\infty} |\hat{\psi}(\omega + 2k\sigma)|^2 = \frac{1}{\sqrt{2\pi}} \sum_{k=-\infty}^{\infty} \hat{S}(\omega + 2k\sigma) = \frac{1}{2\sigma};$$

hence $\{\psi(t-t_n)\}_{n\in\mathbb{Z}}$ is orthonormal in $L^2(\mathbb{R})$. Conversely, let $\{\psi(t-t_n)\}_{n\in\mathbb{Z}}$ be an orthonormal family in $L^2(\mathbb{R})$. Then by Lemma 27.1,

$$\sum_{k=-\infty}^{\infty} |\hat{\psi}(\omega + 2k\sigma)|^2 = 1/2\sigma.$$

Put $\hat{S}(\omega) = \sqrt{2\pi}|\hat{\psi}(\omega)|^2$. Clearly $\hat{S} \in L^1(\mathbb{R})$ with $\hat{S} > 0$ a.e. Moreover,

$$\sum_{k=-\infty}^{\infty} \hat{S}(\omega + 2k\sigma) = \sqrt{2\pi} \sum_{k=-\infty}^{\infty} |\hat{\psi}(\omega + 2k\sigma)|^2 = \frac{\sqrt{\pi}}{\sqrt{2}\sigma},$$

which, in view of Lemma 27.2, implies that $S$ is a sampling function with respect to the sequence $\{t_n\}_{n\in\mathbb{Z}}$. In fact, $S$ can be defined directly by setting

$$S(t) = \int_{-\infty}^{\infty} \psi(x)\overline{\psi}(x-t)dx,$$

which is well defined as can be seen by applying the Cauchy-Schwarz inequality. Moreover,

$$S(t_n) = \int_{-\infty}^{\infty} \psi(x)\overline{\psi}(x-t_n)dx = \delta_{n,0} \quad \text{for all } n$$

and

$$\hat{S}(\omega) = \sqrt{2\pi}|\hat{\psi}(\omega)|^2.$$

■

**Theorem 27.4** *Let $t_n = n\pi/\sigma$, $n \in \mathbb{Z}$ and $\hat{\psi} \in L^2[-\sigma,\sigma]$.*

*(a) Then $\{\psi(t-t_n)\}_{n\in\mathbb{Z}}$ is an orthonormal basis of $B^2_\sigma$ if and only if $|\hat{\psi}(\omega)| = 1/\sqrt{2\sigma}$, a.e.*

*(b) In addition, let $\psi(t)$ be a sampling function with respect to the sequence $\{t_n\}_{n\in\mathbb{Z}}$. Then $\{\psi(t-t_n)\}_{n\in\mathbb{Z}}$ is an orthonormal basis of $B^2_\sigma$ if and only if $\sigma = \pi$ and $\psi(t) = \sin\pi t/\pi t$.*

**Proof.** (a) The proof is straightforward (see [2, p. 132]).
(b) The "if" part is trivial. Now let $\psi \in B^2_\sigma$ be a sampling function with respect to the sequence $\{t_n\}_{n\in\mathbb{Z}}$. By Lemma 27.2, we must have

$$\tilde{\Psi}(\omega) = \sum_{k=-\infty}^{\infty} \hat{\psi}(\omega + 2k\sigma) = \frac{\sqrt{\pi}}{\sqrt{2}\sigma},$$

where $\tilde{\Psi}$ is a periodic function with period $2\sigma$. Since $\{\psi(t-t_n)\}_{n\in\mathbb{Z}}$ is an orthogonal basis of $B^2_\sigma$, the support of $\hat{\psi}$ must be contained in $[-\sigma,\sigma]$ and $\hat{\psi}(\omega) > 0$ a.e thereon. Therefore, $\hat{\psi}(\omega) = \sqrt{\pi}/(\sqrt{2}\sigma)$ a.e on $(-\sigma,\sigma)$, which implies that $\psi(t) = \sin\sigma t/\sigma t$. But from part (a) in order for $\{\psi(t-t_n)\}_{n\in\mathbb{Z}}$ to be orthonormal, we must also have $|\hat{\psi}(\omega)|^2 = 1/2\sigma$. Thus, $\sigma = \pi$. ■

## 27.4 $B$-splines and Orthonormal Sets in the Paley–Wiener Space

In this section we accomplish the main aim of this article, which is to show how the cardinal $B$-splines can be used to generate orthonormal sets and frames in the Paley–Wiener space $B^2_\sigma$, for some $\sigma$.

We define the cardinal $B$-spline, $\phi_n$, of order $n$ as follows:

$$\phi_0(\omega) = \chi_{(-\sigma/2,\sigma/2)}(\omega)\,, \qquad \phi_1(\omega) = \sqrt{2\pi}(\phi_0 * \phi_0)(\omega),$$

and

$$\phi_n(\omega) = \sqrt{(2\pi)}(\phi_{n-1} * \phi_0)(\omega), \quad n = 1, 2, \dots ,$$

where $\chi_A$ is the characteristic function of $A$. It is easy to verify that

$$\phi_n(\omega) > 0 \text{ in } (-(n+1)\sigma/2, (n+1)\sigma/2)\,,$$

and

$$\hat{\phi}_n(t) = (2\pi)^{n/2}\left[\hat{\phi}_0(t)\right]^{n+1} = (\sigma^{n+1}/\sqrt{2\pi})\left(\frac{\sin(\sigma t/2)}{\sigma t/2}\right)^{n+1},$$

where $\hat{\phi}_0(t) = \frac{\sigma}{\sqrt{2\pi}}\left[\sin(\sigma t/2)/(\sigma t/2)\right]$.

By applying multiresolution analysis techniques to $\phi_n(\omega)$, for fixed $n$, a system of orthonormal wavelet basis, known as *Battle-Lemarié wavelets*, can be constructed [2, p. 146 ] on the frequency line. Now we show how the cardinal $B$-splines can be used to generate orthonormal sets in the Paley–Wiener space. But first, let us recall that the set $\{\phi(t - k\pi/\sigma)\}_{k=-\infty}^{\infty}$ is orthonormal in $B^2_\sigma$, where

$$\phi(t) = \sqrt{\sigma/\pi}\left(\frac{\sin \sigma t}{\sigma t}\right).$$

In terms of Young's function, $\phi$ can be rewritten as

$$\phi(t) = \sqrt{\sigma/\pi}\,\mathbf{Y}_1(\sigma t)/(\sigma t)\,.$$

In the next theorem we show, among other things, that there is another orthonormal system in $B^2_\sigma$ that is generated by Young's function of order 3/2.

**Theorem 27.5** *(a) Let $\phi_n(w)$ be the cardinal $B$-spline of order $n$ defined above. Fix $n$ and define $\psi_n(t)$ by*

$$\hat{\psi}_n(\omega) = \frac{1}{\sigma^{(n+1)/2}}\sqrt{\phi_n(\omega)},$$

*and set $\psi_{n,k}(t) = \psi_n\left(t - \frac{2k\pi}{\sigma}\right)$, $k = 0, \pm 1, \pm 2, \dots$ . Then $\{\psi_{n,k}(t)\}_{k=-\infty}^{\infty}$ is an orthonormal set in the Paley-Wiener space $B^2_{(n+1)\sigma/2}$. In particular, for $n = 1$, we have $\left\{\psi_1\left(t - \frac{2k\pi}{\sigma}\right)\right\}_{k\in\mathbb{Z}}$ is orthonormal in $B^2_\sigma$, with*

$$\psi_1(t) = \sqrt{\frac{\sigma}{2}}\frac{\mathbf{Y}_{3/2}(\sigma t)}{(\sigma t)^{3/2}},$$

*and $\mathbf{Y}_{3/2}(z)$ is Young's function of order $\frac{3}{2}$.*

(b) *The set* $\left\{\psi_{1,k}(t) = \psi_1\left(t - \frac{2k\pi}{\sigma}\right)\right\}_{k\in\mathbb{Z}}$ *is not a basis of* $B^2_\sigma$*; nevertheless, it is a frame in* $B^2_\lambda$ *for any* $0 < \lambda < \sigma/2$.

**Proof.** (a) It is evident that the function

$$S_n(t) = \left(\frac{\sin(\sigma t/2)}{\sigma t/2}\right)^{n+1}$$

is a sampling function with respect to the points $2k\pi/\sigma$, $k \in \mathbb{Z}$. Its Fourier transform $\hat{S}_n(\omega) = \sqrt{2\pi}/\sigma^{n+1}\phi_n(\omega)$ is positive. Therefore, by Theorem 27.3, the function $\psi_n(t)$, defined by

$$\hat{\psi}_n(\omega) = \frac{1}{\sqrt[4]{2\pi}}\sqrt{\hat{S}_n(\omega)} = \frac{1}{\sigma^{(n+1)/2}}\sqrt{\phi_n(\omega)}\,,$$

gives rise to an orthonormal set of functions in $L^2(\mathbb{R})$, each of which is obtained from $\psi_n$ by a translation by an integer multiple of $2\pi/\sigma$. Thus, we have by Theorem 27.3 that $\left\{\psi_{n,k}(t) = \psi_n\left(t - 2k\pi/\sigma\right)\right\}_{k\in\mathbb{Z}}$ is orthonormal in $L^2(\mathbb{R})$. But since the support of $\hat{\psi}_n(\omega)$ is $[-(n+1)\sigma/2, (n+1)\sigma/2]$, it follows that $\psi \in B^2_{(n+1)\sigma/2}$. For $n = 1$, we have

$$\hat{\psi}_1(\omega) = \begin{cases} \frac{1}{\sigma}\sqrt{\sigma - |\omega|} & , \quad |\omega| \le \sigma \\ 0 & , \quad |\omega| \ge \sigma. \end{cases}$$

In view of formula (5) [4, p. 11], we have

$$\psi_1(t) = \frac{1}{3}\sqrt{\frac{2\sigma}{\pi}}\left[{}_1F_1(1;\frac{5}{2};i\sigma t) + {}_1F_1(1;\frac{5}{2};-i\sigma t)\right],$$

where ${}_1F_1$ is the confluent hypergeometric function given by

$${}_1F_1(a;b;z) = \sum_{k=0}^{\infty}\frac{(a)_k z^k}{(b)_k k!}, \text{ with } (a)_k = \frac{\Gamma(a+k)}{\Gamma(a)}.$$

With some easy calculations, we can show that

$$\begin{aligned}
{}_1F_1(1;\frac{5}{2};i\sigma t) + {}_1F_1(1;\frac{5}{2};-i\sigma t) &= \sum_{k=0}^{\infty}\frac{1}{(5/2)_k}[(i)^k + (-i)^k](\sigma t)^k \\
&= 2\sum_{k=0}^{\infty}\frac{(-1)^k(\sigma t)^{2k}}{(5/2)_{2k}} \\
&= \frac{3\sqrt{\pi}}{2}\sum_{k=0}^{\infty}\frac{(-1)^k(\sigma t)^{2k}}{\Gamma(2k+3/2+1)} \\
&= \frac{3\sqrt{\pi}}{2}\frac{\mathbf{Y}_{3/2}(\sigma t)}{(\sigma t)^{3/2}},
\end{aligned}$$

which implies that

$$\psi_1(t) = \sqrt{\frac{\sigma}{2}} \frac{\mathbf{Y}_{3/2}(\sigma t)}{(\sigma t)^{3/2}},$$

where $\mathbf{Y}_{3/2}(2)$ is Young's function of order 3/2 (cf. Equation (27.2) ).
(b) This follows immediately from part (b) of Lemma 27.1. ■

The following corollary is a special case of Corollary 1 in Young's paper [6].

**Corollary 27.6** *For $Y_{3/2}(t)$, we have*

$$\int_0^\infty \frac{Y_{3/2}(x)}{x^{3/2}} dx = \sqrt{\pi}.$$

**Proof.** Let $\sigma = \pi$. Then we have

$$\begin{aligned} \hat{\psi}_1(0) &= \frac{1}{\sqrt{\pi}} = \frac{1}{\sqrt{2\pi}} \int_{-\infty}^{\infty} \psi_1(t) dt \\ &= \frac{1}{2} \int_{-\infty}^{\infty} \frac{Y_{3/2}(\pi t)}{(\pi t)^{3/2}} dt = \frac{1}{\pi} \int_0^\infty \frac{Y_{3/2}(x)}{x^{3/2}} dx. \end{aligned}$$

■

*References*

[1] C. Chui, *An Introduction to Wavelets,* Academic Press, New York (1992).

[2] I. Daubechies, *Ten Lectures on Wavelets,* SIAM Publications, Soc. Indust. Appl. Math., Philadelphia (1992).

[3] R. Duffin and A. Schaeffer, A class of nonharmonic Fourier series, *Trans. Amer. Math. Soc.*, Vol. 72 (1952), pp. 341–366.

[4] A. Erdelyi, W. Magnus, F. Oberhettinger, and F. Tricomi, *Tables of Integral Transforms,* Vol. I, II, McGraw-Hill, New York (1954).

[5] B. Ja. Levin, *Distribution of Zeros of Entire Functions*, Transl. Math. Monographs, Vol. 5, Amer. Math. Soc., Providence, RI (1964).

[6] W.H. Young, On infinite integrals involving a generalization of the sine and cosine functions, *Quart. J. Math.,* Vol. 4 (1912), pp. 161—177.

[7] A. I. Zayed, *Function and Generalized Function Transformations,* CRC Press, Boca Raton, Fl (1996).

[8] A. I. Zayed, *Advances in Shannon's Sampling Theory,* CRC Press, Boca Raton, Fl (1993).

# Chapter 28

# Norms of powers and a central limit theorem for complex-valued probabilities

**Bogdan Baishanski**

## 28.1 Introduction

The title of this article could be: "A New Way of Looking at Some Old Results". As one can expect, the new way of looking leads to new results.

After introducing some notation we present first the results obtained mainly in 1950's about the following problem:

**Problem 28.1** *Let* $f(z) = \sum_{0}^{\infty} a_\nu z^\nu$, $\|f\| = \sum |a_\nu| < \infty$*; under which conditions on* $f$ *is the sequence* $\|f^n\|$ *bounded? (Since*

$$\|f^n\|^{1/n} \to \max\{|f(z)|\colon |z| = 1\},$$

*we need to consider only the case when* $\max\{|f(z)|\colon |z| = 1\} = 1$.*)*

This problem is still open, but two results of sufficient generality which were obtained in '50's have some important applications. People working on these problems became aware only in late '60's of the work done (also in the fifties) on the identical problem for absolutely convergent Fourier series. It was realized that the restriction to the power series is artificial, and that with only trivial modifications everything proved for power series worked also in the case of trigonometric series.

Next, we present results on the asymptotic behavior of $\|f^n\|$; several of these results were obtained by the author and his students, and they were obtained without any application in mind. The methods here are ad hoc, and the proofs, sometimes ingenious, are lengthy and technical; the papers are certainly not pleasant to read. So, when we returned to these topics recently and obtained some new results, we tried to make proofs more transparent. Still, the asymptotic behavior of the norms $\|f^n\|$ remained an isolated topic.

However, the attempt to obtain a more transparent presentation of old results led us to better realize the connections of our topic with probability theory. We explain how our awareness of these connections went through three stages, and led us to change our point of view and see for the first time that our topic was (or should have been) the description of the behavior of $n$-fold convolutions $\underbrace{c * c * \cdots * c}_{n \text{ times}}$ as $n \to \infty$, where $c = (c_\nu)_{-\infty}^{\infty} \in \ell_1$. In other words, our topic was (or should have been) the search for an analogue of the Central Limit Theorem for complex-valued probabilities.

After we presented Theorems 28.11 and 28.12 as a Central Limit Theorem for complex-valued probabilities, we learned from M. Pinsky that these problems have risen in different contexts (numerical analysis, partial differential equations, ...) and that versions of the Central Limit Theorem for signed probabilities already existed ([39], [29], [21], [23]). It turns out, however, that our version of the Central Limit Theorem differs greatly from the other versions. We close the article by pointing out these differences.

All theorems in this article, with the exception of Theorem 28.13 and part of Theorem 28.2, are due to the author and his students. (We need to add, however, that Theorem 28.6 is essentially our rediscovery of the Edgeworth-Esseen expansion.)

Perhaps a better way of organizing this article would have been to recognize here three distinct topics:

1. Theorem 28.1, with its refinements: Theorems 28.4 and 28.5, and – what we mention in passing – its applications (equivalent power series, ...) and relations with more general results (Leibenzon, J. P. Kahane, Beurling-Helson, P. Cohen);

2. the unappealing Theorem 28.7 which suprisingly leads to some neat results (Theorems 28.8, 28.9, 28.10) and is related to mathematical statistics via Theorem 28.6; and

3. the main topic of this article – relation between Theorems 28.2, 28.3, 28.11, 28.12, 28.13.

## 28.2 The Five Parameters

In the case of real-valued real-analytic functions $f$, a strict local maximum at $x_0$ is characterized by two numbers: an integer $q = \min\{n\colon f^{(n)}(x_0) \neq 0\}$ and the real number $f^{(n)}(x_0)$. If $f$ is a complex-valued analytic function of a real variable defined in a neighborhood of $x_0$, and if the modulus of $f$ attains a strict local maximum at $x_0$, we shall use five parameters to characterize that maximum: integers $p$ and $q$, and real numbers $\alpha$, $a$, $b$. The integer $q$ is defined by

$$\log f(x_0 + t) - \log f(x_0) = \sum_{j=1}^{q} A_j t^j + o(t^q), \qquad t \to 0 \tag{28.1}$$

where $Re\, A_j = 0$ for $j = 1, 2, \ldots, q-1$, $Re\, A_q \neq 0$. The remaining four parameters are defined by

$$p = \min\{j : j \geq 2\,,\ A_j \neq 0\}$$
$$\alpha = -iA_1\,, \qquad a = Im\, A_p\,, \qquad b = -Re\, A_q\,.$$

It is easy to see that $\alpha$ is real, $2 \leq p \leq q$, $q$ is even, $b > 0$, and, if $p \neq q$, then $a \neq 0$. The principal role play parameters $p$ and $q$, and the crucial difference is between cases $p = q$ (the regular case) and $p \neq q$. Typically we expect that at a local maximum of $|f|$ we have $\frac{d^2}{dt^2}|f(t)| \neq 0$, in that case $q = 2$, and so the generic case is regular.

In most results presented in this article it is not assumed that $f$ is real-analytic, but (unless $|f|$ is constant) it is assumed that at any point where $|f|$ attains its *absolute* maximum in the circle (or on the line) the condition (28.1) is satisfied, so that at each such point the five parameters are well defined.

## 28.3 Boundedness

### *28.3.1 Power Series*

The problem of boundedness of the sequence $\|f^n\|$ arose in several contexts independently. Karamata raised that problem for the bilinear functions $g(z) = \frac{az+b}{cz+d}$ (normalized so that $g(1) = 1$, $|g(z)| \leq 1$ for $|z| \leq 1$), since he was interested in summability methods defined by matrices $[a_{n\nu}]$ where $g^n(z) = \sum_0^\infty a_{n\nu} z^\nu$. (Such summability methods would contain both the Euler-Knopp and Meyer-König methods.) In a letter to Karamata, Szegö sketched a proof that $\|g^n\| = \mathcal{O}(1)$, $n \to \infty$, assuming that $|g|$ attains the maximum on the circle only at $z = 1$. He expressed coefficients $a_{n\nu}$ by Cauchy's formula, then estimated them by using the saddle point method, choosing as path of integration a circle $|z + \varepsilon| = 1 + \varepsilon$ where $\varepsilon > 0$ was sufficiently small so that $|g(z)| < 1$ at all points of the circle except at $z = 1$.

The most interesting case, when $g$ is an automorphism of the unit disk: $g(z) = e^{\theta i}\frac{z - z_0}{1 - \overline{z_0} z}$, $0 < |z_0| < 1$ was solved in [3], where it was proved that

**Theorem 28.1** *If $B$ is a finite Blaschke product, (with the exception of the trivial case $B(z) = e^{\theta i} z^\nu$), then $\|B^n\| \to \infty$, $n \to \infty$.*

Since this is one of the most important results we present in this article, and since the proof is elementary, we shall sketch the proof.

Let $c_n = 1/\max_\nu |a_{n\nu}|$. Since

$$\|B^n\| = \sum_\nu |a_{n\nu}| \geq c_n \sum_\nu |a_{n\nu}|^2$$
$$= c_n \cdot \frac{1}{2\pi} \int_{-\pi}^{\pi} |B^n(e^{ti})|^2 dt = c_n\,,$$

it is sufficient to show that $\max_\nu |a_{n\nu}| \to 0$, $n \to \infty$. But

$$a_{n\nu} = \frac{1}{2\pi} \int_{-\pi}^{\pi} \exp i(nh(t) - \nu t) dt$$

where $h(t) = \arg B(e^{ti})$. Given $\varepsilon > 0$, we remove around each of finitely many zeroes of $h''$ an interval, so that the total length of removed intervals is $< \varepsilon$. Let $\delta = \min |h''|$ on the remaining intervals. Applying the Van der Corput Lemma:

$$\text{"if } \varphi''(t) \geq \rho > 0 \text{ for } t \in (a, b), \text{then } |\int_a^b \exp(i\varphi(t))dt| \leq \frac{8}{\sqrt{\rho}}\text{"}$$

to each of the remaining intervals we obtain that $|a_{n\nu}| \leq \varepsilon + \frac{C}{\sqrt{n\delta}} < 2\varepsilon$ for $n > N(\varepsilon)$.

Another important result is

**Theorem 28.2** *Let $f$ be analytic on the closed unit disk.* $\max\{|f(z)|\colon |z| = 1\} = 1$. *Then, excluding the trivial case when $f$ is a monomial $cz^k$, the necessary and sufficient condition for the boundedness of the sequence $\|f^n\|$ is that $p = q$ at each point where $|f(e^{ti})| = 1$.*

Sufficiency was proved in [3] by estimating $|a_{n\nu}|$; a different, simpler proof was obtained later by Indlekofer [26] and Newman [32], each of them bases the proof on an inequality, in one case the inequality is

$$\left(\sum |d_\nu|\right)^4 \leq M \sum |d_\nu|^2 \sum (1 + (\nu - c)^2)|d_\nu|^2\,,$$

where $M$ is an absolute constant, and $c$ is an aribtrary real number, in the other case the inequality is slightly different: $1 + (\nu - c)^2$ above is replaced by $(\nu - c)^2$ and it is assumed that at least two terms of sequence $(d_\nu)$ are different from zero.

One part of the necessity condition is given by Theorem 28.1, the other part was proved in [10] where it was shown that if $p \neq q$ at one of the points where $|f(e^{ti})| = 1$, then $\sum_\nu |a_{n\nu}|^2$ tends to zero slower than $\max_\nu |a_{n\nu}|$, which implies that $\sum_\nu |a_{n\nu}| \to \infty$.

Theorem 28.1 implies, as was shown by Turan [37], a somewhat surprising fact that the peripheral convergence is not a conformal invariant, namely that the convergence of the MacLaurin series of $f$ at $z = 1$ does not imply that the MacLaurin series of $g = f \circ \varphi$, where $\varphi(z) = \frac{z-\alpha}{1-\alpha z}$, $0 < \alpha < 1$, also converges at $z = 1$. This observation has led to a number of articles on "equivalent power series" and related topics ([1], [2], [4], [11], [38] and see [26] for further references).

Theorem 28.2 was generalized to several dimensions ([19], [20]), applied to analytic continuation [5], and had other applications ([35], [36]).

### *28.3.2 Trigonometric Series*

In later sixties we became aware of three independent lines of research where essentially the same problems were investigated. One of the three, described in the preceding section, dealt with absolutely convergent power series; the other two dealt with absolutely convergent Fourier series:

$$f(x) = \sum_{-\infty}^{\infty} a_\nu e^{i\nu t}, \quad \|f\| = \sum_{-\infty}^{\infty} |a_\nu|.$$

On one side, since the stability of the difference schemes for the solution of partial differential equations reduces to the boundedness of the sequence $\|f^n\|$, $n = 1, 2, \ldots$, numerical analysts have investigated this topic (see references in [18]). On the other hand, the question of homomorphisms of Banach algebra of absolutely convergent Fourier series into itself can be reduced to the question: for which $f$ is the doubly infinite sequence $\|f^n\|$, $-\infty < n < \infty$, bounded? That question can be reformulated as follows: for which $f$ of constant modulus one is the sequence $\|f^n\|$, $n = 1, 2, \ldots$ bounded? The answer is: Only for $f$ of the form $f(t) = \exp(\alpha + kt)i$, where $\alpha$ is real, and $k$ is an integer ([9], [14]). This result contains as a special case Theorem 1, and it is itself a special case of Beurling-Helson theorem [9] and of the very general Paul Cohen's theorem.

As regards Theorem 28.2, with minimal changes the proofs given in [3], [5], [10], [26], [32] (and described in the previous section) remain valid when instead of requiring that $f$ be analytic in the closed unit disk, we require analycity only on the unit circle, replacing thus power series by trigonometric series.

## 28.4 Asymptotic Behavior

A further, non-trivial generalization of Theorem 28.2 is due to Hedstrom [18]; he does not assume that $f = \sum a_\nu e^{\nu t i}$ is real-analytic, but only that it is absolutely continuous and that $f'$ is of bounded variation. His results are more general in another aspect since he allows wider variety of behavior of $f$ at the points where $|f| = 1$, then it is allowed by (28.1). Also, he obtains precise order of growth for $\|f^n\|$ in all cases he investigates.

Even more precise results were obtained by D. Girard ([14], [16]). He proved the following:

**Theorem 28.3** *If $f$ is absoluely continuous, $2\pi$ periodic function such that $f'$ is of bounded variation and if $|f|$ attains its maximum, equal $1$, only at the point $0$, and $f$ is analytic at $0$, then the asymptotic behavior of $\|f^n\|$ depends only on the parameters $p$, $q$, $a$ and $b$ characterizing the maximum at zero. Namely,*

$$\text{if } p = q \quad \text{then } \|f^n\| \to \frac{1}{2\pi}\|\widehat{F}\|_{L^1} \tag{28.2}$$

*where $\widehat{F}$ is the Fourier transform of* $\exp(A_q t^q)$, *($A_q = -b + ia$) and*

$$\text{if } p \neq q\,, \quad \text{then } \|f^n\| \sim Cn^{\frac{1}{2}(1-p/q)}|a|^{\frac{1}{2}}b^{-\frac{p}{2q}}\,, \tag{28.3}$$

*where $C = \left(\frac{2}{\pi}\right)^{\frac{1}{2}+\delta(p)}\frac{(p(p-1))^{\frac{1}{2}}}{q}\Gamma\left(\frac{p}{2q}\right)$, $\delta(p) = 0$ if $p$ is even, and $= 1$ if $p$ is odd.*

Girard has also indicated how the analycity condition at 0 can be relaxed.

A stronger and more precise result than Theorem 28.1 was obtained by J. P. Kahane [27], [28]. He has shown that if $f$ is real analytic $2\pi$-periodic and $|f| = 1$, then $c\sqrt{n} \le \|f^n\| \le C\sqrt{n}$, with positive constants $c$ and $C$. Hedstrom [18] has relaxed the analycity condition, and, for the special case of an automorphism of the unit disk, Girard has obtained the asymptotic formula [15]:

**Theorem 28.4** *If $0 < \alpha < 1$, $\beta = \frac{1-\alpha}{1+\alpha}$, then*

$$\frac{1}{\sqrt{n}}\left\|\left(\frac{z-\alpha}{1-\alpha z}\right)^n\right\| = \frac{8\sqrt{2}}{\Gamma^2\left(\frac{1}{4}\right)}(1-\beta^2)^{\frac{1}{2}}F\left(\frac{1}{2},\,\frac{3}{4};\,\frac{3}{2};\,1-\beta^2\right) + o(1) \tag{28.4}$$

Recently, in his doctoral thesis, Jan Hlavacek has obtained the following generalization of Theorem 28.4 [22].

**Theorem 28.5** *Let $h$ be real, odd, three times continuously differentiable function, $h(t+2\pi) = h(t) + 2k\pi$ for some integer $k$. Let $h''(t) < 0$ for $t \in (0,\pi)$, and let both $h'''(0)$ and $h'''(\pi)$ be different from zero. Then*

$$\frac{1}{\sqrt{n}}\|\exp(inh(t))\| = \left(\frac{2}{\pi}\right)^{\frac{3}{2}}\int_0^\pi \sqrt{|h''(t)|}dt + \mathcal{O}(n^{-\frac{1}{10}})\,, \quad n \to \infty\,. \tag{28.5}$$

Since $h(t) = arg\frac{e^{ti}-\alpha}{1-\alpha e^{ti}}$ satisfies the conditions of Theorem 28.5, this theorem has as a corollary the Girard's asymptotic formula (28.4). We should mention, however, that the argument of a finite Blaschke product $B$ does not satisfy conditions of Theorem 28.5, unless all zeros of $B$ lie on the same ray, and we do not know whether $\frac{1}{\sqrt{n}}\|B^n\|$ converges in general case.

## 28.5 Asymptotic Series

Girard's result (28.2) has also been improved recently: instead of having just the limit of $\|f^n\|$, we have now the whole asymptotic series if $p = q = 2$ [8] (and also in case $p = q \geq 4$ if $a \neq 0$; see [34]). In [8] we have made an effort to write our proofs in a transparent way, and therefore, we first described the behavior of the coefficients $a_{n\nu}$, and then deduced the behavior of $\sum |a_{n\nu}|$. Both results are quite lengthy, but since they have some interesting consequences, we shall state them here.

**Theorem 28.6** *Let $f \in L^1(-\pi, \pi)$ and let $(a_{n\nu})$, $\nu = 0, \pm 1, \pm 2, \ldots$ be the Fourier coefficients of the powers $f^n$, $n = 1, 2, \ldots$. If*

1. (1) $\sup\{|f(t)| : \epsilon < |t| < \pi\} = M_\epsilon < 1$ *for every* $\epsilon > 0$ *and if*
2. (2) $\log f(t) = i\alpha t + At^2 + \sum_{j=1}^{k} B_j t^{j+2} + o(t^{k+2})$, $t \to 0$ *and*
3. (3) $\Re A \neq 0$

*then the following asymptotic expansion holds uniformly in $\nu$, $-\infty < \nu < \infty$, as $n \to \infty$*

$$a_{n\nu} = \frac{1}{2\sqrt{-A\pi}} \sum_{j=0}^{k} n^{-\frac{j+1}{2}} \exp\left(\frac{\gamma_{n\nu}^2}{4A}\right) Q_j(\gamma_{n\nu}) + o\left(n^{-\frac{k+1}{2}}\right)$$

*where* $\gamma_{n\nu} = \frac{\alpha n - \nu}{\sqrt{n}}$, $Q_0(\gamma) = 1$ *and*

$$Q_j(\gamma) = (-i)^{-j}(2\sqrt{-A})^{-j}$$
$$\sum \frac{1}{s_1! s_2! \ldots s_j!} B_1^{s_1} B_2^{s_2} \ldots B_j^{s_j} (4A)^{-s} H_{2s+j}\left(\frac{\gamma}{2\sqrt{-A}}\right)$$

*the sum being taken over all $j$-tuples $(s_1, s_2, \ldots, s_j)$ such that*

$$s_1 + 2s_2 + \ldots + js_j = j, \quad s = s_1 + s_2 + \ldots + s_j,$$

*$H_m$ is the Hermite polynomial of degree $m$, defined by*

$$H_m(x) = (-1)^m e^{x^2} \frac{d^m}{dx^m} e^{-x^2}$$

*and* $\sqrt{-A}$ *is uniquely determined by the condition* $\operatorname{Re}\sqrt{-A} > 0$.

From the theorem above and from some other estimates for non-central coefficients $a_{n\nu}$, we obtain

**Theorem 28.7** *Let* $f(t) = \sum_{-\infty}^{\infty} a_\nu e^{i\nu t}$ *be twice continuously differentiable,* $f(0) = 1$, $|f(t)| < 1$ *for* $0 < |t| \leq \pi$ *and* $\frac{d^2}{dt^2}|f(t)| \neq 0$ *at* $t = 0$. *Let* $||f|| = \sum |a_\nu|$. *If, in addition, d and h are positive integers such that* $d > 24(h+1)^3 + 1$, *and f is d times differentiable in a neighborhood of zero, then there exists an asymptotic expansion of* $||f^n||$ *in powers of* $n^{-1}$:

$$||f^n|| = c_0 + \frac{c_1}{n} + \frac{c_2}{n^2} + \ldots + \frac{c_h}{n^h} + o(n^{-h}) \text{ as } n \to \infty$$

*The coeficient,* $c_m$ *of this asymptotic expansion depends on parameters* $A, B_1, B_2, \ldots, B_{2m}$ *where these parameters are defined by*

$$\log f(t) = i\alpha t + At^2 + t^2 \sum_{j=1}^{d-2} B_j t^j + o(t^d) \text{ as } t \to 0$$

*An expression for* $c_m$ *may be obtained in the following manner:*

1. *Polynomials* $Q_j$ *are defined by:*

$$Q_j(\gamma) = (-2i\sqrt{-A})^{-j} \sum \frac{1}{s_1! s_2! \ldots s_j!} B_1^{s_1} B_2^{s_2} \ldots B_j^{s_j} (4A)^{-s} H_{2s+j}\left(\frac{\gamma}{2\sqrt{-A}}\right)$$

   *the sum being taken over all j-tuples* $(s_1, s_2, \ldots, s_j)$ *such that*

$$s_1 + 2s_2 + \ldots + js_j = j, s = s_1 + s_2 + \ldots + s_j,$$

   $H_m$ *is the Hermite polynomial of degree m,*

$$H_m(x) = (-1)^m e^{x^2} \frac{d^m}{dx^m} e^{-x^2}$$

2. *Next, the polynomials* $S_j$ *are defined recursively:*

$$S_0 = 1$$

$$S_j(\gamma) = -\frac{1}{2}[S_1 S_{j-1} + S_2 S_{j-2} + \ldots + S_{j-1} S_1] + \frac{1}{2}[Q_0 \overline{Q}_j + Q_1 \overline{Q}_{j-1} + \ldots + Q_j \overline{Q}_0]$$

*3. Finally,*

$$c_k = \frac{1}{2\sqrt{\pi|A|}} \int_{-\infty}^{\infty} S_{2k}(\gamma) \exp\left(\frac{\gamma^2}{4} \operatorname{Re}\left(\frac{1}{A}\right)\right) d\gamma$$

There are two remarks to be made concerning the just quoted theorems. First, the expansion for the coefficients $a_{n\nu}$, as well as the expansion for their absolute values, is in powers of $\frac{1}{\sqrt{n}}$, but the expansion for the sums $\sum_\nu |a_{n\nu}|$ is in powers of $\frac{1}{n}$. Second, if the Fourier coefficients of $f$ are all real and non-negative, the same is true for the Fourier coefficients $a_{n\nu}$ of $f^n$, and so $||f^n|| = \sum_\nu |a_{n\nu}| = \sum a_{n\nu} = f^n(0) = 1$ for every $n$, and we obtain, without any computation, that all the coefficients $c_k$ of the asymptotic expansion must vanish for $k \geq 1$. The surprising fact is that applying our Theorem 28.7, we obtain the same conclusion (i.e $c_k = 0$ for $k \geq 1$) when the Fourier coefficients of $f$ are only assumed to be real.

Moreover, if a few initial moments of the sequence $(a_\nu)$ are real, part of the asymptotic series collapses, and we deduce from Theorem 28.7 the simple and surprising

**Theorem 28.8** *Let* $f(t) = \sum_{-\infty}^{\infty} a_\nu e^{i\nu t}$ *be twice continuously differentiable,* $|f(0)| = 1$, $|f(t)| < 1$ *for* $0 < |t| \leq \pi$ *and* $\frac{d^2}{dt^2}|f(t)| \neq 0$ *at* $t = 0$. *If* $f$ *is infinitely differentiable in a neighborhood of zero and if* $i^j f^{(j)}(0)$ *is real for* $j = 1, 2, \ldots, 2h+2$, *then*

$$||f^n|| = 1 + o(n^{-h}), \quad n \to \infty.$$

It is really striking in how roundabout a way the last theorem was proved, and that one cannot see a more direct approach. However, the author has obtained directly a related result, by proving first the following theorem [6].

**Theorem 28.9** *Let* $f$ *be a real-analytic* $2\pi$*-periodic function,*

$$|f(t)| < f(0) = 1 \text{ for } 0 < |t| \leq \pi$$

*and let*

$$\frac{d^2 f(t)}{dt^2} \neq 0.$$

*Let* $f^n(t) = \sum a_{n\nu} e^{\nu t i}$, *and let the coefficients* $(a_{1\nu})$ *be real. Then there exist constants* $\lambda$ *and* $\mu$, $\lambda < \mu$, *such that if* $\lambda < \lambda_1 < \mu_1 < \mu$, *then there exists* $N$ *with the property that the coefficients* $a_{n\nu}$ *are strictly positive for* $n > N$, $\lambda_1 N < \nu < \mu_1 N$.

The author has shown how the constants $\lambda$ and $\mu$ can be determined. This has been proved using the saddle-point method, similarly to the way results on large deviations are obtained in probability theory. It may be noted that this theorem contains, as a special case, the Frobenius theorem in number theory: if $S$ is a finite set of positive integers which are not all divisible by some integer $k$, $k > 1$, then any sufficiently large integer can be represented as sum of numbers from $S$. From Theorem 9 the author has deduced

**Theorem 28.10** *Under the same assumptions as in Theorem 28.9,*

$$\|f^n\| = 1 + \mathcal{O}(e^{-cn}), \quad n \to \infty \tag{28.6}$$

*for some* $c > 0$.

Moreover, the example $f(t) = \sqrt{\frac{1}{3}(2 + e^{ti})}$, when $\|f^n\| = 1$ for $n$ even, $\|f^n\| \neq 1$ for $n$ odd, shows that one cannot expect to improve (28.6) by obtaining the asymptotic behavior of $\|f^n\| - 1$.

## 28.6 Changing the Question

We were aware at the outset of a connection between our topic and the probability theory. Namely, if the Fourier coefficients of $f(t) = \sum a_\nu e^{\nu ti}$ are non-negative, $\sum a_\nu = 1$, $f^n(t) = \sum a_{n\nu} e^{\nu ti}$ and if the random variable $X_1$ is defined by $\mathrm{Prob}\,(X_1 = \nu) = a_\nu$, then

$$a_{n\nu} = \mathrm{Prob}\,(X_1 + X_2 + \cdots + X_n = \nu)$$

where $X_1, X_2, \ldots, X_n$ are independent identically distributed random variables. However, this relation with probability theory seemed irrelevant, since if $a_\nu$'s are nonnegative, then $\|f^n\| = 1$ for every $n$.

Only after we decided to obtain a more transparent presentation of the old (and the new) results, and looked, therefore, at the behavior of the coefficients $(a_{n\nu})$ themselves, the connection with probability theory became relevant.

After we proved Theorem 28.6, the author, leafing through Petrov's book [33] at a book exhibit, learned about the Edgeworth-Esseen asymptotic series, an important result in mathematical statistics, and realized that the only difference between our theorem and the Edgeworth-Esseen Theorem (in addition to terminology) is that, instead of assuming: $a_\nu \geq 0$ for every $\nu \in Z$, and the set $\{\nu\colon a_\nu \neq 0\}$ is not contained in a proper subgroup of $Z$, we were making weaker assumptions:

1. $\left|\sum a_\nu e^{\nu ti}\right|$ attains its absolute maximum on $[-\pi, \pi]$ only at the point $t = 0$, and

2. $\frac{d^2}{dt^2}\left|\sum a_\nu e^{\nu ti}\right| \neq 0$ at $t = 0$.

Looking at results of probability theory from a purely analytic point of view, we realize that some of these results will remain valid if probabilities are allowed complex-values. Complex-valued probabilities make sense! (Perhaps not for statisticians and probabilists, but certainly for analysts.)

The behavior of $a_{n\nu}$'s in case $a_\nu \geq 0$ has been the central problem of the probability theory (limit theorems for sums of i.d.d. random variables); it has been thoroughly investigated, and we know the possibilities: normal law, other stable laws, no law. We see, therefore, that our new topic – describing behavior of $a_{n\nu}$ in case coefficients are complex-valued – is too general and immense. The natural first steps are to answer the questions:

**Question 28.1** *What are – in case of complex-valued probabilities – the analogues of the normal law?*

**Question 28.2** *What is the Central Limit Theorem for complex-valued probabilities?*

To put the problem in the most convenient form, let us observe that the problem we have considered until now (the behavior of $a_{n\nu}$) has been: If $a = (a_\nu)_{-\infty}^{\infty} \in \ell_1$, describe the behavior of $a^{(n)} = \underbrace{a * a * a * \cdots * a}_{n \text{ times}}$ as $n \to \infty$. The characteristic function of the distribution $\underline{\underline{a}}$ is the $2\pi$-periodic function $f(t) = \sum a_\nu e^{\nu ti}$. Certain aspects will be simplified if we consider the continuous case instead of discrete; functions $\varphi \in L^1$ instead of sequences $a \in \ell_1$. The problem becomes: If $\varphi \in L^1(R)$, describe the behavior of $\varphi^{(n)} = \underbrace{\varphi * \varphi * \varphi * \cdots * \varphi}_{n \text{ times}}$ as $n \to \infty$.

## 28.7 Behavior of Scaled $\varphi^{(n)}$ for Large $n$

The only assumptions we make on the complex-valued function $\varphi$ are:

$$\varphi(x) \to 0 \text{ exponentially as } |x| \to \infty, \tag{28.7}$$

$$\varphi \in L^s(R) \text{ for some } s > 1, \tag{28.8}$$

and

$$|\widehat{\varphi}(t)| < \widehat{\varphi}(0) = 1 \text{ for } t \neq 0. \tag{28.9}$$

Property (28.7) and (28.8) implies that $\widehat{\varphi}$ is real-analytic, so the maximum (28.9) at zero can be characterized by the five parameters; condition (28.8) insures that $\widehat{\varphi}^n$ will be integrable for $n$ sufficiently large, so that the

$n$-fold convolution $\varphi * \varphi * \cdots * \varphi$ can be expressed as the inverse Fourier transform of $\widehat{\varphi}^n$.

We can formulate now

**A Heuristic Principle**

UNDER THE CONDITIONS (28.7), (28.8), (28.9), THE ASYMPTOTIC BEHAVIOR, AS $n \to \infty$, OF THE $n$-FOLD CONVOLUTIONS $\varphi^{(n)}$, CORRECTLY SCALED, DEPENDS ONLY ON THE FIVE PARAMETERS.

The principle is obviously imprecise and vague, but it helps us from the outset to classify the possible behaviors of $\varphi^{(n)}$.

So, assuming this principle, to answer the two questions raised above, we need to investigate the behavior of the $n$-fold convolutions $\varphi^{(n)}$ as $n \to \infty$ only in cases

$$\widehat{\varphi}(x) = \exp(Ax^q)\,, \quad \operatorname{Re} A < 0 \tag{28.10}$$

which is a typical representative when $p = q$, and the case

$$\widehat{\varphi}(x) = \exp(ix^p - x^q)\,, \tag{28.11}$$

This is the typical representative when $p$ is different from $q$, since it can be shown that, without loss of generality, we may assume $\alpha = 0$, $b = 1$, and even $a = 1$. Let

$$\alpha_n = n^{1/q}\,. \tag{28.12}$$

In case $p = q$, we shall consider the scaled $n$-fold convolution:

$$\psi_n(x) = \alpha_n \varphi^{(n)}(\alpha_n x)\,. \tag{28.13}$$

We obtain that if $\varphi$ is defined by (28.11), then

$$\begin{aligned} \psi_n(x) &= \frac{1}{2\pi} \int \widehat{\varphi}^n \left(\frac{\xi}{\alpha_n}\right) e^{i\xi x} d\xi \\ &= \frac{1}{2\pi} \int \exp(i\xi x + A\xi^q) d\xi\,; \end{aligned} \tag{28.14}$$

so that in the very special case (28.11), $\psi_n$ is for every $n$ equal to the inverse Fourier transform of $\exp(Ax^q)$. According to our heuristic principle, we can expect the following result.

**Theorem 28.11** *If the function $\varphi$ satisfies conditions (28.7), (28.8) and (28.9), and is normalized so that $\alpha = 0$, and if the parameters $p$ and $q$ characterizing the behavior of $\widehat{\varphi}$ at the origin are equal, then the scaled $n$-fold convolutions $\psi_n$, defined by (28.13), will converge pointwise and in the $L^1$-norm to the inverse Fourier transform of* $\exp(Ax^q)$.

This theorem is an analogue of the Central Limit Theorem for the complex-valued probabilities in case $p = q$, and the inverse Fourier transforms of

$\exp(Ax^q)$ are analogues of the normal distribution. Theorem 28.11 implies immediately Girard's result (28.2) (or to be quite precise, the analogue of (28.2) in the continuous case). A detailed proof of Theorem 28.11 will be included in [25].

Theorem 28.11 is not suprising, much more interesting will be the case when $p \neq q$. In the case $p = q$, it was obvious that the right scaling factor is given by (28.12). What should be the "right" scaling factor if $p \neq q$: $n^{1/p}$ or $n^{1/q}$ (it will be seen in the next section that some authors have used $n^{1/p}$). We claim that *the right scaling* is given by

$$\psi_n(t) = \alpha_n \sqrt{\lambda_n} \varphi^{(n)}(\alpha_n \lambda_n t), \tag{28.15}$$

where $\alpha_n$ is defined by (28.13) and $\lambda_n = n^{1-p/q}$. This scaling is obviously a trick, and that trick turns out to be the crucial step in the investigation of the case $p \neq q$. Since $\alpha_n \varphi^{(n)}(\alpha_n t) = \frac{1}{2\pi} \int \widehat{\varphi}^n \left(\frac{\xi}{\alpha_n}\right) e^{i\xi t} d\xi$, setting here $t = \lambda_n x$, we get in case $\varphi$ is defined by (28.11), that

$$\begin{aligned} \psi_n(x) &= \frac{\sqrt{\lambda_n}}{2\pi} \int \widehat{\varphi}^n \left(\frac{\xi}{\alpha_n}\right) e^{i\xi\lambda_n x} d\xi \\ &= \frac{\sqrt{\lambda_n}}{2\pi} \int \exp(i\lambda_n(\xi x + \xi^p) - \xi^q) d\xi \,. \end{aligned} \tag{28.16}$$

Here we realize the purpose of the trick: it makes possible to apply now the stationary phase method.

Let us note first one particular result that is obtained by applying the stationary phase method to the integral: if $p$ is even, then $|\psi_n(x)|$ converges, as $n \to \infty$, to the integrable function

$$f(x) = C_1 |x|^{-\beta} \exp(-C_2 |x|^\gamma), \tag{28.17}$$

where $\beta = \frac{p-2}{2(p-1)}$, $\gamma = \frac{q}{p-1}$ and $C_1$, $C_2$ depend only on parameters $p$ and $q$, $C_2 > 0$.

The investigation of $\psi_n$ in the special case (28.13) and our heuristic principle lead us to the following result (conjectured by the author, and proved in a stronger form by Natalia Humphreys [25]):

**Theorem 28.12** *If the function $\varphi$ satisfies conditions (28.7), (28.8), and (28.9), and if the parameters $p$ and $q$ characterizing the behavior of $\widehat{\varphi}$ at the origin are distinct, then the scaled $n$-fold convolutions $\psi_n$, defined in (28.15), will exhibit one of two types of very regular oscillatory divergence, depending on $p$ being even or odd:*

1. *If $p$ is even, then*

$$\psi_n(x) = f(x) u_n(x) + r_n(x), \tag{28.18}$$

*where $f$ is given by (28.17), $|u_n(x)| = 1$ for all $x$ and $n$, the sequence $(u_n)$ satisfies the condition: For every interval $[a, b]$ and every arc $\ell$ of the unit circle*

$$\frac{1}{b-a} m\{x : x \in [a,b],\ u_n(x) \in \ell\} \to \frac{1}{2\pi}(\textit{arclength of } \ell),\ n \to \infty$$

*and $r_n(x)$ tends to zero as $n \to \infty$, both pointwise and in $L^1$-norm.*

2. *If $p$ is odd, then*

$$\psi_n(x) = g(x)t_n(x) + r_n(x)\,, \tag{28.19}$$

*where*

$$g(x) = \begin{cases} 0 & \textit{if } x > 0 \\ 2f(x) & \textit{if } x < 0 \end{cases},$$

*$f$ being defined by (28.17), $-1 \leq t_n(x) \leq 1$, the sequence $(t_n)$ satisfies the condition: For every interval $[a, b]$ and for every subinterval $I$ of $[-1, 1]$*

$$\frac{1}{b-a} m\{x \colon x \in [a,b]\,,\ t_n(x) \in I\} \to \frac{1}{\pi} \int\limits_I \frac{dx}{\sqrt{1-x^2}}\,, \quad n \to \infty\,,$$

*and $r_n(x)$ tends to zero as $n \to \infty$, both pointwise and in $L^1$-norm.*

A few remarks are in order.

1. Even in the special case (28.11), the proof of the fact that $\|r_n\|_{L^1} \to 0$, $n \to \infty$, requires some delicate estimates; this is true even in the simplest case of all, when $\widehat{\varphi}(x) = \exp(ix^2 - x^q)$; in that case, we do not need the stationary phase method, the decomposition (28.18) is obvious, but even in this case it is not easy to establish that $\|r_n\|_{L^1} \to 0$ [7].

2. In the special case (28.11), the explicit expression for $u_n$ and $t_n$ are: $u_n(x) = \exp(iC\lambda_n|x|^\delta)$, $t_n(x) = \cos(C\lambda_n|x|^\delta)$, where $\delta = \frac{p}{p-1}$ and $C$ depends only on $p$ and $q$.

3. The assymetry in case $p$ odd may be suprising. To understand this fact, we look again at the special case; (28.11). We fix $x > 0$. Then the function $t(\xi) = x\xi + \xi^p$ is increasing and taking $t$ as the new integration variable we obtain that

$$\int \exp(i\lambda_n(x\xi + \xi^p) - \xi^q)d\xi = \int e^{i\lambda_n t}\phi(t)dt\,, \tag{28.20}$$

where $\phi$ is an analytic function on the real line. Therefore, the integral on the right, as the Fourier transform of an analytic function, tends to zero exponentially as $|\lambda_n| \to \infty$. We obtain then from (28.16) and (28.20) that $\psi_n(x) \to 0$, $n \to \infty$ for every $x > 0$.

4. From Theorem 28.12 we easily derive the Girard's asymptotic formula (28.3) (more precisely, the analogue of that formula in the continuous case). From (28.15) we have that

$$\|\varphi^{(n)}\|_{L^1} = \sqrt{\lambda_n}\|\psi_n\|_{L^1}\,.$$

In case $p$ even, it follows immediately from (28.18) and the fact that $\|r_n\|_{L^1} \to 0$ that $\|\psi_n\|_{L^1} \to \|f\|_{L^1}$, so we obtain

$$\begin{aligned}&\text{If } p \neq q \text{ and } p \text{ is even, then}\\ &\|\varphi^{(n)}\| \sim \|f\|_{L^1}\sqrt{\lambda_n}\,, \quad n \to \infty\,.\end{aligned} \tag{28.21}$$

In case $p$ odd, we obtain from (28.19) similarly that

$$\begin{aligned}\|\psi_n\|_{L^1} &= \|gt_n\|_{L^1} + o(1) \to \frac{2}{\pi}\|g\|_{L^1}\\ &= \frac{2}{\pi}2\|f\|_{L^1(-\infty,0)}\\ &= \frac{2}{\pi}\|f\|_{L^1}\,,\end{aligned}$$

so that

$$\begin{aligned}&\text{if } p \neq q \text{ and } p \text{ is odd, then}\\ &\|\varphi^{(n)}\| \sim \frac{2}{\pi}\|f\|_{L^1}\sqrt{\lambda_n}\,, \quad n \to \infty\,.\end{aligned} \tag{28.22}$$

Evaluating the integral of $f$ in 28.17and substituting in (28.22) and (28.21) we derive (28.3).

5. Now we can explain why we consider the scaling (28.15) to be not just a useful trick, but to represent the "right", the "natural" scaling. We need a result on probability distributions, but first a terminological note: a probability distribution $F$ is improper if there exists $x_0$ such that $F(x) = 0$ for $x < x_0$ and $F(x) = 1$ for $x > x_0$; $F$ is proper if it is not improper. It is known (see Theorems 1 and 2 on pages 40-42 of [17]) that if $F_n$ is a sequence of probability distributions with the property:

$$\begin{aligned}&\text{there exists a sequence } (\delta_n) \text{ of positive reals}\\ &\text{such that } F_n(\delta_n x) \text{ converges to a proper}\\ &\text{distribution } F(x)\,,\end{aligned} \tag{28.23}$$

then the sequence $(\delta_n)$ is essentially unique. Namely, if $(\eta_n)$ is another sequence of positive reals with the property that $F_n(\eta_n x)$ converges to a proper distribution $G(x)$, then $\eta_n/\delta_n$ converges to some positive constant $c$, and $G(x) = F(cx)$. Now we return to the $n$-fold convolutions $\varphi^{(n)}$ in case $p$ even, more precisely to their absolute values $|\varphi^{(n)}|$, and we define the probability distributions $F_n$ by $F_n(x) = \frac{1}{\|\varphi^{(n)}\|_1} \int_{-\infty}^{x} |\varphi^{(n)}(t)|\, dt$. It follows that

$$F_n(\alpha_n \lambda_n x) = \frac{1}{\|\varphi^{(n)}\|_1} \int_{-\infty}^{x} \sqrt{\lambda_n}|\psi^{(n)}(t)|\, dt.$$

Because of (28.19) and (28.22), we obtain that

$$F_n(\alpha_n \lambda_n x) \to F(x) = \frac{1}{\|f\|_1} \int_{-\infty}^{x} f(t)\, dt\,. \tag{28.24}$$

The distribution $F$ is proper, and we see from (28.18) that (28.21) is satisfied with $\delta_n = \alpha_n \lambda_n$, and therefore, that the scaling (28.15) is essentially unique.

6. If $\varphi$ satisfies only the conditions (28.7) and (28.8), the behavior of $\varphi^{(n)}$ can be analyzed with the use of Theorems 28.11 and 28.12. Namely, in that case, $\widehat{\varphi}$ is analytic function of a real variable, and it is easy to see that $|\widehat{\varphi}|$ attains its absolute maximum in at most finitely many points. Since the behavior of $\varphi^{(n)}$ is superposition of contributions from each of these finitely many points, the main task is to describe the behavior of $\varphi^{(n)}$ in case there is only one such point. Without loss of generality we can assume that point to be the origin, and that $\widehat{\varphi}(0) = 1$; in other words, we can assume that (28.9) holds, and then the behavior of $\varphi^{(n)}$ is described by Theorem 28.11 - if $p = q$, and Theorem 28.12, if $p \neq q$. It should be noted that (28.7) is much stronger condition than what is used in the proof of Theorems 28.11 and 28.12.

7. Since the theme of our workshop is divergence, we should note an interesting feature of oscillatory divergence of type (28.18) and (28.19): if a sequence of functions $\psi_n$ diverges as in (28.18), the same is true for any subsequence of $\psi_n$; similarly for type (28.19).

## 28.8 Another Kind of Central Limit Theorems

In our talk at the 7$^{\text{th}}$ IWAA workshop, we presented the content of the preceding sections; in the discussion following the talk, M. Pinsky told us about the work of K. Hochberg on signed probabilities. (Also a question was raised about the relations of our topic with quantum probability; here the answer is simple: quantum probability and complex-valued probability are generalizations in different directions, they seem to be completely unrelated.) We found out later that several authors ([21], [23], [24], [29], [30]) had realized that signed probabilities make sense analytically, had found a corresponding central limit theorem and used it in numerical analysis or partial differential equations. (The first seems to have been Zhukov [39]).

There is a big difference between the work of mentioned authors on one side, and our results on the other side. To illustrate that difference, we present here a version of the Central Limit Theorem, due to Hochberg [23, Theorem 2]; which we generalize slightly, and change terminology and notation in order to make comparison with Theorems 28.11 and 28.12 easier.

**Theorem 28.13** *If* $\varphi \in L^1(R)$, $|\widehat{\varphi}| \leq 1$ *and if* $\widehat{\varphi}(t) = \exp(ct^k + o(t^k))$, $t \to 0$, *then* $n^{\frac{1}{k}}\varphi^{(n)}(n^{\frac{1}{k}}x)$ *converge, as tempered distributions, to the inverse Fourier transform of* $\exp(ct^k)$. *Moreover, if* $k$ *is even and* $Re\, c < 0$, *then* $n^{\frac{1}{k}}\varphi^{(n)}(n^{\frac{1}{k}}x)$ *converge weakly on the space of Fourier transforms of integrable functions.*

Let us observe that $|\widehat{\varphi}| \leq 1$ implies $Re\, c = 0$ if $k$ is odd, and $Re\, c \leq 0$ if $k$ is even.

The proof of Theorem 28.13 is quite short. Let $\theta$ be any integrable function, then

$$\int \widehat{\theta}(x)\, \sqrt[k]{n}\varphi^{(n)}(x\, \sqrt[k]{n})dx = \int \theta(x)\widehat{\varphi}^n\left(\frac{x}{\sqrt[k]{n}}\right) dx \qquad (28.25)$$
$$\to \int \theta(x) \lim_{n\to\infty} \widehat{\varphi}^n\left(\frac{x}{\sqrt[k]{n}}\right) dx$$

since the assumption $|\widehat{\varphi}| \leq 1$ shows that the integrands in the second integral are dominated by the integrable function $|\theta|$.

Since $\widehat{\varphi}(t) = \exp(ct^k + o(t^k))$, $t \to 0$, we obtain that, for every $x$, $\widehat{\varphi}^n\left(\frac{x}{\sqrt[k]{n}}\right) = \exp(cx^k + o(1))$, $n \to \infty$. It follows then from (28.25) that for any $\theta$ integrable

$$\lim_{n\to\infty} \int \widehat{\theta}(x)\, \sqrt[k]{n}\varphi^{(n)}(x\, \sqrt[k]{n}) = \int \theta(x) \exp(cx^k)dx\,. \qquad (28.26)$$

If $Re\, c < 0$, then $k$ is even, so the inverse Fourier transform of $\exp cx^k$ is an integrable function $G$, and the integral on the right is equal to $\int \widehat{\theta}(x)G(x)\, dx$, which proves the last part of the theorem. The remaining case is when $c$ is purely imaginary. In that case we assume that $\widehat{\theta}$ is an arbitrary function from the Schwartz class $\varphi$, and denote again by $G$ the inverse Fourier transform of $\exp cx^k$, which in this case is a tempered distribution. In that case

$$\int \theta(x) \exp(cx^k)dx = (\widehat{G}, \theta)$$
$$= (G, \widehat{\theta})\,,$$

so that (28.26) implies that, for any $\widehat{\theta} \in \mathcal{S}$,

$$\lim_{x \to \infty} ( \sqrt[k]{n} \varphi^{(n)}(x \sqrt[k]{n}), \widehat{\theta}) = (G, \widehat{\theta})$$

which ends the proof of the theorem.

There seems to be only one connection between Theorems 28.12 and 28.13: obviously, if $\sqrt[q]{n}\varphi^{(n)}(x\sqrt[q]{n})$ converges in $L^1$, then it converges to the same limit weakly on the space of Fourier transforms of integrable functions. There are, however, several important differences:

1. for distributional convergence it is irrelevant whether $|\widehat{\varphi}|$ attains its absolute maximum at only one or at several points, for $L^1$-convergence that is an essential difference;

2. if $p \neq q$, then the parameter $q$ is irrelevant for distributional convergence, but it quite relevant for our description, given in Theorem 28.12;

3. if $p \neq q$, then the scaling we use (which was justified earlier in this article) is different from the scaling used for distributional convergence;

4. the distributional convergence approach has been well motivated by applications in numerical analysis, partial differential equations, ...;

5. that approach is more general;

6. there are aspects of the behavior of the $n$-fold convlutions $\varphi^{(n)}$, which are detected by Theorems 28.11 or 28.12, but not by more general theorems.

*References*

[1] Alpar, L., Remarque sur la sommabilite des series de Taylor sur leurs cercles de convergence I, II, III, Magyar. Tud. Akad. Mat. Kutato Int. Kozl., 3(1958), 141-158; 5(1960), 97-152.

[2] Alpar, L., Sur certaines transformees des series de puissances absolument convergentes sur la frontiere de leur cercle de convergence, Magyar. Tud. Akad. Mat. Kutato Int. Kozl., 7(1962), 287-316.

[3] Bajsanski, B., Sur une classe generale de procedes de sommations du type d'Euler-Borel, Acad. Serbe Sci. Publ. Inst. Math., 1 0(1956), 131-152.

[4] Bajsanski, B., Generalisation d'un theoreme de Carlemann. Acad. Serbe Sci. Publ. Inst. Math., 12(1958), 101-108.

[5] Bajsanski, B., Opsta klasa postupaka zbirljivosti Euler-Borel-ovog tipa i njitiova primena na analiticko produzenje, Acad. Serbe Sci. Zbornik, Inst. Math., 7(1959), 1-36.

[6] Baishanski, B., Positivity zones of n-fold convolutions, in preparation, 1998.

[7] Baishanski, B. Behavior of $n$-fold convolutions of complex-valued functions for large $n$, in preparation.

[8] Baishanski, B. and Snell, M., Norms of powers of absolutely convergent Fourier series, Journal d'Analyse, 1998.

[9] Beurling, A. and Helson, H., Fourier-Stieltjes transforms with bounded powers, Math. Scand. 1(1953), 120-126.

[10] Clunie, J. and Vermes, P., Regular Sonnenschein type summability methods, Acad. Roy. Belg. Bull. Cl. Sci.(5), 45(1959), 930-954.

[11] Clunie, J., On Equivalent Power Series, Acta Math. Acad. Sci. Hung., 18(1967), 165-169.

[12] Cramer, H., Random Variables and Probability Distributions, Cambridge University Press, 1970.

[13] Esseen, C.G., Fourier Analysis of Distribution Fuctions: A mathematical study of the Laplace-Gaussian law, Acta Math. (1945), 77, 1-125.

[14] Girard, D., A General Asymptotic Formula For Analytic Functions, Dissertation, Ohio State University, 1968.

[15] Girard, D., The Behavior of the Norm of an Automorphism of the Unit Disk, Pacific J. Math., 47(1973), 443-456.

[16] Girard, D., The Asymptotic Behavior of Norms of Powers of Absolutely Convergent Fourier Series, Pacific J. Math, 37(1971) 357-387.

[17] Gnedenko, B.V. and Kolmogorov, A.N., Limit distributions for sums of independent random variables, Addison-Wesley Publishing Company, 1968.

[18] Hedstrom, G., Norms of Powers of Absolutely Convergent Fourier Series, Michigan Math. J., 13(1966), 249-259.

[19] Heiberg, C., Norms of Powers of Absolutely Convergent Fourier Series in Two Variables, Dissertation, The Ohio State University, 1971.

[20] Heiberg, C., Norms of Powers of Absolutely Convergent Fourier Series in Several Variables, J. Functional Analysis, 14(1973), 382-400.

[21] Hersch, R., A class of "central limit theorems" for convolution products of generalized functions. Trans. Amer. Math. Soc. 1'40 1969 71-85.

[22] Hlavacek, J., Asymptotic formula for the norms of $\exp(inh(t))$, Dissertation, The Ohio State University, 1998.

[23] Hochberg, K. J., A signed measure on path space related to Wiener measure. Ann. Probab. 6 1978, no. 3, 433–458.

[24] Hochberg, K. J., Central limit theorem for signed distributions. Proc. Amer. Math. Soc. 79 1980, no. 2, 298–302.

[25] Humphreys, N., Central limit theorems for complex-valued probabilities, in preparation, 1998.

[26] Indlekofer, K.H., On Turan's Equivalent Power Series, Studies in Pure Mathematics, 357-379, Birkhauser, 1983.

[27] Kahane,J., Transformees de Fourier des Fonctions Sommables. Proceedings of the International Congress of Mathematicians, 1962, Institut Mittag-Leffler, Djursholm, 1963.

[28] Kahane,J., Sur certaines classes de series de Fourier absolument convergentes, J. Math. Pure Appl., (9), 35(1956), 249-259.

[29] Krylov, V. Ju., A limit theorem (Russian) Dokl. Akad. Nauk SSSR 139 1961 20–23.

[30] Krylov, V. Ju., Some properties of the distribution corresponding to the equation $\frac{\partial u}{\partial t} = (-1)^{q-1}\frac{\partial^{2q}u}{\partial x^{2q}}$. Soviet Math. Dokl. 1 1960, 760–763.

[31] Leibenzon, Z.L., On the ring of functions with an absolutely convergent Fourier series, Uspehi Mat. Nauk, vol. 9, no. 3 (61)(1954), 157-162.

[32] Newman, D.J., Homomorphisms of $l_+$, American Journal of Mathematics, 91(1969), 37-46.

[33] Petrov, V.V., Limit theorems of probability theory: Sequences of independent random variables, Clarendon Press, 1995.

[34] Snell, M., An Asymptotic Series for Norms of Powers, Dissertation, The Ohio State University, 1995.

[35] Tam, L., The General Euler-Borel Summability Method, Dissertation, The Ohio State University, 1990.

[36] Tam, L., A Tauberian theorem for the general Euler-Borel summability method, Canad. J. Math., 44 (1992), no. 5, 1100-1120.

[37] Turan, P., A remark concerning the behaviour of a power series on the periphery of its convergence circle, Publ. Inst. Math. Acad. Serbe Sci., 12(1958), 19-26.

[38] Whitford, L.E., On Boundary Behavior of Power Series, Dissertation, The Ohio State University, 1968.

[39] Zhukov, A. I., Limit theorem for difference operators. (Russian) Uspehi Mat. Nauk 14 1959, no. 3 (87), 129–136.

# Chapter 29

# Quasiasymptotics at zero and nonlinear problems in a framework of Colombeau generalized functions

**S. Pilipović**
**M. Stojanović**

*ABSTRACT. We give a survey of basic properties of the $\mathcal{G}$-quasiasymptotics and its applications to nonlinear problems in a framework of Colombeau algebra of generalized functions. We show that solutions to certain nonlinear systems of PDEs and nonlinear integral equations inherit the $\mathcal{G}$-quasiasymptotics of the initial data, or of the free terms.*

## 29.1 Introduction

In distribution theory, the study of quasiasymptotic properties at zero and at infinity is a well established tool developed by the Moscow school around Vladimirov [33]. They have characterized the Laplace transform of tempered distributions via this notion and gave the applications in mathematical physics.

This approach has been continued by the Novi Sad mathematical group. A survey concerning generalized asymptotic behavior of integral transforms is given in [21], where the quasiasymptotics and the $S$-asymptotics related to the convolution type integal transforms are given. An exhaustive list of pertinent references may be found in [21] and [27].

We refer to [34] for the advantages of these notions in applications to some problems of theoretical physics in the framework of Schwartz distributions. For the relations of ordinary asymptotics and quasiasymptotics we refer to [28] and [34]. The ordinary asymptotics implies the quasiasymptotics but the converse does not hold in general.

The objective of this article is to extend the notion of quasiasymptotics to the Colombeau algebra $\mathcal{G}$ of generalized functions. Note that Colombeau algebra of generalized functions now plays an important role in nonlinear PDEs. A survey of developing of Colombeau theory to nonlinear problems and linear ones is given in [20].

Colombeau algebra allows not only the multiplication and hence discussing nonlinear problems, but also an extension of quasiasymptotics to nonlinear problems. Families of functions which present elements in $\mathcal{G}$ provide a suitable framework for nonlinear operations on distributions and extension from linear to nonlinear problems. In addition, we use the $\mathcal{G}$-quasiasymptotics as a source of new asymptotic properties.

We limit our consideration to the $\mathcal{G}$-quasiasymptotics at zero. Definitions, main characterizations, and applications of the $\mathcal{G}$-quasiasymptotics at infinity are given in [26].

Since a generalized function $g$ is represented with an $\varepsilon$-net $g_\varepsilon$, $\varepsilon \in (0,1)$, of smooth functions with the power order growth with respect to $\varepsilon$, the order growth of $\varepsilon$ reflects in some sense its singularity. But the singularity at zero is precesely characterized throughout the analysis of the behavior of $g_\varepsilon(\varepsilon x)$ as $\varepsilon \to 0$. This has already been found for Schwartz distributions (the quasiasymptotics and quasiasymptotic expansion at zero [27]). Since Colombeau space contains elements which are not distributions ($\delta^2$, for example) we reconsider the concept of quasiasymptotic expansion in $\mathcal{G}$.

The following demands on the quasiasymptotics are involved:

(1) The quasiasymptotics in Colombeau algebra preserves the ordinary quasiasymptotics;

(2) $Cd(f)$ has the quasiasymptotics in Colombeau space if and only if a distribution $f$ has the quasiasymptotics in $\mathcal{D}'$, ($Cd(f)$ denotes a Colombeau generalized function which corresponds to $f \in \mathcal{D}'$);

(3) existence of a generalized function which belongs to $\mathcal{G} \setminus \mathcal{D}'$ such that it has the quasiasymptotics.

We will present a framework in which (1) – (3) can be treated in full generality, with a wealth of applications to certain classes of nonlinear PDEs, with singular initial data and integral equations with non-Lipschitz nonlinearity.

Keeping the $\mathcal{G}$-quasiasymptotics framework on a general level enables us to discuss a variety of consistency results with classical results for the distributional and ordinary quasiasymptotics. These results are coherent with the known ones. For example, the definition of the $\mathcal{G}$-quasiasymptotics at zero restricted to Schwartz distributions gives the well-known notion of the quasiasymptotics at zero and in particular, the value at zero.

Note that in applications arise properties related to the $\mathcal{G}$-quasiasymptotics in Colombeau spaces out of properties of the quasiasymptotics in $\mathcal{D}'$.

Another concept of qualitative analysis, the microlocal analysis, in the framework of Colombeau algebra of generalized functions is developed in [9], [10], [14], [1], [30], [7], [29]. The wave front and the dot wave front set of $G \in \mathcal{G}(X)$ are defined and estimated in [9], [10]. A survey of local and microlocal analysis within Colombeau algebra of generalized functions is given in [29]. Here we only note that the dot wave front set $\dot{\mathrm{WF}}_g$ intrinsically characterizes the singularities of elements in $\mathcal{G}$. This notion is used in the analysis of propagation of singularities for differential operators with coefficients in $\dot{\mathcal{G}}^\infty$; such investigations outlines the fact that the structure of $\mathcal{G}$ is more complex than the structure of usual spaces of functions and distributions.

The plan of exposition is as follows: In the section 29.2 we give a construction of the algebra of generalized functions. Section 29.3 contains the definition of the $\mathcal{G}$-quasiasymptotics at zero, basic properties and examples which show what new notion involves. Final section gives an overview of further application of the $\mathcal{G}$-quasiasymptotics at zero. We give a number of typical applications of the $\mathcal{G}$-quasiasymptotics at zero in Colombeau spaces to nonlinear problems such as nonlinear PDEs with singular data, nonlinear non-Lipschitz integral equations, semilinear hyperbolic systems and a Goursat problem with unbounded gradient for a nonlinear term. It is shown that solutions inherit the quasiasymptotic property of initial data. In all these cases we avoid unbounded gradients for the nonlinear term by special regularization.

## 29.2 Algebra of generalized functions

For PDEs with rough initial data or with discontinuous coefficients the Colombeau algebra of generalized functions is a good framework. Many of them have no solution in a space of distributions but may be solved in $\mathcal{G}$.

There are different approaches to Colombeau theory of generalized functions ( [5], [6]) developed by many authors (cf. [2], [15], [16], [18], [17], [31], [19]). All these approaches take as a starting point an infinite index set $I$ and consider the power set $\mathcal{E}(\Omega) = (C^\infty(\Omega))^I$. It refers to the element $(u_\varepsilon)_{\varepsilon \in I}$ of $\mathcal{E}(\Omega)$ as family of smooth functions which are members of factor algebras (or factor spaces) of families of smooth functions. A generalized version of this approach which is concerned with factor set of arbitrary metrisable spaces is developed in [8]. In [3], [4], are introduced factor spaces of Sobolev spaces and their application to solving nonlinear PDEs.

We recall the simplified version of Colombeau theory from [22]. We will investigate an algebra of generalized functions on an open set $\Omega \subset \mathbf{R}^n$.

Let $V$ be a topological vector space whose topology is given by a countable set of seminorms $\mu_k$, $k \in \mathbf{N}$. In fact we will consider the case $V = C^\infty(\Omega)$, $\Omega$ is an open set in $\mathbf{R}^n$, and a sequence of seminorms is given by

$$\mu_k(f) = \sum_{|\alpha| \leq k} (\sup_{x \in \overline{\Omega}_k} |\partial^\alpha f(x)|), \quad k \in \mathbf{N}_0,$$

where $\Omega_k$ is a sequence of open sets such that $\cup_{k=0}^\infty \Omega_k = \Omega$, $\Omega_k \subset\subset \Omega_{k+1}$, $k \in \mathbf{N}_0$. The uniform structure on $C^\infty(\Omega)$, defined by this sequence of seminorms, does not depend on the choice of the sequence $\Omega_k$.

Then $\mathcal{E}_{M,V}$ is the set of locally bounded functions $R(\varepsilon) = R_\varepsilon : (0,1) \to V$

such that for every $k \in \mathbf{N}$ there exists $a \in \mathbf{R}$ with the property that

$$\mu_k(R_\varepsilon) = \mathcal{O}(\varepsilon^a),$$

where $\mathcal{O}(\varepsilon^a)$ means that the left side is smaller then or equal to $C\varepsilon^a$ for some $C > 0$ and every $\varepsilon \in (0, \varepsilon_0)$, $\varepsilon_0 > 0$.

The space of all elements $H \in \mathcal{E}_{M,V}$ with the property that for any $k \in \mathbf{N}$ and for any $a \in \mathbf{R}$, $\mu_k(H_\varepsilon) = \mathcal{O}(\varepsilon^a)$ is denoted by $\mathcal{N}_V$.

The quotient space $\mathcal{G}_V = \mathcal{E}_{M,V}/\mathcal{N}_V$ is called the polynomial Colombeau generalized extension of $V$.

If $V = \mathbf{C}$, then $\mathcal{G}_V$ is called the algebra of generalized constants and it is denoted by $\bar{\mathbf{C}}$. $\mathcal{E}_{M,V}$ is denoted by $\mathcal{E}^0$ and $\mathcal{N}_V$ by $\mathcal{N}^0$ .

If $V = C^\infty(\Omega)$, then $\mathcal{G}_V$ is called the algebra of generalized functions on $\Omega$ and it is denoted by $\mathcal{G}(\Omega)$; $\mathcal{E}_{M,V}$ is denoted by $\mathcal{E}_M(\Omega)$ and $\mathcal{N}_V$ by $\mathcal{N}(\Omega)$. $\bar{\mathbf{C}}$ can be considered as a subalgebra of $\mathcal{G}(\Omega)$.

After proving that the presheaf $U \mapsto \mathcal{G}(U)$ ($U$ is open in $\mathbf{R}^n$) is a sheaf, one can prove that the above embeddings can be extended to embeddings of $C^\infty(\Omega)$ and $\mathcal{D}'(\Omega)$ into $\mathcal{G}(\Omega)$. The support of a generalized function $H$, $supp_g H$, is defined as the complement of the largest open subset $\Omega'$ such that $H_{|\Omega'} = 0$.

If $G$ is a generalized function with a compact support $K \subset\subset \Omega$ ($G \in \mathcal{G}_c(\Omega)$) and $G_\varepsilon$ is a representative of $G$, then its integral is defined by

$$\int G dx = [\int \psi(x) G_\varepsilon(x) dx],$$

where $\psi \in C_0^\infty(\mathbf{R}^n)$, $\psi = 1$ on $K$. This definition does not depend on $\psi$.

We embed $\mathcal{E}'(\Omega)$ into $\mathcal{G}(\Omega)$ in the following way.

Let $\psi \in C_c^\infty(\mathbf{R}^n) = \mathcal{D}$ and $\phi \in \mathcal{S}(\mathbf{R}^n)$ such that it is even, $\mathcal{F}(\phi) = \hat{\phi} \in \mathcal{D}(\mathbf{R}^n)$ and $\hat{\phi} \equiv 1$ on a neighborhood of zero; $\phi$ is called a vision function. Put $\phi_{\varepsilon^2}(x) = 1/\varepsilon^{2n}\phi(x/\varepsilon^2)$, $x \in \mathbf{R}^n, \varepsilon \in (0,1)$. Then, $N_\varepsilon(x) = (\psi * \phi_{\varepsilon^2}(x) - \psi(x))$ belongs to $\mathcal{N}(\Omega)$, where $*$ is a convolution.

Let $f \in \mathcal{D}'(\Omega)$, $\Omega \in \mathbf{R}^n$. For the sake of the $\mathcal{G}$-quasiasymptotics at zero we will use the embedding $\mathcal{I}_2 f = [(f\kappa_\varepsilon) * \phi_{\varepsilon^2}]$, where $\kappa_\varepsilon$ is a characteristic function of $\Omega_\varepsilon = \{x \in \Omega,\ d(x, Compl.\Omega) > \varepsilon\}$ which transfers more appropriately the asymptotic properties of $f$ (cf. [22]).

For elements in $\mathcal{G}$ the following additional types of equalities are defined. Let $G, F \in \mathcal{G}(\Omega)$. Then:

(i) They are equal in the distribution sense, $F \overset{\mathcal{D}'}{=} G$ if

$$\int (G - F)\psi dx = 0, \text{ for any } \psi \in C_c^\infty(\Omega).$$

(ii) They are associated $G \sim F$ if there exist the representatives $G_\varepsilon$ and $F_\varepsilon$ of $G$ and $F$, respectively, such that

$$\lim_{\varepsilon \to 0} \int (G_\varepsilon(x) - F_\varepsilon(x))\psi(x) dx = 0, \text{ for any } \psi \in C_c^\infty(\Omega).$$

Elements $G, H \in \mathcal{G}(\Omega)$ are $L^p$-associated ($p \in [1,\infty]$) in $\omega \subset \Omega$, if $\lim_{\varepsilon\to 0} ||G_\varepsilon - H_\varepsilon||_{L^p(\omega)} = 0$, where $G_\varepsilon$ and $H_\varepsilon$ are representatives of $G$ and $H$, respectively.

Let $p \in [1,\infty]$ and $g \in L^p_{loc}(\Omega)$. Then $G \in \mathcal{G}(\Omega)$ is $L^p$-associated to $g$ if $||g - G_\varepsilon||_{L^p(\omega)} \to 0$, as $\varepsilon \to 0$, for every $\omega \subset\subset \Omega$ and every representative $G_\varepsilon$ of $G$.

The definitions do not depend on the choice of representatives.

In order to give a better insight into the very reach structure of Colombeau space we give several examples (cf. [15]).

**Example 29.1** *Colombeau space $\mathcal{G}$ allows the existence of various Heaviside functions: $H^n \neq H$ in $\mathcal{G}$, $n \geq 2$.*

*Let $\mathcal{G}_{pc}$ be the algebra of piecewise continuous functions in $\mathcal{G}$ on $\mathbf{R}$. It is not subalgebra of $\mathcal{G}$ since in $\mathcal{G}$, $H^n \neq H$, $n \geq 2$ and in $\mathcal{G}_{pc}$, $H^n = H$.*

*Let us explain $H^n \neq H$ in $\mathcal{G}$. If $H^n = H$ would hold in $\mathcal{G}$, we would have*

$$H^{n-1}H' = \frac{1}{n}H',\ H^n H' = \frac{1}{n}HH',$$

*and*

$$H^{n+1}H' = \frac{1}{n+1}HH',\ \frac{1}{n}HH' = \frac{1}{n+1}HH'.$$

*This is not true in $\mathcal{G}$, because $HH' \neq 0$. In $\mathcal{D}'$, $H^2$ is not defined, but in associated sense in $\mathcal{G}$, we have*

$$H^n \approx H \Rightarrow H^{n-1}\delta \approx \frac{1}{n}\delta.$$

**Example 29.2** *Put $\delta^2 = [\frac{1}{\varepsilon^2}\rho^2(\frac{x}{\varepsilon})]$, $\rho \in C_0^\infty(\mathbf{R})$, $\int \rho = 1$, $\int x^m \rho = 0$, $1 \leq m \leq N$, $N \in \mathbf{N}$. We have $\delta^2 \in \mathcal{G} \setminus \mathcal{D}'$, which means that there does not exist a distribution which determines $\delta^2$ in $\mathcal{G}$; $\delta^2(0)$ is represented by a divergent net of complex numbers.*

**Example 29.3** *In $\mathcal{G}$ we have*

$$x_+ \neq xH \text{ and } x_+ \approx xH.$$

**Example 29.4** *The non-associativity of $C^\infty \cdot \mathcal{D}'$ follows from $\delta 1 = \delta$, $\delta 0 = 0$,*

$$xV_p\frac{1}{x} = 1 \text{ and } 0 = (\delta x)V_p\frac{1}{x} \neq \delta(xV_p\frac{1}{x}) = 1.$$

*In $\mathcal{G}$, $x\delta \neq 0$ but in the associated sense $x\delta \approx 0$.*

## 29.3 $\mathcal{G}$-quasiasymptotics at zero

We denote by $\mathcal{K}$ a set of positive measurable functions defined on $(0,1)$ with the property

$$A^{-1}\varepsilon^p \leq c(\varepsilon) \leq A\varepsilon^{-p},\ \varepsilon \in (0,1)$$

for some $A > 0$ and $p > 0$.

Let $\omega$ denote an open set in $\mathbf{R}^n$ which contains 0.

**Definition 29.1** *([22]) Let $F \in \mathcal{G}(\omega)$. It is said that an $F \in \mathcal{G}(\omega)$ has the $\mathcal{G}$-quasiasymptotic behavior at zero with respect to $c(\varepsilon) \in \mathcal{K}$ if there is $F_\varepsilon$, a representative of $F$, such that for every $\psi \in \mathcal{D}(\omega)$ there is $C_\psi \in \mathbf{C}$ such that*

$$\lim_{\varepsilon \to 0} \left\langle \frac{F_\varepsilon(\varepsilon x)}{c(\varepsilon)}, \psi(x) \right\rangle = C_\psi \tag{29.1}$$

*and $C_\psi \neq 0$.*

It is proved in [22] that when $\mathcal{G}$ has the quasiasymptotics, there exists $g \in \mathcal{D}'(\omega)$ such that $C_\psi = \langle g, \psi \rangle$, $\psi \in \mathcal{D}(\omega)$.

In order to denote the difference between the ordinary and the $\mathcal{G}$-quasiasymptotics at zero we use the notation $f \overset{q.}{\sim} g$ and $f \overset{q.c.}{\sim} g$, respectively.

Main properties of $\mathcal{G}$ quasiasymptotics are given in the following two propositions.

**Proposition 29.1** *[22]*

*(a) Let $R_\varepsilon \in \mathcal{N}(\omega)$. Then, for every $c \in \mathcal{K}$, and every $\psi \in \mathcal{D}(\omega)$,*

$$\lim_{\varepsilon \to 0} \left\langle \frac{R_\varepsilon(\varepsilon x_1, ..., \varepsilon x_n)}{c(\varepsilon)}, \psi(x_1, ..., x_n) \right\rangle = 0.$$

*(b) Let $F \in \mathcal{G}(\omega)$, $c \in \mathcal{K}$ and let (29.1) holds. Then, (29.1) holds for every representative $\tilde{F}_\varepsilon$ of $F$.*

*(c) There exists $g \in \mathcal{D}'(\omega)$ such that*

$$C_\psi = \langle g, \psi \rangle,\ \psi \in \mathcal{D}(\omega).$$

**Proposition 29.2** *[22] Let $f \in \mathcal{E}'(\mathbf{R})$, supp$f = K \ni 0$. The following conditions are equivalent:*

*(a) $f \overset{q}{\sim} g \neq 0$, at zero with respect to $c(\varepsilon)$, in $\mathcal{S}'(\mathbf{R})$.*

*(b) $f \overset{q}{\sim} g \neq 0$, at zero with respect to $c(\varepsilon)$, in $\mathcal{D}'(\mathbf{R})$.*

*(c)*

$$\lim_{\varepsilon\to 0}\frac{(f*\phi_{\varepsilon^2})(\varepsilon x)}{c(\varepsilon)} = g \ \text{in}\ \mathcal{S}',\ g\neq 0,$$

*where $\phi$ is the vision function.*

*(d) For every $\theta\in\mathcal{D}(\mathbf{R})$, $\theta(0)\neq 0$, $f\theta \overset{q}{\sim} \theta(0)g\neq 0$ at zero with respect to $c(\varepsilon)$, in $\mathcal{D}'(\mathbf{R})$.*

**Remark 29.1** *Clearly, if (c) holds then the limit in (c) exists in $\mathcal{D}'$.*

We also define the stronger criterion of quasiasymptotics, the so-called strong $\mathcal{G}$-quasiasymptotics.

**Definition 29.2** *Let $F\in\mathcal{G}(\Omega)$. It is said that $F$ has the strong $\mathcal{G}$-quasiasymptotics at zero with the limit $g\in C^\infty(\Omega)$ with respect to $c(\varepsilon)\in\mathcal{K}$ if there exists $F_\varepsilon$, a representative of $F$, such that for every $K\subset\subset\mathbf{R}$*

$$\lim_{\varepsilon\to 0}\frac{F_\varepsilon(\varepsilon x)}{c(\varepsilon)} = g(x)\ \text{uniformly for}\ x\in K.$$

**Example 29.5** *We will prove that the generalized function $\delta^2=[\frac{1}{\varepsilon^2}\rho^2(\frac{x}{\varepsilon})]$ from Example 29.2 has the quasiasymptotics at zero with respect to $c(\varepsilon)=\varepsilon^{-2}$. It holds*

$$\lim_{\varepsilon\to 0}\langle\frac{\delta^2(\varepsilon x)}{\varepsilon^{-2}},\psi(x)\rangle = \lim_{\varepsilon\to 0}\int\rho^2(x)\psi(x)dx = \langle\rho^2,\psi\rangle = C_\psi,\ C_\psi\in\mathbf{C}. \quad (29.2)$$

*Thus, $[\frac{1}{\varepsilon^2}\rho^2(\frac{x}{\varepsilon})]$ has the quasiasymptotics at zero in the sense of Definition 29.1 with respect to $\varepsilon^{-2}$. The limit distribution in (29.2) is not of the form $Cx^\alpha$, as expected in the theory of ordinary quasiasymptotics.*

**Example 29.6** *The generalized function $[(2+sin\frac{1}{\varepsilon})\rho^2(\frac{x}{\varepsilon})]$ has the quasiasymptotics at zero with respect to $c(\varepsilon)=2+sin\frac{1}{\varepsilon}$, but this is not of the form $c(\varepsilon)=\varepsilon^\alpha L(\varepsilon)$, where $L$ is a Karamata slowly varying function (it is measurable, positive and $\lim_{\varepsilon\to 0}\frac{L(\varepsilon t)}{L(\varepsilon)}=1$ uniformly for $t\in[a,b]\subset(0,\infty)$, $\varepsilon<\varepsilon_0/b$, $\varepsilon_0$ is fixed). Also $\langle\frac{1}{c(\varepsilon)}(2+sin\frac{1}{\varepsilon})\rho^2(x),\psi(x)\rangle\to\langle g(x),\psi(x)\rangle$, $\psi\in\mathcal{D}$ and $g$ is not of the form $Cx^\alpha$, $C\neq 0$.*

**Example 29.7** *The generalized function $\frac{1}{\varepsilon^2}\rho^2(\frac{x}{\varepsilon^2})$ has the quasiasymptotics with the limit $\delta\int\rho^2(t)dt$ with respect to $c(\varepsilon)=\varepsilon^{-1}$ because*

$$\langle\frac{1}{\varepsilon^{-1}}\frac{1}{\varepsilon^2}\rho^2(\frac{\varepsilon x}{\varepsilon^2}),\psi(x)\rangle = \langle\frac{1}{\varepsilon}\rho^2(\frac{x}{\varepsilon}),\psi(x)\langle = \langle\rho^2(t),\psi(\varepsilon t)\rangle = \psi(0)\int\rho^2(t)dt.$$

**Example 29.8** *Let $F, G \in \mathcal{G}(\omega)$. If $F \overset{q.c.}{\sim} g$ at zero with respect to $c(\varepsilon)$ and $G \overset{q.c.}{\sim} g_1$, at zero with respect to $c_1(\varepsilon)$, then it is not true, in general, that $GF \overset{q.c.}{\sim} gg_1$ at zero with respect to $c_2(\varepsilon) = c(\varepsilon)c_1(\varepsilon)$, as it is the case in the classical theory. For example, $\frac{1}{\varepsilon^2}\rho(\frac{-x}{\varepsilon^2}) \overset{q.c.}{\sim} \delta$ with respect to $c(\varepsilon) = \varepsilon^{-1}$ but $\frac{1}{\varepsilon^4}\rho^2(\frac{-x}{\varepsilon^2}) \overset{q.c.}{\sim} (\int \rho^2(t)dt)\delta$ with respect to $c(\varepsilon) = \varepsilon^{-3}$.*

**Example 29.9** *The $\sqrt{\delta}$-generalized function (cf. [11]) determined by*

$$\Theta_{\sqrt{\rho},\varepsilon}(x) = \frac{1}{\sqrt{\varepsilon}}\sqrt{\rho(\frac{x}{\varepsilon})},$$

*is associated with the zero distribution. It has the quasiasymptotics at zero in the sense of Definition 29.1 with respect to $c(\varepsilon) = \varepsilon^{-1/2}$. Thus,*

$$\lim_{\varepsilon\to 0}\langle\frac{\Theta_{\sqrt{\rho},\varepsilon}(x)}{\varepsilon^{-1/2}}, \psi(x)\rangle = \lim_{\varepsilon\to 0}\int \rho(x)\psi(x)dx = C_\psi,\ \psi \in \mathcal{D},\ C_\psi \neq 0.$$

**Example 29.10** *Consider a more general function*

$$\Theta_{\rho,\varepsilon}(x) = \frac{1}{\mu(\varepsilon)}\rho(\frac{x}{\nu(\varepsilon)}),\ (cf. [11]),$$

*where $\rho \in \mathcal{D}$, $supp\rho \in [-1,1]$ and $sup_{x\in\mathbf{R}}\rho(x) > 0$, $\mu(\varepsilon) \to 0$, $\nu(\varepsilon) \to 0$ as $\varepsilon \to 0$. In general, such generalized functions are not the images of distributions in $\mathcal{G}(\mathbf{R})$ under the embedding $\mathcal{I}_2$.*

*Let $c(\varepsilon) = \nu(\varepsilon)(\varepsilon\mu(\varepsilon))^{-1}$. Then,*

$$\lim_{\varepsilon\to 0}\langle\frac{\Theta_{\rho,\varepsilon}(\varepsilon x)}{c(\varepsilon)}, \psi(x)\rangle = \lim_{\varepsilon\to 0}\int \frac{\varepsilon}{\nu(\varepsilon)}\rho(\frac{\varepsilon x}{\nu(\varepsilon)})\psi(x)dx =$$

$$\lim_{t\to 0}\int \rho(t)\psi(\frac{\nu(\varepsilon)}{\varepsilon}t)dt = \int \rho(x)\psi_1(x)dx = C_{\psi_1},\ C_{\psi_1} \in \mathbf{C},\ \psi \in \mathcal{D},$$

*if $\frac{\mu(\varepsilon)}{\varepsilon}$ has a limit as $\varepsilon \to 0$.*

## 29.4 Application of $\mathcal{G}$ quasiasymptotics to generalized solutions

We apply the $\mathcal{G}$-quasiasymptotics to some nonlinear problems such as a system of Volterra integral equation, a Cauchy problem for semilinear hyperbolic system (in particular wave equation and the Euler-Lagrange equation), and a Goursat problem.

It is proved in [15] that if the gradient for a nonlinear term is logaritmically bounded then a semilinear hyperbolic system is solvable in the framework of $\mathcal{G}$. For generalized solutions of a Goursat problem with a bounded gradient for nonlinear term see [12].

We consider the general case, when the gradient for a nonlinear term is not bounded and we find special regularizations, called $h$-regularizations, to avoid it. These regularizations are based on a regularization given for a semilinear hyperbolic system in [13] and adopted for each problem separately.

For a class of $h$-regularized equations the proposed method of finding generalized solutions produces a family of approximate solutions to the classical problem with the bounded gradient. Note that the main problems in our investigations of various types of equations and corresponding families of equations are the estimates with respect to $\varepsilon$ which make this theory a nontrivial generalization of the classical one.

Since all of these problems have their integral form we emphasize the example of a system of Volterra integral equations of the second kind. We recall the regularization for Volterra integral equation because it produces the other ones. Note that systems of integral equations, hyperbolic semilinear differential equations and abstract evolution equations can be considered as Volterra type equations. In particular, we give a special regularization for semilinear hyperbolic system.

### *29.4.1 System of nonlinear Volterra integral equations with non-Lipschitz nonlinearity*

We seek generalized solutions to

$$f^i(x) = g^i(x) + \int_0^x K^i(x, y, f(y))dy, \ x \in I \subset \mathbf{R}, \ i = 1, ...n, \qquad (29.3)$$

where $g \in (\mathcal{D}'(I))^n$ and a non-Lipschitz nonlinearity of $K \in L^\infty_{loc}(I \times I \times \mathbf{R}^{2n})$ by using the technique of cutting and regularizing in order to avoid the problems caused by an unbounded gradient of $K$ and the singularities of $g$.

More precisely, we replace $K(x, y, f)$ by $K_\varepsilon(x, y, f)$ which is smooth, bounded and has a bounded gradient with respect to $f$ for every fixed $\varepsilon$ and converges pointwisely to $K$ as $\varepsilon \to 0$.

Recall, elements of $\mathcal{D}'(\Omega)$ for the sake of the $\mathcal{G}$-quasiasymptotics at zero, are embedded into $\mathcal{G}(\Omega)$ by $\mathcal{I}_2 : \mathcal{D}'(\Omega) \to \mathcal{G}(\Omega)$, by $\mathcal{I}_2 f = [(f\kappa_\varepsilon) * \phi_{\varepsilon^2}]$, $\phi_{\varepsilon^2} = \varepsilon^{-2n}\phi(x/\varepsilon^2)$, $x \in \mathbf{R}^n$, is a mollifier which corresponds to a vision function $\phi$.

For a given $g = (g^1, ..., g^n) \in (\mathcal{D}'(I))^n$, we put $g^i_{\varepsilon^2} = (g^i\kappa_\varepsilon) * \phi_{\varepsilon^2}$, $i = 1, ..., n$, where $\kappa_\varepsilon \in C_0^\infty(I)$, $\kappa_\varepsilon = \begin{cases} 1 \text{ on } I_{2\varepsilon} \\ 0 \text{ on } I \setminus I_\varepsilon, \end{cases}$ $I = (-a, a), I_{j\varepsilon} =$

$(-a+j\varepsilon, a-j\varepsilon)$, $j=1,2$.

Assume that $K^i$ in (29.3) is an $L^\infty_{loc}$-function on $I\times I\times\mathbf{R}^{2n}$, $i=1,...,n$. Our aim is to make a regularization $K^i_\varepsilon$ of $K^i$, $i=1,...,n$ such that a family of solutions (determining a generalized solution to the regularized system) appropriately approximates the solution to (29.3) in the case when $g\in(\mathcal{D}'(I))^n$ and $K$ is not of Lipschitz class.

Let $B_r=\{(x,y); x\in\omega_r, y\in\omega_r\}$, where $\omega_r=\{x\in I; |x|<r, d(x, Compl.(I))\geq 1/r\}, r\in\mathbf{N}$.

Let $h$ be such that

$$h(\varepsilon)=\mathcal{O}(|\log\varepsilon|^{1/2}),\ \varepsilon\to 0.$$

We will use the notation $K(x,y,f)=K(x,y,u,v)$, where $f=u+iv$.

Let

$$\bar{K}^i_\varepsilon(x,y,u,v)=\begin{cases} K^i(x,y,u,v), & \text{if } (x,y)\in B_{h(\varepsilon)}, |K^i(x,y,u,v)|\leq h(\varepsilon),\\ 0, & \text{otherwise},\end{cases}$$

for $\varepsilon\in(0,1)$, $r\in\mathbf{N}$, $i=1,...,n$,

$$K^i_\varepsilon(x,y,u,v)=(\bar{K}^i_\varepsilon(\cdot)*(h(\varepsilon)^{2n+2}\theta(h(\varepsilon)\ \cdot)))(x,y,u,v)= \tag{29.4}$$

$$h(\varepsilon)^{2n+2}\int\limits_{B_r\times\mathbf{R}^{2n}}\bar{K}^i_\varepsilon(\xi,\tau,\eta,\nu)\theta(h(\varepsilon)(x-\xi),h(\varepsilon)(y-\tau),h(\varepsilon)(u-\eta),$$

$$h(\varepsilon)(v-\nu))d\xi d\tau d\eta d\nu,\ (x,y,u,v)\in I\times I\times\mathbf{R}^{2n},\ k=1,...,n,\ \varepsilon\in(0,1),$$

where $\theta\in C_0^\infty(\mathbf{R}^{2+2n})$ such that $\theta=1$ on $\{x; |x|\leq 1/2\}$, $\theta(x)=0$ on $\{x; |x|\geq 1\}$ and $\int\theta(x)dx=1$.

We have

$$|\partial/\partial u_j K^i_\varepsilon(x,y,u,v)|=|\partial/\partial u_j\int\limits_{B_{h(\varepsilon)}\times\mathbf{R}^{2n}}h(\varepsilon)^{2n+2}$$

$$\bar{K}^i_\varepsilon(\xi,\tau,\mu,\nu)\theta(h(\varepsilon)(x-\xi),h(\varepsilon)(y-\tau),h(\varepsilon)(u-\eta),h(\varepsilon)(v-\nu))d\xi d\tau d\eta d\nu|$$

$$=h(\varepsilon)|\int\limits_{\mathbf{R}^{2n+2}}\bar{K}^i_\varepsilon(x-\frac{1}{h(\varepsilon)}\xi,y-\frac{1}{h(\varepsilon)}\tau,u_1-\frac{1}{h(\varepsilon)}\eta_1,...,u_n-\frac{1}{h(\varepsilon)}\eta_n,...,v_1$$

$$-\frac{1}{h(\varepsilon)}\nu_1,...,v_n-\frac{1}{h(\varepsilon)}\nu_n)\partial/\partial u_j\theta(\xi,\tau,\eta_1,...,\eta_n,\nu_1,...,\nu_n)$$

$$d\xi d\tau d\eta_1,...,d\eta_n d\nu_1,...,d\nu_n|\leq Ch(\varepsilon)^2.$$

The same holds for $\partial K_\varepsilon^i/\partial v_j$. Thus, there exists $\varepsilon_0 > 0$ such that

$$|K_\varepsilon| \le C|\log\varepsilon|^{1/2},\ |\nabla_{(u,v)}K_\varepsilon| \le C|\log\varepsilon|,\ \text{on } I\times I\times\mathbf{R}^{2n},\ \varepsilon<\varepsilon_0. \tag{29.5}$$

We relate to (29.3), the following family of equations

$$F_\varepsilon^i(x) = g_\varepsilon^i(x) + \int_0^x K_\varepsilon^i(x,y,F_\varepsilon(y))dy + d_\varepsilon^i(x),\ x\in I,\ \varepsilon\in(0,\varepsilon_0), i=1,\dots,n, \tag{29.6}$$

where $g_\varepsilon^i = g^i * \phi_{\varepsilon^2}$, $K_\varepsilon^i$ are defined by (29.4), $F_\varepsilon = (F_\varepsilon^1,\dots,F_\varepsilon^n)$, and $d_\varepsilon^i \in \mathcal{N}(I)$, $i=1,\dots,n$.

We call (29.6) an $\mathcal{E}_M$-regularized Volterra system of equations which is related to (29.3) and the corresponding equation in $\mathcal{G}(I)$,

$$[F_\varepsilon^i(x)] = [g_\varepsilon^i(x)] + [\int_0^x K_\varepsilon^i(x,y,F_\varepsilon(y))dy],\ i=1,\dots,n, \tag{29.7}$$

is called a regularized Volterra system of equations.

**Lemma 29.3** *([24])*

*(a) Let $F_\varepsilon \in (\mathcal{E}_M(I))^n$. Then*

$$I \ni x \mapsto \int_0^x K_\varepsilon^i(x,y,F_\varepsilon(y))dy \in (\mathcal{E}_M(I))^n,\ i=1,\dots,n.$$

*(b) Let $F_\varepsilon \in (\mathcal{N}(I))^n$. Then*

$$I \ni x \mapsto \int_0^x K_\varepsilon^i(x,y,F_\varepsilon(y))dy \in (\mathcal{N}(I))^n,\ i=1,\dots,n.$$

*(c) Let $F_\varepsilon$, $\tilde F_\varepsilon \in (\mathcal{E}_M(I))^n$ such that $F_\varepsilon - \tilde F_\varepsilon \in \mathcal{N}(I)$. Then,*

$$I \ni x \mapsto \int_0^x (K_\varepsilon^i(x,y,F_\varepsilon(y)) - K_\varepsilon^i(x,y,\tilde F(y))dy \in \mathcal{N}(I),\ i=1,\dots,n.$$

**Theorem 29.4** *([24]) The regularized Volterra system of equation (29.7) has a unique solution $[F_\varepsilon] \in (\mathcal{G}(I))^n$.*

There exists $L^\infty$-association between the solution to the regularized system (29.6) and the local exact solution to (29.3) in the case when $g$ is continuous and $K$ is of $C^1$ class. This means that our approximate family of solutions converges to the local exact solution. If $K$ is of an appropriate Lipschitz class and $g \in (\mathcal{D}'(I))^n$, then the generalized solution is represented by a distribution. Thus, if $g$ has a support which consists of a finite number of points we have a delta wave as a solution (cf. [15], p. 127).

**Theorem 29.5** *[24]*

(a) *Let in (29.3), $K = (K^1, ..., K^n) \in (\mathbf{C}^1(I^2 \times \mathbf{R}^{2n}))^n$ and $g = (g^1, ..., g^n) \in (\mathbf{C}(I))^n$. There exist $T > 0$ and a unique solution $f$ to (29.3) in $(\mathbf{C}((-T,T)))^n$, such that the solution $[F_\varepsilon]$ to the regularized system (29.7) is $L^\infty$-associated in $(\mathcal{G}((-T,T)))^n$ to $f$.*

(b) *Let $I = (-J, J)$, $J > 0$ and let $K \in (C^1(I^2 \times \mathbf{R}^{2n}))^n$ such that*

$$\sup_{\substack{x,y \in I^2 \\ u,v \in \mathbf{R}^{2n}}} \{|\nabla K(x,y,u,v)|\} < \infty, \tag{29.8}$$

*and $g \in (L^p(I))^n$, $p \geq 1$. Then, there exists $f \in (L^p(I))^n$ such that the solution $[F_\varepsilon]$ to (29.7) is $L^p$-associated to $f$.*

**Remark 29.2** *(i) Concerning the assertion in (b) instead of the assumptions $K \in (C^1(I^2 \times \mathbf{R}^{2n}))^n$ and (29.8) we can assume that $K \in (C(I^2 \times \mathbf{R}^{2n}))^n$, and that it is globally of Lipschitz class with respect to $u$ and $v$.*

*(ii) One can formulate additional assumptions on $K$ which imply that $f$ is an $(L^p)^n$-solution to (29.3). For example, if $K$ satisfies the assumptions of (b) (or the first part of this remark),*

$$|K(x,y,u,v)| \leq h(\varepsilon), \text{ for } (x,y) \in B_\varepsilon \text{ and } (u,v) \in \mathbf{R}^{2n},\ i = 1, ..., n$$

*and $K \in (L^p(I^2 \times \mathbf{R}^{2n}))^n$, then $\bar{K}^i_\varepsilon = K^i$ and $K^i_\varepsilon \to K^i$ in the $L^p(I^2 \times \mathbf{R}^{2n})$-norm, $i = 1, ..., n$. By using estimates as in the proof of assertion (b) one can prove that $f$ is an $L^p$ solution to (29.3).*

*(iii) It is an open problem whether assumptions on $K$ in (ii) and assumption $g \in (\mathcal{D}'(I))^n$ imply the existence of an $f \in (\mathcal{D}'(I))^n$ such that the solution $[F_\varepsilon]$ to (29.7) is associated to $f$.*

We will prove that the quasiasimptotic behavior at zero of $g$ determines the $\mathcal{G}$-quasiasymptotic behavior at zero of the solution to (29.7) and consequently to (29.3). Let $g^i \in \mathcal{D}'(I)$, $K^i \in L^\infty_{loc}(I \times I \times \mathbf{R}^{2n})$, in (29.7) $i = 1, ..., n$. Then, the solution to (29.7) belongs to $(\mathcal{G}(I))^n$ and the following theorem holds.

**Theorem 29.6** *Let $c(\varepsilon) \in \mathcal{K}$, $\lim_{\varepsilon \to 0} \dfrac{\varepsilon}{c(\varepsilon)} = 0$ and suppose that*

$\lim_{\varepsilon \to 0} \dfrac{g^i(\varepsilon x)}{c(\varepsilon)}$ *exists in $\mathcal{D}'(I)$, $i = 1, 2, ..., n$. Let $g^i_\varepsilon$ in (29.7) and (29.6) be* *determined by* $g_\varepsilon^i = (g^i\kappa_\varepsilon)*\phi_{\varepsilon^2}$. *Then, the solution* $[F(x)] = [F_\varepsilon^1(x), ..., F_\varepsilon^n(x)]$ *to (29.7) has the* $\mathcal{G}$*-quasiasymptotics at zero with respect to* $c(\varepsilon)$, *i.e.*

$$\lim_{\varepsilon\to 0}\left\langle \frac{F_\varepsilon^i(\varepsilon x)}{c(\varepsilon)}, \psi(x)\right\rangle = C_\psi^i,\ C_\psi^i \in \mathbf{C},\ \psi\in\mathcal{D}(I),\ i=1,...,n,$$

*provided one of the following two conditions holds:*

($a$) $\varepsilon|\log\varepsilon|^{1/2}/c(\varepsilon)\to 0,$ *as* $\varepsilon\to 0$;

($b$) *For every* $[-J,J]\subset I$

$$\frac{\varepsilon|\log\varepsilon|}{c(\varepsilon)}\sup_{x\in J}|g(\varepsilon x)|\to 0,\ \ as\ \varepsilon\to 0.$$

**Proof.** Let $\psi\in\mathcal{D}(I)$, $\operatorname{supp}\psi\subset[-J,J]\subset I$. Then, by (29.6),

$$\left\langle \frac{F_\varepsilon^i(\varepsilon x)}{c(\varepsilon)}, \psi(x)\right\rangle = \left\langle \frac{g_\varepsilon^i(\varepsilon x)}{c(\varepsilon)}, \psi(x)\right\rangle +$$

$$\int\int_0^\varepsilon \psi(x)\frac{1}{c(\varepsilon)}K_\varepsilon^i(\varepsilon x, y, F_\varepsilon(y))dydx,\ i=1,...,n,\ \varepsilon\in(0,\varepsilon_0). \tag{29.9}$$

In case ($a$) the assertion follows because of (29.5). Let us consider case ($b$). Let in (29.6) the variable $x$ be replaced by $\varepsilon x$. We use:

$$|K_\varepsilon^i(x,y,F(y))| \le |K_\varepsilon^i(x,y,0)| + |F(y)||\nabla_{(u,v)}K_\varepsilon^i(x,y,\sigma F(y))|, 0<\sigma<1, \tag{29.10}$$

and obtain

$$|F_\varepsilon(\varepsilon x)| \le |g_\varepsilon(\varepsilon x)| + \int_0^{\varepsilon x}(|K_\varepsilon(\varepsilon x,y,0)| + |F_\varepsilon(y)||\nabla_{(u,v)}K_\varepsilon(\varepsilon x,y,\sigma F_\varepsilon(y)|)dy.$$

Applying the Gronwall inequality and (29.5) we obtain

$$\sup_{x\in J}|F_\varepsilon(\varepsilon x)| \le (\sup_{x\in J}|g_\varepsilon(\varepsilon x)| + \sup_{x,y\in J\times J}|K(x,y,0)|(\varepsilon sT))\exp(C\varepsilon x|log\varepsilon|).$$

Thus,

$$\sup_{x\in J}|F_\varepsilon(\varepsilon x)| \le C(\sup_{x\in J}|g_\varepsilon(\varepsilon x)| + \varepsilon). \tag{29.11}$$

Applying (29.11) and (29.10) we obtain

$$\int_0^{\varepsilon x}|K_\varepsilon(\varepsilon x,y,F_\varepsilon(y))|dy \le C\varepsilon(1 + (\sup_{x\in J}|g(\varepsilon x)| + \varepsilon)|\log\varepsilon|),\ x\in J.$$

Noting assumptions (*b*) it follows that the integral in (29.9) tends to zero as $\varepsilon \to 0$ and

$$\lim_{\varepsilon\to 0}\left\langle \frac{F^i_\varepsilon(\varepsilon x)}{c(\varepsilon)}, \psi(x)\right\rangle = \lim_{\varepsilon\to 0}\left\langle \frac{g^i_\varepsilon(\varepsilon x)}{c(\varepsilon)}, \psi(x)\right\rangle, \ i = 1, ..., n.$$

■

**Remark 29.3** *([24]) A similar proof holds true for equation (29.3) with assumptions on* $g \in (C(I))^n$ *and* $K \in (C^1(I \times I \times \mathbf{R}^{2n}))^n$ *given in Theorem 29.5. Thus, the corresponding assertion holds for (29.3).*

The above proof and the appropriate assertions can be applied to a semilinear hyperbolic system because it has the same integral form due to characteristic curves.

## *29.4.2 Semilinear hyperbolic system*

We recall a regularization for a semilinear strictly hyperbolic $(n \times n)$-system in two independent variables, $(x, t) \in \mathbf{R}^2$,

$$(\partial_t + \Lambda(x,t)\partial_x)u(x,t) = F(x,t,u(x,t)) \tag{29.12}$$

$$u(x,0) = (u_1(x,0), ..., u_n(x,0)) = (a_1(x), ..., a_n(x)) \in (\mathcal{G}(\mathbf{R}))^n,$$

where $\Lambda(x,t)$ is a diagonal matrix with the real distinct smooth functions on the diagonal, and let $(x,t,u) \mapsto F(x,t,u)$ be a smooth function on $\mathbf{R}^2 \times \mathbf{C}^n$ with an unbounded gradient $i = 1, ..., n$. The smoothness of $F^i$, $i = 1, ..., n$ implies slightly different regularization of $F$.

Fix a decreasing function $h : (0,1) \to (0,\infty)$ such that

$$h(\varepsilon) = \mathcal{O}(C|log\varepsilon|^{1/2}),\ h(\varepsilon) \to 0,\ \text{as } \varepsilon \to 0. \tag{29.13}$$

Let $B_r$, be the cube $|x| \leq r$, $|t| \leq r$, $|u| \leq r$, and $u = (u_1, v_1, ..., u_n, v_n)$. We denote by $\varepsilon_i$ a decreasing sequence of positive numbers such that $h(\varepsilon_{i+1}) = i$, $i \in \mathbf{N}$. We have $h(\varepsilon) \geq i - 1$ for $\varepsilon_{i+1} \leq \varepsilon < \varepsilon_i$. Let for $i \in \mathbf{N}$

$$S_i = B_i \cap \{(x,t,u,v), |F(x,t,u,v)| \leq i-1\}$$

$$\cap\{(x,t,u,v), |\nabla_{(u,v)}F(x,t,u,v)| \leq i-1\}.$$

Let $\chi_i$ be the characteristic function for $S_i$, $i \in \mathbf{N}$, and $\alpha \in C_0^\infty(\mathbf{R})$, $\int \alpha(t)dt = 1$, $\alpha \geq 1$, $\alpha = 1$ in a neighborhood of $t = 0$ and $\alpha_\varepsilon = 1/\varepsilon\alpha(t/\varepsilon)$, $t \in \mathbf{R}$. Put

$$\chi_{h(\varepsilon)} = (\chi_i * \alpha_{h(\varepsilon)^{-1}}),\ \varepsilon \in [\varepsilon_{i+1}, \varepsilon_i),\ i \in \mathbf{N},\ \ F^k_{h(\varepsilon)} = F^k\chi_{h(\varepsilon)},\ \varepsilon \in (0, \varepsilon_1),$$

$k=1,...,n$. Then, there exists a constant $C>0$ such that

$$||F_{h(\varepsilon)}||_{L^\infty(\mathbf{R}^{2n+2})} \leq Ch(\varepsilon), \quad ||\nabla_{(u,v)}F_{h(\varepsilon)}||_{L^\infty(\mathbf{R}^{2n+2})} \leq Ch(\varepsilon)^2. \tag{29.14}$$

The following is the regularized system obtained with the aid of $h$:

$$(\partial_t + \Lambda(x,t)\partial_x)u_{h(\varepsilon)}(x,t) = F_{h(\varepsilon)}(x,t,u_{h(\varepsilon)}(x,t)), \; u_{h(\varepsilon)}(x,0) = A(x). \tag{29.15}$$

Recall, $\mathcal{O}_M(\mathbf{C}^p)$ denotes the subspace of functions $f$ of $C^\infty$ $(\mathbf{C}^p)$ for which $f^{(\alpha)}$ is slowly increasing at infinity for every $\alpha \in \mathbf{N}_0^{2p}$, where we identify $f(z_1,\ldots,z_p)$ with $f(x_1,y_1\ldots,x_p,y_p)$ and $f^{(\alpha)}$ denotes $\dfrac{\partial^{|\alpha|}f}{\partial x_1^{\alpha_1}\partial y_1^{\alpha_2}\ldots\partial x_p^{\alpha_{2p-1}}\partial y_p^{\alpha_{2p}}}$.

**Proposition 29.7** *( [13]) (i) Assume that every component of the mapping $u \mapsto F(x,t,u)$ belongs to $\mathcal{O}_M(\mathbf{C}^n)$ and has uniform bounds for $(x,t) \in K \subset\subset \mathbf{R}^2$. Then the regularized system (29.15) has a unique solution in $(\mathcal{G}(\mathbf{R}^2))^n$ whenever the initial data are in $(\mathcal{G}(\mathbf{R}))^n$.*

*(ii) Let the initial data $(a_1,...,a_n)$ in (29.12) belong to $(C(\mathbf{R}))^n$ and $K_0 = [-a,a]$, $a>0$. The solution $u_{h(\varepsilon)}$ to the regularized system (29.15) is $L^\infty$-associated with the continuous local solution $u$ to (29.12) in $K_{T_0}$, for some $T_0>0$.*

**Proposition 29.8** *([23]) (i) Assume that the conditions of Proposition 29.7 (i) are satisfied and that for every characteristic curve $\gamma$ to (29.15) (which pass through $\varepsilon x$ at $\tau = \varepsilon t$),*

$$\lim_{\varepsilon\to 0} \frac{a_{i,\varepsilon}(\gamma_i(\varepsilon x, \varepsilon t, 0))}{c(\varepsilon)} \textit{ exists in } \mathcal{D}'(\mathbf{R}^2),\ i=1,...,n,$$

*where $c(\varepsilon) \in \mathcal{K}$, and one of the following conditions hold*

$$\lim_{\varepsilon\to 0} \frac{\varepsilon|\log\varepsilon|^{1/2}}{c(\varepsilon)} = 0,$$

$$\frac{\varepsilon|\log\varepsilon|}{c(\varepsilon)} \sup_{(x,t)\in K} |a_\varepsilon(\gamma_i(\varepsilon x,\varepsilon t,0))| \to 0, \textit{ for every } K\subset\subset \mathbf{R}^2,\ i=1,...,n.$$

*Then, the solution $([u_{1,h(\varepsilon)}(x,t)],...,[u_{n,h(\varepsilon)}(x,t)])$ to (29.15) has the $\mathcal{G}$-quasiasymptotic behavior at zero with respect to $c(\varepsilon)$, i. e.*

$$\lim_{\varepsilon\to 0}\left\langle \frac{u_{i,h(\varepsilon)}(\varepsilon x,\varepsilon t)}{c(\varepsilon)}, \psi(x,t)\right\rangle = C_{i,\psi},\ C_{i,\psi}\in\mathbf{C},\ \psi\in\mathcal{D}(\mathbf{R}^2),\ i=1,...,n.$$

*(ii) Assume that the conditions of Proposition 29.7 (ii) hold and that*

$$\lim_{\varepsilon\to 0+} \frac{a_i(\gamma_i(\varepsilon x,\varepsilon t,0))}{c(\varepsilon)} \textit{ exists in } \mathcal{D}'((-J,J)\times(-T_0,T_0))$$

*for some $T_0 > 0$, where $c(\varepsilon) \in \mathcal{K}$. Then, the local continuous solution to (29.12) has the same quasiasymptotics at $(0,0)$ as those of the generalized solution to (29.15) where $A(x) = [(A_{1\varepsilon}(x), ..., A_{n\varepsilon}(x))]$, $A_{i\varepsilon}(x) = (a_i\beta_\varepsilon) * \phi_{\varepsilon^2}$, $i = 1, ..., n$ and $\beta_\varepsilon$ is the characteristic function of $(-J+\varepsilon, J+\varepsilon)$.*

A nonlinear wave equation and the Euler-Lagrange equation can be transformed to semilinear hyperbolic system by corresponding substitution (cf. [23]). They have an integral form. Hence, we consider them as examples.

### *29.4.3 Nonlinear wave equation*

Nonlinear wave equation

$$u_{tt}(x,t) - \triangle u(x,t) = f(x,t,u(x,t))$$

$$u_{|t=0} = \tilde{u}_0(x),\ u_{t|t=0} = \tilde{u}_1(x),\ \tilde{u}_0, \tilde{u}_1 \in \mathcal{D}'(\mathbf{R})$$

where $f$ is $C^\infty$ function with unbounded gradient such that $f(x,t,0) = 0$ has an equivalent system of integral form

$$\begin{aligned} u_1(x,t) &= a_1(x) + \int_0^t f(u_2(x-t+s,s))ds \\ u_2(x,t) &= a_2(x) + \int_0^t u_1(x+t-s,s)ds. \end{aligned}$$

obtained by substitution $u_2 = u$, $u_1 = u_t - u_x$, where $a_1(x) = \tilde{u}_1(x) - \tilde{u}_{0x}(x), a_2(x) = \tilde{u}_0(x)$ and the characteristic curves are the straight lines $\gamma_1(s) \equiv x - t + s = 0$, $\gamma_2(s) \equiv x + t - s = 0$.

Since (29.13) and (29.14) hold, we obtain

$$|\nabla f_{h(\varepsilon)}| \leq C|log\varepsilon| \text{ for some } C > 0. \tag{29.16}$$

The regularized system is

$$(\partial_t + \partial_x)u_{1,h(\varepsilon)} = f_{h(\varepsilon)}(u_{2,h(\varepsilon)}) + d_{1,\varepsilon}(x,t) \tag{29.17}$$

$$(\partial_t - \partial_x)u_{2,h(\varepsilon)} = u_{1,h(\varepsilon)}$$

$$u_{i,\varepsilon}(x,0) = a_{i,\varepsilon}(x) + d_{2i,\varepsilon},\ i = 1,2$$

where

$$a_{i\varepsilon} = a_i\zeta_\varepsilon * \phi_{\varepsilon^2}(x),\ i = 1,2, \tag{29.18}$$

$\zeta_\varepsilon$ is the characteristic function of $(-1/\varepsilon, 1/\varepsilon)$, $d_{1,\varepsilon} \in \mathcal{N}(\mathbf{R}^2)$, $d_{2i,\varepsilon} \in \mathcal{N}(\mathbf{R})$. We call (29.17) and the corresponding equation in $(\mathcal{G}(\mathbf{R})^2)^2$ the $h$-regularized systems to nonlinear wave equation.

According to the classical theory, in this case there exists a unique solution (cf. [32]). We can prove the following proposition.

**Proposition 29.9** *([23]) Let $a_1, a_2 \in \mathcal{D}'(\mathbf{R})$ and $c(\varepsilon) \in \mathcal{K}$. Assume*

$$\lim_{\varepsilon \to 0} \frac{a_2(\varepsilon x + \varepsilon t)}{c(\varepsilon)}, \text{ exists in } \mathcal{D}'(\mathbf{R}), \tag{29.19}$$

$$\lim_{\varepsilon \to 0} \frac{\varepsilon |\log \varepsilon|}{c(\varepsilon)} \sup_{(x,t) \in K} \{|a_{1,\varepsilon}(\varepsilon x - \varepsilon t)| + |a_{2,\varepsilon}(\varepsilon x + \varepsilon t)|\} = 0 \text{ for every } K \subset\subset \mathbf{R}^2, \tag{29.20}$$

*($a_{i\varepsilon}$ is given by (29.18), $i = 1, 2$). Then, $[u_{2,h(\varepsilon)}(x,t)])$ to the h-regularized system has the $\mathcal{G}$-quasiasymptotics at zero with respect to $c(\varepsilon)$, i. e.*

$$\lim_{\varepsilon \to 0} \left\langle \frac{u_{2,\varepsilon}(\varepsilon x, \varepsilon t)}{c(\varepsilon)}, \psi(x,t) \right\rangle = C_{2,\psi}, \ C_{2,\psi} \in \mathbf{C}, \ \psi \in \mathcal{D}(\mathbf{R}^2).$$

**Remark 29.4** *In fact we have proved that the generalized solution $u_\varepsilon = u_{2\varepsilon}$ to the h-regularized wave equation has the corresponding the $\mathcal{G}$-quasi-asymptotics.*

**Proof.** We note that (29.19) holds with $a_{2,\varepsilon}$ instead of $a_2$.

Let $\psi \in \mathcal{D}(\mathbf{R}^2)$, $supp\psi \subset\subset K_T$ for a sufficiently large compact set $K_0 \subset \mathbf{R}$ and $T > 0$. Due to $f(x,t,0) = 0$, we have

$$|u_{1,\varepsilon}(\varepsilon x, \varepsilon t)| \leq |a_{1,\varepsilon}(\varepsilon x - \varepsilon t)| + \sup_{x \in \mathbf{R}} \{|f'_{h(\varepsilon)}|\} \int_0^{\varepsilon t} |u_{2,\varepsilon}(\varepsilon x - \varepsilon t + \tau, \tau)| d\tau,$$

$$|u_{2,\varepsilon}(\varepsilon x, \varepsilon t)| \leq |a_{2,\varepsilon}(\varepsilon x + \varepsilon t)| + \int_0^{\varepsilon t} |u_{1,\varepsilon}(\varepsilon x + \varepsilon t - \tau, \tau)| d\tau,$$

$\varepsilon \in (0, \varepsilon_0)$, for suitable $\varepsilon_0$. The Gronwall inequality and (29.16) imply

$$\sup_{(x,t) \in K_T} \sqrt{|u_{1,\varepsilon}(\varepsilon x, \varepsilon t)|^2 + |u_{2,\varepsilon}(\varepsilon x, \varepsilon t)|^2} \tag{29.21}$$

We apply these estimates in the examination of the $\mathcal{G}$-quasiasymptotics of $u_{2,\varepsilon}$,

$$\lim_{\varepsilon \to 0} \left\langle \frac{u_{2,\varepsilon}(\varepsilon x, \varepsilon t)}{c(\varepsilon)}, \psi(x,t) \right\rangle = \lim_{\varepsilon \to 0} \left\langle \frac{a_2(\varepsilon x + \varepsilon t)}{c(\varepsilon)}, \psi(x,y) \right\rangle +$$

$$\left\langle \frac{\int_0^{ct} (u_{1,\varepsilon}(\varepsilon x + \varepsilon t - \tau, \tau)}{c(\varepsilon)}, \psi(x,t) \right\rangle.$$

By assumptions in (29.19) and (29.20) and by (29.21) it follows

$$\lim_{\varepsilon \to 0} \left\langle \frac{u_{2,\varepsilon}(\varepsilon x, \varepsilon t)}{c(\varepsilon)}, \psi(x,t) \right\rangle = C_{2,\psi}, \ C_{2,\psi} \in \mathbf{C}, \ \psi \in \mathcal{D}(\mathbf{R}^2).$$

■

### *29.4.4 Euler-Lagrange equation*

As another example of a semilinear hyperbolic system we consider the Euler-Lagrange equation with one space variable

$$\ddot{x} + V'(x) = 0, \; x_{t_0} = x_0, \; \dot{x}(t_0) = \dot{x}_0, \; x_0, \dot{x}_0 \in \mathbf{C}. \tag{29.22}$$

Let $V$ be a $C^\infty$ function with unbounded second derivative in (29.22), then we find a regularization $V_{h(\varepsilon)} = V\chi_{h(\varepsilon)}$, ($\chi_\varepsilon$ is defined in 29.4.2) such that

$$|V''_{h(\varepsilon)}(x)| \leq C|log\varepsilon|^{3/2}, \text{ for some } C > 0.$$

Setting $x = u_2$, $u_1 = \dot{u}^2$, $x \in \mathbf{R}$, we transform the equation into regularized system

$$\frac{d}{dt}u_1 = -[V'_{h(\varepsilon)}(u_{2\varepsilon})], \; \frac{d}{dt}u_2 = u_1, \; u_2(0) = [x_{0\varepsilon}], \; u_1(0) = [\dot{x}_{0\varepsilon}]$$

which corresponds to the $h$-regularized equation

$$\ddot{x} + [V'_{h(\varepsilon)}(x)] = 0, \; x(0) = [x_{0\varepsilon}], \; \dot{x}(0) = [\dot{x}_{0\varepsilon}]. \tag{29.23}$$

**Proposition 29.10** *([23]) Let $V \in C^\infty(\mathbf{R})$, $V'(0) = 0$, $c(\varepsilon) \in \mathcal{K}$ and let $\lim_{\varepsilon\to 0} \frac{|x^i_{0\varepsilon}|}{c(\varepsilon)}$, $i = 1, 2$. Then, the solution $[u_{2,h(\varepsilon)}(t)] = [x_\varepsilon(t)]$, to (29.23) has the $\mathcal{G}$-quasiasymptotics at zero with respect to $c(\varepsilon)$, i. e.*

$$\lim_{\varepsilon\to 0} \left\langle \frac{x_\varepsilon(\varepsilon t)}{c(\varepsilon)}, \psi(t) \right\rangle = C_\psi, \; C_\psi \in \mathbf{C}, \; \psi \in \mathcal{D}(\mathbf{R}).$$

### *29.4.5 Goursat problem*

.

Let $\phi$ and $\psi$ be generalized functions of one variable (for instance, distributions) and $F$ be a given nonlinear function. Similarly, by the technique of regularization given for the Volterra equation it is proved in [25] that

$$\begin{array}{l} \partial_x\partial_y U(x,y) = F(\cdot, U(x,y)) \\ U_{|\{y=0\}} = \phi(x) \\ U_{|\{x=0\}} = \psi(y) \end{array} \tag{29.24}$$

admits the $\mathcal{G}$-quasiasymptotics at zero of the solution under some constraints on $c(\varepsilon)$ with the unbounded gradient.

The integral form of (29.24) is given by

$$U_{h(\varepsilon)}(\varepsilon x, \varepsilon y) = U_{0\varepsilon}(\varepsilon x, \varepsilon y) + \int_0^{\varepsilon x}\int_0^{\varepsilon y} F_{h(\varepsilon)}(\xi, \eta, U_{h(\varepsilon)}(\xi, \eta)) d\xi d\eta, \tag{29.25}$$

where

$$U_{0\varepsilon}(\varepsilon x, \varepsilon y) = \phi_\varepsilon(\varepsilon x) + \psi_\varepsilon(\varepsilon y) - \phi_\varepsilon(0).$$

Using the regularization for the Volterra equation we obtain the regularized system

$$\partial_x \partial_y [U_{h(\varepsilon)}(x,y)] = [F_{h(\varepsilon)}(x,y,[U_{h(\varepsilon)}(x,y)])], \tag{29.26}$$

$$[U_{h(\varepsilon)|\{y=0\}}] = \phi(x)$$

$$[U_{h(\varepsilon)|\{x=0\}}] = \psi(y).$$

The $h$-regularized system has $[U_{h(\varepsilon)}(x,y)]$ as a unique solution in $\mathcal{G}(\mathbf{R}^2)$ (cf. [25]). Thus, the following theorem holds.

**Proposition 29.11** *([25]) Let* $c(\varepsilon) \in \mathcal{K}$, $\lim_{\varepsilon\to 0} \dfrac{\varepsilon^2 |log\varepsilon|}{c(\varepsilon)} = 0$ *and let* $\lim\limits_{\varepsilon\to 0} \dfrac{U_{0\varepsilon}}{c(\varepsilon)}$ *exists in* $\mathcal{D}'(\omega)$ *where* $\omega$ *is an open set and* $(0,0) \in \omega$*. Then, the solution* $[U_{h(\varepsilon)}(x,y)]$ *to (29.26) has the quasiasymptotics at zero with respect to* $c(\varepsilon)$*, i. e.*

$$\lim_{\varepsilon\to 0} \left\langle \frac{U_{h\varepsilon}(\varepsilon x, \varepsilon y)}{c(\varepsilon)}, \psi(x,y) \right\rangle = C_\psi,\ C_\psi \in \mathbf{C},\ \psi \in \mathcal{D}(\omega),\ i = 1, ..., n.$$

**Proof.** Let $(x,y) \in K$. According to a special regularization we have the following constraint

$$\begin{array}{l} F_{h(\varepsilon)} \in C^\infty(\mathbf{R}^3, \mathbf{R}), \\ \forall K \subset\subset \mathbf{R}^2, \sup_{\substack{(x,y)\in K \\ z\in\mathbf{R}}} |\partial_z F_{h(\varepsilon)}(x,y,z)| < C|log\varepsilon|,\ \varepsilon \in (0,1). \end{array} \tag{29.27}$$

Let $\kappa_{h\varepsilon} = 1 + \sup_{\substack{(x,y)\in K \\ t\in\mathbf{R}}} |\partial_z F_{h(\varepsilon)}(\xi,\eta,t)|$. For all $(\xi,\eta) \in K$ the following holds

$$|F_{h(\varepsilon)}(\xi,\eta,U_\varepsilon(\xi,\eta))| \le |F_{h(\varepsilon)}(\xi,\eta,0)| + \kappa_{h\varepsilon}|U_\varepsilon(\xi,\eta)|.$$

Setting this in (29.25) we obtain

$$|U_{h\varepsilon}(\varepsilon x, \varepsilon y)| \le |U_{0\varepsilon}(\varepsilon x, \varepsilon y)| + \int_0^{\varepsilon x}\int_0^{\varepsilon y} (|F_{h(\varepsilon)}(\xi,\eta,0)|$$

$$+\kappa_{h\varepsilon} \sup_{(\xi,\eta)\in K} |U_{h\varepsilon}(\xi,\eta)|) d\xi d\eta$$

and

$$|U_{h\varepsilon}(\varepsilon x, \varepsilon y)| \leq |U_{0\varepsilon}(\varepsilon x, \varepsilon y)| + \sup_{(\xi,\eta)\in K} (|F_{h(\varepsilon)}(\xi,\eta,0)|)(\varepsilon^2)|xy|$$

$$+ \int_0^{\varepsilon x}\int_0^{\varepsilon y} \kappa_{h\varepsilon}|U_{h\varepsilon}(\xi,\eta)|d\xi d\eta.$$

The Gronwall inequality and the estimate for $\kappa_{h\varepsilon}$ yield

$$|U_{h\varepsilon}(\varepsilon x, \varepsilon y)| \leq C(|U_{0\varepsilon}(\varepsilon x, \varepsilon y)| + \varepsilon^2). \tag{29.28}$$

Using (29.27) and (29.28) we obtain

$$\int_0^{\varepsilon x}\int_0^{\varepsilon y} |F_{h(\varepsilon)}(\xi,\eta,U_{h\varepsilon}(\xi,\eta))|d\xi d\eta \leq \sup_{(\xi,\eta)\in K} (|F_{h(\varepsilon)}(\xi,\eta,0)|)(\varepsilon^2)|xy|$$

$$+|\kappa_{h\varepsilon}| \sup_{(\xi,\eta)\in K} |U_{h\varepsilon}(\xi,\eta)|(\varepsilon)^2|xy| \leq C\varepsilon^2\left(1 + |log\varepsilon|(|U_{0\varepsilon}(\varepsilon x,\varepsilon y)| + \varepsilon^2)\right)$$

$$\leq C\varepsilon^2(1 + |U_{0\varepsilon}(\varepsilon x,\varepsilon y)||log\varepsilon| + \varepsilon^2|log\varepsilon|).$$

Thus, for $(x,y) \in K$,

$$\int_0^{\varepsilon x}\int_0^{\varepsilon y} |F_{h(\varepsilon)}(\xi,\eta,U_{h\varepsilon}(\xi,\eta))|d\xi d\eta \leq C\varepsilon^2(1 + |U_{0\varepsilon}(\varepsilon x,\varepsilon y)||log\varepsilon| + \varepsilon^2|log\varepsilon|). \tag{29.29}$$

Then, we have in (29.25)

$$\left\langle \frac{U_{h\varepsilon}(\varepsilon x,\varepsilon y)}{c(\varepsilon)}, \psi(x,y)\right\rangle = \left\langle \frac{U_{0\varepsilon}(\varepsilon x,\varepsilon y)}{c(\varepsilon)}, \psi(x,y)\right\rangle$$

$$+ \int_{(x,y)\in I\times I} \psi(x,y)\left(\frac{1}{c(\varepsilon)}\int_0^{\varepsilon x}\int_0^{\varepsilon y} F_{h(\varepsilon)}(\xi,\eta,U_{h\varepsilon}(\xi,\eta))d\xi d\eta\right) dxdy.$$

Since (29.29) and $\lim_{\varepsilon\to 0} \frac{\varepsilon^2|log\varepsilon|}{c(\varepsilon)} = 0$ hold by assumption, the integral part tends to zero, and the assertion of Proposition follows. ■

Consequently, we obtain the same result for (29.24).

## *REFERENCES*

[1] Aragona J., Biagioni H. A., Intrinsic Definition of the Colombeau Algebra of Generalized Functions, Anal. Math. 17(2), 75-132(1991).

[2] Biagioni H. A., A Nonlinear Theory of Generalized Functions, Springer Verlag, Berlin-Heildelberg-New York, 1990.

[3] Biagioni H. A., Oberguggenberger M., Generalized Solutions to the Korteweg-de Vries and the Regularized Long-wave Equations, SIAM J. Math. Anal. 23(4), 923-940(1992).

[4] Biagioni H. A., Iorio R. J., Generalized Solutions to the Kuramoto-Sivashinski Equation, Proc. Int. Conf. Gen. Funct., N. Sad, 1996, Editor: Integral Transforms and Special Functions.

[5] Colombeau J. F., Elementary Introduction to New Generalized Functions, North Holland, 1985.

[6] Colombeau J. F., New Generalized Functions and Multiplications of Distributions, North Holland, 1982.

[7] Colombeau J. F., Meril A., Generalized Functions and Multiplication of Distributions on $C^\infty$-manifolds, J. Math. Anal. Appl. 186(2), 357-363(1994).

[8] Delcroix A., Scarpalesos D., Topology on Asymptotic Algebras of Generalized Functions and Applications, Preprint.

[9] Djapić N., Pilipović S., Colombeau Generalized Functions on a Manifold-the Micro Local Properties, Indag. Math. N. S. 7(3), 293-309(1996).

[10] Djapić N., Pilipović S., Scarpalesos D., Intrinsic Microlocal Characterization of Colombeau Generalized Functions, J. Anal. Math.(1998)(to appear).

[11] Hermann R., Oberguggenberger M., Generalized Functions, Calculus of Variations and Nonlinear ODEs, Preprint.

[12] Marty J. A., Nuiro S. P., Valmorin V., Algèbres Différentielles et Problème de Goursat non Linéaire à Données Irrégulières, J. Math. Pure Applic, Preprint.

[13] Nedeljkov M., Pilipović S., Generalized Solutions to a Semilinear Hyperbolic System with a non-Lipschitz Nonlinearity, Mh. Math. 125, 255-261(1998).

[14] Nedeljkov M., Pilipović S., Hypoelliptic Differential Operators with Generalized Constant Coefficients, Proc. Edinburgh Math. Soc. 41, 47-60(1998).

[15] Oberguggenberger M., Multiplication of Distributions and Applications to Partial Differential Equations, Pitman. Res. Not. Math. 259, Longman Sci. Techn. Essex, 1992.

[16] Oberguggenberger M., Nonlinear Theories of Generalized Functions, Proc. Conf. on Appl. Nonst. Anal. to Anal. Func. Anal. and Prob., Blaubeuren, July 19-25, 1992.

[17] Oberguggenberger M., Contributions of Nonstandard Analysis to PDEs, Proc. Coll. on Nonst. Math., Aveiro, Portugal 1994.

[18] Oberguggenberger M., Generalized Solutions to Semilinear Hyperbolic System, Mh. Math., 103, 133-144(1987).

[19] Pilipović S., Colombeau Generalized Functions and Pseudo-differential Operators, Lecture Notes of Tokyo University, Tokyo, 1993.

[20] Pilipović S., Linear Equations with Singular Coefficients and Nonlinear Equations with non-Lipschitz Nonlinearities, Proc. Int. Conf. Gen. Funct., N. Sad, 1996, Editor: Integral Transforms and Special Functions.

[21] Pilipović S., Generalized Asymptotics, Math. Moravica, Proc. $5^{th}$ IWWA, 169-181(1997).

[22] Pilipović S., Stojanović M., On the Behavior of Colombeau Generalized Functions at a Point-Application to Semilinear Systems, Publ. Math. Debreceen, 51/1-2, 111-126(1997).

[23] Pilipović S., Stojanović M., $\mathcal{G}$-quasiasymptotics of a Solution to Semilinear Hyperbolic System with one Space Variable, Preprint.

[24] Pilipović S., Stojanović M., Generalized Solutions to Nonlinear Volterra Integral Equations with non-Lipschitz nonlinearity, Nonlin. Anal. Th. Meth.& Appl. (1998) (to appear).

[25] Pilipović S., Stojanović M., Generalized Goursat Problem and a Quasiasymptotics of a Solution, Facta Univ., Niš, (1998) (to appear).

[26] Pilipović S., Stojanović M., Note on $\mathcal{G}$-quasiasymptotics at Infinity, Preprint.

[27] Pilipović S., Stanković B., Takači A., Asymptotic Behavior and Stieltjes Transformation of Distributions, Teubner, Band 116, Leipzig 1990.

[28] Pilipović S., On the Behavior of a Distribution at Origin, Math. Nachr. 141, 27-32(1989).

[29] Pilipović S., Local and Microlocal Analysis in the Space of Colombeau Generalized Functions, Workshop on Nonlin. Th. Gen. Funct., E-Schredinger-I, Viena, Proc. (ed. M. Oberggugenberger)(1988) (to appear).

[30] Roever J. W., Damsma M., Colombeau Algebras on a $C^\infty$-manifold, Indag. Math. (N.S.), 341-358(1991).

[31] Scarpalezos D., Colombeau Generalized Functions: Topological Structures, Microlocal Properties. A simplified point of view, Paris, 1995. Preprint.

[32] Struwe H., Semilinear Wave Equations, Bull. Amer. Math. Soc., 1, 26(1992).

[33] Vladimirov V. S., Droshshinov Yu. N., Zavyalov B. I., Tauberian Theorems for Generalized Functions, Kluwer Academic Publishers, Maine, 1988.

[34] Zavyalov B. I. On the Asymptotic Properties of Functions Holomorphic in Tubular Cones, Mat. Sbornik, 136(178), 1(5), 97-114(1988)(in Russian).